Trapezoid: A four-sided figure with one pair of parallel sides

Area: $A = \frac{1}{2}h(b_1 + b_2)$

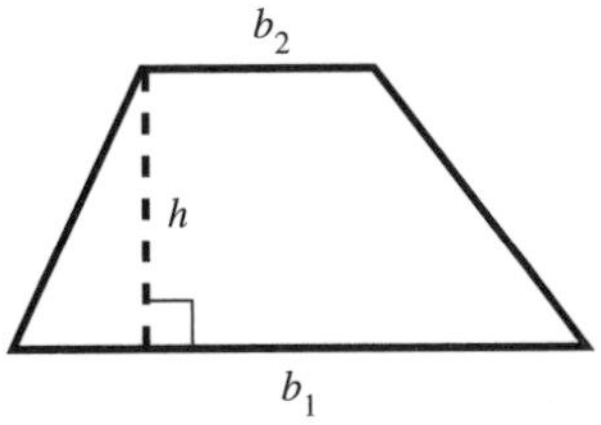

Parallelogram: A four-sided figure with opposite sides parallel

Area: $A = bh$

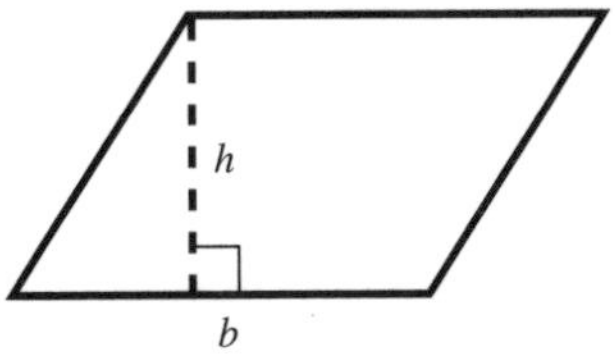

Rectangle: A four-sided figure with four right angles

Area: $A = LW$

Perimeter: $P = 2L + 2W$

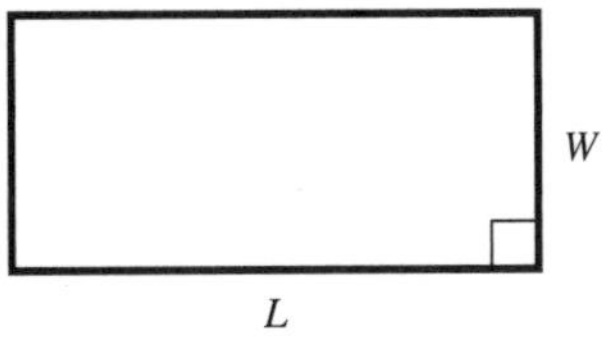

Rhombus: A four-sided figure with four equal sides

Perimeter: $P = 4a$

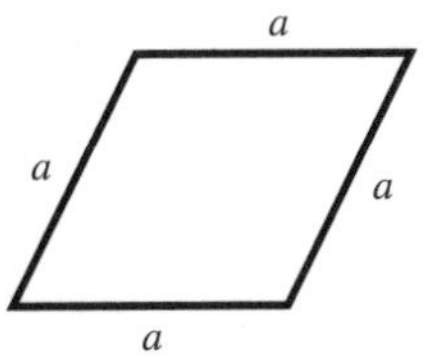

Square: A four-sided figure with four equal sides and four right angles

Area: $A = s^2$

Perimeter: $P = 4s$

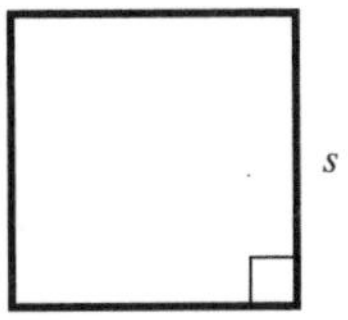

Circle

Area: $A = \pi r^2$

Circumference: $C = 2\pi r$

Diameter: $d = 2r$

Value of pi: $\pi \approx 3.14$

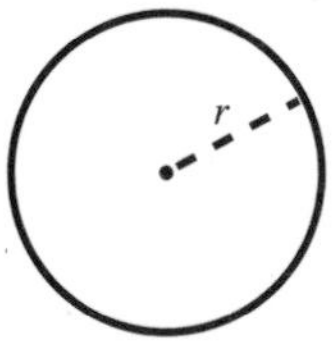

Sphere

Volume: $V = \frac{4}{3}\pi r^3$

Surface Area: $S = 4\pi r^2$

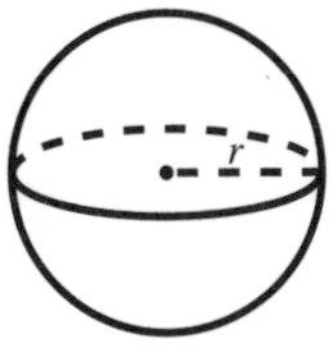

Right Circular Cone

Volume: $V = \frac{1}{3}\pi r^2 h$

Lateral Surface Area: $S = \pi r\sqrt{r^2 + h^2}$

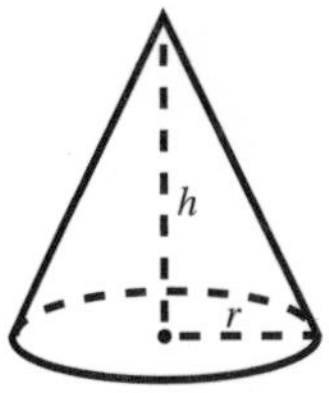

Right Circular Cylinder

Volume: $V = \pi r^2 h$

Lateral Surface Area: $S = 2\pi rh$

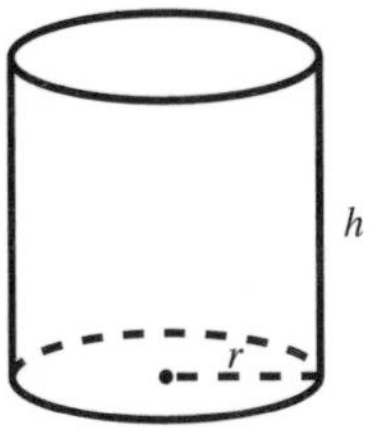

Rectangular Solid

Volume: $V = LWH$

Surface Area:
$A = 2LW + 2WH + 2LH$

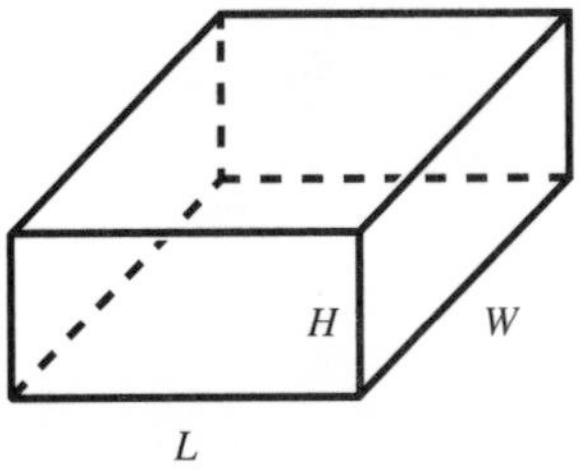

Elementary Algebra

Elementary Algebra

fifth edition

Mark Dugopolski
Southeastern Louisiana University

McGraw Hill Higher Education

Boston Burr Ridge, IL Dubuque, IA Madison, WI New York San Francisco St. Louis
Bangkok Bogotá Caracas Kuala Lumpur Lisbon London Madrid Mexico City
Milan Montreal New Delhi Santiago Seoul Singapore Sydney Taipei Toronto

The McGraw-Hill Companies

Higher Education

ELEMENTARY ALGEBRA, FIFTH EDITION

Published by McGraw-Hill, a business unit of The McGraw-Hill Companies, Inc., 1221 Avenue of the Americas, New York, NY 10020.

This book is printed on acid-free paper.

1 2 3 4 5 6 7 8 9 0 VNH/VNH 0 9 8 7 6 5 4
1 2 3 4 5 6 7 8 9 0 VNH/VNH 0 9 8 7 6 5 4

ISBN 0–07–293466–2
ISBN 0–07–302236–5 (Annotated Instructor's Edition)

Publisher, Mathematics and Statistics: *William K. Barter*
Publisher, Developmental Mathematics: *Elizabeth J. Haefele*
Director of Development: *David Dietz*
Senior Developmental Editor: *Randy Welch*
Executive Marketing Manager: *Michael Weitz*
Marketing Manager: *Steven R. Stembridge*
Senior Project Manager: *Vicki Krug*
Lead Production Supervisor: *Sandy Ludovissy*
Senior Media Project Manager: *Sandra M. Schnee*
Lead Media Technology Producer: *Jeff Huettman*
Designer: *Rick D. Noel*
Cover/Interior Designer: *Elise Lansdon/Lansdon Design*
(USE) Cover Image: *©Photonica, J61-284-3, Pattern with Star Shape, Shigeru Tanaka*
Lead Photo Research Coordinator: *Carrie K. Burger*
Photo Research: *Pam Carley*
Supplement Producer: *Brenda A. Ernzen*
Compositor: *Interactive Composition Corporation*
Typeface: *10.5/12 Times Roman*
Printer: *Von Hoffmann Corporation*

Photo Credits
Page 69: © Vol. 141/Corbis; p. 76: Alex Rodriguez © Reuters/Corbis; p. 145: © George Disario/Corbis; p. 164: © Vol. 166/Corbis; p. 188: © Ann M. Job/AP/Wide World Photos; p. 228 top right: © Gary Conner/PhotoEdit; p. 234: © Michael Keller/Corbis; p. 247: © Richard T. Nowitz/Corbis; p. 287: © DV169/Digital Vision; p. 547: © DV49/Digital Vision. All other photos © PhotoDisc/Getty.

Library of Congress Cataloging-in-Publication Data

Dugopolski, Mark.
Elementary algebra / Mark Dugopolski. — 5th ed.
p. cm.
Includes index.
ISBN 0–07–293466–2 (student ed. : acid-free paper) — ISBN 0–07–302236–5 (annotated instructor's ed. : acid-free paper)
1. Algebra—Textbooks. I. Title.

QA152.3.D839 2006
512.9—dc22 2004019510
CIP

www.mhhe.com

In loving memory of my parents,
Walter and Anne Dugopolski

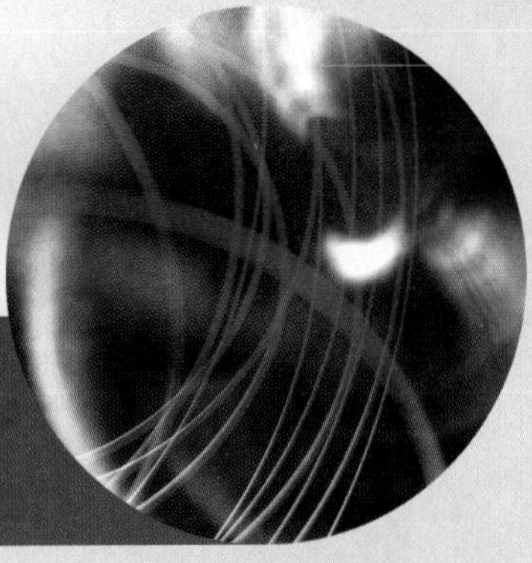

Contents

Chapter 5

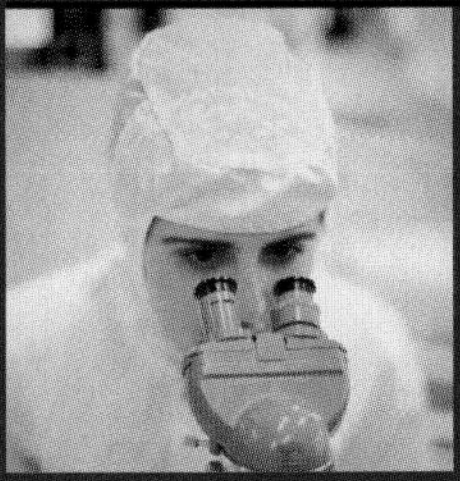

Polynomials and Exponents 287

Chapter 6

Factoring 349

Chapter 7

Rational Expressions 405

Preface

FROM THE AUTHOR

I would like to thank the many students and faculty who have used my books over the years. You have provided me with excellent feedback that has assisted me in writing a better, more student-focused book in each edition. Your comments are always taken seriously, and I have adjusted my focus on each revision to satisfy your needs.

In this edition in particular, you told me that you wanted a fresh, **new page layout** that was clean and open so that students could read it easily and without distraction. Last fall a group of 15 of you met in Boston to choose this design for my books. I hope that you like our implementation of your feedback, and I will always welcome your feedback on both the content and appearance of the future editions of my texts.

Additionally, you asked me to add **more exercises** that fall on the **easier side of the difficulty spectrum.** I took special time and effort to add the types of exercises you requested in these revisions, and I hope that they will assist your students to succeed in the course through ample practice of the fundamentals before moving on to exercises of medium difficulty.

Understandable Explanations

I originally undertook the task of writing my own book for the elementary algebra course so I could explain mathematical concepts to students in language they would understand. Most books claim to do this, but my experience with a variety of texts had proven otherwise. What students and faculty will find in my book are **short, precise explanations** of terms and concepts that are written in **understandable language.**

For example, when I introduce the Commutative Property of Addition, I make the concrete analogy that "the price of a hamburger plus a Coke is the same as the price of a Coke plus a hamburger," a mathematical fact in their daily lives that students can readily grasp. Math doesn't need to remain a mystery to students, and students reading my book will find other analogies like this one that connect abstractions to everyday experiences.

Detailed Examples Keyed to Exercises

My experience as a teacher has taught me two things about examples: they need to be detailed, and they need to help students do their homework. As a result, users of my book will find abundant examples with every step carefully laid out and explained where necessary so that students can follow along in class if the instructor is demonstrating an example on the board. Students will also be able to read them on their own later when they're ready to do the exercise sets.

I have also introduced a **double cross-referencing** system between my examples and exercise sets so that no matter which one students start with, they'll see the connection to the other. All examples in this edition now refer to specific exercises by ending with a phrase such as "Now do Exercises 11–18" so that students will have the opportunity for immediate practice of that concept. If students work an exercise and find they are stumped on how to finish it, they'll see that for that group of exercises they're directed to a specific example to follow as a model. Either way, students will find my book's examples give them the guidance they need to succeed in the course.

Varied Exercises and Applications

A third goal of mine in writing this book was to give students **more variety** in the kinds of exercises they perform than I found in other books. Students won't find an intimidating page of endless drills in my book, but instead will see exercises in manageable groups with specific goals. They will also be able to augment their math proficiency using different formats (true/false, written response, multiple choice) and different methods (discussion, collaboration, calculators). Not only is there an abundance of skill-building exercises, I have also researched a wide variety of **realistic applications** using **real data** so that those "dreaded word problems" will be seen as a useful and practical extension of what students have learned.

Finally, and new to this edition, every chapter ends with **critical thinking exercises** that go beyond numerical computation and call on students to employ their intuitive problem-solving skills to find the answers to mathematical puzzles in **fun and innovative** ways. With all of these resources to choose from, I am sure that instructors will be comfortable adapting my book to fit their course, and that students will appreciate having a text written for their level and to stimulate their interest.

Listening to Student and Instructor Concerns

McGraw-Hill has given me a wonderful resource for making my textbook more responsive to the immediate concerns of students and faculty. In addition to sending my manuscript out for review by instructors at many different colleges, several times a year McGraw-Hill holds symposia and focus groups with math instructors where the emphasis is *not* on selling products but instead on the **publisher listening** to the needs of faculty and their students. These encounters have provided me with a wealth of ideas on how to improve my chapter organization, make the page layout of my books more readable, and fine-tune exercises in every chapter. Consequently, students and faculty will feel comfortable using my book because it incorporates their specific suggestions and anticipates their needs. These events have particularly helped me in the shaping of the Fifth Edition.

Improvements in the Fifth Edition

- After consulting with a panel of experienced instructors on the **optimal page layout** to use with their students, the entire textbook was redesigned to be more open, easy to read, and easy for students to follow.
- All chapters now end in an exercise section called **"Critical Thinking: For Individual or Group Work,"** which focuses on intuitive problem-solving skills.

These exercises go beyond routine algebraic skills to give students the opportunity to think creatively to solve puzzles and challenges.

- **New exercises** have been added or updated throughout the text. Most of the added exercises fall on the easier side of the difficulty spectrum to give students more practice with the fundamentals.
- Every example is now keyed to specific exercises with the advice **"Now do Exercises . . . ,"** so that students will quickly see which exercises they should do to reinforce the concepts presented in the example.
- While interval and set notation are both introduced, **interval notation** is used as the preferred method due to instructor feedback.
- All the **"Math at Work"** features have been completely rewritten to show how math is used in different professions rather than focus on specific individuals.
- **Teaching Tips** have been added in the margins of the Annotated Instructor's Edition to provide instructors with practical classroom advice on how to present specific topics.
- In Chapter 7, all **answers for rational expressions** have been made consistent so that the denominator is always factored and the numerator is not.
- **Notes on Collaborative Learning** are now included in the Annotated Instructor's Edition.
- A new **Final Exam Review** has been added as an appendix to give students extra practice at the end of the course and to alleviate their math anxiety when preparing for the final exam.
- A new appendix on **Sets** has been added to give students an optional review of this topic.
- The **index** has been expanded to include more entries based on student and instructor requests.

Acknowledgments

I would like to extend my appreciation to the people at McGraw-Hill for their wholehearted support in producing the new editions of my books. My thanks go to Liz Haefele, Publisher, for being an energetic champion behind her authors and books; to David Dietz, Director of Development, for making the revision process work like a well-oiled machine; to Randy Welch, Senior Developmental Editor, for his advice on shaping the new editions; to Vicki Krug, Senior Project Manager, for expertly overseeing the many details of the production process; to Rick Noel, Designer, for the wonderful new design of my texts; to Carrie Burger, Lead Photo Research Coordinator, for her aid in picking out excellent photos; to Hal Whipple, for checking the accuracy of my texts; to Brenda Ernzen, Supplements Producer, for producing top-notch print supplements; and to Jeff Huettman, Lead Media Technology Producer, and Sandy Schnee, Senior Media Project Manager, for shepherding the development of high-quality media supplements that accompany my textbook. To all of them, my many thanks for their efforts to make my books best-sellers when there are many good books for faculty to choose from.

I sincerely appreciate the efforts of the reviewers who made many helpful suggestions to improve my series of books. I would like to extend special thanks to Mitch Levy for his many detailed and valuable suggestions for improving my exercise sets, and to Richard Maurer for keeping a user diary and alerting me to specific areas that he and his students have identified where I could make things more clear and precise.

Elise Adamson, *Wayland Baptist University*

Ebrahim Ahmadizadeh, *Northampton Community College*

W. Todd Ashby, *Charleston Southern University*

Viola Lee Bean, *Boise State University*

Monika Bender, *Central Texas College*

Mary Kay Best, *Coastal Bend College–Beeville*

Steve Boettcher, *Estrella Mountain Community College*

Annette M. Burden, *Youngstown State University*

Gail Burkett, *Palm Beach Community College*

Linda Clay, *Albuquerque Technical Vocational Institute*

John F. Close, *Salt Lake Community College*

Vivian Dennis-Monzingo, *Eastfield College*

Donna Densmore, *Bossier Parish Community College*

Mark deSaint-Rat, *Miami University–Middletown*

Lenore Desilets, *De Anza College*

William A. Echols, *Houston Community College*

Mike Everett, *Santa Ana College*

Pat Foard, *South Plains College*

Linda Franko, *Cuyahoga Community College*

Joseph Fritzsche, *University of Phoenix*

Corinna Goehring, *Jackson State Community College*

Wael Hassinan, *University of Phoenix*

Steven Hatfield, *Marshall University*

Erin Hines, *College of the Redwoods–Eureka*

Laura L. Hoye, *Trident Technical College*

Matthew Hudock, *St. Philip's College*

Barbara Hughes, *San Jacinto College–Pasadena*

Linda Hurst, *Central Texas College*

Domingo Javier-Litong, *Houston Community College*

Laura Kalbaugh, *Wake Technical Community College*

Krystyna Karminska, *Thomas Nelson Community College*

Joselle D. Kehoe, *DeVry Institute of Technology*

Tor Kwembe, *Chicago State University*

Suzann Kyriazopoulous, *DeVry University–Chicago Campus*

Angela Lawrenz, *Blinn College*

Sheila Ledford, *Coastal Georgia Community College*

Mitchel Levy, *Broward Community College*

Charyl Link, *Kansas City Kansas Community College*

Frederick Lippman, *Shasta College*

Sergio Loch, *Grand View College*

Carol Marinas, *Barry University*

Richard Maurer, *University of Phoenix*

Robert McCoy, *University of Alaska–Anchorage*

David Meredith, *San Francisco State University*

Margaret Michener, *University of Nebraska–Kearney*

Pam Miller, *Phoenix College*

Barbara Miller, *Lexington Community College*

Juan Molina, *Austin Community College*
Joyce Nemeth, *Broward Community College*
Thomas Notermann, *Devry University–Tinley Park*
Kim Nunn, *Northeast State Technical Community College*
Charles Odion, *Houston Community College*
Michele Olsen, *College of the Redwoods*
Frank Pecchioni, *Jefferson Community College*
Joanne Peeples, *El Paso Community College*
Avis Proctor, *Broward Community College*
Togba Sapolucia, *Houston Community College*
E. Jenell Sargent, *Tennessee State University–Nashville*
Paula Schornick, *Seminole State College*
Patty Schovanec, *Texas Tech University*
Mohsen Shirani, *Tennessee State University–Nashville*
Jefferson Shirley, *De Anza College*
Julia Simms, *Southern Illinois University–Edwardsville*
Donald W. Solomon, *University of Wisconsin–Milwaukee*
Sandra L. Spain, *Thomas Nelson Community College*
Brian Stewart, *Tarrant County College–Southeast*
Jo Temple, *Texas Tech University*
Burnette Thompson, *Houston Community College–Northwest*
Timothy Thompson, *Oregon Institute of Technology*
Lourdes Triana, *Humboldt State University*
David Turner, *Faulkner University*
Vivian Turner, *Rochester College*
Emmanuel Ekwere Usen, *Houston Community College*
Nina Verheyn, *Kilgore College*
Paul Visintainer, *Augusta Technical College*
Robert Vogeler, *North Idaho College*
Pam Wahl, *Middlesex Community College*
Brenda Weaver, *Stillman College*
Marjorie Whitmore, *Northwest Arkansas Community College*
Joel D. Williams, *Houston Community College*
Walter Wooden, *Broward Community College*
Kevin Yokoyama, *College of the Redwoods*
Vivian Zabrocki, *Montana State University–Billings*
Jane R. Zegestowsky, *Penn State–Abington*
Limin Zhang, *Columbia Basin College*
Deborah Zopf, *Henry Ford Community College*

I also want to express my sincere appreciation to my wife, Cheryl, for her invaluable patience and support.

Mark Dugopolski
Ponchatoula, Louisiana

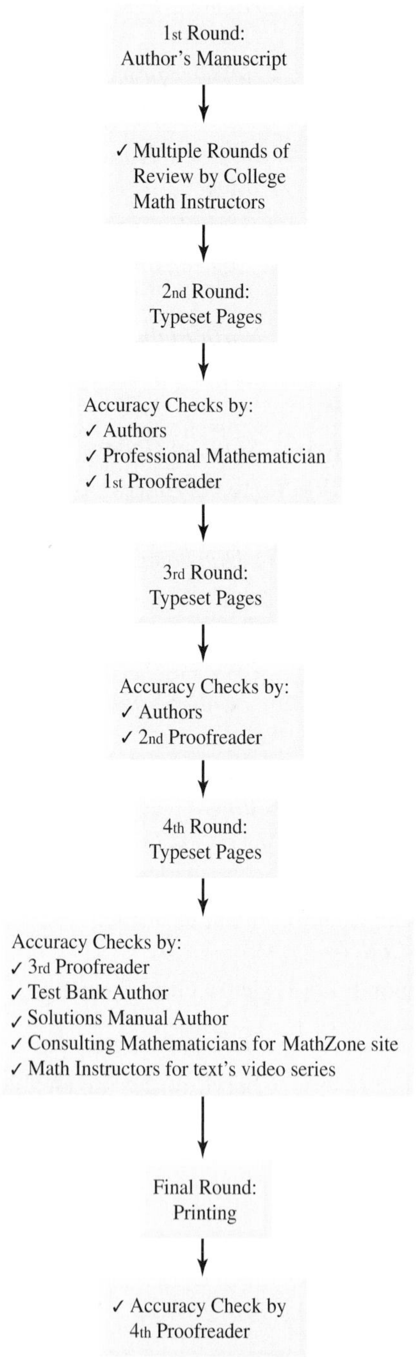

A COMMITMENT TO ACCURACY

You have a right to expect an accurate textbook, and McGraw-Hill invests considerable time and effort to make sure that we deliver one. Listed below are the many steps we take to make sure this happens.

OUR ACCURACY VERIFICATION PROCESS

First Round

Step 1: Numerous **college math instructors** review the manuscript and report on any errors that they may find, and the authors make these corrections in their final manuscript.

Second Round

Step 2: Once the manuscript has been typeset, the **authors** check their manuscript against the first page proofs to ensure that all illustrations, graphs, examples, exercises, solutions, and answers have been correctly laid out on the pages, and that all notation is correctly used.

Step 3: An outside, **professional mathematician** works through every example and exercise in the page proofs to verify the accuracy of the answers.

Step 4: A **proofreader** adds a triple layer of accuracy assurance in the first pages by hunting for errors, then a second, corrected round of page proofs is produced.

Third Round

Step 5: The **author team** reviews the second round of page proofs for two reasons: 1) to make certain that any previous corrections were properly made, and 2) to look for any errors they might have missed on the first round.

Step 6: A **second proofreader** is added to the project to examine the new round of page proofs to double check the author team's work and to lend a fresh, critical eye to the book before the third round of paging.

Fourth Round

Step 7: A **third proofreader** inspects the third round of page proofs to verify that all previous corrections have been properly made and that there are no new or remaining errors.

Step 8: Meanwhile, in partnership with **independent mathematicians,** the text accuracy is verified from a variety of fresh perspectives:

- The **test bank author** checks for consistency and accuracy as they prepare the computerized test item file.
- The **solutions manual author** works every single exercise and verifies their answers, reporting any errors to the publisher.
- A **consulting group of mathematicians,** who write material for the text's MathZone site, notifies the publisher of any errors they encounter in the page proofs.
- A video production company employing **expert math instructors** for the text's videos will alert the publisher of any errors they might find in the page proofs.

Final Round

Step 9: The **project manager,** who has overseen the book from the beginning, performs a **fourth proofread** of the textbook during the printing process, providing a final accuracy review.

⇒ What results is a mathematics textbook that is as accurate and error-free as is humanly possible, and our authors and publishing staff are confident that our many layers of quality assurance have produced textbooks that are the leaders of the industry for their integrity and correctness.

Guided Tour: Features and Supplements

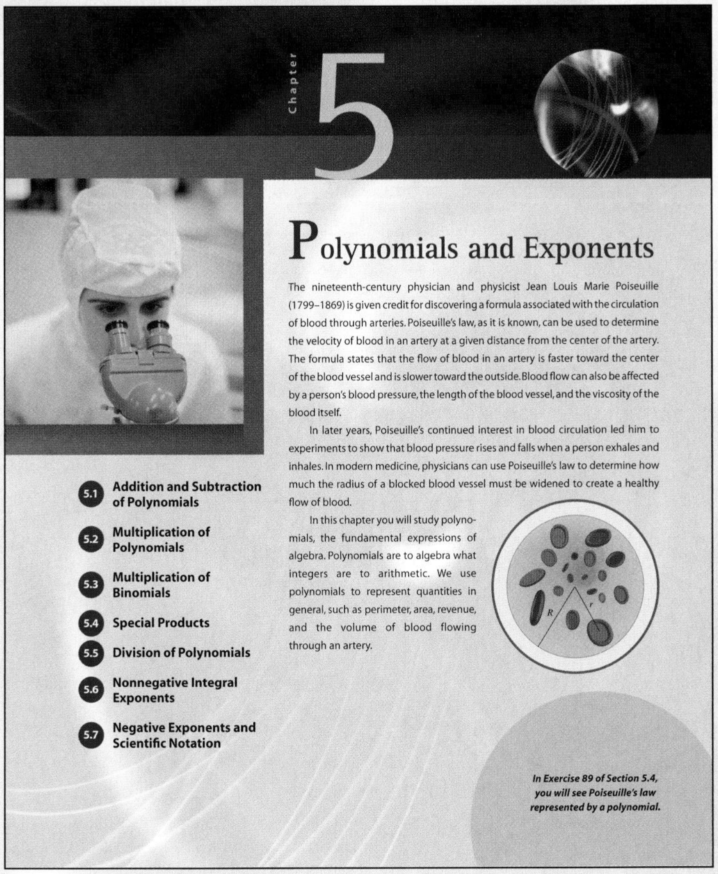

Chapter 5

Polynomials and Exponents

The nineteenth-century physician and physicist Jean Louis Marie Poiseuille (1799–1869) is given credit for discovering a formula associated with the circulation of blood through arteries. Poiseuille's law, as it is known, can be used to determine the velocity of blood in an artery at a given distance from the center of the artery. The formula states that the flow of blood in an artery is faster toward the center of the blood vessel and is slower toward the outside. Blood flow can also be affected by a person's blood pressure, the length of the blood vessel, and the viscosity of the blood itself.

In later years, Poiseuille's continued interest in blood circulation led him to experiments to show that blood pressure rises and falls when a person exhales and inhales. In modern medicine, physicians can use Poiseuille's law to determine how much the radius of a blocked blood vessel must be widened to create a healthy flow of blood.

In this chapter you will study polynomials, the fundamental expressions of algebra. Polynomials are to algebra what integers are to arithmetic. We use polynomials to represent quantities in general, such as perimeter, area, revenue, and the volume of blood flowing through an artery.

In Exercise 89 of Section 5.4, you will see Poiseuille's law represented by a polynomial.

Chapter Opener

Each chapter opener features a real-world situation that can be modeled using mathematics. The application then refers students to a specific exercise in the chapter's exercise sets.

89. ***Poiseuille's law.*** According to the nineteenth-century physician Poiseuille, the velocity (in centimeters per second) of blood r centimeters from the center of an artery of radius R centimeters is given by

$$v = k(R - r)(R + r),$$

where k is a constant. Rewrite the formula using a special product rule.

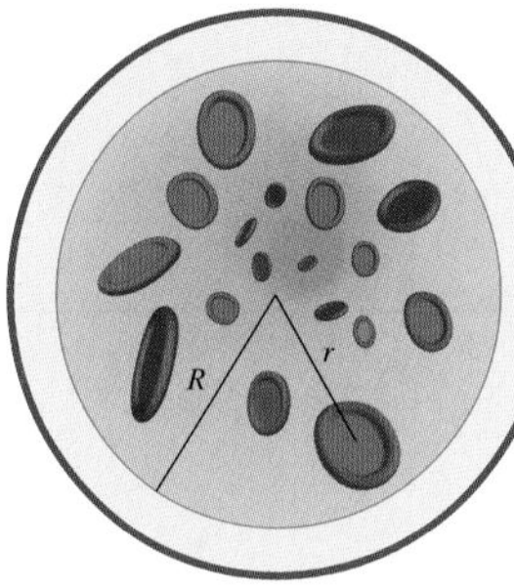

Figure for Exercise 89

EXAMPLE 4

Higher powers of a binomial
Expand each binomial.

a) $(x + 4)^3$ **b)** $(y - 2)^4$

Solution

a) $(x + 4)^3 = (x + 4)^2(x + 4)$
$= (x^2 + 8x + 16)(x + 4)$
$= (x^2 + 8x + 16)x + (x^2 + 8x + 16)4$
$= x^3 + 8x^2 + 16x + 4x^2 + 32x + 64$
$= x^3 + 12x^2 + 48x + 64$

b) $(y - 2)^4 = (y - 2)^2(y - 2)^2$
$= (y^2 - 4y + 4)(y^2 - 4y + 4)$
$= (y^2 - 4y + 4)(y^2) + (y^2 - 4y + 4)(-4y) + (y^2 - 4y + 4)(4)$
$= y^4 - 4y^3 + 4y^2 - 4y^3 + 16y^2 - 16y + 4y^2 - 16y + 16$
$= y^4 - 8y^3 + 24y^2 - 32y + 16$

Now do Exercises 47–54

Examples
Examples refer directly to exercises, and those exercises in turn refer back to that example. This **double cross-referencing** helps students connect examples to exercises no matter which one they start with.

Expand each binomial. See Example 4.

47. $(x + 1)^3$
48. $(y - 1)^3$
49. $(2a - 3)^3$
50. $(3w - 1)^3$
51. $(a - 3)^4$
52. $(2b + 1)^4$
53. $(a + b)^4$
54. $(2a - 3b)^4$

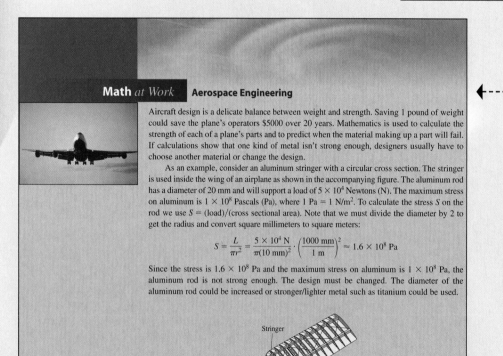

Math *at Work* **Aerospace Engineering**

Aircraft design is a delicate balance between weight and strength. Saving 1 pound of weight could save the plane's operators \$5000 over 20 years. Mathematics is used to calculate the strength of each of a plane's parts and to predict when the material making up a part will fail. If calculations show that one kind of metal isn't strong enough, designers usually have to choose another material or change the design.

As an example, consider an aluminum stringer with a circular cross section. The stringer is used inside the wing of an airplane as shown in the accompanying figure. The aluminum rod has a diameter of 20 mm and will support a load of 5×10^4 Newtons (N). The maximum stress on aluminum is 1×10^8 Pascals (Pa), where 1 Pa = 1 N/m^2. To calculate the stress S on the rod we use S = (load)/(cross sectional area). Note that we must divide the diameter by 2 to get the radius and convert square millimeters to square meters:

$$S = \frac{L}{\pi r^2} = \frac{5 \times 10^4 \text{ N}}{\pi(10 \text{ mm})^2} \cdot \left(\frac{1000 \text{ mm}}{1 \text{ m}}\right)^2 \approx 1.6 \times 10^8 \text{ Pa}$$

Since the stress is 1.6×10^8 Pa and the maximum stress on aluminum is 1×10^8 Pa, the aluminum rod is not strong enough. The design must be changed. The diameter of the aluminum rod could be increased or stronger/lighter metal such as titanium could be used.

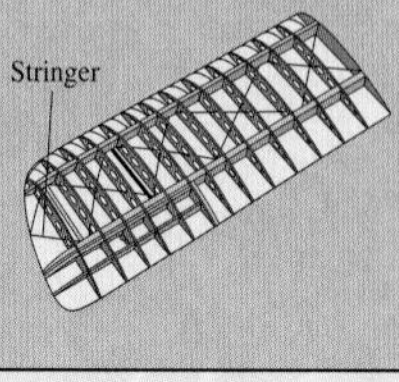

Math at Work
The Math at Work feature appears in each chapter to reinforce the book's theme of real applications in the everyday world of work.

Converting to Scientific Notation

To convert a positive number to scientific notation, we just reverse the strategy for converting from scientific notation.

Strategy for Converting to Scientific Notation

1. Count the number of places (n) that the decimal must be moved so that it will follow the first nonzero digit of the number.
2. If the original number was larger than 10, use 10^n.
3. If the original number was smaller than 1, use 10^{-n}.

Remember that the scientific notation for a number larger than 10 will have a positive power of 10 and the scientific notation for a number between 0 and 1 will have a negative power of 10.

Strategy Boxes

The strategy boxes provide a handy reference for students to use when they review key concepts and techniques to prepare for tests and homework.

Margin Notes

Margin notes include **Helpful Hints,** which give advice on the topic they're adjacent to; **Study Tips,** which give more general advice in improving study habits; **Calculator Close-Ups,** which provide advice on using calculators to verify students' work; and **Teaching Tips,** which are especially helpful in programs with new instructors who are looking for alternate ways to explain and reinforce material.

Helpful Hint

The exponent rules in this section apply to expressions that involve only multiplication and division. This is not too surprising since exponents, multiplication, and division are closely related. Recall that $a^3 = a \cdot a \cdot a$ and $a \div b = a \cdot b^{-1}$.

Study Tip

Remember that everything we do in solving problems is based on principles (which are also called rules, theorems, and definitions). These principles justify the steps we take. Be sure that you understand the reasons. If you just memorize procedures without understanding, you will soon forget the procedures.

Calculator Close-Up

With a calculator's built-in scientific notation, some parentheses can be omitted as shown below. Writing out the powers of 10 can lead to errors.

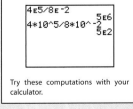

```
4E5/8E-2
                 5E6
4*10^5/8*10^-2
                 5E2
```

Try these computations with your calculator.

Teaching Tip Many students have trouble converting 24×10^{14} to scientific notation. Be sure they write $2.4 \times 10^1 \times 10^{14}$ and then add exponents.

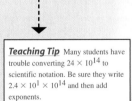

Warm-Ups ▼

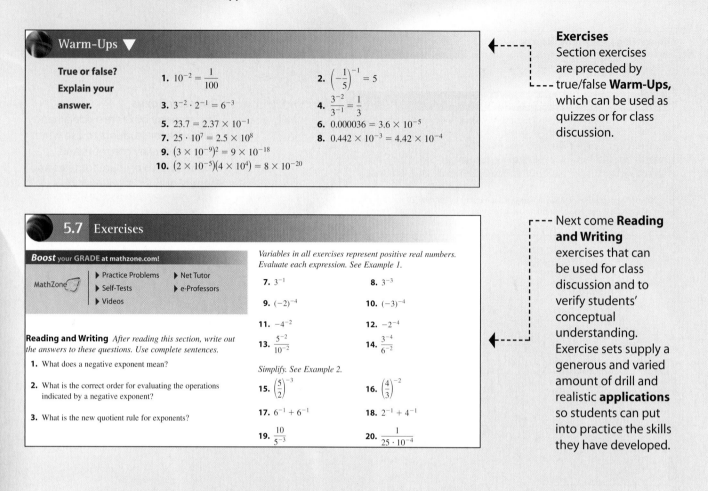

True or false? Explain your answer.

1. $10^{-2} = \frac{1}{100}$

2. $\left(-\frac{1}{5}\right)^{-1} = 5$

3. $3^{-2} \cdot 2^{-1} = 6^{-3}$

4. $\frac{3^{-2}}{3^{-1}} = \frac{1}{3}$

5. $23.7 = 2.37 \times 10^{-1}$

6. $0.000036 = 3.6 \times 10^{-5}$

7. $25 \cdot 10^{7} = 2.5 \times 10^{8}$

8. $0.442 \times 10^{-3} = 4.42 \times 10^{-4}$

9. $(3 \times 10^{-9})^2 = 9 \times 10^{-18}$

10. $(2 \times 10^{-5})(4 \times 10^{4}) = 8 \times 10^{-20}$

Exercises
Section exercises are preceded by true/false **Warm-Ups,** which can be used as quizzes or for class discussion.

5.7 Exercises

Boost your GRADE at mathzone.com!

MathZone
- Practice Problems
- Self-Tests
- Videos
- Net Tutor
- e-Professors

Reading and Writing *After reading this section, write out the answers to these questions. Use complete sentences.*

1. What does a negative exponent mean?

2. What is the correct order for evaluating the operations indicated by a negative exponent?

3. What is the new quotient rule for exponents?

Variables in all exercises represent positive real numbers. Evaluate each expression. See Example 1.

7. 3^{-1}

8. 3^{-3}

9. $(-2)^{-4}$

10. $(-3)^{-4}$

11. -4^{-2}

12. -2^{-4}

13. $\frac{5^{-2}}{10^{-2}}$

14. $\frac{3^{-4}}{6^{-2}}$

Simplify. See Example 2.

15. $\left(\frac{5}{2}\right)^{-3}$

16. $\left(\frac{4}{3}\right)^{-2}$

17. $6^{-1} + 6^{-1}$

18. $2^{-1} + 4^{-1}$

19. $\frac{10}{5^{-3}}$

20. $\frac{1}{25 \cdot 10^{-4}}$

Next come **Reading and Writing** exercises that can be used for class discussion and to verify students' conceptual understanding. Exercise sets supply a generous and varied amount of drill and realistic **applications** so students can put into practice the skills they have developed.

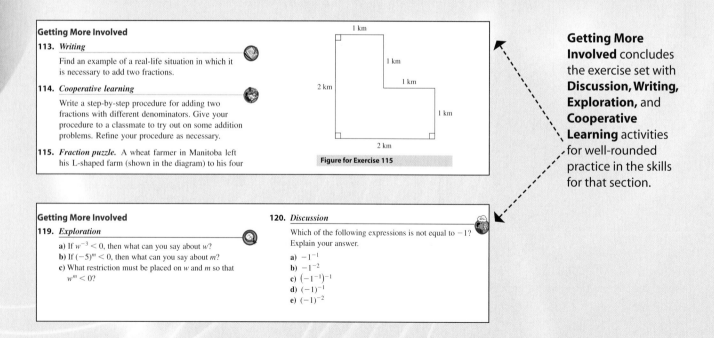

Getting More Involved

113. *Writing*

Find an example of a real-life situation in which it is necessary to add two fractions.

114. *Cooperative learning*

Write a step-by-step procedure for adding two fractions with different denominators. Give your procedure to a classmate to try out on some addition problems. Refine your procedure as necessary.

115. ***Fraction puzzle.*** A wheat farmer in Manitoba left his L-shaped farm (shown in the diagram) to his four

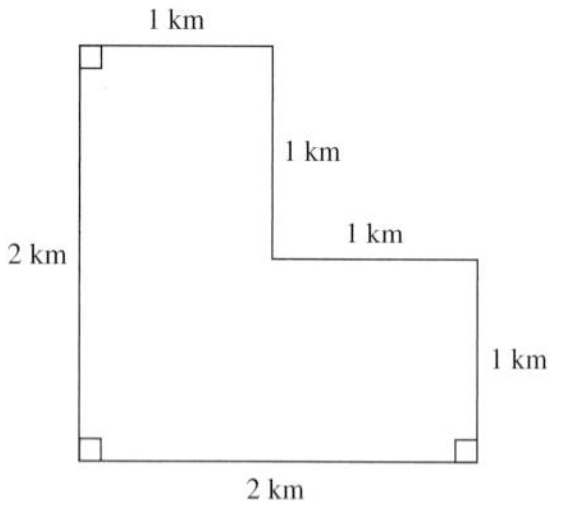

Figure for Exercise 115

Getting More Involved concludes the exercise set with **Discussion, Writing, Exploration,** and **Cooperative Learning** activities for well-rounded practice in the skills for that section.

Getting More Involved

119. *Exploration*

a) If $w^{-3} < 0$, then what can you say about w?
b) If $(-5)^m < 0$, then what can you say about m?
c) What restriction must be placed on w and m so that $w^m < 0$?

120. *Discussion*

Which of the following expressions is not equal to -1? Explain your answer.

a) -1^{-1}
b) -1^{-2}
c) $(-1^{-1})^{-1}$
d) $(-1)^{-1}$
e) $(-1)^{-2}$

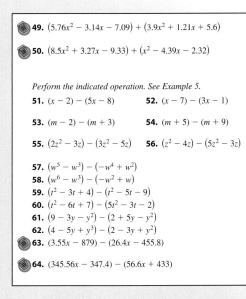

49. $(5.76x^2 - 3.14x - 7.09) + (3.9x^2 + 1.21x + 5.6)$

50. $(8.5x^2 + 3.27x - 9.33) + (x^2 - 4.39x - 2.32)$

Perform the indicated operation. See Example 5.

51. $(x - 2) - (5x - 8)$
52. $(x - 7) - (3x - 1)$

53. $(m - 2) - (m + 3)$
54. $(m + 5) - (m + 9)$

55. $(2z^2 - 3z) - (3z^2 - 5z)$
56. $(z^2 - 4z) - (5z^2 - 3z)$

57. $(w^5 - w^3) - (-w^4 + w^2)$
58. $(w^6 - w^3) - (-w^2 + w)$
59. $(t^2 - 3t + 4) - (t^2 - 5t - 9)$
60. $(t^2 - 6t + 7) - (5t^2 - 3t - 2)$
61. $(9 - 3y - y^2) - (2 + 5y - y^2)$
62. $(4 - 5y + y^3) - (2 - 3y + y^2)$
63. $(3.55x - 879) - (26.4x - 455.8)$

64. $(345.56x - 347.4) - (56.6x + 433)$

Calculator Exercises

Optional calculator exercises provide students with the opportunity to use scientific or graphing calculators to solve various problems.

Collaborative Activities

In addition to the cooperative learning activities in the section exercises, an extensive Collaborative Activity concludes each chapter so that students can solve math problems in a group setting.

Collaborative Activities

Grouping: Two students per group
Topic: Multiplying Polynomials

Area as a Model of Binomial Multiplication

Drawings and diagrams are often used to illustrate mathematical ideas. In this activity we use areas of rectangles to illustrate multiplication of binomials.

Example. The product $15 \cdot 13$ is the area of a 15 by 13 rectangle. Rewrite the product as $(10 + 5)(10 + 3)$ and make the following drawing.

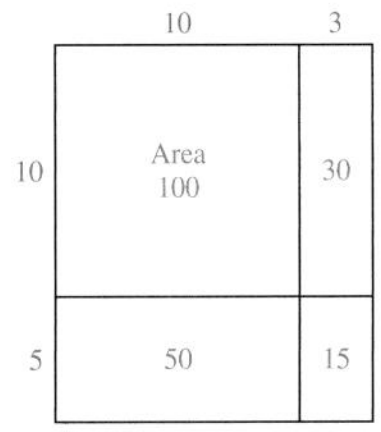

Note that the areas of the four regions are the four parts of FOIL:

$$\begin{aligned}(10 + 5)(10 + 3) &= \overset{\text{First}}{10^2} + \overset{\text{Outer}}{10 \cdot 3} + \overset{\text{Inner}}{5 \cdot 10} + \overset{\text{Last}}{5 \cdot 3} \\ &= 100 + 30 + 50 + 15 \\ &= 195\end{aligned}$$

Exercises

1. Make a drawing using graph paper to illustrate the product $12 \cdot 13$ as the product of two binomials $(10 + 2)(10 + 3)$. Have each square of your graph paper represent one unit. One partner should draw the rectangle and find the total of the areas of the four regions that correspond to FOIL while the other should find $12 \cdot 13$ in the "usual" way. Check your answers.
2. We have been using 10 for the first term of our binomials, but now we will use x. Choose any positive integer for x and draw an $x + 2$ by $x + 4$ rectangle on your graph paper. Have one partner draw the rectangle and find the total of the areas of the four regions that correspond to FOIL while the other will find $(x + 2)(x + 4)$ in the "usual" way. Check your answers.
3. Suppose that the area of the $x + 2$ by $x + 4$ rectangle drawn in the last exercise is actually 120. Does your rectangle have an area of 120? If not, then draw an $x + 2$ by $x + 4$ rectangle with an area of 120. What is x? Is x an integer?
4. Is there an $x + 3$ by $x + 6$ rectangle for which the area is 120 and x is a positive integer?
5. Is there an $x + 7$ by $x + 14$ rectangle for which the area is 120 and x is a positive integer?
6. How many ways are there to partition a 10 by 12 rectangle into four regions using sides $x + a$ and $x + b$, where x, a, and b are positive integers?

Chapter 5 Wrap-Up

Summary

Polynomials		Examples
Term	A number or the product of a number and one or more variables raised to powers	$5x^3$, $-4x$, 7
Polynomial	A single term or a finite sum of terms	$2x^5 - 9x^2 + 11$
Degree of a polynomial	The highest degree of any of the terms	Degree of $2x - 9$ is 1. Degree of $5x^3 - x^2$ is 3.
Naming a Polynomial	A polynomial can be named with a letter such as P or $P(x)$ (function notation).	$P = x^2 - 1$ $P(x) = x^2 - 1$

Wrap-Up
The extensive and varied review in the chapter Wrap-Up will help students prepare for tests. First comes the **Summary** with key terms and concepts illustrated by examples, then **Enriching Your Mathematical Word Power** enables students to test their recall of new terminology in a multiple choice format.

Enriching Your Mathematical Word Power

For each mathematical term, choose the correct meaning.

1. term
a. an expression containing a number or the product of a number and one or more variables
b. the amount of time spent in this course
c. a word that describes a number
d. a variable

c. an equation that has only one solution
d. a polynomial that has one term

6. FOIL
a. a method for adding polynomials
b. first, outer, inner, last
c. an equation with no solution
d. a polynomial with five terms

Review Exercises

Next come **Review Exercises,** which are first linked back to the section of the chapter that they review, and then the exercises are mixed without section references in the **Miscellaneous** section.

5.1 *Perform the indicated operations.*

1. $(2w - 6) + (3w + 4)$

2. $(1 - 3y) + (4y - 6)$

3. $(x^2 - 2x - 5) - (x^2 + 4x - 9)$

4. $(3 - 5x - x^2) - (x^2 - 7x + 8)$

5. $(5 - 3w + w^2) + (w^2 - 4w - 9)$

6. $(-2t^2 + 3t - 4) + (t^2 - 7t + 2)$

7. $(4 - 3m - m^2) - (m^2 - 6m + 5)$

8. $(n^3 - n^2 + 9) - (n^4 - n^3 + 5)$

20. $(x + 2)(x^2 - 2x + 4)$

21. $(x^2 - 2x + 4)(3x - 2)$

22. $(5x + 3)(x^2 - 5x + 4)$

5.3 *Perform the indicated operations.*

23. $(q - 6)(q + 8)$

24. $(w + 5)(w + 12)$

90. $\frac{a^{10}}{a^{-4}}$

91. $\frac{a^3}{a^{-7}}$

92. $\frac{b^{-2}}{b^{-6}}$

Miscellaneous

Perform the indicated operations.

115. $(x + 3)(x + 7)$

116. $(k + 5)(k + 4)$

117. $(t - 3y)(t - 4y)$

Chapter 5 Test

Perform the indicated operations.

1. $(7x^3 - x^2 - 6) + (5x^2 + 2x - 5)$

2. $(x^2 - 3x - 5) - (2x^2 + 6x - 7)$

3. $\dfrac{6y^3 - 9y^2}{-3y}$

4. $(x - 2) \div (2 - x)$

5. $(x^3 - 2x^2 - 4x + 3) \div (x - 3)$

6. $3x^2(5x^3 - 7x^2 + 4x - 1)$

Find the products.

7. $(x + 5)(x - 2)$

8. $(3a - 7)(2a + 5)$

9. $(a - 7)^2$

10. $(4x + 3y)^2$

11. $(b - 3)(b + 3)$

12. $(3t^2 - 7)(3t^2 + 7)$

23. $\dfrac{6t^{-7}}{2t^9}$

24. $\dfrac{w^{-6}}{w^{-4}}$

25. $(-3s^{-3}t^2)^{-2}$

26. $(-2x^{-6}y)^3$

Study Tip

Before you take an in-class exam on this chapter, work the sample test given here. Set aside one hour to work this test and use the answers in the back of this book to grade yourself. Even though your instructor might not ask exactly the same questions, you will get a good idea of your test readiness.

Convert to scientific notation.

27. 5,433,000

28. 0.0000065

Chapter Test
The test gives students additional practice to make sure they're ready for the real thing, with **all** answers provided at the back of the book and **all** solutions available in the Student's Solutions Manual.

Making Connections | A Review of Chapters 1–5

Evaluate each arithmetic expression.

1. $-16 \div (-2)$ **2.** $-16 \div \left(-\dfrac{1}{2}\right)$

3. $(-5)^2 - 3(-5) + 1$ **4.** $-5^2 - 4(-5) + 3$

5. $2^{15} \div 2^{10}$ **6.** $2^6 - 2^5$

7. $-3^2 \cdot 4^2$ **8.** $(-3 \cdot 4)^2$

9. $\left(\dfrac{1}{2}\right)^3 + \dfrac{1}{2}$ **10.** $\left(\dfrac{2}{3}\right)^2 - \dfrac{1}{3}$

11. $(5 + 3)^2$ **12.** $5^2 + 3^2$

13. $3^{-1} + 2^{-1}$ **14.** $2^{-2} - 3^{-2}$

15. $(30 - 1)(30 + 1)$ **16.** $(30 - 1) \div (1 - 30)$

Perform the indicated operations.

17. $(x + 3)(x + 5)$ **18.** $x + 3(x + 5)$

19. $-5t^3v \cdot 3t^2v^6$ **20.** $(-10t^3v^2) \div (-2t^2v)$

21. $(x^2 + 8x + 15) + (x + 5)$

22. $(x^2 + 8x + 15) - (x + 5)$

23. $(x^2 + 8x + 15) \div (x + 5)$

24. $(x^2 + 8x + 15)(x + 5)$

25. $(-6y^3 + 8y^2) \div (-2y^2)$

26. $(18y^4 - 12y^3 + 3y^2) \div (3y^2)$

Solve each equation.

27. $2x + 1 = 0$ **28.** $x - 7 = 0$

29. $\dfrac{3}{4}x - 3 = \dfrac{1}{2}$ **30.** $\dfrac{x}{2} - \dfrac{3}{4} = \dfrac{1}{8}$

31. $2(x - 3) = 3(x - 2)$

32. $2(3x - 3) = 3(2x - 2)$

Solve.

33. Find the x-intercept for the line $y = 2x + 1$.

34. Find the y-intercept for the line $y = x - 7$.

35. Find the slope of the line $y = 2x + 1$.

36. Find the slope of the line that goes through (0, 0) and $\left(\frac{1}{2}, \frac{1}{3}\right)$.

37. If $y = \frac{3}{4}x - 3$ and y is $\frac{1}{2}$, then what is x?

38. Find y if $y = \frac{x}{2} - \frac{3}{4}$ and x is $\frac{1}{2}$.

Solve the problem.

39. ***Average cost.*** Pineapple Recording plans to spend \$100,000 to record a new CD by the Woozies and \$2.25 per CD to manufacture the disks. The polynomial $2.25n + 100{,}000$ represents the total cost in dollars for recording and manufacturing n disks. Find an expression that represents the average cost per disk by dividing the total cost by n. Find the average cost per disk for $n = 1000$, 100,000, and 1,000,000. What happens to the large initial investment of \$100,000 if the company sells one million CDs?

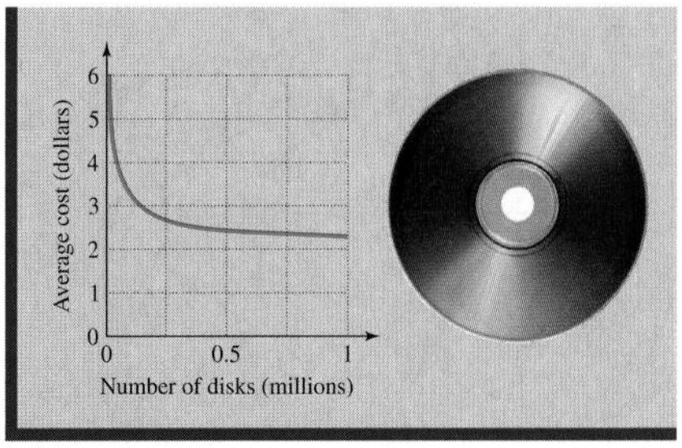

Figure for Exercise 39

The **Making Connections** feature following the Chapter Test is a cumulative review of all chapters up to and including the one just finished, helping to tie the course concepts together for students on a regular basis.

Critical **Thinking** | For Individual or Group Work | Chapter 5

These exercises can be solved by a variety of techniques, which may or may not require algebra. So be creative and think critically. Explain all answers. Answers are in the Instructor's Edition of this text.

1. ***Counting cubes.*** What is the total number of cubes that are in each of the following diagrams?

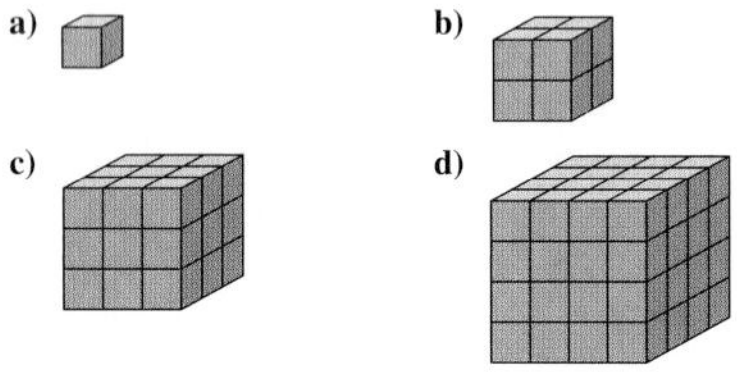

2. ***More cubes.*** Imagine a large cube that is made up of 125 small cubes like those in the previous exercise. What is the total number of cubes that could be found in this arrangement?

3. ***Timely coincidence.*** Starting at 8 A.M. determine the number of times in the next 24 hours for which the hour and minute hands on a clock coincide?

Photo for Exercise 3

4. ***Chess board.*** There are 64 squares on a square chess board. How many squares are neither diagonal squares nor edge squares?

Photo for Exercise 4

5. ***Last digit.*** Find the last digit in 3^{9999}.

6. ***Reconciling remainders.*** Find a positive integer smaller than 500 that has a remainder of 3 when divided by 5, a remainder of 6 when divided by 9, and a remainder of 8 when divided by 11.

7. ***Exact sum.*** Find this sum exactly:

$$\frac{1}{2} + \frac{1}{2^2} + \frac{1}{2^3} + \frac{1}{2^4} + \cdots + \frac{1}{2^{19}}$$

8. ***Ten-digit number.*** Find a 10-digit number whose first digit is the number of 1's in the 10-digit number, whose second digit is the number of 2's in the 10-digit number, whose third digit is the number of 3's in the 10-digit number, and so on. The ninth digit must be the number of nines in the 10-digit number and the tenth digit must be the number of zeros in the 10-digit number.

Critical Thinking
New to this edition, the Critical Thinking section that concludes every chapter encourages students to think creatively to solve unique and intriguing problems and puzzles.

SUPPLEMENTS FOR INSTRUCTORS

Annotated Instructor's Edition

This version of the student text contains **answers** to all odd- and even-numbered exercises in addition to helpful **teaching tips.** The answers are printed on the same page as the exercises themselves so that there is no need to consult a separate appendix or answer key.

Instructor's Testing and Resource CD

The cross-platform CD-ROM provides a wealth of resources for the instructor. Supplements featured on this CD-ROM include a **computerized test bank** utilizing Brownstone Diploma® **algorithm-based** testing software to quickly create customized exams. This user-friendly program enables instructors to search for questions by topic, format, or difficulty level; edit existing questions or add new ones; and scramble questions and answer keys for multiple versions of the same test.

Instructor's Solutions Manual

This supplement contains detailed solutions to all exercises in the text and is prepared by Mark Dugopolski. The methods used to solve the problems in the manual are the same as those used to solve the examples in the textbook.

MathZone www.mathzone.com*

*Web-based product also available on CD-ROM

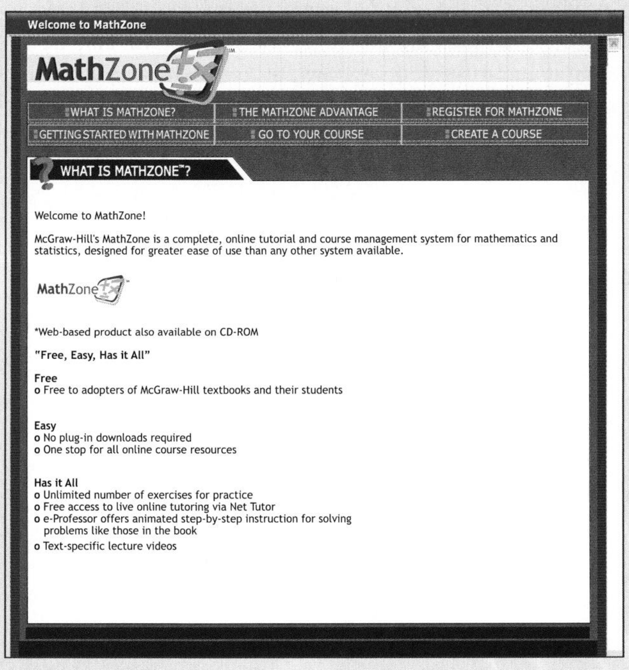
Welcome to MathZone
MathZone
WHAT IS MATHZONE?
THE MATHZONE ADVANTAGE
REGISTER FOR MATHZONE
GETTING STARTED WITH MATHZONE
GO TO YOUR COURSE
CREATE A COURSE
WHAT IS MATHZONE™?
Welcome to MathZone!
McGraw-Hill's MathZone is a complete, online tutorial and course management system for mathematics and statistics, designed for greater ease of use than any other system available.
MathZone
*Web-based product also available on CD-ROM
"Free, Easy, Has it All"
Free
o Free to adopters of McGraw-Hill textbooks and their students
Easy
o No plug-in downloads required
o One stop for all online course resources
Has it All
o Unlimited number of exercises for practice
o Free access to live online tutoring via Net Tutor
o e-Professor offers animated step-by-step instruction for solving problems like those in the book
o Text-specific lecture videos

McGraw-Hill's **MathZone** is a complete, online tutorial and course management system for mathematics and statistics, designed for greater ease of use than any other system available. Free on adoption of a McGraw-Hill title, instructors can create and share courses and assignments with colleagues and adjuncts in a matter of a few clicks of the mouse. All assignments, questions, e-Professors, online tutoring, and video lectures are directly tied to text-specific materials in Dugopolski, *Elementary Algebra, 5th Edition.* MathZone courses are customized to your textbook, but you can edit questions and algorithms, import your own content, and create announcements and due dates for assignments. MathZone has automatic grading and reporting of easy-to-assign algorithmically generated homework, quizzing, and testing. All student activity within MathZone is automatically recorded and available to you through a fully integrated grade book that can be downloaded to Excel.

ALEKS® (**A**ssessment and **LE**arning in **K**nowledge **S**paces) is an artificial intelligence-based system for individualized math learning, available over the Web. ALEKS delivers precise, qualitative diagnostic assessments of students' math knowledge, guides them in the selection of appropriate new study material, and records their progress toward mastery of curricular goals in a robust classroom management system. See page xxx for more details regarding ALEKS.

PageOut

PageOut is McGraw-Hill's unique, intuitive tool enabling instructors to create a full-featured, professional quality course website *without* being a technical expert. With PageOut you can post your syllabus online, assign content from the Dugopolski MathZone site, add links to important off-site resources, and maintain student results in the online grade book. PageOut is free for every McGraw-Hill Higher Education user and, if you're short on time, we even have a team ready to help you create your site. Contact your McGraw-Hill representative for further information.

SUPPLEMENTS FOR STUDENTS

Student's Solutions Manual

This supplement, prepared by Mark Dugopolski, contains complete worked-out solutions to all odd-numbered exercises in the textbook and all odd- and even-numbered problems for the Chapter Tests and Making Connections. Solutions for Critical Thinking are in the Instructor's Solutions Manual only. The methods used to solve the problems in the manual are the same as those used to solve the examples in the textbook. This tool can be an invaluable aid to students who want to check their work and improve their grades by comparing their own solutions to those found in the manual and finding specific areas where they can do better.

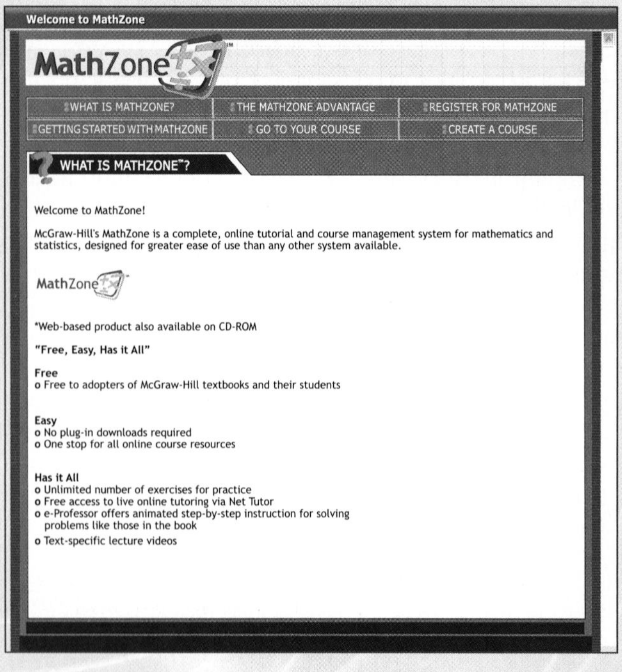

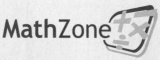

MathZone www.mathzone.com*

*Web-based product also available on CD-ROM

McGraw-Hill's MathZone is a powerful new online tutorial for homework, quizzing, testing, and interactive applications. MathZone offers:

- **Practice exercises** based on the text and generated in an unlimited number for as much practice as needed to master any topic you study.
- **Videos** of classroom instructors giving lectures and showing you how to solve exercises from the text.

- **e-Professors** to take you through animated, step-by-step instructions (delivered via on-screen text and synchronized audio) for solving problems in the book, enabling you to digest each step at your own pace.
- **NetTutor,** which offers live, personalized tutoring via the Internet.
- Every assignment, question, e-Professor, and video lecture is derived directly from Dugopolski, *Elementary Algebra, 5th Edition.*

NetTutor

Also available separately from MathZone, NetTutor is a revolutionary system that enables students to interact with a live tutor over the Web by using NetTutor's Web-based, graphical chat capabilities. Students can also submit questions and receive answers, browse previously answered questions, and view previous live chat sessions. NetTutor can be accessed on the text's MathZone site through the Student Edition.

ALEKS® v2.0

ALEKS® (**A**ssessment and **LE**arning in **K**nowledge **S**paces) is an artificial intelligence-based system for individualized math learning, available over the Web. ALEKS delivers precise, qualitative diagnostic assessments of students' math knowledge, guides them in the selection of appropriate new study material, and records their progress toward mastery of curricular goals in a robust classroom management system. See page xxx for more details regarding ALEKS.

Dugopolski Video Series

The video series is available on DVD and VHS tape and features an instructor introducing topics and working through selected odd-numbered exercises from the text, explaining how to complete them step by step. The DVDs are **closed-captioned** for the hearing-impaired and also **subtitled in Spanish.**

Math for the Anxious: Building Basic Skills,
by Rosanne Proga

Math for the Anxious: Building Basic Skills is written to provide a practical approach to the problem of math anxiety. By combining strategies for success with a pain-free introduction to basic math content, students will overcome their anxiety and find greater success in their math courses.

ALEKS® v2.0

ALEKS is an artificial intelligence-based system for individualized math learning, available for Higher Education from McGraw-Hill over the World Wide Web.

ALEKS delivers precise assessments of math knowledge, guides the student in the selection of appropriate new study material, and records student progress toward mastery of goals.

ALEKS interacts with a student much as a skilled human tutor would, moving between explanation and practice as needed, correcting and analyzing errors, defining terms and changing topics on request. By accurately assessing a student's knowledge, ALEKS can focus clearly on what the student is ready to learn next, helping to master the course content more quickly and easily.

ALEKS is:

- **A comprehensive course management system.** It tells the instructor exactly what students know and don't know.
- **Artificial intelligence.** It totally individualizes assessment and learning.
- **Customizable.** ALEKS can be set to cover the material in your course.
- **Web-based.** It uses a standard browser for easy Internet access.
- **Inexpensive.** There are no setup fees or site license fees.

ALEKS 2.0 adds the following new features:

- **Automatic Textbook Integration**
- **New Instructor Module**
- **Instructor-Created Quizzes**
- **New Message Center**

ALEKS maintains the features that have made it so popular including:

- **Web-Based Delivery** No complicated network or lab setup
- **Immediate Feedback** for students in learning mode
- **Integrated Tracking of Student Progress and Activity**
- **Individualized Instruction** which gives students problems they are ***Ready to Learn***

For more information please contact your McGraw-Hill Sales Representative or visit ALEKS at http://www.highedmath.aleks.com.

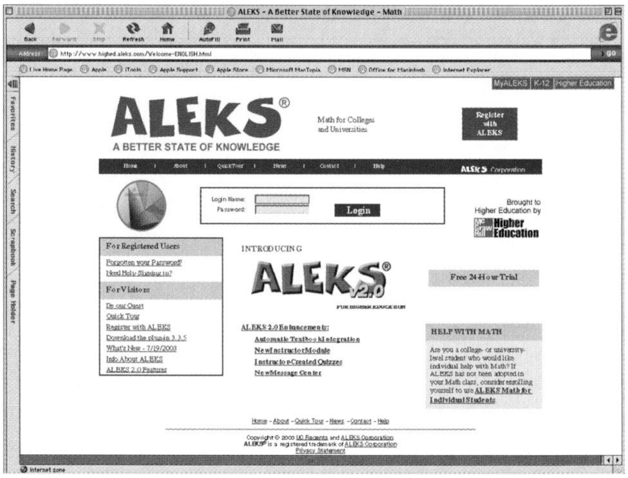

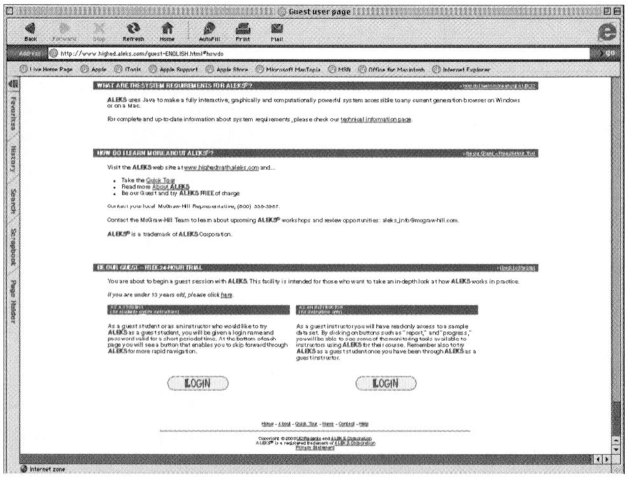

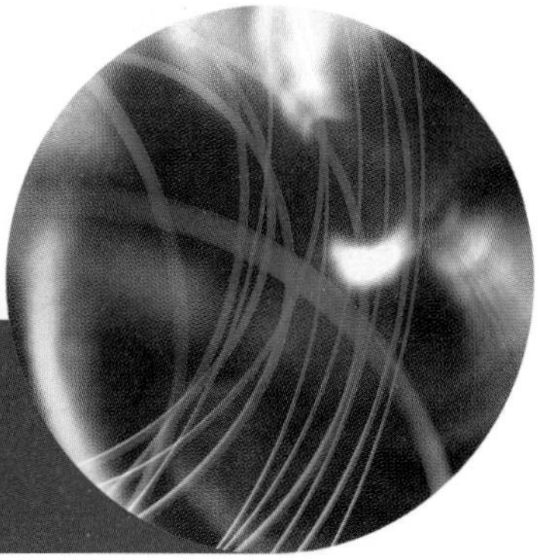

Applications Index

Chemistry

Construction

Consumer Applications

Environment

Investment

Science

Sports

Statistics/Demographics

Chapter 1

Real Numbers and Their Properties

It has been said that baseball is the "great American pastime." All of us who have played the game or who have only been spectators believe we understand the game. But do we realize that a pitcher must aim for an invisible three-dimensional target that is about 20 inches wide by 23 inches high by 17 inches deep and that a pitcher must throw so that the batter has difficulty hitting the ball? A curve ball may deflect 14 inches to skim over the outside corner of the plate, or a knuckle ball can break 11 inches off center when it is 20 feet from the plate and then curve back over the center of the plate.

The batter is trying to hit a rotating ball that can travel up to 120 miles per hour and must make split-second decisions about shifting his weight, changing his stride, and swinging the bat. The size of the bat each batter uses depends on his strengths, and pitchers in turn try to capitalize on a batter's weaknesses.

Millions of baseball fans enjoy watching this game of strategy and numbers. Many watch their favorite teams at the local ballparks, while others cheer for the home team on television. Of course, baseball fans are always interested in which team is leading the division and the number of games that their favorite team is behind the leader. Finding the number of games behind for each team in the division involves both arithmetic and algebra. Algebra provides the formula for finding games behind, and arithmetic is used to do the computations.

In Exercise 103 of Section 1.6 we will find the number of games behind for each team in the American League East.

1.1 The Real Numbers

In this Section

- The Integers
- The Rational Numbers
- The Number Line
- The Real Numbers
- Intervals of Real Numbers
- Absolute Value

The numbers that we use in algebra are called the real numbers. We start the discussion of the real numbers with some simpler sets of numbers.

The Integers

The most fundamental collection or **set** of numbers is the set of **counting numbers** or **natural numbers.** Of course, these are the numbers that we use for counting. The set of natural numbers is written in symbols as follows.

The Natural Numbers

$$\{1, 2, 3, \ldots\}$$

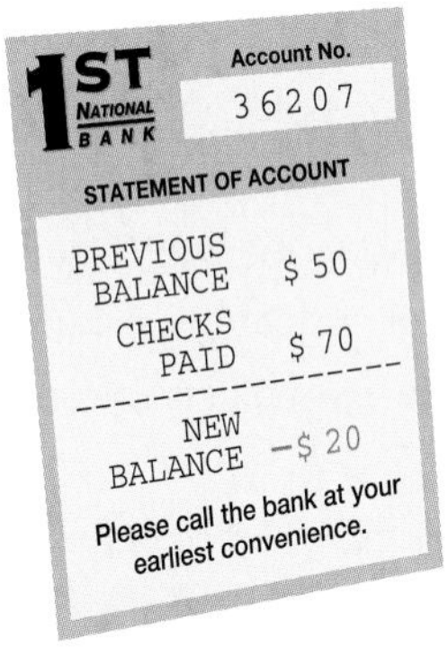

Figure 1.1

Braces, { }, are used to indicate a set of numbers. The three dots after 1, 2, and 3, which are read "and so on," mean that the pattern continues without end. There are infinitely many natural numbers.

The natural numbers, together with the number 0, are called the **whole numbers.** The set of whole numbers is written as follows.

The Whole Numbers

$$\{0, 1, 2, 3, \ldots\}$$

Although the whole numbers have many uses, they are not adequate for indicating losses or debts. A debt of \$20 can be expressed by the negative number -20 (negative twenty). See Fig. 1.1. When a thermometer reads 10 degrees below zero on a Fahrenheit scale, we say that the temperature is $-10°$F. See Fig. 1.2. The whole numbers together with the negatives of the counting numbers form the set of **integers.**

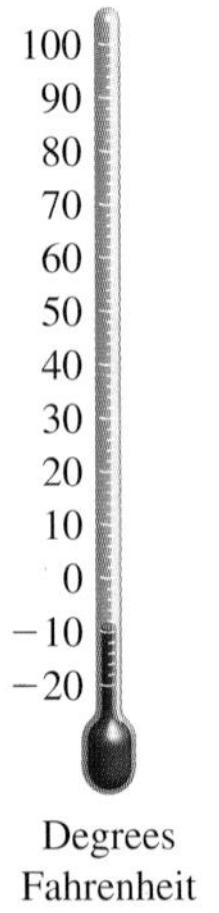

Figure 1.2

The Integers

$$\{\ldots, -3, -2, -1, 0, 1, 2, 3, \ldots\}$$

The Rational Numbers

A **rational number** is any number that can be expressed as a ratio (or quotient) of two integers. The set of rational numbers includes both the positive and negative fractions. We cannot list the rational numbers as easily as we listed the numbers in the other sets we have been discussing. So we write the set of rational numbers in symbols using **set-builder notation** as follows.

The Rational Numbers

$$\left\{\frac{a}{b} \,\middle|\, a \text{ and } b \text{ are integers, with } b \neq 0\right\}$$

↑ The set of ↑ such that ↑ conditions

Helpful Hint

Rational numbers are used for ratios. For example, if 2 out of 5 students surveyed attend summer school, then the ratio of students who attend summer school to the total number surveyed is 2/5. Note that the ratio 2/5 does not tell how many were surveyed or how many attend summer school.

We read this notation as "the set of numbers of the form $\frac{a}{b}$ such that a and b are integers, with $b \neq 0$." We rule out $b = 0$ because division by zero does not make sense and it is not a defined operation. Note how we use the letters a and b to represent numbers here. A letter used to represent some numbers is called a **variable.**

Examples of rational numbers are

$$\frac{3}{1},\quad \frac{5}{4},\quad -\frac{7}{10},\quad \frac{0}{6},\quad \frac{5}{1},\quad -\frac{77}{3},\quad \text{and}\quad \frac{-3}{-6}.$$

Note that we usually use simpler forms for some of these rational numbers. For instance, $\frac{3}{1} = 3$ and $\frac{0}{6} = 0$. The integers are rational numbers because any integer can be written with a denominator of 1.

If you divide the denominator into the numerator, then you can convert a rational number to decimal form. As a decimal, every rational number either repeats indefinitely $\left(\frac{1}{3} = 0.\overline{3} = 0.333\ldots\right)$ or terminates $\left(\frac{1}{8} = 0.125\right)$. The line over the 3 indicates that it repeats forever. The part that repeats can have more digits than the display of your calculator. In this case you will have to divide by hand to do the conversion. For example, try converting $\frac{11}{17}$ to a repeating decimal.

The Number Line

The number line is a diagram that helps us visualize numbers and their relationships to each other. A number line is like the scale on the thermometer in Fig. 1.2. To construct a number line, we draw a straight line and label any convenient point with the number 0. Now we choose any convenient length and use it to locate other points. Points to the right of 0 correspond to the positive numbers, and points to the left of 0 correspond to the negative numbers. Zero is neither positive nor negative. The number line is shown in Fig. 1.3.

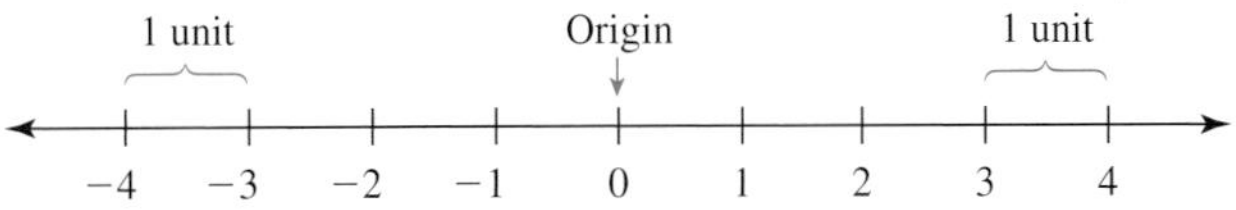

Figure 1.3

The numbers corresponding to the points on the line are called the **coordinates** of the points. The distance between two consecutive integers is called a **unit** and is the same for any two consecutive integers. The point with coordinate 0 is called the **origin.** The numbers on the number line increase in size from left to right. *When we compare the size of any two numbers, the larger number lies to the right of the smaller on the number line.* Zero is larger than any negative number and smaller than any positive number.

EXAMPLE 1

Comparing numbers on a number line

Determine which number is the larger in each given pair of numbers.

a) $-3, 2$ **b)** $0, -4$ **c)** $-2, -1$

Solution

a) The larger number is 2, because 2 lies to the right of -3 on the number line. In fact, any positive number is larger than any negative number.

b) The larger number is 0, because 0 lies to the right of -4 on the number line.

c) The larger number is -1, because -1 lies to the right of -2 on the number line.

Now do Exercises 7–16

The set of integers is illustrated or *graphed* in Fig. 1.4 by drawing a point for each integer. The three dots to the right and left below the number line and the blue arrows indicate that the numbers go on indefinitely in both directions.

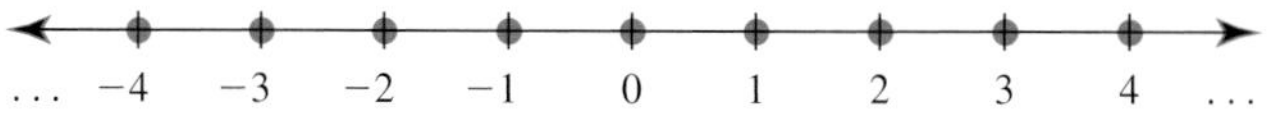

Figure 1.4

EXAMPLE 2

Graphing numbers on a number line

List the numbers described, and graph the numbers on a number line.

a) The whole numbers less than 4

b) The integers between 3 and 9

c) The integers greater than -3

Solution

a) The whole numbers less than 4 are 0, 1, 2, and 3. These numbers are shown in Fig. 1.5.

Figure 1.5

b) The integers between 3 and 9 are 4, 5, 6, 7, and 8. Note that 3 and 9 are not considered to be *between* 3 and 9. The graph is shown in Fig. 1.6.

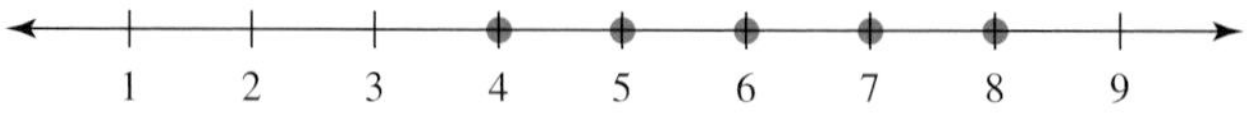

Figure 1.6

c) The integers greater than -3 are -2, -1, 0, 1, and so on. To indicate the continuing pattern, we use three dots on the graph shown in Fig. 1.7.

Figure 1.7

Now do Exercises 17–26

Study Tip

Start a personal library. This book as well as other books that you study from should be the basis for your library. You can also add books to your library at garage-sale prices when your bookstore sells its old texts. If you need to reference some material in the future, it is much easier to use a familiar book.

The Real Numbers

For every rational number there is a point on the number line. For example, the number $\frac{1}{2}$ corresponds to a point halfway between 0 and 1 on the number line, and $-\frac{5}{4}$ corresponds to a point one and one-quarter units to the left of 0, as shown in Fig. 1.8. Since there is a correspondence between numbers and points on the number line, the points are often referred to as numbers.

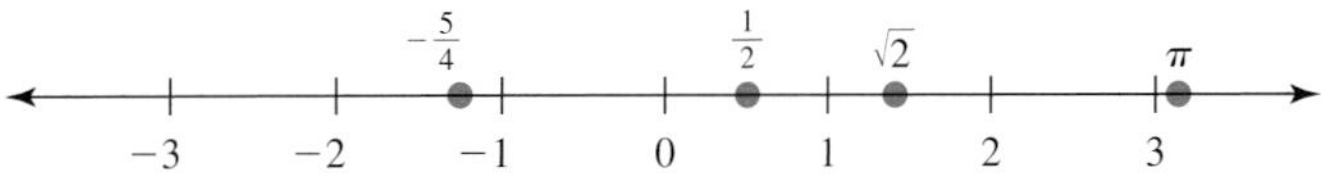

Figure 1.8

Calculator Close-Up

A calculator can give rational approximations for irrational numbers such as $\sqrt{2}$ and π.

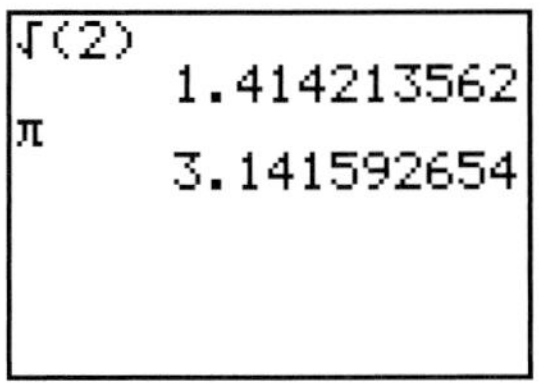

The calculator screens in this text may differ from the screen of the calculator model you use. If so, you may have to consult your manual to get the desired results.

The set of numbers that corresponds to *all* points on a number line is called the set of **real numbers** or R. A graph of the real numbers is shown on a number line by shading all points as in Fig. 1.9. All rational numbers are real numbers, but there are points on the number line that do not correspond to rational numbers. Those real numbers that are not rational are called **irrational.** An irrational number cannot be written as a ratio of integers. It can be shown that numbers such as $\sqrt{2}$ (the square root of 2) and π (Greek letter pi) are irrational. The number $\sqrt{2}$ is a number that can be multiplied by itself to obtain $2 (\sqrt{2} \cdot \sqrt{2} = 2)$. The number π is the ratio of the circumference and diameter of any circle. Irrational numbers are not as easy to represent as rational numbers. That is why we use symbols such as $\sqrt{2}$, $\sqrt{3}$, and π for irrational numbers. When we perform computations with irrational numbers, we sometimes use rational approximations for them. For example, $\sqrt{2} \approx 1.414$ and $\pi \approx 3.14$. The symbol $\approx$ means "is approximately equal to." Note that not all square roots are irrational. For example, $\sqrt{9} = 3$, because $3 \cdot 3 = 9$. We will deal with irrational numbers in greater depth when we discuss roots in Chapter 8.

Figure 1.9

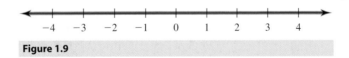

Figure 1.10 summarizes the sets of numbers that make up the real numbers, and shows the relationships between them.

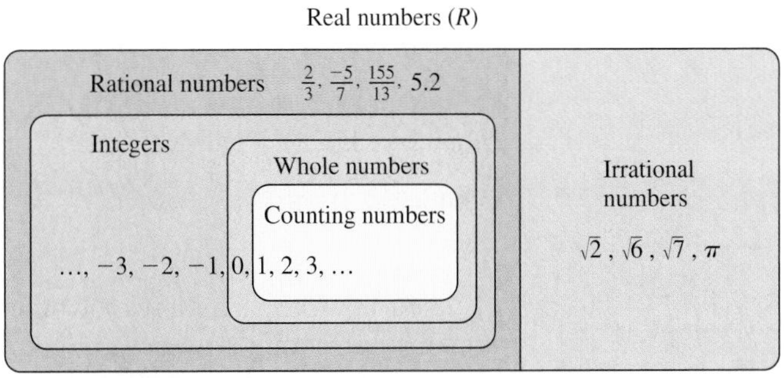

Figure 1.10

EXAMPLE 3

Types of numbers

Determine whether each statement is true or false.

a) Every rational number is an integer.

b) Every counting number is an integer.

c) Every irrational number is a real number.

Solution

a) False. For example, $\frac{1}{2}$ is a rational number that is not an integer.

b) True, because the integers consist of the counting numbers, the negatives of the counting numbers, and zero.

c) True, because the rational numbers together with the irrational numbers form the real numbers.

Now do Exercises 27–38

Intervals of Real Numbers

Retailers often have a sale for a certain *interval* of time. Between 6 A.M. and 8 A.M. you get a 20% discount. A **bounded** or finite interval of real numbers is the set of real numbers that are between two real numbers, which are called the **endpoints** of the interval. The endpoints may or may not belong to an interval. **Interval notation** is used to represent intervals of real numbers. In interval notation, parentheses are used to indicate that the endpoints do not belong to the interval and brackets indicate that the endpoints do belong to the interval. The following box shows the four types of finite intervals for two real numbers a and b, where a is less than b.

Finite Intervals

Verbal Description	Interval Notation	Graph
The set of real numbers between a and b	(a, b)	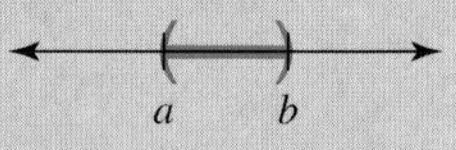
The set of real numbers between a and b inclusive	$[a, b]$	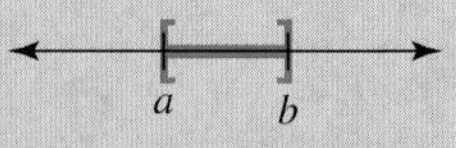
The set of real numbers greater than a and less than or equal to b	$(a, b]$	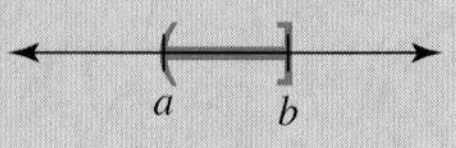
The set of real numbers greater than or equal to a and less than b	$[a, b)$	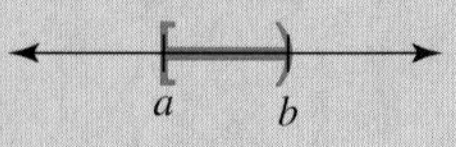

The interval [2, 5)

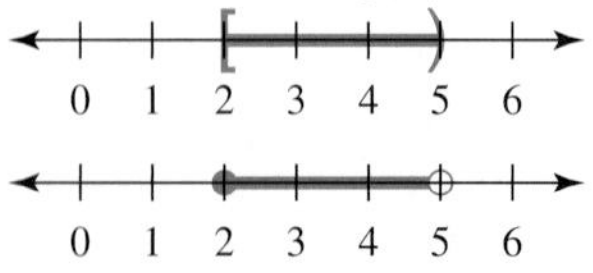

Figure 1.11

Note how the parentheses and brackets are used on the graph and in the interval notation. It is also common to draw the graph of an interval of real numbers using an open circle for an endpoint that does not belong to the interval and a closed circle for an endpoint that belongs to the interval. For example, see the graphs of the interval [2, 5) in Fig. 1.11.

In this text, graphs of intervals will be drawn with parentheses and brackets so that they agree with interval notation.

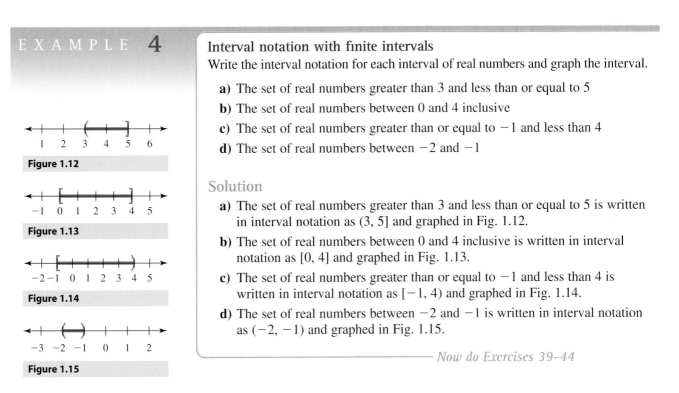

EXAMPLE 4

Interval notation with finite intervals

Write the interval notation for each interval of real numbers and graph the interval.

a) The set of real numbers greater than 3 and less than or equal to 5

b) The set of real numbers between 0 and 4 inclusive

c) The set of real numbers greater than or equal to -1 and less than 4

d) The set of real numbers between -2 and -1

Solution

a) The set of real numbers greater than 3 and less than or equal to 5 is written in interval notation as $(3, 5]$ and graphed in Fig. 1.12.

b) The set of real numbers between 0 and 4 inclusive is written in interval notation as $[0, 4]$ and graphed in Fig. 1.13.

c) The set of real numbers greater than or equal to -1 and less than 4 is written in interval notation as $[-1, 4)$ and graphed in Fig. 1.14.

d) The set of real numbers between -2 and -1 is written in interval notation as $(-2, -1)$ and graphed in Fig. 1.15.

Now do Exercises 39–44

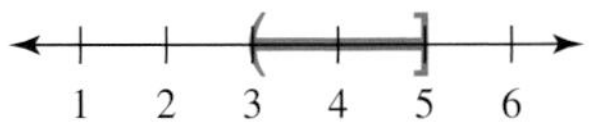

Figure 1.12

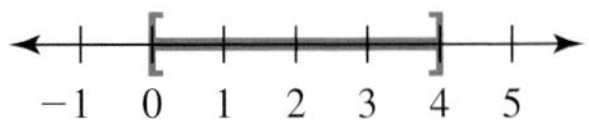

Figure 1.13

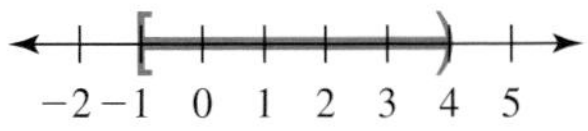

Figure 1.14

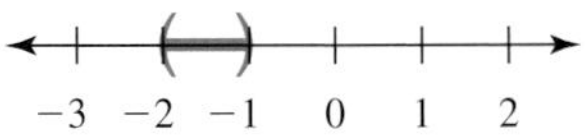

Figure 1.15

Some sales never end. After 8 A.M. all merchandise is 10% off. An **unbounded** or **infinite interval** of real numbers is missing at least one endpoint. It may extend infinitely far to the right or left on the number line. In this case the infinity symbol ∞ is used as an endpoint in the interval notation. Note that parentheses are always used next to ∞ or $-\infty$ in interval notation. The following box shows the five types of infinite intervals for a real number a.

Infinite Intervals

Verbal Description	Interval Notation	Graph
The set of real numbers greater than a	(a, ∞)	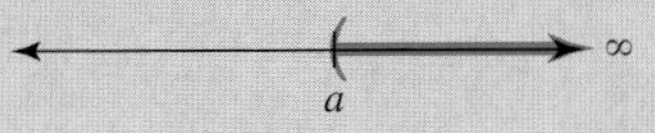
The set of real numbers greater than or equal to a	$[a, \infty)$	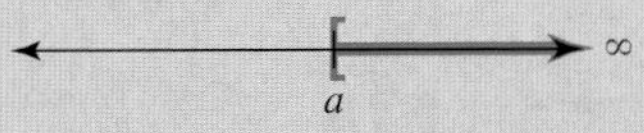
The set of real numbers less than a	$(-\infty, a)$	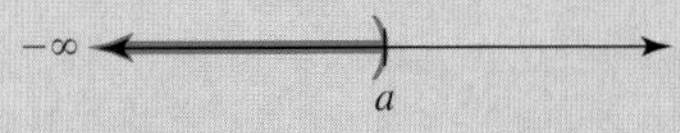
The set of real numbers less than or equal to a	$(-\infty, a]$	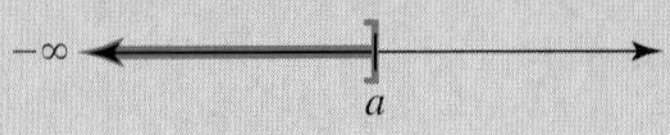
The set of all real numbers	$(-\infty, \infty)$	

EXAMPLE 5

Interval notation with infinite intervals

Write each interval of real numbers in interval notation and graph it.

a) The set of real numbers greater than or equal to 3

b) The set of real numbers less than -2

c) The set of real numbers greater than 2.5

Solution

a) The set of real numbers greater than or equal to 3 is written in interval notation as $[3, \infty)$ and graphed in Fig. 1.16.

Figure 1.16

b) The set of real numbers less than -2 is written in interval notation as $(-\infty, -2)$ and graphed in Fig. 1.17.

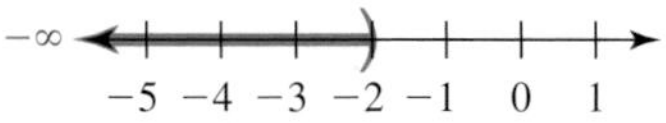

Figure 1.17

c) The set of real numbers greater than 2.5 is written in interval notation as $(2.5, \infty)$ and graphed in Fig. 1.18.

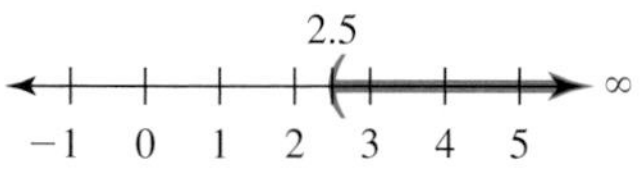

Figure 1.18

Now do Exercises 45–50

Absolute Value

The concept of absolute value will be used to define the basic operations with real numbers in Section 1.3. The **absolute value** of a number is the number's distance from 0 on the number line. For example, the numbers 5 and -5 are both five units away from 0 on the number line. So the absolute value of each of these numbers is 5. See Fig. 1.19. We write $|a|$ for "the absolute value of a." So

$$|5| = 5 \quad \text{and} \quad |-5| = 5.$$

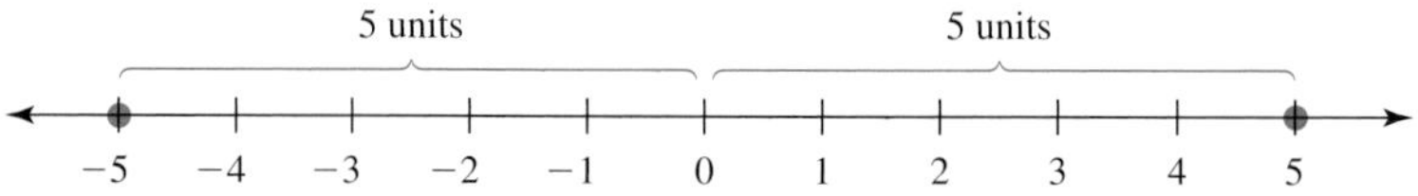

Figure 1.19

The notation $|a|$ represents distance, and distance is never negative. So $|a|$ is greater than or equal to zero for any real number a.

EXAMPLE 6

Finding absolute value

Evaluate.

a) $|3|$ **b)** $|-3|$ **c)** $|0|$

d) $\left|\frac{2}{3}\right|$ **e)** $|-0.39|$

Solution

a) $|3| = 3$ because 3 is three units away from 0.

b) $|-3| = 3$ because -3 is three units away from 0.

c) $|0| = 0$ because 0 is zero units away from 0.

Study Tip

Exercise sets are designed to gradually increase in difficulty. So start from the beginning and work lots of exercises. If you get stuck, go back and study the corresponding examples. If you are still stuck, move ahead to a new type of exercise.

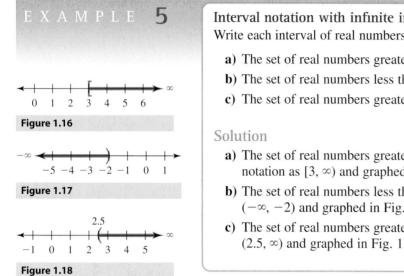

d) $\left|\frac{2}{3}\right| = \frac{2}{3}$

e) $|-0.39| = 0.39$

Now do Exercises 51–58

Two numbers that are located on opposite sides of zero and have the same absolute value are called **opposites** of each other. The numbers 5 and -5 are opposites of each other. We say that the opposite of 5 is -5 and the opposite of -5 is 5. The symbol "$-$" is used to indicate "opposite" as well as "negative." When the negative sign is used before a number, it should be read as "negative." When it is used in front of parentheses or a variable, it should be read as "opposite." For example, $-(5) = -5$ means "the opposite of 5 is negative 5," and $-(-5) = 5$ means "the opposite of negative 5 is 5." Zero does not have an opposite in the same sense as nonzero numbers. Zero is its own opposite. We read $-(0) = 0$ as the "the opposite of zero is zero."

In general, $-a$ means "the opposite of a." If a is positive, $-a$ is negative. If a is negative, $-a$ is positive. Opposites have the following property.

Opposite of an Opposite

For any real number a,

$$-(-a) = a.$$

Remember that we have defined $|a|$ to be the distance between 0 and a on the number line. Using opposites, we can give a symbolic definition of absolute value.

Absolute Value

$$|a| = \begin{cases} a & \text{if } a \text{ is positive or zero} \\ -a & \text{if } a \text{ is negative} \end{cases}$$

EXAMPLE 7

Using the symbolic definition of absolute value

Evaluate.

a) $|8|$

b) $|0|$

c) $|-8|$

Solution

a) If a is positive, then $|a| = a$. Since 8 is greater than 0, $|8| = 8$.

b) If a is 0, then $|a| = a$. So $|0| = 0$.

c) If a is negative, then $|a| = -a$. So $|-8| = -(-8) = 8$.

Now do Exercises 59–64

Warm-Ups

True or false? Explain your answer.

1. The natural numbers and the counting numbers are the same.
2. The number 8,134,562,877,565 is a counting number.
3. Zero is a counting number.
4. Zero is not a rational number.
5. The opposite of negative 3 is positive 3.
6. The absolute value of 4 is -4.
7. $-(-9) = 9$
8. The real number π is in the interval (3, 4).
9. Negative 6 is greater than negative 3.
10. Negative 5 is between 4 and 6.

1.1 Exercises

Reading and Writing *After reading this section write out the answers to these questions. Use complete sentences.*

1. What are the integers?
2. What are the rational numbers?
3. What is the difference between a rational and an irrational number?
4. What is a number line?
5. How do you know that one number is larger than another?
6. What is the ratio of the circumference and diameter of any circle?

Determine which number is the larger in each given pair of numbers. See Example 1.

7. $-3, 6$
8. $7, -10$
9. $0, -6$
10. $-8, 0$
11. $-3, -2$
12. $-5, -8$
13. $-12, -15$
14. $-13, -7$
15. $-2.9, -2.1$
16. $2.1, 2.9$

List the numbers described and graph them on a number line. See Example 2.

17. The counting numbers smaller than 6
18. The natural numbers larger than 4
19. The whole numbers smaller than 5
20. The integers between -3 and 3
21. The whole numbers between -5 and 5
22. The integers smaller than -1

23. The counting numbers larger than -4

24. The natural numbers between -5 and 7

25. The integers larger than $\frac{1}{2}$

26. The whole numbers smaller than $\frac{7}{4}$

Determine whether each statement is true or false. Explain your answer. See Example 3.

27. Every integer is a rational number.
28. Every counting number is a whole number.
29. Zero is a counting number.
30. Every whole number is a counting number.
31. The ratio of the circumference and diameter of a circle is an irrational number.
32. Every rational number can be expressed as a ratio of integers.
33. Every whole number can be expressed as a ratio of integers.
34. Some of the rational numbers are integers.
35. Some of the integers are natural numbers.
36. There are infinitely many rational numbers.
37. Zero is an irrational number.
38. Every irrational number is a real number.

Write each interval of real numbers in interval notation and graph it. See Example 4.

39. The set of real numbers between 0 and 1

40. The set of real numbers between 2 and 6

41. The set of real numbers between -2 and 2 inclusive

42. The set of real numbers between -3 and 4 inclusive

43. The set of real numbers greater than 0 and less than or equal to 5

44. The set of real numbers greater than or equal to -1 and less than 6

Write each interval of real numbers in interval notation and graph it. See Example 5.

45. The set of real numbers greater than 4

46. The set of real numbers greater than 2

47. The set of real numbers less than or equal to -1

48. The set of real numbers less than or equal to -4

49. The set of real numbers greater than or equal to 0

50. The set of real numbers greater than or equal to 6

Determine the values of the following. See Examples 6 and 7.

51. $|-6|$ **52.** $|4|$
53. $|0|$ **54.** $|2|$
55. $|7|$ **56.** $|-7|$
57. $|-9|$ **58.** $|-2|$
59. $|-45|$ **60.** $|-30|$
61. $\left|\frac{3}{4}\right|$ **62.** $\left|-\frac{1}{2}\right|$
63. $|-5.09|$ **64.** $|0.00987|$

Select the smaller number in each given pair of numbers.

65. $-16, 9$ **66.** $-12, -7$
67. $-\frac{5}{2}, -\frac{9}{4}$ **68.** $\frac{5}{8}, \frac{6}{7}$
69. $|-3|, 2$ **70.** $|-6|, 0$
71. $|-4|, 3$ **72.** $|5|, -4$

Which number in each given pair has the larger absolute value?

73. $-5, -9$ **74.** $-12, -8$
75. $16, -9$ **76.** $-12, 7$

Determine which number in each pair is closer to 0 on the number line.

77. $-4, -5$ **78.** $-8.1, 7.9$
79. $-2.01, -1.99$ **80.** $2.01, 1.99$

81. $-75, 74$ **82.** $-75, -74$

What is the distance on the number line between 0 and each of the following numbers?

83. 5.25 **84.** 4.2 **85.** -40

86. -33 **87.** $-\frac{1}{2}$ **88.** $-\frac{1}{3}$

Consider the following nine integers:

$$-4, -3, -2, -1, 0, 1, 2, 3, 4$$

89. Which of these integers has an absolute value equal to 3?

90. Which of these integers has an absolute value equal to 0?

91. Which of these integers has an absolute value greater than 2?

92. Which of these integers has an absolute value greater than 1?

93. Which of these integers has an absolute value less than 2?

94. Which of these integers has an absolute value less than 4?

Write the interval notation for the interval of real numbers shown in each graph.

95. 

96. 

97. 

98. 

99. 

100. 

True or false? Explain your answer.

101. If we add the absolute values of -3 and -5, we get 8.

102. If we multiply the absolute values of -2 and 5, we get 10.

103. The absolute value of any negative number is greater than 0.

104. The absolute value of any positive number is less than 0.

105. The absolute value of -9 is larger than the absolute value of 6.

106. The absolute value of 12 is larger than the absolute value of -11.

Getting More Involved

107. ***Writing***

Find a real-life question for which the answer is a rational number that is not an integer.

108. ***Exploration***

a) Find a rational number between $\frac{1}{3}$ and $\frac{1}{4}$.
b) Find a rational number between -3.205 and -3.114.
c) Find a rational number between $\frac{2}{3}$ and 0.6667.
d) Explain how to find a rational number between any two given rational numbers.

109. ***Discussion***

Suppose that a is a negative real number. Determine whether each of the following is positive or negative, and explain your answer.

a) $-a$ **b)** $|-a|$ **c)** $-|a|$ **d)** $-(-a)$ **e)** $-|-a|$

110. ***Discussion***

Determine whether each number listed in the table below is a member of each set listed on the side of the table. For example, $\frac{1}{2}$ is a real number and a rational number. So check marks are placed in those two cells of the table.

	$\frac{1}{2}$	-2	π	$\sqrt{3}$	$\sqrt{9}$	6	0	$-\frac{7}{3}$
Real	✓							
Irrational								
Rational	✓							
Integer								
Whole								
Counting								

1.2 Fractions

In this Section

- Equivalent Fractions
- Multiplying Fractions
- Dividing Fractions
- Adding and Subtracting Fractions
- Fractions, Decimals, and Percents
- Applications

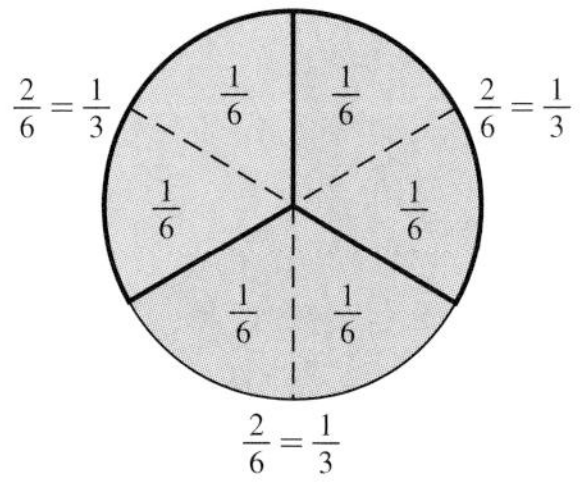

Figure 1.20

In this section and Sections 1.3 and 1.4 we will discuss operations performed with real numbers. We begin by reviewing operations with fractions. Note that this section on fractions is not an entire arithmetic course. We are simply reviewing selected fraction topics that will be used in this text.

Equivalent Fractions

If a pizza is cut into 3 equal pieces and you eat 2, you have eaten $\frac{2}{3}$ of the pizza. If the pizza is cut into 6 equal pieces and you eat 4, you have still eaten 2 out of every 3 pieces. So the fraction $\frac{4}{6}$ is considered **equal** or **equivalent** to $\frac{2}{3}$. See Fig. 1.20. Every fraction can be written in infinitely many equivalent forms. Consider the following equivalent forms of $\frac{2}{3}$:

$$\frac{2}{3} = \frac{4}{6} = \frac{6}{9} = \frac{8}{12} = \frac{10}{15} = \ldots$$

The three dots mean "and so on."

Notice that each equivalent form of $\frac{2}{3}$ can be obtained by multiplying the numerator (top number) and denominator (bottom number) of $\frac{2}{3}$ by a nonzero number. For example,

$$\frac{2}{3} = \frac{2 \cdot 5}{3 \cdot 5} = \frac{10}{15}.$$ The raised dot indicates multiplication.

Converting a fraction into an equivalent fraction with a larger denominator is called **building up** the fraction. A fraction can be built up by multiplying the numerator and denominator by the same number, or by using multiplication of fractions (which will be discussed later in this section). For example, to build up $\frac{2}{3}$ to a denominator of 15 we could multiply $\frac{2}{3}$ by the number 1, using $\frac{5}{5}$ as an equivalent form of 1:

$$\frac{2}{3} = \frac{2}{3} \cdot 1 = \frac{2}{3} \cdot \frac{5}{5} = \frac{10}{15}$$

Building Up Fractions

If $b \neq 0$ and $c \neq 0$, then

$$\frac{a}{b} = \frac{a \cdot c}{b \cdot c}.$$

Multiplying the numerator and denominator of a fraction by a nonzero number changes the fraction's appearance but not its value.

EXAMPLE 1

Building up fractions

Build up each fraction so that it is equivalent to the fraction with the indicated denominator.

a) $\frac{3}{4} = \frac{?}{28}$ **b)** $\frac{5}{3} = \frac{?}{30}$

Solution

a) Because $4 \cdot 7 = 28$, we multiply both the numerator and denominator by 7:

$$\frac{3}{4} = \frac{3 \cdot 7}{4 \cdot 7} = \frac{21}{28}$$

b) Because $3 \cdot 10 = 30$, we multiply both the numerator and denominator by 10:

$$\frac{5}{3} = \frac{5 \cdot 10}{3 \cdot 10} = \frac{50}{30}$$

Now do Exercises 7–18

The method for building up fractions shown in Example 1 will be used again on rational expressions in Chapter 7. So it is good to use this method and show the details. The same goes for the method of reducing fractions that is coming next.

Converting a fraction to an equivalent fraction with a smaller denominator is called **reducing** the fraction. For example, to reduce $\frac{10}{15}$, we *factor* 10 as $2 \cdot 5$ and 15 as $3 \cdot 5$, and then divide out the *common factor* 5:

$$\frac{10}{15} = \frac{2 \cdot \cancel{5}}{3 \cdot \cancel{5}} = \frac{2}{3}$$

The fraction $\frac{2}{3}$ cannot be reduced further because the numerator 2 and the denominator 3 have no factors (other than 1) in common. So we say that $\frac{2}{3}$ is in **lowest terms.**

Reducing Fractions

If $b \neq 0$ and $c \neq 0$, then

$$\frac{a \cdot c}{b \cdot c} = \frac{a}{b}.$$

Dividing the numerator and denominator of a fraction by a nonzero number changes the fraction's appearance but not its value.

EXAMPLE 2

Reducing fractions

Reduce each fraction to lowest terms.

a) $\frac{15}{24}$ **b)** $\frac{42}{30}$

c) $\frac{13}{26}$ **d)** $\frac{35}{7}$

Calculator Close-Up

To reduce a fraction to lowest terms using a graphing calculator, display the fraction and use the fraction feature.

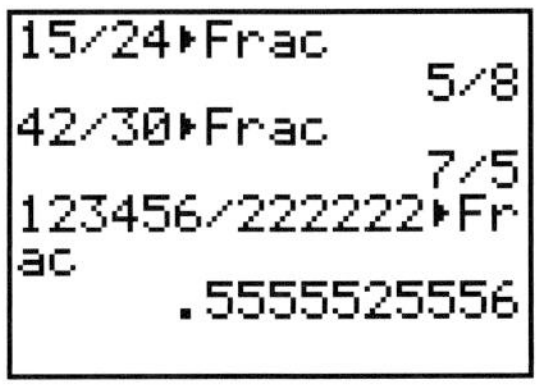

If the fraction is too complicated, the calculator will return a decimal equivalent instead of reducing it.

Solution

For each fraction, factor the numerator and denominator and then divide by the common factor:

a) $\dfrac{15}{24} = \dfrac{\cancel{3} \cdot 5}{\cancel{3} \cdot 8} = \dfrac{5}{8}$ **b)** $\dfrac{42}{30} = \dfrac{7 \cdot \cancel{6}}{5 \cdot \cancel{6}} = \dfrac{7}{5}$

c) $\dfrac{13}{26} = \dfrac{1 \cdot \cancel{13}}{2 \cdot \cancel{13}} = \dfrac{1}{2}$ The number 1 in the numerator is essential.

d) $\dfrac{35}{7} = \dfrac{5 \cdot \cancel{7}}{1 \cdot \cancel{7}} = \dfrac{5}{1} = 5$

Now do Exercises 19–34

Strategy for Obtaining Equivalent Fractions

Equivalent fractions can be obtained by multiplying or dividing the numerator and denominator by the same nonzero number.

Multiplying Fractions

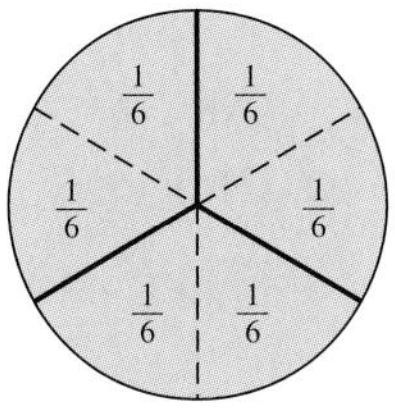

Figure 1.21

Suppose a pizza is cut into three equal pieces. If you eat $\frac{1}{2}$ of one piece, you have eaten $\frac{1}{6}$ of the pizza. See Fig. 1.21. You can obtain $\frac{1}{6}$ by multiplying $\frac{1}{2}$ and $\frac{1}{3}$:

$$\frac{1}{2} \cdot \frac{1}{3} = \frac{1 \cdot 1}{2 \cdot 3} = \frac{1}{6}$$

This example illustrates the definition of multiplication of fractions. To multiply two fractions, we multiply their numerators and multiply their denominators.

Multiplication of Fractions

If $b \neq 0$ and $d \neq 0$, then

$$\frac{a}{b} \cdot \frac{c}{d} = \frac{a \cdot c}{b \cdot d}.$$

EXAMPLE 3

Multiplying fractions

Find the product, $\frac{2}{3} \cdot \frac{5}{8}$.

Solution

Multiply the numerators and the denominators:

$$\frac{2}{3} \cdot \frac{5}{8} = \frac{10}{24}$$

$$= \frac{\cancel{2} \cdot 5}{\cancel{2} \cdot 12} \quad \text{Factor the numerator and denominator.}$$

$$= \frac{5}{12} \quad \text{Divide out the common factor 2.}$$

Now do Exercises 35–40

It is usually easier to reduce before multiplying, as shown in Example 4.

EXAMPLE 4

Reducing before multiplying

Find the indicated products.

a) $\frac{1}{3} \cdot \frac{3}{4}$ **b)** $\frac{4}{5} \cdot \frac{15}{22}$

Solution

a) $\frac{1}{3} \cdot \frac{3}{4} = \frac{1}{\cancel{3}} \cdot \frac{\cancel{3}}{4} = \frac{1}{4}$

b) Factor the numerators and denominators, and then divide out the common factors before multiplying:

$$\frac{4}{5} \cdot \frac{15}{22} = \frac{2 \cdot \cancel{2}}{\cancel{5}} \cdot \frac{3 \cdot \cancel{5}}{\cancel{2} \cdot 11} = \frac{6}{11}$$

Now do Exercises 41–46

Calculator Close-Up

A graphing calculator can multiply fractions and get fractional answers using the fraction feature. Note how a mixed number is written on a graphing calculator.

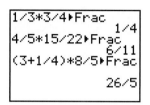

```
1/3*3/4▸Frac
             1/4
4/5*15/22▸Frac
            6/11
(3+1/4)*8/5▸Frac
            26/5
```

Dividing Fractions

Suppose that a pizza is cut into three pieces. If one piece is divided between two people $\left(\frac{1}{3} \div 2\right)$, then each of these two people gets $\frac{1}{6}$ of the pizza. Of course $\frac{1}{3}$ times $\frac{1}{2}$ is also $\frac{1}{6}$. So dividing by 2 is equivalent to multiplying by $\frac{1}{2}$. In symbols:

$$\frac{1}{3} \div 2 = \frac{1}{3} \div \frac{2}{1} = \frac{1}{3} \cdot \frac{1}{2} = \frac{1}{6}$$

The pizza example illustrates the general rule for dividing fractions.

Division of Fractions

If $b \neq 0$, $c \neq 0$, and $d \neq 0$, then

$$\frac{a}{b} \div \frac{c}{d} = \frac{a}{b} \cdot \frac{d}{c}.$$

In general if $m \div n = p$, then n is called the **divisor** and p (the result of the division) is called the **quotient** of m and n. We also refer to $m \div n$ and $\frac{m}{n}$ as the quotient of m and n. So in words, *to find the quotient of two fractions we invert the divisor and multiply.*

EXAMPLE 5

Dividing fractions

Find the indicated quotients.

a) $\frac{1}{3} \div \frac{7}{6}$ **b)** $\frac{2}{3} \div 5$

c) $\frac{3}{8} \div \frac{3}{2}$

Calculator Close-Up

When the divisor is a fraction on a graphing calculator, it must be in parentheses. A different result is obtained without using parentheses. Note that when the divisor is a whole number, parentheses are not necessary.

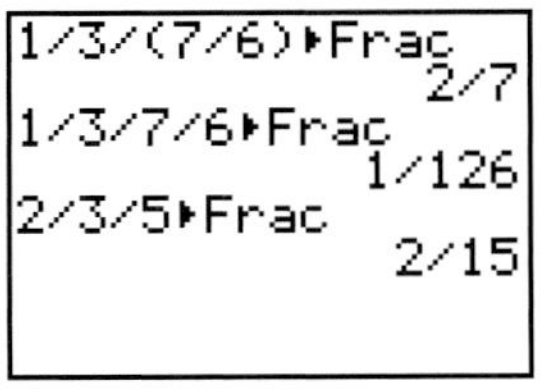

Try these computations on your calculator.

Solution

In each case we invert the divisor (the number on the right) and multiply.

a) $\frac{1}{3} \div \frac{7}{6} = \frac{1}{3} \cdot \frac{6}{7}$ Invert the divisor.

$= \frac{1}{\not{3}} \cdot \frac{2 \cdot \not{3}}{7}$ Reduce.

$= \frac{2}{7}$ Multiply.

b) $\frac{2}{3} \div 5 = \frac{2}{3} \div \frac{5}{1} = \frac{2}{3} \cdot \frac{1}{5} = \frac{2}{15}$

c) $\frac{3}{8} \div \frac{3}{2} = \frac{3}{8} \cdot \frac{2}{3} = \frac{\not{3} \cdot 1}{4 \cdot \not{2}} \cdot \frac{\not{2}}{\not{3}} = \frac{1}{4}$

Now do Exercises 47–56

Adding and Subtracting Fractions

To understand addition and subtraction of fractions, again consider the pizza that is cut into six equal pieces as shown in Fig. 1.22. If you eat $\frac{3}{6}$ and your friend eats $\frac{2}{6}$, together you have eaten $\frac{5}{6}$ of the pizza. Similarly, if you remove $\frac{1}{6}$ from $\frac{6}{6}$ you have $\frac{5}{6}$ left. To add or subtract fractions with identical denominators, we add or subtract their numerators and write the result over the common denominator.

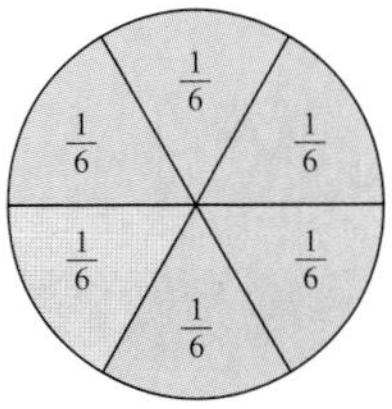

$\frac{3}{6} + \frac{2}{6} = \frac{5}{6}$

Figure 1.22

Addition and Subtraction of Fractions

If $b \neq 0$, then

$$\frac{a}{b} + \frac{c}{b} = \frac{a+c}{b} \quad \text{and} \quad \frac{a}{b} - \frac{c}{b} = \frac{a-c}{b}.$$

An **improper fraction** is a fraction in which the numerator is larger than the denominator. For example, $\frac{7}{6}$ is an improper fraction. A **mixed number** is a natural number plus a fraction, with the plus sign removed. For example, $1\frac{1}{6}$ (or $1 + \frac{1}{6}$) is a mixed number. Since $1 + \frac{1}{6} = \frac{6}{6} + \frac{1}{6} = \frac{7}{6}$, we have $1\frac{1}{6} = \frac{7}{6}$.

EXAMPLE 6

Adding and subtracting fractions

Perform the indicated operations.

a) $\frac{1}{7} + \frac{2}{7}$

b) $\frac{7}{10} - \frac{3}{10}$

Helpful Hint

A good way to remember that you need common denominators for addition is to think of a simple example. If you own 1/3 share of a car wash and your spouse owns 1/3, then together you own 2/3 of the business.

Solution

a) $\frac{1}{7} + \frac{2}{7} = \frac{3}{7}$

b) $\frac{7}{10} - \frac{3}{10} = \frac{4}{10} = \frac{\not{2} \cdot 2}{\not{2} \cdot 5} = \frac{2}{5}$

Now do Exercises 57–60

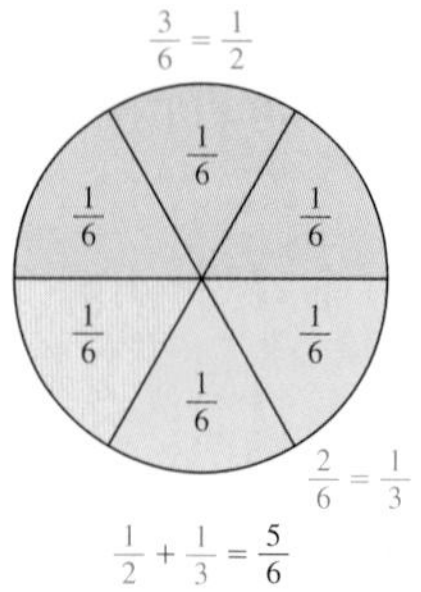

Figure 1.23

To add or subtract fractions with different denominators, we must convert them to equivalent fractions with the same denominator and then add or subtract. For example, to add $\frac{1}{2}$ and $\frac{1}{3}$, we build up each fraction to a denominator of 6. See Fig. 1.23. Since $\frac{1}{2} = \frac{3}{6}$ and $\frac{1}{3} = \frac{2}{6}$, we have

$$\frac{1}{2} + \frac{1}{3} = \frac{3}{6} + \frac{2}{6} = \frac{5}{6}.$$

The smallest number that is a multiple of the denominators of two or more fractions is called the **least common denominator (LCD).** So 6 is the LCD for $\frac{1}{2}$ and $\frac{1}{3}$. Note that we obtained the LCD 6 by examining Fig. 1.23. We must have a more systematic way.

The procedure for finding the LCD is based on factors. For example, to find the LCD for the denominators 6 and 9, factor 6 and 9 as $6 = 2 \cdot 3$ and $9 = 3 \cdot 3$. To obtain a multiple of both 6 and 9 the number must have two 3's as factors and one 2. So the LCD for 6 and 9 is $2 \cdot 3 \cdot 3$ or 18. If any number is omitted from $2 \cdot 3 \cdot 3$, we will not have a multiple of both 6 and 9. So each factor found in either 6 or 9 appears in the LCD the maximum number of times that it appears in either 6 or 9. The general strategy follows.

Helpful Hint

The *least* common denominator is *greater than* or equal to all of the denominators, because they must all divide into the LCD.

Strategy for Finding the LCD

1. Factor each denominator completely.
2. The LCD contains each distinct factor the maximum number of times that it occurs in any of the denominators.

Note that a **prime number** is a number 2 or larger that has no factors other than itself and 1. If a denominator is prime (such as 2, 3, 5, 7, 11) then we do not factor it. A number is **factored completely** when it is written as a product of prime numbers.

EXAMPLE 7

Adding and subtracting fractions

Perform the indicated operations.

a) $\frac{3}{4} + \frac{1}{6}$ **b)** $\frac{1}{3} - \frac{1}{12}$

c) $\frac{7}{12} + \frac{5}{18}$ **d)** $2\frac{1}{3} + \frac{5}{9}$

Solution

a) First factor the denominators as $4 = 2 \cdot 2$ and $6 = 2 \cdot 3$. Since 2 occurs twice in 4 and once in 6, it appears twice in the LCD. Since 3 appears once in 6 and not at all in 4, it appears once in the LCD. So the LCD is $2 \cdot 2 \cdot 3$

Study Tip

Read the material in the text before it is discussed in class, even if you do not totally understand it. The classroom discussion will be the second time you have seen the material and it will be easier to question points that you do not understand.

or 12. Now build up each denominator to 12:

$$\frac{3}{4}+\frac{1}{6}=\frac{3\cdot 3}{4\cdot 3}+\frac{1\cdot 2}{6\cdot 2}\quad\text{Build up each denominator to 12.}$$
$$=\frac{9}{12}+\frac{2}{12}\quad\text{Simplify.}$$
$$=\frac{11}{12}\quad\text{Add.}$$

b) The denominators are 12 and 3. Factor 12 as $12 = 2 \cdot 6 = 2 \cdot 2 \cdot 3$. Since 3 is a prime number we do not factor it. Since 2 occurs twice in 12 and not at all in 3, it appears twice in the LCD. Since 3 occurs once in 3 and once in 12, 3 appears once in the LCD. The LCD is $2 \cdot 2 \cdot 3$ or 12. So we must build up $\frac{1}{3}$ to have a denominator of 12:

$$\frac{1}{3}-\frac{1}{12}=\frac{1\cdot 4}{3\cdot 4}-\frac{1}{12}\quad\text{Build up the first fraction to the LCD.}$$
$$=\frac{4}{12}-\frac{1}{12}\quad\text{Simplify.}$$
$$=\frac{3}{12}\quad\text{Subtract.}$$
$$=\frac{1}{4}\quad\text{Reduce to lowest terms.}$$

c) Since $12 = 2 \cdot 6 = 2 \cdot 2 \cdot 3$ and $18 = 2 \cdot 9 = 2 \cdot 3 \cdot 3$, the factor 2 appears twice in the LCD and the factor 3 appears twice in the LCD. So the LCD is $2 \cdot 2 \cdot 3 \cdot 3$ or 36:

$$\frac{7}{12}+\frac{5}{18}=\frac{7\cdot 3}{12\cdot 3}+\frac{5\cdot 2}{18\cdot 2}\quad\text{Build up each denominator to 36.}$$
$$=\frac{21}{36}+\frac{10}{36}\quad\text{Simplify.}$$
$$=\frac{31}{36}\quad\text{Add.}$$

d) To perform addition with the mixed number $2\frac{1}{3}$, first convert it into an improper fraction: $2\frac{1}{3} = 2 + \frac{1}{3} = \frac{6}{3} + \frac{1}{3} = \frac{7}{3}$.

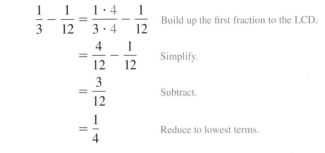

$$2\frac{1}{3}+\frac{5}{9}=\frac{7}{3}+\frac{5}{9}\quad\text{Write } 2\tfrac{1}{3} \text{ as an improper fraction.}$$
$$=\frac{7\cdot 3}{3\cdot 3}+\frac{5}{9}\quad\text{The LCD is 9.}$$
$$=\frac{21}{9}+\frac{5}{9}\quad\text{Simplify.}$$
$$=\frac{26}{9}\quad\text{Add.}$$

Note that $\frac{1}{3} + \frac{5}{9} = \frac{3}{9} + \frac{5}{9} = \frac{8}{9}$. Then add on the 2 to get $2\frac{8}{9}$, which is the same as $\frac{26}{9}$.

Now do Exercises 61–72

Helpful Hint

Recall the *place value* for decimal numbers:

```
tenths
 ┌hundredths
 │┌thousandths
 ││┌ten thousandths
0.2635
```

So $0.2635 = \frac{2635}{10{,}000}$.

Fractions, Decimals, and Percents

In the decimal number system, fractions with a denominator of 10, 100, 1000, and so on are written as decimal numbers. For example,

$$\frac{3}{10} = 0.3, \quad \frac{25}{100} = 0.25, \quad \text{and} \quad \frac{5}{1000} = 0.005.$$

Fractions with a denominator of 100 are often written as percents. Think of the percent symbol (%) as representing the denominator of 100. For example,

$$\frac{25}{100} = 25\%, \quad \frac{5}{100} = 5\%, \quad \text{and} \quad \frac{300}{100} = 300\%.$$

Example 8 illustrates further how to convert from any one of the forms (fraction, decimal, percent) to the others.

EXAMPLE 8

Changing forms

Convert each given fraction, decimal, or percent into its other two forms.

a) $\frac{1}{5}$ **b)** 6% **c)** 0.1

Solution

a) $\frac{1}{5} = \frac{1 \cdot 20}{5 \cdot 20} = \frac{20}{100} = 20\%$ and $\frac{1}{5} = \frac{1 \cdot 2}{5 \cdot 2} = \frac{2}{10} = 0.2$

So $\frac{1}{5} = 0.2 = 20\%$. Note that a fraction can also be converted to a decimal by dividing the denominator into the numerator with long division.

b) $6\% = \frac{6}{100} = 0.06$ and $\frac{6}{100} = \frac{\not{2} \cdot 3}{\not{2} \cdot 50} = \frac{3}{50}$

So $6\% = 0.06 = \frac{3}{50}$.

c) $0.1 = \frac{1}{10} = \frac{1 \cdot 10}{10 \cdot 10} = \frac{10}{100} = 10\%$

So $0.1 = \frac{1}{10} = 10\%$.

Now do Exercises 73–84

Calculator Close-Up

A calculator can convert fractions to decimals and decimals to fractions. The calculator shown here converts the terminating decimal 0.333333333333 into 1/3 even though 1/3 is a repeating decimal with infinitely many threes after the decimal point.

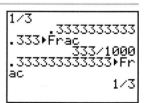

```
1/3
          .3333333333
.333▸Frac
             333/1000
.333333333333▸Fr
ac
                  1/3
```

Applications

The dimensions for lumber used in construction are usually given in fractions. For example, a 2 × 4 stud used for framing a wall is actually $1\frac{1}{2}$ in. by $3\frac{1}{2}$ in. by $92\frac{5}{8}$ in. A 2 × 12 floor joist is actually $1\frac{1}{2}$ in. by $11\frac{1}{2}$ in.

Math *at Work* Stock Price Analysis

Stock market analysts use mathematics daily to evaluate the potential success of a stock based on its financial statements and its current performance. Each analyst has a philosophy of investing. If an analyst is working for a mutual fund that specializes in retirement investing for clients with a lengthy time horizon, the analyst may recommend higher-risk stocks. If the client base is older and has a shorter time horizon, the analyst may recommend more secure investments.

There are hundreds of ratios and formulas that a stock market analyst uses to estimate the value of a stock. Two popular ones are the capital asset pricing model (CAPM) and the price/earnings ratio (P/E). The CAPM is used to assess the price of a stock in relation to general movements in the stock market whereas the P/E ratio is used to compare the price of one stock to others in the same industry.

Using CAPM a stock's price P is determined by $P = A + BM$, where A is the stock's variance, B is the stock's fluctuation in relation to the market, and M is the market level. For example, a stock trading at \$10.50 on the New York Stock Exchange has a variance of 3.24 and fluctuation of 0.001058 using the Dow Jones Industrial Average. If the Dow is at 9242, then $P = 3.24 + 0.001058(9242) \approx 13.02$. So the stock is worth \$13.02 and is a good buy at \$10.50. If the company has earned \$1.53 per share, then $\text{P/E} = 10.50/1.53 \approx 6.9$. If other stocks in the same industry have higher P/E ratios, then this stock is a good buy.

Since there are hundreds of ways to analyze a stock and all analysts have access to the same data, the analysts must decide which data are most important. The analyst must also look beyond data and formulas to determine whether to buy a stock.

EXAMPLE 9 Framing a two-story house

In framing a two-story house, a carpenter uses a 2×4 shoe, a wall stud, two 2×4 plates, then 2×12 floor joists, and a $\frac{3}{4}$-in. plywood floor, before starting the second level. Use the dimensions in Fig. 1.24 to find the total height of the framing shown.

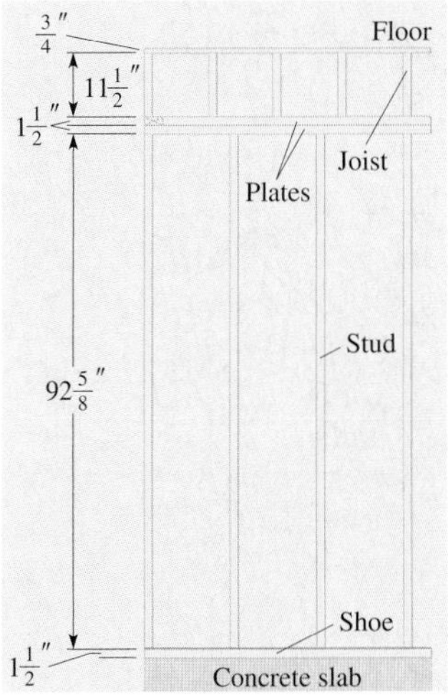

Figure 1.24

Solution

We can find the total height using multiplication and addition:

$$
\begin{aligned}
3 \cdot 1\frac{1}{2} + 92\frac{5}{8} + 11\frac{1}{2} + \frac{3}{4} &= 4\frac{1}{2} + 92\frac{5}{8} + 11\frac{1}{2} + \frac{3}{4} \\
&= 4\frac{4}{8} + 92\frac{5}{8} + 11\frac{4}{8} + \frac{6}{8} \\
&= 107\frac{19}{8} \\
&= 107 + \frac{16}{8} + \frac{3}{8} = 107 + 2 + \frac{3}{8} = 109\frac{3}{8}
\end{aligned}
$$

The total height of the framing shown is $109\frac{3}{8}$ in.

Now do Exercises 109–112

Warm-Ups

True or false? Explain your answer.

1. Every fraction is equal to infinitely many equivalent fractions.
2. The fraction $\frac{8}{12}$ is equivalent to the fraction $\frac{4}{6}$.
3. The fraction $\frac{8}{12}$ reduced to lowest terms is $\frac{4}{6}$.
4. $\frac{1}{2} \cdot \frac{2}{3} = \frac{1}{3}$
5. $\frac{1}{2} \cdot \frac{3}{5} = \frac{3}{10}$
6. $\frac{1}{2} \cdot \frac{6}{5} = \frac{6}{10}$
7. $\frac{1}{2} \div 3 = \frac{1}{6}$
8. $5 \div \frac{1}{2} = 10$
9. $\frac{1}{2} + \frac{1}{4} = \frac{2}{6}$
10. $2 - \frac{1}{2} = \frac{3}{2}$

1.2 Exercises

Reading and Writing *After reading this section write out the answers to these questions. Use complete sentences.*

1. What are equivalent fractions?
2. How can you find all fractions that are equivalent to a given fraction?
3. What does it mean to reduce a fraction to lowest terms?
4. For which operations with fractions are you required to have common denominators? Why?
5. How do you convert a fraction to a decimal?
6. How do you convert a percent to a fraction?

Build up each fraction or whole number so that it is equivalent to the fraction with the indicated denominator. See Example 1.

7. $\frac{3}{4} = \frac{?}{8}$
8. $\frac{5}{7} = \frac{?}{21}$
9. $\frac{8}{3} = \frac{?}{12}$
10. $\frac{7}{2} = \frac{?}{8}$
11. $5 = \frac{?}{2}$
12. $9 = \frac{?}{3}$
13. $\frac{3}{4} = \frac{?}{100}$
14. $\frac{1}{2} = \frac{?}{100}$
15. $\frac{3}{10} = \frac{?}{100}$
16. $\frac{2}{5} = \frac{?}{100}$
17. $\frac{5}{3} = \frac{?}{42}$
18. $\frac{5}{7} = \frac{?}{98}$

Reduce each fraction to lowest terms. See Example 2.

19. $\frac{3}{6}$
20. $\frac{2}{10}$
21. $\frac{12}{18}$
22. $\frac{30}{40}$
23. $\frac{15}{5}$
24. $\frac{39}{13}$
25. $\frac{50}{100}$
26. $\frac{5}{1000}$
27. $\frac{200}{100}$
28. $\frac{125}{100}$
29. $\frac{18}{48}$
30. $\frac{34}{102}$
31. $\frac{26}{42}$
32. $\frac{70}{112}$
33. $\frac{84}{91}$
34. $\frac{121}{132}$

Find each product. See Examples 3 and 4.

35. $\frac{2}{3} \cdot \frac{5}{9}$
36. $\frac{1}{8} \cdot \frac{1}{8}$
37. $\frac{1}{3} \cdot 15$
38. $\frac{1}{4} \cdot 16$
39. $\frac{3}{4} \cdot \frac{14}{15}$
40. $\frac{5}{8} \cdot \frac{12}{35}$
41. $\frac{2}{5} \cdot \frac{35}{26}$
42. $\frac{3}{10} \cdot \frac{20}{21}$
43. $\frac{1}{2} \cdot \frac{6}{5}$

44. $\frac{1}{2} \cdot \frac{3}{5}$

45. $\frac{1}{2} \cdot \frac{1}{3}$

46. $\frac{3}{16} \cdot \frac{1}{7}$

Find each quotient. See Example 5.

47. $\frac{3}{4} \div \frac{1}{4}$

48. $\frac{2}{3} \div \frac{1}{2}$

49. $\frac{1}{3} \div 5$

50. $\frac{3}{5} \div 3$

51. $5 \div \frac{5}{4}$

52. $8 \div \frac{2}{3}$

53. $\frac{6}{10} \div \frac{3}{4}$

54. $\frac{2}{3} \div \frac{10}{21}$

55. $\frac{3}{16} \div \frac{5}{2}$

56. $\frac{1}{8} \div \frac{5}{16}$

Find each sum or difference. See Examples 6 and 7.

57. $\frac{1}{4} + \frac{1}{4}$

58. $\frac{1}{10} + \frac{1}{10}$ $= \frac{2}{10} = \frac{1}{5}$

59. $\frac{5}{12} - \frac{1}{12}$

60. $\frac{17}{14} - \frac{5}{14}$

61. $\frac{1}{2} - \frac{1}{4}$

62. $\frac{1}{3} + \frac{1}{6}$ $\frac{2}{6} + \frac{1}{6} = \frac{3}{6} = \frac{1}{2}$

63. $\frac{1}{3} + \frac{1}{4}$

64. $\frac{1}{2} + \frac{3}{5}$

65. $\frac{3}{4} - \frac{2}{3}$

66. $\frac{4}{5} - \frac{3}{4}$

67. $\frac{1}{6} + \frac{5}{8}$

68. $\frac{3}{4} + \frac{1}{6}$

69. $\frac{5}{24} - \frac{1}{18}$

70. $\frac{3}{16} - \frac{1}{20}$

71. $3\frac{5}{6} + \frac{5}{16}$

72. $5\frac{3}{8} - \frac{15}{16}$

Convert each given fraction, decimal, or percent into its other two forms. See Example 8.

73. $\frac{3}{5}$

74. $\frac{19}{20}$

75. 9%

76. 60%

77. 0.08

78. 0.4

79. $\frac{3}{4}$

80. $\frac{5}{8}$

81. 2%

82. 120%

83. 0.01

84. 0.005

Perform the indicated operations.

85. $\frac{3}{8} \div \frac{1}{8}$

86. $\frac{7}{8} \div \frac{3}{14}$

87. $\frac{3}{4} \cdot \frac{28}{21}$

88. $\frac{5}{16} \cdot \frac{3}{10}$

89. $\frac{7}{12} + \frac{5}{32}$

90. $\frac{2}{15} + \frac{8}{21}$

91. $\frac{5}{24} - \frac{1}{15}$

92. $\frac{9}{16} - \frac{1}{12}$

93. $3\frac{1}{8} + \frac{15}{16}$

94. $5\frac{1}{4} - \frac{9}{16}$

95. $7\frac{2}{3} \cdot 2\frac{1}{4}$

96. $6\frac{1}{2} \div \frac{7}{2}$

97. $\frac{1}{2} + \frac{1}{3} + \frac{1}{4}$

98. $\frac{1}{2} + \frac{1}{3} - \frac{1}{6}$

99. $\frac{1}{2} \cdot \frac{1}{2} \cdot \frac{1}{2}$

100. $\frac{2}{3} \cdot \frac{2}{3} \cdot \frac{2}{3}$

Fill in the blank so that each equation is correct.

101. $\frac{1}{4} + \frac{}{\_} = \frac{5}{8}$

102. $\frac{1}{3} + \frac{}{\_} = \frac{4}{9}$

103. $\frac{5}{16} - \frac{}{\_} = \frac{1}{8}$

104. $\frac{3}{5} - \frac{}{\_} = \frac{1}{10}$

105. $\frac{4}{9} \cdot \frac{}{\_} = \frac{8}{27}$

106. $\frac{3}{8} \cdot \frac{}{\_} = \frac{3}{4}$

107. $\frac{2}{3} \div \frac{}{\_} = \frac{4}{3}$

108. $\frac{1}{15} \div \frac{}{\_} = \frac{1}{5}$

Solve each problem. See Example 9.

109. ***Inheritance.*** Marie is entitled to one-sixth of an estate because of one relationship to the deceased and one-thirty-second of the estate because of another relationship to the deceased. What is the total portion of the estate that she will receive?

110. ***Diversification.*** Helen has $\frac{1}{5}$ of her portfolio in U.S. stocks, $\frac{1}{8}$ of her portfolio in European stocks, and $\frac{1}{10}$ of her portfolio in Japanese stocks. The remainder is invested in municipal bonds. What fraction of her portfolio is invested in municipal bonds? What percent is invested in municipal bonds?

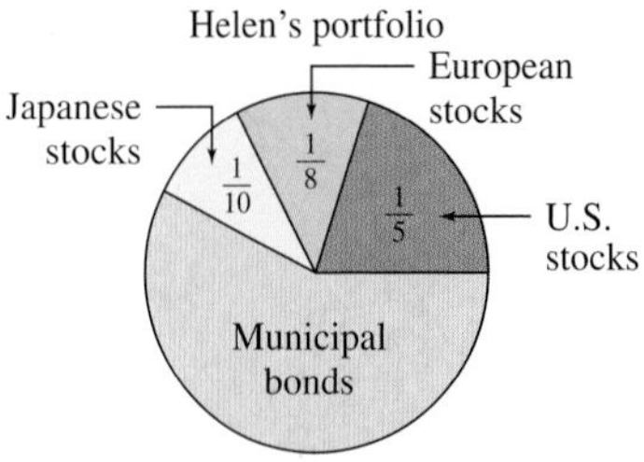

Figure for Exercise 110

111. ***Concrete patio.*** A contractor plans to pour a concrete rectangular patio.

a) Use the table to find the approximate volume of concrete in cubic yards for a 9 ft by 12 ft patio that is 4 inches thick.

b) Find the exact volume of concrete in cubic feet and cubic yards for a patio that is $12\frac{1}{2}$ feet long, $8\frac{3}{4}$ feet wide, and 4 inches thick.

Concrete required for 4 in. thick patio

L (ft)	W (ft)	V (yd^3)
16	14	2.8
14	10	1.7
12	9	1.3
10	8	1.0

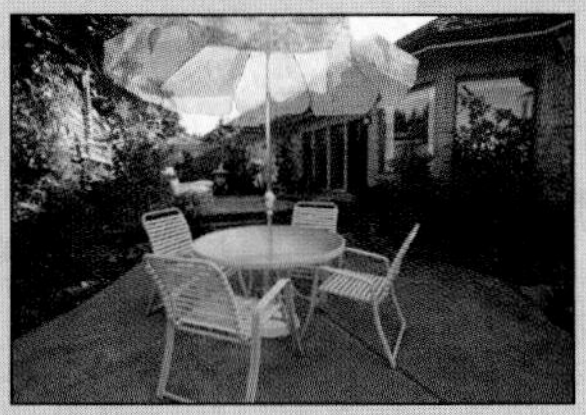

Figure for Exercise 111

112. ***Bundle of studs.*** A lumber yard receives 2×4 studs in a bundle that contains 25 rows (or layers) of studs with 20 studs in each row. A 2×4 stud is actually $1\frac{1}{2}$ in. by $3\frac{1}{2}$ in. by $92\frac{5}{8}$ in. Find the cross-sectional area of a bundle in square inches. Find the volume of a bundle in cubic feet. (The formula $V = LWH$ gives the volume of a rectangular solid.)

Getting More Involved

113. *Writing*

Find an example of a real-life situation in which it is necessary to add two fractions.

114. *Cooperative learning*

Write a step-by-step procedure for adding two fractions with different denominators. Give your procedure to a classmate to try out on some addition problems. Refine your procedure as necessary.

115. ***Fraction puzzle.*** A wheat farmer in Manitoba left his L-shaped farm (shown in the diagram) to his four daughters. Divide the property into four pieces so that each piece is exactly the same size and shape.

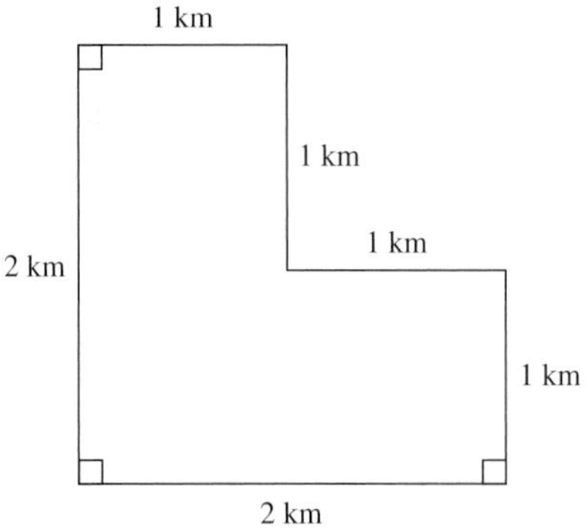

Figure for Exercise 115

1.3 Addition and Subtraction of Real Numbers

In this Section

- **Addition of Two Negative Numbers**
- **Addition of Numbers with Unlike Signs**
- **Subtraction of Signed Numbers**

In arithmetic we add and subtract only positive numbers and zero. In Section 1.1 we introduced the concept of absolute value of a number. Now we will use absolute value to extend the operations of addition and subtraction to the real numbers. We will work only with rational numbers in this chapter. You will learn to perform operations with irrational numbers in Chapter 8.

Addition of Two Negative Numbers

A good way to understand positive and negative numbers is to *think of the positive numbers as assets and the negative numbers as debts.* For this illustration we can think of assets simply as cash. For example, if you have \$3 and \$5 in cash, then your total cash is \$8. You get the total by adding two positive numbers.

Think of debts as unpaid bills such as the electric bill or the phone bill. If you have debts of \$70 and \$80, then your total debt is \$150. You can get the total debt by adding negative numbers:

$$\underset{\$70\text{ debt}}{\underset{\uparrow}{(-70)}} \quad \underset{\text{plus}}{\underset{\uparrow}{+}} \quad \underset{\$80\text{ debt}}{\underset{\uparrow}{(-80)}} \quad = \quad \underset{\$150\text{ debt}}{\underset{\uparrow}{-150}}$$

We think of this addition as adding the absolute values of -70 and -80 $(70 + 80 = 150)$, and then putting a negative sign on that result to get -150. These examples illustrate the following rule.

Sum of Two Numbers with Like Signs

To find the sum of two numbers with the same sign, add their absolute values. The sum has the same sign as the given numbers.

EXAMPLE 1

Adding numbers with like signs

Perform the indicated operations.

a) $23 + 56$ **b)** $(-12) + (-9)$ **c)** $(-3.5) + (-6.28)$ **d)** $\left(-\frac{1}{2}\right) + \left(-\frac{1}{4}\right)$

Solution

a) The sum of two positive numbers is a positive number: $23 + 56 = 79$.

b) The absolute values of -12 and -9 are 12 and 9, and $12 + 9 = 21$. So

$$(-12) + (-9) = -21.$$

c) Add the absolute values of -3.5 and -6.28, and put a negative sign on the sum. Remember to line up the decimal points when adding decimal numbers:

$$\begin{array}{r} 3.50 \\ \underline{6.28} \\ 9.78 \end{array}$$

So $(-3.5) + (-6.28) = -9.78$.

d) $\left(-\frac{1}{2}\right) + \left(-\frac{1}{4}\right) = \left(-\frac{2}{4}\right) + \left(-\frac{1}{4}\right) = -\frac{3}{4}$

Now do Exercises 7–14

Study Tip

Exchange phone numbers, cellular phone numbers, pager numbers, and e-mail addresses with several students in your class. If you miss class and you can't reach your instructor, then you will have someone who can tell you the assignments. If you are stuck on a problem, you can contact a classmate for help.

Addition of Numbers with Unlike Signs

If you have a debt of \$5 and have only \$5 in cash, then your debts equal your assets (in absolute value), and your net worth is \$0. **Net worth** is the total of debts and assets. Symbolically,

$$\underset{\$5\text{ debt}}{\underset{\uparrow}{-5}} \quad + \quad \underset{\$5\text{ cash}}{\underset{\uparrow}{5}} \quad = \quad \underset{\text{Net worth}}{\underset{\uparrow}{0.}}$$

For any number a, a and its opposite, $-a$, have a sum of zero. For this reason, a and $-a$ are called **additive inverses** of each other. Note that the words "negative," "opposite," and "additive inverse" are often used interchangeably.

Additive Inverse Property

For any number a,

$$a + (-a) = 0 \quad \text{and} \quad (-a) + a = 0.$$

EXAMPLE 2

Finding the sum of additive inverses

Evaluate.

a) $34 + (-34)$ **b)** $-\frac{1}{4} + \frac{1}{4}$ **c)** $2.97 + (-2.97)$

Solution

a) $34 + (-34) = 0$

b) $-\frac{1}{4} + \frac{1}{4} = 0$

c) $2.97 + (-2.97) = 0$

Now do Exercises 15–18

Helpful Hint

We use the illustrations with debts and assets to make the rules for adding signed numbers understandable. However, in the end the carefully written rules tell us exactly how to perform operations with signed numbers, and we must obey the rules.

To understand the sum of a positive and a negative number that are not additive inverses of each other, consider the following situation. If you have a debt of \$6 and \$10 in cash, you may have \$10 in hand, but your net worth is only \$4. Your assets exceed your debts (in absolute value), and you have a positive net worth. In symbols,

$$-6 + 10 = 4.$$

Note that to get 4, we actually subtract 6 from 10.

If you have a debt of \$7 but have only \$5 in cash, then your debts exceed your assets (in absolute value). You have a negative net worth of $-\$2$. In symbols,

$$-7 + 5 = -2.$$

Note that to get the 2 in the answer, we subtract 5 from 7.

As you can see from these examples, the sum of a positive number and a negative number (with different absolute values) may be either positive or negative. These examples help us to understand the rule for adding numbers with unlike signs and different absolute values.

Sum of Two Numbers with Unlike Signs (and Different Absolute Values)

To find the sum of two numbers with unlike signs (and different absolute values), subtract their absolute values.

- The answer is positive if the number with the larger absolute value is positive.
- The answer is negative if the number with the larger absolute value is negative.

EXAMPLE 3

Adding numbers with unlike signs

Evaluate.

a) $-5 + 13$ **b)** $6 + (-7)$ **c)** $-6.4 + 2.1$

d) $-5 + 0.09$ **e)** $\left(-\frac{1}{3}\right) + \left(\frac{1}{2}\right)$ **f)** $\frac{3}{8} + \left(-\frac{5}{6}\right)$

Calculator Close-Up

Your calculator can add signed numbers. Most calculators have a key for subtraction and a different key for the negative sign.

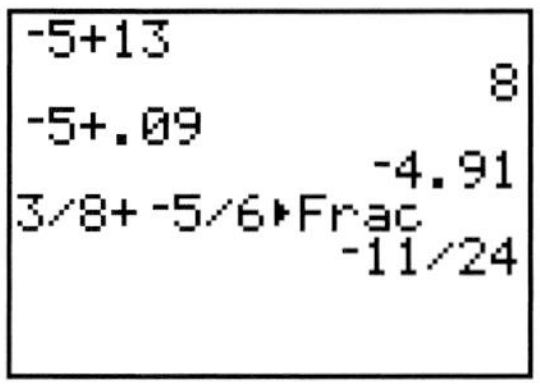

You should do the exercises in this section by hand and then check with a calculator.

Study Tip

The keys to success are desire and discipline. You must want success and you must discipline yourself to do what it takes to get success. There are a lot of things that you can't do anything about, but you can learn to be disciplined. Set your goals, make plans, and schedule your time. Before you know it you will have the discipline that is necessary for success.

Solution

a) The absolute values of -5 and 13 are 5 and 13. Subtract them to get 8. Since the number with the larger absolute value is 13 and it is positive, the result is positive:

$$-5 + 13 = 8$$

b) The absolute values of 6 and -7 are 6 and 7. Subtract them to get 1. Since -7 has the larger absolute value, the result is negative:

$$6 + (-7) = -1$$

c) Line up the decimal points and subtract 2.1 from 6.4.

$$\begin{array}{r} 6.4 \\ -2.1 \\ \hline 4.3 \end{array}$$

Since 6.4 is larger than 2.1, and 6.4 has a negative sign, the sign of the answer is negative. So $-6.4 + 2.1 = -4.3$.

d) Line up the decimal points and subtract 0.09 from 5.00.

$$\begin{array}{r} 5.00 \\ -0.09 \\ \hline 4.91 \end{array}$$

Since 5.00 is larger than 0.09, and 5.00 has the negative sign, the sign of the answer is negative. So $-5 + 0.09 = -4.91$.

e) $\left(-\frac{1}{3}\right) + \left(\frac{1}{2}\right) = \left(-\frac{2}{6}\right) + \left(\frac{3}{6}\right) = \frac{1}{6}$

f) $\frac{3}{8} + \left(-\frac{5}{6}\right) = \frac{9}{24} + \left(-\frac{20}{24}\right) = -\frac{11}{24}$

Now do Exercises 19–28

Subtraction of Signed Numbers

Each subtraction problem with signed numbers is solved by doing an equivalent addition problem. So before attempting subtraction of signed numbers be sure that you understand addition of signed numbers.

Now think of subtraction as removing debts or assets, and think of addition as receiving debts or assets. If you have \$100 in cash and \$30 is taken from you, your resulting net worth is the same as if you have \$100 cash and a phone bill for \$30 arrives in the mail. In symbols,

$$\underset{}{100} \quad \underset{\text{Remove}}{\overset{-}{\uparrow}} \quad \underset{\text{Cash}}{\overset{30}{\uparrow}} \quad = \quad 100 \quad \underset{\text{Receive}}{\overset{+}{\uparrow}} \quad \underset{\text{Debt}}{\overset{(-30).}{\uparrow}}$$

Removing cash is equivalent to receiving a debt.

Suppose you have \$15 but owe a friend \$5. Your net worth is only \$10. If the debt of \$5 is canceled or forgiven, your net worth will go up to \$15, the same as if you received \$5 in cash. In symbols,

$$10 \quad \underset{\text{Remove}}{\overset{-}{\uparrow}} \quad \underset{\text{Debt}}{\overset{(-5)}{\uparrow}} \quad = \quad 10 \quad \underset{\text{Receive}}{\overset{+}{\uparrow}} \quad \underset{\text{Cash}}{\overset{5.}{\uparrow}}$$

Removing a debt is equivalent to receiving cash.

Notice that each subtraction problem is equivalent to an addition problem in which we add the opposite of what we want to subtract. In other words, *subtracting a number is the same as adding its opposite.*

Subtraction of Real Numbers

For any real numbers a and b,

$$a - b = a + (-b).$$

EXAMPLE 4

Subtracting signed numbers

Perform each subtraction.

a) $-5 - 3$

b) $5 - (-3)$

c) $-5 - (-3)$

d) $\frac{1}{2} - \left(-\frac{1}{4}\right)$

e) $-3.6 - (-5)$

f) $0.02 - 8$

Solution

To do *any* subtraction, we can change it to addition of the opposite.

a) $-5 - 3 = -5 + (-3) = -8$

b) $5 - (-3) = 5 + (3) = 8$

c) $-5 - (-3) = -5 + 3 = -2$

d) $\frac{1}{2} - \left(-\frac{1}{4}\right) = \frac{2}{4} + \frac{1}{4} = \frac{3}{4}$

e) $-3.6 - (-5) = -3.6 + 5 = 1.4$

f) $0.02 - 8 = 0.02 + (-8) = -7.98$

Now do Exercises 29–56

Warm-Ups ▼

True or false? Explain your answer.

1. $-9 + 8 = -1$
2. $(-2) + (-4) = -6$
3. $0 - 7 = -7$
4. $5 - (-2) = 3$
5. $-5 - (-2) = -7$
6. The additive inverse of -3 is 0.
7. If b is a negative number, then $-b$ is a positive number.
8. The sum of a positive number and a negative number is a negative number.
9. The result of a subtracted from b is the same as b plus the opposite of a.
10. If a and b are negative numbers, then $a - b$ is a negative number.

1.3 Exercises

Reading and Writing *After reading this section write out the answers to these questions. Use complete sentences.*

1. What operations did we study in this section?
2. How do you find the sum of two numbers with the same sign?
3. When can we say that two numbers are additive inverses of each other?
4. What is the sum of two numbers with opposite signs and the same absolute value?
5. How do we find the sum of two numbers with unlike signs?
6. What is the relationship between subtraction and addition?

Perform the indicated operation. See Example 1.

7. $3 + 10$ **8.** $81 + 19$

9. $(-3) + (-10)$ **10.** $(-81) + (-19)$

11. $-0.25 + (-0.9)$ **12.** $-0.8 + (-2.35)$

13. $\left(-\frac{1}{3}\right) + \left(-\frac{1}{6}\right)$ **14.** $\frac{2}{3} + \frac{1}{12}$

Evaluate. See Examples 2 and 3.

15. $-8 + 8$ **16.** $20 + (-20)$

17. $-\frac{17}{50} + \frac{17}{50}$ **18.** $\frac{12}{13} + \left(-\frac{12}{13}\right)$

19. $-7 + 9$ **20.** $10 + (-30)$

21. $7 + (-13)$ **22.** $-8 + 20$

23. $8.6 + (-3)$ **24.** $-9.5 + 12$

25. $3.9 + (-6.8)$ **26.** $-5.24 + 8.19$

27. $\frac{1}{4} + \left(-\frac{1}{2}\right)$ **28.** $-\frac{2}{3} + 2$

Fill in the parentheses to make each statement correct. See Example 4.

29. $8 - 2 = 8 + (?)$

30. $3.5 - 1.2 = 3.5 + (?)$

31. $4 - 12 = 4 + (?)$

32. $\frac{1}{2} - \frac{5}{6} = \frac{1}{2} + (?)$

33. $-3 - (-8) = -3 + (?)$

34. $-9 - (-2.3) = -9 + (?)$

35. $8.3 - (-1.5) = 8.3 + (?)$

36. $10 - (-6) = 10 + (?)$

Perform the indicated operation. See Example 4.

37. $6 - 10$
38. $3 - 19$
39. $-3 - 7$
40. $-3 - 12$
41. $5 - (-6)$
42. $5 - (-9)$
43. $-6 - 5$
44. $-3 - 6$
45. $\frac{1}{4} - \frac{1}{2}$
46. $\frac{2}{5} - \frac{2}{3}$
47. $\frac{1}{2} - \left(-\frac{1}{4}\right)$
48. $\frac{2}{3} - \left(-\frac{1}{6}\right)$
49. $10 - 3$
50. $13 - 3$
51. $1 - 0.07$
52. $0.03 - 1$
53. $7.3 - (-2)$
54. $-5.1 - 0.15$
55. $-0.03 - 5$
56. $0.7 - (-0.3)$

Perform the indicated operations. Do not use a calculator.

57. $-5 + 8$
58. $-6 + 10$
59. $-6 + (-3)$
60. $(-13) + (-12)$
61. $-80 - 40$
62. $44 - (-15)$
63. $61 - (-17)$
64. $-19 - 13$
65. $(-12) + (-15)$
66. $-12 + 12$
67. $13 + (-20)$
68. $15 + (-39)$
69. $-102 - 99$
70. $-94 - (-77)$
71. $-161 - 161$
72. $-19 - 88$
73. $-16 + 0.03$
74. $0.59 + (-3.4)$
75. $0.08 - 3$
76. $1.8 - 9$
77. $-3.7 + (-0.03)$
78. $0.9 + (-1)$
79. $-2.3 - (-6)$
80. $-7.08 - (-9)$
81. $\frac{3}{4} + \left(-\frac{3}{5}\right)$
82. $-\frac{1}{3} + \frac{3}{5}$
83. $-\frac{1}{12} - \left(-\frac{3}{8}\right)$
84. $-\frac{1}{17} - \left(-\frac{1}{17}\right)$

Fill in the parentheses so that each equation is correct.

85. $-5 + (\quad) = 8$
86. $-9 + (\quad) = 22$
87. $12 + (\quad) = 2$
88. $13 + (\quad) = -4$
89. $10 - (\quad) = -4$
90. $14 - (\quad) = -8$
91. $6 - (\quad) = 10$
92. $3 - (\quad) = 15$
93. $-4 - (\quad) = -1$
94. $-11 - (\quad) = 2$

Use a calculator to perform the indicated operations.

95. $45.87 + (-49.36)$
96. $-0.357 + (-3.465)$
97. $0.6578 + (-1)$
98. $-2.347 + (-3.5)$
99. $-3.45 - 45.39$
100. $9.8 - 9.974$
101. $-5.79 - 3.06$
102. $0 - (-4.537)$

Solve each problem.

103. ***Overdrawn.*** Willard opened his checking account with a deposit of \$97.86. He then wrote checks and had other charges as shown in his account register. Find his current balance.

Deposit		97.86
Wal-Mart	27.89	
Kmart	42.32	
ATM cash	25.00	
Service charge	3.50	
Check printing	8.00	

Figure for Exercise 103

104. ***Net worth.*** Melanie's house is worth \$125,000, but she still owes \$78,422 on her mortgage. She has \$21,236 in a savings account and has \$9,477 in credit card debt. She owes \$6,131 to the credit union and figures that her cars and other household items are worth a total of \$15,000. What is Melanie's net worth?

105. ***Falling temperatures.*** At noon the temperature in Montreal was 5°C. By midnight the mercury had fallen 12°. What was the temperature at midnight?

106. ***Bitter cold.*** The overnight low temperature in Milwaukee was −13°F for Monday night. The temperature went up 20° during the day on Tuesday and then fell 15° to reach Tuesday night's overnight low temperature.

a) What was the overnight low Tuesday night?

b) Judging from the accompanying graph, was the average low for the week above or below 0°F?

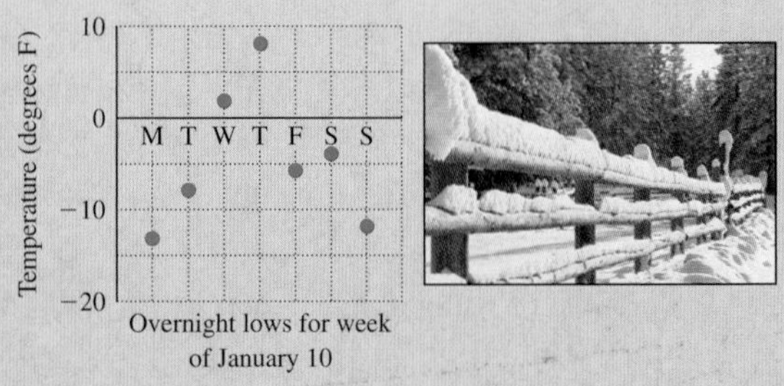

Figure for Exercise 106

Getting More Involved

107. *Writing*

What does absolute value have to do with adding signed numbers? Can you add signed numbers without using absolute value?

108. *Discussion*

Why do we learn addition of signed numbers before subtraction?

109. *Discussion*

Aimee and Joni are traveling south in separate cars on Interstate 5 near Stockton. While they are speaking to each other on cellular telephones, Aimee gives her location as mile marker x and Joni gives her location as mile marker y. Which of the following expressions gives the distance between them? Explain your answer.

a) $y - x$ **b)** $x - y$
c) $|x - y|$ **d)** $|y - x|$
e) $|x| + |y|$

1.4 Multiplication and Division of Real Numbers

In this Section

- Multiplication of Real Numbers
- Division of Real Numbers
- Division by Zero

In this section we will complete the study of the four basic operations with real numbers.

Multiplication of Real Numbers

The result of multiplying two numbers is referred to as the **product** of the numbers. The numbers multiplied are called **factors.** In algebra we use a raised dot between the factors to indicate multiplication, or we place symbols next to one another to indicate multiplication. Thus $a \cdot b$ or ab are both referred to as the product of a and b. When multiplying numbers, we may enclose them in parentheses to make the meaning clear. To write 5 times 3, we may write it as $5 \cdot 3$, 5(3), (5)3, or (5)(3). In multiplying a number and a variable, no sign is used between them. Thus $5x$ is used to represent the product of 5 and x.

Multiplication is just a short way to do repeated additions. Adding together five 3's gives

$$3 + 3 + 3 + 3 + 3 = 15.$$

So we have the multiplication fact $5 \cdot 3 = 15$. Adding together five -3's gives

$$(-3) + (-3) + (-3) + (-3) + (-3) = -15.$$

So we should have $5(-3) = -15$. We can think of $5(-3) = -15$ as saying that taking on five debts of \$3 each is equivalent to a debt of \$15. Losing five debts of \$3 each is equivalent to gaining \$15, so we should have $(-5)(-3) = 15$.

These examples illustrate the rule for multiplying signed numbers.

Helpful Hint

The product of two numbers with like signs is positive, but the product of three numbers with like signs can be positive or negative. For example,

$$2 \cdot 2 \cdot 2 = 8$$

and

$$(-2)(-2)(-2) = -8.$$

Product of Signed Numbers

To find the product of two nonzero real numbers, multiply their absolute values.

- The product is *positive* if the numbers have *like* signs.
- The product is *negative* if the numbers have *unlike* signs.

EXAMPLE 1

Multiplying signed numbers

Evaluate each product.

a) $(-2)(-3)$ **b)** $3(-6)$ **c)** $-5 \cdot 10$

d) $\left(-\frac{1}{3}\right)\left(-\frac{1}{2}\right)$ **e)** $(-0.02)(0.08)$ **f)** $(-300)(-0.06)$

Calculator Close-Up

Try finding the products in Example 1 with your calculator.

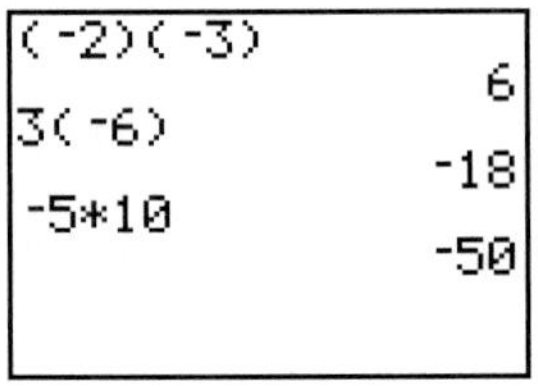

Solution

a) First find the product of the absolute values:

$$|-2| \cdot |-3| = 2 \cdot 3 = 6$$

Because -2 and -3 have the same sign, we get $(-2)(-3) = 6$.

b) First find the product of the absolute values:

$$|3| \cdot |-6| = 3 \cdot 6 = 18$$

Because 3 and -6 have unlike signs, we get $3(-6) = -18$.

c) $-5 \cdot 10 = -50$ Unlike signs, negative result

d) $\left(-\frac{1}{3}\right)\left(-\frac{1}{2}\right) = \frac{1}{6}$ Like signs, positive result

e) When multiplying decimals, we total the number of decimal places in the factors to get the number of decimal places in the product. Thus

$$(-0.02)(0.08) = -0.0016.$$

f) $(-300)(-0.06) = 18$

Now do Exercises 7–18

Division of Real Numbers

We say that $10 \div 5 = 2$ because $2 \cdot 5 = 10$. This example illustrates how division is defined in terms of multiplication.

Division of Real Numbers

If a, b, and c are any real numbers with $b \neq 0$, then

$$a \div b = c \qquad \text{provided that} \qquad c \cdot b = a.$$

Using the definition of division, we get

$$10 \div (-2) = -5$$

because $(-5)(-2) = 10$;

$$-10 \div 2 = -5$$

because $(-5)(2) = -10$; and

$$-10 \div (-2) = 5$$

because $(5)(-2) = -10$. From these examples we see that the rule for dividing signed numbers is similar to that for multiplying signed numbers.

Division of Signed Numbers

To find the quotient of two nonzero real numbers, divide their absolute values.

- The quotient is *positive* if the two numbers have *like* signs.
- The quotient is *negative* if the two numbers have *unlike* signs.

Zero divided by any nonzero real number is zero.

EXAMPLE 2

Dividing signed numbers

Evaluate.

a) $(-8) \div (-4)$ **b)** $(-8) \div 8$ **c)** $8 \div (-4)$

d) $-4 \div \frac{1}{3}$ **e)** $-2.5 \div 0.05$ **f)** $0 \div (-6)$

Solution

a) $(-8) \div (-4) = 2$ Same sign, positive result

b) $(-8) \div 8 = -1$ Unlike signs, negative result

c) $8 \div (-4) = -2$

d) $-4 \div \frac{1}{3} = -4 \cdot \frac{3}{1}$ Invert and multiply.

$= -4 \cdot 3$

$= -12$

e) $-2.5 \div 0.05 = \frac{-2.5}{0.05}$ Write in fraction form.

$= \frac{-2.5 \cdot 100}{0.05 \cdot 100}$ Multiply by 100 to eliminate the decimals.

$= \frac{-250}{5}$ Simplify.

$= -50$ Divide.

f) $0 \div (-6) = 0$

Now do Exercises 19–62

Helpful Hint

Do not use negative numbers in long division. To find $-378 \div 7$, divide 378 by 7:

$$\begin{array}{r} 54 \\ 7\overline{)378} \\ \underline{35} \\ 28 \\ \underline{28} \\ 0 \end{array}$$

Since a negative divided by a positive is negative

$$-378 \div 7 = -54.$$

Division can also be indicated by a fraction bar. For example,

$$24 \div 6 = \frac{24}{6} = 4.$$

If signed numbers occur in a fraction, we use the rules for dividing signed numbers. For example,

$$\frac{-9}{3} = -3, \quad \frac{9}{-3} = -3, \quad \frac{-1}{2} = \frac{1}{-2} = -\frac{1}{2}, \quad \text{and} \quad \frac{-4}{-2} = 2.$$

Study Tip

If you don't know how to get started on the exercises, go back to the examples. Cover the solution in the text with a piece of paper and see if you can solve the example. After you have mastered the examples, then try the exercises again.

Note that if one negative sign appears in a fraction, the fraction has the same value whether the negative sign is in the numerator, in the denominator, or in front of the fraction. If the numerator and denominator of a fraction are both negative, then the fraction has a positive value.

Division by Zero

Why do we exclude division by zero from the definition of division? If we write $10 \div 0 = c$, we need to find a number c such that $c \cdot 0 = 10$. This is impossible. If we write $0 \div 0 = c$, we need to find a number c such that $c \cdot 0 = 0$. In fact, $c \cdot 0 = 0$ is true for any value of c. Having $0 \div 0$ equal to any number would be confusing in doing computations. Thus $a \div b$ is defined only for $b \neq 0$. Quotients such as

$$8 \div 0, \quad 0 \div 0, \quad \frac{8}{0}, \quad \text{and} \quad \frac{0}{0}$$

are said to be **undefined.**

Warm-Ups ▼

True or false? Explain your answer.

1. The product of 7 and y is written as $7y$.
2. The product of -2 and 5 is 10.
3. The quotient of x and 3 can be written as $x \div 3$ or $\frac{x}{3}$.
4. $0 \div 6$ is undefined.
5. $(-9) \div (-3) = 3$
6. $6 \div (-2) = -3$
7. $\left(-\frac{1}{2}\right)\left(-\frac{1}{2}\right) = \frac{1}{4}$
8. $(-0.2)(0.2) = -0.4$
9. $\left(-\frac{1}{2}\right) \div \left(-\frac{1}{2}\right) = 1$
10. $\frac{0}{0} = 0$

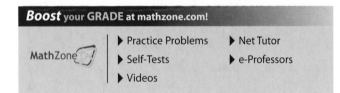

1.4 Exercises

Reading and Writing *After reading this section write out the answers to these questions. Use complete sentences.*

1. What operations did we study in this section?
2. What is a product?
3. How do you find the product of two signed numbers?
4. What is the relationship between division and multiplication?
5. How do you find the quotient of nonzero real numbers?
6. Why is division by zero undefined?

Evaluate. See Example 1.

7. $-3 \cdot 9$
8. $6(-4)$
9. $(-12)(-11)$
10. $(-9)(-15)$
11. $-\frac{3}{4} \cdot \frac{4}{9}$
12. $\left(-\frac{2}{3}\right)\left(-\frac{6}{7}\right)$
13. $0.5(-0.6)$
14. $(-0.3)(0.3)$
15. $(-12)(-12)$
16. $(-11)(-11)$
17. $-3 \cdot 0$
18. $0(-7)$

Evaluate. See Example 2.

19. $8 \div (-8)$
20. $-6 \div 2$
21. $(-90) \div (-30)$
22. $(-20) \div (-40)$
23. $\frac{44}{-66}$
24. $\frac{-33}{-36}$
25. $\left(-\frac{2}{3}\right) \div \left(-\frac{4}{5}\right)$
26. $-\frac{1}{3} \div \frac{4}{9}$
27. $\frac{-125}{0}$
28. $-37 \div 0$
29. $0 \div \left(-\frac{1}{3}\right)$
30. $0 \div 43.568$
31. $40 \div (-0.5)$
32. $3 \div (-0.1)$
33. $-0.5 \div (-2)$
34. $-0.75 \div (-0.5)$

Perform the indicated operations.

35. $(25)(-4)$
36. $(5)(-4)$
37. $(-3)(-9)$
38. $(-51) \div (-3)$
39. $-9 \div 3$
40. $86 \div (-2)$
41. $20 \div (-5)$
42. $(-8)(-6)$
43. $(-6)(5)$
44. $(-18) \div 3$
45. $(-57) \div (-3)$
46. $(-30)(4)$
47. $(0.6)(-0.3)$
48. $(-0.2)(-0.5)$
49. $(-0.03)(-10)$
50. $(0.05)(-1.5)$
51. $(-0.6) \div (0.1)$
52. $8 \div (-0.5)$
53. $(-0.6) \div (-0.4)$
54. $(-63) \div (-0.9)$
55. $-\frac{12}{5}\left(-\frac{55}{6}\right)$
56. $-\frac{9}{10} \cdot \frac{4}{3}$
57. $-2\frac{3}{4} \div 8\frac{1}{4}$
58. $-9\frac{1}{2} \div \left(-3\frac{1}{6}\right)$

Use a calculator to perform the indicated operations. Round approximate answers to two decimal places.

59. $(0.45)(-365)$
60. $8.5 \div (-0.15)$
61. $(-52) \div (-0.034)$
62. $(-4.8)(5.6)$

Fill in the parentheses so that each equation is correct.

63. $-5 \cdot (\quad) = 60$
64. $-9 \cdot (\quad) = 54$
65. $12 \cdot (\quad) = -96$
66. $11 \cdot (\quad) = -44$
67. $24 \div (\quad) = -4$
68. $51 \div (\quad) = -17$
69. $-36 \div (\quad) = 36$
70. $-48 \div (\quad) = 6$
71. $-40 \div (\quad) = -8$
72. $-13 \div (\quad) = -1$

Perform the indicated operations. Use a calculator to check.

73. $(-4)(-4)$
74. $-4 - 4$
75. $-4 + (-4)$
76. $-4 \div (-4)$
77. $-4 + 4$
78. $-4 \cdot 4$
79. $-4 - (-4)$
80. $0 \div (-4)$
81. $0.1 - 4$
82. $(0.1)(-4)$
83. $(-4) \div (0.1)$
84. $-0.1 - 4$
85. $(-0.1)(-4)$
86. $-0.1 + 4$
87. $|-0.4|$
88. $|0.4|$
89. $\frac{-0.06}{0.3}$
90. $\frac{2}{-0.04}$
91. $\frac{3}{-0.4}$
92. $\frac{-1.2}{-0.03}$
93. $-\frac{1}{5} + \frac{1}{6}$
94. $-\frac{3}{5} - \frac{1}{4}$
95. $\left(-\frac{3}{4}\right)\left(\frac{2}{15}\right)$
96. $-1 \div \left(-\frac{1}{4}\right)$

Use a calculator to perform the indicated operations. Round approximate answers to three decimal places.

97. $\frac{45.37}{6}$
98. $(-345) \div (28)$
99. $(-4.3)(-4.5)$
100. $\frac{-12.34}{-3}$
101. $\frac{0}{6.345}$
102. $0 \div (34.51)$
103. $199.4 \div 0$
104. $\frac{23.44}{0}$

Getting More Involved

105. ***Discussion***

If you divide \$0 among five people, how much does each person get? If you divide \$5 among zero people, how much does each person get? What do these questions illustrate?

106. ***Discussion***

What is the difference between the non-negative numbers and the positive numbers?

107. ***Writing***

Why do we learn multiplication of signed numbers before division?

108. ***Writing***

Try to rewrite the rules for multiplying and dividing signed numbers without using the idea of absolute value. Are your rewritten rules clearer than the original rules?

1.5 Exponential Expressions and the Order of Operations

In this Section

- **Arithmetic Expressions**
- **Exponential Expressions**
- **The Order of Operations**

In Sections 1.3 and 1.4 you learned how to perform operations with a pair of real numbers to obtain a third real number. In this section you will learn to evaluate expressions involving several numbers and operations.

Arithmetic Expressions

The result of writing numbers in a meaningful combination with the ordinary operations of arithmetic is called an **arithmetic expression** or simply an **expression.** Consider the expressions

$$(3 + 2) \cdot 5 \quad \text{and} \quad 3 + (2 \cdot 5).$$

The parentheses are used as **grouping symbols** and indicate which operation to perform first. Because of the parentheses, these expressions have different values:

$$(3 + 2) \cdot 5 = 5 \cdot 5 = 25$$
$$3 + (2 \cdot 5) = 3 + 10 = 13$$

Absolute value symbols and fraction bars are also used as grouping symbols. The numerator and denominator of a fraction are treated as if each is in parentheses.

EXAMPLE 1

Using grouping symbols

Evaluate each expression.

a) $(3 - 6)(3 + 6)$

b) $|3 - 4| - |5 - 9|$

c) $\dfrac{4 - (-8)}{5 - 9}$

Solution

a) $(3 - 6)(3 + 6) = (-3)(9)$ Evaluate within parentheses first.

$= -27$ Multiply.

b) $|3 - 4| - |5 - 9| = |-1| - |-4|$ Evaluate within absolute value symbols.

$= 1 - 4$ Find the absolute values.

$= -3$ Subtract.

c) $\dfrac{4 - (-8)}{5 - 9} = \dfrac{12}{-4}$ Evaluate the numerator and denominator.

$= -3$ Divide.

Now do Exercises 7–18

Calculator Close-Up

One advantage of a graphing calculator is that you can enter an entire expression on its display and then evaluate it. If your calculator does not allow built-up form for fractions, then you must use parentheses around the numerator and denominator as shown here.

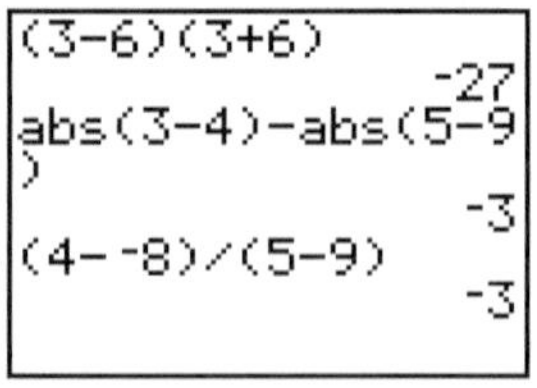

Exponential Expressions

An arithmetic expression with repeated multiplication can be written by using exponents. For example,

$$2 \cdot 2 \cdot 2 = 2^3 \quad \text{and} \quad 5 \cdot 5 = 5^2.$$

Study Tip

If you need help, do not hesitate to get it. Math has a way of building upon the past. What you learn today will be used tomorrow, and what you learn tomorrow will be used the day after. If you don't straighten out problems immediately, then you can get hopelessly lost. If you are having trouble, see your instructor to find out what help is available.

The 3 in 2^3 is the number of times that 2 occurs in the product $2 \cdot 2 \cdot 2$, while the 2 in 5^2 is the number of times that 5 occurs in $5 \cdot 5$. We read 2^3 as "2 cubed" or "2 to the third power." We read 5^2 as "5 squared" or "5 to the second power." In general, an expression of the form a^n is called an **exponential expression** and is defined as follows.

Exponential Expression

For any counting number n,

$$a^n = \underbrace{a \cdot a \cdot a \cdot \ldots \cdot a}_{n \text{ factors}}.$$

We call a the **base** and n the **exponent.**

The expression a^n is read "a to the nth power." If the exponent is 1, it is usually omitted. For example, $9^1 = 9$.

EXAMPLE 2

Using exponential notation

Write each product as an exponential expression.

a) $6 \cdot 6 \cdot 6 \cdot 6 \cdot 6$ **b)** $(-3)(-3)(-3)(-3)$ **c)** $\frac{3}{2} \cdot \frac{3}{2} \cdot \frac{3}{2}$

Solution

a) $6 \cdot 6 \cdot 6 \cdot 6 \cdot 6 = 6^5$

b) $(-3)(-3)(-3)(-3) = (-3)^4$

c) $\frac{3}{2} \cdot \frac{3}{2} \cdot \frac{3}{2} = \left(\frac{3}{2}\right)^3$

Now do Exercises 19–26

EXAMPLE 3

Writing an exponential expression as a product

Write each exponential expression as a product without exponents.

a) y^6 **b)** $(-2)^4$ **c)** $\left(\frac{5}{4}\right)^3$ **d)** $(-0.1)^2$

Solution

a) $y^6 = y \cdot y \cdot y \cdot y \cdot y \cdot y$

b) $(-2)^4 = (-2)(-2)(-2)(-2)$

c) $\left(\frac{5}{4}\right)^3 = \frac{5}{4} \cdot \frac{5}{4} \cdot \frac{5}{4}$

d) $(-0.1)^2 = (-0.1)(-0.1)$

Now do Exercises 27–34

To evaluate an exponential expression, write the base as many times as indicated by the exponent, then multiply the factors from left to right.

EXAMPLE 4

Evaluating exponential expressions

Evaluate.

a) 3^3 **b)** $(-2)^3$ **c)** $\left(\frac{2}{3}\right)^4$ **d)** $(0.4)^2$

Calculator Close-Up

You can use the power key for any power. Most calculators also have an x^2 key that gives the second power. Note that parentheses must be used when raising a fraction to a power.

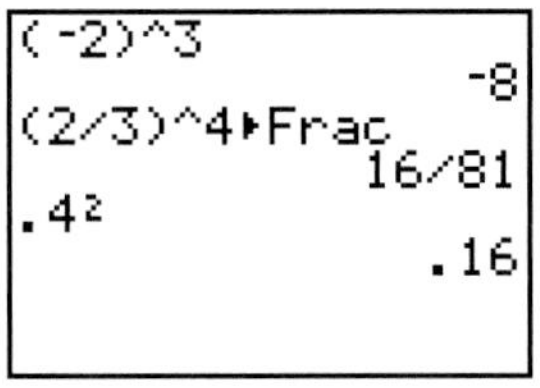

Solution

a) $3^3 = 3 \cdot 3 \cdot 3 = 9 \cdot 3 = 27$

b) $(-2)^3 = (-2)(-2)(-2)$
$= 4(-2)$
$= -8$

c) $$\left(\frac{2}{3}\right)^4 = \frac{2}{3} \cdot \frac{2}{3} \cdot \frac{2}{3} \cdot \frac{2}{3} = \frac{4}{9} \cdot \frac{2}{3} \cdot \frac{2}{3} = \frac{8}{27} \cdot \frac{2}{3} = \frac{16}{81}$$

d) $(0.4)^2 = (0.4)(0.4) = 0.16$

Now do Exercises 35–50

CAUTION Note that $3^3 \neq 9$. We do not multiply the exponent and the base when evaluating an exponential expression.

Be especially careful with exponential expressions involving negative numbers. An exponential expression with a negative base is written with parentheses around the base as in $(-2)^4$:

$$(-2)^4 = (-2)(-2)(-2)(-2) = 16$$

To evaluate $-(2^4)$, use the base 2 as a factor four times, then find the opposite:

$$-(2^4) = -(2 \cdot 2 \cdot 2 \cdot 2) = -(16) = -16$$

We often omit the parentheses in $-(2^4)$ and simply write -2^4. So

$$-2^4 = -(2^4) = -16.$$

To evaluate $-(-2)^4$, use the base -2 as a factor four times, then find the opposite:

$$-(-2)^4 = -(16) = -16$$

EXAMPLE 5

Evaluating exponential expressions involving negative numbers

Evaluate.

a) $(-10)^4$ **b)** -10^4

c) $-(-0.5)^2$ **d)** $-(5-8)^2$

Solution

a) $(-10)^4 = (-10)(-10)(-10)(-10)$ Use -10 as a factor four times.

$= 10{,}000$

b) $-10^4 = -(10^4)$ Rewrite using parentheses.

$= -(10{,}000)$ Find 10^4.

$= -10{,}000$ Then find the opposite of 10,000.

c) $-(-0.5)^2 = -(-0.5)(-0.5)$ Use -0.5 as a factor two times.

$= -(0.25)$

$= -0.25$

d) $-(5-8)^2 = -(-3)^2$ Evaluate within parentheses first.

$= -(9)$ Square -3 to get 9.

$= -9$ Take the opposite of 9 to get -9.

Now do Exercises 51–58

Helpful Hint

"Please Excuse My Dear Aunt Sally" (PEMDAS) is often used as a memory aid for the order of operations. Do Parentheses, Exponents, Multiplication and Division, then Addition and Subtraction. Multiplication and division have equal priority. The same goes for addition and subtraction.

The Order of Operations

When we evaluate expressions, operations within grouping symbols are always performed first. For example,

$$(3+2)\cdot 5 = (5)\cdot 5 = 25 \quad \text{and} \quad (2\cdot 3)^2 = 6^2 = 36.$$

To make expressions look simpler, we often omit some or all parentheses. In this case, we must agree on the order in which to perform the operations. We agree to do multiplication before addition and exponential expressions before multiplication. So

$$3 + 2\cdot 5 = 3 + 10 = 13 \quad \text{and} \quad 2\cdot 3^2 = 2\cdot 9 = 18.$$

We state the complete **order of operations** in the following box.

Order of Operations

If an expression contains no grouping symbols, evaluate it using the following order. If an expression contains operations within grouping symbols, evaluate the expressions within grouping symbols first, using the following order.

1. Evaluate each exponential expression (in order from left to right).

2. Perform multiplication and division (in order from left to right).

3. Perform addition and subtraction (in order from left to right).

Multiplication and division have equal priority in the order of operations. If both appear in an expression, they are performed in order from left to right. The same holds for addition and subtraction. For example,

$$8 \div 4 \cdot 3 = 2 \cdot 3 = 6 \qquad \text{and} \qquad 9 - 3 + 5 = 6 + 5 = 11.$$

EXAMPLE 6

Using the order of operations

Evaluate each expression.

a) $2^3 \cdot 3^2$ **b)** $2 \cdot 5 - 3 \cdot 4 + 4^2$ **c)** $2 \cdot 3 \cdot 4 - 3^3 + \frac{8}{2}$

Calculator Close-Up

Most calculators follow the same order of operations shown here. Evaluate these expressions with your calculator.

```
2^3*3²
                72
2*5-3*4+4²
                14
2*3*4-3^3+8/2
                 1
```

Solution

a) $2^3 \cdot 3^2 = 8 \cdot 9$ Evaluate exponential expressions before multiplying.

$= 72$

b) $2 \cdot 5 - 3 \cdot 4 + 4^2 = 2 \cdot 5 - 3 \cdot 4 + 16$ Exponential expressions first

$= 10 - 12 + 16$ Multiplication second

$= 14$ Addition and subtraction from left to right

c) $2 \cdot 3 \cdot 4 - 3^3 + \frac{8}{2} = 2 \cdot 3 \cdot 4 - 27 + \frac{8}{2}$ Exponential expressions first

$= 24 - 27 + 4$ Multiplication and division second

$= 1$ Addition and subtraction from left to right

Now do Exercises 59–70

When grouping symbols are used, we perform operations within grouping symbols first. The order of operations is followed within the grouping symbols.

EXAMPLE 7

Grouping symbols and the order of operations

Evaluate.

a) $3 - 2(7 - 2^3)$ **b)** $3 - |7 - 3 \cdot 4|$ **c)** $\frac{9 - 5 + 8}{-5^2 - 3(-7)}$

Solution

a) $3 - 2(7 - 2^3) = 3 - 2(7 - 8)$ Evaluate within parentheses first.

$= 3 - 2(-1)$

$= 3 - (-2)$ Multiply.

$= 5$ Subtract.

b) $3 - |7 - 3 \cdot 4| = 3 - |7 - 12|$ Evaluate within the absolute value symbols first.

$= 3 - |-5|$

$= 3 - 5$ Evaluate the absolute value.

$= -2$ Subtract.

c) $\frac{9 - 5 + 8}{-5^2 - 3(-7)} = \frac{12}{-25 + 21} = \frac{12}{-4} = -3$ Numerator and denominator are treated as if in parentheses.

Now do Exercises 71–84

When grouping symbols occur within grouping symbols, we evaluate within the innermost grouping symbols first and then work outward. In this case, brackets [ ] can be used as grouping symbols along with parentheses to make the grouping clear.

EXAMPLE 8

Grouping within grouping

Evaluate each expression.

a) $6 - 4[5 - (7 - 9)]$ **b)** $-2|3 - (9 - 5)| - |-3|$

Solution

a) $6 - 4[5 - (7 - 9)] = 6 - 4[5 - (-2)]$ Innermost parentheses first

$= 6 - 4[7]$ Next evaluate within the brackets.

$= 6 - 28$ Multiply.

$= -22$ Subtract.

b) $-2|3 - (9 - 5)| - |-3| = -2|3 - 4| - |-3|$ Innermost grouping first

$= -2|-1| - |-3|$ Evaluate within the first absolute value.

$= -2 \cdot 1 - 3$ Evaluate absolute values.

$= -2 - 3$ Multiply.

$= -5$ Subtract.

Now do Exercises 85–92

Calculator Close-Up

Graphing calculators can handle grouping symbols within grouping symbols. Since parentheses must occur in pairs, you should have the same number of left parentheses as right parentheses. You might notice other grouping symbols on your calculator, but they may or may not be used for grouping. See your manual.

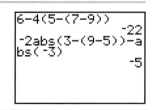

```
6-4(5-(7-9))
                 -22
-2abs(3-(9-5))-a
bs(-3)
                  -5
```

Warm-Ups ▼

True or false? Explain your answer.

1. $(-3)^2 = -6$

2. $5 - 3 \cdot 2 = 4$

3. $(5 - 3)2 = 4$

4. $|5 - 6| = |5| - |6|$

5. $5 + 6 \cdot 2 = (5 + 6) \cdot 2$

6. $(2 + 3)^2 = 2^2 + 3^2$

7. $5 - 3^3 = 8$

8. $(5 - 3)^3 = 8$

9. $6 - \frac{6}{2} = \frac{0}{2}$

10. $\frac{6 - 6}{2} = 0$

1.5 Exercises

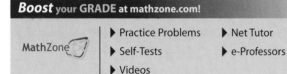

Reading and Writing *After reading this section write out the answers to these questions. Use complete sentences.*

1. What is an arithmetic expression?
2. What is the purpose of grouping symbols?
3. What is an exponential expression?
4. What is the difference between -3^6 and $(-3)^6$?
5. What is the purpose of the order of operations?
6. What were the different types of grouping symbols used in this section?

Evaluate each expression. See Example 1.

7. $(4-3)(5-9)$ **8.** $(5-7)(-2-3)$

9. $|3+4|-|-2-4|$ **10.** $|-4+9|+|-3-5|$

11. $\dfrac{7-(-9)}{3-5}$ **12.** $\dfrac{-8+2}{-1-1}$

13. $(-6+5)(7)$ **14.** $-6+(5\cdot 7)$

15. $(-3-7)-6$ **16.** $-3-(7-6)$

17. $-16\div(8\div 2)$ **18.** $(-16\div 8)\div 2$

Write each product as an exponential expression. See Example 2.

19. $4\cdot 4\cdot 4\cdot 4$ **20.** $1\cdot 1\cdot 1\cdot 1\cdot 1$

21. $(-5)(-5)(-5)(-5)$ **22.** $(-7)(-7)(-7)$

23. $(-y)(-y)(-y)$ **24.** $x\cdot x\cdot x\cdot x\cdot x$

25. $\dfrac{3}{7}\cdot\dfrac{3}{7}\cdot\dfrac{3}{7}\cdot\dfrac{3}{7}\cdot\dfrac{3}{7}$ **26.** $\dfrac{y}{2}\cdot\dfrac{y}{2}\cdot\dfrac{y}{2}\cdot\dfrac{y}{2}$

Write each exponential expression as a product without exponents. See Example 3.

27. 5^3 **28.** $(-8)^4$

29. b^2 **30.** $(-a)^5$

31. $\left(-\dfrac{1}{2}\right)^5$

32. $\left(-\dfrac{13}{12}\right)^3$

33. $(0.22)^4$

34. $(1.25)^6$

Evaluate each exponential expression. See Examples 4 and 5.

35. 3^4 **36.** 5^3 **37.** 0^9

38. 0^{12} **39.** $(-5)^4$ **40.** $(-2)^5$

41. $(-6)^3$ **42.** $(-12)^2$ **43.** $(10)^5$

44. $(-10)^6$ **45.** $(-0.1)^3$ **46.** $(-0.2)^2$

47. $\left(\dfrac{1}{2}\right)^3$ **48.** $\left(\dfrac{2}{3}\right)^3$ **49.** $\left(-\dfrac{1}{2}\right)^2$

50. $\left(-\dfrac{2}{3}\right)^2$ **51.** -8^2 **52.** -7^2

53. -8^4 **54.** -7^4

55. $-(7-10)^3$ **56.** $-(6-9)^4$

57. $(-2^2)-(3^2)$ **58.** $(-3^4)-(-5^2)$

Evaluate each expression. See Example 6.

59. $3^2\cdot 2^2$ **60.** $5\cdot 10^2$

61. $-3\cdot 2+4\cdot 6$ **62.** $-5\cdot 4-8\cdot 3$

63. $(-3)^3+2^3$ **64.** $3^2-5(-1)^3$

65. $-21+36\div 3^2$ **66.** $-18-9^2\div 3^3$

67. $-3\cdot 2^3-5\cdot 2^2$ **68.** $2\cdot 5-3^2+4\cdot 0$

69. $\dfrac{-8}{2}+2\cdot 3\cdot 5-2^3$ **70.** $-4\cdot 2\cdot 6-\dfrac{12}{3}+3^3$

Evaluate each expression. See Example 7.

71. $(-3+4^2)(-6)$ **72.** $-3\cdot(2^3+4)\cdot 5$

73. $(-3\cdot 2+6)^3$ **74.** $5-2(-3+2)^3$

75. $2-5(3-4\cdot 2)$ **76.** $(3-7)(4-6\cdot 2)$

77. $3-2\cdot|5-6|$ **78.** $3-|6-7\cdot 3|$

79. $(3^2-5)\cdot|3\cdot 2-8|$

80. $|4-6\cdot 3|+|6-9|$

81. $\dfrac{3-4\cdot 6}{7-10}$ **82.** $\dfrac{6-(-8)^2}{-3-(-1)}$

83. $\dfrac{7 - 9 - 3^2}{9 - 7 - 3}$ **84.** $\dfrac{3^2 - 2 \cdot 4}{-30 + 2 \cdot 4^2}$

Evaluate each expression. See Example 8.

85. $3 + 4[9 - 6(2 - 5)]$

86. $9 + 3[5 - (3 - 6)^2]$

87. $6^2 - [(2 + 3)^2 - 10]$

88. $3[(2 - 3)^2 + (6 - 4)^2]$

89. $4 - 5 \cdot |3 - (3^2 - 7)|$

90. $2 + 3 \cdot |4 - (7^2 - 6^2)|$

91. $-2|3 - (7 - 3)| - |-9|$

92. $[3 - (2 - 4)][3 + |2 - 4|]$

Evaluate each expression. Use a calculator to check.

93. $1 + 2^3$ **94.** $(1 + 2)^3$

95. $(-2)^2 - 4(-1)(3)$ **96.** $(-2)^2 - 4(-2)(-3)$

97. $4^2 - 4(1)(-3)$ **98.** $3^2 - 4(-2)(3)$

99. $(-11)^2 - 4(5)(0)$ **100.** $(-12)^2 - 4(3)(0)$

101. $-5^2 - 3 \cdot 4^2$ **102.** $-6^2 - 5(-3)^2$

103. $[3 + 2(-4)]^2$ **104.** $[6 - 2(-3)]^2$

105. $|-1| - |-1|$ **106.** $4 - |1 - 7|$

107. $\dfrac{4 - (-4)}{-2 - 2}$ **108.** $\dfrac{3 - (-7)}{3 - 5}$

109. $3(-1)^2 - 5(-1) + 4$

110. $-2(1)^2 - 5(1) - 6$

111. $5 - 2^2 + 3^4$ **112.** $5 + (-2)^2 - 3^2$

113. $-2 \cdot |9 - 6^2|$ **114.** $8 - 3|5 - 4^2 + 1|$

115. $-3^2 - 5[4 - 2(4 - 9)]$

116. $-2[(3 - 4)^3 - 5] + 7$

117. $1 - 5|5 - (9 + 1)|$

118. $|6 - 3 \cdot 7| + |7 - (5 - 2)|$

Use a calculator to evaluate each expression. Round approximate answers to four decimal places.

119. $3.2^2 - 4(3.6)(-2.2)$

120. $(-4.5)^2 - 4(-2.8)(-4.6)$

121. $(5.63)^3 - [4.7 - (-3.3)^2]$

122. $9.8^3 - [1.2 - (4.4 - 9.6)^2]$

123. $\dfrac{3.44 - (-8.32)}{6.89 - 5.43}$

124. $\dfrac{-4.56 - 3.22}{3.44 - (-6.26)}$

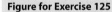

Solve each problem.

125. ***Population of the United States.*** In 2004 the population of the United States was 294.4 million (U.S. Census Bureau, www.census.gov). If the population continues to grow at an annual rate of 1.05%, then the population in the year 2015 will be $294.4(1.0105)^{11}$ million.

a) Evaluate the expression to find the predicted population in 2015 to the nearest tenth of a million people.

b) Use the accompanying graph to estimate the year in which the population will reach 350 million people.

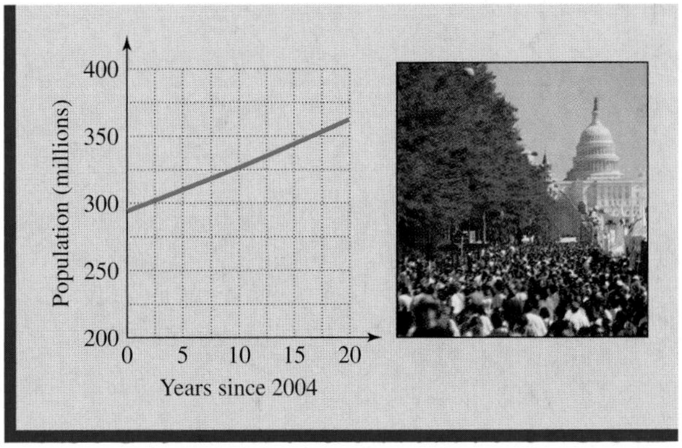

Figure for Exercise 125

126. ***Population of Mexico.*** In 2004 the population of Mexico was 106.5 million. If Mexico's population continues to grow at an annual rate of 1.43%, then the population in 2015 will be $106.5(1.0143)^{11}$ million.

a) Find the predicted population in 2015 to the nearest tenth of a million people.

b) Use the result of Exercise 125 to determine whether United States or Mexico will have the greater increase in population between 2004 and 2015.

Getting More Involved

127. ***Discussion***

How do the expressions $(-5)^3$, $-(5^3)$, -5^3, $-(-5)^3$, and $-1 \cdot 5^3$ differ?

128. ***Discussion***

How do the expressions $(-4)^4$, $-(4^4)$, -4^4, $-(-4)^4$, and $-1 \cdot 4^4$ differ?

1.6 Algebraic Expressions

In this Section

- Identifying Algebraic Expressions
- Translating Algebraic Expressions
- Evaluating Algebraic Expressions
- Equations
- Applications

In Section 1.5 you studied arithmetic expressions. In this section you will study expressions that are more general—expressions that involve variables.

Identifying Algebraic Expressions

Since variables (or letters) are used to represent numbers, we can use variables in arithmetic expressions. The result of combining numbers and variables with the ordinary operations of arithmetic (in some meaningful way) is called an **algebraic expression** or simply an **expression.** For example,

$$x + 2, \qquad \pi r^2, \qquad b^2 - 4ac, \qquad \text{and} \qquad \frac{a - b}{c - d}$$

are algebraic expressions.

Expressions are often named by the last operation to be performed in the expression. For example, the expression $x + 2$ is a **sum** because the only operation in the expression is addition. The expression $a - bc$ is referred to as a **difference** because subtraction is the last operation to be performed. The expression $3(x - 4)$ is a **product,** while $\frac{3}{x - 4}$ is a **quotient.** The expression $(a + b)^2$ is a **square** because the addition is performed before the square is found.

EXAMPLE 1

Naming expressions

Identify each expression as either a sum, difference, product, quotient, or square.

a) $3(x + 2)$ **b)** $b^2 - 4ac$

c) $\frac{a - b}{c - d}$ **d)** $(a - b)^2$

Solution

a) In $3(x + 2)$ we add before we multiply. So this expression is a product.

b) By the order of operations the last operation to perform in $b^2 - 4ac$ is subtraction. So this expression is a difference.

c) The last operation to perform in this expression is division. So this expression is a quotient.

d) In $(a - b)^2$ we subtract before we square. This expression is a square.

Now do Exercises 7–18

Helpful Hint

Sum, difference, product, and quotient are nouns. They are used as names for expressions. Add, subtract, multiply, and divide are verbs. They indicate an action to perform.

Translating Algebraic Expressions

Algebra is useful because it can be used to solve problems. Since problems are often communicated verbally, we must be able to translate verbal expressions into algebraic expressions and translate algebraic expressions into verbal expressions. Consider the following examples of verbal expressions and their corresponding algebraic expressions.

Verbal Expressions and Corresponding Algebraic Expressions

Verbal Expression	Algebraic Expression
The sum of $5x$ and 3	$5x + 3$
The product of 5 and $x + 3$	$5(x + 3)$
The sum of 8 and $\frac{x}{3}$	$8 + \frac{x}{3}$
The quotient of $8 + x$ and 3	$\frac{8 + x}{3}$, $(8 + x)/3$, or $(8 + x) \div 3$
The difference of 3 and x^2	$3 - x^2$
The square of $3 - x$	$(3 - x)^2$

Note that the word "difference" must be used carefully. To be consistent, we say that the difference between a and b is $a - b$. So the difference between 10 and 12 is $10 - 12$ or -2. However, outside of a textbook most people would say that the difference in age between a 10-year-old and a 12-year-old is 2, not -2. Users of the English language do not follow precise rules like we follow in mathematics. Of course, in mathematics we must make our mathematics and our English sentences perfectly clear. So we try to avoid using "difference" in an ambiguous or vague manner. (We will study verbal and algebraic expressions further in Section 2.5.)

Example 2 shows how the terms sum, difference, product, quotient, and square are used to describe expressions.

EXAMPLE 2

Algebraic expressions to verbal expressions

Translate each algebraic expression into a verbal expression. Use the word sum, difference, product, quotient, or square.

a) $\frac{3}{x}$ **b)** $2y + 1$ **c)** $3x - 2$ **d)** $(a - b)(a + b)$ **e)** $(a + b)^2$

Solution

a) The quotient of 3 and x

b) The sum of $2y$ and 1

c) The difference of $3x$ and 2

d) The product of $a - b$ and $a + b$

e) The square of the sum $a + b$

Now do Exercises 19–28

EXAMPLE 3

Verbal expressions to algebraic expressions

Translate each verbal expression into an algebraic expression.

a) The quotient of $a + b$ and 5

b) The difference of x^2 and y^2

c) The product of π and r^2

d) The square of the difference $x - y$

Solution

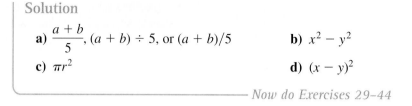

a) $\frac{a+b}{5}$, $(a+b) \div 5$, or $(a+b)/5$ **b)** $x^2 - y^2$

c) πr^2 **d)** $(x - y)^2$

Now do Exercises 29–44

Study Tip

Get to class early so that you are relaxed and ready to go when class starts. Collect your thoughts and get your questions ready. If your instructor arrives early, you might be able to get your questions answered before class. Take responsibility for your education. For many come to learn, but not all learn.

Evaluating Algebraic Expressions

The value of an algebraic expression depends on the values given to the variables. For example, the value of $x - 2y$ when $x = -2$ and $y = -3$ is found by replacing x and y by -2 and -3, respectively:

$$x - 2y = -2 - 2(-3) = -2 - (-6) = 4$$

If $x = 1$ and $y = 2$, the value of $x - 2y$ is found by replacing x by 1 and y by 2, respectively:

$$x - 2y = 1 - 2(2) = 1 - 4 = -3$$

Note that we use the order of operations when evaluating an algebraic expression.

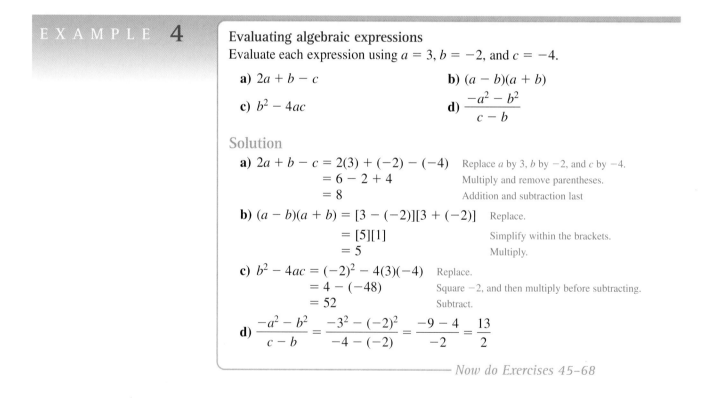

EXAMPLE 4

Evaluating algebraic expressions

Evaluate each expression using $a = 3$, $b = -2$, and $c = -4$.

a) $2a + b - c$ **b)** $(a - b)(a + b)$

c) $b^2 - 4ac$ **d)** $\frac{-a^2 - b^2}{c - b}$

Solution

a) $2a + b - c = 2(3) + (-2) - (-4)$ Replace a by 3, b by -2, and c by -4.
$= 6 - 2 + 4$ Multiply and remove parentheses.
$= 8$ Addition and subtraction last

b) $(a - b)(a + b) = [3 - (-2)][3 + (-2)]$ Replace.
$= [5][1]$ Simplify within the brackets.
$= 5$ Multiply.

c) $b^2 - 4ac = (-2)^2 - 4(3)(-4)$ Replace.
$= 4 - (-48)$ Square -2, and then multiply before subtracting.
$= 52$ Subtract.

d) $\frac{-a^2 - b^2}{c - b} = \frac{-3^2 - (-2)^2}{-4 - (-2)} = \frac{-9 - 4}{-2} = \frac{13}{2}$

Now do Exercises 45–68

Mathematical notation is readily available in scientific word processors. However, on Internet pages or in email, multiplication is often written with a star (*), fractions are written with a slash (/), and exponents with a caret (^). For example, $\frac{x+y}{2x^3}$ is written as $(x + y)/(2*x\text{^}3)$. If the numerator or denominator contain more than one symbol it is best to enclose them in parentheses to avoid confusion. An expression such as $1/2x$ is confusing. If your class evaluates it for $x = 4$, some students will probably assume that it is $1/(2x)$ and get $1/8$, and some will assume that it is $(1/2)x$ and get 2.

Equations

An **equation** is a statement of equality of two expressions. For example,

$$11 - 5 = 6, \qquad x + 3 = 9, \qquad 2x + 5 = 13, \qquad \text{and} \qquad \frac{x}{2} - 4 = 1$$

are equations. In an equation involving a variable, any number that gives a true statement when we replace the variable by the number is said to **satisfy** the equation and is called a **solution** or **root** to the equation. For example, 6 is a solution to $x + 3 = 9$ because $6 + 3 = 9$ is true. Because $5 + 3 = 9$ is false, 5 is not a solution to the equation $x + 3 = 9$. We have **solved** an equation when we have found all solutions to the equation. You will learn how to solve certain equations in Chapter 2.

EXAMPLE 5

Satisfying an equation

Determine whether the given number is a solution to the equation following it.

a) $6,\ 3x - 7 = 9$

b) $-3,\ \dfrac{2x - 4}{5} = -2$

c) $-5,\ -x - 2 = 3(x + 6)$

Study Tip

Ask questions in class. If you don't ask questions, then the instructor might believe that you have total understanding. When one student has a question, there are usually several who have the same question but do not speak up. Asking questions not only helps you to learn, but it keeps the classroom more lively and interesting.

Solution

a) Replace x by 6 in the equation $3x - 7 = 9$:

$$3(6) - 7 = 9$$
$$18 - 7 = 9$$
$$11 = 9 \quad \text{False}$$

The number 6 is not a solution to the equation $3x - 7 = 9$.

b) Replace x by -3 in the equation $\dfrac{2x - 4}{5} = -2$:

$$\frac{2(-3) - 4}{5} = -2$$
$$\frac{-10}{5} = -2$$
$$-2 = -2 \quad \text{True}$$

The number -3 is a solution to the equation.

c) Replace x by -5 in $-x - 2 = 3(x + 6)$:

$$-(-5) - 2 = 3(-5 + 6)$$
$$5 - 2 = 3(1)$$
$$3 = 3 \quad \text{True}$$

The number -5 is a solution to the equation $-x - 2 = 3(x + 6)$.

Now do Exercises 69–84

Just as we translated verbal expressions into algebraic expressions, we can translate verbal sentences into algebraic equations. In an algebraic equation we use the equality symbol (=). Equality is indicated in words by phrases such as "is equal to," "is the same as," or simply "is."

EXAMPLE 6

Writing equations

Translate each sentence into an equation.

a) The sum of x and 7 is 12.

b) The product of 4 and x is the same as the sum of y and 5.

c) The quotient of $x + 3$ and 5 is equal to -1.

Solution

a) $x + 7 = 12$ **b)** $4x = y + 5$ **c)** $\frac{x + 3}{5} = -1$

Now do Exercises 85–92

Applications

Algebraic expressions are used to describe or **model** real-life situations. We can evaluate an algebraic expression for many values of a variable to get a collection of data. A graph (picture) of this data can give us useful information. For example, a forensic scientist can use a graph to estimate the length of a person's femur from the person's height.

EXAMPLE 7

Reading a graph

A forensic scientist uses the expression $69.1 + 2.2F$ as an estimate of the height in centimeters of a male with a femur of length F centimeters (National Space Biomedical Research Institute, www.nsbri.org).

a) If the femur of a male skeleton measures 50.6 cm, then what was the person's height?

b) Use the graph shown in Fig. 1.25 to estimate the length of a femur for a person who is 150 cm tall.

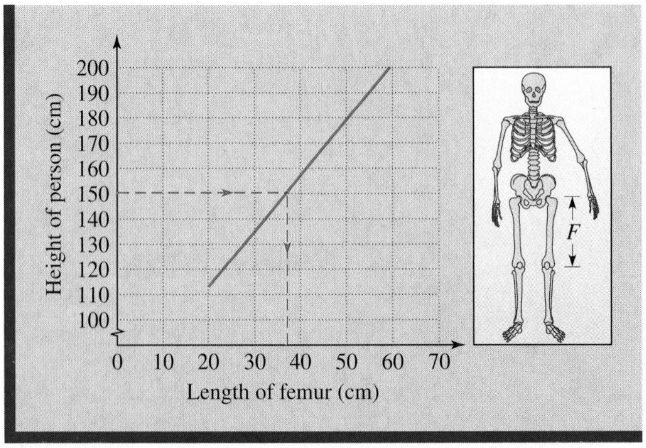

Figure 1.25

Study Tip

Find a group of students to work with outside of class. Don't just settle for answers. Make sure that everyone in the group understands the solution to a problem. You will find that you really understand a concept when you can explain it to someone else.

Solution

a) To find the height of the person, we use $F = 50.6$ in the expression $69.1 + 2.2F$:

$$69.1 + 2.2(50.6) \approx 180.4$$

So the person was approximately 180.4 cm tall.

b) To find the length of a femur for a person who is 150 cm tall, first locate 150 cm on the height scale of the graph in Fig. 1.25. Now draw a horizontal line to the graph and then a vertical line down to the length scale. So the length of a femur for a person who is 150 cm tall is approximately 36 cm.

Now do Exercises 101–106

Warm-Ups ▼

True or false? Explain your answer.

1. The expression $2x + 3y$ is referred to as a sum.
2. The expression $5(y - 9)$ is a difference.
3. The expression $2(x + 3y)$ is a product.
4. The expression $\frac{x}{2} + \frac{y}{3}$ is a quotient.
5. The expression $(a - b)(a + b)$ is a product of a sum and a difference.
6. If x is -2, then the value of $2x + 4$ is 8.
7. If $a = -3$, then $a^3 - 5 = 22$.
8. The number 5 is a solution to the equation $2x - 3 = 13$.
9. The product of $x + 3$ and 5 is $(x + 3)5$.
10. The expression $2(x + 7)$ should be read as "the sum of 2 times x plus 7."

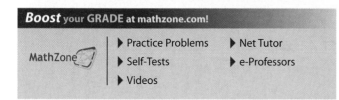

1.6 Exercises

Reading and Writing *After reading this section write out the answers to these questions. Use complete sentences.*

1. What is an algebraic expression?
2. What is the difference between an algebraic expression and an arithmetic expression?
3. How can you tell whether an algebraic expression should be referred to as a sum, difference, product, quotient, or square?
4. How do you evaluate an algebraic expression?
5. What is an equation?
6. What is a solution to an equation?

Identify each expression as a sum, difference, product, quotient, square, or cube. See Example 1.

7. $a^3 - 1$
8. $b(b - 1)$

9. $(w - 1)^3$

10. $m^2 + n^2$

11. $3x + 5y$

12. $\frac{a - b}{b - a}$

13. $\frac{u}{v} - \frac{v}{u}$

14. $(s - t)^2$

15. $3(x + 5y)$

16. $a - \frac{a}{2}$

17. $\left(\frac{2}{z}\right)^2$

18. $(2q - p)^3$

Use the term sum, difference, product, quotient, square, or cube to translate each algebraic expression into a verbal expression. See Example 2.

19. $x^2 - a^2$

20. $a^3 + b^3$

21. $(x - a)^2$

22. $(a + b)^3$

23. $\frac{x - 4}{2}$

24. $2(x - 3)$

25. $\frac{x}{2} - 4$

26. $2x - 3$

27. $(ab)^3$

28. a^3b^3

Translate each verbal expression into an algebraic expression. Do not simplify. See Example 3.

29. The sum of 8 and y

30. The sum of $8x$ and $3y$

31. The product of $5x$ and z

32. The product of $x + 9$ and $x + 12$

33. The difference of 8 and $7x$

34. The difference of a^3 and b^3

35. The quotient of 6 and $x + 4$

36. The quotient of $x - 7$ and $7 - x$

37. The square of $a + b$

38. The cube of $x - y$

39. The sum of the cube of x and the square of y

40. The quotient of the square of a and the cube of b

41. The product of 5 and the square of m

42. The difference of the square of m and the square of n

43. The square of the sum of s and t

44. The cube of the difference of a and b

Evaluate each expression using $a = -1$, $b = 2$, and $c = -3$. See Example 4.

45. $-(a - b)$

46. $b - a$

47. $-b^2 + 7$

48. $-c^2 - b^2$

49. $c^2 - 2c + 1$

50. $b^2 - 2b + 4$

51. $a^3 - b^3$

52. $b^3 - c^3$

53. $(a - b)(a + b)$

54. $(a - c)(a + c)$

55. $b^2 - 4ac$

56. $a^2 - 4bc$

57. $\frac{a - c}{a - b}$

58. $\frac{b - c}{b + a}$

59. $\frac{2}{a} + \frac{6}{b} - \frac{9}{c}$

60. $\frac{c}{a} + \frac{6}{b} - \frac{b}{a}$

61. $a \div |-a|$

62. $|a| \div a$

63. $|b| - |a|$

64. $|c| + |b|$

65. $-|-a - c|$

66. $-|-a - b|$

67. $(3 - |a - b|)^2$

68. $(|b + c| - 2)^3$

Determine whether the given number is a solution to the equation following it. See Example 5.

69. $2, 3x + 7 = 13$

70. $-1, -3x + 7 = 10$

71. $-2, \frac{3x - 4}{2} = 5$

72. $-3, \frac{-2x + 9}{3} = 5$

73. $-2, -x + 4 = 6$

74. $-9, -x + 3 = 12$

75. $4, 3x - 7 = x + 1$

76. $5, 3x - 7 = 2x + 1$

77. $3, -2(x - 1) = 2 - 2x$

78. $-8, x - 9 = -(9 - x)$

79. $1, x^2 + 3x - 4 = 0$

80. $-1, x^2 + 5x + 4 = 0$

81. $8, \frac{x}{x - 8} = 0$

82. $3, \frac{x - 3}{x + 3} = 0$

83. $-6, \frac{x + 6}{x + 6} = 1$

84. $9, \frac{9}{x - 9} = 0$

Translate each sentence into an equation. See Example 6.

85. The sum of $5x$ and $3x$ is $8x$.

86. The sum of $\frac{y}{2}$ and 3 is 7.

87. The product of 3 and $x + 2$ is equal to 12.

88. The product of -6 and $7y$ is equal to 13.

89. The quotient of x and 3 is the same as the product of x and 5.

90. The quotient of $x + 3$ and $5y$ is the same as the product of x and y.

91. The square of the sum of a and b is equal to 9.

92. The sum of the squares of a and b is equal to the square of c.

Fill in the tables with the appropriate values for the given expressions.

93.

x	$2x - 3$
-2	
-1	
0	
1	
2	

94.

x	$-\frac{1}{2}x + 4$
-4	
-2	
0	
2	
4	

95.

a	a^2	a^3	a^4
2			
$\frac{1}{2}$			
10			
0.1			

96.

b	$\frac{1}{b}$	$\frac{1}{b^2}$	$\frac{1}{b^3}$
3			
$\frac{1}{3}$			
10			
0.1			

Use a calculator to find the value of $b^2 - 4ac$ for each of the following choices of a, b, and c.

97. $a = 4.2$, $b = 6.7$, $c = 1.8$

98. $a = -3.5$, $b = 9.1$, $c = 3.6$

99. $a = -1.2$, $b = 3.2$, $c = 5.6$

100. $a = 2.4$, $b = -8.5$, $c = -5.8$

Solve each problem. See Example 7.

101. ***Forensics.*** A forensic scientist uses the expression $81.7 + 2.4T$ to estimate the height in centimeters of a male with a tibia of length T centimeters. If a male skeleton has a tibia of length 36.5 cm, then what was the height of the person? Use the accompanying graph to estimate the length of a tibia for a male with a height of 180 cm.

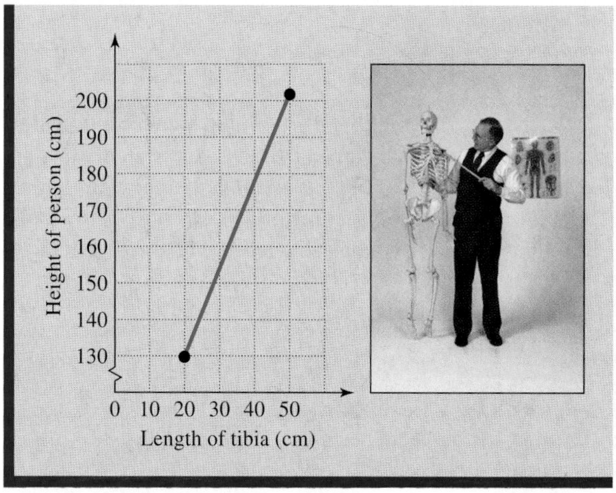

Figure for Exercise 101

102. ***Forensics.*** A forensic scientist uses the expression $72.6 + 2.5T$ to estimate the height in centimeters of a female with a tibia of length T centimeters. If a female skeleton has a tibia of length 32.4 cm, then what was the height of the person? Find the length of your tibia in centimeters, and use the expression from this exercise or the previous exercise to estimate your height.

103. ***Games behind.*** In baseball a team's standing is measured by its percentage of wins and by the number of games it

	W	L	Pct	GB
NY Yankees	101	61	0.623	–
Boston	95	67	0.586	?
Toronto	86	76	0.531	?
Baltimore	71	91	0.438	?
Tampa Bay	63	99	0.389	?

Table for Exercise 103

is behind the leading team in its division. The expression

$$\frac{(X - x) + (y - Y)}{2}$$

gives the number of games behind for a team with x wins and y losses, where the division leader has X wins and Y losses. The table shown on the previous page gives the won-lost records for the American League East at the end of 2001 (www.espn.com). Fill in the column for the games behind (GB).

104. ***Fly ball.*** The approximate distance in feet that a baseball travels when hit at an angle of 45° is given by the expression

$$\frac{(v_0)^2}{32}$$

where v_0 is the initial velocity in feet per second. If Barry Bonds of the Giants hits a ball at a 45° angle with an initial velocity of 120 feet per second, then how far will the ball travel? Use the accompanying graph to estimate the initial velocity for a ball that has traveled 370 feet.

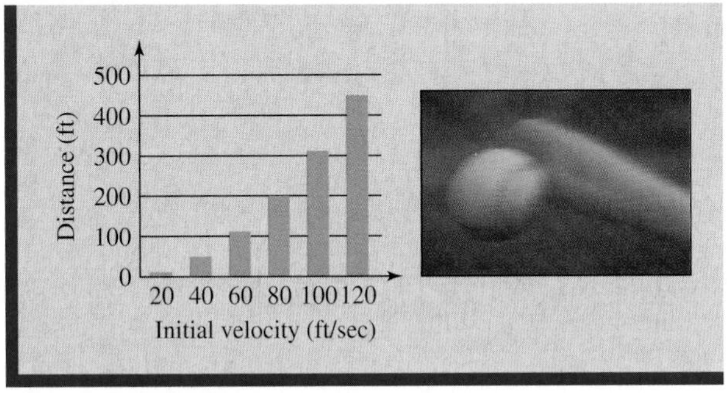

Figure for Exercise 104

105. ***Football field.*** The expression $2L + 2W$ gives the perimeter of a rectangle with length L and width W. What is the perimeter of a football field with length 100 yards and width 160 feet?

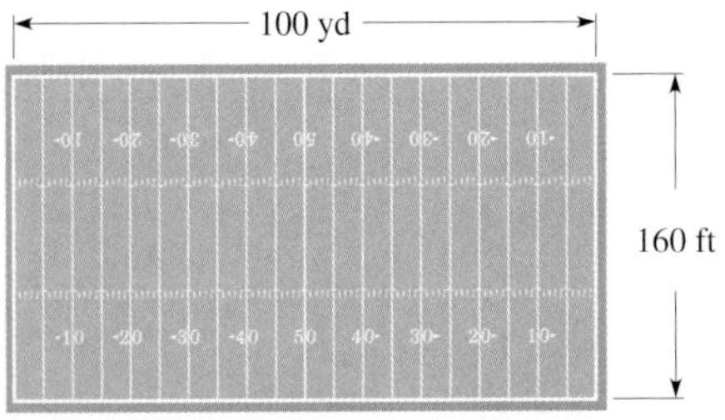

Figure for Exercise 105

106. ***Crop circles.*** The expression πr^2 gives the area of a circle with radius r. How many square meters of wheat were destroyed when an alien ship made a crop circle of diameter 25 meters in the wheat field at the Southwind Ranch? Find π on your calculator.

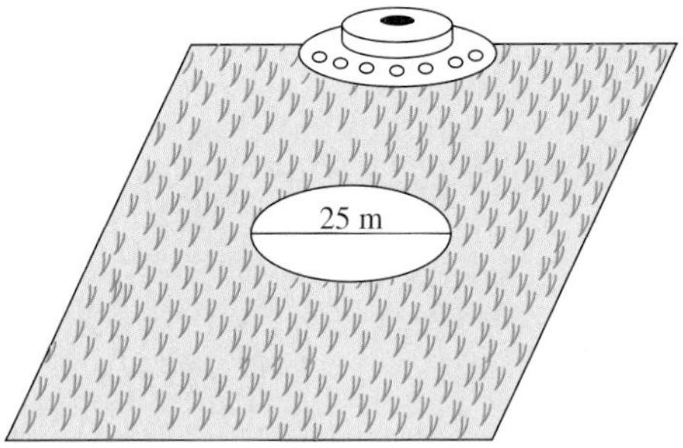

Figure for Exercise 106

Getting More Involved

107. ***Writing***

Explain why the square of the sum of two numbers is different from the sum of the squares of two numbers.

108. ***Cooperative learning***

The sum of the integers from 1 through n is $\frac{n(n+1)}{2}$. The sum of the squares of the integers from 1 through n is $\frac{n(n+1)(2n+1)}{6}$. The sum of the cubes of the integers from 1 through n is $\frac{n^2(n+1)^2}{4}$. Use the appropriate expressions to find the following values.

a) The sum of the integers from 1 through 50

b) The sum of the squares of the integers from 1 through 40

c) The sum of the cubes of the integers from 1 through 30

d) The square of the sum of the integers from 1 through 20

e) The cube of the sum of the integers from 1 through 10

1.7 Properties of the Real Numbers

In this Section

- The Commutative Properties
- The Associative Properties
- The Distributive Property
- The Identity Properties
- The Inverse Properties
- Multiplication Property of Zero
- Applications

Everyone knows that the price of a hamburger plus the price of a Coke is the same as the price of a Coke plus the price of a hamburger. But do you know that this example illustrates the commutative property of addition? The properties of the real numbers are commonly used by anyone who performs the operations of arithmetic. In algebra we must have a thorough understanding of these properties.

The Commutative Properties

We get the same result whether we evaluate $3 + 5$ or $5 + 3$. This example illustrates the commutative property of addition. The fact that $4 \cdot 6$ and $6 \cdot 4$ are equal illustrates the commutative property of multiplication.

Commutative Property of Addition

For any real numbers a and b,

$$a + b = b + a.$$

Commutative Property of Multiplication

For any real numbers a and b,

$$ab = ba.$$

EXAMPLE 1

The commutative property of addition

Use the commutative property of addition to rewrite each expression.

a) $2 + (-10)$ **b)** $8 + x^2$ **c)** $2y - 4x$

Solution

a) $2 + (-10) = -10 + 2$

b) $8 + x^2 = x^2 + 8$

c) $2y - 4x = 2y + (-4x) = -4x + 2y$

Now do Exercises 7–12

EXAMPLE 2

The commutative property of multiplication

Use the commutative property of multiplication to rewrite each expression.

a) $n \cdot 3$ **b)** $(x + 2) \cdot 3$ **c)** $5 - yx$

Solution

a) $n \cdot 3 = 3 \cdot n = 3n$ **b)** $(x + 2) \cdot 3 = 3(x + 2)$

c) $5 - yx = 5 - xy$

Now do Exercises 13–18

Addition and multiplication are commutative operations, but what about subtraction and division? Since $5 - 3 = 2$ and $3 - 5 = -2$, subtraction is not commutative. To see that division is not commutative, try dividing \$8 among 4 people and \$4 among 8 people.

The Associative Properties

Consider the computation of $2 + 3 + 6$. Using the order of operations, we add 2 and 3 to get 5 and then add 5 and 6 to get 11. If we add 3 and 6 first to get 9 and then add 2 and 9, we also get 11. So

$$(2 + 3) + 6 = 2 + (3 + 6).$$

We get the same result for either order of addition. This property is called the **associative property of addition.** The commutative and associative properties of addition are the reason that a hamburger, a Coke, and French fries cost the same as French fries, a hamburger, and a Coke.

We also have an **associative property of multiplication.** Consider the following two ways to find the product of 2, 3, and 4:

$$(2 \cdot 3)4 = 6 \cdot 4 = 24$$

$$2(3 \cdot 4) = 2 \cdot 12 = 24$$

We get the same result for either arrangement.

Helpful Hint

In arithmetic we would probably write $(2 + 3) + 7 = 12$ without thinking about the associative property. In algebra, we need the associative property to understand that

$$\begin{aligned}(x + 3) + 7 &= x + (3 + 7)\\ &= x + 10.\end{aligned}$$

Associative Property of Addition

For any real numbers a, b, and c,

$$(a + b) + c = a + (b + c).$$

Associative Property of Multiplication

For any real numbers a, b, and c,

$$(ab)c = a(bc).$$

EXAMPLE 3

Using the properties of multiplication

Use the commutative and associative properties of multiplication and exponential notation to rewrite each product.

a) $(3x)(x)$ **b)** $(xy)(5yx)$

Solution

a) $(3x)(x) = 3(x \cdot x) = 3x^2$

b) The commutative and associative properties of multiplication allow us to rearrange the multiplication in any order. We generally write numbers before variables, and we usually write variables in alphabetical order:

$$(xy)(5yx) = 5xxyy = 5x^2y^2$$

Now do Exercises 19–24

Study Tip

Find out what help is available at your school. Accompanying this text are video tapes, solution manuals, and a computer tutorial. Around most campuses you will find tutors available for hire, but most schools have a math lab where you can get help for free. Some schools even have free one-on-one tutoring available through special programs.

Consider the expression

$$3 - 9 + 7 - 5 - 8 + 4 - 13.$$

According to the accepted order of operations, we could evaluate this by computing from left to right. However, using the definition of subtraction, we can rewrite this expression as addition:

$$3 + (-9) + 7 + (-5) + (-8) + 4 + (-13)$$

The commutative and associative properties of addition allow us to add these numbers in any order we choose. It is usually faster to add the positive numbers, add the negative numbers, and then combine those two totals:

$$3 + 7 + 4 + (-9) + (-5) + (-8) + (-13) = 14 + (-35) = -21$$

Note that by performing the operations in this manner, we must subtract only once. There is no need to rewrite this expression as we have done here. We can sum the positive numbers and the negative numbers from the original expression and then combine their totals.

EXAMPLE 4

Using the properties of addition

Evaluate.

a) $3 - 7 + 9 - 5$

b) $4 - 5 - 9 + 6 - 2 + 4 - 8$

Solution

a) First add the positive numbers and the negative numbers:

$$\begin{aligned} 3 - 7 + 9 - 5 &= 12 + (-12) \\ &= 0 \end{aligned}$$

b) $$\begin{aligned} 4 - 5 - 9 + 6 - 2 + 4 - 8 &= 14 + (-24) \\ &= -10 \end{aligned}$$

Now do Exercises 25–34

It is certainly not essential that we evaluate the expressions of Example 4 as shown. We get the same answer by adding and subtracting from left to right. However, in algebra, just getting the answer is not always the most important point. Learning new methods often increases understanding.

Even though addition is associative, subtraction is not an associative operation. For example, $(8 - 4) - 3 = 1$ and $8 - (4 - 3) = 7$. So

$$(8 - 4) - 3 \neq 8 - (4 - 3).$$

We can also use a numerical example to show that division is not associative. For instance, $(16 \div 4) \div 2 = 2$ and $16 \div (4 \div 2) = 8$. So

$$(16 \div 4) \div 2 \neq 16 \div (4 \div 2).$$

The Distributive Property

If four men and five women pay \$3 each for a movie, there are two ways to find the total amount spent:

$$3(4 + 5) = 3 \cdot 9 = 27$$

$$3 \cdot 4 + 3 \cdot 5 = 12 + 15 = 27$$

Helpful Hint

To visualize the distributive property, we can determine the number of circles shown here in two ways:

o o o o   o o o o o
o o o o   o o o o o
o o o o   o o o o o

There are $3 \cdot 9$ or 27 circles, or there are $3 \cdot 4$ circles in the first group and $3 \cdot 5$ circles in the second group for a total of 27 circles.

Since we get \$27 either way, we can write

$$3(4 + 5) = 3 \cdot 4 + 3 \cdot 5.$$

We say that the multiplication by 3 is *distributed* over the addition. This example illustrates the **distributive property.**

Consider the following expressions involving multiplication and subtraction:

$$5(6 - 4) = 5 \cdot 2 = 10$$

$$5 \cdot 6 - 5 \cdot 4 = 30 - 20 = 10$$

Since both expressions have the same value, we can write

$$5(6 - 4) = 5 \cdot 6 - 5 \cdot 4.$$

Multiplication by 5 is distributed over each number in the parentheses. This example illustrates that multiplication distributes over subtraction.

Distributive Property

For any real numbers a, b, and c,

$$a(b + c) = ab + ac \quad \text{and} \quad a(b - c) = ab - ac.$$

We can use the distributive property to remove parentheses. If we start with $4(x + 3)$ and write

$$4(x + 3) = 4x + 4 \cdot 3 = 4x + 12,$$

we are using it to multiply 4 and $x + 3$ or to remove the parentheses. We wrote the product $4(x + 3)$ as the sum $4x + 12$.

EXAMPLE 5

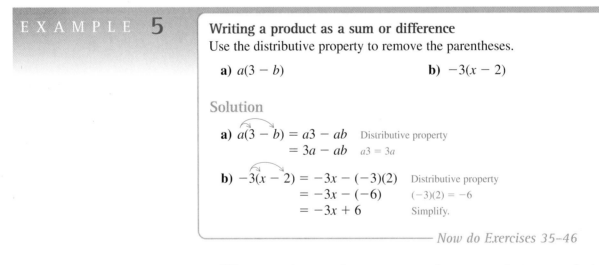

Writing a product as a sum or difference

Use the distributive property to remove the parentheses.

a) $a(3 - b)$ **b)** $-3(x - 2)$

Solution

a) $a(3 - b) = a3 - ab$ Distributive property

$= 3a - ab$ $a3 = 3a$

b) $-3(x - 2) = -3x - (-3)(2)$ Distributive property

$= -3x - (-6)$ $(-3)(2) = -6$

$= -3x + 6$ Simplify.

Now do Exercises 35–46

When we write a number or an expression as a product, we are **factoring.** If we start with $3x + 15$ and write

$$3x + 15 = 3x + 3 \cdot 5 = 3(x + 5),$$

we are using the distributive property to factor $3x + 15$. We factored out the common factor 3.

EXAMPLE 6

Writing a sum or difference as a product

Use the distributive property to factor each expression.

a) $7x - 21$ **b)** $5a + 5$

Solution

a) $7x - 21 = 7x - 7 \cdot 3$ Write 21 as $7 \cdot 3$.

$= 7(x - 3)$ Distributive property

b) $5a + 5 = 5a + 5 \cdot 1$ Write 5 as $5 \cdot 1$.

$= 5(a + 1)$ Factor out the common factor 5.

Now do Exercises 47–58

Study Tip

Don't cram for a test. Some students try to cram weeks of work into one "all-nighter." These same students are seen frantically paging through the text up until the moment that the test papers are handed out. These practices create a lot of test anxiety and will only make you sick. Start studying for a test several days in advance, and get a good night's sleep before a test. If you keep up with homework, then there will be no need to cram.

The Identity Properties

The numbers 0 and 1 have special properties. Multiplication of a number by 1 does not change the number, and addition of 0 to a number does not change the number. That is why 1 is called the **multiplicative identity** and 0 is called the **additive identity.**

Additive Identity Property

For any real number a,

$$a + 0 = 0 + a = a.$$

Multiplicative Identity Property

For any real number a,

$$a \cdot 1 = 1 \cdot a = a.$$

The Inverse Properties

The idea of additive inverses was introduced in Section 1.3. Every real number a has an **additive inverse** or **opposite,** $-a$, such that $a + (-a) = 0$. Every nonzero real number a also has a **multiplicative inverse** or **reciprocal,** written $\frac{1}{a}$, such that $a \cdot \frac{1}{a} = 1$. Note that the sum of additive inverses is the additive identity and that the product of multiplicative inverses is the multiplicative identity.

Additive Inverse Property

For any real number a, there is a unique number $-a$ such that

$$a + (-a) = 0.$$

Multiplicative Inverse Property

For any nonzero real number a, there is a unique number $\frac{1}{a}$ such that

$$a \cdot \frac{1}{a} = 1.$$

We are already familiar with multiplicative inverses for rational numbers. For example, the multiplicative inverse of $\frac{2}{3}$ is $\frac{3}{2}$ because

$$\frac{2}{3} \cdot \frac{3}{2} = \frac{6}{6} = 1.$$

EXAMPLE 7

Multiplicative inverses

Find the multiplicative inverse of each number.

a) 5 **b)** 0.3

c) $-\frac{3}{4}$ **d)** 1.7

Solution

a) The multiplicative inverse of 5 is $\frac{1}{5}$ because

$$5 \cdot \frac{1}{5} = 1.$$

b) To find the reciprocal of 0.3, we first write 0.3 as a ratio of integers:

$$0.3 = \frac{3}{10}$$

The multiplicative inverse of 0.3 is $\frac{10}{3}$ because

$$\frac{3}{10} \cdot \frac{10}{3} = 1.$$

c) The reciprocal of $-\frac{3}{4}$ is $-\frac{4}{3}$ because

$$\left(-\frac{3}{4}\right)\left(-\frac{4}{3}\right) = 1.$$

d) First convert 1.7 to a ratio of integers:

$$1.7 = 1\frac{7}{10} = \frac{17}{10}$$

The multiplicative inverse is $\frac{10}{17}$.

Now do Exercises 59–70

Calculator Close-Up

You can find multiplicative inverses with a calculator as shown here.

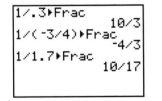

```
1/.3▸Frac
             10/3
1/(-3/4)▸Frac
             -4/3
1/1.7▸Frac
            10/17
```

When the divisor is a fraction, it must be in parentheses.

Study Tip

When you get a test back, do not simply file it in your notebook or the waste basket. While the material is fresh in your mind, rework all problems that you missed. Ask questions about anything that you don't understand and save your test for future reference.

Multiplication Property of Zero

Zero has a property that no other number has. Multiplication involving zero always results in zero.

Multiplication Property of Zero

For any real number a,

$$0 \cdot a = 0 \qquad \text{and} \qquad a \cdot 0 = 0.$$

EXAMPLE 8

Identifying the properties

Name the property that justifies each equation.

a) $5 \cdot 7 = 7 \cdot 5$

b) $4 \cdot \frac{1}{4} = 1$

c) $1 \cdot 864 = 864$

d) $6 + (5 + x) = (6 + 5) + x$

e) $3x + 5x = (3 + 5)x$

f) $6 + (x + 5) = 6 + (5 + x)$

g) $\pi x^2 + \pi y^2 = \pi(x^2 + y^2)$

h) $325 + 0 = 325$

i) $-3 + 3 = 0$

j) $455 \cdot 0 = 0$

Solution

a) Commutative property of multiplication

b) Multiplicative inverse property

c) Multiplicative identity property

d) Associative property of addition

e) Distributive property

f) Commutative property of addition

g) Distributive property

h) Additive identity property

i) Additive inverse property

j) Multiplication property of 0

Now do Exercises 71–90

Applications

Reciprocals are important in problems involving work. For example, if you wax one car in 3 hours, then your rate is $\frac{1}{3}$ of a car per hour. If you can wash one car in 12 minutes $\left(\frac{1}{5}\text{ of an hour}\right)$, then you are washing cars at the rate of 5 cars per hour. In general, if you can complete a task in x hours, then your rate is $\frac{1}{x}$ tasks per hour.

EXAMPLE 9

Washing rates

A car wash has two machines. The old machine washes one car in 0.1 hour, while the new machine washes one car in 0.08 hour. If both machines are operating, then at what rate (in cars per hour) are the cars being washed?

Solution

The old machine is working at the rate of $\frac{1}{0.1}$ cars per hour, and the new machine is working at the rate of $\frac{1}{0.08}$ cars per hour. Their rate working together is the sum of their individual rates:

$$\frac{1}{0.1} + \frac{1}{0.08} = 10 + 12.5 = 22.5$$

So working together, the machines are washing 22.5 cars per hour.

Now do Exercises 105–108

Helpful Hint

When machines or people are working together, we can add their rates provided they do not interfere with each other's work. If operating both car wash machines causes a traffic jam, then the rate together might not be 22.5 cars per hour.

Warm-Ups

True or false? Explain your answer.

1. $24 \div (4 \div 2) = (24 \div 4) \div 2$

2. $1 \div 2 = 2 \div 1$

3. $6 - 5 = -5 + 6$

4. $9 - (4 - 3) = (9 - 4) - 3$

5. Multiplication is a commutative operation.

6. $5x + 5 = 5(x + 1)$ for any value of x.

7. The multiplicative inverse of 0.02 is 50.

8. $-3(x - 2) = -3x + 6$ for any value of x.

9. $3x + 2x = (3 + 2)x$ for any value of x.

10. The additive inverse of 0 is 0.

1.7 Exercises

Reading and Writing *After reading this section write out the answers to these questions. Use complete sentences.*

1. What is the difference between the commutative property of addition and the associative property of addition?

2. Which property involves two different operations?

3. What is factoring?

4. Which two numbers play a prominent role in the properties studied here?

5. What is the purpose of studying the properties of real numbers?

6. What is the relationship between rate and time?

Use the commutative property of addition to rewrite each expression. See Example 1.

7. $9 + r$ **8.** $t + 6$ **9.** $3(2 + x)$

10. $P(1 + rt)$ **11.** $4 - 5x$ **12.** $b - 2a$

Use the commutative property of multiplication to rewrite each expression. See Example 2.

13. $x \cdot 6$ **14.** $y \cdot (-9)$ **15.** $(x - 4)(-2)$

16. $a(b + c)$ **17.** $4 - y \cdot 8$ **18.** $z \cdot 9 - 2$

Use the commutative and associative properties of multiplication and exponential notation to rewrite each product. See Example 3.

19. $(4w)(w)$ **20.** $(y)(2y)$ **21.** $3a(ba)$

22. $(x \cdot x)(7x)$ **23.** $(x)(9x)(xz)$ **24.** $y(y \cdot 5)(wy)$

Evaluate by finding first the sum of the positive numbers and then the sum of the negative numbers. See Example 4.

25. $8 - 4 + 3 - 10$

26. $-3 + 5 - 12 + 10$

27. $8 - 10 + 7 - 8 - 7$

28. $6 - 11 + 7 - 9 + 13 - 2$

29. $-4 - 11 + 7 - 8 + 15 - 20$

30. $-8 + 13 - 9 - 15 + 7 - 22 + 5$

31. $-3.2 + 2.4 - 2.8 + 5.8 - 1.6$

32. $5.4 - 5.1 + 6.6 - 2.3 + 9.1$

33. $3.26 - 13.41 + 5.1 - 12.35 - 5$

34. $5.89 - 6.1 + 8.58 - 6.06 - 2.34$

Use the distributive property to remove the parentheses. See Example 5.

35. $3(x - 5)$
36. $4(b - 1)$
37. $a(2 + t)$
38. $b(a + w)$
39. $-3(w - 6)$
40. $-3(m - 5)$
41. $-4(5 - y)$
42. $-3(6 - p)$
43. $-1(a - 7)$
44. $-1(c - 8)$
45. $-1(t + 4)$
46. $-1(x + 7)$

Use the distributive property to factor each expression. See Example 6.

47. $2m + 12$
48. $3y + 6$
49. $4x - 4$
50. $6y + 6$
51. $4y - 16$
52. $5x + 15$
53. $4a + 8$
54. $7a - 35$
55. $x + xy$
56. $a - ab$
57. $6a - 2b$
58. $8a + 2c$

Find the multiplicative inverse (reciprocal) of each number. See Example 7.

59. $\frac{1}{2}$
60. $\frac{1}{3}$
61. -5
62. -6
63. 7
64. 8
65. 1
66. -1
67. -0.25
68. 0.75
69. 2.5
70. 3.5

Name the property that justifies each equation. See Example 8.

71. $3 \cdot x = x \cdot 3$
72. $x + 5 = 5 + x$
73. $2(x - 3) = 2x - 6$
74. $a(bc) = (ab)c$
75. $-3(xy) = (-3x)y$
76. $3(x + 1) = 3x + 3$
77. $4 + (-4) = 0$
78. $1.3 + 9 = 9 + 1.3$
79. $x^2 \cdot 5 = 5x^2$
80. $0 \cdot \pi = 0$
81. $1 \cdot 3y = 3y$
82. $(0.1)(10) = 1$
83. $2a + 5a = (2 + 5)a$
84. $3 + 0 = 3$
85. $-7 + 7 = 0$
86. $1 \cdot b = b$
87. $(2346)0 = 0$
88. $4x + 4 = 4(x + 1)$
89. $ay + y = y(a + 1)$
90. $ab + bc = b(a + c)$

Complete each equation, using the property named.

91. $a + y =$ ____, commutative property of addition
92. $6x + 6 =$ ____, distributive property
93. $5(aw) =$ ____, associative property of multiplication
94. $x + 3 =$ ____, commutative property of addition
95. $\frac{1}{2}x + \frac{1}{2} =$ ____, distributive property
96. $-3(x - 7) =$ ____, distributive property
97. $6x + 15 =$ ____, distributive property
98. $(x + 6) + 1 =$ ____, associative property of addition
99. $4(0.25) =$ ____, multiplicative inverse property
100. $-1(5 - y) =$ ____, distributive property
101. $0 = 96($____$)$, multiplication property of zero
102. $3 \cdot ($____$) = 3$, multiplicative identity property
103. $0.33($____$) = 1$, multiplicative inverse property
104. $-8(1) =$ ____, multiplicative identity property

Solve each problem. See Example 9.

105. ***Laying bricks.*** A bricklayer lays one brick in 0.04 hour, while his apprentice lays one brick in 0.05 hour.

a) If both are working, then at what combined rate (in bricks per hour) are they laying bricks?

b) Which person is working faster?

Figure for Exercise 105

106. ***Recovering golf balls.*** Susan and Joan are diving for golf balls in a large water trap. Susan recovers a golf ball every 0.016 hour while Joan recovers a ball every 0.025 hour. If both are working, then at what rate (in golf balls per hour) are they recovering golf balls?

107. ***Population explosion.*** In 2004 the population of the earth was increasing by one person every 0.4308 second (U.S. Census Bureau, www.census.gov).

a) At what rate in people per second is the population of the earth increasing?

b) At what rate in people per week is the population of the earth increasing?

108. ***Farmland conversion.*** The amount of farmland in the United States is decreasing by one acre every 0.00876 hours as farmland is being converted to nonfarm use (American Farmland Trust, www.farmland.org). At what rate in acres per day is the farmland decreasing?

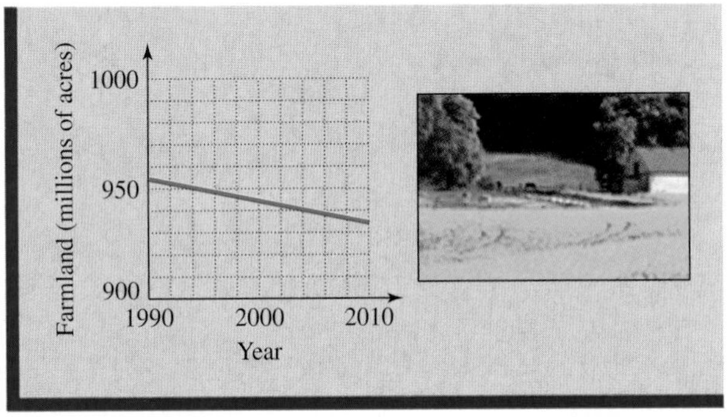

Figure for Exercise 108

Getting More Involved

109. *Writing*

The perimeter of a rectangle is the sum of twice the length and twice the width. Write in words another way to find the perimeter that illustrates the distributive property.

110. *Discussion*

Eldrid bought a loaf of bread for \$1.69 and a gallon of milk for \$2.29. Using a tax rate of 5%, he correctly figured that the tax on the bread would be 8 cents and the tax on the milk would be 11 cents, for a total of \$4.17. However, at the cash register he was correctly charged \$4.18. How could this happen? Which property of the real numbers is in question in this case?

111. *Exploration*

Determine whether each of the following pairs of tasks are "commutative." That is, does the order in which they are performed produce the same result?

a) Put on your coat; put on your hat.

b) Put on your shirt; put on your coat.

Find another pair of "commutative" tasks and another pair of "noncommutative" tasks.

1.8 Using the Properties to Simplify Expressions

In this Section

- Using the Properties in Computation
- Like Terms
- Combining Like Terms
- Products and Quotients
- Removing Parentheses

The properties of the real numbers can be helpful when we are doing computations. In this section we will see how the properties can be applied in arithmetic and algebra.

Using the Properties in Computation

The properties of the real numbers can often be used to simplify computations. For example, to find the product of 26 and 200, we can write

$$\begin{aligned}(26)(200) &= (26)(2 \cdot 100)\\ &= (26 \cdot 2)(100)\\ &= 52 \cdot 100\\ &= 5200\end{aligned}$$

It is the associative property that allows us to multiply 26 by 2 to get 52, then multiply 52 by 100 to get 5200.

EXAMPLE 1

Using the properties

Use the appropriate property to aid you in evaluating each expression.

a) $347 + 35 + 65$ **b)** $3 \cdot 435 \cdot \frac{1}{3}$ **c)** $6 \cdot 28 + 4 \cdot 28$

Study Tip

Being a full-time student is a full-time job. A successful student spends from two to four hours studying outside of class for every hour spent in the classroom. It is rare to find a person who can handle two full-time jobs and it is just as rare to find a successful full-time student who also works full time.

Solution

a) Notice that the sum of 35 and 65 is 100. So apply the associative property as follows:

$$347 + (35 + 65) = 347 + 100$$
$$= 447$$

b) Use the commutative and associative properties to rearrange this product. We can then do the multiplication quickly:

$$3 \cdot 435 \cdot \frac{1}{3} = 435\left(3 \cdot \frac{1}{3}\right) \quad \text{Commutative and associative properties}$$
$$= 435 \cdot 1 \quad \text{Multiplicative inverse property}$$
$$= 435 \quad \text{Multiplicative identity property}$$

c) Use the distributive property to rewrite this expression.

$$6 \cdot 28 + 4 \cdot 28 = (6 + 4)28$$
$$= 10 \cdot 28$$
$$= 280$$

Now do Exercises 7–22

Like Terms

An expression containing a number or the product of a number and one or more variables raised to powers is called a **term.** For example,

$$-3, \quad 5x, \quad -3x^2y, \quad a, \quad \text{and} \quad -abc$$

are terms. The number preceding the variables in a term is called the **coefficient.** In the term $5x$, the coefficient of x is 5. In the term $-3x^2y$ the coefficient of x^2y is -3. In the term a, the coefficient of a is 1 because $a = 1 \cdot a$. In the term $-abc$ the coefficient of abc is -1 because $-abc = -1 \cdot abc$. If two terms contain the same variables with the same exponents, they are called **like terms.** For example, $3x^2$ and $-5x^2$ are like terms, but $3x^2$ and $-5x^3$ are not like terms.

Combining Like Terms

Using the distributive property on an expression involving the sum of like terms allows us to combine the like terms as shown in Example 2.

EXAMPLE 2

Combining like terms

Use the distributive property to perform the indicated operations.

a) $3x + 5x$ **b)** $-5xy - (-4xy)$

EXAMPLE 6

Simplifying quotients

Simplify.

a) $\dfrac{10x}{5}$ **b)** $\dfrac{4x + 8}{2}$

Solution

a) Since dividing by 5 is equivalent to multiplying by $\frac{1}{5}$, we have

$$\frac{10x}{5} = \frac{1}{5}(10x) = \left(\frac{1}{5} \cdot 10\right)x = (2)x = 2x.$$

Note that you can simply divide 10 by 5 to get 2.

b) Since dividing by 2 is equivalent to multiplying by $\frac{1}{2}$, we have

$$\frac{4x + 8}{2} = \frac{1}{2}(4x + 8) = 2x + 4.$$

Note that both 4 and 8 are divided by 2.

Now do Exercises 59–70

CAUTION It is not correct to divide only one term in the numerator by the denominator. For example,

$$\frac{4 + 7}{2} \neq 2 + 7$$

because $\frac{4+7}{2} = \frac{11}{2}$ and $2 + 7 = 9$.

Removing Parentheses

Multiplying a number by -1 merely changes the sign of the number. For example,

$$(-1)(7) = -7 \quad \text{and} \quad (-1)(-8) = 8.$$

So -1 times a number is the *opposite* of the number. Using variables, we write

$$(-1)x = -x \quad \text{or} \quad -1(y + 5) = -(y + 5).$$

When a minus sign appears in front of a sum, we can change the minus sign to -1 and use the distributive property. For example,

$$\begin{aligned} -(w + 4) &= -1(w + 4) \\ &= (-1)w + (-1)4 && \text{Distributive property} \\ &= -w + (-4) && \textit{Note: } -1 \cdot w = -w, -1 \cdot 4 = -4 \\ &= -w - 4 \end{aligned}$$

Note how the minus sign in front of the parentheses caused all of the signs to change: $-(w + 4) = -w - 4$. As another example, consider the following:

$$\begin{aligned} -(x - 3) &= -1(x - 3) \\ &= (-1)x - (-1)3 \\ &= -x - (-3) \\ &= -x + 3 \end{aligned}$$

Study Tip

Take notes in class. Write down everything you can. As soon as possible after class, rewrite your notes. Fill in details and make corrections. Make a note of examples and exercises in the text that are similar to examples in your notes. If your instructor takes the time to work an example in class, it is a good bet that your instructor expects you to understand the concepts involved.

Calculator Close-Up

A negative sign in front of parentheses changes the sign of every term inside the parentheses.

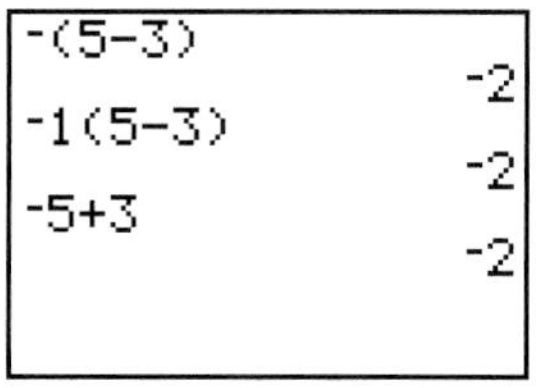

CAUTION When removing parentheses preceded by a minus sign, you must change the sign of *every* term within the parentheses.

EXAMPLE 7

Removing parentheses

Simplify each expression.

a) $5 - (x + 3)$ **b)** $3x - 6 - (2x - 4)$ **c)** $-6x - (-x + 2)$

Solution

a) $5 - (x + 3) = 5 - x - 3$ Change the sign of each term in parentheses.

$= 5 - 3 - x$ Commutative property of addition

$= 2 - x$ Combine like terms.

b) $3x - 6 - (2x - 4) = 3x - 6 - 2x + 4$ Remove parentheses and change signs.

$= 3x - 2x - 6 + 4$ Commutative property of addition

$= x - 2$ Combine like terms.

c) $-6x - (-x + 2) = -6x + x - 2$ Remove parentheses and change signs.

$= -5x - 2$ Combine like terms.

Now do Exercises 71–78

The commutative and associative properties of addition allow us to rearrange the terms so that we may combine the like terms. However, it is not necessary to actually write down the rearrangement. We can identify the like terms and combine them without rearranging.

EXAMPLE 8

Simplifying algebraic expressions

Simplify.

a) $(-2x + 3) + (5x - 7)$ **b)** $-3x + 6x + 5(4 - 2x)$

c) $-2x(3x - 7) - (x - 6)$ **d)** $x - 0.02(x + 500)$

Solution

a) $(-2x + 3) + (5x - 7) = 3x - 4$ Combine like terms.

b) $-3x + 6x + 5(4 - 2x) = -3x + 6x + 20 - 10x$ Distributive property

$= -7x + 20$ Combine like terms.

c) $-2x(3x - 7) - (x - 6) = -6x^2 + 14x - x + 6$ Distributive property

$= -6x^2 + 13x + 6$ Combine like terms.

d) $x - 0.02(x + 500) = 1x - 0.02x - 10$ Distributive property

$= 0.98x - 10$ Combine like terms.

Now do Exercises 79–96

Warm-Ups ▼

True or false? Explain your answer.

A statement involving variables should be marked true only if it is true for all values of the variable.

1. $3(x + 6) = 3x + 18$

2. $-3x + 9 = -3(x + 9)$

3. $-1(x - 4) = -x + 4$

4. $3a + 4a = 7a$

5. $(3a)(4a) = 12a$

6. $3(5 \cdot 2) = 15 \cdot 6$

7. $x + x = x^2$

8. $x \cdot x = 2x$

9. $3 + 2x = 5x$

10. $-(5x - 2) = -5x + 2$

1.8 Exercises

Reading and Writing *After reading this section write out the answers to these questions. Use complete sentences.*

1. What are like terms?

2. What is the coefficient of a term?

3. What can you do to like terms that you cannot do to unlike terms?

4. What operations can you perform with unlike terms?

5. What is the difference between a positive sign preceding a set of parentheses and a negative sign preceding a set of parentheses?

6. What happens when a number is multiplied by -1?

Use the appropriate properties to evaluate the expressions. See Example 1.

7. $35(200)$

8. $15(300)$

9. $\frac{4}{3}(0.75)$

10. $5(0.2)$

11. $256 + 78 + 22$

12. $12 + 88 + 376$

13. $35 \cdot 3 + 35 \cdot 7$

14. $98 \cdot 478 + 2 \cdot 478$

15. $18 \cdot 4 \cdot 2 \cdot \frac{1}{4}$

16. $19 \cdot 3 \cdot 2 \cdot \frac{1}{3}$

17. $(120)(300)$

18. $150 \cdot 200$

19. $12 \cdot 375(-6 + 6)$

20. $354^2(-2 \cdot 4 + 8)$

21. $78 + 6 + 8 + 4 + 2$

22. $-47 + 12 - 6 - 12 + 6$

Combine like terms where possible. See Examples 2 and 3.

23. $5w + 6w$

24. $4a + 10a$

25. $4x - x$

26. $a - 6a$

27. $2x - (-3x)$

28. $2b - (-5b)$

29. $-3a - (-2a)$

30. $-10m - (-6m)$

31. $-a - a$

32. $a - a$

33. $10 - 6t$

34. $9 - 4w$

35. $3x^2 + 5x^2$

36. $3r^2 + 4r^2$

37. $-4x + 2x^2$

38. $6w^2 - w$

39. $5mw^2 - 12mw^2$

40. $4ab^2 - 19ab^2$

41. $\frac{1}{3}a + \frac{1}{2}a$

42. $\frac{3}{5}b - b$

Simplify the following products or quotients. See Examples 4–6.

43. $3(4h)$

44. $2(5h)$

45. $6b(-3)$

46. $-3m(-1)$

47. $(-3m)(3m)$

48. $(2x)(-2x)$

49. $(-3d)(-4d)$

50. $(-5t)(-2t)$

51. $(-y)(-y)$

52. $y(-y)$

53. $-3a(5b)$

54. $-7w(3r)$

55. $-3a(2 + b)$

56. $-2x(3 + y)$

57. $-k(1-k)$

58. $-t(t-1)$

59. $\frac{3y}{3}$

60. $\frac{-9t}{9}$

61. $\frac{-15y}{5}$

62. $\frac{-12b}{2}$

63. $2\left(\frac{y}{2}\right)$

64. $6\left(\frac{m}{3}\right)$

65. $8y\left(\frac{y}{4}\right)$

66. $10\left(\frac{2a}{5}\right)$

67. $\frac{6a-3}{3}$

68. $\frac{-8x+6}{2}$

69. $\frac{-9x+6}{-3}$

70. $\frac{10-5x}{-5}$

Simplify each expression. See Example 7.

71. $x-(3x-1)$

72. $4x-(2x-5)$

73. $5-(y-3)$

74. $8-(m-6)$

75. $2m+3-(m+9)$

76. $7-8t-(2t+6)$

77. $-3-(-w+2)$

78. $-5x-(-2x+9)$

Simplify the following expressions by combining like terms. See Example 8.

79. $3x+5x+6+9$

80. $2x+6x+7+15$

81. $(-2x+3)+(7x-4)$

82. $(-3x+12)+(5x-9)$

83. $3a-7-(5a-6)$

84. $4m-5-(m-2)$

85. $2(a-4)-3(-2-a)$

86. $2(w+6)-3(-w-5)$

87. $3x(2x-3)+5(2x-3)$

88. $2a(a-5)+4(a-5)$

89. $-b(2b-1)-4(2b-1)$

90. $-2c(c-8)-3(c-8)$

91. $-5m+6(m-3)+2m$

92. $-3a+2(a-5)+7a$

93. $5-3(x+2)-6$

94. $7+2(k-3)-k+6$

95. $x-0.05(x+10)$

96. $x-0.02(x+300)$

Simplify each expression.

97. $3x-(4-x)$

98. $2+8x-11x$

99. $y-5-(-y-9)$

100. $a-(b-c-a)$

101. $7-(8-2y-m)$

102. $x-8-(-3-x)$

103. $\frac{1}{2}(10-2x)+\frac{1}{3}(3x-6)$

104. $\frac{1}{2}(x-20)-\frac{1}{5}(x+15)$

105. $\frac{1}{2}(3a+1)-\frac{1}{3}(a-5)$

106. $\frac{1}{4}(6b+2)-\frac{2}{3}(3b-2)$

107. $0.2(x+3)-0.05(x+20)$

108. $0.08x+0.12(x+100)$

109. $2k+1-3(5k-6)-k+4$

110. $2w-3+3(w-4)-5(w-6)$

111. $-3m-3[2m-3(m+5)]$

112. $6h+4[2h-3(h-9)-(h-1)]$

Solve each problem.

113. ***Married filing jointly.*** The value of the expression

$$7820+0.25(x-56{,}800)$$

is the 2003 federal income tax for a married couple filing jointly with a taxable income of x dollars, where x is over \$56,800 but not over \$114,650 (Internal Revenue Service, www.irs.gov).

a) Simplify the expression.

b) Use the expression to find the amount of tax for a couple with a taxable income of \$80,000.

c) Use the accompanying graph to estimate the 2003 federal income tax for a couple with a taxable income of \$200,000

d) Use the accompanying graph to estimate the taxable income for a couple who paid \$80,000 in federal income tax.

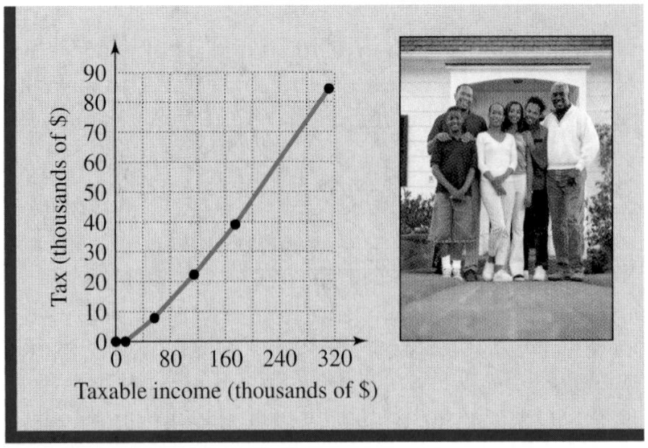

Figure for Exercise 113

114. ***Marriage penalty eliminated.*** The value of the expression

$$3910 + 0.25(x - 28{,}400)$$

is the 2003 federal income tax for a single taxpayer with taxable income of x dollars, where x is over \$28,400 but not over \$68,800.

a) Simplify the expression.

b) Find the amount of tax for a single taxpayer with taxable income of \$40,000.

c) Who pays more, a married couple with a joint taxable income of \$80,000 or two single taxpayers with taxable incomes of \$40,000 each? See Exercise 113.

115. ***Perimeter of a corral.*** The perimeter of a rectangular corral that has width x feet and length $x + 40$ feet is $2(x) + 2(x + 40)$. Simplify the expression for the perimeter. Find the perimeter if $x = 30$ feet.

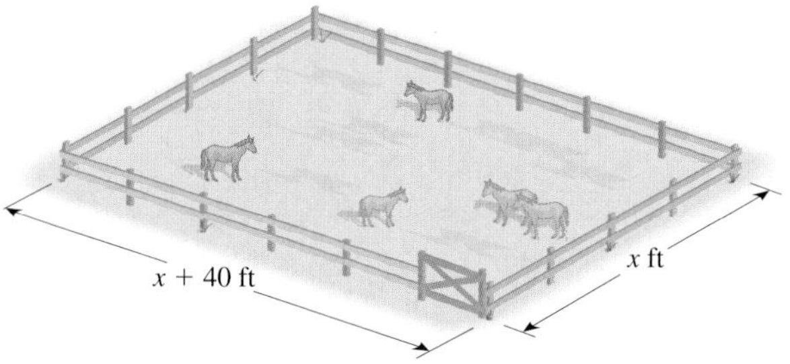

Figure for Exercise 115

Getting More Involved

116. ***Discussion***

What is wrong with the way in which each of the following expressions is simplified?

a) $4(2 + x) = 8 + x$

b) $4(2x) = 8 \cdot 4x = 32x$

c) $\dfrac{4 + x}{2} = 2 + x$

d) $5 - (x - 3) = 5 - x - 3 = 2 - x$

117. ***Discussion***

An instructor asked his class to evaluate the expression $1/2x$ for $x = 5$. Some students got 0.1; others got 2.5. Which answer is correct and why?

Collaborative Activities

Grouping: Four students per group

Topic: Signed numbers

Walking the Number Line

This activity will help you understand adding and subtracting integers. There are four roles that will be rotated in your group; positive sign holder, negative sign holder, problem reader, and number-line walker.

Preparation. Using 13 note cards or pieces of paper write the integers from -6 to $+6$, one number to a card. Also make a card with a positive sign on it and another card with a negative sign. Place the 13 number cards in order on the floor about a step apart to create a number line.

Assume Your Position. After choosing your roles, have the positive sign holder stand at the positive end of the 13-card number line after $+6$, facing to the center. The negative sign holder will stand at the negative end of the line after -6, facing the center. The walker will be near the line ready to start. The problem reader is close by ready to read the first problem.

Rules. The number line walker will stand on the first number that the problem reader reads. The walker will face the positive sign holder if the second number is positive or face the negative sign holder if the second number is negative. For addition the walker will walk forward the number of spaces of the second number and for subtraction the walker will walk backward the number of spaces of the second number.

Examples. The reader reads $-1 + (-3)$. The walker begins at -1, faces the negative sign holder, then walks forward three steps. The walker should be on -4, which is the correct result for $-1 + (-3)$.

The reader reads $1 - (-4)$. The walker begins at 1, faces the negative sign holder, then walks backward 4 steps. The walker should be on 5, which is the correct result of $1 - (-4)$.

Exercises. Try the following exercises. After each two exercises, rotate the roles. Be sure to check that the walker is on the correct result.

1. $-2 + 3$

2. $-2 + (-3)$

3. $2 + (-3)$

4. $2 - (-3)$

5. $2 + 3$

6. $2 - 3$

7. $-2 + (-3)$

8. $-2 - (-3)$

Extension. Make up your own problems using integers between -6 and $+6$ and walk out the results on the number line.

Chapter 1 Wrap-Up

Summary

The Real Numbers		**Examples**
Counting or natural numbers	$\{1, 2, 3, \ldots\}$	
Whole numbers	$\{0, 1, 2, 3, \ldots\}$	
Integers	$\{\ldots, -3, -2, -1, 0, 1, 2, 3, \ldots\}$	
Rational numbers	$\left\{\frac{a}{b} \,\middle|\, a \text{ and } b \text{ are integers with } b \neq 0\right\}$	$\frac{3}{2}$, 5, −6, 0
Irrational numbers	$\{x \mid x \text{ is a real number that is not rational}\}$	$\sqrt{2}, \sqrt{3}, \pi$
Real numbers	The set of real numbers consists of all rational numbers together with all irrational numbers.	
Intervals of real numbers	If a is less than b, then the set of real numbers between a and b is written as (a, b). The set of real numbers between a and b inclusive is written as $[a, b]$.	The notation (1, 9) represents the real numbers between 1 and 9. The notation [1, 9] represents the real numbers between 1 and 9 inclusive.

Fractions		**Examples**
Reducing fractions	$\frac{a \cdot c}{b \cdot c} = \frac{a}{b}$	$\frac{4}{6} = \frac{2 \cdot 2}{2 \cdot 3} = \frac{2}{3}$
Building up fractions	$\frac{a}{b} = \frac{a \cdot c}{b \cdot c}$	$\frac{3}{8} = \frac{3 \cdot 5}{8 \cdot 5} = \frac{15}{40}$
Multiplying fractions	$\frac{a}{b} \cdot \frac{c}{d} = \frac{ac}{bd}$	$\frac{2}{3} \cdot \frac{4}{5} = \frac{8}{15}$
Dividing fractions	$\frac{a}{b} \div \frac{c}{d} = \frac{a}{b} \cdot \frac{d}{c}$	$\frac{2}{3} \div \frac{4}{5} = \frac{2}{3} \cdot \frac{5}{4} = \frac{10}{12} = \frac{5}{6}$

Adding or subtracting fractions	$\frac{a}{b} + \frac{c}{b} = \frac{a+c}{b}$ $\frac{a}{b} - \frac{c}{b} = \frac{a-c}{b}$	$\frac{1}{5} + \frac{2}{5} = \frac{3}{5}$ $\frac{3}{5} - \frac{2}{5} = \frac{1}{5}$
Least common denominator	The smallest number that is a multiple of all denominators.	$\frac{1}{4} + \frac{1}{6} = \frac{3}{12} + \frac{2}{12} = \frac{5}{12}$
Operations with Real Numbers		**Examples**
Absolute value	$\lvert a \rvert = \begin{cases} a & \text{if } a \text{ is positive or zero} \\ -a & \text{if } a \text{ is negative} \end{cases}$	$\lvert 3 \rvert = 3$, $\lvert 0 \rvert = 0$ $\lvert -3 \rvert = 3$
Sum of two numbers with like signs	Add their absolute values. The sum has the same sign as the given numbers.	$-3 + (-4) = -7$
Sum of two numbers with unlike signs (and different absolute values)	Subtract the absolute values of the numbers. The answer is positive if the number with the larger absolute value is positive. The answer is negative if the number with the larger absolute value is negative.	$-4 + 7 = 3$ $-7 + 4 = -3$
Sum of opposites	The sum of any number and its opposite is 0.	$-6 + 6 = 0$
Subtraction of signed numbers	$a - b = a + (-b)$ Subtract any number by adding its opposite.	$3 - 5 = 3 + (-5) = -2$ $4 - (-3) = 4 + 3 = 7$
Product or quotient	Like signs ↔ Positive result Unlike signs ↔ Negative result	$(-3)(-2) = 6$ $(-8) \div 2 = -4$
Definition of exponents	For any counting number n, $a^n = \underbrace{a \cdot a \cdot a \cdot \ldots \cdot a}_{n \text{ factors}}$.	$2^3 = 2 \cdot 2 \cdot 2 = 8$
Order of operations	No parentheses or absolute value present: 1. Exponential expressions 2. Multiplication and division 3. Addition and subtraction With parentheses or absolute value: First evaluate within each set of parentheses or absolute value, using the order of operations.	 $5 + 2^3 = 13$ $2 + 3 \cdot 5 = 17$ $4 + 5 \cdot 3^2 = 49$ $(2 + 3)(5 - 7) = -10$ $2 + 3 \lvert 2 - 5 \rvert = 11$

Properties of the Real Numbers		Examples
	For any real numbers a, b, and c	
Commutative property of Addition Multiplication	 $a + b = b + a$ $a \cdot b = b \cdot a$	 $5 + 7 = 7 + 5$ $6 \cdot 3 = 3 \cdot 6$
Associative property of Addition Multiplication	 $a + (b + c) = (a + b) + c$ $a \cdot (b \cdot c) = (a \cdot b) \cdot c$	 $1 + (2 + 3) = (1 + 2) + 3$ $2(3 \cdot 4) = (2 \cdot 3)4$
Distributive properties	$a(b + c) = ab + ac$ $a(b - c) = ab - ac$	$2(3 + x) = 6 + 2x$ $-2(x - 5) = -2x + 10$
Additive identity property	$a + 0 = a$ and $0 + a = a$ Zero is the additive identity.	$5 + 0 = 0 + 5 = 5$
Multiplicative identity property	$1 \cdot a = a$ and $a \cdot 1 = a$ One is the multiplicative identity.	$7 \cdot 1 = 1 \cdot 7 = 7$
Additive inverse property	For any real number a, there is a number $-a$ (additive inverse or opposite) such that $a + (-a) = 0$ and $-a + a = 0$.	$3 + (-3) = 0$ $-3 + 3 = 0$
Multiplicative inverse property	For any nonzero real number a there is a number $\frac{1}{a}$ (multiplicative inverse or reciprocal) such that $a \cdot \frac{1}{a} = 1$ and $\frac{1}{a} \cdot a = 1$.	$3 \cdot \frac{1}{3} = 1$ $\frac{1}{3} \cdot 3 = 1$
Multiplication property of 0	$a \cdot 0 = 0$ and $0 \cdot a = 0$	$5 \cdot 0 = 0$ $0(-7) = 0$

Enriching Your Mathematical Word Power

For each mathematical term, choose the correct meaning.

1. like terms
a. terms that are identical
b. the terms of a sum
c. terms that have the same variables with the same exponents
d. terms with the same variables

2. equivalent fractions
a. identical fractions
b. fractions that represent the same number
c. fractions with the same denominator
d. fractions with the same numerator

3. variable
a. a letter that is used to represent some numbers
b. the letter x
c. an equation with a letter in it
d. not the same

4. reducing
a. less than
b. losing weight
c. making equivalent
d. dividing out common factors

5. lowest terms
a. numerator is smaller than the denominator
b. no common factors
c. the best interest rate
d. when the numerator is 1

6. additive inverse
a. the number -1
b. the number 0
c. the opposite of addition
d. opposite

7. order of operations
a. the order in which operations are to be performed in the absence of grouping symbols
b. the order in which the operations were invented
c. the order in which operations are written
d. a list of operations in alphabetical order

8. least common denominator
a. the smallest divisor of all denominators
b. the denominator that appears the least
c. the smallest identical denominator
d. the least common multiple of the denominators

9. absolute value
a. definite value
b. positive number
c. distance from 0 on the number line
d. the opposite of a number

10. natural numbers
a. the counting numbers
b. numbers that are not irrational
c. the nonnegative numbers
d. numbers that we find in nature

Review Exercises

1.1 *Which of the numbers* $-\sqrt{5}$, -2, 0, 1, 2, 3.14, π, *and* 10 *are*

1. whole numbers?

2. natural numbers?

3. integers?

4. rational numbers?

5. irrational numbers?

6. real numbers?

Study Tip

Note how the review exercises are arranged according to the sections in this chapter. If you are having trouble with a certain type of problem, refer back to the appropriate section for examples and explanations.

True or false? Explain your answer.

7. Every whole number is a rational number.

8. Zero is not a rational number.

9. The counting numbers between -4 and 4 are -3, -2, -1, 0, 1, 2, and 3.

10. There are infinitely many integers.

11. The set of counting numbers smaller than the national debt is infinite.

12. The decimal number 0.25 is a rational number.

13. Every integer greater than -1 is a whole number.

14. Zero is the only number that is neither rational nor irrational.

Graph each set of numbers.

15. The set of integers between -3 and 3

16. The set of natural numbers between -3 and 3

17. The set of real numbers between -1 and 4

18. The set of real numbers between -2 and 3 inclusive

Write the interval notation for each interval of real numbers.

19. The set of real numbers between 4 and 6 inclusive

20. The set of real numbers greater than 2 and less than 5

21. The set of real numbers greater than or equal to -30

22. The set of real numbers less than 50

1.2 *Perform the indicated operations.*

23. $\frac{1}{3} + \frac{3}{8}$

24. $\frac{2}{3} - \frac{1}{4}$

25. $\frac{3}{5} \cdot 10$

26. $\frac{3}{5} \div 10$

27. $\frac{2}{5} \cdot \frac{15}{14}$

28. $7 \div \frac{1}{2}$

29. $4 + \frac{2}{3}$

30. $\frac{7}{12} - \frac{1}{4}$

31. $\frac{1}{2} + \frac{1}{3} + \frac{1}{4}$

32. $\frac{3}{4} \div 9$

1.3 *Evaluate.*

33. $-5 + 7$

34. $-9 + (-4)$

35. $35 - 48$

36. $-3 - 9$

37. $-12 + 5$

38. $-12 - 5$

39. $-12 - (-5)$

40. $-9 - (-9)$

41. $-0.05 + 12$

42. $-0.03 + (-2)$

43. $-0.1 - (-0.05)$

44. $-0.3 + 0.3$

45. $\frac{1}{3} - \frac{1}{2}$

46. $-\frac{2}{3} + \frac{1}{4}$

47. $-\frac{1}{3} + \left(-\frac{2}{5}\right)$

48. $\frac{1}{3} - \left(-\frac{1}{4}\right)$

1.4 *Evaluate.*

49. $(-3)(5)$

50. $(-9)(-4)$

51. $(-8) \div (-2)$

52. $50 \div (-5)$

53. $\frac{-20}{-4}$

54. $\frac{30}{-5}$

55. $\left(-\frac{1}{2}\right)\left(-\frac{1}{3}\right)$

56. $8 \div \left(-\frac{1}{3}\right)$

57. $-0.09 \div 0.3$

58. $4.2 \div (-0.3)$

59. $(0.3)(-0.8)$

60. $0 \div (-0.0538)$

61. $(-5)(-0.2)$

62. $\frac{1}{2}(-12)$

1.5 *Evaluate.*

63. $3 + 7(9)$

64. $(3 + 7)9$

65. $(3 + 4)^2$

66. $3 + 4^2$

67. $3 + 2 \cdot |5 - 6 \cdot 4|$

68. $3 - (8 - 9)$

69. $(3 - 7) - (4 - 9)$

70. $3 - 7 - 4 - 9$

71. $-2 - 4(2 - 3 \cdot 5)$

72. $3^2 - 7 + 5^2$

73. $3^2 - (7 + 5)^2$

74. $|4 - 6 \cdot 3| - |7 - 9|$

75. $\frac{-3 - 5}{2 - (-2)}$

76. $\frac{1 - 9}{4 - 6}$

77. $\frac{6 + 3}{3} - 5 \cdot 4 + 1$

78. $\frac{2 \cdot 4 + 4}{3} - 3(1 - 2)$

1.6 *Let $a = -1$, $b = -2$, and $c = 3$. Find the value of each algebraic expression.*

79. $b^2 - 4ac$

80. $a^2 - 4b$

81. $(c - b)(c + b)$

82. $(a + b)(a - b)$

83. $a^2 + 2ab + b^2$

84. $a^2 - 2ab + b^2$

85. $a^3 - b^3$

86. $a^3 + b^3$

87. $\frac{b + c}{a + b}$

88. $\frac{b - c}{2b - a}$

89. $|a - b|$

90. $|b - a|$

91. $(a + b)c$

92. $ac + bc$

Determine whether the given number is a solution to the equation following it.

93. $4,\ 3x - 2 = 10$

94. $1,\ 5(x + 3) = 20$

95. $-6,\ \frac{3x}{2} = 9$

96. $-30,\ \frac{x}{3} - 4 = 6$

97. $15,\ \frac{x + 3}{2} = 9$

98. $1,\ \frac{12}{2x + 1} = 4$

99. $4,\ -x - 3 = 1$

100. $7,\ -x + 1 = 6$

1.7 *Name the property that justifies each statement.*

101. $a(x + y) = ax + ay$

102. $3(4y) = (3 \cdot 4)y$

103. $(0.001)(1000) = 1$

104. $xy = yx$

105. $0 + y = y$

106. $325 \cdot 1 = 325$

107. $3 + (2 + x) = (3 + 2) + x$

108. $2x - 6 = 2(x - 3)$

109. $5 \cdot 200 = 200 \cdot 5$

110. $3 + (x + 2) = (x + 2) + 3$

111. $-50 + 50 = 0$

112. $43 \cdot 59 \cdot 82 \cdot 0 = 0$

113. $12 \cdot 1 = 12$

114. $3x + 1 = 1 + 3x$

1.8 *Simplify by combining like terms.*

115. $3a + 7 - (4a - 5)$

116. $2m + 6 - (m - 2)$

117. $2a(3a - 5) + 4a$

118. $3a(a - 5) + 5a(a + 2)$

119. $3(t - 2) - 5(3t - 9)$

120. $2(m + 3) - 3(3 - m)$

121. $0.1(a + 0.3) - (a + 0.6)$

122. $0.1(x + 0.3) - (x - 0.9)$

123. $0.05(x - 20) - 0.1(x + 30)$

124. $0.02(x - 100) + 0.2(x - 50)$

125. $5 - 3x(-5x - 2) + 12x^2$

126. $7 - 2x(3x - 7) - x^2$

127. $-(a - 2) - 2 - a$

128. $-(w - y) - 3(y - w)$

129. $x(x + 1) + 3(x - 1)$

130. $y(y - 2) + 3(y + 1)$

Miscellaneous

Evaluate each expression. Use a calculator to check.

131. $752(-13) + 752(13)$

132. $75 - (-13)$

133. $|15 - 23|$

134. $4^2 - 6^2$

135. $-6^2 + 3(5)$

136. $(0.03)(-200)$

137. $\frac{2}{5} + \frac{1}{10}$

138. $\frac{2 + 1}{5 + 10}$

139. $(0.05) \div (-0.1)$

140. $(4 - 9)^2 + (2 \cdot 3 - 1)^2$

141. $2\left(-\frac{1}{2}\right)^2 + \left(-\frac{1}{2}\right) - 1$

142. $\left(-\frac{6}{7}\right)\left(\frac{21}{26}\right)$

Simplify each expression if possible.

143. $\frac{2x + 4}{2}$

144. $4(2x)$

145. $4 + 2x$

146. $4(2 + x)$

147. $4 \cdot \frac{x}{2}$

148. $4 - (x - 2)$

149. $-4(x - 2)$

150. $(4x)(2x)$

151. $4x + 2x$

152. $2 + (x + 4)$

153. $4 \cdot \frac{x}{4}$

154. $4 \cdot \frac{3x}{2}$

155. $2 \cdot x \cdot 4$

156. $4 - 2(2 - x)$

157. $2(x - 4) - x(x - 4)$

158. $-x(2 - x) - 2(2 - x)$

159. $\frac{1}{2}(x - 4) - \frac{1}{4}(x - 2)$

160. $\frac{1}{4}(x + 2) - \frac{1}{2}(x - 4)$

Fill in the tables with the appropriate values for the given expressions.

161.

x	$-\frac{1}{3}x + 1$
-6	
-3	
0	
3	
6	

162.

x	$\frac{1}{2}x + 3$
-4	
-2	
0	
2	
4	

163.

a	a^2	a^3	a^4
5			
-4			

164.

b	$\frac{1}{b}$	$\frac{1}{b^2}$	$\frac{1}{b^3}$
-3			
$-\frac{1}{2}$			

Solve each problem.

165. ***Telemarketing.*** Brenda and Nicki sell memberships in an automobile club over the telephone. Brenda sells one membership every 0.125 hour, and Nicki sells one membership every 0.1 hour. At what rate (in memberships per hour) are the memberships being sold when both are working?

166. ***High-income bracket.*** The expression

$$90{,}514.5 + 0.35(x - 311{,}950)$$

represents the amount for the 2003 federal income tax in dollars for a single taxpayer with x dollars of taxable income, where x is over \$311,950 (www.irs.gov).

a) Simplify the expression.

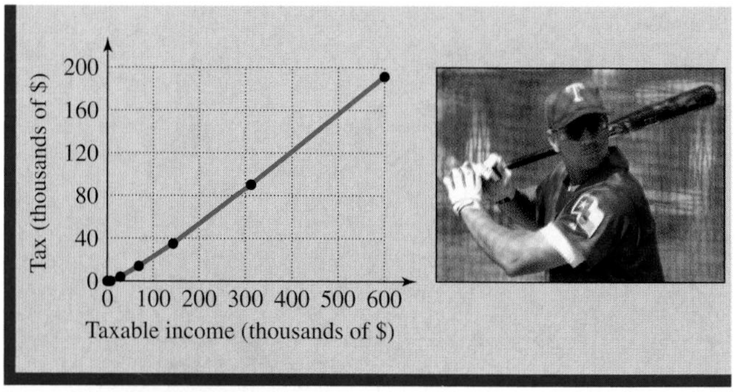

Figure for Exercise 166

b) Use the graph in the accompanying figure to estimate the amount of tax for a single taxpayer with a taxable income of \$450,000.

c) Find the amount of tax for MLB player Alex Rodriguez for 2003. At \$22 million he was the highest paid baseball player that year (www.usatoday.com).

Chapter 1 Test

Which of the numbers $-3, -\sqrt{3}, -\frac{1}{4}, 0, \sqrt{5}, \pi,$ *and* 8 *are*

1. Whole numbers?

2. Integers?

3. Rational numbers?

4. Irrational numbers?

Evaluate each expression.

5. $6 + 3(-9)$

6. $(-2)^2 - 4(-2)(-1)$

7. $\dfrac{-3^2 - 9}{3 - 5}$

8. $-5 + 6 - 12 + 4$

9. $0.05 - 1$

10. $(5 - 9)(5 + 9)$

11. $(878 + 89) + 11$

12. $6 + |3 - 5(2)|$

13. $8 - 3|7 - 10|$

14. $(839 + 974)[3(-4) + 12]$

15. $974(7) + 974(3)$

16. $-\frac{2}{3} + \frac{3}{8}$

17. $(-0.05)(400)$

18. $\left(-\frac{3}{4}\right)\left(\frac{2}{9}\right)$

19. $13 \div \left(-\frac{1}{3}\right)$

Study Tip

Before you take an in-class exam on this chapter, work the sample test given here. Set aside one hour to work this test and use the answers in the back of this book to grade yourself. Even though your instructor might not ask exactly the same questions, you will get a good idea of your test readiness.

Graph each set of numbers.

20. The set of whole numbers less than 5

21. The set of real numbers less than or equal to 4

Write the interval notation for each interval of real numbers.

22. The real numbers greater than 2

23. The real numbers greater than or equal to 3 and less than 9

Identify the property that justifies each equation.

24. $2(x + 7) = 2x + 14$

25. $48 \cdot 1000 = 1000 \cdot 48$

26. $2 + (6 + x) = (2 + 6) + x$

27. $-348 + 348 = 0$

28. $1 \cdot (-6) = -6$

29. $0 \cdot 388 = 0$

Use the distributive property to write each sum or difference as a product.

30. $3x + 30$

31. $7w - 7$

Simplify each expression.

32. $6 + 4x + 2x$

33. $6 + 4(x - 2)$

34. $5x - (3 - 2x)$

35. $x + 10 - 0.1(x + 25)$

36. $2a(4a - 5) - 3a(-2a - 5)$

37. $\dfrac{6x + 12}{6}$

38. $8 \cdot \dfrac{t}{2}$

39. $(-9xy)(-6xy)$

40. $\frac{1}{2}(3x + 2) - \frac{1}{4}(3x - 2)$

Evaluate each expression if $a = -2$, $b = 3$, *and* $c = 4$.

41. $b^2 - 4ac$

42. $\dfrac{a - b}{b - c}$

43. $(a - c)(a + c)$

Determine whether the given number is a solution to the equation following it.

44. -2, $3x - 4 = 2$

45. 13, $\dfrac{x + 3}{8} = 2$

46. -3, $-x + 5 = 8$

Solve each problem.

47. Burke and Nora deliver pizzas for Godmother's Pizza. Burke averages one delivery every 0.25 hour, and Nora averages one delivery every 0.2 hour. At what rate (in deliveries per hour) are the deliveries made when both are working?

48. A forensic scientist uses the expression $80.405 + 3.660R - 0.06(A - 30)$ to estimate the height in centimeters for a male with a radius (bone in the forearm) of length R centimeters and age A in years, where A is over 30. Simplify the expression. Use the expression to estimate the height of an 80-year-old male with a radius of length 25 cm.

Critical Thinking | For Individual or Group Work | Chapter 1

These exercises can be solved by a variety of techniques, which may or may not require algebra. So be creative and think critically. Explain all answers. Answers are in the Instructor's Edition of this text.

1. ***Dividing evenly.*** Suppose that you have a three-ounce glass, a five-ounce glass, and an eight-ounce glass, as shown in the accompanying figure. The two smaller glasses are empty, but the largest glass contains eight ounces of milk. How can you divide the milk into two equal parts by using only these three glasses as measuring devices?

Figure for Exercise 1

2. ***Totaling one hundred.*** Start with the sequence of digits 123456789. Place any number of plus or minus signs between the digits in the sequence so that the value of the resulting expression is 100. For example, we could write

$$123 - 45 + 6 + 78 - 9,$$

but the value is not 100.

3. ***More hundreds.*** We can easily find an expression whose value is 6 using only 2's. For example, $2^2 + 2 = 6$. Find an expression whose value is 100 using only 3's. Only 4's, and so on.

4. ***Forming triangles.*** It is possible to draw three straight lines through a capital M to form nine nonoverlapping triangles. Try it.

5. ***The right time.*** Starting at 12 noon determine the number of times in the next 24 hours for which the hour and minute hands on a clock form a right angle?

Photo for Exercise 5

6. ***Perfect power.*** One is the smallest positive integer that is a perfect square, a perfect cube, and a perfect fifth power. What is the next larger positive integer that is a perfect square, a perfect cube, and a perfect fifth power?

7. ***Summing the digits.*** The sum of all of the digits that are used in writing the integers from 29 through 32 is

$$2 + 9 + 3 + 0 + 3 + 1 + 3 + 2$$

or 23. Find the sum of all of the digits that are used in writing the integers from 1 through 1000 without using a calculator.

8. ***Integral rectangles.*** Find all rectangles whose sides are integers and the numerical value for the area is equal to the numerical value for the perimeter.

Chapter 2

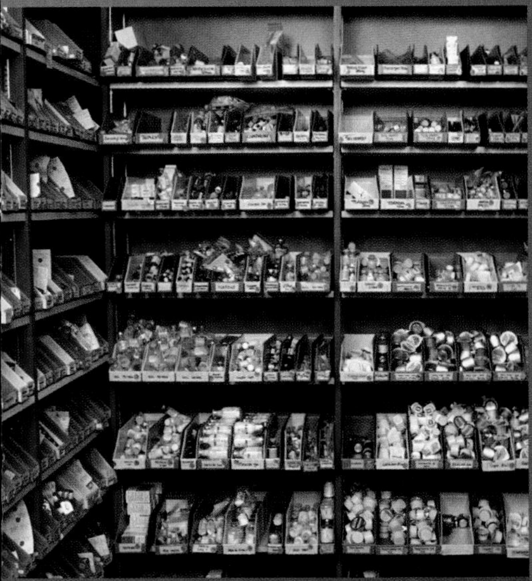

Linear Equations and Inequalities in One Variable

Some ancient peoples chewed on leaves to cure their headaches. Thousands of years ago, the Egyptians used honey, salt, cedar oil, and sycamore bark to cure illnesses. Currently, some of the indigenous people of North America use black birch as a pain reliever.

Today, we are grateful for modern medicine and the seemingly simple cures for illnesses. From our own experiences we know that just the right amount of a drug can work wonders but too much of a drug can do great harm. Even though physicians often prescribe the same drug for children and adults, the amount given must be tailored to the individual. The portion of a drug given to children is usually reduced on the basis of factors such as the weight and height of the child. Likewise, older adults frequently need a lower dosage of medication than what would be prescribed for a younger, more active person.

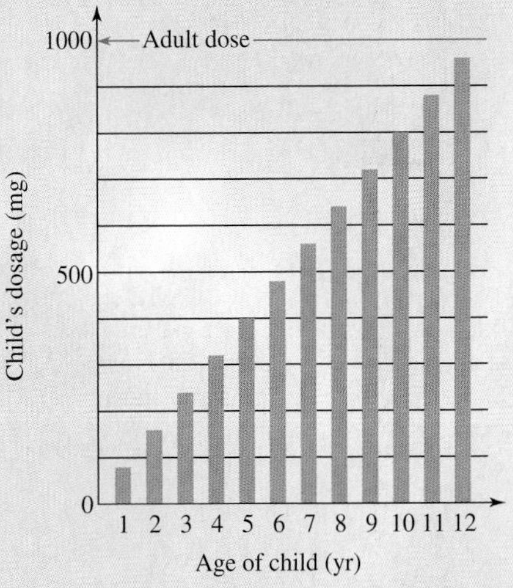

Various algebraic formulas have been developed for determining the proper dosage for a child and an older adult.

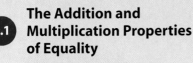

In Exercises 97 and 98 of Section 2.4 you will see two formulas that are used to determine a child's dosage by using the adult dosage and the child's age.

2.1 The Addition and Multiplication Properties of Equality

In this Section

- The Addition Property of Equality
- The Multiplication Property of Equality
- Variables on Both Sides
- Applications

In Section 1.6, an **equation** was defined as a statement that two expressions are equal. A **solution** to an equation is a number that can be used in place of the variable to make the equation a true statement. The **solution set** is the set of all solutions to an equation. Equations with the same solution set are **equivalent equations.** To **solve** an equation means to find all solutions to the equation. In this section you will learn systematic procedures for solving equations.

The Addition Property of Equality

If two workers have equal salaries and each gets a $1000 raise, then they will have equal salaries after the raise. If two people are the same age now, then in 5 years they will still be the same age. If you add the same number to two equal quantities, the results will be equal. This idea is called the *addition property of equality:*

The Addition Property of Equality

Adding the same number to both sides of an equation does not change the solution to the equation. In symbols, if $a = b$, then

$$a + c = b + c.$$

EXAMPLE 1

Adding the same number to both sides

Solve $x - 3 = -7$.

Solution

We can remove the 3 from the left side of the equation by adding 3 to each side of the equation:

$$x - 3 = -7$$

$$x - 3 + 3 = -7 + 3 \quad \text{Add 3 to each side.}$$

$$x + 0 = -4 \quad \text{Simplify each side.}$$

$$x = -4 \quad \text{Zero is the additive identity.}$$

Since -4 satisfies the last equation, it should also satisfy the original equation because all of the previous equations are equivalent. Check that -4 satisfies the original equation by replacing x by -4:

$$x - 3 = -7 \quad \text{Original equation}$$

$$-4 - 3 = -7 \quad \text{Replace } x \text{ by } -4.$$

$$-7 = -7 \quad \text{Simplify.}$$

Since $-4 - 3 = -7$ is correct, $\{-4\}$ is the solution set to the equation.

Now do Exercises 7–16

Helpful Hint

Think of an equation like a balance scale. To keep the scale in balance, what you add to one side you must also add to the other side.

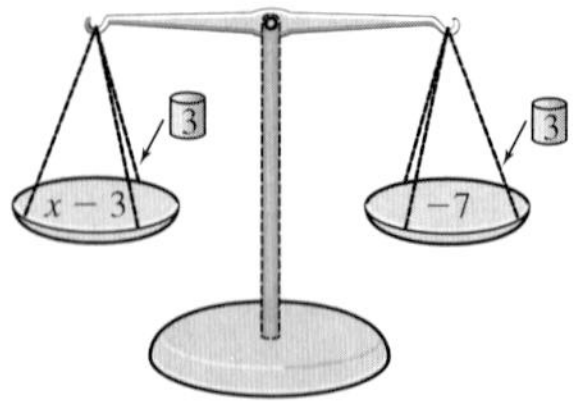

Note that enclosing the solutions to an equation in braces is not absolutely necessary. It is simply a formal way of saying "This is my final answer."

The equations that we work with in this section and Sections 2.2 and 2.3 are called linear equations.

Linear Equation

A **linear equation in one variable** x is an equation that can be written in the form

$$ax + b = 0,$$

where a and b are real numbers and $a \neq 0$.

An equation such as $2x + 3 = 0$ is a linear equation. We also refer to equations such as

$$x + 8 = 0, \quad 3x = 7, \quad 2x + 5 = 9 - 5x, \quad \text{and} \quad 3 + 5(x - 1) = -7 + x$$

as linear equations, because these equations could be written in the form $ax + b = 0$ using the properties of equality.

In Example 1, we used addition to isolate the variable on the left-hand side of the equation. Once the variable is isolated, we can determine the solution to the equation. Because subtraction is defined in terms of addition, we can also use subtraction to isolate the variable.

EXAMPLE 2

Subtracting the same number from both sides

Solve $9 + x = -2$.

Solution

We can remove the 9 from the left side by adding -9 to each side or by subtracting 9 from each side of the equation:

$$9 + x = -2$$
$$9 + x - 9 = -2 - 9 \quad \text{Subtract 9 from each side.}$$
$$x = -11 \quad \text{Simplify each side.}$$

Check that -11 satisfies the original equation by replacing x by -11:

$$9 + x = -2 \quad \text{Original equation}$$
$$9 + (-11) = -2 \quad \text{Replace } x \text{ by } -11.$$

Since $9 + (-11) = -2$ is correct, $\{-11\}$ is the solution set to the equation.

Now do Exercises 17–26

Our goal in solving equations is to isolate the variable. In Examples 1 and 2, the variable was isolated on the left side of the equation. In Example 3, we isolate the variable on the right side of the equation.

EXAMPLE 3

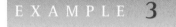

Isolating the variable on the right side

Solve $\frac{1}{2} = -\frac{1}{4} + y$.

Solution

We can remove $-\frac{1}{4}$ from the right side by adding $\frac{1}{4}$ to both sides of the equation:

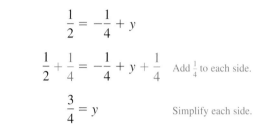

$$\frac{1}{2} = -\frac{1}{4} + y$$

$$\frac{1}{2} + \frac{1}{4} = -\frac{1}{4} + y + \frac{1}{4} \quad \text{Add } \tfrac{1}{4} \text{ to each side.}$$

$$\frac{3}{4} = y \quad \text{Simplify each side.}$$

Check that $\frac{3}{4}$ satisfies the original equation by replacing y by $\frac{3}{4}$:

$$\frac{1}{2} = -\frac{1}{4} + y \quad \text{Original equation}$$

$$\frac{1}{2} = -\frac{1}{4} + \frac{3}{4} \quad \text{Replace } y \text{ by } \tfrac{3}{4}.$$

$$\frac{1}{2} = \frac{2}{4} \quad \text{Simplify.}$$

Since $\frac{1}{2} = \frac{2}{4}$ is correct, $\left\{\frac{3}{4}\right\}$ is the solution set to the equation.

Now do Exercises 27–34

Study Tip

Don't simply work exercises to get answers. Keep reminding yourself of what it is that you are doing. Look for the big picture. What properties are you using? What does a solution mean? Is it reasonable? Does it check?

The Multiplication Property of Equality

To isolate a variable that is involved in a product or a quotient, we need the multiplication property of equality.

The Multiplication Property of Equality

Multiplying both sides of an equation by the same nonzero number does not change the solution to the equation. In symbols, if $a = b$ and $c \neq 0$, then

$$ac = bc.$$

We specified that $c \neq 0$ in the multiplication property of equality because multiplying by 0 can change the solution to an equation. For example, $x = 4$ is satisfied only by 4, but $0 \cdot x = 0 \cdot 4$ is true for any real number x.

In Example 4 we use the multiplication property of equality to solve an equation.

EXAMPLE 4

Multiplying both sides by the same number

Solve $\frac{z}{2} = 6$.

Solution

We isolate the variable z by multiplying each side of the equation by 2.

$$\frac{z}{2} = 6 \quad \text{Original equation}$$
$$2 \cdot \frac{z}{2} = 2 \cdot 6 \quad \text{Multiply each side by 2.}$$
$$1z = 12 \quad \text{Because } 2 \cdot \frac{z}{2} = 2 \cdot \frac{1}{2}z = 1z$$
$$z = 12 \quad \text{Multiplicative identity}$$

Because $\frac{12}{2} = 6$, $\{12\}$ is the solution set to the equation.

Now do Exercises 35–42

Because dividing by a number is the same as multiplying by its reciprocal, the multiplication property of equality allows us to divide each side of the equation by any nonzero number.

EXAMPLE 5

Dividing both sides by the same number

Solve $-5w = 30$.

Solution

Since w is multiplied by -5, we can isolate w by multiplying by $-\frac{1}{5}$ or by dividing each side by -5:

$$-5w = 30 \quad \text{Original equation}$$
$$\frac{-5w}{-5} = \frac{30}{-5} \quad \text{Divide each side by } -5.$$
$$1 \cdot w = -6 \quad \text{Because } \frac{-5}{-5} = 1$$
$$w = -6 \quad \text{Multiplicative identity}$$

Because $-5(-6) = 30$, $\{-6\}$ is the solution set to the equation.

Now do Exercises 43–52

In Example 6, the coefficient of the variable is a fraction. We could divide each side by the coefficient as we did in Example 5, but it is easier to multiply each side by the reciprocal of the coefficient.

EXAMPLE 6

Multiplying by the reciprocal

Solve $\frac{4}{5}p = 40$.

Solution

Multiply each side by $\frac{5}{4}$, the reciprocal of $\frac{4}{5}$, to isolate p on the left side.

$$\frac{4}{5}p = 40$$

$$\frac{5}{4} \cdot \frac{4}{5}p = \frac{5}{4} \cdot 40 \quad \text{Multiply each side by } \tfrac{5}{4}.$$

$$1 \cdot p = 50 \quad \text{Multiplicative inverses}$$

$$p = 50 \quad \text{Multiplicative identity}$$

Because $\frac{4}{5} \cdot 50 = 40$, we can be sure that the solution set is $\{50\}$.

Now do Exercises 53–60

Helpful Hint

You could solve this equation by multiplying each side by 5 to get $4p = 200$, and then dividing each side by 4 to get $p = 50$.

If the coefficient of the variable is an integer, we usually divide each side by that integer, as we did in solving $-5w = 30$ in Example 5. Of course we could also solve that equation by multiplying each side by $-\frac{1}{5}$. If the coefficient of the variable is a fraction, we usually multiply each side by the reciprocal of the fraction as we did in solving $\frac{4}{5}p = 40$ in Example 6. Of course we could also solve that equation by dividing each side by $\frac{4}{5}$. If $-x$ appears in an equation, we can multiply by -1 to get x or divide by -1 to get x, because $-1(-x) = x$ and $\frac{-x}{-1} = x$.

EXAMPLE 7

Multiplying by −1

Solve $-h = 12$.

Solution

Multiply each side by -1 to get h on the left side.

$$-h = 12$$

$$-1(-h) = -1 \cdot 12$$

$$h = -12$$

Since $-(-12) = 12$, the solution set is $\{-12\}$.

Now do Exercises 61–68

Variables on Both Sides

In Example 8, the variable occurs on both sides of the equation. Because the variable represents a real number, we can still isolate the variable by using the addition property of equality. Note that it does not matter whether the variable ends up on the right side or the left side.

EXAMPLE 8

Subtracting an algebraic expression from both sides

Solve $-9 + 6y = 7y$.

Solution

The expression $6y$ can be removed from the left side of the equation by subtracting $6y$ from both sides.

$$-9 + 6y = 7y$$

$$-9 + 6y - 6y = 7y - 6y \quad \text{Subtract } 6y \text{ from each side.}$$

$$-9 = y \quad \text{Simplify each side.}$$

Check by replacing y by -9 in the original equation:

$$-9 + 6(-9) = 7(-9)$$

$$-63 = -63$$

The solution set to the equation is $\{-9\}$.

Now do Exercises 69–76

Helpful Hint

It does not matter whether the variable ends up on the left or right side of the equation. Whether we get $y = -9$ or $-9 = y$ we can still conclude that the solution is -9.

Applications

In Example 9, we use the multiplication property of equality in an applied situation.

EXAMPLE 9

Comparing populations

In the 2000 census, Georgia had $\frac{2}{3}$ as many people as Illinois (U.S. Bureau of Census, www.census.gov). If the population of Georgia was 8 million, then what was the population of Illinois?

Solution

If p represents the population of Illinois, then $\frac{2}{3}p$ represents the population of Georgia. Since the population of Georgia was 8 million we can write the equation $\frac{2}{3}p = 8$. To find p, solve the equation:

$$\frac{2}{3}p = 8$$

$$\frac{3}{2} \cdot \frac{2}{3}p = \frac{3}{2} \cdot 8 \quad \text{Multiply each side by } \frac{3}{2}.$$

$$p = 12 \quad \text{Simplify.}$$

So the population of Illinois was 12 million in 2000.

Now do Exercises 97–100

Warm-Ups

True or false? Explain your answer.

1. The solution to $x - 5 = 5$ is 10.
2. The equation $\frac{x}{2} = 4$ is equivalent to the equation $x = 8$.
3. To solve $\frac{3}{4}y = 12$, we should multiply each side by $\frac{3}{4}$.
4. The equation $\frac{x}{7} = 4$ is equivalent to $\frac{1}{7}x = 4$.
5. Multiplying each side of an equation by any real number will result in an equation that is equivalent to the original equation.
6. To isolate t in $2t = 7 + t$, subtract t from each side.
7. To solve $\frac{2r}{3} = 30$, we should multiply each side by $\frac{3}{2}$.
8. Adding any real number to both sides of an equation will result in an equation that is equivalent to the original equation.
9. The equation $5x = 0$ is equivalent to $x = 0$.
10. The solution to $2x - 3 = x + 1$ is 4.

2.1 Exercises

Reading and Writing *After reading this section, write out the answers to these questions. Use complete sentences.*

1. What does the addition property of equality say?
2. What are equivalent equations?
3. What is the multiplication property of equality?
4. What is a linear equation in one variable?
5. How can you tell if your solution to an equation is correct?
6. To obtain an equivalent equation, what are you not allowed to do to both sides of the equation?

Solve each equation. Show your work and check your answer. See Example 1.

7. $x - 6 = -5$ **8.** $x - 7 = -2$

9. $-13 + x = -4$ **10.** $-8 + x = -12$

11. $y - \frac{1}{2} = \frac{1}{2}$ **12.** $y - \frac{1}{4} = \frac{1}{2}$

13. $w - \frac{1}{3} = \frac{1}{3}$ **14.** $w - \frac{1}{3} = \frac{1}{2}$

15. $a - 0.2 = -0.08$ **16.** $b - 1 = -0.03$

Solve each equation. Show your work and check your answer. See Example 2.

17. $x + 3 = -6$ **18.** $x + 4 = -3$

19. $12 + x = -7$ **20.** $19 + x = -11$

21. $t + \frac{1}{2} = \frac{3}{4}$ **22.** $t + \frac{1}{3} = 1$

23. $\frac{1}{19} + m = \frac{1}{19}$ **24.** $\frac{1}{3} + n = \frac{1}{2}$

25. $a + 0.05 = 6$ **26.** $b + 4 = -0.7$

Solve each equation. Show your work and check your answer. See Example 3.

27. $2 = x + 7$

28. $3 = x + 5$

29. $-13 = y - 9$

30. $-14 = z - 12$

31. $0.5 = -2.5 + x$

32. $0.6 = -1.2 + x$

33. $\frac{1}{8} = -\frac{1}{8} + r$

34. $\frac{1}{6} = -\frac{1}{6} + h$

Solve each equation. Show your work and check your answer. See Example 4.

35. $\frac{x}{2} = -4$

36. $\frac{x}{3} = -6$

37. $0.03 = \frac{y}{60}$

38. $0.05 = \frac{y}{80}$

39. $\frac{a}{2} = \frac{1}{3}$

40. $\frac{b}{2} = \frac{1}{5}$

41. $\frac{1}{6} = \frac{c}{3}$

42. $\frac{1}{12} = \frac{d}{3}$

Solve each equation. Show your work and check your answer. See Example 5.

43. $-3x = 15$

44. $-5x = -20$

45. $20 = 4y$

46. $18 = -3a$

47. $2w = 2.5$

48. $-2x = -5.6$

49. $5 = 20x$

50. $-3 = 27d$

51. $5x = \frac{3}{4}$

52. $3x = -\frac{2}{3}$

Solve each equation. Show your work and check your answer. See Example 6.

53. $\frac{3}{2}x = -3$

54. $\frac{2}{3}x = -8$

55. $90 = \frac{3y}{4}$

56. $14 = \frac{7y}{8}$

57. $-\frac{3}{5}w = -\frac{1}{3}$

58. $-\frac{5}{2}t = -\frac{3}{5}$

59. $\frac{2}{3} = -\frac{4x}{3}$

60. $\frac{1}{14} = -\frac{6p}{7}$

Solve each equation. Show your work and check your answer. See Example 7.

61. $-x = 8$

62. $-x = 4$

63. $-y = -\frac{1}{3}$

64. $-y = -\frac{7}{8}$

65. $3.4 = -z$

66. $4.9 = -t$

67. $-k = -99$

68. $-m = -17$

Solve each equation. Show your work and check your answer. See Example 8.

69. $4x = 3x - 7$

70. $3x = 2x + 9$

71. $9 - 6y = -5y$

72. $12 - 18w = -17w$

73. $-6x = 8 - 7x$

74. $-3x = -6 - 4x$

75. $\frac{1}{2}c = 5 - \frac{1}{2}c$

76. $-\frac{1}{2}h = 13 - \frac{3}{2}h$

Use the appropriate property of equality to solve each equation.

77. $12 = x + 17$

78. $-3 = x + 6$

79. $\frac{3}{4}y = -6$

80. $\frac{5}{9}z = -10$

81. $-3.2 + x = -1.2$

82. $t - 3.8 = -2.9$

83. $2a = \frac{1}{3}$

84. $-3w = \frac{1}{2}$

85. $-9m = 3$

86. $-4h = -2$

87. $-b = -44$

88. $-r = 55$

89. $\frac{2}{3}x = \frac{1}{2}$

90. $\frac{3}{4}x = \frac{1}{3}$

91. $-5x = 7 - 6x$

92. $-\frac{1}{2} + 3y = 4y$

93. $\frac{5a}{7} = -10$

94. $\frac{7r}{12} = -14$

95. $\frac{1}{2}v = -\frac{1}{2}v + \frac{3}{8}$

96. $\frac{1}{3}s + \frac{7}{9} = \frac{4}{3}s$

Solve each problem by writing and solving an equation. See Example 9.

97. ***Births to teenagers.*** In 2000 there were 48.5 births per 1000 females 15 to 19 years of age (National Center for Health Statistics, www.cdc.gov/nchs). This birth rate is $\frac{4}{5}$ of the birth rate for teenagers in 1991.

a) Write an equation and solve it to find the birth rate for teenagers in 1991.

b) Use the accompanying graph to estimate the birth rate to teenagers in 1996.

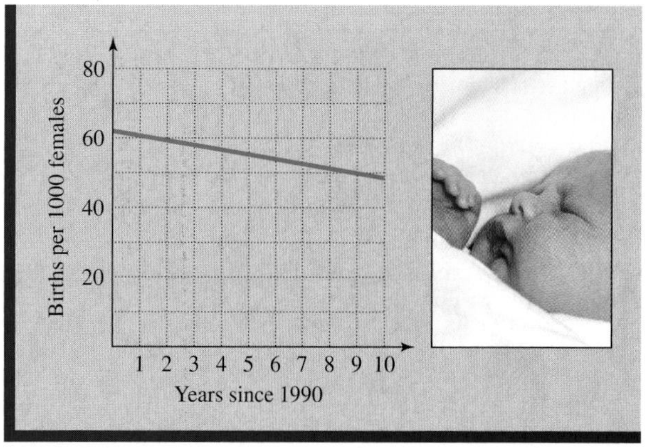

Figure for Exercise 97

98. ***World grain demand.*** Freeport McMoRan projects that in 2010 world grain supply will be 1.8 trillion metric tons and the supply will be only $\frac{3}{4}$ of world grain demand. What will world grain demand be in 2010?

99. ***Advancers and decliners.*** On Thursday, $\frac{2}{3}$ of the stocks traded on the New York Stock Exchange advanced in price. If 1918 stocks advanced, then how many stocks were traded on that day?

100. ***Births in the United States.*** In 2000, one-third of all births in the United States were to unmarried women (National Center for Health Statistics, www.cdc.gov/nchs). If there were 1,352,938 births to unmarried women, then how many births were there in 2000?

Photo for Exercise 98

2.2 Solving General Linear Equations

In this Section

- **Equations of the Form $ax + b = 0$**
- **Equations of the Form $ax + b = cx + d$**
- **Equations with Parentheses**
- **Applications**

All of the equations that we solved in Section 2.1 required only a single application of a property of equality. In this section you will solve equations that require more than one application of a property of equality.

Equations of the Form $ax + b = 0$

To solve an equation of the form $ax + b = 0$ we might need to apply both the addition property of equality and the multiplication property of equality.

EXAMPLE 1

Using the addition and multiplication properties of equality

Solve $3r - 5 = 0$.

Solution

To isolate r, first add 5 to each side, then divide each side by 3.

$$3r - 5 = 0 \quad \text{Original equation}$$

$$3r - 5 + 5 = 0 + 5 \quad \text{Add 5 to each side.}$$

$$3r = 5 \quad \text{Combine like terms.}$$

$$\frac{3r}{3} = \frac{5}{3} \quad \text{Divide each side by 3.}$$

$$r = \frac{5}{3} \quad \text{Simplify.}$$

Checking $\frac{5}{3}$ in the original equation gives

$$3 \cdot \frac{5}{3} - 5 = 5 - 5 = 0.$$

So $\left\{\frac{5}{3}\right\}$ is the solution set to the equation.

Now do Exercises 5–12

Helpful Hint

If we divide each side by 3 first, we must divide each term on the left side by 3 to get $r - \frac{5}{3} = 0$. Then add $\frac{5}{3}$ to each side to get $r = \frac{5}{3}$. Although we get the correct answer, we usually save division to the last step so that fractions do not appear until necessary.

CAUTION In solving $ax + b = 0$ we usually use the addition property of equality first and the multiplication property last. Note that this is the reverse of the order of operations (multiplication before addition), because we are undoing the operations that are done in the expression $ax + b$.

EXAMPLE 2

Using the addition and multiplication properties of equality

Solve $-\frac{2}{3}x + 8 = 0$.

Solution

To isolate x, first subtract 8 from each side, then multiply each side by $-\frac{3}{2}$.

$$-\frac{2}{3}x + 8 = 0 \quad \text{Original equation}$$

$$-\frac{2}{3}x + 8 - 8 = 0 - 8 \quad \text{Subtract 8 from each side.}$$

$$-\frac{2}{3}x = -8 \quad \text{Combine like terms.}$$

$$-\frac{3}{2}\left(-\frac{2}{3}x\right) = -\frac{3}{2}(-8) \quad \text{Multiply each side by } -\tfrac{3}{2}.$$

$$x = 12 \quad \text{Simplify.}$$

Checking 12 in the original equation gives

$$-\frac{2}{3}(12) + 8 = -8 + 8 = 0.$$

So $\{12\}$ is the solution set to the equation.

Now do Exercises 13–20

Equations of the Form $ax + b = cx + d$

In solving equations our goal is to isolate the variable. We use the addition property of equality to eliminate unwanted terms. Note that it does not matter whether the variable ends up on the right or left side. For some equations we will perform fewer steps if we isolate the variable on the right side.

EXAMPLE 3

Isolating the variable on the right side

Solve $3w - 8 = 7w$.

Solution

To eliminate the $3w$ from the left side, we can subtract $3w$ from both sides.

$$3w - 8 = 7w \quad \text{Original equation}$$

$$3w - 8 - 3w = 7w - 3w \quad \text{Subtract } 3w \text{ from each side.}$$

$$-8 = 4w \quad \text{Simplify each side.}$$

$$\frac{-8}{4} = \frac{4w}{4} \quad \text{Divide each side by 4.}$$

$$-2 = w \quad \text{Simplify.}$$

Study Tip

Talk to your classmates. Discuss new terms and ideas. How does this lesson fit in with the last lesson? Form a study group. Does your college have a learning lab where you can study together?

To check, replace w with -2 in the original equation:

$$3w - 8 = 7w \quad \text{Original equation}$$
$$3(-2) - 8 = 7(-2)$$
$$-14 = -14$$

Since -2 satisfies the original equation, the solution set is $\{-2\}$.

Now do Exercises 21–28

You should solve the equation in Example 3 by isolating the variable on the left side to see that it takes more steps. In Example 4, it is simplest to isolate the variable on the left side.

EXAMPLE 4

Isolating the variable on the left side

Solve $\frac{1}{2}b - 8 = 12$.

Solution

To eliminate the 8 from the left side, we add 8 to each side.

$$\frac{1}{2}b - 8 = 12 \quad \text{Original equation}$$
$$\frac{1}{2}b - 8 + 8 = 12 + 8 \quad \text{Add 8 to each side.}$$
$$\frac{1}{2}b = 20 \quad \text{Simplify each side.}$$
$$2 \cdot \frac{1}{2}b = 2 \cdot 20 \quad \text{Multiply each side by 2.}$$
$$b = 40 \quad \text{Simplify.}$$

To check, replace b with 40 in the original equation:

$$\frac{1}{2}b - 8 = 12 \quad \text{Original equation}$$
$$\frac{1}{2}(40) - 8 = 12$$
$$12 = 12$$

Since 40 satisfies the original equation, the solution set is $\{40\}$.

Now do Exercises 29–36

In Example 5 both sides of the equation contain two terms.

EXAMPLE 5

Study Tip

Take good notes. Note taking helps you to concentrate in class and provides a source for review. Rewrite your notes after class and fill in anything that is missing. Practice solving the problems that were demonstrated in class.

Solving $ax + b = cx + d$

Solve $2m - 4 = 4m - 10$.

Solution

First, we decide to isolate the variable on the left side. So we must eliminate the 4 from the left side and eliminate $4m$ from the right side:

$$\begin{aligned} 2m - 4 &= 4m - 10 & \\ 2m - 4 + 4 &= 4m - 10 + 4 & &\text{Add 4 to each side.} \\ 2m &= 4m - 6 & &\text{Simplify each side.} \\ 2m - 4m &= 4m - 6 - 4m & &\text{Subtract } 4m \text{ from each side.} \\ -2m &= -6 & &\text{Simplify each side.} \\ \frac{-2m}{-2} &= \frac{-6}{-2} & &\text{Divide each side by } -2. \\ m &= 3 & &\text{Simplify.} \end{aligned}$$

To check, replace m by 3 in the original equation:

$$\begin{aligned} 2m - 4 &= 4m - 10 & &\text{Original equation} \\ 2 \cdot 3 - 4 &= 4 \cdot 3 - 10 & \\ 2 &= 2 & \end{aligned}$$

Since 3 satisfies the original equation, the solution set is $\{3\}$.

Now do Exercises 37–44

Equations with Parentheses

Equations that contain parentheses or like terms on the same side should be simplified as much as possible before applying any properties of equality.

EXAMPLE 6

Simplifying before using properties of equality

Solve $2(q - 3) + 5q = 8(q - 1)$.

Solution

First remove parentheses and combine like terms on each side of the equation.

$$\begin{aligned} 2(q - 3) + 5q &= 8(q - 1) & &\text{Original equation} \\ 2q - 6 + 5q &= 8q - 8 & &\text{Distributive property} \\ 7q - 6 &= 8q - 8 & &\text{Combine like terms.} \\ 7q - 6 + 6 &= 8q - 8 + 6 & &\text{Add 6 to each side.} \\ 7q &= 8q - 2 & &\text{Combine like terms.} \\ 7q - 8q &= 8q - 2 - 8q & &\text{Subtract } 8q \text{ from each side.} \\ -q &= -2 & \\ -1(-q) &= -1(-2) & &\text{Multiply each side by } -1. \\ q &= 2 & &\text{Simplify.} \end{aligned}$$

Calculator Close-Up

You can check an equation by entering the equation on the home screen as shown here. The equal sign is in the TEST menu.

When you press ENTER, the calculator returns the number 1 if the equation is true or 0 if the equation is false. Since the calculator shows a 1, we can be sure that 2 is the solution.

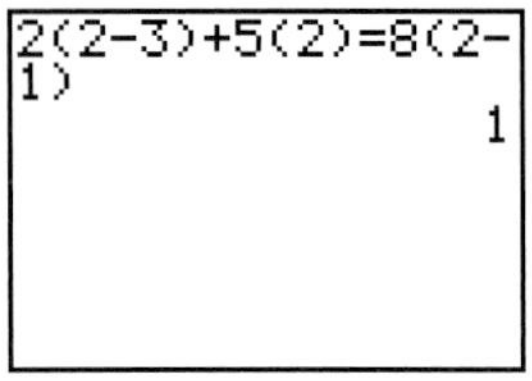

To check, we replace q by 2 in the original equation and simplify:

$$2(q - 3) + 5q = 8(q - 1) \quad \text{Original equation}$$
$$2(2 - 3) + 5(2) = 8(2 - 1) \quad \text{Replace } q \text{ by 2.}$$
$$2(-1) + 10 = 8(1)$$
$$8 = 8$$

Because both sides have the same value, the solution set is $\{2\}$.

Now do Exercises 45–52

Linear equations can vary greatly in appearance, but there is a strategy that you can use for solving any of them. The following strategy summarizes the techniques that we have been using in the examples. Keep it in mind when you are solving linear equations.

Strategy for Solving Equations

1. Remove parentheses by using the distributive property and then combine like terms to simplify each side as much as possible.
2. Use the addition property of equality to get like terms from opposite sides onto the same side so that they may be combined.
3. The multiplication property of equality is generally used last.
4. Check that the solution satisfies the original equation.

Applications

Linear equations occur in business situations where there is a fixed cost and a per item cost. A mail order company might charge \$3 plus \$2 per CD for shipping and handling. A lawyer might charge \$300 plus \$65 per hour for handling your lawsuit. AT&T might charge 5 cents per minute plus \$2.95 for long distance calls. Example 7 illustrates the kind of problem that can be solved in this situation.

EXAMPLE 7

Long distance charges

With AT&T's One Rate plan you are charged 5 cents per minute plus \$2.95 for long distance service for one month. If a long distance bill is \$4.80, then what is the number of minutes used?

Solution

Let x represent the number of minutes of calls in the month. At \$0.05 per minute, the cost for x minutes is the product $0.05x$ dollars. Since there is a fixed cost of \$2.95, an expression for the total cost is $0.05x + 2.95$ dollars. Since the total cost is \$4.80, we have $0.05x + 2.95 = 4.80$. Solve this equation to find x.

$$0.05x + 2.95 = 4.80$$
$$0.05x + 2.95 - 2.95 = 4.80 - 2.95 \quad \text{Subtract 2.95 from each side.}$$
$$0.05x = 1.85 \quad \text{Simplify.}$$
$$\frac{0.05x}{0.05} = \frac{1.85}{0.05} \quad \text{Divide each side by 0.05.}$$
$$x = 37 \quad \text{Simplify.}$$

So the bill is for 37 minutes.

Now do Exercises 93–100

Warm-Ups ▼

True or false? Explain your answer.

1. The solution to $4x - 3 = 3x$ is 3.
2. The equation $2x + 7 = 8$ is equivalent to $2x = 1$.
3. To solve $3x - 5 = 8x + 7$, you should add 5 to each side and subtract $8x$ from each side.
4. To solve $5 - 4x = 9 + 7x$, you should subtract 9 from each side and then subtract $7x$ from each side.
5. Multiplying each side of an equation by the same nonzero real number will result in an equation that is equivalent to the original equation.
6. To isolate y in $3y - 7 = 6$, divide each side by 3 and then add 7 to each side.
7. To solve $\frac{3w}{4} = 300$, we should multiply each side by $\frac{4}{3}$.
8. The equation $-n = 9$ is equivalent to $n = -9$.
9. The equation $-y = -7$ is equivalent to $y = 7$.
10. The solution to $7x = 5x$ is 0.

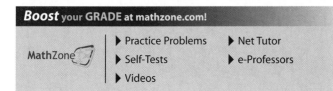

2.2 Exercises

Reading and Writing *After reading this section, write out the answers to these questions. Use complete sentences.*

1. What properties of equality do you apply to solve $ax + b = 0$?
2. Which property of equality is usually applied last?
3. What property of equality is used to solve $-x = 8$?
4. What is usually the first step in solving a linear equation involving parentheses?

Solve each equation. Show your work and check your answer. See Examples 1 and 2.

5. $5a - 10 = 0$
6. $8y + 24 = 0$
7. $-3y - 6 = 0$
8. $-9w - 54 = 0$

9. $3x - 2 = 0$

10. $5y + 1 = 0$

11. $2p + 5 = 0$

12. $9z - 8 = 0$

13. $\frac{1}{2}w - 3 = 0$

14. $\frac{3}{8}t + 6 = 0$

15. $-\frac{2}{3}x + 8 = 0$

16. $-\frac{1}{7}z - 5 = 0$

17. $-m + \frac{1}{2} = 0$

18. $-y - \frac{3}{4} = 0$

19. $3p + \frac{1}{2} = 0$

20. $9z - \frac{1}{4} = 0$

Solve each equation. See Examples 3 and 4.

21. $6x - 8 = 4x$

22. $9y + 14 = 2y$

23. $4z = 5 - 2z$

24. $3t = t - 3$

25. $4a - 9 = 7$

26. $7r + 5 = 47$

27. $9 = -6 - 3b$

28. $13 = 3 - 10s$

29. $\frac{1}{2}w - 4 = 13$

30. $\frac{1}{3}q + 13 = -5$

31. $6 - \frac{1}{3}d = \frac{1}{3}d$

32. $9 - \frac{1}{2}a = \frac{1}{4}a$

33. $2w - 0.4 = 2$

34. $10h - 1.3 = 6$

35. $x = 3.3 - 0.1x$

36. $y = 2.4 - 0.2y$

Solve each equation. See Example 5.

37. $3x - 3 = x + 5$

38. $9y - 1 = 6y + 5$

39. $4 - 7d = 13 - 4d$

40. $y - 9 = 12 - 6y$

41. $c + \frac{1}{2} = 3c - \frac{1}{2}$

42. $x - \frac{1}{4} = \frac{1}{2} - x$

43. $\frac{2}{3}a - 5 = \frac{1}{3}a + 5$

44. $\frac{1}{2}t - 3 = \frac{1}{4}t - 9$

Solve each equation. See Example 6.

45. $5(a - 1) + 3 = 28$

46. $2(w + 4) - 1 = 1$

47. $2 - 3(q - 1) = 10 - (q + 1)$

48. $-2(y - 6) = 3(7 - y) - 5$

49. $2(x - 1) + 3x = 6x - 20$

50. $3 - (r - 1) = 2(r + 1) - r$

51. $2\left(y - \frac{1}{2}\right) = 4\left(y - \frac{1}{4}\right) + y$

52. $\frac{1}{2}(4m - 6) = \frac{2}{3}(6m - 9) + 3$

Solve each linear equation. Show your work and check your answer.

53. $2x = \frac{1}{3}$

54. $3x = \frac{6}{11}$

55. $5t = -2 + 4t$

56. $8y = 6 + 7y$

57. $3x - 7 = 0$

58. $5x + 4 = 0$

59. $-x + 6 = 5$

60. $-x - 2 = 9$

61. $-9 - a = -3$

62. $4 - r = 6$

63. $2q + 5 = q - 7$

64. $3z - 6 = 2z - 7$

65. $-3x + 1 = 5 - 2x$

66. $5 - 2x = 6 - x$

67. $-12 - 5x = -4x + 1$

68. $-3x - 4 = -2x + 8$

69. $3x + 0.3 = 2 + 2x$

70. $2y - 0.05 = y + 1$

71. $k - 0.6 = 0.2k + 1$

72. $2.3h + 6 = 1.8h - 1$

73. $0.2x - 4 = 0.6 - 0.8x$

74. $0.3x = 1 - 0.7x$

75. $-3(k - 6) = 2 - k$

76. $-2(h - 5) = 3 - h$

77. $2(p + 1) - p = 36$

78. $3(q + 1) - q = 23$

79. $7 - 3(5 - u) = 5(u - 4)$

80. $v - 4(4 - v) = -2(2v - 1)$

81. $4(x + 3) = 12$

82. $5(x - 3) = -15$

83. $\frac{w}{5} - 4 = -6$

84. $\frac{q}{2} + 13 = -22$

85. $\frac{2}{3}y - 5 = 7$

86. $\frac{3}{4}u - 9 = -6$

87. $4 - \frac{2n}{5} = 12$

88. $9 - \frac{2m}{7} = 19$

89. $-\frac{1}{3}p - \frac{1}{2} = \frac{1}{2}$

90. $-\frac{3}{4}z - \frac{2}{3} = \frac{1}{3}$

91. $3.5x - 23.7 = -38.75$

92. $3(x - 0.87) - 2x = 4.98$

Solve each problem. See Example 7.

93. ***The practice.*** A lawyer charges \$300 plus \$65 per hour for a divorce. If the total charge for Bill's divorce was \$1405, then for what number of hours did the lawyer work on the case?

94. ***The plumber.*** Tamika paid \$165 to her plumber for a service call. If her plumber charges \$45 plus \$40 per hour for a service call, then for how many hours did the plumber work?

95. ***Celsius temperature.*** If the air temperature in Quebec is 68° Fahrenheit, then the solution to the equation $\frac{9}{5}C + 32 = 68$ gives the Celsius temperature of the air. Find the Celsius temperature.

96. ***Fahrenheit temperature.*** Water boils at 212°F.

a) Use the accompanying graph to determine the Celsius temperature at which water boils.

b) Find the Fahrenheit temperature of hot tap water at 70°C by solving the equation

$$70 = \frac{5}{9}(F - 32).$$

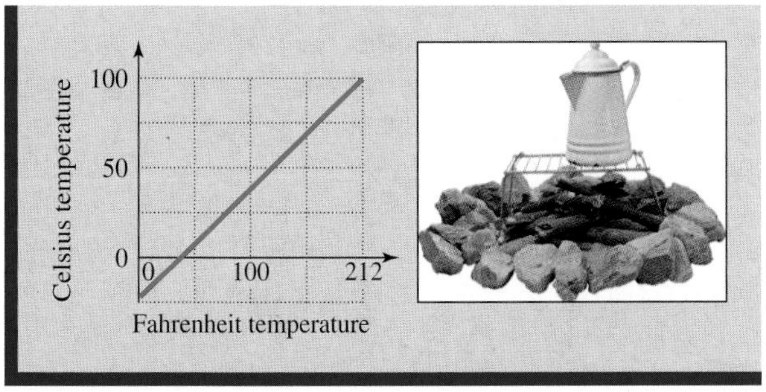

Figure for Exercise 96

97. ***Rectangular patio.*** If a rectangular patio has a length that is 3 feet longer than its width and a perimeter of 42 feet, then the width can be found by solving the equation $2x + 2(x + 3) = 42$. What is the width?

98. ***Perimeter of a triangle.*** The perimeter of the triangle shown in the accompanying figure is 12 meters. Determine the values of x, $x + 1$, and $x + 2$ by solving the equation

$$x + (x + 1) + (x + 2) = 12.$$

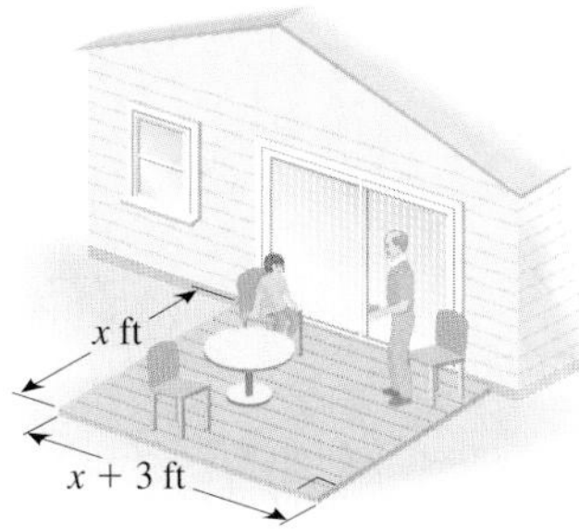

Figure for Exercise 97

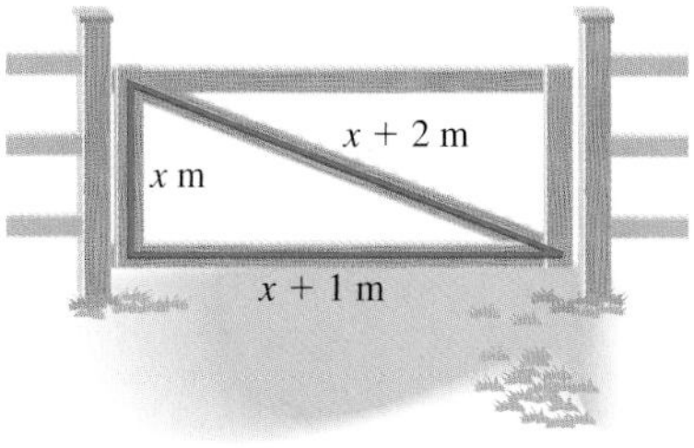

Figure for Exercise 98

99. ***Cost of a car.*** Jane paid 9% sales tax and a \$150 title and license fee when she bought her new Saturn for a total of \$16,009.50. If x represents the price of the car, then x satisfies $x + 0.09x + 150 = 16{,}009.50$. Find the price of the car by solving the equation.

100. ***Cost of labor.*** An electrician charged Eunice \$29.96 for a service call plus \$39.96 per hour for a total of \$169.82 for installing her electric dryer. If n represents the number of hours for labor, then n satisfies

$$39.96n + 29.96 = 169.82.$$

Find n by solving this equation.

2.3 More Equations

In this Section

- Equations Involving Fractions
- Equations Involving Decimals
- Simplifying the Process
- Identities
- Conditional Equations
- Inconsistent Equations

In this section we will solve more equations of the type that we solved in Sections 2.1 and 2.2. However, some equations in this section will contain fractions or decimal numbers. Some equations will have infinitely many solutions and some will have no solution.

Equations Involving Fractions

We solved some equations involving fractions in Sections 2.1 and 2.2. Here, we will solve equations with fractions by eliminating all fractions in the first step. All of the fractions will be eliminated if we multiply each side by the least common denominator.

EXAMPLE 1

Multiplying by the least common denominator

Solve $\frac{y}{2} - 1 = \frac{y}{3} + 1$.

Solution

The least common denominator (LCD) for the denominators 2 and 3 is 6. Since both 2 and 3 divide into 6 evenly, multiplying each side by 6 will eliminate the fractions:

$$6\left(\frac{y}{2} - 1\right) = 6\left(\frac{y}{3} + 1\right) \quad \text{Multiply each side by 6.}$$

$$6 \cdot \frac{y}{2} - 6 \cdot 1 = 6 \cdot \frac{y}{3} + 6 \cdot 1 \quad \text{Distributive property}$$

$$3y - 6 = 2y + 6 \quad \text{Simplify: } 6 \cdot \tfrac{y}{2} = 3y$$

$$3y = 2y + 12 \quad \text{Add 6 to each side.}$$

$$y = 12 \quad \text{Subtract } 2y \text{ from each side.}$$

Check 12 in the original equation:

$$\frac{12}{2} - 1 = \frac{12}{3} + 1$$

$$5 = 5$$

Since 12 satisfies the original equation, the solution set is $\{12\}$.

Now do Exercises 7–22

Helpful Hint

Note that the fractions in Example 1 will be eliminated if you multiply each side of the equation by any number divisible by both 2 and 3. For example, multiplying by 24 yields

$$12y - 24 = 8y + 24$$
$$4y = 48$$
$$y = 12.$$

Equations Involving Decimals

When an equation involves decimal numbers, we can work with the decimal numbers or we can eliminate all of the decimal numbers by multiplying both sides by 10, or 100, or 1000, and so on. Multiplying a decimal number by 10 moves the decimal point one place to the right. Multiplying by 100 moves the decimal point two places to the right, and so on.

EXAMPLE 2

An equation involving decimals

Solve $0.3p + 8.04 = 12.6$.

Solution

The largest number of decimal places appearing in the decimal numbers of the equation is two (in the number 8.04). Therefore we multiply each side of the

Helpful Hint

After you have used one of the properties of equality on each side of an equation, be sure to simplify all expressions as much as possible before using another property of equality. This step is like making sure that all of the injured football players are removed from the field before proceeding to the next play.

equation by 100 because multiplying by 100 moves decimal points two places to the right:

$0.3p + 8.04 = 12.6$	Original equation
$100(0.3p + 8.04) = 100(12.6)$	Multiplication property of equality
$100(0.3p) + 100(8.04) = 100(12.6)$	Distributive property
$30p + 804 = 1260$	
$30p + 804 - 804 = 1260 - 804$	Subtract 804 from each side.
$30p = 456$	
$\frac{30p}{30} = \frac{456}{30}$	Divide each side by 30.
$p = 15.2$	

You can use a calculator to check that

$$0.3(15.2) + 8.04 = 12.6.$$

The solution set is $\{15.2\}$.

Now do Exercises 23–32

EXAMPLE 3

Another equation with decimals

Solve $0.5x + 0.4(x + 20) = 13.4$.

Solution

First use the distributive property to remove the parentheses:

$0.5x + 0.4(x + 20) = 13.4$	Original equation
$0.5x + 0.4x + 8 = 13.4$	Distributive property
$10(0.5x + 0.4x + 8) = 10(13.4)$	Multiply each side by 10.
$5x + 4x + 80 = 134$	Simplify.
$9x + 80 = 134$	Combine like terms.
$9x + 80 - 80 = 134 - 80$	Subtract 80 from each side.
$9x = 54$	Simplify.
$x = 6$	Divide each side by 9.

Check 6 in the original equation:

$0.5(6) + 0.4(6 + 20) = 13.4$	Replace x by 6.
$3 + 0.4(26) = 13.4$	
$3 + 10.4 = 13.4$	

Since both sides of the equation have the same value, the solution set is $\{6\}$.

Now do Exercises 33–36

CAUTION If you multiply each side by 10 in Example 3 before using the distributive property, be careful how you handle the terms in parentheses:

$$10 \cdot 0.5x + 10 \cdot 0.4(x + 20) = 10 \cdot 13.4$$
$$5x + 4(x + 20) = 134$$

It is not correct to multiply 0.4 by 10 *and also* to multiply $x + 20$ by 10.

Simplifying the Process

It is very important to develop the skill of solving equations in a systematic way, writing down every step as we have been doing. As you become more skilled at solving equations, you will probably want to simplify the process a bit. One way to simplify the process is by writing only the result of performing an operation on each side. Another way is to isolate the variable on the side where the variable has the larger coefficient, when the variable occurs on both sides. We use these ideas in Example 4 and in future examples in this text.

EXAMPLE 4

Simplifying the process

Solve each equation.

a) $2a - 3 = 0$ **b)** $2k + 5 = 3k + 1$

Study Tip

I hear and I forget; I see and I remember; I do and I understand. There is no substitute for doing exercises, lots of exercises.

Solution

a) Add 3 to each side, then divide each side by 2:

$$2a - 3 = 0$$
$$2a = 3 \quad \text{Add 3 to each side.}$$
$$a = \frac{3}{2} \quad \text{Divide each side by 2.}$$

Check that $\frac{3}{2}$ satisfies the original equation. The solution set is $\left\{\frac{3}{2}\right\}$.

b) For this equation we can get a single k on the right by subtracting $2k$ from each side. (If we subtract $3k$ from each side, we get $-k$, and then we need another step.)

$$2k + 5 = 3k + 1$$
$$5 = k + 1 \quad \text{Subtract } 2k \text{ from each side.}$$
$$4 = k \quad \text{Subtract 1 from each side.}$$

Check that 4 satisfies the original equation. The solution set is $\{4\}$.

Now do Exercises 37–52

Identities

It is easy to find equations that are satisfied by any real number that we choose as a replacement for the variable. For example, the equations

$$x \div 2 = \frac{1}{2}x, \qquad x + x = 2x, \qquad \text{and} \qquad x + 1 = x + 1$$

are satisfied by all real numbers. The equation

$$\frac{5}{x} = \frac{5}{x}$$

is satisfied by any real number except 0 because division by 0 is undefined.

Identity

An equation that is satisfied by every real number for which both sides are defined is called an **identity.**

We cannot recognize that the equation in Example 5 is an identity until we have simplified each side.

EXAMPLE 5

Solving an identity

Solve $7 - 5(x - 6) + 4 = 3 - 2(x - 5) - 3x + 28$.

Solution

We first use the distributive property to remove the parentheses:

$$7 - 5(x - 6) + 4 = 3 - 2(x - 5) - 3x + 28$$
$$7 - 5x + 30 + 4 = 3 - 2x + 10 - 3x + 28$$
$$41 - 5x = 41 - 5x \quad \text{Combine like terms.}$$

This last equation is true for any value of x because the two sides are identical. So the solution set to the original equation is the set of all real numbers or R.

Now do Exercises 53–54

Study Tip

Life is a game that holds many rewards for those who compete. Winning is never an accident. To win you must know the rules and have a game plan.

CAUTION If you get an equation in which both sides are identical, as in Example 5, there is no need to continue to simplify the equation. If you do continue, you will eventually get $0 = 0$, from which you can still conclude that the equation is an identity.

Conditional Equations

The statement $2x + 4 = 10$ is true only on condition that we choose $x = 3$. The equation $x^2 = 4$ is satisfied only if we choose $x = 2$ or $x = -2$. These equations are called conditional equations.

Conditional Equation

A **conditional equation** is an equation that is satisfied by at least one real number but is not an identity.

Every equation that we solved in Sections 2.1 and 2.2 is a conditional equation.

Inconsistent Equations

It is easy to find equations that are false no matter what number we use to replace the variable. Consider the equation

$$x = x + 1.$$

If we replace x by 3, we get $3 = 3 + 1$, which is false. If we replace x by 4, we get $4 = 4 + 1$, which is also false. Clearly, there is no number that will satisfy $x = x + 1$. Other examples of equations with no solutions include

$$x = x - 2, \qquad x - x = 5, \qquad \text{and} \qquad 0 \cdot x + 6 = 7.$$

Inconsistent Equation

An equation that has no solution is called an **inconsistent equation.**

The solution set to an inconsistent equation has no members. The set with no members is called the **empty set** and it is denoted by the symbol $\varnothing$.

EXAMPLE 6

Solving an inconsistent equation

Solve $2 - 3(x - 4) = 4(x - 7) - 7x$.

Solution

Use the distributive property to remove the parentheses:

$2 - 3(x - 4) = 4(x - 7) - 7x$	The original equation
$2 - 3x + 12 = 4x - 28 - 7x$	Distributive property
$14 - 3x = -28 - 3x$	Combine like terms on each side.
$14 - 3x + 3x = -28 - 3x + 3x$	Add $3x$ to each side.
$14 = -28$	Simplify.

The last equation is not true for any x. So the solution set to the original equation is the empty set, $\varnothing$. The equation is inconsistent.

Now do Exercises 55–72

Keep the following points in mind in solving equations.

Summary: Identities and Inconsistent Equations

1. An equation that is equivalent to an equation in which both sides are identical is an identity. The equation is satisfied by all real numbers for which both sides are defined.
2. An equation that is equivalent to an equation that is always false is inconsistent. The equation has no solution. The solution set is the empty set, $\varnothing$.

Warm-Ups ▼

True or false? Explain your answer.

1. To solve $\frac{1}{2}x - \frac{1}{3} = x + \frac{1}{6}$ multiply each side by 6.
2. The equation $\frac{1}{2}x - \frac{1}{3} = x + \frac{1}{6}$ is equivalent to $3x - 2 = 6x + 1$.
3. The equation $0.2x + 0.03x = 8$ is equivalent to $20x + 3x = 8$.
4. The solution set to $3h + 8 = 0$ is $\left\{\frac{8}{3}\right\}$.
5. The equation $5a + 3 = 0$ is an inconsistent equation.
6. The equation $2t = t$ is a conditional equation.
7. The equation $w - 0.1w = 0.9w$ is an identity.
8. All real numbers satisfy the equation $1 \div x = \frac{1}{x}$.
9. The equation $\frac{x}{x} = 1$ is an identity.
10. The equation $x - x = 99$ has no solution.

2.3 Exercises

Reading and Writing *After reading this section, write out the answers to these questions. Use complete sentences.*

1. What is the usual first step when solving an equation involving fractions?
2. What is a good first step for solving an equation involving decimals?
3. What is an identity?
4. What is a conditional equation?
5. What is an inconsistent equation?
6. What is the solution set to an inconsistent equation?

Solve each equation by first eliminating the fractions. See Example 1.

7. $\frac{x}{4} - \frac{3}{10} = 0$
8. $\frac{x}{15} + \frac{1}{6} = 0$
9. $3x - \frac{1}{6} = \frac{1}{2}$
10. $5x + \frac{1}{2} = \frac{3}{4}$
11. $\frac{x}{2} + 3 = x - \frac{1}{2}$
12. $13 - \frac{x}{2} = x - \frac{1}{2}$
13. $\frac{x}{2} + \frac{x}{3} = 20$
14. $\frac{x}{2} - \frac{x}{3} = 5$
15. $\frac{w}{2} + \frac{w}{4} = 12$
16. $\frac{a}{4} - \frac{a}{2} = -5$
17. $\frac{3z}{2} - \frac{2z}{3} = -10$
18. $\frac{3m}{4} + \frac{m}{2} = -5$
19. $\frac{1}{3}p - 5 = \frac{1}{4}p$
20. $\frac{1}{2}q - 6 = \frac{1}{5}q$
21. $\frac{1}{6}v + 1 = \frac{1}{4}v - 1$
22. $\frac{1}{15}k + 5 = \frac{1}{6}k - 10$

Solve each equation by first eliminating the decimal numbers. See Examples 2 and 3.

23. $x - 0.2x = 72$

24. $x - 0.1x = 63$

25. $0.3x + 1.2 = 0.5x$

26. $0.4x - 1.6 = 0.6x$

27. $0.02x - 1.56 = 0.8x$

28. $0.6x + 10.4 = 0.08x$

29. $0.1a - 0.3 = 0.2a - 8.3$

30. $0.5b + 3.4 = 0.2b + 12.4$

31. $0.05r + 0.4r = 27$

32. $0.08t + 28.3 = 0.5t - 9.5$

33. $0.05y + 0.03(y + 50) = 17.5$

34. $0.07y + 0.08(y - 100) = 44.5$

35. $0.1x + 0.05(x - 300) = 105$

36. $0.2x - 0.05(x - 100) = 35$

Solve each equation. If you feel proficient enough, try simplifying the process, as described in Example 4.

37. $2x - 9 = 0$

38. $3x + 7 = 0$

39. $-2x + 6 = 0$

40. $-3x - 12 = 0$

41. $\frac{z}{5} + 1 = 6$

42. $\frac{s}{2} + 2 = 5$

43. $\frac{c}{2} - 3 = -4$

44. $\frac{b}{3} - 4 = -7$

45. $3 = t + 6$

46. $-5 = y - 9$

47. $5 + 2q = 3q$

48. $-4 - 5p = -4p$

49. $8x - 1 = 9 + 9x$

50. $4x - 2 = -8 + 5x$

51. $-3x + 1 = -1 - 2x$

52. $-6x + 3 = -7 - 5x$

Solve each equation. Identify each as a conditional equation, an inconsistent equation, or an identity. See Examples 5 and 6.

53. $x + x = 2x$

54. $2x - x = x$

55. $a - 1 = a + 1$

56. $r + 7 = r$

57. $3y + 4y = 12y$

58. $9t - 8t = 7$

59. $-4 + 3(w - 1) = w + 2(w - 2) - 1$

60. $4 - 5(w + 2) = 2(w - 1) - 7w - 4$

61. $3(m + 1) = 3(m + 3)$

62. $5(m - 1) - 6(m + 3) = 4 - m$

63. $x + x = 2$

64. $3x - 5 = 0$

65. $2 - 3(5 - x) = 3x$

66. $3 - 3(5 - x) = 0$

67. $(3 - 3)(5 - z) = 0$

68. $(2 \cdot 4 - 8)p = 0$

69. $\frac{0}{x} = 0$

70. $\frac{2x}{2} = x$

71. $x \cdot x = x^2$

72. $\frac{2x}{2x} = 1$

Solve each equation.

73. $3x - 5 = 2x - 9$

74. $5x - 9 = x - 4$

75. $x + 2(x + 4) = 3(x + 3) - 1$

76. $u + 3(u - 4) = 4(u - 5)$

77. $23 - 5(3 - n) = -4(n - 2) + 9n$

78. $-3 - 4(t - 5) = -2(t + 3) + 11$

79. $0.05x + 30 = 0.4x - 5$

80. $x - 0.08x = 460$

81. $-\frac{2}{3}a + 1 = 2$

82. $-\frac{3}{4}t = \frac{1}{2}$

83. $\frac{y}{2} + \frac{y}{6} = 20$

84. $\frac{3w}{5} - 1 = \frac{w}{2} + 1$

85. $0.09x - 0.2(x + 4) = -1.46$

86. $0.08x + 0.5(x + 100) = 73.2$

87. $436x - 789 = -571$

88. $0.08x + 4533 = 10x + 69$

89. $\frac{x}{344} + 235 = 292$

90. $34(x - 98) = \frac{x}{2} + 475$

Solve each problem.

91. ***Sales commission.*** Danielle sold her house through an agent who charged 8% of the selling price. After the commission was paid, Danielle received \$117,760. If x is the selling price, then x satisfies

$$x - 0.08x = 117{,}760.$$

Solve this equation to find the selling price.

92. ***Raising rabbits.*** Before Roland sold two female rabbits, half of his rabbits were female. After the sale, only one-third of his rabbits were female. If x represents his original number of rabbits, then

$$\frac{1}{2}x - 2 = \frac{1}{3}(x - 2).$$

Solve this equation to find the number of rabbits that he had before the sale.

93. ***Eavesdropping.*** Reginald overheard his boss complaining that his federal income tax for 2003 was \$60,531.

a) Use the accompanying graph to estimate his boss's taxable income for 2003.

b) Find his boss's exact taxable income for 2003 by solving the equation

$$39{,}096.50 + 0.33(x - 174{,}700) = 60{,}531.$$

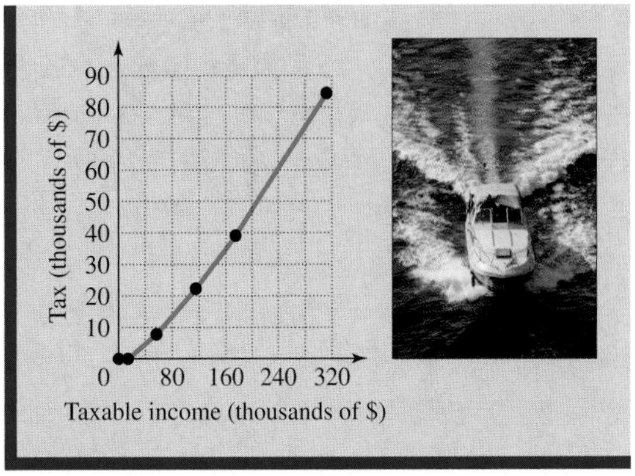

Figure for Exercise 93

94. ***Federal taxes.*** According to Bruce Harrell, CPA, the federal income tax for a class C corporation is found by solving a linear equation. The reason for the equation is that the amount x of federal tax is deducted before the state tax is figured, and the amount of state tax is deducted before the federal tax is figured. To find the amount of federal tax for a corporation with a taxable income of \$200,000, for which the federal tax rate is 25% and the state tax rate is 10%, Bruce must solve

$$x = 0.25[200{,}000 - 0.10(200{,}000 - x)].$$

Solve the equation for Bruce.

2.4 Formulas

In this Section

- **Solving for a Variable**
- **Finding the Value of a Variable**

In this section, you will learn to rewrite formulas using the same properties of equality that we used to solve equations. You will also learn how to find the value of one of the variables in a formula when we know the value of all of the others.

Solving for a Variable

Most drivers know the relationship between distance, rate, and time. For example, if you drive 70 mph for 3 hours, then you will travel 210 miles. At 60 mph a 300-mile trip will take 5 hours. If a 400-mile trip took 8 hours, then you averaged 50 mph. The relationship between distance D, rate R, and time T is expressed by the formula

$$D = R \cdot T.$$

A **formula** or **literal equation** is an equation involving two or more variables.

To find the time for a 300-mile trip at 60 mph, you are using the formula in the form $T = \frac{D}{R}$. The process of rewriting a formula for one variable in terms of the others is called **solving for a certain variable.** To solve for a certain variable, we use the same techniques that we use in solving equations.

EXAMPLE 1

Solving for a certain variable

Solve the formula $D = RT$ for T.

Solution

Since T is multiplied by R, dividing each side of the equation by R will isolate T:

$$D = RT \quad \text{Original formula}$$

$$\frac{D}{R} = \frac{R \cdot T}{R} \quad \text{Divide each side by } R.$$

$$\frac{D}{R} = T \quad \text{Divide out (or cancel) the common factor } R.$$

$$T = \frac{D}{R} \quad \text{It is customary to write the single variable on the left.}$$

Now do Exercises 7–18

The formula $C = \frac{5}{9}(F - 32)$ is used to find the Celsius temperature for a given Fahrenheit temperature. If we solve this formula for F, then we have a formula for finding Fahrenheit temperature for a given Celsius temperature.

EXAMPLE 2

Solving for a certain variable

Solve the formula $C = \frac{5}{9}(F - 32)$ for F.

Solution

We could apply the distributive property to the right side of the equation, but it is simpler to proceed as follows:

$$C = \frac{5}{9}(F - 32)$$

$$\frac{9}{5}C = \frac{9}{5} \cdot \frac{5}{9}(F - 32) \quad \text{Multiply each side by } \tfrac{9}{5}\text{, the reciprocal of } \tfrac{5}{9}.$$

$$\frac{9}{5}C = F - 32 \quad \text{Simplify.}$$

$$\frac{9}{5}C + 32 = F - 32 + 32 \quad \text{Add 32 to each side.}$$

$$\frac{9}{5}C + 32 = F \quad \text{Simplify.}$$

The formula is usually written as $F = \frac{9}{5}C + 32$.

Now do Exercises 19–24

When solving for a variable that appears more than once in the equation, we must combine the terms to obtain a single occurrence of the variable. *When a formula has been solved for a certain variable, that variable will not occur on both sides of the equation.*

EXAMPLE 3

Solving for a variable that appears on both sides

Solve $5x - b = 3x + d$ for x.

Solution

First get all terms involving x onto one side and all other terms onto the other side:

$$5x - b = 3x + d \quad \text{Original formula}$$

$$5x - 3x - b = d \quad \text{Subtract } 3x \text{ from each side.}$$

$$5x - 3x = b + d \quad \text{Add } b \text{ to each side.}$$

$$2x = b + d \quad \text{Combine like terms.}$$

$$x = \frac{b + d}{2} \quad \text{Divide each side by 2.}$$

The formula solved for x is $x = \frac{b + d}{2}$.

Now do Exercises 25–32

In Chapter 3, it will be necessary to solve an equation involving x and y for y.

EXAMPLE 4

Solving for y

Solve $x + 2y = 6$ for y. Write the answer in the form $y = mx + b$, where m and b are fixed real numbers.

Solution

$$x + 2y = 6 \quad \text{Original equation}$$

$$2y = 6 - x \quad \text{Subtract } x \text{ from each side.}$$

$$\frac{1}{2} \cdot 2y = \frac{1}{2}(6 - x) \quad \text{Multiply each side by } \tfrac{1}{2}.$$

$$y = 3 - \frac{1}{2}x \quad \text{Distributive property}$$

$$y = -\frac{1}{2}x + 3 \quad \text{Rearrange to get } y = mx + b \text{ form.}$$

Now do Exercises 33–42

Helpful Hint

If we simply wanted to solve $x + 2y = 6$ for y, we could have written

$$y = \frac{6 - x}{2} \text{ or } y = \frac{-x + 6}{2}.$$

However, in Example 4 we requested the form $y = mx + b$. This form is a popular form that we will study in detail in Chapter 3.

Notice that in Example 4 we multiplied each side of the equation by $\frac{1}{2}$, and so we multiplied each term on the right-hand side by $\frac{1}{2}$. Instead of multiplying by $\frac{1}{2}$, we could have divided each side of the equation by 2. We would then divide each term on the right side by 2. This idea is illustrated in Example 5.

EXAMPLE 5

Solving for y

Solve $2x - 3y = 9$ for y. Write the answer in the form $y = mx + b$, where m and b are real numbers. (When we study lines in Chapter 3 you will see that $y = mx + b$ is the slope-intercept form of the equation of a line.)

Solution

$$2x - 3y = 9 \quad \text{Original equation}$$

$$-3y = -2x + 9 \quad \text{Subtract } 2x \text{ from each side.}$$

$$\frac{-3y}{-3} = \frac{-2x + 9}{-3} \quad \text{Divide each side by } -3.$$

$$y = \frac{-2x}{-3} + \frac{9}{-3} \quad \text{By the distributive property, each term is divided by } -3.$$

$$y = \frac{2}{3}x - 3 \quad \text{Simplify.}$$

Now do Exercises 43–54

Even though we wrote $y = \frac{2}{3}x - 3$ in Example 5, the equation is still considered to be in the form $y = mx + b$ because we could have written $y = \frac{2}{3}x + (-3)$.

Finding the Value of a Variable

In many situations we know the values of all variables in a formula except one. We use the formula to determine the unknown value.

EXAMPLE 6

Finding the value of a variable in a formula

If $2x - 3y = 9$, find y when $x = 6$.

Solution

Method 1: First solve the equation for y. Because we have already solved this equation for y in Example 5 we will not repeat that process in this example. We have

$$y = \frac{2}{3}x - 3.$$

Now replace x by 6 in this equation:

$$y = \frac{2}{3}(6) - 3$$

$$= 4 - 3 = 1$$

So when $x = 6$, we have $y = 1$.

Method 2: First replace x by 6 in the original equation, then solve for y:

$$2x - 3y = 9 \quad \text{Original equation}$$

$$2 \cdot 6 - 3y = 9 \quad \text{Replace } x \text{ by 6.}$$

$$12 - 3y = 9 \quad \text{Simplify.}$$

$$-3y = -3 \quad \text{Subtract 12 from each side.}$$

$$y = 1 \quad \text{Divide each side by } -3.$$

So when $x = 6$, we have $y = 1$.

Now do Exercises 63–72

If we had to find the value of y for many different values of x, it would be best to solve the equation for y, then insert the various values of x. Method 1 of Example 6 would be the better method. If we must find only one value of y, it does not matter which method we use. When doing the exercises corresponding to this example, you should try both methods.

The next example involves the simple interest formula $I = Prt$, where I is the amount of interest, P is the principal or the amount invested, r is the annual interest rate, and t is the time in years. The interest rate is generally expressed as a percent. When using a rate in computations, you must convert it to a decimal.

EXAMPLE 7

Using the simple interest formula

If the simple interest is \$120, the principal is \$400, and the time is 2 years, find the rate.

Solution

First, solve the formula $I = Prt$ for r, then insert values of P, I, and t:

$Prt = I$	Simple interest formula
$\frac{Prt}{Pt} = \frac{I}{Pt}$	Divide each side by Pt.
$r = \frac{I}{Pt}$	Simplify.
$r = \frac{120}{400 \cdot 2}$	Substitute the values of I, P, and t.
$r = 0.15$	Simplify.
$r = 15\%$	Move the decimal point two places to the right.

Now do Exercises 73–76

Helpful Hint

All interest computation is based on simple interest. However, depositors do not like to wait two years to get interest as in Example 7. More often the time is $\frac{1}{12}$ year or $\frac{1}{365}$ year. Simple interest computed every month is said to be compounded monthly. Simple interest computed every day is said to be compounded daily.

In solving a geometric problem, it is always helpful to draw a diagram, as we do in Example 8.

EXAMPLE 8

Using a geometric formula

The perimeter of a rectangle is 36 feet. If the width is 6 feet, then what is the length?

Solution

First, put the given information on a diagram as shown in Fig. 2.1. Substitute the given values into the formula for the perimeter of a rectangle found inside the front cover of this book, and then solve for L. (We could solve for L first and then insert the given values.)

$P = 2L + 2W$	Perimeter of a rectangle
$36 = 2L + 2 \cdot 6$	Substitute 36 for P and 6 for W.
$36 = 2L + 12$	Simplify.
$24 = 2L$	Subtract 12 from each side.
$12 = L$	Divide each side by 2.

Check: If $L = 12$ and $W = 6$, then $P = 2(12) + 2(6) = 36$ feet. So we can be certain that the length is 12 feet.

Now do Exercises 77–80

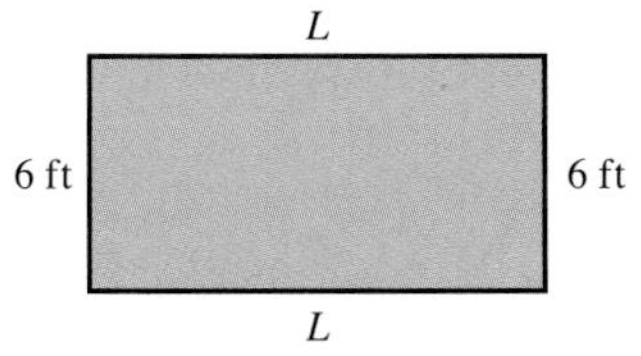

Figure 2.1

If L is the list price or original price of an item and r is the rate of discount, then the amount of discount is rL, the product of the rate and the list price. The sale price S is the list price minus the amount of discount. So $S = L - rL$. The rate of discount is generally expressed as a percent. In computations, rates must be written as decimals or fractions.

EXAMPLE 9

Finding the original price

What was the original price of a stereo that sold for \$560 after a 20% discount?

Solution

Express 20% as the decimal 0.20 or 0.2 and use the formula $S = L - rL$:

$$\text{Selling price} = \text{list price} - \text{amount of discount}$$

$$560 = L - 0.2L$$

$$10(560) = 10(L - 0.2L) \quad \text{Multiply each side by 10.}$$

$$5600 = 10L - 2L \quad \text{Remove the parentheses.}$$

$$5600 = 8L \quad \text{Combine like terms.}$$

$$\frac{5600}{8} = \frac{8L}{8} \quad \text{Divide each side by 8.}$$

$$700 = L$$

Since 20% of \$700 is \$140 and \$700 − \$140 = \$560, we can be sure that the original price was \$700. Note that if the discount is 20%, then the selling price is 80% of the list price. So we could have started with the equation $560 = 0.80L$.

Now do Exercises 81–86

Study Tip

Don't wait for inspiration to strike, it probably won't. Algebra is learned one tiny step at a time. So do lots of exercises and keep taking those tiny steps.

Warm-Ups ▼

True or false? Explain your answer.

1. If we solve $D = R \cdot T$ for T, we get $T \cdot R = D$.
2. If we solve $a - b = 3a - m$ for a, we get $a = 3a - m + b$.
3. Solving $A = LW$ for L, we get $L = \frac{W}{A}$.
4. Solving $D = RT$ for R, we get $R = \frac{d}{t}$.
5. The perimeter of a rectangle is the product of its length and width.
6. The volume of a shoe box is the product of its length, width, and height.
7. The sum of the length and width of a rectangle is one-half of its perimeter.
8. Solving $y - x = 5$ for y gives us $y = x + 5$.
9. If $x = -1$ and $y = -3x + 6$, then $y = 3$.
10. The circumference of a circle is the product of its diameter and the number π.

2.4 Exercises

Boost your GRADE at mathzone.com!

MathZone
- Practice Problems
- Self-Tests
- Videos
- Net Tutor
- e-Professors

Reading and Writing *After reading this section, write out the answers to these questions. Use complete sentences.*

1. What is a formula?

2. What is a literal equation?

3. What does it mean to solve a formula for a certain variable?

4. How do you solve a formula for a variable that appears on both sides?

5. What are the two methods shown for finding the value of a variable in a formula?

6. What formula expresses the perimeter of a rectangle in terms of its length and width?

Solve each formula for the specified variable. See Examples 1 and 2.

7. $D = RT$ for R

8. $A = LW$ for W

9. $C = \pi D$ for D

10. $F = ma$ for a

11. $I = Prt$ for P

12. $I = Prt$ for t

13. $F = \frac{9}{5}C + 32$ for C

14. $y = \frac{3}{4}x - 7$ for x

15. $A = \frac{1}{2}bh$ for h

16. $A = \frac{1}{2}bh$ for b

17. $P = 2L + 2W$ for L

18. $P = 2L + 2W$ for W

19. $A = \frac{1}{2}(a + b)$ for a

20. $A = \frac{1}{2}(a + b)$ for b

21. $S = P + Prt$ for r

22. $S = P + Prt$ for t

23. $A = \frac{1}{2}h(a + b)$ for a

24. $A = \frac{1}{2}h(a + b)$ for b

Solve each equation for x. See Example 3.

25. $5x + a = 3x + b$

26. $2c - x = 4x + c - 5b$

27. $4(a + x) - 3(x - a) = 0$

28. $-2(x - b) - (5a - x) = a + b$

29. $3x - 2(a - 3) = 4x - 6 - a$

30. $2(x - 3w) = -3(x + w)$

31. $3x + 2ab = 4x - 5ab$

32. $x - a = -x + a + 4b$

Solve each equation for y. See Examples 4 and 5.

33. $x + y = -9$

34. $3x + y = -5$

35. $x + y - 6 = 0$

36. $4x + y - 2 = 0$

37. $2x - y = 2$

38. $x - y = -3$

39. $3x - y + 4 = 0$

40. $-2x - y + 5 = 0$

41. $x + 2y = 4$

42. $3x + 2y = 6$

43. $2x - 2y = 1$

44. $3x - 2y = -6$

45. $y + 2 = 3(x - 4)$

46. $y - 3 = -3(x - 1)$

47. $y - 1 = \frac{1}{2}(x - 2)$

48. $y - 4 = -\frac{2}{3}(x - 9)$

49. $\frac{1}{2}x - \frac{1}{3}y = -2$

50. $\frac{x}{2} + \frac{y}{4} = \frac{1}{2}$

51. $y - 2 = \frac{3}{2}(x + 3)$

52. $y + 4 = \frac{2}{3}(x - 2)$

53. $y - \frac{1}{2} = -\frac{1}{4}\left(x - \frac{1}{2}\right)$

54. $y + \frac{1}{2} = -\frac{1}{3}\left(x + \frac{1}{2}\right)$

Fill in the tables using the given formulas.

55. $y = -3x + 30$

x	y
−10	
0	
10	
20	
30	

56. $y = 4x - 20$

x	y
−10	
−5	
0	
5	
10	

57. $F = \frac{9}{5}C + 32$

C	F
−10	
−5	
0	
40	
100	

58. $C = \frac{5}{9}(F - 32)$

F	C
−40	
14	
32	
59	
86	

59. $T = \frac{400}{R}$

R (mph)	T (hr)
10	
20	
40	
80	
100	

60. $R = \frac{100}{T}$

T (hr)	R (mph)
1	
5	
20	
50	
100	

61. $S = \frac{n(n+1)}{2}$

n	S
1	
2	
3	
4	
5	

62. $S = \frac{n(n+1)(2n+1)}{6}$

n	S
1	
2	
3	
4	
5	

For each equation that follows, find y given that x = 2. See Example 6.

63. $y = 3x - 4$

64. $y = -2x + 5$

65. $3x - 2y = -8$

66. $4x + 6y = 8$

67. $\frac{3x}{2} - \frac{5y}{3} = 6$

68. $\frac{2y}{5} - \frac{3x}{4} = \frac{1}{2}$

69. $y - 3 = \frac{1}{2}(x - 6)$

70. $y - 6 = -\frac{3}{4}(x - 2)$

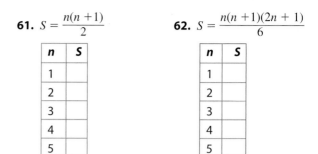

71. $y - 4.3 = 0.45(x - 8.6)$

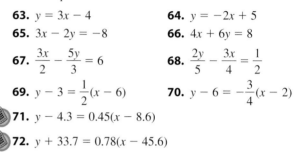

72. $y + 33.7 = 0.78(x - 45.6)$

Solve each of the following problems. Some geometric formulas that may be helpful can be found inside the front cover of this text. See Examples 7–9.

73. ***Finding the rate.*** If the simple interest on \$5000 for 3 years is \$600, then what is the rate?

74. ***Finding the rate.*** Wayne paid \$420 in simple interest on a loan of \$1000 for 7 years. What was the rate?

75. ***Finding the time.*** Kathy paid \$500 in simple interest on a loan of \$2500. If the annual interest rate was 5%, then what was the time?

76. ***Finding the time.*** Robert paid \$240 in simple interest on a loan of \$1000. If the annual interest rate was 8%, then what was the time?

77. ***Finding the length.*** The area of a rectangle is 28 square yards. The width is 4 yards. Find the length.

78. ***Finding the width.*** The area of a rectangle is 60 square feet. The length is 4 feet. Find the width.

79. ***Finding the length.*** If it takes 600 feet of wire fencing to fence a rectangular feed lot that has a width of 75 feet, then what is the length of the lot?

80. ***Finding the depth.*** If it takes 500 feet of fencing to enclose a rectangular lot that is 104 feet wide, then how deep is the lot?

81. ***Finding MSRP.*** What was the manufacturer's suggested retail price (MSRP) for a Lexus SC 430 that sold for \$54,450 after a 10% discount?

82. ***Finding MSRP.*** What was the MSRP for a Hummer H1 that sold for \$107,272 after an 8% discount?

83. ***Finding the original price.*** Find the original price if there is a 15% discount and the sale price is \$255.

84. ***Finding the list price.*** Find the list price if there is a 12% discount and the sale price is \$4400.

85. ***Rate of discount.*** Find the rate of discount if the discount is \$40 and the original price is \$200.

86. ***Rate of discount.*** Find the rate of discount if the discount is \$20 and the original price is \$250.

87. ***Width of a football field.*** The perimeter of a football field in the NFL, excluding the end zones, is 920 feet. How wide is the field?

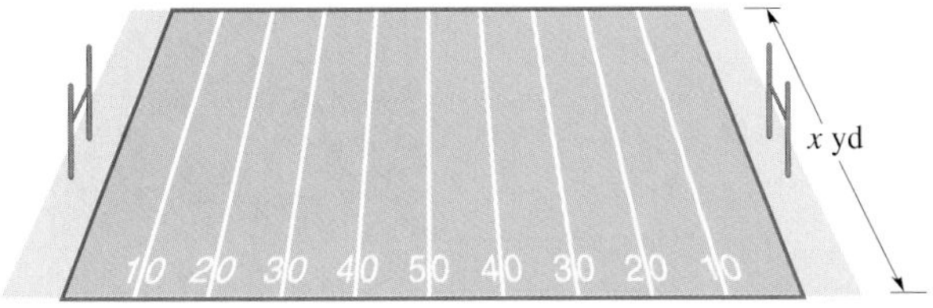

Figure for Exercise 87

88. ***Perimeter of a frame.*** If a picture frame is 16 inches by 20 inches, then what is its perimeter?

89. ***Volume of a box.*** A rectangular box measures 2 feet wide, 3 feet long, and 4 feet deep. What is its volume?

90. ***Volume of a refrigerator.*** The volume of a rectangular refrigerator is 20 cubic feet. If the top measures 2 feet by 2.5 feet, then what is the height?

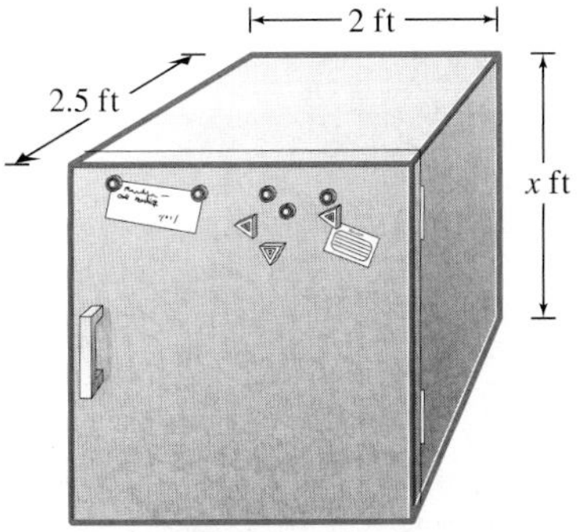

Figure for Exercise 90

91. ***Radius of a pizza.*** If the circumference of a pizza is 8π inches, then what is the radius?

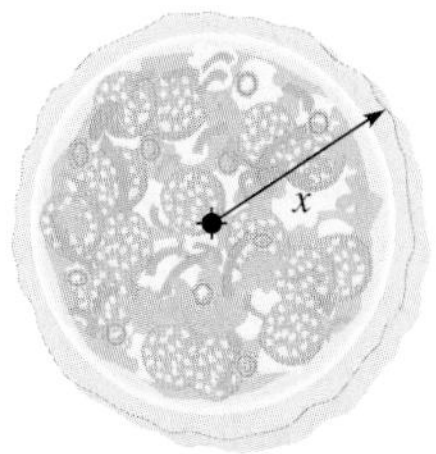

Figure for Exercise 91

92. ***Diameter of a circle.*** If the circumference of a circle is 4π meters, then what is the diameter?

93. ***Height of a banner.*** If a banner in the shape of a triangle has an area of 16 square feet with a base of 4 feet, then what is the height of the banner?

Figure for Exercise 93

94. ***Length of a leg.*** If a right triangle has an area of 14 square meters and one leg is 4 meters in length, then what is the length of the other leg?

95. ***Length of the base.*** A trapezoid with height 20 inches and lower base 8 inches has an area of 200 square inches. What is the length of its upper base?

96. ***Height of a trapezoid.*** The end of a flower box forms the shape of a trapezoid. The area of the trapezoid is 300 square

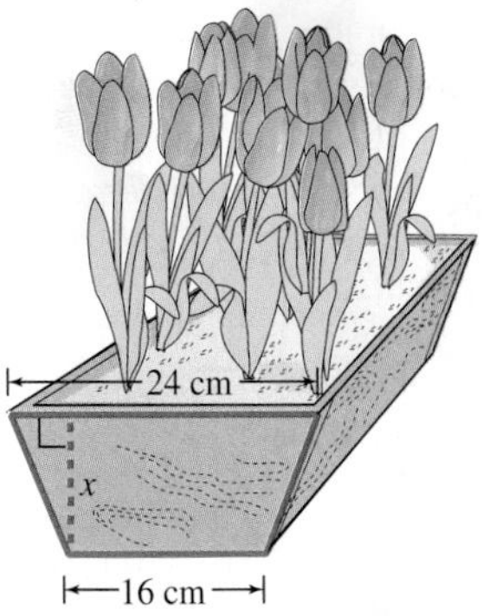

Figure for Exercise 96

centimeters. The bases are 16 centimeters and 24 centimeters in length. Find the height.

97. ***Fried's rule.*** Doctors often prescribe the same drugs for children as they do for adults. The formula $d = 0.08aD$ (Fried's rule) is used to calculate the child's dosage d, where a is the child's age and D is the adult dosage. If a doctor prescribes 1000 milligrams of acetaminophen for an adult, then how many milligrams would the doctor prescribe for an eight-year-old child? Use the bar graph to determine the age at which a child would get the same dosage as an adult.

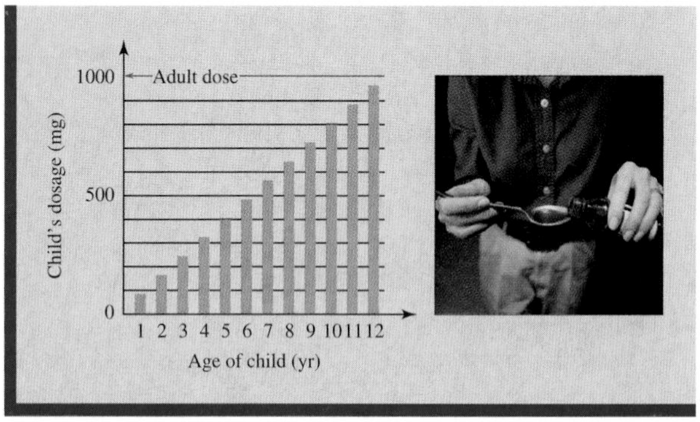

Figure for Exercise 97

98. ***Cowling's rule.*** Cowling's rule is another method for determining the dosage of a drug to prescribe to a child. For this rule, the formula

$$d = \frac{D(a + 1)}{24}$$

gives the child's dosage d, where D is the adult dosage and a is the age of the child in years. If the adult dosage of a drug is 600 milligrams and a doctor uses this formula to determine that a child's dosage is 200 milligrams, then how old is the child?

99. ***Administering Vancomycin.*** A patient is to receive 750 mg of the antibiotic Vancomycin. However, Vancomycin comes in a solution containing 1 gram (available dose) of Vancomycin per 5 milliliters (quantity) of solution. Use the formula

$$\text{Amount} = \frac{\text{desired dose}}{\text{available dose}} \times \text{quantity}$$

to find the amount of this solution that should be administered to the patient.

100. ***International communications.*** The global investment in telecom infrastructure since 1990 can be modeled by the formula

$$I = 7.5T + 115,$$

where I is in billions of dollars and t is the number of years since 1990 (*Fortune,* www.fortune.com).

a) Use the formula to find the global investment in 2000.

b) Use the accompanying graph to estimate the year in which the global investment will reach \$250 billion.

c) Use the formula to find the year in which the global investment will reach \$250 billion.

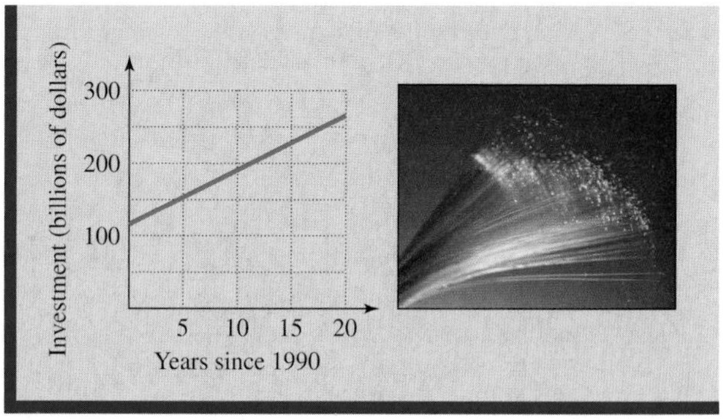

Figure for Exercise 100

101. ***The 2.4-meter rule.*** A 2.4-meter sailboat is a one-person boat that is about 13 feet in length, has a displacement of about 550 pounds, and a sail area of about 81 square feet. To compete in the 2.4-meter class, a boat must satisfy the formula

$$2.4 = \frac{L + 2D - F\sqrt{S}}{2.37},$$

where L = length, F = freeboard, D = girth, and S = sail area. Solve the formula for L.

Photo for Exercise 101

2.5 Translating Verbal Expressions into Algebraic Expressions

In this Section

- **Writing Algebraic Expressions**
- **Pairs of Numbers**
- **Consecutive Integers**
- **Using Formulas**
- **Writing Equations**

You translated some verbal expressions into algebraic expressions in Section 1.6; in this section you will study translating in more detail.

Writing Algebraic Expressions

The following box contains a list of some frequently occurring verbal expressions and their equivalent algebraic expressions.

Translating Words into Algebra

	Verbal Phrase	Algebraic Expression
Addition:	The sum of a number and 8	$x + 8$
	Five is added to a number	$x + 5$
	Two more than a number	$x + 2$
	A number increased by 3	$x + 3$
Subtraction:	Four is subtracted from a number	$x - 4$
	Three less than a number	$x - 3$
	The difference between 7 and a number	$7 - x$
	A number decreased by 2	$x - 2$
Multiplication:	The product of 5 and a number	$5x$
	Twice a number	$2x$
	One-half of a number	$\frac{1}{2}x$
	Five percent of a number	$0.05x$
Division:	The ratio of a number to 6	$\frac{x}{6}$
	The quotient of 5 and a number	$\frac{5}{x}$
	Three divided by some number	$\frac{3}{x}$

EXAMPLE 1

Writing algebraic expressions

Translate each verbal expression into an algebraic expression.

a) The sum of a number and 9

b) Eighty percent of a number

c) A number divided by 4

d) The result of a number subtracted from 5

e) Three less than a number

Solution

a) If x is the number, then the sum of x and 9 is $x + 9$.

b) If w is the number, then eighty percent of the number is $0.80w$.

c) If y is the number, then the number divided by 4 is $\frac{y}{4}$.

d) If z is the number, then the result of subtracting z from 5 is $5 - z$.

e) If a is the number, then 3 less than a is $a - 3$.

Now do Exercises 7–18

Helpful Hint

We know that x and $10 - x$ have a sum of 10 for any value of x. We can easily check that fact by adding:

$$x + 10 - x = 10$$

In general it is not true that x and $x - 10$ have a sum of 10, because

$$x + x - 10 = 2x - 10.$$

For what value of x is the sum of x and $x - 10$ equal to 10?

Pairs of Numbers

There is often more than one unknown quantity in a problem, but a relationship between the unknown quantities is given. For example, if one unknown number is 5 more than another unknown number, we can use

$$x \quad \text{and} \quad x + 5,$$

to represent them. Note that x and $x + 5$ can also be used to represent two unknown numbers that differ by 5, for if two numbers differ by 5, one of the numbers is 5 more than the other.

How would you represent two numbers that have a sum of 10? If one of the numbers is 2, the other is certainly $10 - 2$, or 8. Thus if x is one of the numbers, then $10 - x$ is the other. The expressions

$$x \quad \text{and} \quad 10 - x$$

have a sum of 10 for any value of x.

EXAMPLE 2

Algebraic expressions for pairs of numbers

Write algebraic expressions for each pair of numbers.

a) Two numbers that differ by 12 **b)** Two numbers with a sum of -8

Solution

a) The expressions x and $x - 12$ represent two numbers that differ by 12. We can check by subtracting:

$$x - (x - 12) = x - x + 12 = 12$$

Of course, x and $x + 12$ also differ by 12 because $x + 12 - x = 12$.

b) The expressions x and $-8 - x$ have a sum of -8. We can check by addition:

$$x + (-8 - x) = x - 8 - x = -8$$

Now do Exercises 19–30

Pairs of numbers occur in geometry in discussing measures of angles. You will need the following facts about degree measures of angles.

Degree Measures of Angles

Two angles are called **complementary** if the sum of their degree measures is 90°.

Two angles are called **supplementary** if the sum of their degree measures is 180°.

The sum of the degree measures of the three angles of any triangle is 180°.

For complementary angles, we use x and $90 - x$ for their degree measures. For supplementary angles, we use x and $180 - x$. Complementary angles that share a common side form a right angle. Supplementary angles that share a common side form a straight angle or straight line.

EXAMPLE 3

Degree measures

Write algebraic expressions for each pair of angles shown.

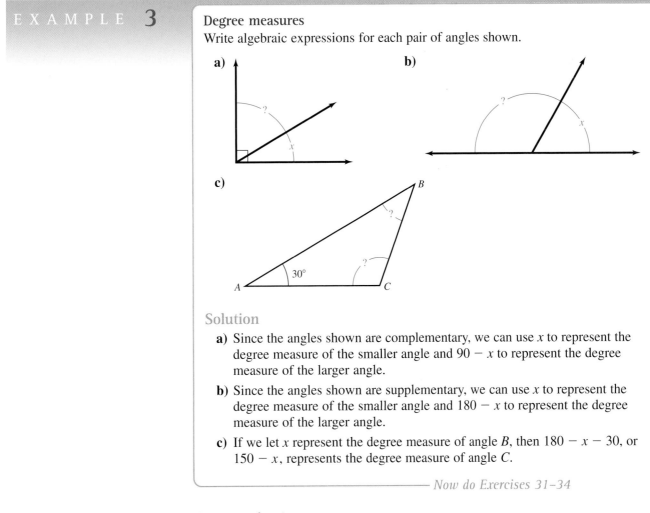

a)

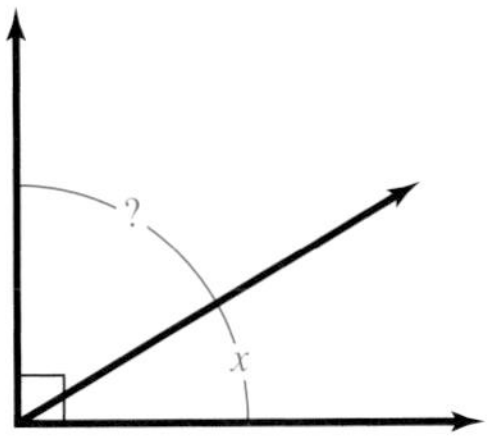

b)

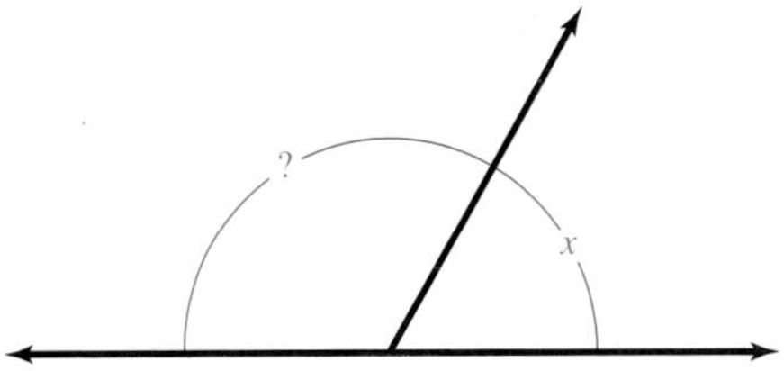

c)

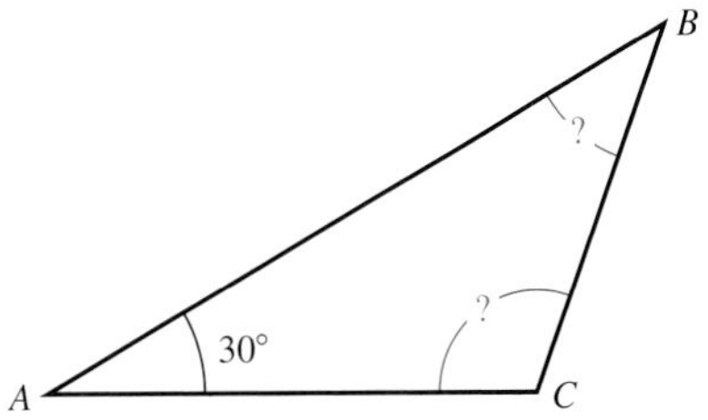

Solution

a) Since the angles shown are complementary, we can use x to represent the degree measure of the smaller angle and $90 - x$ to represent the degree measure of the larger angle.

b) Since the angles shown are supplementary, we can use x to represent the degree measure of the smaller angle and $180 - x$ to represent the degree measure of the larger angle.

c) If we let x represent the degree measure of angle B, then $180 - x - 30$, or $150 - x$, represents the degree measure of angle C.

Now do Exercises 31–34

Consecutive Integers

Note that each integer is one larger than the previous integer. For example, if $x = 5$, then $x + 1 = 6$ and $x + 2 = 7$. So if x is an integer, then x, $x + 1$, and $x + 2$ represent three consecutive integers. Each even (or odd) integer is two larger than the previous even (or odd) integer. For example, if $x = 6$, then $x + 2 = 8$, and $x + 4 = 10$. If $x = 7$, then $x + 2 = 9$, and $x + 4 = 11$. So x, $x + 2$, and $x + 4$ represent three consecutive even integers if x is even and three consecutive odd integers if x is odd.

CAUTION The expressions x, $x + 1$, and $x + 3$ do not represent three consecutive odd integers no matter what x represents.

EXAMPLE 4

Expressions for integers

Write algebraic expressions for the following unknown integers.

a) Two consecutive integers, the smallest of which is w.

b) Three consecutive even integers, the smallest of which is z.

c) Four consecutive odd integers, the smallest of which is y.

Solution

a) Each integer is 1 larger than the preceding integer. So if w represents the smallest of two consecutive integers, then w and $w + 1$ represent the integers.

b) Each even integer is 2 larger than the preceding even integer. So if z represents the smallest of three consecutive even integers, then z, $z + 2$, and $z + 4$ represent the three consecutive even integers.

c) Each odd integer is 2 larger than the preceding odd integer. So if y represents the smallest of four consecutive odd integers, then y, $y + 2$, $y + 4$, and $y + 6$ represent the four consecutive odd integers.

Now do Exercises 35–42

Using Formulas

In writing expressions for unknown quantities, we often use standard formulas such as those given inside the front cover of this book.

EXAMPLE 5

Writing algebraic expressions using standard formulas

Find an algebraic expression for

a) the distance if the rate is 30 miles per hour and the time is T hours.

b) the discount if the rate is 40% and the original price is p dollars.

Solution

a) Using the formula $D = RT$, we have $D = 30T$. So $30T$ is an expression that represents the distance in miles.

b) Since the discount is the rate times the original price, an algebraic expression for the discount is $0.40p$ dollars.

Now do Exercises 43–66

Writing Equations

To solve a problem using algebra, we describe or **model** the problem with an equation. In this section we write the equations only, and in Section 2.6 we write and solve them. Sometimes we must write an equation from the information given in the problem and sometimes we use a standard model to get the equation. Some standard models are shown in the following box.

Study Tip

It is a good idea to work with others, but don't be misled. Working a problem with help is not the same as working a problem on your own. In the end, mathematics is personal. Make sure that you can do it. Give yourself a weekly quiz by working 10 exercises randomly selected from the material covered since the last test.

Uniform Motion Model

Distance = Rate · Time $\quad D = R \cdot T$

Percentage Models

What number is 5% of 40? $\quad x = 0.05 \cdot 40$

Ten is what percent of 80? $\quad 10 = x \cdot 80$

Twenty is 4% of what number? $\quad 20 = 0.04 \cdot x$

Selling Price and Discount Model

Discount = Rate of discount · Original price

Selling Price = Original price − Discount

Real Estate Commission Model

Commission = Rate of commission · Selling price

Amount for owner = Selling price − Commission

Geometric Models for Perimeter

Perimeter of any figure = the sum of the lengths of the sides

Rectangle: $P = 2L + 2W$ $\quad$ Square: $P = 4s$

Geometric Models for Area

Rectangle: $A = LW$ $\quad$ Square: $A = s^2$

Parallelogram: $A = bh$ $\quad$ Triangle: $A = \frac{1}{2}bh$

More geometric formulas can be found inside the front cover of this text.

EXAMPLE 6

Writing equations

Identify the variable and write an equation that describes each situation.

a) Find two numbers that have a sum of 14 and a product of 45.

b) A coat is on sale for 25% off the list price. If the sale price is \$87, then what is the list price?

c) What percent of 8 is 2?

d) The value of x dimes and $x - 3$ quarters is \$2.05.

Helpful Hint

At this point we are simply learning to write equations that model certain situations. Don't worry about solving these equations now. In Section 2.6 we will solve problems by writing an equation and solving it.

Solution

a) Let x = one of the numbers and $14 - x$ = the other number. Since their product is 45, we have

$$x(14 - x) = 45.$$

b) Let $x =$ the list price and $0.25x =$ the amount of discount. We can write an equation expressing the fact that the selling price is the list price minus the discount:

$$\text{List price} - \text{discount} = \text{selling price}$$
$$x - 0.25x = 87$$

c) If we let x represent the percentage, then the equation is $x \cdot 8 = 2$, or $8x = 2$.

d) The value of x dimes at 10 cents each is $10x$ cents. The value of $x - 3$ quarters at 25 cents each is $25(x - 3)$ cents. We can write an equation expressing the fact that the total value of the coins is 205 cents:

$$\text{Value of dimes} + \text{value of quarters} = \text{total value}$$
$$10x + 25(x - 3) = 205$$

Now do Exercises 67–92

CAUTION The value of the coins in Example 6(d) is either 205 cents or 2.05 dollars. If the total value is expressed in dollars, then all of the values must be expressed in dollars. So we could also write the equation as

$$0.10x + 0.25(x - 3) = 2.05.$$

Warm-Ups ▼

True or false? Explain your answer.

1. For any value of x, the numbers x and $x + 6$ differ by 6.
2. For any value of a, a and $10 - a$ have a sum of 10.
3. If Jack ran at x miles per hour for 3 hours, he ran $3x$ miles.
4. If Jill ran at x miles per hour for 10 miles, she ran for $10x$ hours.
5. If the realtor gets 6% of the selling price and the house sells for x dollars, the owner gets $x - 0.06x$ dollars.
6. If the owner got \$50,000 and the realtor got 10% of the selling price, the house sold for \$55,000.
7. Three consecutive odd integers can be represented by x, $x + 1$, and $x + 3$.
8. The value in cents of n nickels and d dimes is $0.05n + 0.10d$.
9. If the sales tax rate is 5% and x represents the price of the goods purchased, then the total bill is $1.05x$.
10. If the length of a rectangle is 4 feet more than the width w, then the perimeter is $w + (w + 4)$ feet.

2.5 Exercises

Reading and Writing *After reading this section, write out the answers to these questions. Use complete sentences.*

1. What are the different ways of verbally expressing the operation of addition?

2. How can you algebraically express two numbers using only one variable?

3. What are complementary angles?

4. What are supplementary angles?

5. What is the relationship between distance, rate, and time?

6. What is the difference between expressing consecutive even integers and consecutive odd integers algebraically?

Translate each verbal expression into an algebraic expression. See Example 1.

7. The sum of a number and 3
8. Two more than a number
9. Three less than a number
10. Four subtracted from a number
11. The product of a number and 5
12. Five divided by some number
13. Ten percent of a number
14. Eight percent of a number
15. The ratio of a number and 3
16. The quotient of 12 and a number
17. One-third of a number
18. Three-fourths of a number

Write algebraic expressions for each pair of numbers. See Example 2.

19. Two numbers with a difference of 15
20. Two numbers that differ by 9
21. Two numbers with a sum of 6
22. Two numbers with a sum of 5
23. Two numbers with a sum of -4
24. Two numbers with a sum of -8
25. Two numbers such that one is 3 larger than the other

26. Two numbers such that one is 8 smaller than the other

27. Two numbers such that one is 5% of the other

28. Two numbers such that one is 40% of the other

29. Two numbers such that one is 30% more than the other

30. Two numbers such that one is 20% smaller than the other

Each of the following figures shows a pair of angles. Write algebraic expressions for the degree measures of each pair of angles. See Example 3.

31.

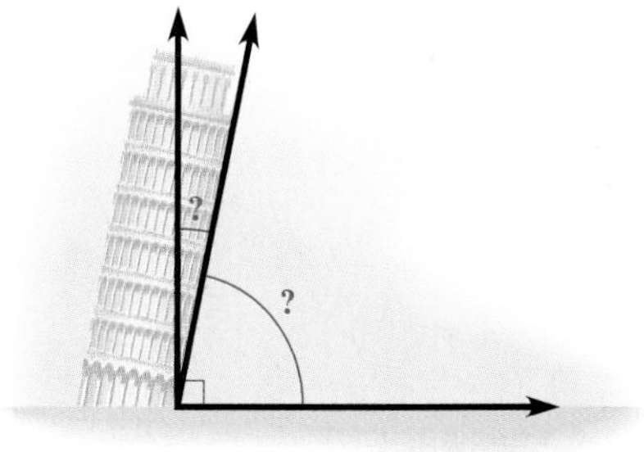

Figure for Exercise 31

32.

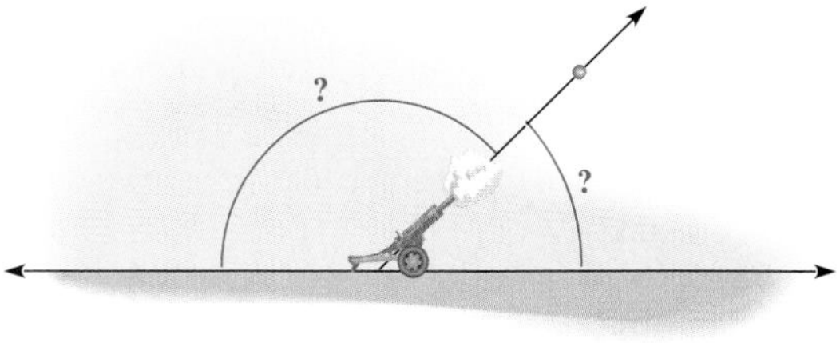

Figure for Exercise 32

33.

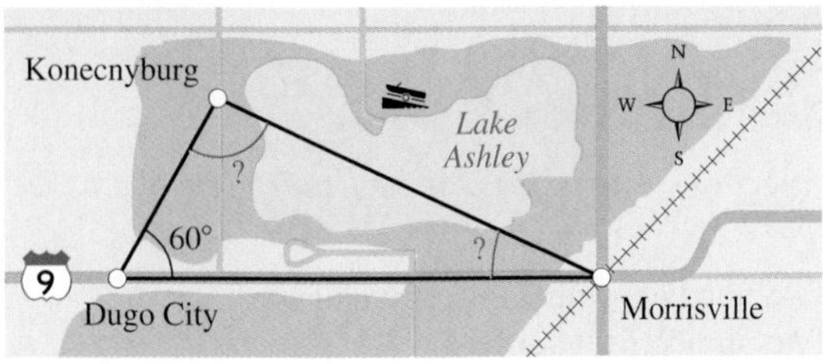

Figure for Exercise 33

34.

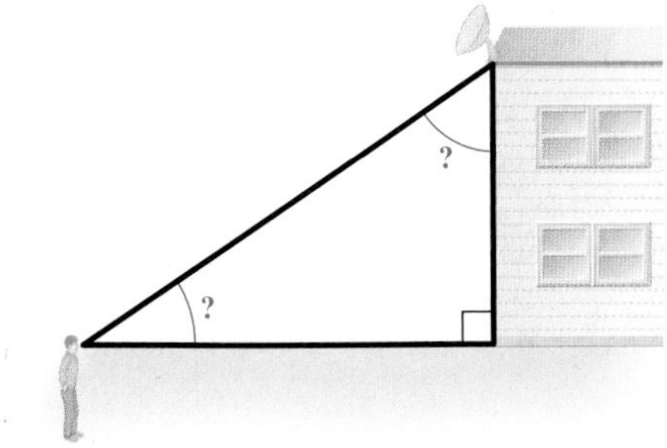

Figure for Exercise 34

Write algebraic expressions for the following unknown integers. See Example 4.

35. Two consecutive even integers, the smallest of which is n

36. Two consecutive odd integers, the smallest of which is x

37. Two consecutive integers

38. Three consecutive even integers

39. Three consecutive odd integers

40. Three consecutive integers

41. Four consecutive even integers

42. Four consecutive odd integers

Find an algebraic expression for the quantity in italics using the given information. See Example 5.

43. The *distance,* given that the rate is x miles per hour and the time is 3 hours

44. The *distance,* given that the rate is $x + 10$ miles per hour and the time is 5 hours

45. The *discount,* given that the rate is 25% and the original price is q dollars

46. The *discount,* given that the rate is 10% and the original price is t yen

47. The *time,* given that the distance is x miles and the rate is 20 miles per hour

48. The *time,* given that the distance is 300 kilometers and the rate is $x + 30$ kilometers per hour

49. The *rate,* given that the distance is $x - 100$ meters and the time is 12 seconds

50. The *rate,* given that the distance is 200 feet and the time is $x + 3$ seconds

51. The *area* of a rectangle with length x meters and width 5 meters

52. The *area* of a rectangle with sides b yards and $b - 6$ yards

53. The *perimeter* of a rectangle with length $w + 3$ inches and width w inches

54. The *perimeter* of a rectangle with length r centimeters and width $r - 1$ centimeters

55. The *width* of a rectangle with perimeter 300 feet and length x feet

56. The *length* of a rectangle with area 200 square feet and width w feet

57. The *length* of a rectangle, given that its width is x feet and its length is 1 foot longer than twice the width

58. The *length* of a rectangle, given that its width is w feet and its length is 3 feet shorter than twice the width

59. The *area* of a rectangle, given that the width is x meters and the length is 5 meters longer than the width

60. The *perimeter* of a rectangle, given that the length is x yards and the width is 10 yards shorter

61. The *simple interest,* given that the principal is $x + 1000$, the rate is 18%, and the time is 1 year

62. The *simple interest,* given that the principal is $3x$, the rate is 6%, and the time is 1 year

63. The *price per pound* of peaches, given that x pounds sold for \$16.50

64. The *rate per hour* of a mechanic who gets \$480 for working x hours

65. The *degree measure* of an angle, given that its complementary angle has measure x degrees

66. The *degree measure* of an angle, given that its supplementary angle has measure x degrees

Identify the variable and write an equation that describes each situation. Do not solve the equation. See Example 6.

67. Two numbers differ by 5 and have a product of 8.

68. Two numbers differ by 6 and have a product of -9.

69. Herman's house sold for x dollars. The real estate agent received 7% of the selling price and Herman received \$84,532.

70. Gwen sold her car on consignment for x dollars. The saleswoman's commission was 10% of the selling price and Gwen received \$6570.

71. What percent of 500 is 100?

72. What percent of 40 is 120?

73. The value of x nickels and $x + 2$ dimes is \$3.80.

74. The value of d dimes and $d - 3$ quarters is \$6.75.

75. The sum of a number and 5 is 13.

76. Twelve subtracted from a number is -6.

77. The sum of three consecutive integers is 42.

78. The sum of three consecutive odd integers is 27.

79. The product of two consecutive integers is 182.

80. The product of two consecutive even integers is 168.

81. Twelve percent of Harriet's income is \$3000.

82. If 9% of the members buy tickets, then we will sell 252 tickets to this group.

83. Thirteen is 5% of what number?

84. Three hundred is 8% of what number?

85. The length of a rectangle is 5 feet longer than the width, and the area is 126 square feet.

86. The length of a rectangle is 1 yard shorter than twice the width, and the perimeter is 298 yards.

87. The value of n nickels and $n - 1$ dimes is 95 cents.

88. The value of q quarters, $q + 1$ dimes, and $2q$ nickels is 90 cents.

89. The measure of an angle is 38° smaller than the measure of its supplementary angle.

90. The measure of an angle is 16° larger than the measure of its complementary angle.

91. ***Target heart rate.*** For a cardiovascular workout, fitness experts recommend that you reach your target heart rate and stay at that rate for at least 20 minutes (HealthStatus, www.healthstatus.com). To find your target heart rate, find the sum of your age and your resting heart rate, then subtract that sum from 220. Find 60% of that result and add it to your resting heart rate.

a) Write an equation with variable r expressing the fact that the target heart rate for 30-year-old Bob is 144.

b) Judging from the accompanying graph, does the target heart rate for a 30-year-old increase or decrease as the resting heart rate increases?

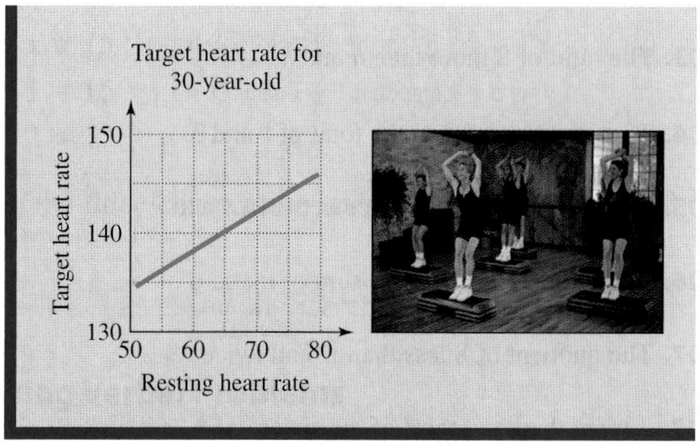

Figure for Exercise 91

92. ***Adjusting the saddle.*** The saddle height on a bicycle should be 109% of the rider's inside leg measurement L (www.harriscyclery.com). See the figure on the next page. Write an equation expressing the fact that the saddle height for Brenda is 36 in.

Warm-Ups

True or false? Explain your answer.

1. The first step in solving a word problem is to write the equation.
2. You should always write down what the variable represents.
3. Diagrams and tables are used as aids in solving problems.
4. To represent two consecutive odd integers, we use x and $x + 1$.
5. If $5x$ is 2 miles more than $3(x + 20)$, then $5x + 2 = 3(x + 20)$.
6. We can represent two numbers with a sum of 6 by x and $6 - x$.
7. Two numbers that differ by 7 can be represented by x and $x + 7$.
8. The degree measures of two complementary angles can be represented by x and $90 - x$.
9. The degree measures of two supplementary angles can be represented by x and $x + 180$.
10. If x is half as large as $x + 50$, then $2x = x + 50$.

2.6 Exercises

Reading and Writing *After reading this section, write out the answers to these questions. Use complete sentences.*

1. What types of problems are discussed in this section?
2. Why do we solve number problems?
3. What is uniform motion?
4. What are supplementary angles?
5. What are complementary angles?
6. What should you always do when solving a geometric problem?

Show a complete solution to each problem. See Example 1.

7. ***Consecutive integers.*** Find three consecutive integers whose sum is 141.
8. ***Consecutive even integers.*** Find three consecutive even integers whose sum is 114.
9. ***Consecutive odd integers.*** Two consecutive odd integers have a sum of 152. What are the integers?
10. ***Consecutive odd integers.*** Four consecutive odd integers have a sum of 120. What are the integers?
11. ***Consecutive integers.*** Find four consecutive integers whose sum is 194.
12. ***Consecutive even integers.*** Find four consecutive even integers whose sum is 340.

Show a complete solution to each problem. See Examples 2 and 3.

13. ***Olympic swimming.*** If an Olympic swimming pool is twice as long as it is wide and the perimeter is 150 meters, then what are the length and width?

Study Tip

Don't spend too much time on a single problem. If you get stuck on a problem, look at some examples in the text, move on to the next problem, or get help. It is often helpful to work some other problems and then come back to that one pesky problem.

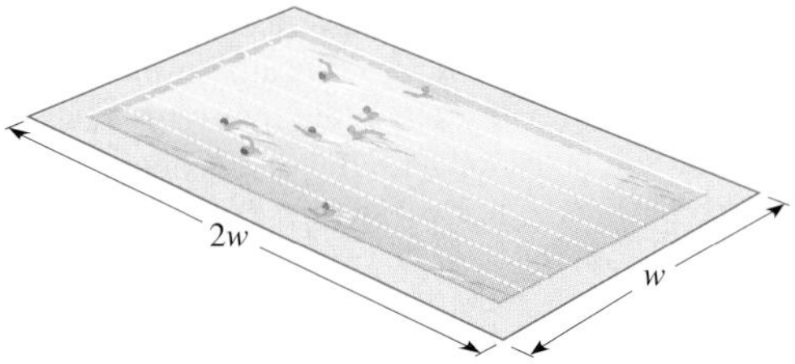

Figure for Exercise 13

14. ***Wimbledon tennis.*** If the perimeter of a tennis court is 228 feet and the length is 6 feet longer than twice the width, then what are the length and width?

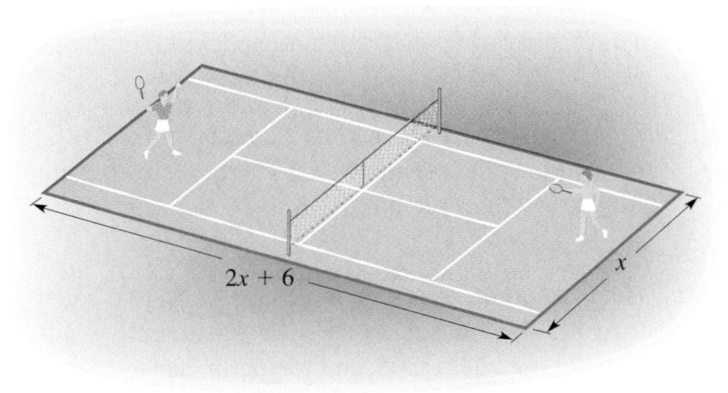

Figure for Exercise 14

15. ***Framed.*** Julia framed an oil painting that her uncle gave her. The painting was 4 inches longer than it was wide, and it took 176 inches of frame molding. What were the dimensions of the picture?

16. ***Industrial triangle.*** Geraldo drove his truck from Indianapolis to Chicago, then to St. Louis, and then back to Indianapolis. He observed that the second side of his triangular route was 81 miles short of being twice as long as the first side and that the third side was 61 miles longer than the first side. If he traveled a total of 720 miles, then how long is each side of this triangular route?

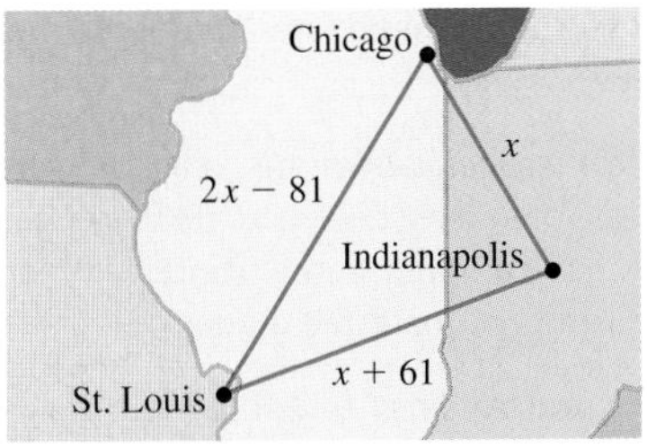

Figure for Exercise 16

17. ***Triangular banner.*** A banner in the shape of an isosceles triangle has a base that is 5 inches shorter than either of the equal sides. If the perimeter of the banner is 34 inches, then what is the length of the equal sides?

18. ***Border paper.*** Dr. Good's waiting room is 8 feet longer than it is wide. When Vincent wallpapered Dr. Good's waiting room, he used 88 feet of border paper. What are the dimensions of Dr. Good's waiting room?

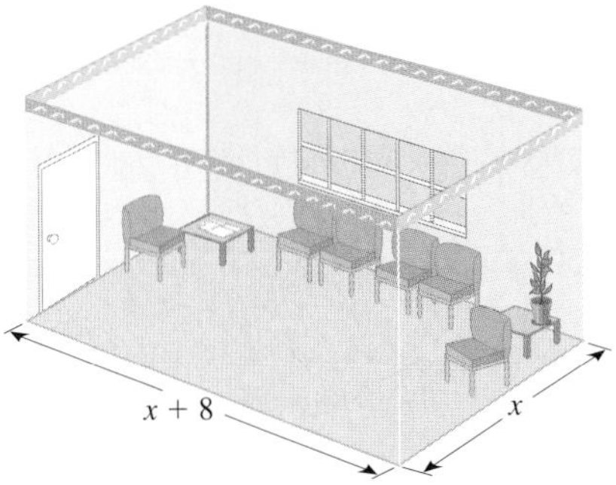

Figure for Exercise 18

19. ***Ramping up.*** A civil engineer is planning a highway overpass as shown in the figure. Find the degree measure of the angle marked w.

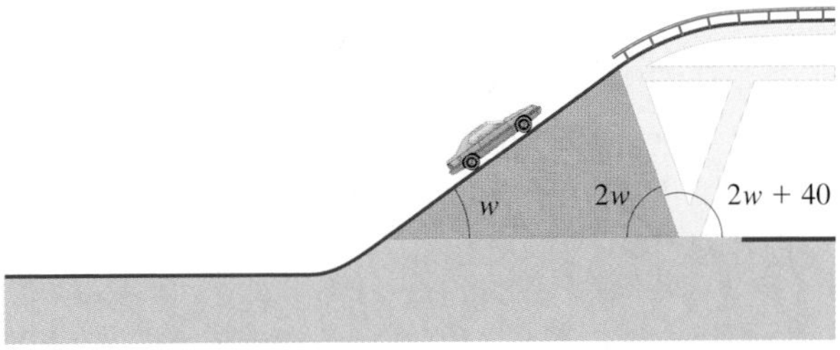

Figure for Exercise 19

20. ***Ramping down.*** For the other side of the overpass, the engineer has drawn the plans shown in the figure. Find the degree measure of the angle marked z.

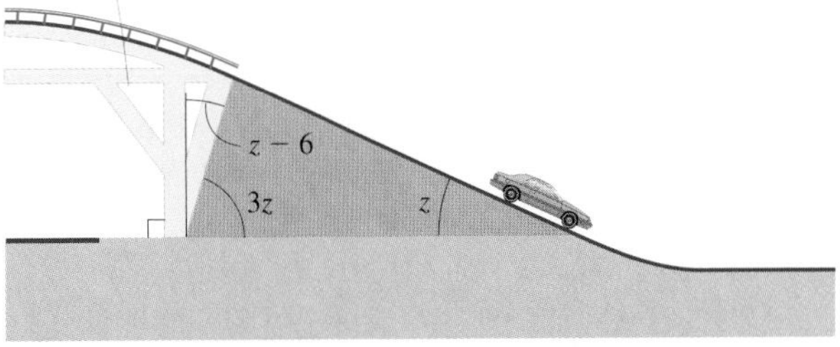

Figure for Exercise 20

Perform the indicated unit conversions. See Example 4. Round approximate answers to the nearest hundredth. Answers can vary slightly depending on the conversion factors used.

21. Convert 96 feet to inches.
22. Convert 33 yards to feet.
23. Convert 14.22 miles to kilometers.
24. Convert 33.6 kilometers to miles
25. Convert 13.5 centimeters to inches.
26. Convert 42.1 inches to centimeters.
27. Convert 14.2 ounces to grams.
28. Convert 233 grams to ounces.
29. Convert 40 miles per hour to feet per second.

30. Convert 200 feet per second to miles per hour.

31. Convert 500 feet per second to kilometers per hour.

32. Convert 230 yards per second to miles per minute.

Show a complete solution to each problem. See Examples 5 and 6.

33. ***Highway miles.*** Bret drove for 4 hours on the freeway, then decreased his speed by 20 miles per hour and drove for 5 more hours on a country road. If his total trip was 485 miles, then what was his speed on the freeway?

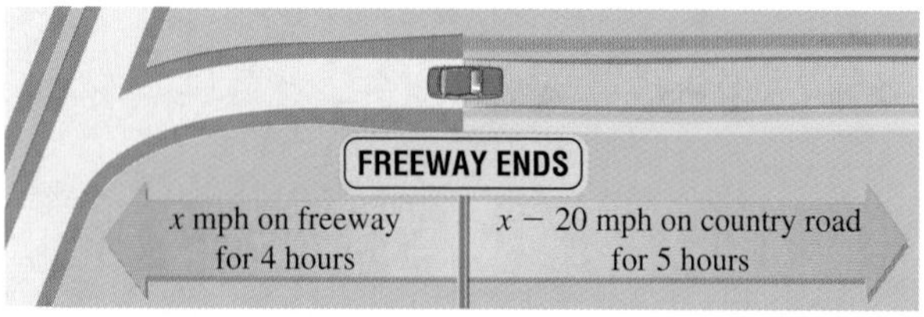

Figure for Exercise 33

34. ***Walking and running.*** On Saturday morning, Lynn walked for 2 hours and then ran for 30 minutes. If she ran twice as fast as she walked and she covered 12 miles altogether, then how fast did she walk?

35. ***Driving all night.*** Kathryn drove her rig 5 hours before dawn and 6 hours after dawn. If her average speed was 5 miles per hour more in the dark and she covered 630 miles altogether, then what was her speed after dawn?

36. ***Commuting to work.*** On Monday, Roger drove to work in 45 minutes. On Tuesday he averaged 12 miles per hour more, and it took him 9 minutes less to get to work. How far does he travel to work?

37. ***Head winds.*** A jet flew at an average speed of 640 mph from Los Angeles to Chicago. Because of head winds the jet averaged only 512 mph on the return trip, and the return trip took 48 minutes longer. How many hours was the flight from Chicago to Los Angeles? How far is it from Chicago to Los Angeles?

38. ***Ride the Peaks.*** Penny's bicycle trip from Colorado Springs to Pikes Peak took 1.5 hours longer than the return trip to Colorado Springs. If she averaged 6 mph on the way to Pikes Peak and 15 mph for the return trip, then how long was the ride from Colorado Springs to Pikes Peak?

Solve each problem.

39. ***Perimeter of a frame.*** The perimeter of a rectangular frame is 64 in. If the width of the frame is 8 in. less than the length, then what are the length and width of the frame?

40. ***Perimeter of a box.*** The width of a rectangular box is 20% of the length. If the perimeter is 192 cm, then what are the length and width of the box?

41. ***Isosceles triangle.*** An isosceles triangle has two equal sides. If the shortest side of an isosceles triangle is 2 ft less than one of the equal sides and the perimeter is 13 ft, then what are the lengths of the sides?

42. ***Scalene triangle.*** A scalene triangle has three unequal sides. The perimeter of a scalene triangle is 144 m. If the first side is twice as long as the second side and the third side is 24 m longer than the second side, then what are the measures of the sides?

43. ***Angles of a scalene triangle.*** The largest angle in a scalene triangle is six times as large as the smallest. If the middle angle is twice the smallest, then what are the degree measures of the three angles?

44. ***Angles of a right triangle.*** If one of the acute angles in a right triangle is 38°, then what are the degree measures of all three angles?

45. ***Angles of an isosceles triangle.*** One of the equal angles in an isosceles triangle is four times as large as the smallest angle in the triangle. What are the degree measures of the three angles?

46. ***Angles of an isosceles triangle.*** The measure of one of the equal angles in an isosceles triangle is 10° larger than twice the smallest angle in the triangle. What are the degree measures of the three angles?

47. ***Super Bowl score.*** The 1977 Super Bowl was played in the Rose Bowl in Pasadena. In that football game the Oakland Raiders scored 18 more points than the Minnesota

Vikings. If the total number of points scored was 46, then what was the final score for the game?

48. ***Top payrolls.*** Payrolls for the three highest paid baseball teams (the Yankees, Mets, and Braves) for 2003 totaled \$376 million (www.usatoday.com). If the team payroll for the Yankees was \$36 million greater than the payroll for the Mets and the payroll for the Mets was \$11 million greater than the payroll for the Braves, then what was the 2003 payroll for each team?

49. ***Idabel to Lawton.*** Before lunch, Sally drove from Idabel to Ardmore, averaging 50 mph. After lunch she continued on to Lawton, averaging 53 mph. If her driving time after lunch was 1 hour less than her driving time before lunch and the total trip was 256 miles, then how many hours did she drive before lunch? How far is it from Ardmore to Lawton?

50. ***Norfolk to Chadron.*** On Monday, Chuck drove from Norfolk to Valentine, averaging 47 mph. On Tuesday, he continued on to Chadron, averaging 69 mph. His driving time on Monday was 2 hours longer than his driving time on Tuesday. If the total distance from Norfolk to Chadron is 326 miles, then how many hours did he drive on Monday? How far is it from Valentine to Chadron?

51. ***Golden oldies.*** Joan Crawford, John Wayne, and James Stewart were born in consecutive years (*Doubleday Almanac*). Joan Crawford was the oldest of the three, and James Stewart was the youngest. In 1950, after all three had their birthdays, the sum of their ages was 129. In what years were they born?

52. ***Leading men.*** Bob Hope was born 2 years after Clark Gable and 2 years before Henry Fonda (*Doubleday Almanac*). In 1951, after all three of them had their birthdays, the sum of their ages was 144. In what years were they born?

53. ***Trimming a garage door.*** A carpenter used 30 ft of molding in three pieces to trim a garage door. If the long piece was 2 ft longer than twice the length of each shorter piece, then how long was each piece?

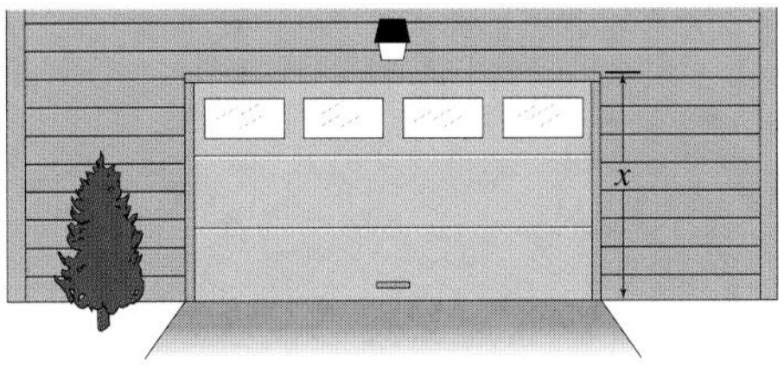

Figure for Exercise 53

54. ***Fencing dog pens.*** Clint is constructing two adjacent rectangular dog pens. Each pen will be three times as long as it is wide, and the pens will share a common long side. If Clint has 65 ft of fencing, what are the dimensions of each pen?

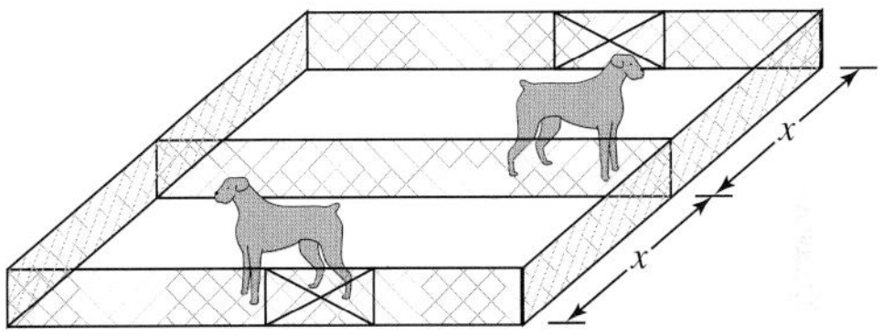

Figure for Exercise 54

2.7 Discount, Investment, and Mixture Applications

In this Section

- **Discount Problems**
- **Commission Problems**
- **Investment Problems**
- **Mixture Problems**

In this section, we continue our study of applications of algebra. The problems in this section involve percents.

Discount Problems

When an item is sold at a discount, the amount of the discount is usually described as being a percentage of the original price. The percentage is called the **rate of discount.** Multiplying the rate of discount and the original price gives the amount of the discount.

EXAMPLE 1

Finding the original price

Ralph got a 12% discount when he bought his new 2002 Corvette Coupe. If the amount of his discount was \$6606, then what was the original price of the Corvette?

Solution

Let x represent the original price. The discount is found by multiplying the 12% rate of discount and the original price:

$$\text{Rate of discount} \cdot \text{original price} = \text{amount of discount}$$

$$0.12x = 6606$$

$$x = \frac{6606}{0.12} \quad \text{Divide each side by 0.12.}$$

$$x = 55{,}050$$

To check, find 12% of \$55,050. Since $0.12 \cdot 55{,}050 = 6606$, the original price of the Corvette was \$55,050.

Now do Exercises 7–8

EXAMPLE 2

Finding the original price

When Susan bought her new car, she also got a discount of 12%. She paid \$17,600 for her car. What was the original price of Susan's car?

Helpful Hint

To get familiar with the problem, guess that the original price was \$30,000. Then her discount is 0.12(30,000) or \$3600. The price she paid would be 30,000 − 3600 or \$26,400, which is incorrect.

Solution

Let x represent the original price for Susan's car. The amount of discount is 12% of x, or $0.12x$. We can write an equation expressing the fact that the original price minus the discount is the price Susan paid.

$$\text{Original price} - \text{discount} = \text{sale price}$$

$$x - 0.12x = 17{,}600$$

$$0.88x = 17{,}600 \quad 1.00x - 0.12x = 0.88x$$

$$x = \frac{17{,}600}{0.88} \quad \text{Divide each side by 0.88.}$$

$$x = 20{,}000$$

Check: 12% of \$20,000 is \$2400, and \$20,000 − \$2400 = \$17,600. The original price of Susan's car was \$20,000.

Now do Exercises 9–10

Commission Problems

A salesperson's commission for making a sale is often a percentage of the selling price. **Commission problems** are very similar to other problems involving percents. The commission is found by multiplying the rate of commission and the selling price.

EXAMPLE 3

Real estate commission

Sarah is selling her house through a real estate agent whose commission rate is 7%. What should the selling price be so that Sarah can get the $83,700 she needs to pay off the mortgage?

Solution

Let x be the selling price. The commission is 7% of x (not 7% of $83,700). Sarah receives the selling price less the sales commission:

$$\text{Selling price} - \text{commission} = \text{Sarah's share}$$

$$x - 0.07x = 83{,}700$$

$$0.93x = 83{,}700 \quad 1.00x - 0.07x = 0.93x$$

$$x = \frac{83{,}700}{0.93}$$

$$x = 90{,}000$$

Check: 7% of $90,000 is $6300, and $90,000 − $6300 = $83,700. So the house should sell for $90,000.

Now do Exercises 11–14

Investment Problems

The interest on an investment is a percentage of the investment, just as the sales commission is a percentage of the sale amount. However, in **investment problems** we must often account for more than one investment at different rates. So it is a good idea to make a table, as in Example 4.

EXAMPLE 4

Diversified investing

Ruth Ann invested some money in a certificate of deposit with an annual yield of 9%. She invested twice as much in a mutual fund with an annual yield of 10%. Her interest from the two investments at the end of the year was $232. How much was invested at each rate?

Helpful Hint

To get familiar with the problem, guess that she invested $1000 at 9% and $2000 at 10%. Then her interest in one year would be

$$0.09(1000) + 0.10(2000)$$

or $290, which is close but incorrect.

Solution

When there are many unknown quantities, it is often helpful to identify them in a table. Since the time is 1 year, the amount of interest is the product of the interest rate and the amount invested.

	Interest rate	Amount invested	Interest for 1 year
CD	9%	x	$0.09x$
Mutual fund	10%	$2x$	$0.10(2x)$

Since the total interest from the investments was \$232, we can write the following equation:

$$\text{CD interest} + \text{mutual fund interest} = \text{total interest}$$

$$0.09x + 0.10(2x) = 232$$

$$0.09x + 0.20x = 232$$

$$0.29x = 232$$

$$x = \frac{232}{0.29}$$

$$x = 800$$

$$2x = 1600$$

To check, we find the total interest:

$$0.09(800) + 0.10(1600) = 72 + 160$$

$$= 232$$

So Ruth Ann invested \$800 at 9% and \$1600 at 10%.

Now do Exercises 15–18

Study Tip

Finding out what happened in class and attending class are not the same. Attend every class and use class as a learning time. Take notes, ask questions, and make sure that you are learning in the classroom.

Mixture Problems

Mixture problems are concerned with the result of mixing two quantities, each of which contains another substance. Notice how similar the following mixture problem is to the last investment problem.

EXAMPLE 5

Mixing milk

How many gallons of milk containing 4% butterfat must be mixed with 80 gallons of 1% milk to obtain 2% milk?

Helpful Hint

To get familiar with the problem, guess that we need 100 gal of 4% milk. Mixing that with 80 gal of 1% milk would produce 180 gal of 2% milk. Now the two milks separately have

$$0.04(100) + 0.01(80)$$

or 4.8 gal of fat. Together the amount of fat is 0.02(180) or 3.6 gal. Since these amounts are not equal, our guess is incorrect.

Solution

It is helpful to draw a diagram and then make a table to classify the given information.

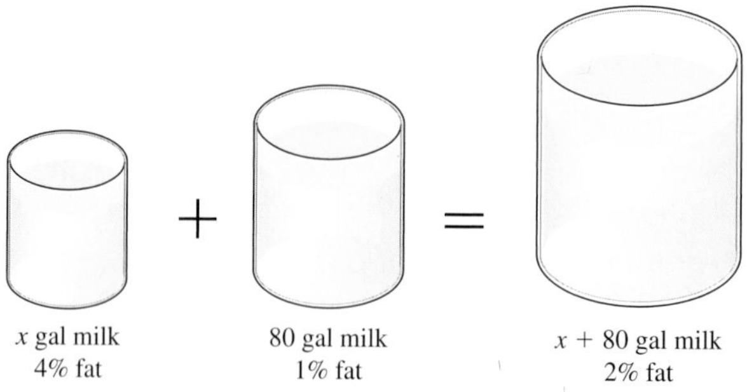

	Percentage of fat	Amount of milk	Amount of fat
4% milk	4%	x	$0.04x$
1% milk	1%	80	$0.01(80)$
2% milk	2%	$x + 80$	$0.02(x + 80)$

The equation expresses the fact that the total fat from the first two types of milk is the same as the fat in the mixture:

$$\begin{aligned}
\text{Fat in 4\% milk} + \text{fat in 1\% milk} &= \text{fat in 2\% milk} \\
0.04x + 0.01(80) &= 0.02(x + 80) \\
0.04x + 0.8 &= 0.02x + 1.6 && \text{Simplify.} \\
100(0.04x + 0.8) &= 100(0.02x + 1.6) && \text{Multiply each side by 100.} \\
4x + 80 &= 2x + 160 && \text{Distributive property.} \\
2x + 80 &= 160 && \text{Subtract } 2x \text{ from each side.} \\
2x &= 80 && \text{Subtract 80 from each side.} \\
x &= 40 && \text{Divide each side by 2.}
\end{aligned}$$

To check, calculate the total fat:

$$2\% \text{ of 120 gallons} = 0.02(120) = 2.4 \text{ gallons of fat}$$
$$0.04(40) + 0.01(80) = 1.6 + 0.8 = 2.4 \text{ gallons of fat}$$

So we mix 40 gallons of 4% milk with 80 gallons of 1% milk to get 120 gallons of 2% milk.

Now do Exercises 19–22

Study Tip

Don't expect to understand a new topic the first time that you see it. Learning mathematics takes time, patience, and repetition. Keep reading the text, asking questions, and working problems. Someone once said, "All mathematics is easy once you understand it."

In mixture problems, the solutions might contain fat, alcohol, salt, or some other substance. We always assume that the substance neither appears nor disappears in the process. For example, if there are 3 grams of salt in one glass of water and 2 grams in another, then there are exactly 5 grams in a mixture of the two.

Warm-Ups ▼

True or false? Explain your answer.

1. If Jim gets a 12% commission for selling a $1000 Wonder Vac, then his commission is $120.
2. If Bob earns a 5% commission on an $80,000 motorhome sale, then Bob earns $400.
3. If Sue gets a 20% discount on a TV with a list price of x dollars, then Sue pays $0.8x$ dollars.
4. If you get a 6% discount on a car that has an MSRP of x dollars, then your discount is $0.6x$ dollars.
5. If the original price is w and the discount is 8%, then the selling price is $w - 0.08w$.
6. If x is the selling price and the commission is 8% of the selling price, then the commission is $0.08x$.
7. If you need $40,000 for your house and the agent gets 10% of the selling price, then the agent gets $4000, and the house sells for $44,000.
8. If you mix 10 liters of a 20% acid solution with x liters of a 30% acid solution, then the total amount of acid is $2 + 0.3x$ liters.
9. A 10% acid solution mixed with a 14% acid solution results in a 24% acid solution.
10. If a TV costs x dollars and sales tax is 5%, then the total bill is $1.05x$ dollars.

2.7 Exercises

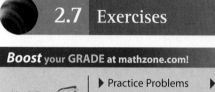

Reading and Writing *After reading this section, write out the answers to these questions. Use complete sentences.*

1. What types of problems are discussed in this section?

2. What is the difference between discount and rate of discount?

3. What is the relationship between discount, original price, rate of discount, and sale price?

4. What do mixture problems and investment problems have in common?

5. Why do we make a table when solving certain problems.

6. What is the relationship between amount of interest, amount invested, and interest rate?

Show a complete solution to each problem. See Examples 1 and 2.

7. ***Close-out sale.*** At a 25% off sale, Jose saved \$80 on a 19-inch Panasonic TV. What was the original price of the television.

8. ***Nice tent.*** A 12% discount on a Walrus tent saved Melanie \$75. What was the original price of the tent?

9. ***Circuit city.*** After getting a 20% discount, Robert paid \$320 for a Pioneer CD player for his car. What was the original price of the CD player?

10. ***Chrysler Sebring.*** After getting a 15% discount on the price of a new Chrysler Sebring convertible, Helen paid \$27,000. What was the original price of the convertible?

Show a complete solution to each problem. See Example 3.

11. ***Selling price of a home.*** Kirk wants to get \$115,000 for his house. The real estate agent gets a commission equal to 8% of the selling price for selling the house. What should the selling price be?

Photo for Exercise 11

12. ***Horse trading.*** Gene is selling his palomino at an auction. The auctioneer's commission is 10% of the selling price. If Gene still owes \$810 on the horse, then what must the horse sell for so that Gene can pay off his loan?

13. ***Sales tax collection.*** Merilee sells tomatoes at a roadside stand. Her total receipts including the 7% sales tax were \$462.24. What amount of sales tax did she collect?

14. ***Toyota Corolla.*** Gwen bought a new Toyota Corolla. The selling price plus the 8% state sales tax was \$15,714. What was the selling price?

Show a complete solution to each problem. See Example 4.

15. ***Wise investments.*** Wiley invested some money in the Berger 100 Fund and \$3000 more than that amount in the Berger 101 Fund. For the year he was in the fund, the 100 Fund paid 18% simple interest and the 101 Fund paid 15% simple interest. If the income from the two investments totaled \$3750 for one year, then how much did he invest in each fund?

16. ***Loan shark.*** Becky lent her brother some money at 8% simple interest, and she lent her sister twice as much at twice the interest rate. If she received a total of 20 cents interest, then how much did she lend to each of them?

17. ***Investing in bonds.*** David split his \$25,000 inheritance between Fidelity Short-Term Bond Fund with an annual yield of 5% and T. Rowe Price Tax-Free Short-Intermediate Fund with an annual yield of 4%. If his total income for one year on the two investments was \$1140, then how much did he invest in each fund?

18. ***High-risk funds.*** Of the \$50,000 that Natasha pocketed on her last real estate deal, \$20,000 went to charity. She invested part of the remainder in Dreyfus New Leaders Fund with an annual yield of 16% and the rest in Templeton Growth Fund with an annual yield of 25%. If she made \$6060 on these investments in one year, then how much did she invest in each fund?

Show a complete solution to each problem. See Example 5.

19. ***Mixing milk.*** How many gallons of milk containing 1% butterfat must be mixed with 30 gallons of milk containing 3% butterfat to obtain a mixture containing 2% butterfat?

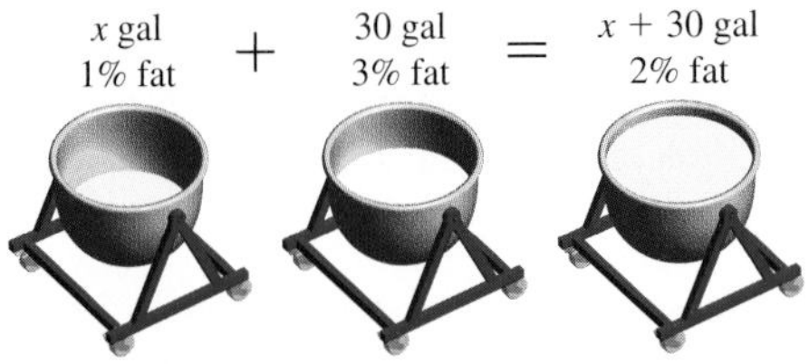

Figure for Exercise 19

20. ***Acid solutions.*** How many gallons of a 5% acid solution should be mixed with 30 gallons of a 10% acid solution to obtain a mixture that is 8% acid?

21. ***Alcohol solutions.*** Gus has on hand a 5% alcohol solution and a 20% alcohol solution. He needs 30 liters of a 10% alcohol solution. How many liters of each solution should he mix together to obtain the 30 liters?

22. ***Adjusting antifreeze.*** Angela needs 20 quarts of 50% antifreeze solution in her radiator. She plans to obtain this by mixing some pure antifreeze with an appropriate amount of a 40% antifreeze solution. How many quarts of each should she use?

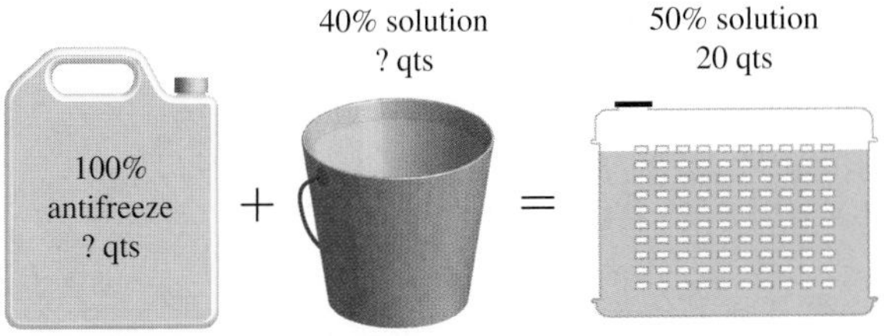

Figure for Exercise 22

Solve each problem.

23. ***Registered voters.*** If 60% of the registered voters of Lancaster County voted in the November election and 33,420 votes were cast, then how many registered voters are there in Lancaster County?

Photo for Exercise 23

24. ***Tough on crime.*** In a random sample of voters, 594 respondents said that they favored passage of a \$33 billion crime bill. If the number in favor of the crime bill was 45% of the number of voters in the sample, then how many voters were in the sample?

25. ***Ford Taurus.*** At an 8% sales tax rate, the sales tax on Peter's new Ford Taurus was \$1200. What was the price of the car?

26. ***Taxpayer blues.*** Last year, Faye paid 24% of her income to taxes. If she paid \$9600 in taxes, then what was her income?

27. ***Making a profit.*** A retail store buys shirts for \$8 and sells them for \$14. What percent increase is this?

28. ***Monitoring AIDS.*** If 28 new AIDS cases were reported in Landon County this year and 35 new cases were reported last year, then what percent decrease in new cases is this?

29. ***High school integration.*** Wilson High School has 400 students, of whom 20% are African American. The school board plans to merge Wilson High with Jefferson High. This one school will then have a student population that is 44% African American. If Jefferson currently has a student population that is 60% African American, then how many students are at Jefferson?

30. ***Junior high integration.*** The school board plans to merge two junior high schools into one school of 800 students in which 40% of the students will be Caucasian. One of the schools currently has 58% Caucasian students; the other has only 10% Caucasian students. How many students are in each of the two schools?

31. ***Hospital capacity.*** When Memorial Hospital is filled to capacity, it has 18 more people in semiprivate rooms (two patients to a room) than in private rooms. The room rates are \$200 per day for a private room and \$150 per day for a semiprivate room. If the total receipts for rooms is

$17,400 per day when all are full, then how many rooms of each type does the hospital have?

32. ***Public relations.*** Memorial Hospital is planning an advertising campaign. It costs the hospital $3000 each time a television ad is aired and $2000 each time a radio ad is aired. The administrator wants to air 60 more television ads than radio ads. If the total cost of airing the ads is $580,000, then how many ads of each type will be aired?

33. ***Mixed nuts.*** Cashews sell for $4.80 per pound, and pistachios sell for $6.40 per pound. How many pounds of pistachios should be mixed with 20 pounds of cashews to get a mixture that sells for $5.40 per pound?

34. ***Premium blend.*** Premium coffee sells for $6.00 per pound, and regular coffee sells for $4.00 per pound. How many pounds of each type of coffee should be blended to obtain 100 pounds of a blend that sells for $4.64 per pound?

35. ***Nickels and dimes.*** Candice paid her library fine with 10 coins consisting of nickels and dimes. If the fine was $0.80, then how many of each type of coin did she use?

36. ***Dimes and quarters.*** Jeremy paid for his breakfast with 36 coins consisting of dimes and quarters. If the bill was $4.50, then how many of each type of coin did he use?

37. ***Cooking oil.*** Crisco Canola Oil is 7% saturated fat. Crisco blends corn oil that is 14% saturated fat with Crisco Canola Oil to get Crisco Canola and Corn Oil, which is 11% saturated fat. How many gallons of corn oil must Crisco mix with 600 gallons of Crisco Canola Oil to get Crisco Canola and Corn Oil?

38. ***Chocolate ripple.*** The Delicious Chocolate Shop makes a dark chocolate that is 35% fat and a white chocolate that is 48% fat. How many kilograms of dark chocolate should be mixed with 50 kilograms of white chocolate to make a ripple blend that is 40% fat?

39. ***Hawaiian Punch.*** Hawaiian Punch is 10% fruit juice. How much water would you have to add to one gallon of Hawaiian Punch to get a drink that is 6% fruit juice?

40. ***Diluting wine.*** A restaurant manager has 2 liters of white wine that is 12% alcohol. How many liters of white grape juice should he add to get a drink that is 10% alcohol?

41. ***Bargain hunting.*** A smart shopper bought 5 pairs of shorts and 8 tops for a total of $108. If the price of a pair of shorts was twice the price of a top, then what was the price of each type of clothing?

42. ***VCRs and CDs.*** The manager of a stereo shop placed an order for $10,710 worth of VCRs at $120 each and CD players at $150 each. If the number of VCRs she ordered was three times the number of CD players, then how many of each did she order?

2.8 Inequalities

In Chapter 1, we defined inequality in terms of the number line. One number is greater than another number if it lies to the right of the other number on the number line. In this section you will study inequality in greater depth.

In this Section

- **Basic Ideas**
- **Graphing Inequalities**
- **Graphing Compound Inequalities**
- **Checking Inequalities**
- **Writing Inequalities**

Helpful Hint

A good way to learn inequality symbols is to notice that the inequality symbol always points at the smaller number. This observation will help you read an inequality such as $-2 < x$. Reading right to left, we say that x is greater than -2. It is usually easier to understand an inequality if you read the variable first.

Basic Ideas

The symbols used to express inequality and their meanings are given in the following box.

Inequality Symbols

Symbol	Meaning
$<$	Is less than
$\le$	Is less than or equal to
$>$	Is greater than
$\ge$	Is greater than or equal to

The statement $a < b$ means that a is to the left of b on the number line as shown in Fig. 2.4. The statement $c > d$ means that c is to the right of d on the number line, as shown in Fig. 2.5. Of course, $a < b$ has the same meaning as $b > a$. The statement $a \leq b$ means that either a is to the left of b or a corresponds to the same point as b on the number line. The statement $a \leq b$ has the same meaning as the statement $b \geq a$.

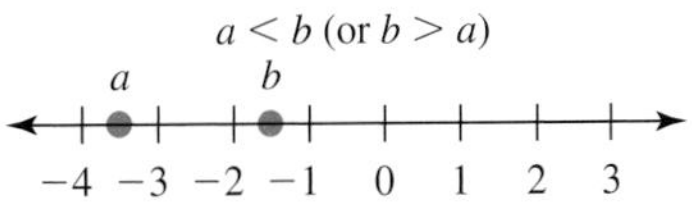

Figure 2.4

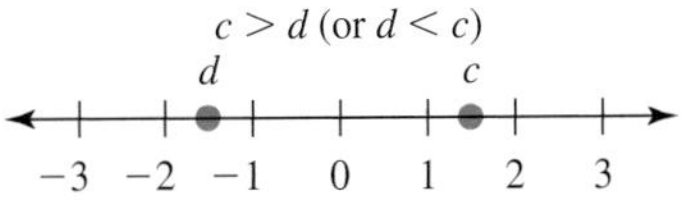

Figure 2.5

EXAMPLE 1

Verifying inequalities

Determine whether each of the following statements is correct.

a) $3 < 4$ **b)** $-1 < -2$ **c)** $-2 \leq 0$

d) $0 \geq 0$ **e)** $2(-3) + 8 > 9$ **f)** $(-2)(-5) \leq 10$

Calculator Close-Up

A graphing calculator can determine whether an inequality is correct. Use the inequality symbols from the TEST menu to enter the inequality.

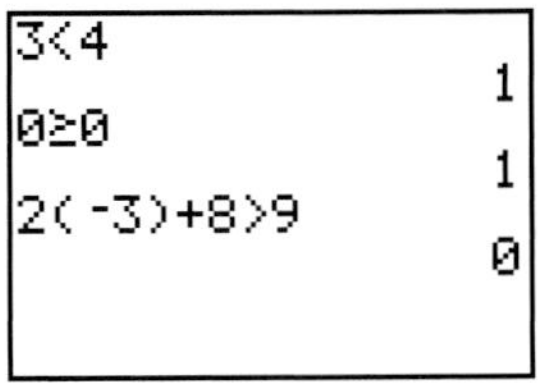

When ENTER is pressed, the calculator returns a 1 if the inequality is correct or a 0 if the inequality is incorrect.

Solution

a) Locate 3 and 4 on the number line shown in Fig. 2.6. Because 3 is to the left of 4 on the number line, $3 < 4$ is correct.

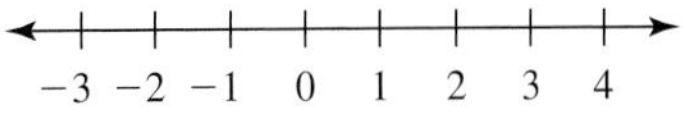

Figure 2.6

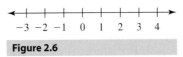

b) Locate -1 and -2 on the number line shown in Fig. 2.6. Because -1 is to the right of -2, on the number line, $-1 < -2$ is not correct.

c) Because -2 is to the left of 0 on the number line, $-2 \leq 0$ is correct.

d) Because 0 is equal to 0, $0 \geq 0$ is correct.

e) Simplify the left side of the inequality to get $2 > 9$, which is not correct.

f) Simplify the left side of the inequality to get $10 \leq 10$, which is correct.

Now do Exercises 7–20

Graphing Inequalities

If a is a fixed real number, then any real number x located to the right of a on the number line satisfies $x > a$. The set of real numbers located to the right of a on the number line is the solution set to $x > a$. This solution set is written in set-builder notation as $\{x \mid x > a\}$, or more simply in interval notation as (a, ∞). We **graph the inequality** by graphing the solution set (a, ∞). Recall from Chapter 1 that a bracket means that an endpoint is included in an interval and a parenthesis means that an endpoint is not included in an interval.

EXAMPLE 2

Graphing inequalities

State the solution set to each inequality in interval notation and sketch its graph.

a) $x < 5$ **b)** $-2 < x$ **c)** $x \geq 10$

Solution

a) All real numbers less than 5 satisfy $x < 5$. The solution set is the interval $(-\infty, 5)$ and the graph of the solution set is shown in Fig. 2.7.

b) The inequality $-2 < x$ indicates that x is greater than -2. The solution set is the interval $(-2, \infty)$ and the graph of the inequality is shown in Fig. 2.8.

c) All real numbers greater than or equal to 10 satisfy $x \geq 10$. The solution set is the interval $[10, \infty)$ and the graph is shown in Fig. 2.9.

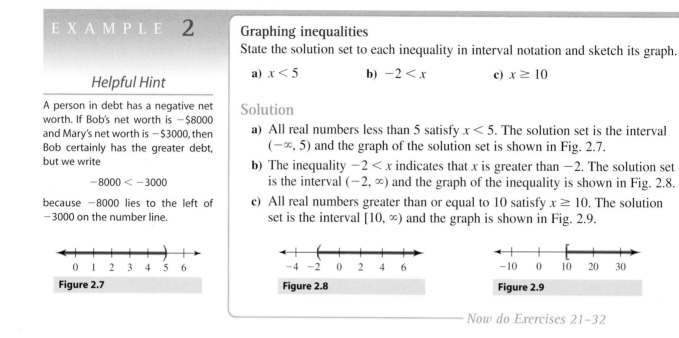

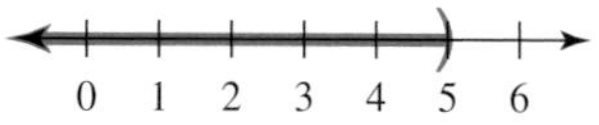

Figure 2.7

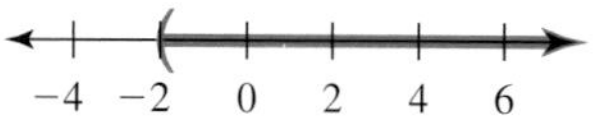

Figure 2.8

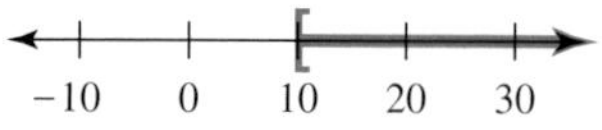

Figure 2.9

Now do Exercises 21–32

Helpful Hint

A person in debt has a negative net worth. If Bob's net worth is $-\$8000$ and Mary's net worth is $-\$3000$, then Bob certainly has the greater debt, but we write

$$-8000 < -3000$$

because -8000 lies to the left of -3000 on the number line.

Graphing Compound Inequalities

A statement involving more than one inequality is a **compound inequality.** We will study one type of compound inequality here and see other types in Section 8.1.

If a and b are real numbers and $a < b$, then the compound inequality

$$a < x < b$$

means that $a < x$ *and* $x < b$. Reading x first makes $a < x < b$ clearer:

"x is greater than a *and* x is less than b."

If x is greater than a and less than b, then x is between a and b. So the solution set to $a < x < b$ is the interval (a, b).

EXAMPLE 3

Graphing compound inequalities

State the solution set to each inequality in interval notation and sketch its graph.

a) $2 < x < 3$ **b)** $-2 \leq x < 1$

Solution

a) All real numbers between 2 and 3 satisfy $2 < x < 3$. The solution set is the interval $(2, 3)$ and the graph of the solution set is shown in Fig. 2.10.

b) The real numbers that satisfy $-2 \leq x < 1$ are between -2 and 1, including -2 but not including 1. So the solution set is the interval $[-2, 1)$ and the graph of this compound inequality is shown in Fig. 2.11.

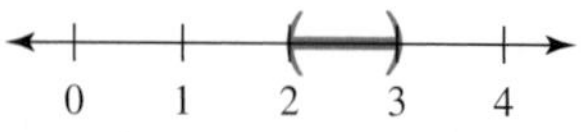

Figure 2.10

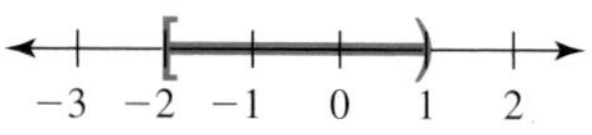

Figure 2.11

Now do Exercises 33–40

Study Tip

No two students learn in the same way or at the same speed. No one can tell you exactly how to study and learn. Learning is personal. You must discover what it takes for you to learn mathematics and then do whatever it takes.

CAUTION We write $a < x < b$ only if $a < b$, and we write $a > x > b$ only if $a > b$. Similar rules hold for $\leq$ and $\geq$. So $4 < x < 9$ and $-6 \geq x \geq -8$ are correct uses of this notation, but $5 < x < 2$ is not correct. Also, the inequalities should *not* point in opposite directions as in $5 < x > 7$.

Checking Inequalities

In Examples 2 and 3 we determined the solution sets to some inequalities. In Section 2.9 more complicated inequalities will be solved by using steps similar to those used for solving equations. In Example 4, we determine whether a given number satisfies an inequality of the type that we will be solving in Section 2.9.

EXAMPLE 4

Checking inequalities

Determine whether the given number satisfies the inequality following it.

a) $0,\ 2x - 3 \leq -5$ **b)** $-4,\ x - 5 > 2x + 1$ **c)** $\frac{13}{3},\ 6 < 3x - 5 < 14$

Calculator Close-Up

To check 13/3 in

$$6 < 3x - 5 < 14$$

we check each part of the compound inequality separately.

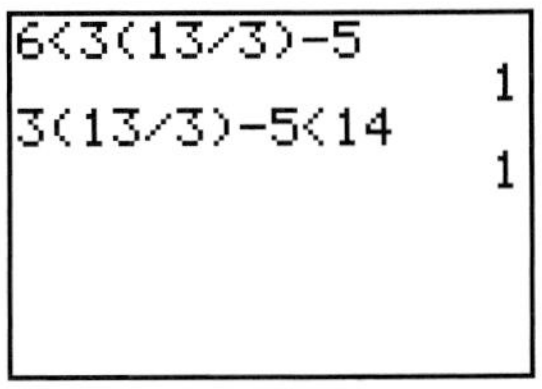

Because both parts of the compound inequality are correct, 13/3 satisfies the compound inequality.

Solution

a) Replace x by 0 in the inequality and simplify:

$$2x - 3 \leq -5$$
$$2 \cdot 0 - 3 \leq -5$$
$$-3 \leq -5 \quad \text{Incorrect}$$

Since this last inequality is incorrect, 0 is not a solution to the inequality.

b) Replace x by -4 and simplify:

$$x - 5 > 2x + 1$$
$$-4 - 5 > 2(-4) + 1$$
$$-9 > -7 \quad \text{Incorrect}$$

Since this last inequality is incorrect, -4 is not a solution to the inequality.

c) Replace x by $\frac{13}{3}$ and simplify:

$$6 < 3x - 5 < 14$$
$$6 < 3 \cdot \frac{13}{3} - 5 < 14$$
$$6 < 13 - 5 < 14$$
$$6 < 8 < 14 \quad \text{Correct}$$

Since 8 is greater than 6 and less than 14, this inequality is correct. So $\frac{13}{3}$ satisfies the original inequality.

Now do Exercises 51–68

Writing Inequalities

Inequalities occur in applications, just as equations do. Certain verbal phrases indicate inequalities. For example, if you must be at least 18 years old to vote, then you can vote if you are 18 or older. The phrase "at least" means "greater than or equal to." If an elevator has a capacity of at most 20 people, then it can hold 20 people or fewer. The phrase "at most" means "less than or equal to."

Math *at Work* Body Mass Index

Medical professionals say that two-thirds of all Americans are overweight and excess weight has about the same effect on life expectancy as smoking. How can you tell if you are overweight or normal? Body mass index (BMI) can help you decide. To determine BMI divide your weight in kilograms by the square of your height in meters. Don't know your weight and height in the metric system? Then use the formula BMI $= 703W/H^2$, where W is your weight in pounds and H is your height in inches.

If $23 < \text{BMI} < 25$, then you are probably not overweight. If BMI ≥ 26, then you are probably overweight and are statistically likely to have a lower life expectancy. According to the National Heart, Lung, and Blood Institute, you are overweight if $25 < \text{BMI} < 29.9$ and obese if BMI ≥ 30. If your BMI is between 17 and 22, your life span might be longer than average. Men are usually happy with a BMI between 23 and 25 and women like to see their BMI between 20 and 22. However, BMI does not distinguish between muscle and fat and can wrongly suggest that a person with a short muscular build is overweight. Also, the BMI does not work well for children, because normal varies with age.

If you want to learn more about body mass index or don't want to do the calculations yourself, then check out any of the numerous websites that discuss BMI and even have online BMI calculators. Just do a search for body mass index.

EXAMPLE 5

Writing inequalities

Write an inequality that describes each situation.

a) Lois plans to spend at most \$500 on a washing machine including the 9% sales tax.

b) The length of a certain rectangle must be 4 meters longer than the width, and the perimeter must be at least 120 meters.

c) Fred made a 76 on the midterm exam. To get a B, the average of his midterm and his final exam must be between 80 and 90.

Solution

a) If x is the price of the washing machine, then $0.09x$ is the amount of sales tax. Since the total must be less than or equal to \$500, the inequality is

$$x + 0.09x \leq 500.$$

b) If W represents the width of the rectangle, then $W + 4$ represents the length. Since the perimeter $(2W + 2L)$ must be greater than or equal to 120, the inequality is

$$2(W) + 2(W + 4) \geq 120.$$

c) If we let x represent Fred's final exam score, then his average is $\frac{x + 76}{2}$. To indicate that the average is between 80 and 90, we use the compound inequality

$$80 < \frac{x + 76}{2} < 90.$$

Now do Exercises 77–89

CAUTION In Example 4(b) you are given that L is 4 meters longer than W. So $L = W + 4$, and you can use $W + 4$ in place of L. If you knew only that L was longer than W, then you would know only that $L > W$.

Warm-Ups ▼

True or false? Explain your answer.

1. $-2 \leq -2$
2. $-5 < 4 < 6$
3. $-3 < 0 < -1$
4. The inequalities $7 < x$ and $x > 7$ have the same graph.
5. The graph of $x < -3$ includes the point at -3.
6. The number 5 satisfies the inequality $x > 2$.
7. The number -3 is a solution to $-2 < x$.
8. The number 4 satisfies the inequality $2x - 1 < 4$.
9. The number 0 is a solution to the inequality $2x - 3 \leq 5x - 3$.
10. The inequalities $2x - 1 < x$ and $x < 2x - 1$ have the same solutions.

2.8 Exercises

Boost your GRADE at mathzone.com!

MathZone
- Practice Problems
- Self-Tests
- Videos
- Net Tutor
- e-Professors

Reading and Writing *After reading this section, write out the answers to these questions. Use complete sentences.*

1. What are the inequality symbols used in this section?
2. What different looking inequality means the same as $a < b$?
3. How do you know when to use a bracket and when to use a parenthesis when graphing an inequality on a number line?
4. What is a compound inequality?
5. What is the meaning of the compound inequality $a < b < c$?
6. What is the difference between "at most" and "at least?"

Determine whether each of the following statements is correct. See Example 1.

7. $-3 < 5$
8. $-6 < 0$
9. $4 \leq 4$
10. $-3 \geq -3$
11. $-6 > -5$
12. $-2 < -9$
13. $-4 \leq -3$
14. $-5 \geq -10$
15. $(-3)(4) - 1 < 0 - 3$
16. $2(4) - 6 \leq -3(5) + 1$
17. $-4(5) - 6 \geq 5(-6)$
18. $4(8) - 30 > 7(5) - 2(17)$
19. $7(4) - 12 \leq 3(9) - 2$
20. $-3(4) + 12 \leq 2(3) - 6$

State the solution set to each inequality in interval notation and sketch its graph. See Examples 2 and 3.

21. $x \leq 3$
22. $x \leq -7$
23. $x > -2$

24. $x > 4$

25. $-1 > x$

26. $0 > x$

27. $-2 \leq x$

28. $-5 \geq x$

29. $x \geq \frac{1}{2}$

30. $x \geq -\frac{2}{3}$

31. $x \leq 5.3$

32. $x \leq -3.4$

33. $-3 < x < 1$

34. $0 < x < 5$

35. $3 \leq x \leq 7$

36. $-3 \leq x \leq -1$

37. $-5 \leq x < 0$

38. $-2 < x \leq 2$

39. $40 < x \leq 100$

40. $0 \leq x < 600$

For each graph, write the corresponding inequality and the solution set to the inequality using interval notation.

41. 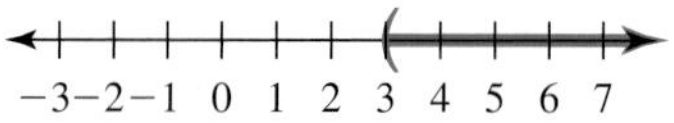

42. 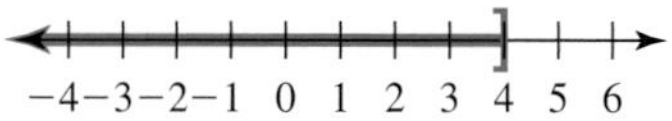

43. 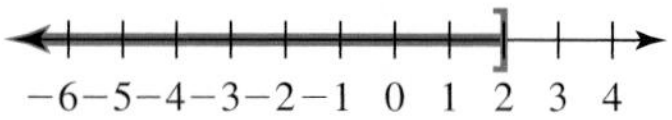

44. 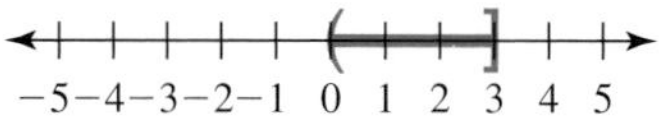

45. 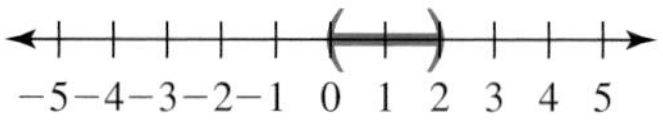

46. 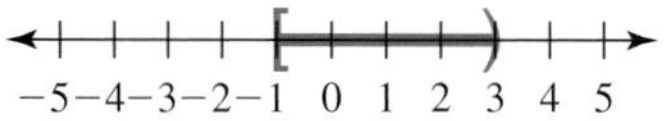

47. 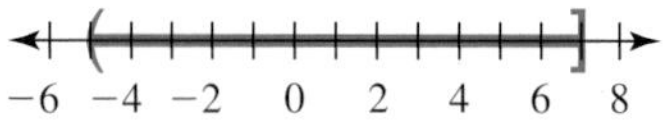

48. 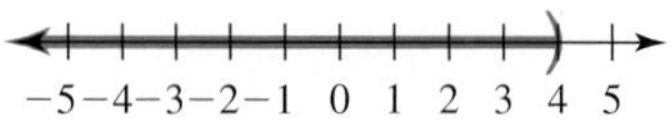

49. 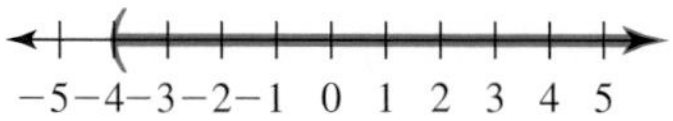

50. 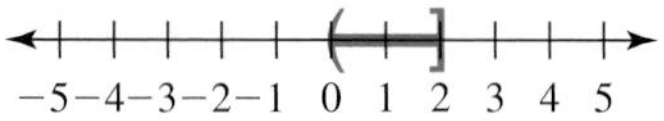

Determine whether the given number satisfies the inequality following it. See Example 4.

51. $-9, -x > 3$
52. $5, -3 < -x$
53. $-2, 5 \leq x$
54. $4, 4 \geq x$
55. $-6, 2x - 3 > -11$
56. $4, 3x - 5 < 7$
57. $3, -3x + 4 > -7$
58. $-4, -5x + 1 > -5$
59. $0, 3x - 7 \leq 5x - 7$
60. $0, 2x + 6 \geq 4x - 9$
61. $2.5, -10x + 9 \leq 3(x + 3)$
62. $1.5, 2x - 3 \leq 4(x - 1)$
63. $-7, -5 < x < 9$
64. $-9, -6 \leq x \leq 40$
65. $-2, -3 \leq 2x + 5 \leq 9$
66. $-5, -3 < -3x - 7 \leq 8$
67. $-3.4, -4.25x - 13.29 < 0.89$
68. $4.8, 3.25x - 14.78 \leq 1.3$

For each inequality, determine which of the numbers -5.1, 0, *and* 5.1 *satisfies the inequality.*

69. $x > -5$

70. $x \le 0$

71. $5 < x$

72. $-5 > x$

73. $5 < x < 7$

74. $5 < -x < 7$

75. $-6 < -x < 6$

76. $-5 \le x - 0.1 \le 5$

Write an inequality to describe each situation. See Example 5.

77. ***Sales tax.*** At an 8% sales tax rate, Susan paid more than \$1500 sales tax when she purchased her new Camaro. Let p represent the price of the Camaro.

78. ***Internet shopping.*** Carlos paid less than \$1000 including \$40 for shipping and 9% sales tax when he bought his new computer. Let p represent the price of the computer.

79. ***Fine dining.*** At Burger Brothers the price of a hamburger is twice the price of an order of French fries, and the price of a Coke is \$0.25 more than the price of the fries. Burger Brothers advertises that you can get a complete meal (burger, fries, and Coke) for under \$2.00. Let p represent the price of an order of fries.

80. ***Cats and dogs.*** Willow Creek Kennel boards only cats and dogs. One Friday night there were twice as many dogs as cats in the kennel and at least 30 animals spent the night there. Let d represent the number of dogs.

81. ***Barely passing.*** Travis made 44 and 72 on the first two tests in algebra and has one test remaining. The average on the three tests must be at least 60 for Travis to pass the course. Let s represent his score on the last test.

82. ***Ace the course.*** Florence made 87 on her midterm exam in psychology. The average of her midterm and her final must be at least 90 to get an A in the course. Let s represent her score on the final.

83. ***Coast to coast.*** On Howard's recent trip from Bangor to San Diego, he drove for 8 hours each day and traveled between 396 and 453 miles each day. Let R represent his average speed for each day.

84. ***Mother's Day present.*** Bart and Betty are looking at color televisions that range in price from \$399.99 to \$579.99. Bart can afford more than Betty and has agreed to spend \$100 more than Betty when they purchase this gift for their mother. Let b represent Betty's portion of the gift.

85. ***Positioning a ladder.*** Write an inequality in the variable x for the degree measure of the angle at the base of the ladder shown in the figure, given that the angle at the base must be between 60° and 70°.

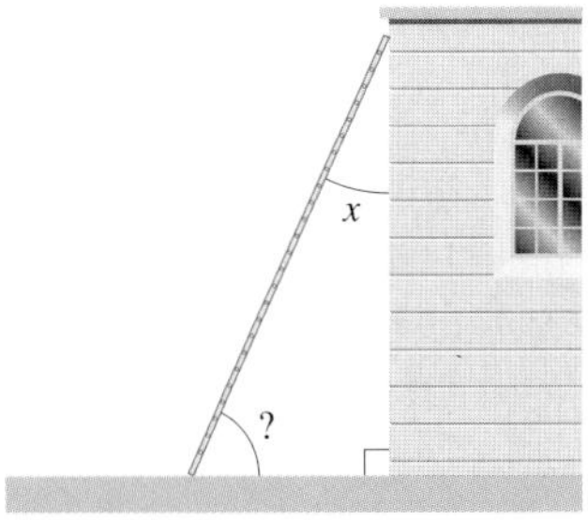

Figure for Exercise 85

86. ***Building a ski ramp.*** Write an inequality in the variable x for the degree measure of the smallest angle of the triangle shown in the figure, given that the degree measure of the smallest angle is at most 30°.

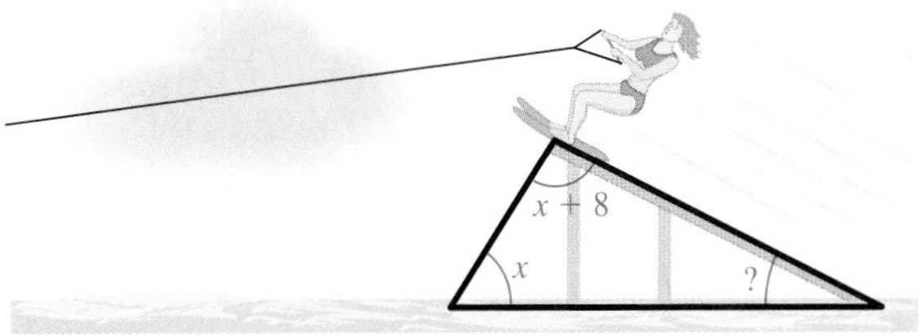

Figure for Exercise 86

87. ***Maximum girth.*** United Parcel Service defines the girth of a box as the sum of the length, twice the width, and twice the height. The maximum girth that UPS will ship is 130 in.

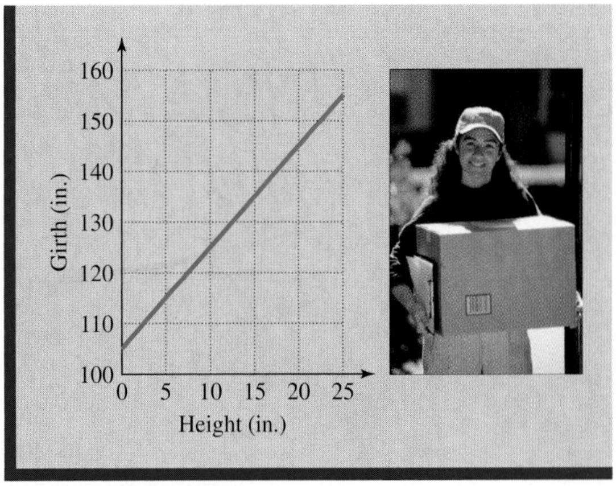

Figure for Exercise 87

a) If a box has a length of 45 in. and a width of 30 in., then what inequality must be satisfied by the height?

b) The accompanying graph shows the girth of a box with a length of 45 in., a width of 30 in., and height of h in. Use the graph to estimate the maximum height that is allowed for this box.

88. ***Batting average.*** Near the end of the season a professional baseball player has 93 hits in 317 times at bat for an average of 93/317 or 0.293. He gets a \$1 million bonus if his season average is over 0.300. He estimates that he will bat 20 more times before the season ends. Let x represent the number of hits in the last 20 at bats of the season.

a) Write an inequality that must be satisfied for him to get the bonus.

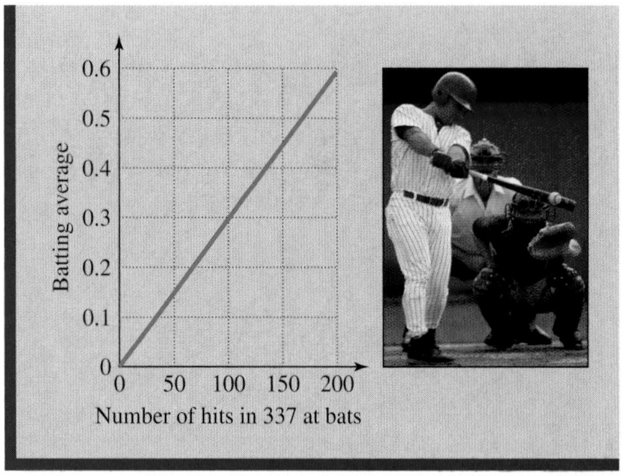

Figure for Exercise 88

b) Use the accompanying graph to estimate the number of hits in 337 at bats that will put his average over 0.300.

Solve.

89. ***Bicycle gear ratios.*** The gear ratio r for a bicycle is defined by the formula

$$r = \frac{Nw}{n},$$

where N is the number of teeth on the chainring (by the pedal), n is the number of teeth on the cog (by the wheel), and w is the wheel diameter in inches (*Cycling,* Burkett and Darst). The following chart gives uses for the various gear ratios.

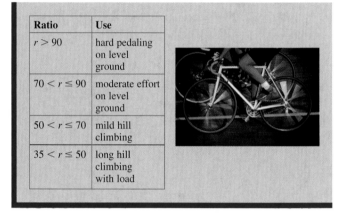

Ratio	Use
$r > 90$	hard pedaling on level ground
$70 < r \le 90$	moderate effort on level ground
$50 < r \le 70$	mild hill climbing
$35 < r \le 50$	long hill climbing with load

A bicycle with a 27-inch diameter wheel has 50 teeth on the chainring and 17 teeth on the cog. Find the gear ratio and indicate what this gear ratio is good for.

2.9 Solving Inequalities and Applications

In this Section

- **Rules for Inequalities**
- **Solving Inequalities**
- **Applications of Inequalities**

To solve equations, we write a sequence of equivalent equations that ends in a very simple equation whose solution is obvious. In this section you will learn that the procedure for solving inequalities is the same. However, the rules for performing operations on each side of an inequality are slightly different from the rules for equations.

Rules for Inequalities

Equivalent inequalities are inequalities that have exactly the same solutions. Inequalities such as $x > 3$ and $x + 2 > 5$ are equivalent because any number that is larger than 3 certainly satisfies $x + 2 > 5$ and any number that satisfies $x + 2 > 5$ must certainly be larger than 3.

We can get equivalent inequalities by performing operations on each side of an inequality just as we do for solving equations. If we start with the inequality $6 < 10$ and add 2 to each side, we get the true statement $8 < 12$. Examine the results of performing the same operation on each side of $6 < 10$.

Helpful Hint

You can think of an inequality like a seesaw that is out of balance.

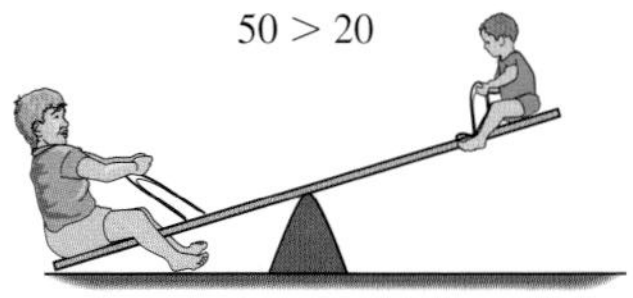

If the same weight is added to or subtracted from each side, it will remain in the same state of imbalance.

Perform these operations on each side:

	Add 2	Subtract 2	Multiply by 2	Divide by 2
Start with $6 < 10$	$8 < 12$	$4 < 8$	$12 < 20$	$3 < 5$

All of the resulting inequalities are correct. Now if we repeat these operations using -2, we get the following results.

Perform these operations on each side:

	Add -2	Subtract -2	Multiply by -2	Divide by -2
Start with $6 < 10$	$4 < 8$	$8 < 12$	$-12 > -20$	$-3 > -5$

Notice that the direction of the inequality symbol is the same for all of the results except the last two. When we multiplied each side by -2 and when we divided each side by -2, we had to reverse the inequality symbol to get a correct result. These tables illustrate the rules for solving inequalities.

Study Tip

Get in the habit of checking your work and having confidence in your answers. The answers to the odd-numbered exercises are in the back of this book, but you should look in the answer section only after you have checked on your own. You will not always have an answer section available.

Addition Property of Inequality

If we add the same number to each side of an inequality we get an equivalent inequality. If $a < b$, then $a + c < b + c$.

The addition property of inequality also allows us to subtract the same number from each side of an inequality because subtraction is defined in terms of addition.

Helpful Hint

Changing the signs of numbers, changes their relative position on the number line. For example, 3 lies to the left of 5 on the number line, but -3 lies to the right of -5. So $3 < 5$, but $-3 > -5$. Since multiplying and dividing by a negative cause sign changes, these operations reverse the inequality.

Multiplication Property of Inequality

If we multiply each side of an inequality by the same *positive* number, we get an equivalent inequality. If $a < b$ and $c > 0$, then $ac < bc$. If we multiply each side of an inequality by the same *negative* number and *reverse the inequality symbol,* we get an equivalent inequality. If $a < b$ and $c < 0$, then $ac > bc$.

The multiplication property of inequality also enables us to divide each side of an inequality by a nonzero number because division is defined in terms of multiplication. So if we multiply or divide each side by a negative number, the inequality symbol is reversed.

EXAMPLE 1

Writing equivalent inequalities

Write the appropriate inequality symbol in the blank so that the two inequalities are equivalent.

a) $x + 3 > 9$, x _____ 6 **b)** $-2x \le 6$, x _____ -3

Solution

a) If we subtract 3 from each side of $x + 3 > 9$, we get the equivalent inequality $x > 6$.

b) If we divide each side of $-2x \le 6$ by -2, we get the equivalent inequality $x \ge -3$.

Now do Exercises 7–14

CAUTION We use the properties of inequality just as we use the properties of equality. However, when we multiply or divide each side by a negative number, we must reverse the inequality symbol.

Solving Inequalities

To solve inequalities, we use the properties of inequality to isolate x on one side.

EXAMPLE 2

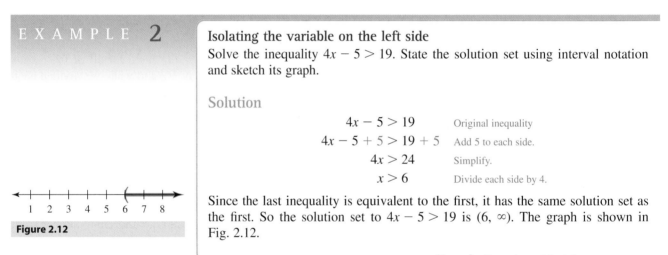

Isolating the variable on the left side

Solve the inequality $4x - 5 > 19$. State the solution set using interval notation and sketch its graph.

Solution

$$4x - 5 > 19 \quad \text{Original inequality}$$
$$4x - 5 + 5 > 19 + 5 \quad \text{Add 5 to each side.}$$
$$4x > 24 \quad \text{Simplify.}$$
$$x > 6 \quad \text{Divide each side by 4.}$$

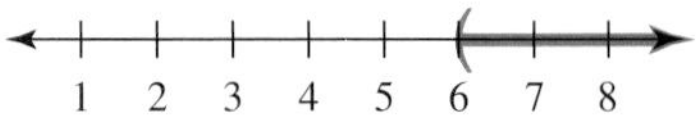

Figure 2.12

Since the last inequality is equivalent to the first, it has the same solution set as the first. So the solution set to $4x - 5 > 19$ is $(6, \infty)$. The graph is shown in Fig. 2.12.

Now do Exercises 15–16

Calculator Close-Up

You can use the TABLE feature of a graphing calculator to numerically support the solution to the inequality $4x - 5 > 19$ in Example 2. Use the Y = key to enter the equation $y_1 = 4x - 5$.

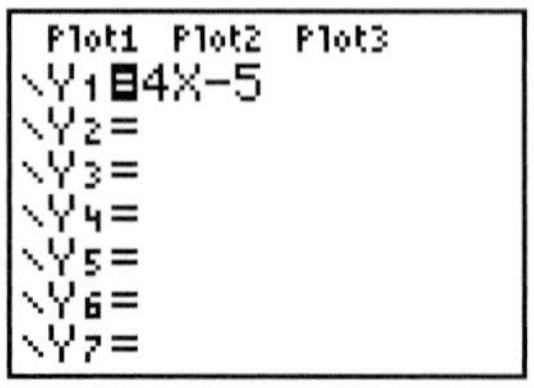

Next, use TBLSET to set the table so that the values of x start at 4.5 and the change in x is 0.5.

Finally, press TABLE to see lists of x-values and the corresponding y-values.

Notice that when x is larger than 6, y_1 (or $4x - 5$) is larger than 19. The table verifies or supports the algebraic solution, but it should not replace the algebraic method.

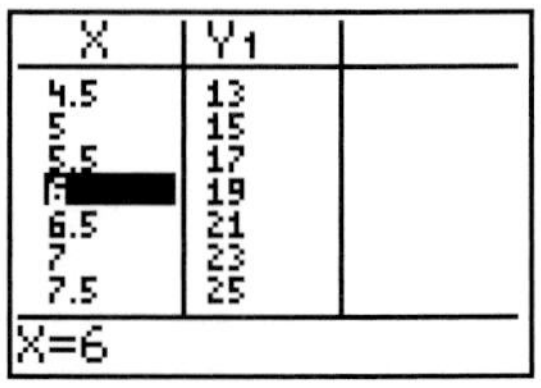

Remember that $5 < x$ is equivalent to $x > 5$. So the variable can be isolated on the right side of an inequality as shown in Example 3.

EXAMPLE 3

Isolating the variable on the right side

Solve the inequality $5x - 2 \le 7x - 5$. State the solution set using interval notation and sketch its graph.

Solution

$$\begin{aligned} 5x - 2 &\le 7x - 5 && \text{Original inequality} \\ 5x - 2 - 5x &\le 7x - 5 - 5x && \text{Subtract } 5x \text{ from each side.} \\ -2 &\le 2x - 5 && \text{Simplify.} \\ 3 &\le 2x && \text{Add 5 to each side.} \\ \frac{3}{2} &\le x && \text{Divide each side by 2.} \end{aligned}$$

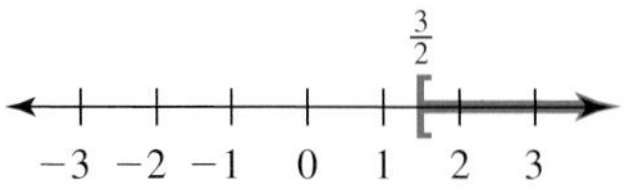

Figure 2.13

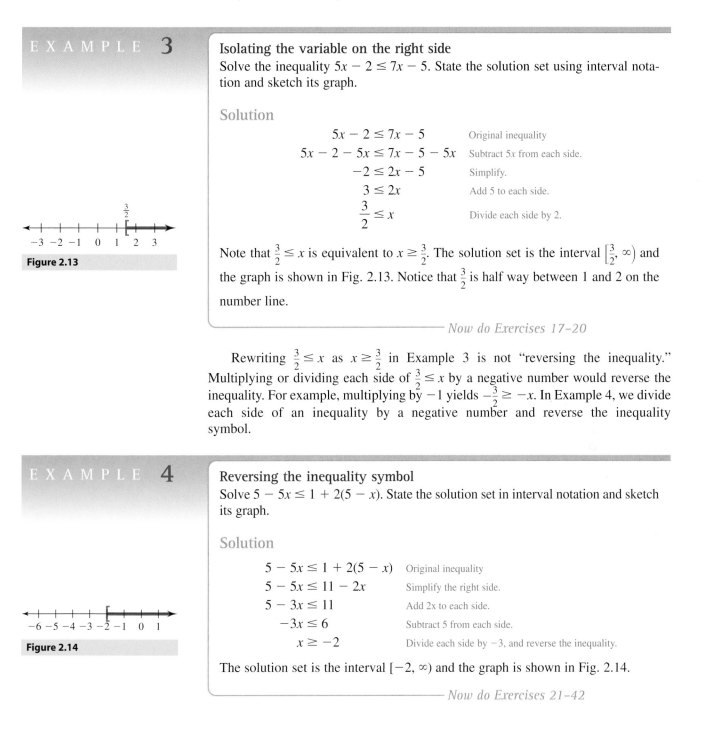

Note that $\frac{3}{2} \le x$ is equivalent to $x \ge \frac{3}{2}$. The solution set is the interval $\left[\frac{3}{2}, \infty\right)$ and the graph is shown in Fig. 2.13. Notice that $\frac{3}{2}$ is half way between 1 and 2 on the number line.

Now do Exercises 17–20

Rewriting $\frac{3}{2} \le x$ as $x \ge \frac{3}{2}$ in Example 3 is not "reversing the inequality." Multiplying or dividing each side of $\frac{3}{2} \le x$ by a negative number would reverse the inequality. For example, multiplying by -1 yields $-\frac{3}{2} \ge -x$. In Example 4, we divide each side of an inequality by a negative number and reverse the inequality symbol.

EXAMPLE 4

Reversing the inequality symbol

Solve $5 - 5x \le 1 + 2(5 - x)$. State the solution set in interval notation and sketch its graph.

Solution

$$\begin{aligned} 5 - 5x &\le 1 + 2(5 - x) && \text{Original inequality} \\ 5 - 5x &\le 11 - 2x && \text{Simplify the right side.} \\ 5 - 3x &\le 11 && \text{Add 2x to each side.} \\ -3x &\le 6 && \text{Subtract 5 from each side.} \\ x &\ge -2 && \text{Divide each side by } -3\text{, and reverse the inequality.} \end{aligned}$$

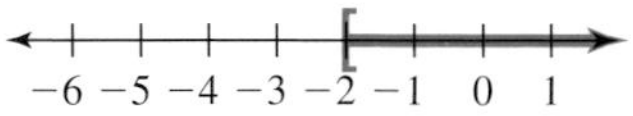

Figure 2.14

The solution set is the interval $[-2, \infty)$ and the graph is shown in Fig. 2.14.

Now do Exercises 21–42

We can use the rules for solving inequalities on the compound inequalities that we studied in Section 2.8.

EXAMPLE 5

Solving a compound inequality

Solve $-9 \le \frac{2x}{3} - 7 < 5$. State the solution set in interval notation and sketch its graph.

Solution

$$-9 \le \frac{2x}{3} - 7 < 5 \quad \text{Original inequality}$$

$$-9 + 7 \le \frac{2x}{3} - 7 + 7 < 5 + 7 \quad \text{Add 7 to each part.}$$

$$-2 \le \frac{2x}{3} < 12 \quad \text{Simplify.}$$

$$\frac{3}{2}(-2) \le \frac{3}{2} \cdot \frac{2x}{3} < \frac{3}{2} \cdot 12 \quad \text{Multiply each part by } \frac{3}{2}.$$

$$-3 \le x < 18 \quad \text{Simplify.}$$

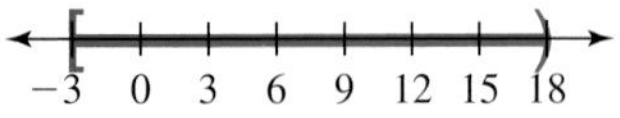

Figure 2.15

Since the last compound inequality is equivalent to the first, the solution set is $[-3, 18)$. The graph is shown in Fig. 2.15.

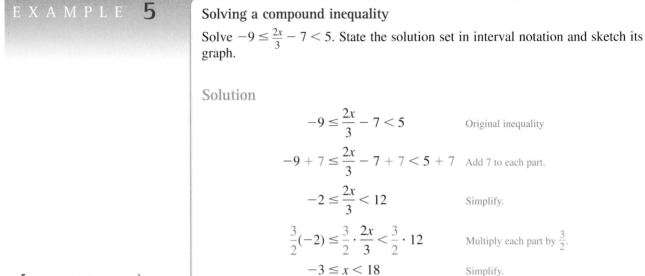

Now do Exercises 43–46

CAUTION There are many negative numbers in Example 5, but the inequality was not reversed, since we did not multiply or divide by a negative number. An inequality is reversed only if you multiply or divide by a negative number.

EXAMPLE 6

Reversing inequality symbols in a compound inequality

Solve $-3 \le 5 - x \le 5$. State the solution set in interval notation and sketch its graph.

Solution

$$-3 \le 5 - x \le 5 \quad \text{Original inequality}$$

$$-3 - 5 \le 5 - x - 5 \le 5 - 5 \quad \text{Subtract 5 from each part.}$$

$$-8 \le -x \le 0 \quad \text{Simplify.}$$

$$(-1)(-8) \ge (-1)(-x) \ge (-1)(0) \quad \text{Multiply each part by } -1,$$

$$8 \ge x \ge 0 \quad \text{reversing the inequality symbols.}$$

It is customary to write $8 \ge x \ge 0$ with the smallest number on the left:

$$0 \le x \le 8$$

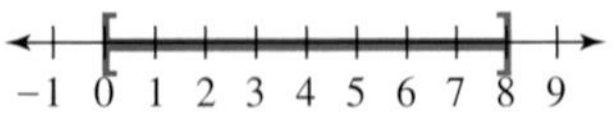

Figure 2.16

Since the last compound inequality is equivalent to the first, the solution set is $[0, 8]$. The graph is shown in Fig. 2.16.

Now do Exercises 47–56

Applications of Inequalities

Example 7 shows how inequalities can be used in applications.

EXAMPLE 7

Averaging test scores

Mei Lin made a 76 on the midterm exam in history. To get a B, the average of her midterm and her final exam must be between 80 and 90. For what range of scores on the final exam will she get a B?

Solution

Let x represent the final exam score. Her average is then $\frac{x + 76}{2}$. The inequality expresses the fact that the average must be between 80 and 90:

$$80 < \frac{x + 76}{2} < 90$$

$$2(80) < 2\left(\frac{x + 76}{2}\right) < 2(90) \quad \text{Multiply each part by 2.}$$

$$160 < x + 76 < 180 \quad \text{Simplify.}$$

$$160 - 76 < x + 76 - 76 < 180 - 76 \quad \text{Subtract 76 from each part.}$$

$$84 < x < 104 \quad \text{Simplify.}$$

The last inequality indicates that Mei Lin's final exam score must be between 84 and 104.

Now do Exercises 63–78

Helpful Hint

Remember that all inequality symbols in a compound inequality must point in the same direction. We usually have them all point to the left so that the numbers are increasing in size as you go from left to right in the inequality.

Warm-Ups ▼

True or false? Explain your answer.

1. The inequality $2x > 18$ is equivalent to $x > 9$.
2. The inequality $x - 5 > 0$ is equivalent to $x < 5$.
3. We can divide each side of an inequality by any real number.
4. The inequality $-2x \le 6$ is equivalent to $-x \le 3$.
5. The statement "x is at most 7" is written as $x < 7$.
6. "The sum of x and $0.05x$ is at least 76" is written as $x + 0.05x \ge 76$.
7. The statement "x is not more than 85" is written as $x < 85$.
8. The inequality $-3 > x > -9$ is equivalent to $-9 < x < -3$.
9. If x is the sale price of Glen's truck, the sales tax rate is 8%, and the title fee is \$50, then the total that he pays is $1.08x + 50$ dollars.
10. If the selling price of the house, x, less the sales commission of 6% must be at least \$60,000, then $x - 0.06x \le 60{,}000$.

2.9 Exercises

Reading and Writing *After reading this section, write out the answers to these questions. Use complete sentences.*

1. What are equivalent inequalities?

2. What is the addition property of inequality?

3. What is the multiplication property of inequality?

4. What similarities are there between solving equations and solving inequalities?

5. How do we solve compound inequalities?

6. How do you know when to reverse the direction of an inequality symbol?

Write the appropriate inequality symbol in the blank so that the two inequalities are equivalent. See Example 1.

7. $x + 7 > 0$
$x \_\_ -7$

8. $x - 6 < 0$
$x \_\_ 6$

9. $9 \le 3w$
$w \_\_ 3$

10. $10 \ge 5z$
$z \_\_ 2$

11. $-4k < -4$
$k \_\_ 1$

12. $-9t > 27$
$t \_\_ -3$

13. $-\frac{1}{2}y \ge 4$
$y \_\_ -8$

14. $-\frac{1}{3}x \le 4$
$x \_\_ -12$

Solve each inequality. State the solution set in interval notation and sketch its graph. See Examples 2–4.

15. $x + 3 > 0$

16. $x + 9 \le -8$

17. $-3 < w - 1$

18. $9 > w - 12$

19. $8 > 2b$

20. $35 < 7b$

21. $-8z \le 4$

22. $-4y \ge -10$

23. $3y - 2 < 7$

24. $2y - 5 > -9$

25. $3 - 9z \le 6$

26. $5 - 6z \ge 13$

27. $6 > -r + 3$

28. $6 \le 12 - r$

29. $5 - 4p > -8 - 3p$

30. $7 - 9p > 11 - 8p$

31. $-\frac{5}{6}q \ge -20$

32. $-\frac{2}{3}q \geq -4$

33. $1 - \frac{1}{4}t \geq \frac{1}{8}$

34. $\frac{1}{6} - \frac{1}{3}t > 0$

35. $0.1x + 0.35 > 0.2$

36. $1 - 0.02x \leq 0.6$

37. $2x + 5 < x - 6$

38. $3x - 4 < 2x + 9$

39. $x - 4 < 2(x + 3)$

40. $2x + 3 < 3(x - 5)$

41. $0.52x - 35 < 0.45x + 8$

42. $8455(x - 3.4) > 4320$

Solve each compound inequality. State the solution set in interval notation and sketch its graph. See Examples 5 and 6.

43. $5 < x - 3 < 7$

44. $2 < x - 5 < 6$

45. $3 < 2v + 1 < 10$

46. $-3 < 3v + 4 < 7$

47. $-4 \leq 5 - k \leq 7$

48. $2 \leq 3 - k \leq 8$

49. $-2 < 7 - 3y \leq 22$

50. $-1 \leq 1 - 2y < 3$

51. $5 < \frac{2u}{3} - 3 < 17$

52. $-4 < \frac{3u}{4} - 1 < 11$

53. $-2 < \frac{4m - 4}{3} \leq \frac{2}{3}$

54. $0 \leq \frac{3 - 2m}{2} < 9$

55. $0.02 < 0.54 - 0.0048x < 0.05$

56. $0.44 < \frac{34.55 - 22.3x}{124.5} < 0.76$

Solve each inequality. State the solution set in interval notation and sketch its graph.

57. $\frac{1}{2}x - 1 \leq 4 - \frac{1}{3}x$

58. $\frac{y}{4} - \frac{5}{12} \geq \frac{y}{3} + \frac{1}{4}$

59. $\frac{1}{2}\left(x - \frac{1}{4}\right) > \frac{1}{4}\left(6x - \frac{1}{2}\right)$

60. $-\frac{1}{2}\left(z - \frac{2}{5}\right) < \frac{2}{3}\left(\frac{3}{4}z - \frac{6}{5}\right)$

61. $\frac{1}{3} < \frac{1}{4}x - \frac{1}{6} < \frac{7}{12}$

62. $-\frac{3}{5} < \frac{1}{5} - \frac{2}{15}w < -\frac{1}{3}$

Solve each of the following problems by using an inequality. See Example 7.

63. ***Boat storage.*** The length of a rectangular boat storage shed must be 4 meters more than the width, and the perimeter must be at least 120 meters. What is the range of values for the width?

64. ***Fencing a garden.*** Elka is planning a rectangular garden that is to be twice as long as it is wide. If she can afford to buy at most 180 feet of fencing, then what are the possible values for the width?

Photo for Exercise 64

65. ***Car shopping.*** Harold Ivan is shopping for a new car. In addition to the price of the car, there is a 5% sales tax and a \$144 title and license fee. If Harold Ivan decides that he will spend less than \$9970 total, then what is the price range for the car?

66. ***Car selling.*** Ronald wants to sell his car through a broker who charges a commission of 10% of the selling price. Ronald still owes \$11,025 on the car. Ronald must get enough to at least pay off the loan. What is the range of the selling price?

67. ***Microwave oven.*** Sherie is going to buy a microwave in a city with an 8% sales tax. She has at most \$594 to spend. In what price range should she look?

68. ***Dining out.*** At Burger Brothers the price of a hamburger is twice the price of an order of French fries, and the price of a Coke is \$0.40 more than the price of the fries. Burger Brothers advertises that you can get a complete meal (burger, fries, and Coke) for under \$4.00. What is the price range of an order of fries?

69. ***Averaging test scores.*** Tilak made 44 and 72 on the first two tests in algebra and has one test remaining. For Tilak to pass the course, the average on the three tests must be at least 60. For what range of scores on his last test will Tilak pass the course?

70. ***Averaging income.*** Helen earned \$400 in January, \$450 in February, and \$380 in March. To pay all of her bills, she must average at least \$430 per month. For what income in April would her average for the four months be at least \$430?

71. ***Going for a C.*** Professor Williams gives only a midterm exam and a final exam. The semester average is computed by taking $\frac{1}{3}$ of the midterm exam score plus $\frac{2}{3}$ of the final exam score. To get a C, Stacy must have a semester average between 70 and 79 inclusive. If Stacy scored only 48 on the midterm, then for what range of scores on the final exam will Stacy get a C?

72. ***Different weights.*** Professor Williamson counts his midterm as $\frac{2}{3}$ of the grade and his final as $\frac{1}{3}$ of the grade. Wendy scored only 48 on the midterm. What range of scores on the final exam would put Wendy's average between 70 and 79 inclusive? Compare to the previous exercise.

73. ***Average driving speed.*** On Halley's recent trip from Bangor to San Diego, she drove for 8 hours each day and traveled between 396 and 453 miles each day. In what range was her average speed for each day of the trip?

74. ***Driving time.*** On Halley's trip back to Bangor, she drove at an average speed of 55 mph every day and traveled between 330 and 495 miles per day. In what range was her daily driving time?

75. ***Sailboat navigation.*** As the sloop sailed north along the coast, the captain sighted the lighthouse at points A and B as shown in the figure. If the degree measure of the angle at the lighthouse is less than 30°, then what are the possible values for x?

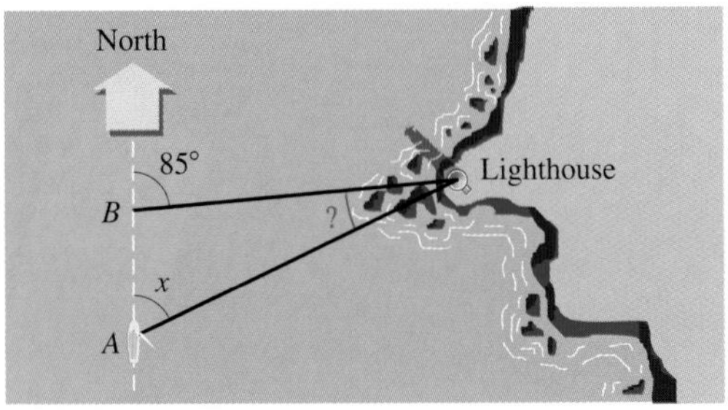

Figure for Exercise 75

76. ***Flight plan.*** A pilot started at point A and flew in the direction shown in the diagram for some time. At point B she made a 110° turn to end up at point C, due east of where she started. If the measure of angle C is less than 85°, then what are the possible values for x?

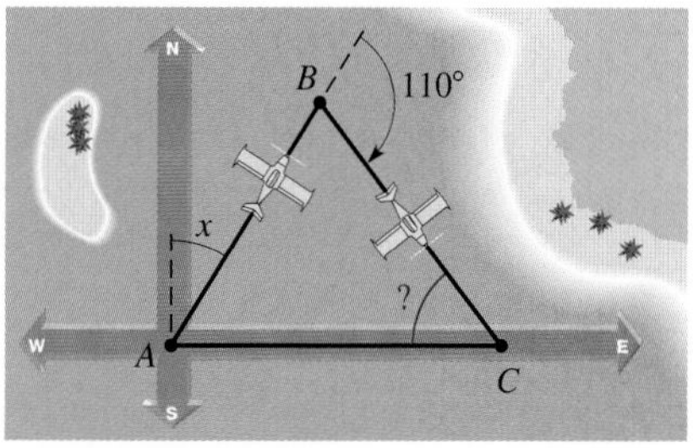

Figure for Exercise 76

77. ***Bicycle gear ratios.*** The gear ratio r for a bicycle is defined by the formula

$$r = \frac{Nw}{n},$$

where N is the number of teeth on the chainring (by the pedal), n is the number of teeth on the cog (by the wheel), and w is the wheel diameter in inches (www.sheldonbrown.com/gears).

a) If the wheel has a diameter of 27 in. and there are 12 teeth on the cog, then for what number of teeth on the chainring is the gear ratio between 60 and 80?

b) If a bicycle has 48 teeth on the chainring and 17 teeth on the cog, then for what diameter wheel is the gear ratio between 65 and 70?

c) If a bicycle has a 26-in. diameter wheel and 40 teeth on the chainring, then for what number of teeth on the cog is the gear ratio less than 75?

78. ***Virtual demand.*** The weekly demand (the number bought by consumers) for the Acme Virtual Pet is given by the formula

$$d = 9000 - 60p$$

where p is the price each in dollars.

a) What is the demand when the price is \$30 each?

b) In what price range will the demand be above 6000?

Collaborative Activities

Grouping: Three students per group

Topic: Expressions, Equations, and Inequalities

Expression-Equation-Inequality

In this activity you will practice your skills with simplifying expressions, solving equations, and solving and graphing inequalities. The roles are simplifier, solver, and grapher. The simplifier will simplify all expressions, the solver will solve all equations, and the grapher will first solve and then graph all inequalities.

Sorting Out the Work. Each of the following exercises contains one expression to be simplified, one equation to be solved, and one inequality to be graphed, not necessarily in that order. For each exercise your group will first sort out the three parts: give the expression to the simplifier, the equation to the solver, and the inequality to the grapher. After each person has completed his or her assigned task, pass the paper to the left for checking. Take on the role of the paper that you now have and proceed to the next exercise.

1. a) $-2x \geq 20$

b) $2(x - 4) + x - 10$

c) $3x - 5 = 4x + 1$

2. a) $\frac{1}{2}x + 3 + \frac{7}{2}x - 5$

b) $4(x + 3) = 3(1 + x)$

c) $5x - 7 > 8x + 5$

3. a) $4(2x - 5) \leq 5x - 1$

b) $3(x + 2) - (12 - x)$

c) $\frac{5}{3}x + 4 = \frac{2}{3}x$

4. a) $5x - 6 + x = 9 + 4x - 1$

b) $\frac{2(5x + 1)}{3} > 2$

c) $5x - x + 2x$

Creating Your Own. Each student now creates and solves the type of problem he or she worked last, using the terms given here. For example, given the terms $4y$, $2y$, and -4 you could make up the expression $4y - 2y + (-4)$, the equation $4y - 2y = -4$, and the inequality $4y + 2y < -4$. Pass your work to the person on your left to check. Then proceed to the next exercise as before.

5. $6x, 2x, 3$

6. $-10x, 5x, 20, -15$

7. $3y, -7y, 5, -4$

8. $-4w, 7, -6w, 22$

Chapter 2 Wrap-Up

Summary

Equations		**Examples**
Linear equation	An equation of the form $ax + b = 0$ with $a \neq 0$	$3x + 7 = 0$
Identity	An equation that is satisfied by every number for which both sides are defined	$x + x = 2x$
Conditional equation	An equation that has at least one solution but is not an identity	$5x - 10 = 0$
Inconsistent equation	An equation that has no solution	$x = x + 1$
Equivalent equations	Equations that have exactly the same solutions	$2x + 1 = 5$ $2x = 4$
Properties of equality	If the same number is added to or subtracted from each side of an equation, the resulting equation is equivalent to the original equation.	$x - 5 = -9$ $x = -4$
	If each side of an equation is multiplied or divided by the same nonzero number, the resulting equation is equivalent to the original equation.	$9x = 27$ $x = 3$
Solving equations	**1.** Remove parentheses by using the distributive property and then combine like terms to simplify each side as much as possible. **2.** Use the addition property of equality to get like terms from opposite sides onto the same side so that they may be combined. **3.** The multiplication property of equality is generally used last. **4.** Check that the solution satisfies the original equation.	$2(x - 3) = -7 + 3(x - 1)$ $2x - 6 = -10 + 3x$ $-x - 6 = -10$ $-x = -4$ $x = 4$ *Check:* $2(4 - 3) = -7 + 3(4 - 1)$ $2 = 2$

Applications

Steps in solving applied problems	**1.** Read the problem. **2.** If possible, draw a diagram to illustrate the problem. **3.** Choose a variable and write down what it represents.

4. Represent any other unknowns in terms of that variable.
5. Write an equation that describes the situation.
6. Solve the equation.
7. Answer the original question.
8. Check your answer by using it to solve the original problem (not the equation).

Inequalities		Examples
Properties of inequality	Addition, subtraction, multiplication, and division may be performed on each side of an inequality, just as we do in solving equations, with one exception. When multiplying or dividing by a negative number, the inequality symbol is reversed.	$-3x + 1 > 7$ $-3x > 6$ $x < -2$

Enriching Your Mathematical Word Power

For each mathematical term, choose the correct meaning.

1. linear equation
a. an equation in which the terms are in line
b. an equation of the form $ax + b = 0$ where $a \neq 0$
c. the equation $a = b$
d. an equation of the form $a^2 + b^2 = c^2$

2. identity
a. an equation that is satisfied by all real numbers
b. an equation that is satisfied by every real number
c. an equation that is identical
d. an equation that is satisfied by every real number for which both sides are defined

3. conditional equation
a. an equation that has at least one real solution
b. an equation that is correct
c. an equation that is satisfied by at least one real number but is not an identity
d. an equation that we are not sure how to solve

4. inconsistent equation
a. an equation that is wrong
b. an equation that is only sometimes consistent
c. an equation that has no solution
d. an equation with two variables

5. equivalent equations
a. equations that are identical
b. equations that are correct
c. equations that are equal
d. equations that have the same solution

6. formula
a. an equation
b. a type of race car
c. a process
d. an equation involving two or more variables

7. literal equation
a. a formula
b. an equation with words
c. a false equation
d. a fact

8. complementary angles
a. angles that compliment each other
b. angles whose degree measures total 90°
c. angles whose degree measures total 180°
d. angles with the same vertex

9. supplementary angles
a. angles with soft flexible sides
b. angles whose degree measures total 90°
c. angles whose degree measures total 180°
d. angles that form a square

10. uniform motion
a. movement of an army
b. movement in a straight line
c. consistent motion
d. motion at a constant rate

Review Exercises

2.1 *Solve each equation and check your answer.*

1. $x - 23 = 12$

2. $14 = 18 + y$

3. $\frac{2}{3}u = -4$

4. $-\frac{3}{8}r = 15$

5. $-5y = 35$

6. $-12 = 6h$

7. $6m = 13 + 5m$

8. $19 - 3n = -2n$

Study Tip

Note how the review exercises are arranged according to the sections in this chapter. If you are having trouble with a certain type of problem, refer back to the appropriate section for examples and explanations.

2.2 *Solve each equation and check your answer.*

9. $2x - 5 = 9$

10. $5x - 8 = 38$

11. $3p - 14 = -4p$

12. $36 - 9y = 3y$

13. $2z + 12 = 5z - 9$

14. $15 - 4w = 7 - 2w$

15. $2(h - 7) = -14$

16. $2(t - 7) = 0$

17. $3(w - 5) = 6(w + 2) - 3$

18. $2(a - 4) + 4 = 5(9 - a)$

2.3 *Solve each equation. Identify each equation as a conditional equation, an inconsistent equation, or an identity.*

19. $2(x - 7) - 5 = 5 - (3 - 2x)$

20. $2(x - 7) + 5 = -(9 - 2x)$

21. $2(w - w) = 0$

22. $2y - y = 0$

23. $\frac{3r}{3r} = 1$

24. $\frac{3t}{3} = 1$

25. $\frac{1}{2}a - 5 = \frac{1}{3}a - 1$

26. $\frac{1}{2}b - \frac{1}{2} = \frac{1}{4}b$

27. $0.06q + 14 = 0.3q - 5.2$

28. $0.05(z + 20) = 0.1z - 0.5$

29. $0.05(x + 100) + 0.06x = 115$

30. $0.06x + 0.08(x + 1) = 0.41$

Solve each equation.

31. $2x + \frac{1}{2} = 3x + \frac{1}{4}$

32. $5x - \frac{1}{3} = 6x - \frac{1}{2}$

33. $\frac{x}{2} - \frac{3}{4} = \frac{x}{6} + \frac{1}{8}$

34. $\frac{1}{3} - \frac{x}{5} = \frac{1}{2} - \frac{x}{10}$

35. $\frac{5}{6}x = -\frac{2}{3}$

36. $-\frac{2}{3}x = \frac{3}{4}$

37. $-\frac{1}{2}(x - 10) = \frac{3}{4}x$

38. $-\frac{1}{3}(6x - 9) = 23$

39. $3 - 4(x - 1) + 6 = -3(x + 2) - 5$

40. $6 - 5(1 - 2x) + 3 = -3(1 - 2x) - 1$

41. $5 - 0.1(x - 30) = 18 + 0.05(x + 100)$

42. $0.6(x - 50) = 18 - 0.3(40 - 10x)$

2.4 *Solve each equation for x.*

43. $ax + b = 0$

44. $mx + e = t$

45. $ax - 2 = b$

46. $b = 5 - x$

47. $LWx = V$

48. $3xy = 6$

49. $2x - b = 5x$

50. $t - 5x = 4x$

Solve each equation for y. Write the answer in the form $y = mx + b$, where m and b are real numbers.

51. $5x + 2y = 6$

52. $5x - 3y + 9 = 0$

53. $y - 1 = -\frac{1}{2}(x - 6)$

54. $y + 6 = \frac{1}{2}(x + 8)$

55. $\frac{1}{2}x + \frac{1}{4}y = 4$

56. $-\frac{x}{3} + \frac{y}{2} = 1$

Find the value of y in each formula if $x = -3$.

57. $y = 3x - 4$

58. $2x - 3y = -7$

59. $5xy = 6$

60. $3xy - 2x = -12$

61. $y - 3 = -2(x - 4)$

62. $y + 1 = 2(x - 5)$

Fill in the tables using the given formulas.

63. $y = -5x + 10$

x	y
−1	
0	
1	
2	
3	

64. $y = 2x - 4$

x	y
0	
1	
2	
3	
4	

65. $y = \frac{2}{3}x - 1$

x	y
−3	
0	
3	
6	

66. $y = 10x + 100$

x	y
−20	
−10	
0	
10	

2.5 *Translate each verbal expression into an algebraic expression.*

67. The sum of a number and 9

68. The product of a number and 7

69. Two numbers that differ by 8

70. Two numbers with a sum of 12

71. Sixty-five percent of a number

72. One half of a number

Identify the variable, and write an equation that describes each situation. Do not solve the equation.

73. One side of a rectangle is 5 feet longer than the other, and the area is 98 square feet.

74. One side of a rectangle is one foot longer than twice the other side, and the perimeter is 56 feet.

75. By driving 10 miles per hour slower than Jim, Barbara travels the same distance in 3 hours as Jim does in 2 hours.

76. Gladys and Ned drove 840 miles altogether, with Gladys averaging 5 miles per hour more in her 6 hours at the wheel than Ned did in his 5 hours at the wheel.

77. The sum of three consecutive even integers is 90.

78. The sum of two consecutive odd integers is 40.

79. The three angles of a triangle have degree measures of t, $2t$, and $t - 10$.

80. Two complementary angles have degree measures p and $3p - 6$.

2.6–7 *Solve each problem.*

81. ***Odd integers.*** If the sum of three consecutive odd integers is 237, then what are the integers?

82. ***Even integers.*** Find two consecutive even integers that have a sum of 450.

83. ***Driving to the shore.*** Lawanda and Betty both drive the same distance to the shore. By driving 15 miles per hour faster than Betty, Lawanda can get there in 3 hours while Betty takes 4 hours. How fast does each of them drive?

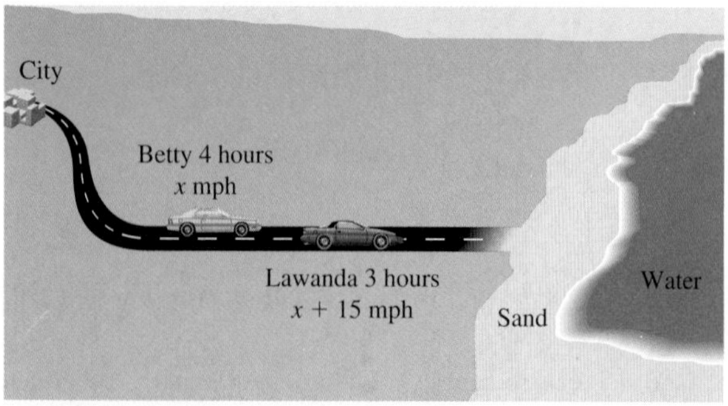

Figure for Exercise 83

84. ***Rectangular lot.*** The length of a rectangular lot is 50 feet more than the width. If the perimeter is 500 feet, then what are the length and width?

85. ***Combined savings.*** Wanda makes \$6000 more per year than her husband does. Wanda saves 10% of her income for retirement, and her husband saves 6%. If together they save \$5400 per year, then how much does each of them make per year?

86. ***Layoffs looming.*** American Products plans to lay off 10% of its employees in its aerospace division and 15% of its employees in its agricultural division. If altogether 12% of the 3000 employees in these two divisions will be laid off, then how many employees are in each division?

2.8 *Determine whether the given number is a solution to the inequality following it.*

87. $3, -2x + 5 \le x - 6$

88. $-2, 5 - x > 4x + 3$

89. $-1, -2 \le 6 + 4x < 0$

90. $0, 4x + 9 \ge 5(x - 3)$

For each graph write the corresponding inequality and the solution set to the inequality using interval notation.

91.
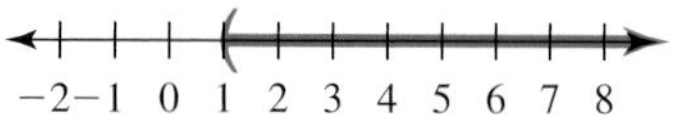

92.
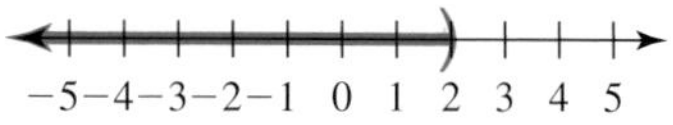

93.
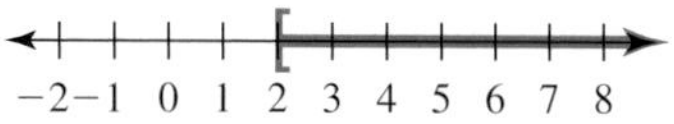

94.
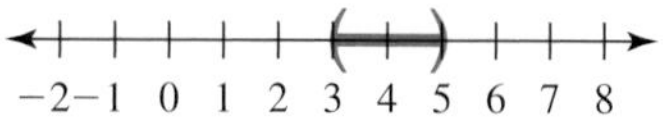

95.
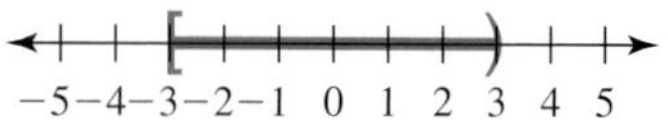

96.
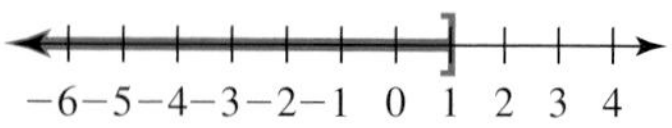

97.
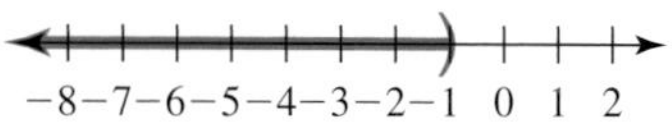

98.
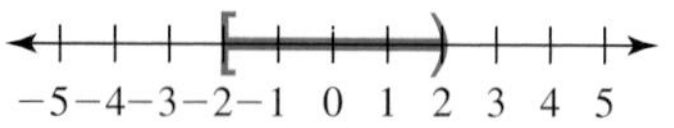

2.9 *Solve each inequality. State the solution set in interval notation and sketch its graph.*

99. $x + 2 > 1$

100. $x - 3 > 7$

101. $3x - 5 < x + 1$

102. $5x - 5 > 9 - 2x$

103. $-\frac{3}{4}x \geq 3$

104. $-\frac{2}{3}x \leq 10$

105. $3 - 2x < 11$

106. $5 - 3x > 35$

107. $-3 < 2x - 1 < 9$

108. $2 \leq 3x + 2 < 8$

109. $0 \leq 1 - 2x < 5$

110. $-5 < 3 - 4x \leq 7$

111. $-1 \leq \frac{2x - 3}{3} \leq 1$

112. $-3 < \frac{4 - x}{2} < 2$

113. $\frac{1}{3} < \frac{1}{3} + \frac{x}{2} < \frac{5}{6}$

114. $-\frac{3}{8} \leq -\frac{1}{4}x + \frac{1}{8} < \frac{5}{8}$

Miscellaneous

Use an equation, inequality, or formula to solve each problem.

115. ***Long-term yields.*** The accompanying graph shows the *yield curve* for U.S. Treasury Bonds on February 20, 2002 (Bloomberg, www.bloomberg.com). The annual yield on a 30-year treasury bond was 5.375%. Use the simple interest formula to find the amount of interest earned during the first year on a 30-year bond of $10,000.

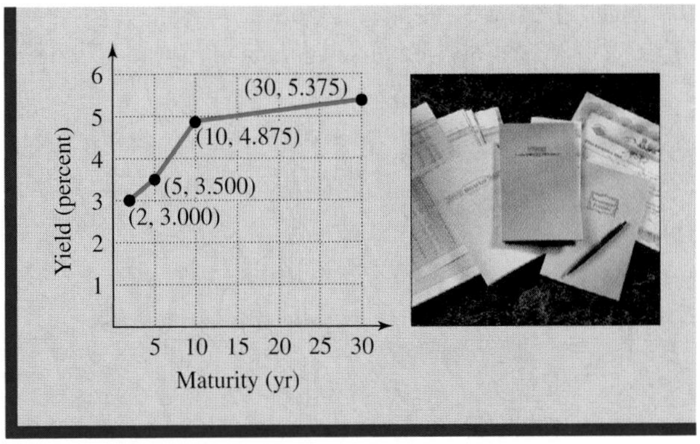

Figure for Exercise 115

116. ***Short-term yields.*** On February 20, 2002 the annual yield on a two-year treasury bond was 3.000%. Use the simple interest formula to find the amount of interest earned during the first year on a two-year treasury bond of $10,000.

117. ***Combined videos.*** The owners of ABC Video discovered that they had no movies in common with XYZ Video and bought XYZ's entire stock. Although XYZ had 200 titles, they had no children's movies, while 60% of ABC's titles were children's movies. If 40% of the movies in the combined stock are children's movies, then how many movies did ABC have before the merger?

118. ***Living comfortably.*** Gary has figured that he needs to take home $30,400 a year to live comfortably. If the government gets 24% of Gary's income, then what must his income be for him to live comfortably?

119. ***Bracing a gate.*** The diagonal brace on a rectangular gate forms an angle with the horizontal side with degree measure x and an angle with the vertical side with degree measure $2x - 3$. Find x.

120. ***Digging up the street.*** A contractor wants to install a pipeline connecting point A with point C on opposite sides of a road as shown in the figure. To save money, the contractor has decided to lay the pipe to point B and then under the road to point C. Find the measure of the angle marked x in the figure.

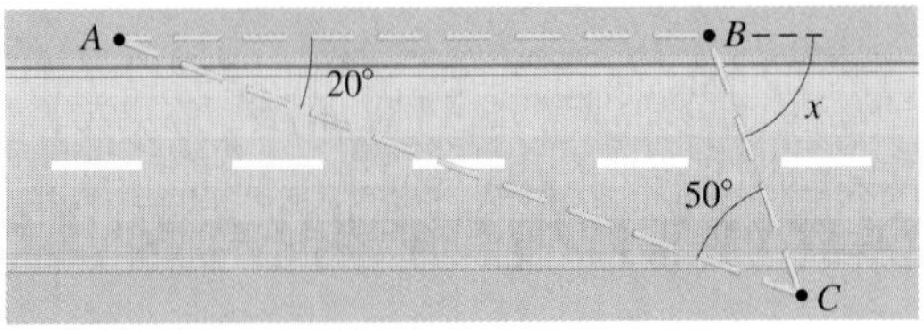

Figure for Exercise 120

121. ***Perimeter of a triangle.*** One side of a triangle is 1 foot longer than the shortest side, and the third side is twice as long as the shortest side. If the perimeter is less than 25 feet, then what is the range of the length of the shortest side?

122. ***Restricted hours.*** Alana makes \$5.80 per hour working in the library. To keep her job, she must make at least \$116 per week; but to keep her scholarship, she must not earn more than \$145 per week. What is the range of the number of hours per week that she may work?

Chapter 2 Test

Solve each equation.

1. $-10x - 6 + 4x = -4x + 8$

2. $5(2x - 3) = x + 3$

3. $-\frac{2}{3}x + 1 = 7$

4. $x + 0.06x = 742$

5. $x - 0.03x = 0.97$

6. $6x - 7 = 0$

7. $\frac{1}{2}x - \frac{1}{3} = \frac{1}{4}x + \frac{1}{6}$

8. $2(x + 6) = 2x - 5$

9. $x + 7x = 8x$

Study Tip

Determine your readiness for an in-class test by working this sample test without looking at examples or notes. Give yourself one hour and grade your paper using the answers in the back of this book.

Solve for the indicated variable.

10. $2x - 3y = 9$ for y

11. $m = aP - w$ for a

For each graph write the corresponding inequality and the solution set to the inequality using interval notation.

12.

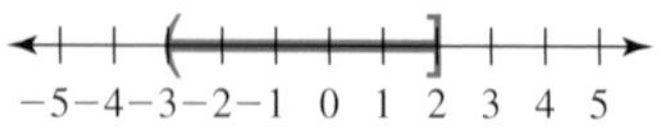

13.

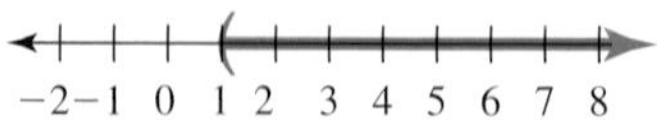

Solve each inequality. State the solution set in interval notation and sketch its graph.

14. $4 - 3(w - 5) < -2w$

15. $1 < \frac{1 - 2x}{3} < 5$

16. $1 < 3x - 2 < 7$

17. $-\frac{2}{3}y < 4$

Write a complete solution to each problem.

18. The perimeter of a rectangle is 72 meters. If the width is 8 meters less than the length, then what is the width of the rectangle?

19. If the area of a triangle is 54 square inches and the base is 12 inches, then what is the height?

20. How many liters of a 20% alcohol solution should Maria mix with 50 liters of a 60% alcohol solution to obtain a 30% solution?

21. Brandon gets a 40% discount on loose diamonds where he works. The cost of the setting is \$250. If he plans to spend at most \$1450, then what is the price range (list price) of the diamonds that he can afford?

22. If the degree measure of the smallest angle of a triangle is one-half of the degree measure of the second largest angle and one-third of the degree measure of the largest angle, then what is the degree measure of each angle?

Making Connections | A Review of Chapters 1–2

Simplify each expression.

1. $3x + 5x$

2. $3x \cdot 5x$

3. $\frac{4x + 2}{2}$

4. $5 - 4(3 - x)$

5. $3x + 8 - 5(x - 1)$

6. $(-6)^2 - 4(-3)2$

7. $3^2 \cdot 2^3$

8. $4(-7) - (-6)(3)$

9. $-2x \cdot x \cdot x$

10. $(-1)(-1)(-1)(-1)(-1)$

Perform the following operations.

11. $\frac{1}{2} + \frac{1}{6}$

12. $\frac{1}{2} - \frac{1}{3}$

13. $\frac{5}{3} \cdot \frac{1}{15}$

14. $\frac{2}{3} \cdot \frac{5}{6}$

15. $6 \cdot \left(\frac{5}{3} + \frac{1}{2}\right)$

16. $15\left(\frac{2}{3} - \frac{2}{15}\right)$

17. $4 \cdot \left(\frac{x}{2} + \frac{1}{4}\right)$

18. $12\left(\frac{5}{6}x - \frac{3}{4}\right)$

Find the solution set to each equation or inequality.

19. $x - \frac{1}{2} = \frac{1}{6}$

20. $x + \frac{1}{3} = \frac{1}{2}$

21. $x - \frac{1}{2} > \frac{1}{6}$

22. $x + \frac{1}{3} \le \frac{1}{2}$

23. $\frac{3}{5}x = \frac{1}{15}$

24. $\frac{3}{2}x = \frac{5}{6}$

25. $-\frac{3}{5}x \le \frac{1}{15}$

26. $-\frac{3}{2}x > \frac{5}{6}$

27. $\frac{5}{3}x + \frac{1}{2} = 1$

28. $\frac{2}{3}x - \frac{2}{15} = 2$

29. $\frac{x}{2} + \frac{1}{4} = \frac{1}{2}$

30. $\frac{5}{6}x - \frac{3}{4} = \frac{5}{12}$

31. $3x + 5x = 8$

32. $3x + 5x = 8x$

33. $3x + 5x = 7x$

34. $3x + 5 = 8$

35. $3x + 5x > 7x$

36. $3x + 5x > 8x$

37. $3x + 1 = 7$

38. $5 - 4(3 - x) = 1$

39. $3x + 8 = 5(x - 1)$

40. $x - 0.05x = 190$

Solve the problem.

41. ***Linear Depreciation.*** In computing income taxes, a company is allowed to depreciate a \$20,000 computer system over five years. Using *linear depreciation,* the value V of the computer system at any year t from 0 through 5 is given by

$$V = C - \frac{(C - S)}{5}t,$$

where C is the initial cost of the system and S is the scrap value of the system.

a) What is the value of the computer system after two years if its scrap value is \$4000?

b) If the value of the system after three years is claimed to be \$14,000, then what is the scrap value of the company's system?

c) If the accompanying graph models the depreciation of the system, then what is the scrap value of the system?

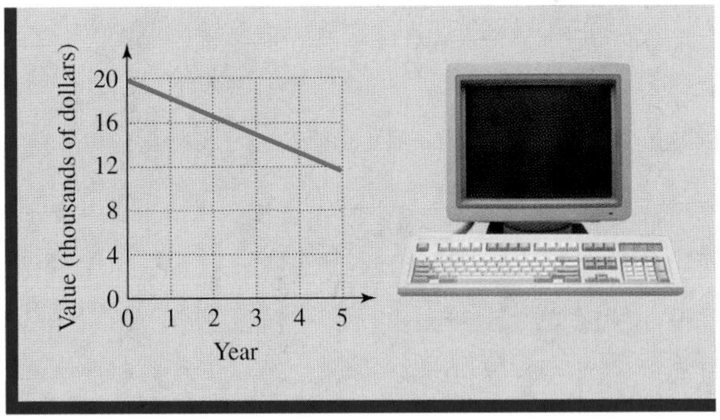

Figure for Exercise 41

Study Tip

Don't wait until the final exam to review material. Do some review on a regular basis. The Making Connections exercises on this page can be used to review, compare, and contrast different concepts that you have studied. A good time to work these exercises is between a test and the start of new material.

Critical **Thinking** | For Individual or Group Work | Chapter 2

These exercises can be solved by a variety of techniques, which may or may not require algebra. So be creative and think critically. Explain all answers. Answers are in the Instructor's Edition of this text.

1. ***Visible squares.*** How many squares are visible in each of the following diagrams?

a)

b)

c)
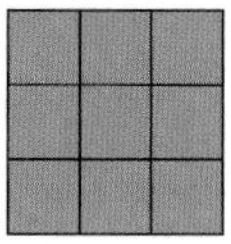

d)
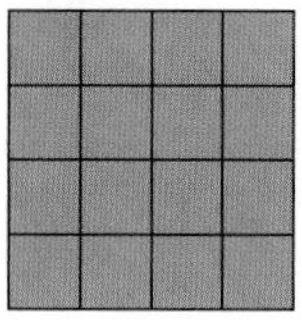

2. ***Baker's dilemma.*** A baker needs 8 cups of flour. He sends his apprentice to the flour bin with a scoop that holds 6 cups and a scoop that holds 11 cups. How can the apprentice measure 8 cups of flour with these scoops?

Photo for Exercise 2

3. ***Totaling one hundred.*** Start with the sequence of digits 987654321. Place any number of plus or minus signs between the digits in the sequence so that the value of the resulting expression is 100. For example,

$$98 + 7 - 6 - 5 + 4 + 3 - 2 + 1 = 100.$$

4. ***Four threes.*** Check out these equations:

$$\frac{3 \cdot 3}{3 \cdot 3} = 1, \frac{3}{3} + \frac{3}{3} = 2, (3 - 3)3 + 3 = 3.$$

Using exactly four 3's write arithmetic expressions whose values are 4, 5, 6, and so on. How far can you go?

5. ***Palindrome time.*** A palindrome is a sequence of words or numbers the reads the same forward or backward. For example, "A TOYOTA" is a palindrome and 14341 is a palindromic number. How many times per day does a digital clock display a palindromic number? Of course the answer depends on the format in which the digital clock displays the time. First, state precisely the type of digital clock display you are using, then count the palindromic numbers for that type of display.

6. ***Reversible products.*** Find the product of 32 and 46. Now reverse the digits and find the product of 23 and 64. The products are the same. Does this happen with any pair of two-digit numbers? Find two other pairs of two-digit numbers (with different digits) that have this property.

7. ***Running late.*** Alice, Bea, Carl, and Don all have an 8 o'clock class. Alice's watch is 8 minutes fast, but she thinks it is 4 minutes slow. Bea's watch is 8 minutes slow, but she thinks it is 8 minutes fast. Carl's watch is 4 minutes slow, but he thinks it is 8 minutes fast. Don's watch is 4 minutes fast, but he thinks it is 8 minutes slow. Each student leaves so they will get to class at exactly 8 o'clock. Each student assumes the correct time is what they think it is by their watch. Who is late to class and by how much?

8. ***Automorphic numbers.*** Automorphic numbers are integers whose squares end in the given integer. Since $1^2 = 1$ and $6^2 = 36$, both 1 and 6 are automorphic. Find the next four automorphic numbers.

Chapter 3

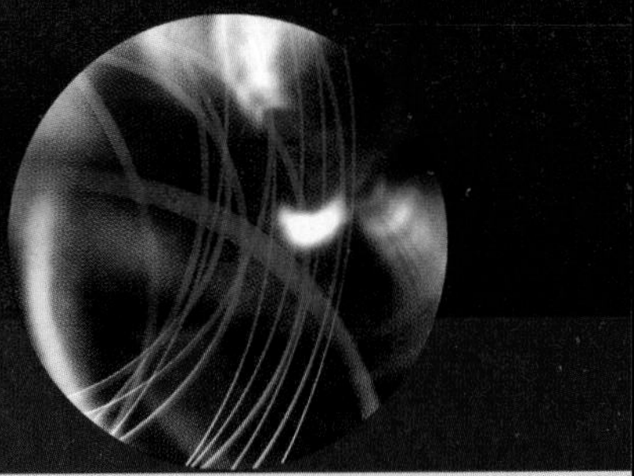

Linear Equations in Two Variables and Their Graphs

- **3.1 Graphing Lines in the Coordinate Plane**
- **3.2 Slope**
- **3.3 Equations of Lines in Slope-Intercept Form**
- **3.4 The Point-Slope Form**
- **3.5 Variation**

If you pick up any package of food and read the label, you will find a long list that usually ends with some mysterious looking names. Many of these strange elements are food additives. A food additive is a substance or a mixture of substances other than basic foodstuffs that is present in food as a result of production, processing, storage, or packaging. They can be natural or synthetic and are categorized in many ways: preservatives, coloring agents, processing aids, and nutritional supplements, to name a few.

Food additives have been around since prehistoric humans discovered that salt would help to preserve meat. Today, food additives can include simple ingredients such as red color from Concord grape skins, calcium, or an enzyme. Throughout the centuries there have been lively discussions on what is healthy to eat. At the present time the food industry is working to develop foods that have less cholesterol, fats, and other unhealthy ingredients.

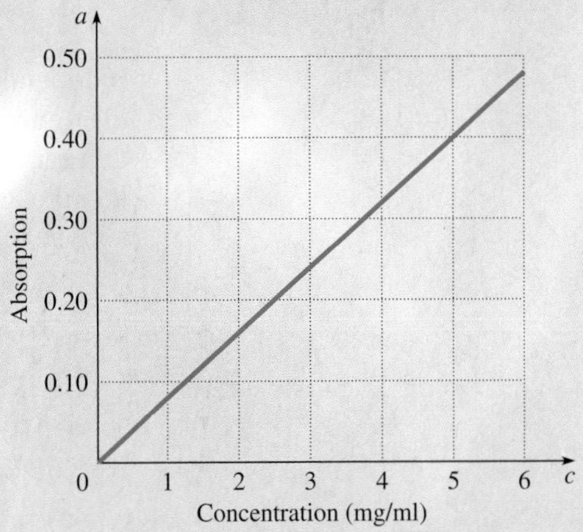

Although they frequently have different viewpoints, the food industry and the Food and Drug Administration (FDA) are working to provide consumers with information on a healthier diet. Recent developments such as the synthetically engineered tomato stirred great controversy, even though the FDA declared the tomato safe to eat.

In Exercise 91 of Section 3.4 you will see how a food chemist uses a linear equation in testing the concentration of an enzyme in a fruit juice.

3.1 Graphing Lines in the Coordinate Plane

In this Section

- Ordered Pairs
- The Rectangular Coordinate System
- Plotting Points
- Graphing a Linear Equation
- Graphing a Line Using Intercepts
- Applications

In Chapter 1 you learned to graph numbers on a number line. We also used number lines to illustrate the solution to inequalities in Chapter 2. In this section you will learn to graph pairs of numbers in a coordinate system made up of a pair of number lines. We will use this coordinate system to illustrate the solution to equations and inequalities in two variables.

Ordered Pairs

The equation $y = 2x - 1$ is an equation in two variables. This equation is satisfied if we choose a value for x and a value for y that make it true. If we choose $x = 2$ and $y = 3$, then $y = 2x - 1$ becomes

$$\overset{y}{\downarrow}\qquad\overset{x}{\downarrow}$$
$$3 = 2(2) - 1.$$
$$3 = 3$$

Because the last statement is true, we say that the pair of numbers 2 and 3 **satisfies the equation** or is a **solution to the equation.** We use the **ordered pair** (2, 3) to represent $x = 2$ and $y = 3$. The format is to always write the value for x first and the value for y second. The numbers in an ordered pair are called **coordinates.** In the pair (2, 3) the first coordinate or x-coordinate is 2 and the second coordinate or y-coordinate is 3. Note that the ordered pair (3, 2) does not satisfy the equation $y = 2x - 1$, because for $x = 3$ and $y = 2$ we have

$$2 \neq 2(3) - 1.$$

Helpful Hint

In this chapter you will be doing a lot of graphing. Using graph paper will help you understand the concepts and help you recognize errors. For your convenience, a page of graph paper can be found on page 231 of this text. Make as many copies of it as you wish.

The variable corresponding to the first coordinate of an ordered pair is called the **independent variable** and the variable corresponding to the second coordinate is called the **dependent variable.** We think of the value for the first coordinate as being selected arbitrarily and the value for the second coordinate as being determined from the first coordinate by a rule such as $y = 2x - 1$. Of course, if the ordered pair must satisfy a simple equation, then we can find either coordinate when given the other coordinate.

EXAMPLE 1

Finding solutions to an equation

Each of the ordered pairs below is missing one coordinate. Complete each ordered pair so that it satisfies the equation $y = -3x + 4$.

a) (2,) **b)** (, −5) **c)** (0,)

Solution

a) The x-coordinate of (2,) is 2. Let $x = 2$ in the equation $y = -3x + 4$:

$$\begin{aligned} y &= -3 \cdot 2 + 4 \\ &= -6 + 4 \\ &= -2 \end{aligned}$$

The ordered pair (2, −2) satisfies the equation.

b) The y-coordinate of (, −5) is −5. Let $y = -5$ in the equation $y = -3x + 4$:

$$-5 = -3x + 4$$
$$-9 = -3x$$
$$3 = x$$

The ordered pair (3, −5) satisfies the equation.

c) Replace x by 0 in the equation $y = -3x + 4$:

$$y = -3 \cdot 0 + 4 = 4$$

So (0, 4) satisfies the equation.

Now do Exercises 7–16

The Rectangular Coordinate System

We use the **rectangular** (or **Cartesian**) **coordinate system** to get a visual image of ordered pairs of real numbers. The rectangular coordinate system consists of two number lines drawn at a right angle to one another, intersecting at zero on each number line, as shown in Fig. 3.1. The plane containing these number lines is called the **coordinate plane.** On the horizontal number line the positive numbers are to the right of zero, and on the vertical number line the positive numbers are above zero.

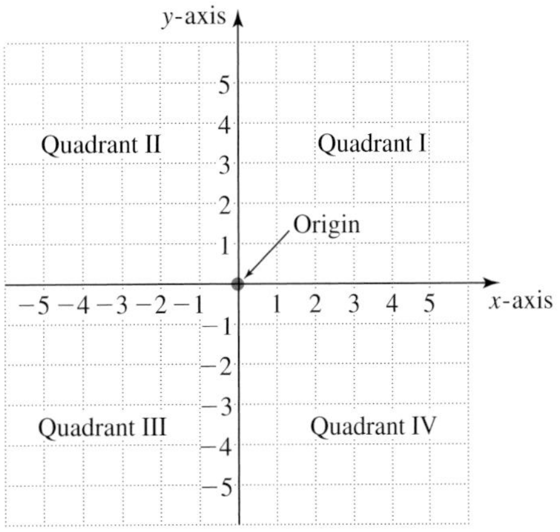

Figure 3.1

The horizontal number line is called the ***x*-axis,** and the vertical number line is called the ***y*-axis.** The point at which they intersect is called the **origin.** The two number lines divide the plane into four regions called **quadrants.** They are numbered as shown in Fig. 3.1. The quadrants do not include any points on the axes.

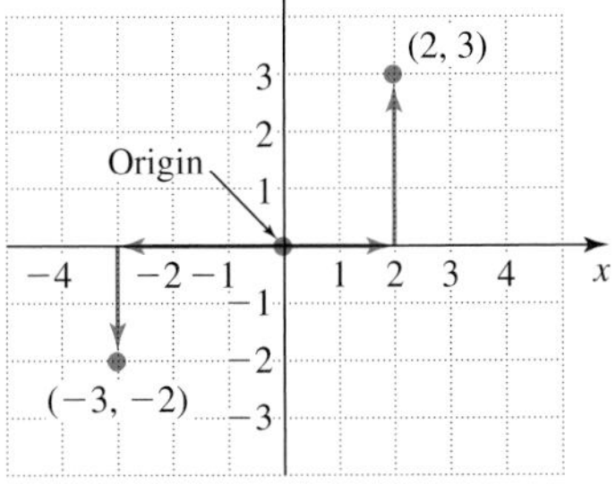

Figure 3.2

Plotting Points

Just as every real number corresponds to a point on the number line, *every pair of real numbers corresponds to a point in the rectangular coordinate system.* For example, the point corresponding to the pair (2, 3) is found by starting at the origin and moving two units to the right and then three units up. The point corresponding to the pair (−3, −2) is found by starting at the origin and moving three units to the left and then two units down. Both of these points are shown in Fig. 3.2.

When we locate a point in the rectangular coordinate system, we are **plotting** or **graphing** the point. Because ordered pairs of numbers correspond to points in the coordinate plane, we frequently refer to an ordered pair as a point.

EXAMPLE 2

Plotting points

Plot the points (2, 5), (−1, 4), (−3, −4), and (3, −2).

Solution

To locate (2, 5), start at the origin, move two units to the right, and then move up five units. To locate (−1, 4), start at the origin, move one unit to the left, and then move up four units. All four points are shown in Fig. 3.3.

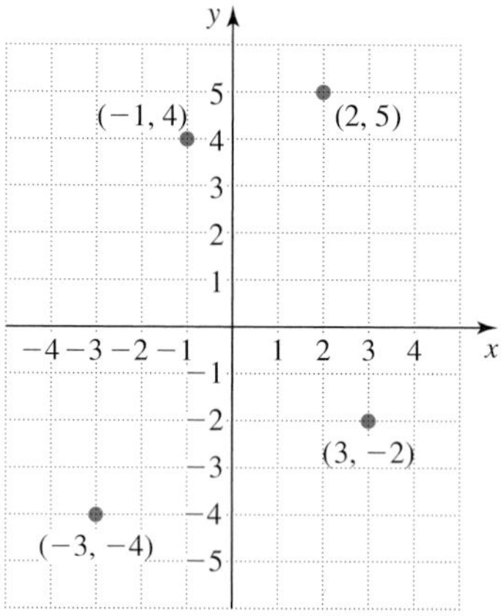

Figure 3.3

Now do Exercises 17–32

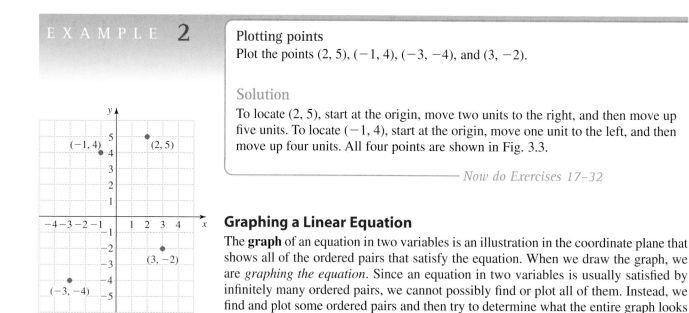

Graphing a Linear Equation

The **graph** of an equation in two variables is an illustration in the coordinate plane that shows all of the ordered pairs that satisfy the equation. When we draw the graph, we are *graphing the equation.* Since an equation in two variables is usually satisfied by infinitely many ordered pairs, we cannot possibly find or plot all of them. Instead, we find and plot some ordered pairs and then try to determine what the entire graph looks like from the few ordered pairs that we have plotted.

EXAMPLE 3

Graphing an equation

Graph the equation $y = 2x - 1$ in the coordinate plane.

Solution

To find ordered pairs that satisfy $y = 2x - 1$, we arbitrarily select some x-coordinates and calculate the corresponding y-coordinates:

If $x = -3$, then $y = 2(-3) - 1 = -7$.
If $x = -2$, then $y = 2(-2) - 1 = -5$.
If $x = -1$, then $y = 2(-1) - 1 = -3$.
If $x = 0$, then $y = 2(0) - 1 = -1$.
If $x = 1$, then $y = 2(1) - 1 = 1$.
If $x = 2$, then $y = 2(2) - 1 = 3$.
If $x = 3$, then $y = 2(3) - 1 = 5$.

We can make a table for these results as follows:

x	−3	−2	−1	0	1	2	3
$y = 2x - 1$	−7	−5	−3	−1	1	3	5

Calculator Close-Up

You can make a table of values for x and y with a graphing calculator. Enter the equation $y = 2x - 1$ using $Y =$ and then press TABLE.

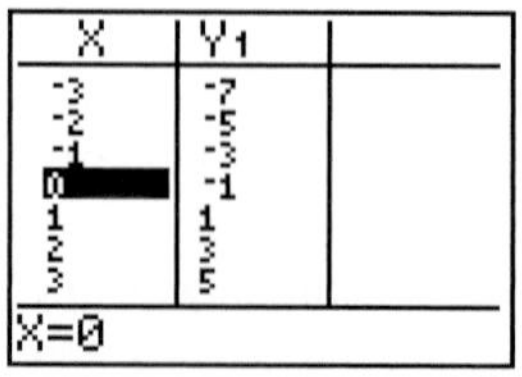

The ordered pairs $(-3, -7)$, $(-2, -5)$, $(-1, -3)$, $(0, -1)$, $(1, 1)$, $(2, 3)$, and $(3, 5)$ are graphed in Fig. 3.4. Notice that the points lie in a straight line. If we choose any real number for x and find the point that satisfies $y = 2x - 1$, we get another point on this line. Likewise, any point on this line satisfies the equation. So the graph of $y = 2x - 1$ is the straight line in Fig. 3.5. The arrows on the ends of the line indicate that it goes indefinitely in both directions.

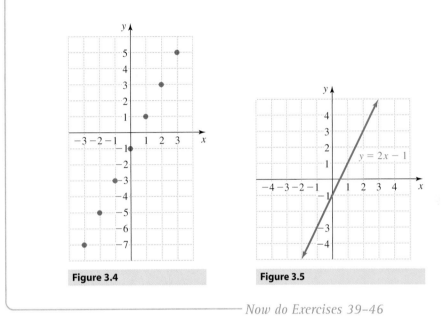

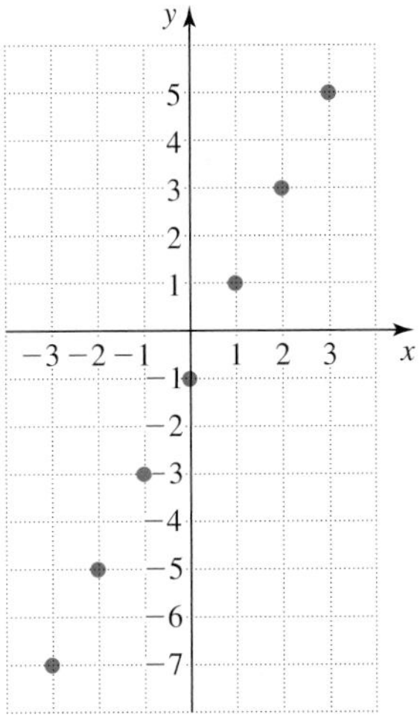

Figure 3.4

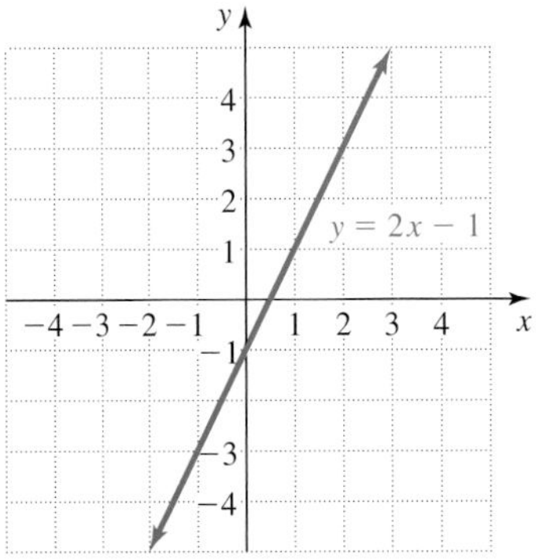

Figure 3.5

Now do Exercises 39–46

Because its graph is a straight line, the equation $y = 2x - 1$ in Example 3 is called a *linear equation in two variables.*

Linear Equation in Two Variables

A **linear equation in two variables** is an equation that can be written in the form

$$Ax + By = C,$$

where A, B, and C are real numbers, with A and B not both equal to zero.

Equations such as

$$x - y = 5, \quad y = 2x + 3, \quad 2x - 5y - 9 = 0, \quad \text{and} \quad x = 8$$

are linear equations because they could all be rewritten in the form $Ax + By = C$. The graph of any linear equation is a straight line.

EXAMPLE 4

Graphing an equation

Graph the equation $3x + y = 2$. Plot at least five points.

Solution

It is easier to make a table of ordered pairs if the equation is solved for y. So subtract $3x$ from each side to get $y = -3x + 2$. Now select some values for x and then calculate the corresponding y-coordinates:

If $x = -2$, then $y = -3(-2) + 2 = 8$.
If $x = -1$, then $y = -3(-1) + 2 = 5$.
If $x = 0$, then $y = -3(0) + 2 = 2$.
If $x = 1$, then $y = -3(1) + 2 = -1$.
If $x = 2$, then $y = -3(2) + 2 = -4$.

The following table shows these five ordered pairs:

x	-2	-1	0	1	2
$y = -3x + 2$	8	5	2	-1	-4

The graph of the line through these points is shown in Fig. 3.6.

Now do Exercises 47–50

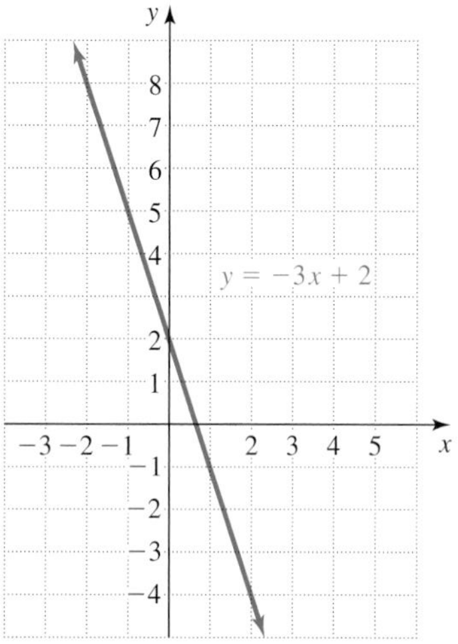

Figure 3.6

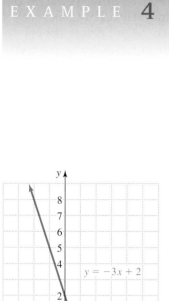

Calculator Close-Up

To graph $y = -3x + 2$, enter the equation using the Y = key:

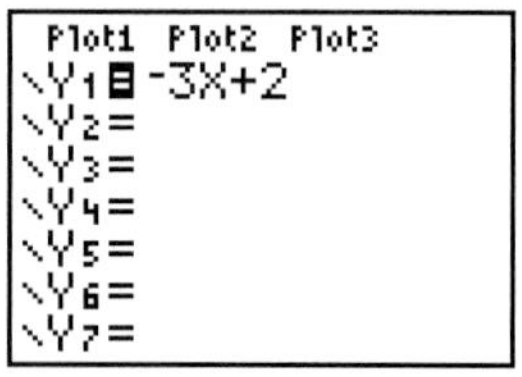

Next, set the viewing window (WINDOW) to get the desired view of the graph. Xmin and Xmax indicate the minimum and maximum x-values used for the graph; likewise for Ymin and Ymax. Xscl and Yscl (scale) give

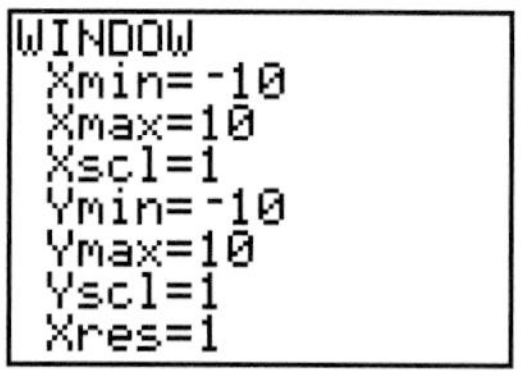

the distance between tick marks on the respective axes.

Press GRAPH to get the graph:

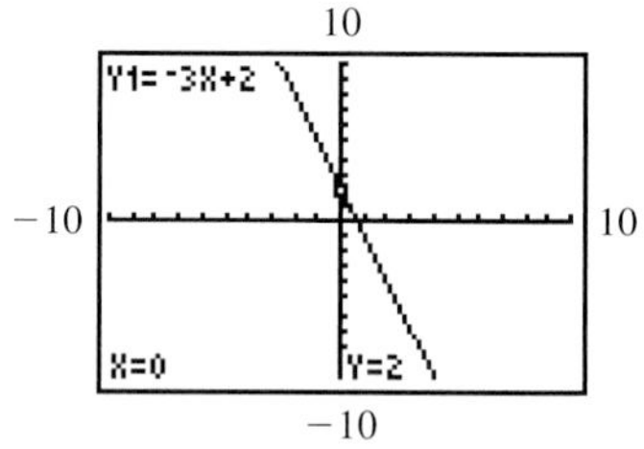

Even though the graph is not really "straight," it is consistent with the graph of $y = -3x + 2$ in Fig. 3.6.

In the linear equation $Ax + By = C$ either A or B could be zero. For example, $0 \cdot x + y = 4$ is a linear equation. Since x is multiplied by 0 this equation is usually written simply as $y = 4$. To graph $y = 4$ in the coordinate plane, we must understand that it comes from $0 \cdot x + y = 4$.

EXAMPLE 5

Horizontal and vertical lines

Graph each linear equation.

a) $y = 4$ **b)** $x = 3$

Solution

a) The equation $y = 4$ is a simplification of $0 \cdot x + y = 4$. So if y is replaced with 4, then we can use any real number for x. For example, $(-1, 4)$ satisfies $0 \cdot x + y = 4$ because $0(-1) + 4 = 4$ is correct. The following table shows five ordered pairs that satisfy $y = 4$.

x	−2	−1	0	1	2
y = 4	4	4	4	4	4

Figure 3.7 shows a horizontal line through these points.

b) The equation $x = 3$ is a simplification of $x + 0 \cdot y = 3$. So if x is replaced with 3, then we can use any real number for y. For example, $(3, -2)$ satisfies $x + 0 \cdot y = 3$ because $3 + 0(-2) = 3$ is correct. The following table shows five ordered pairs that satisfy $x = 3$.

x = 3	3	3	3	3	3
y	−2	−1	0	1	2

Figure 3.8 shows a vertical line through these points.

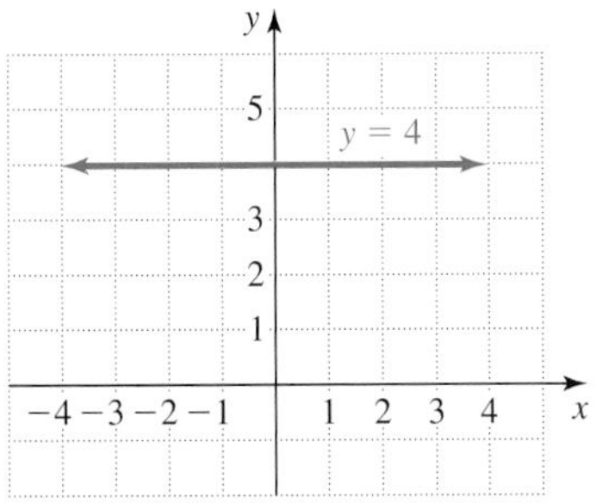

Figure 3.7

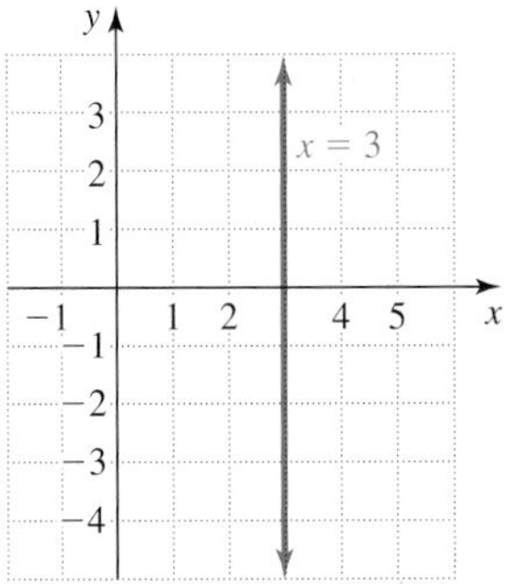

Figure 3.8

Now do Exercises 51–62

Calculator Close-Up

You cannot graph the vertical line $x = 3$ on most graphing calculators. The only equations that can be graphed are ones in which y is written in terms of x.

CAUTION If an equation such as $x = 3$ is discussed in the context of equations in two variables, then we assume that it is a simplified form of $x + 0 \cdot y = 3$, and there are infinitely many ordered pairs that satisfy the equation. If the equation $x = 3$ is discussed in the context of equations in a single variable, then $x = 3$ has only one solution, 3.

All of the equations we have considered so far have involved single-digit numbers. If an equation involves large numbers, then we must change the scale on the x-axis, the y-axis, or both to accommodate the numbers involved. The change of scale is arbitrary, and the graph will look different for different scales.

EXAMPLE 6

Adjusting the scale

Graph the equation $y = 20x + 500$. Plot at least five points.

Solution

The following table shows five ordered pairs that satisfy the equation.

x	-20	-10	0	10	20
$y = 20x + 500$	100	300	500	700	900

To fit these points onto a graph, we change the scale on the x-axis to let each division represent 10 units and change the scale on the y-axis to let each division represent 200 units. The graph is shown in Fig. 3.9.

Now do Exercises 75–80

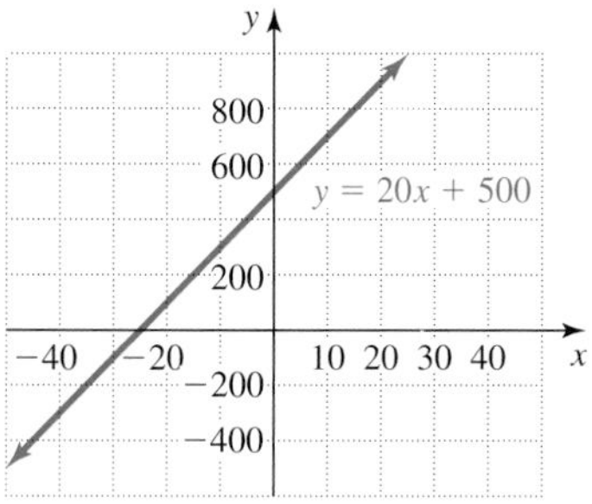

Figure 3.9

Graphing a Line Using Intercepts

We know that the graph of a linear equation is a straight line. Because it takes only two points to determine a line, we can graph a linear equation using only two points. The two points that are the easiest to locate are usually the points where the line crosses the axes. The point where the graph crosses the x-axis is the ***x*-intercept.** The y-coordinate of the x-intercept is zero. The point where the graph crosses the y-axis is the ***y*-intercept.** The x-coordinate of the y-intercept is zero.

EXAMPLE 7

Graphing a line using intercepts

Graph the equation $2x - 3y = 6$ by using the x- and y-intercepts.

Helpful Hint

You can find the intercepts for $2x - 3y = 6$ using the *cover-up method.* Cover up $-3y$ with your pencil, then solve $2x = 6$ mentally to get $x = 3$ and an x-intercept of (3, 0). Now cover up $2x$ and solve $-3y = 6$ to get $y = -2$ and a y-intercept of $(0, -2)$.

Solution

To find the x-intercept, let $y = 0$ in the equation $2x - 3y = 6$:

$$2x - 3 \cdot 0 = 6$$
$$2x = 6$$
$$x = 3$$

The x-intercept is (3, 0). To find the y-intercept, let $x = 0$ in $2x - 3y = 6$:

$$2 \cdot 0 - 3y = 6$$
$$-3y = 6$$
$$y = -2$$

The y-intercept is $(0, -2)$. Locate the intercepts and draw a line through them as shown in Fig. 3.10. To check, find one additional point that satisfies the equation, say (6, 2), and see whether the line goes through that point.

Calculator Close-Up

To check the result in Example 6, graph $y = (2/3)x - 2$:

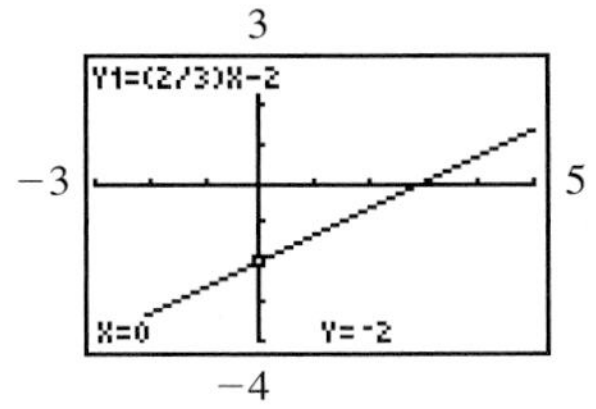

Since the calculator graph appears to be the same as the graph in Fig. 3.10, it supports the conclusion that Fig. 3.10 is correct.

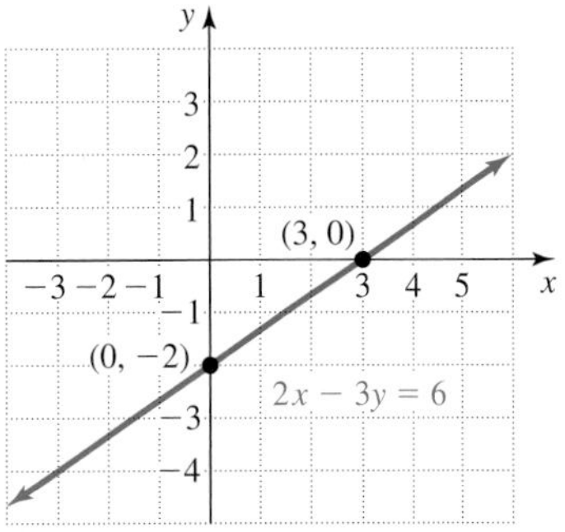

Figure 3.10

Now do Exercises 81–88

Applications

Linear equations occur in many real-life situations. If the cost of plans for a house is \$475 for one copy plus \$30 for each additional copy, then $C = 475 + 30x$, where x is the number of additional copies. If you have \$1000 budgeted for landscaping with trees at \$50 each and bushes at \$20 each, then $50t + 20b = 1000$, where t is the number of trees and b is the number of bushes. In the next example we see a linear equation that models ticket demand.

EXAMPLE 8

Ticket demand

The demand for tickets to see the Ice Gators play hockey can be modeled by the equation $d = 8000 - 100p$, where d is the number of tickets sold and p is the price per ticket in dollars.

a) How many tickets will be sold at \$20 per ticket?

b) Find the intercepts and interpret them.

c) Graph the linear equation.

d) What happens to the demand as the price increases?

Solution

a) If tickets are \$20 each, then $d = 8000 - 100 \cdot 20 = 6000$. So at \$20 per ticket, the demand will be 6000 tickets.

b) Replace d with 0 in the equation $d = 8000 - 100p$ and solve for p:

$$0 = 8000 - 100p$$
$$100p = 8000 \quad \text{Add } 100p \text{ to each side.}$$
$$p = 80 \quad \text{Divide each side by 100.}$$

If $p = 0$, then $d = 8000 - 100 \cdot 0 = 8000$. So the intercepts are (0, 8000) and (80, 0). If the tickets are free, the demand will be 8000 tickets. At \$80 per ticket, no tickets will be sold.

c) Graph the line using the intercepts (0, 8000) and (80, 0) as shown in Fig. 3.11. The line is graphed in the first quadrant only, because negative values for demand or price are meaningless.

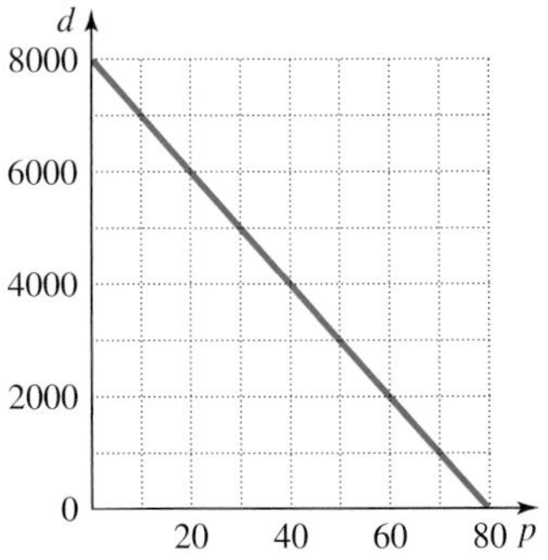

Figure 3.11

d) When the tickets are free, the demand is high. As the price increases, the demand goes down. At \$80 per ticket, there will be no demand.

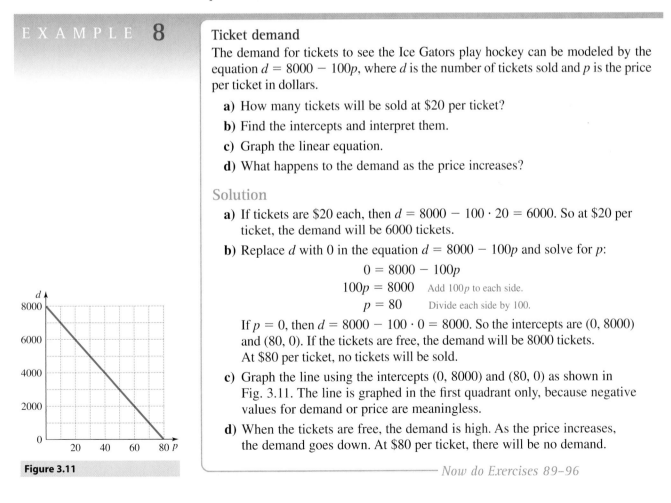

Now do Exercises 89–96

Math *at Work* Predicting the Future

No one knows what the future may bring, but everyone plans for and tries to predict the future. Stock market analysts predict the profits of companies, pollsters predict the outcomes of elections, and urban planners predict sizes of cities. These predictions of the future are often based on the trends of the past.

Consider the accompanying table, which shows the population of the United States in millions for each census year from 1950 through 2000. It certainly appears that the population is going up and it would be a safe bet to predict that the population in 2010 will be somewhat larger than 279 million. We get a different perspective if we look at the accompanying graph of the population data. Not only does the graph show an increasing population, it shows the population increasing in a linear manner. Now we can make a prediction based on the line that appears to fit the data. The equation of this line, the *regression line,* is $y = 2.47x - 4666$, where x is the year and y is the population. The equation of the regression line can be found with a computer or graphing calculator. Now if $x = 2010$, then $y = 2.47(2010) - 4666 \approx 299$. So we can predict 299 million people in 2010.

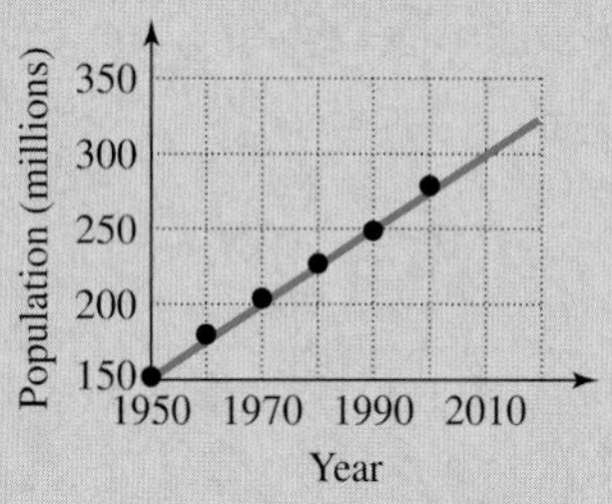

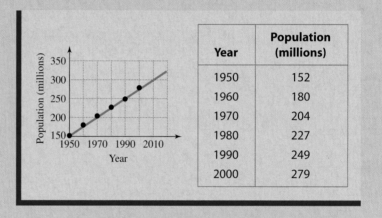

Year	Population (millions)
1950	152
1960	180
1970	204
1980	227
1990	249
2000	279

Warm-Ups ▼

True or false? Explain your answer.

1. The point (2, 4) satisfies the equation $2y - 3x = -8$.
2. If (1, 5) satisfies an equation, then (5, 1) also satisfies the equation.
3. The origin is in quadrant I.
4. The point (4, 0) is on the y-axis.
5. The graph of $x + 0 \cdot y = 9$ is the same as the graph of $x = 9$.
6. The graph of $x = -5$ is a vertical line.
7. The graph of $0 \cdot x + y = 6$ is a horizontal line.
8. The y-intercept for the line $x + 2y = 5$ is (5, 0).
9. The point (5, −3) is in quadrant II.
10. The point (−349, 0) is on the x-axis.

3.1 Exercises

Reading and Writing *After reading this section, write out the answers to these questions. Use complete sentences.*

1. What is an ordered pair?

2. What is the rectangular coordinate system?

3. What name is given to the point of intersection of the x-axis and the y-axis?

4. What is the graph of an equation?

5. What is a linear equation in two variables?

6. What are intercepts?

Complete each ordered pair so that it satisfies the given equation. See Example 1.

7. $y = 3x + 9$: (0,), (, 24), (2,)

8. $y = 2x + 5$: (8,), (−1,), (, −1)

9. $y = -3x - 7$: (0,), $\left(\frac{1}{3}, \quad\right)$, (, −5)

10. $y = -5x - 3$: (−1,), $\left(-\frac{1}{2}, \quad\right)$, (, −2)

11. $y = 1.2x + 54.3$: (0,), (10,), (, 54.9)

12. $y = 1.8x + 22.6$: (1,), (−10,), (, 22.6)

13. $2x - 3y = 6$: (3,), (, −2), (12,)

14. $3x + 5y = 0$: (−5,), (, −3), (10,)

15. $0 \cdot y + x = 5$: (, −3), (, 5), (, 0)

16. $0 \cdot x + y = -6$: (3,), (−1,), (4,)

Plot the points on a rectangular coordinate system. See Example 2.

17. (1, 5)

18. (4, 3)

19. (−2, 1)

20. (−3, 5)

21. $\left(3, -\frac{1}{2}\right)$

22. $\left(2, -\frac{1}{3}\right)$

23. (−2, −4)

24. (−3, −5)

25. (0, 3)

26. (0, −2)

27. (−3, 0)

28. (5, 0)

29. $(\pi, 1)$

30. $(-2, \pi)$

31. (1.4, 4)

32. (−3, 0.4)

Use the given equations to find the missing coordinates in the following tables.

33. $y = -2x + 5$

x	y
−2	
0	
2	
	−3
	−7

34. $y = -x + 4$

x	y
−2	
0	
2	
	0
	−2

35. $y = \frac{1}{3}x + 2$

x	y
−6	
−3	
	2
	3

36. $y = -\frac{1}{2}x + 1$

x	y
−2	
−1	
	1
	$\frac{1}{2}$

37. $y - 20x = 400$

x	y
−30	
	0
−10	
0	
	600

38. $200x + y = 50$

x	y
$-\frac{1}{2}$	
	100
0	
	0
$\frac{1}{2}$	

Graph each equation. Plot at least five points for each equation. Use graph paper. See Examples 3–5. If you have a graphing calculator, use it to check your graphs when possible.

Can use 3

39. $y = x + 1$

40. $y = x - 1$

41. $y = 2x + 1$

42. $y = 3x - 1$

43. $y = 3x - 2$

44. $y = 2x + 3$

45. $y = x$

46. $y = -x$

47. $y = 1 - x$

48. $y = 2 - x$

49. $y = -2x + 3$

50. $y = -3x + 2$

51. $y = -3$

52. $y = 2$

53. $x = 2$

54. $x = -4$

55. $2x + y = 5$

56. $3x + y = 5$

57. $x + 2y = 4$

58. $x - 2y = 6$

59. $x - 3y = 6$

60. $x + 4y = 5$

61. $y = 0.36x + 0.4$

62. $y = 0.27x - 0.42$

For each point, name the quadrant in which it lies or the axis on which it lies.

63. $(-3, 45)$

64. $(-33, 47)$

65. $(-3, 0)$

66. $(0, -9)$

67. $(-2.36, -5)$

68. $(89.6, 0)$

69. $(3.4, 8.8)$

70. $(4.1, 44)$

71. $\left(-\frac{1}{2}, 50\right)$

72. $\left(-6, -\frac{1}{2}\right)$

73. $(0, -99)$

74. $(\pi, 0)$

Graph each equation. Plot at least five points for each equation. Use graph paper. See Example 6. If you have a graphing calculator, use it to check your graphs.

75. $y = x + 1200$

76. $y = 2x - 3000$

77. $y = 50x - 2000$

78. $y = -300x + 4500$

79. $y = -400x + 2000$

80. $y = 500x + 3$

For each equation, state the x-intercept and y-intercept. Then graph the equation using the intercepts and a third point. See Example 7.

81. $3x + 2y = 6$

82. $2x + y = 6$

83. $x - 4y = 4$

84. $-2x + y = 4$

85. $y = \frac{3}{4}x - 9$

86. $y = -\frac{1}{2}x + 5$

87. $\frac{1}{2}x + \frac{1}{4}y = 1$

88. $\frac{1}{3}x - \frac{1}{2}y = 3$

Solve each problem. See Example 8.

89. ***Percentage of full benefit.*** The age at which you retire affects your Social Security benefits. The accompanying graph gives the percentage of full benefit for each age from 62 through 70, based on current legislation and retirement after the year 2005 (Source: Social Security Administration). What percentage of full benefit does a person receive if that person retires at age 63? At what age will a retiree receive the full benefit? For what ages do you receive more than the full benefit?

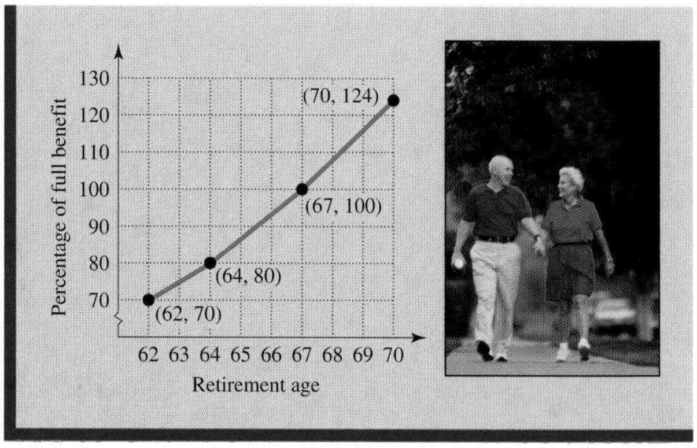

Figure for Exercise 89

90. ***Heel motion.*** When designing running shoes, Chris Edington studies the motion of a runner's foot. The following data gives the coordinates of the heel (in centimeters) at intervals of 0.05 millisecond during one cycle of level treadmill running at 3.8 meters per second (*Sagittal Plane Kinematics, Milliron and Cavanagh*):

(31.7, 5.7), (48.0, 5.7), (68.3, 5.8), (88.9, 6.9),
(107.2, 13.3), (119.4, 24.7), (127.2, 37.8),
(125.7, 52.0), (116.1, 60.2), (102.2, 59.5),
(88.7, 50.2), (73.9, 35.8), (52.6, 20.6),
(29.6, 10.7), (22.4, 5.9)

Graph these ordered pairs to see the heel motion.

91. ***Medicaid spending.*** The cost in billions of dollars for federal Medicaid (health care for the poor) can be modeled by the equation

$$C = 3.2n + 65.3,$$

where n is the number of years since 1990 (Health Care Financing Administration, www.hcfa.gov).

a) What was the cost of federal Medicaid in 2000?
b) In what year will the cost reach \$150 billion?
c) Graph the equation for n ranging from 0 through 20.

92. ***Dental services.*** The national cost C in billions of dollars for dental services can be modeled by the linear equation

$$C = 2.85n + 30.52,$$

where n is the number of years since 1990 (Health Care Financing Administration, www.hcfa.gov).

a) Find and interpret the C-intercept for the line.
b) Find and interpret the n-intercept for the line.
c) Graph the line for n ranging from 0 through 20.
d) If this trend continues, then in what year will the cost of dental services reach 100 billion?

93. ***Hazards of depth.*** The accompanying table shows the depth below sea level and atmospheric pressure (*Encyclopedia of Sports Science,* 1997). The equation

$$A = 0.03d + 1$$

expresses the atmospheric pressure in terms of the depth d.

a) Find the atmospheric pressure at the depth where nitrogen narcosis begins.
b) Find the maximum depth for intermediate divers.
c) Graph the equation for d ranging from 0 to 250 feet.

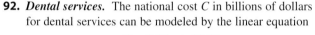

Depth (ft)	Atmospheric Pressure (atm)	Comments
21	1.63	Bends are a danger
60	2.8	Maximum for beginners
100		Nitrogen narcosis begins
	4.9	Maximum for intermediate
200	7.0	Severe nitrogen narcosis
250	8.5	Extremely dangerous depth

Figure for Exercise 93

94. ***Demand equation.*** Helen's Health Foods usually sells 400 cans of ProPac Muscle Punch per week when the price is \$5 per can. After experimenting with prices for some time, Helen has determined that the weekly demand can be found by using the equation

$$d = 600 - 40p,$$

where d is the number of cans and p is the price per can.

a) Will Helen sell more or less Muscle Punch if she raises her price from \$5?
b) What happens to her sales every time she raises her price by \$1?

c) Graph the equation.

d) What is the maximum price that she can charge and still sell at least one can?

95. ***Advertising blitz.*** Furniture City in Toronto had \$24,000 to spend on advertising a year-end clearance sale. A 30-second radio ad costs \$300, and a 30-second local television ad costs \$400. To model this situation, the advertising manager wrote the equation $300x + 400y = 24{,}000$. What do x and y represent? Graph the equation. How many solutions are there to the equation, given that the number of ads of each type must be a whole number?

96. ***Material allocation.*** A tent maker had 4500 square yards of nylon tent material available. It takes 45 square yards of nylon to make an 8×10 tent and 50 square yards to make a 9×12 tent. To model this situation, the manager wrote the equation $45x + 50y = 4500$. What do x and y represent? Graph the equation. How many solutions are there to the equation, given that the number of tents of each type must be a whole number?

Graphing Calculator Exercises

Graph each straight line on your graphing calculator using a viewing window that shows both intercepts. Answers may vary.

97. $2x + 3y = 1200$

98. $3x - 700y = 2100$

99. $200x - 300y = 6$

100. $300x + 5y = 20$

101. $y = 300x - 1$

102. $y = 300x - 6000$

3.2 Slope

In this Section

- Slope Concepts
- Slope Using Coordinates
- Graphing a Line Given a Point and Its Slope
- Parallel Lines
- Perpendicular Lines
- Interpreting Slope

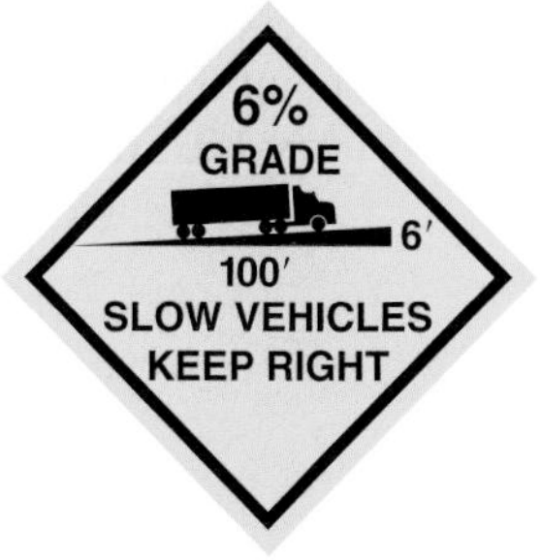

Figure 3.12

In Section 3.1 you learned that the graph of a linear equation is a straight line. In this section, we will continue our study of lines in the coordinate plane.

Slope Concepts

If a highway rises 6 feet in a horizontal run of 100 feet, then the grade is $\frac{6}{100}$ or 6%. See Fig. 3.12. The grade of a road is a measurement of the steepness of the road. It is the rate at which the road is going upward.

The steepness of a line is called the **slope** of the line and it is measured like the grade of a road. As you move from (1, 1) to (4, 3) in Fig. 3.13 the x-coordinate increases by 3 and the y-coordinate increases by 2. The line rises 2 units in a horizontal run of 3 units. So the slope of the line is $\frac{2}{3}$. The slope is the rate at which the y-coordinate is increasing. It increases 2 units for every 3-unit increase in x or it increases $\frac{2}{3}$ of a unit for every 1-unit increase in x. In general, we have the following definition of slope.

Slope

$$\text{Slope} = \frac{\text{change in } y\text{-coordinate}}{\text{change in } x\text{-coordinate}}$$

If we move from the point (4, 3) to the point (1, 1), there is a change of -2 in the y-coordinate and a change of -3 in the x-coordinate. See Fig. 3.14. In this case we get

$$\text{Slope} = \frac{-2}{-3} = \frac{2}{3}.$$

Note that going from (4, 3) to (1, 1) gives the same slope as going from (1, 1) to (4, 3).

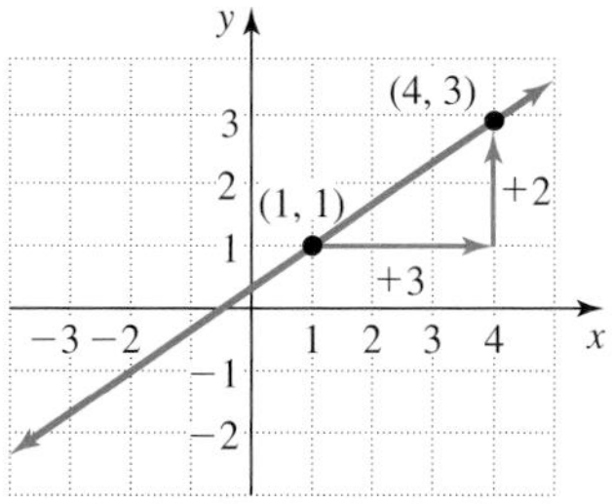

Figure 3.13

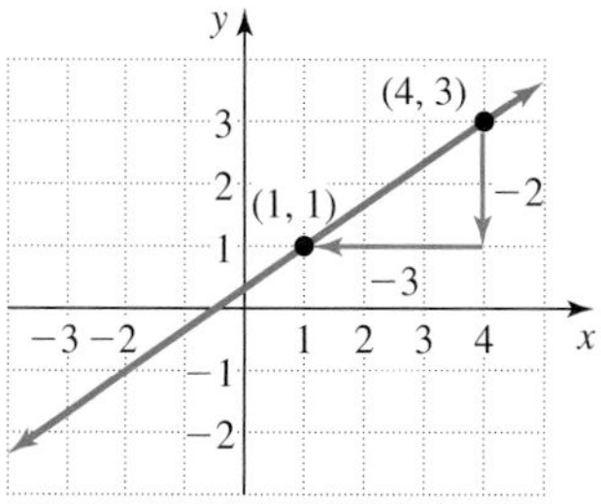

Figure 3.14

We call the change in y-coordinate the **rise** and the change in x-coordinate the **run.** Moving up is a positive rise, and moving down is a negative rise. Moving to the right is a positive run, and moving to the left is a negative run. We usually use the letter m to stand for slope. So we have

$$m = \frac{\text{change in } y}{\text{change in } x} = \frac{\text{rise}}{\text{run}}.$$

EXAMPLE 1

Finding the slope of a line

Find the slopes of the given lines by going from point A to point B.

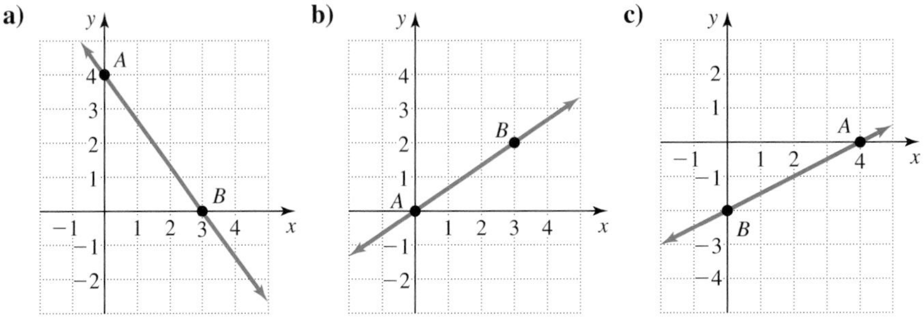

Study Tip

Working problems 1 hour per day every day of the week is better than working problems for 7 hours on one day of the week. It is usually better to spread out your study time than to try and learn everything in one big session.

Solution

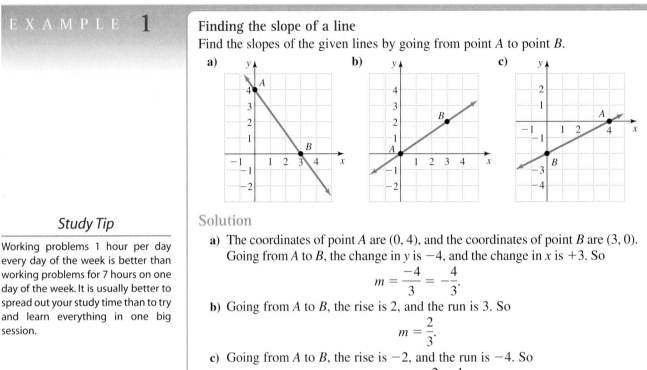

a) The coordinates of point A are $(0, 4)$, and the coordinates of point B are $(3, 0)$. Going from A to B, the change in y is -4, and the change in x is $+3$. So

$$m = \frac{-4}{3} = -\frac{4}{3}.$$

b) Going from A to B, the rise is 2, and the run is 3. So

$$m = \frac{2}{3}.$$

c) Going from A to B, the rise is -2, and the run is -4. So

$$m = \frac{-2}{-4} = \frac{1}{2}.$$

Now do Exercises 7–10

CAUTION The change in y is always in the numerator, and the change in x is always in the denominator.

The ratio of rise to run is the ratio of the lengths of the two legs of any right triangle whose hypotenuse is on the line. As long as one leg is vertical and the other is horizontal, all such triangles for a certain line have the same shape. These triangles are similar triangles. The ratio of the length of the vertical side to the length of the horizontal side for any two such triangles is the same number. So we get the same value for the slope no matter which two points of the line are used to calculate it or in which order the points are used.

EXAMPLE 2

Finding slope

Find the slope of the line shown here using points A and B, points A and C, and points B and C.

Helpful Hint

It is good to think of what the slope represents when x and y are measured quantities rather than just numbers. For example, if the change in y is 50 miles and the change in x is 2 hours, then the slope is 25 mph (or 25 miles per 1 hour). So the slope is the amount of change in y for a change of one in x.

Solution

Using A and B, we get

$$m = \frac{\text{rise}}{\text{run}} = \frac{1}{4}.$$

Using A and C, we get

$$m = \frac{\text{rise}}{\text{run}} = \frac{2}{8} = \frac{1}{4}.$$

Using B and C, we get

$$m = \frac{\text{rise}}{\text{run}} = \frac{1}{4}.$$

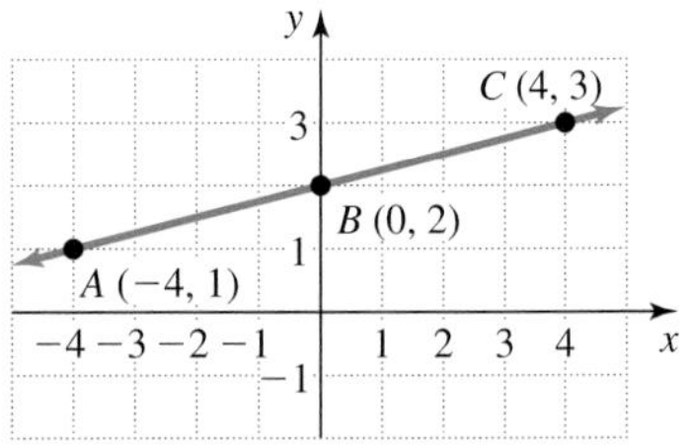

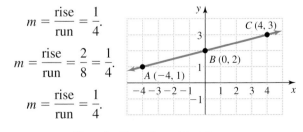

Now do Exercises 11–18

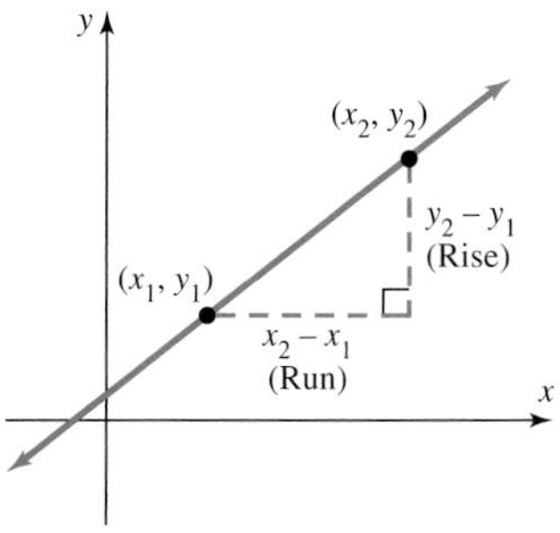

Figure 3.15

Slope Using Coordinates

One way to obtain the rise and run is from a graph. The rise and run can also be found by using the coordinates of two points on the line as shown in Fig. 3.15.

Coordinate Formula for Slope

The slope of the line containing the points (x_1, y_1) and (x_2, y_2) is given by

$$m = \frac{y_2 - y_1}{x_2 - x_1},$$

provided that $x_2 - x_1 \neq 0$.

EXAMPLE 3

Using coordinates to find slope

Find the slope of each of the following lines.

a) The line through (0, 5) and (6, 3)

b) The line through (−3, 4) and (−5, −2)

c) The line through (−4, 2) and the origin

Solution

a) If $(x_1, y_1) = (0, 5)$ and $(x_2, y_2) = (6, 3)$ then

$$m = \frac{y_2 - y_1}{x_2 - x_1}$$
$$= \frac{3 - 5}{6 - 0} = \frac{-2}{6} = -\frac{1}{3}.$$

If $(x_1, y_1) = (6, 3)$ and $(x_2, y_2) = (0, 5)$ then

$$m = \frac{y_2 - y_1}{x_2 - x_1}$$
$$= \frac{5 - 3}{0 - 6} = \frac{2}{-6} = -\frac{1}{3}.$$

Note that it does not matter which point is called (x_1, y_1) and which is called (x_2, y_2). In either case the slope is $-\frac{1}{3}$.

b) Let $(x_1, y_1) = (-3, 4)$ and $(x_2, y_2) = (-5, -2)$:

$$m = \frac{y_2 - y_1}{x_2 - x_1}$$
$$= \frac{-2 - 4}{-5 - (-3)}$$
$$= \frac{-6}{-2} = 3$$

c) Let $(x_1, y_1) = (0, 0)$ and $(x_2, y_2) = (-4, 2)$:

$$m = \frac{2 - 0}{-4 - 0} = \frac{2}{-4} = -\frac{1}{2}$$

Now do Exercises 19–30

Study Tip

Students who have difficulty with algebra often schedule it in a class that meets one day per week so they do not have to see it as often. However, many students do better in classes that meet more often for shorter time periods. So schedule your classes to maximize your chances of success.

CAUTION It does not matter which point is called (x_1, y_1) and which is called (x_2, y_2), but if you divide $y_2 - y_1$ by $x_1 - x_2$, the slope will have the wrong sign.

Because division by zero is undefined, slope is undefined if $x_2 - x_1 = 0$ or $x_2 = x_1$. The x-coordinates of two distinct points on a line are equal only if the points are on a vertical line. *So slope is undefined for vertical lines.* The concept of slope does not exist for a vertical line.

Any two points on a horizontal line have equal y-coordinates. So for points on a horizontal line we have $y_2 - y_1 = 0$. Since $y_2 - y_1$ is in the numerator of the slope formula, *the slope for any horizontal line is zero.* We never refer to a line as having "no slope" because in English no can mean zero or does not exist.

EXAMPLE 4

Slope for vertical and horizontal lines

Find the slope of the line through each pair of points.

a) (2, 1) and (2, −3)

b) (−2, 2) and (4, 2)

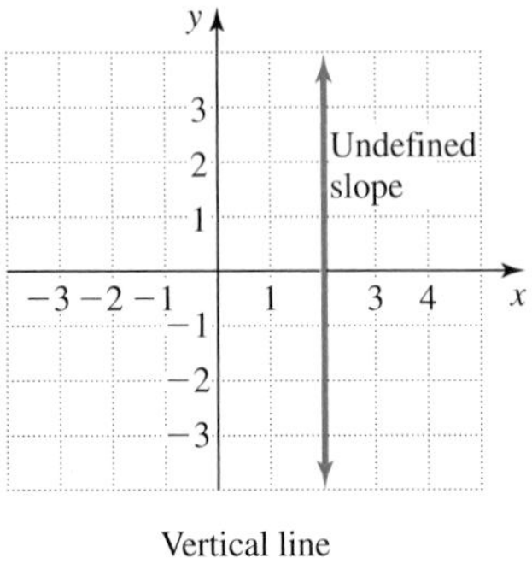

Figure 3.16

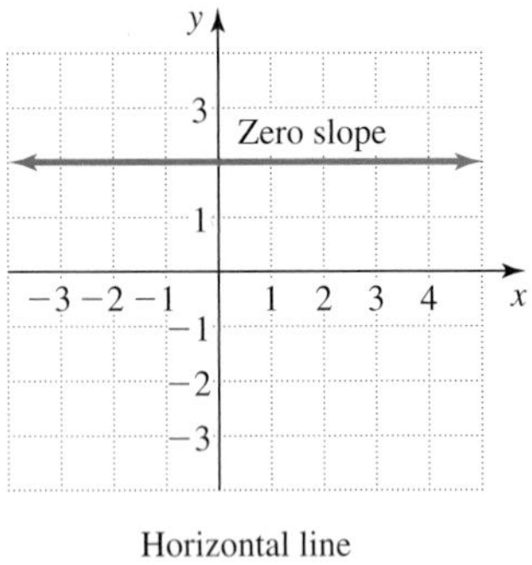

Figure 3.17

Solution

a) The points (2, 1) and (2, −3) are on the vertical line shown in Fig. 3.16. Since slope is undefined for vertical lines, this line does not have a slope. Using the slope formula we get

$$m = \frac{-3 - 1}{2 - 2} = \frac{-4}{0}.$$

Since division by zero is undefined, we can again conclude that slope is undefined for the vertical line through the given points.

b) The points (−2, 2) and (4, 2) are on the horizontal line shown in Fig. 3.17. Using the slope formula we get

$$m = \frac{2 - 2}{-2 - 4} = \frac{0}{-6} = 0.$$

So the slope of the horizontal line through these points is 0.

Now do Exercises 31–34

Note that for a line with *positive slope,* the y-values increase as the x-values increase. For a line with *negative slope,* the y-values decrease as the x-values increase. See Fig. 3.18.

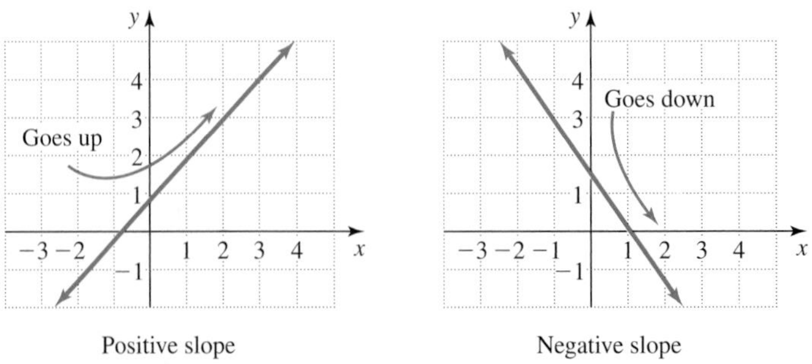

Figure 3.18

Graphing a Line Given a Point and Its Slope

To graph a line from its equation we usually make a table of ordered pairs and then draw a line through the points or we use the intercepts. In Example 5 we will graph a line using one point and the slope. From the slope we find additional points by using the rise and the run.

EXAMPLE 5

Graphing a line given a point and its slope

Graph each line.

a) The line through (2, 1) with slope $\frac{3}{4}$

b) The line through $(-2, 4)$ with slope -3

Solution

a) First locate the point (2, 1). Because the slope is $\frac{3}{4}$, we can find another point on the line by going up three units and to the right four units to get the point (6, 4), as shown in Fig. 3.19. Now draw a line through (2, 1) and (6, 4). Since $\frac{3}{4} = \frac{-3}{-4}$ we could have obtained the second point by starting at (1, 2) and going down 3 units and to the left 4 units.

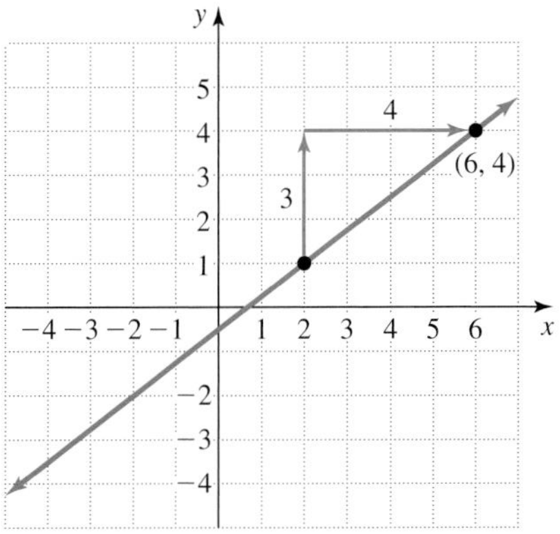

Figure 3.19

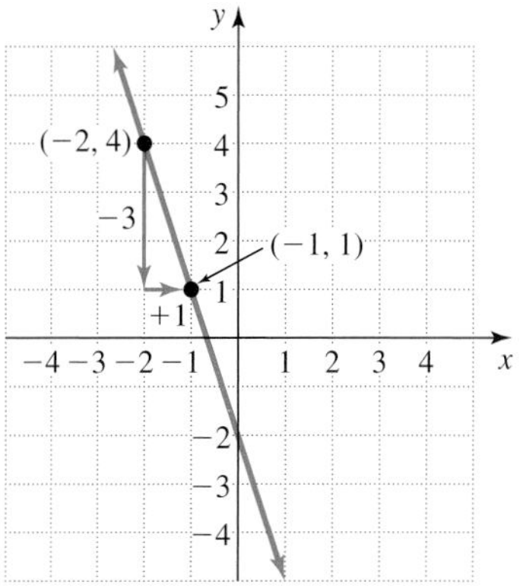

Figure 3.20

b) First locate the point $(-2, 4)$. Because the slope is -3, or $\frac{-3}{1}$, we can locate another point on the line by starting at $(-2, 4)$ and moving down three units and then one unit to the right to get the point $(-1, 1)$. Now draw a line through $(-2, 4)$ and $(-1, 1)$ as shown in Fig. 3.20. Since $\frac{-3}{1} = \frac{3}{-1}$ we could have obtained the second point by starting at $(-2, 4)$ and going up 3 units and to the left 1 unit.

Now do Exercises 37–42

Calculator Close-Up

When we graph a line we usually draw a graph that shows both intercepts, because they are important features of the graph. If the intercepts are not between -10 and 10, you will have to adjust the window to get a good graph. The viewing window that has *x*- and *y*-values ranging from a minimum of -10 to a maximum of 10 is called the *standard viewing window.*

Parallel Lines

Every nonvertical line has a unique slope, but there are infinitely many lines with a given slope. All lines that have a given slope are parallel.

Parallel Lines

Nonvertical lines are parallel if and only if they have equal slopes. Any two vertical lines are parallel to each other.

EXAMPLE 6

Graphing parallel lines

Draw a line through the point $(-2, 1)$ with slope $\frac{1}{2}$ and a line through $(3, 0)$ with slope $\frac{1}{2}$.

Solution

Because slope is the ratio of rise to run, a slope of $\frac{1}{2}$ means that we can locate a second point of the line by starting at $(-2, 1)$ and going up one unit and to the right two units. For the line through $(3, 0)$ we start at $(3, 0)$ and go up one unit and to the right two units. See Fig. 3.21.

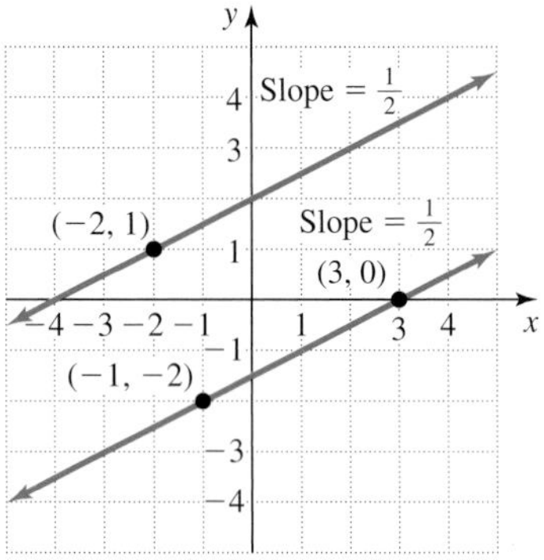

Figure 3.21

Now do Exercises 43–44

Perpendicular Lines

Figure 3.22 shows two right triangles with acute angles of 20° and 70° and legs with lengths a and b ($a > 0$, $b > 0$) positioned along a vertical line. The angle between lines l_1 and l_2 in Fig. 3.22 must be 90° because that angle along with 20° and 70° together form the vertical line. Now the slope of l_1 is $\frac{a}{b}$ and the slope of l_2 is $\frac{-b}{a}$. That is, the slope of one line is the opposite of the reciprocal of the slope of the other.

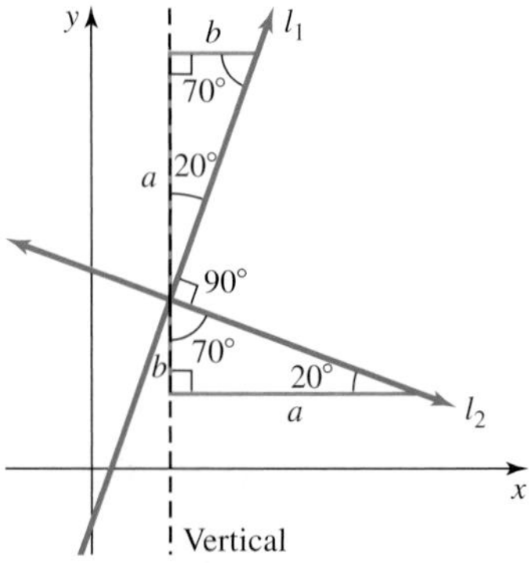

Figure 3.22

This example illustrates the following rule.

Perpendicular Lines

Two lines with slopes m_1 and m_2 are perpendicular if and only if

$$m_1 = -\frac{1}{m_2}.$$

Any vertical line is perpendicular to any horizontal line.

Notice that we cannot compare slopes of horizontal and vertical lines to see if they are perpendicular because slope is not defined for vertical lines.

EXAMPLE 7

Graphing perpendicular lines

Draw two lines through the point $(-1, 2)$, one with slope $-\frac{1}{3}$ and the other with slope 3.

Solution

Because slope is the ratio of rise to run, a slope of $-\frac{1}{3}$ means that we can locate a second point on the line by starting at $(-1, 2)$ and going down one unit and to the right three units. For the line with slope 3, we start at $(-1, 2)$ and go up three units and to the right one unit. See Fig. 3.23.

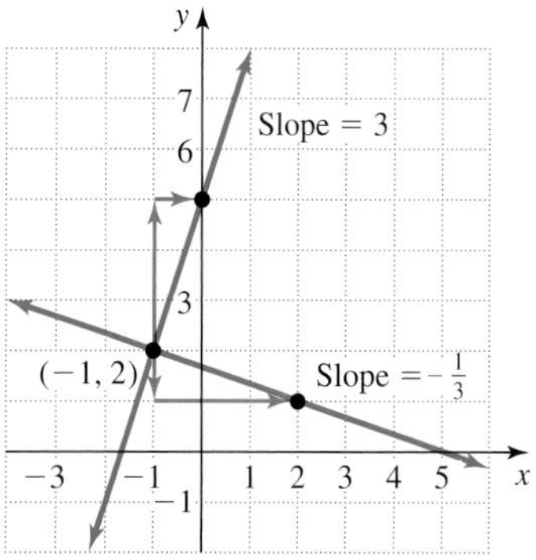

Figure 3.23

Now do Exercises 45–52

Helpful Hint

The relationship between the slopes of perpendicular lines can also be remembered as

$$m_1 \cdot m_2 = -1.$$

For example, lines with slopes -3 and $\frac{1}{3}$ are perpendicular because $-3 \cdot \frac{1}{3} = -1$.

Interpreting Slope

Slope of a line is the ratio of the rise and the run. If the rise is measured in dollars and the run in days, then the slope is measured in dollars per day or dollars/day. The slope of a line is the rate at which the dependent variable is increasing or decreasing.

EXAMPLE 8

Interpreting slope

A car goes from 60 mph to 0 mph in 120 feet after applying the brakes.

a) Find and interpret the slope of the line shown here.

b) What is the velocity at a distance of 80 feet?

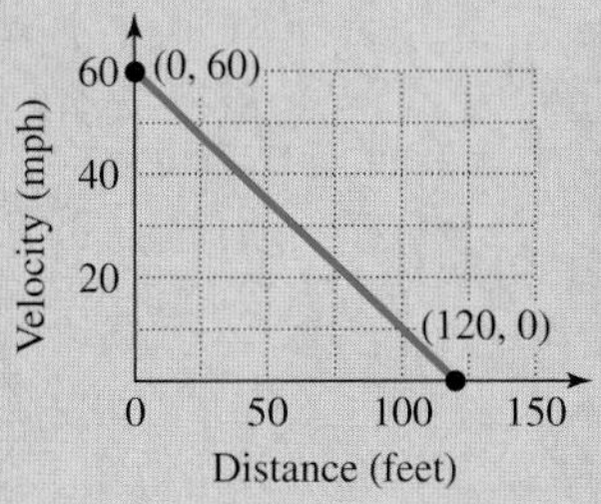

Solution

a) Find the slope of the line through (0, 60) and (120, 0):

$$m = \frac{60 - 0}{0 - 120} = -0.5$$

Because the vertical axis is miles per hour and the horizontal axis is feet, the slope is -0.5 mph/ft, which means the car is losing 0.5 mph of velocity for every foot it travels after the brakes are applied.

b) If the velocity is decreasing 0.5 mph for every foot the car travels, then in 80 feet the velocity goes down 0.5(80) or 40 mph. So the velocity at 80 feet is $60 - 40$ or 20 mph.

Now do Exercises 61–64

EXAMPLE 9

Finding points when given the slope

Assume that the base price of a new Jeep Wrangler is increasing \$300 per year. Find the data that is missing from the table.

Year	Price (dollars)
2001	15,600
2002	
2003	
	18,300
	20,100

Solution

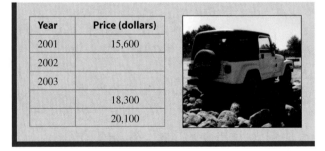

The price in 2002 is \$15,900 and in 2003 it is \$16,200 because the slope is \$300 per year. The rise in price from \$16,200 to \$18,300 is \$2100, which takes 7 years at \$300 per year. So in 2010 the price is \$18,300. The rise from \$18,300 to \$20,100 is \$1800, which takes 6 years at \$300 per year. So in 2016 the price is \$20,100.

Now do Exercises 65–66

Warm-Ups

True or false? Explain your answer.

1. Slope is a measurement of the steepness of a line.
2. Slope is rise divided by run.
3. Every line in the coordinate plane has a slope.
4. The line through the point (1, 1) and the origin has slope 1.
5. Slope can never be negative.
6. A line with slope 2 is perpendicular to any line with slope -2.
7. The slope of the line through (0, 3) and (4, 0) is $\frac{3}{4}$.
8. Two different lines cannot have the same slope.
9. The line through (1, 3) and $(-5, 3)$ has zero slope.
10. Slope can have units such as feet per second.

3.2 Exercises

Boost your GRADE at mathzone.com!

MathZone
- Practice Problems
- Self-Tests
- Videos
- Net Tutor
- e-Professors

Reading and Writing *After reading this section, write out the answers to these questions. Use complete sentences.*

1. What is the slope of a line?
2. What is the difference between rise and run?
3. For which lines is slope undefined?
4. Which lines have zero slope?
5. What is the difference between lines with positive slope and lines with negative slope?
6. What is the relationship between the slopes of perpendicular lines?

In Exercises 7–18, find the slope of each line. See Examples 1 and 2.

7.

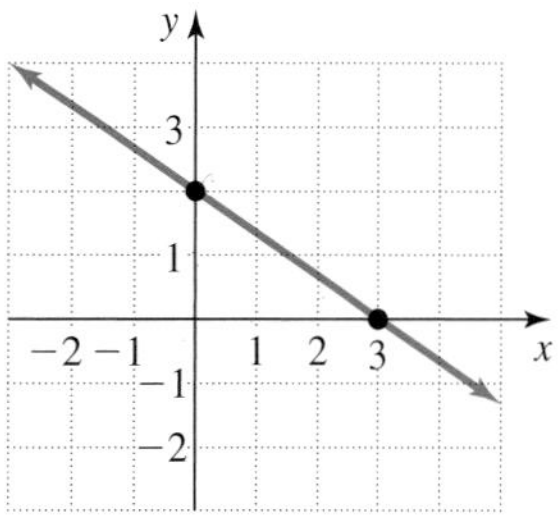

8.

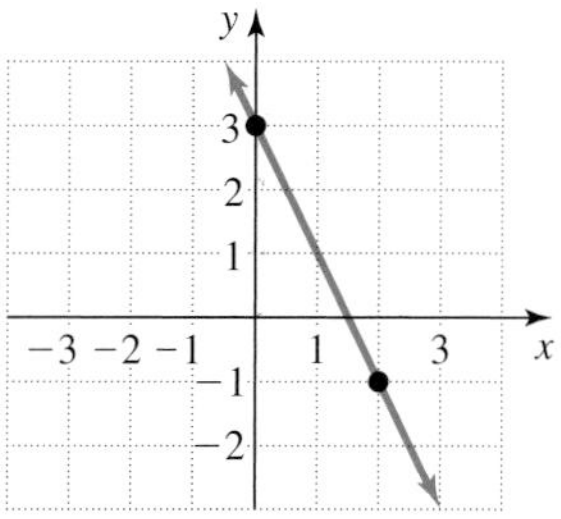

9.

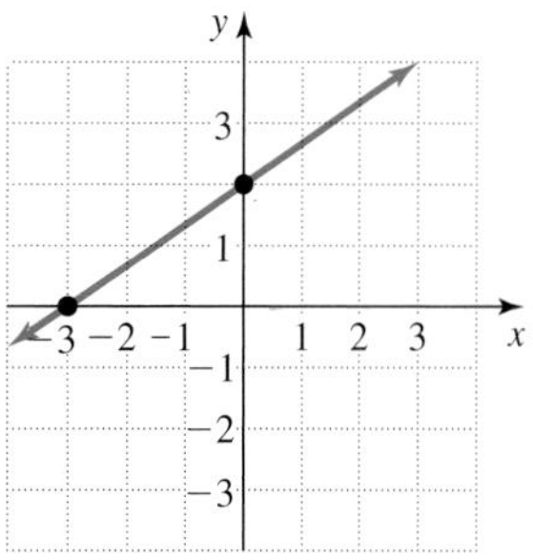

10.

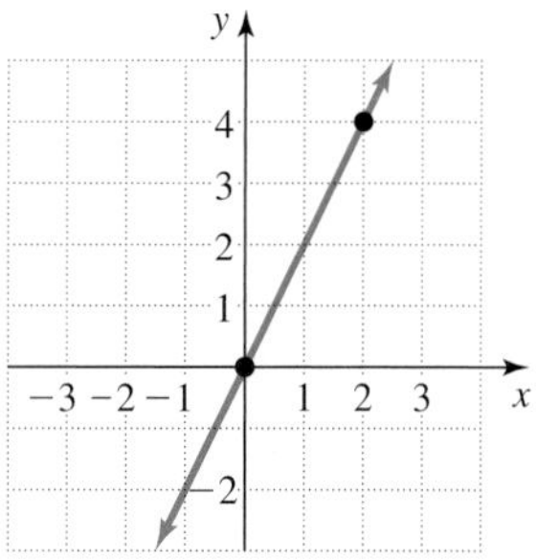

11.

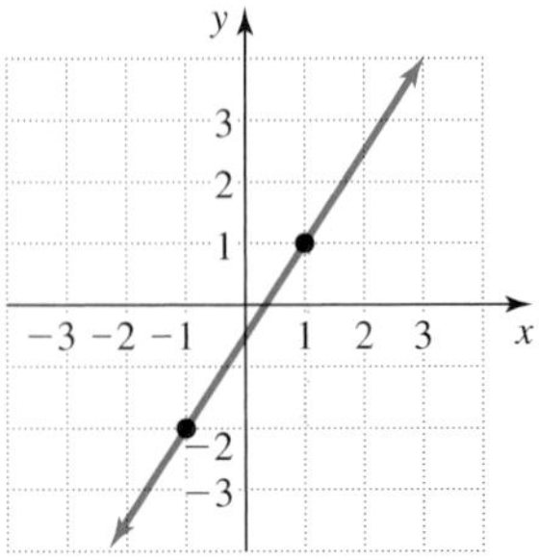

12.

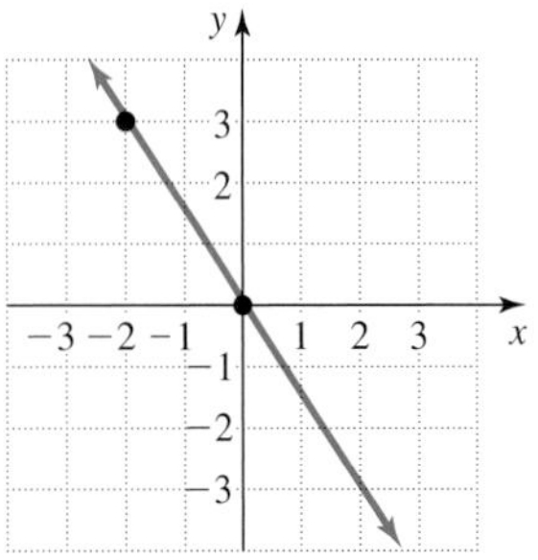

13.

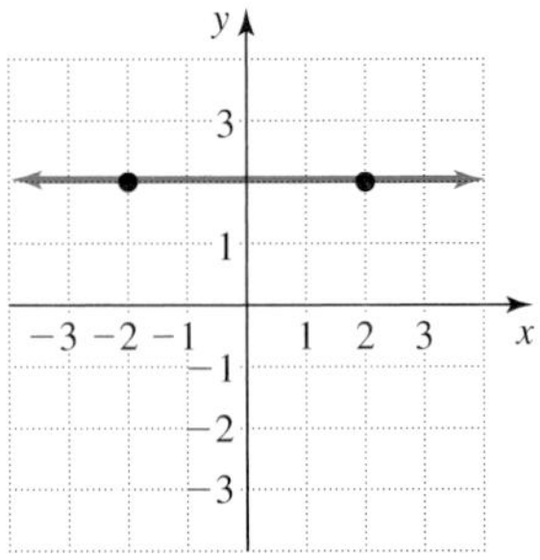

14.

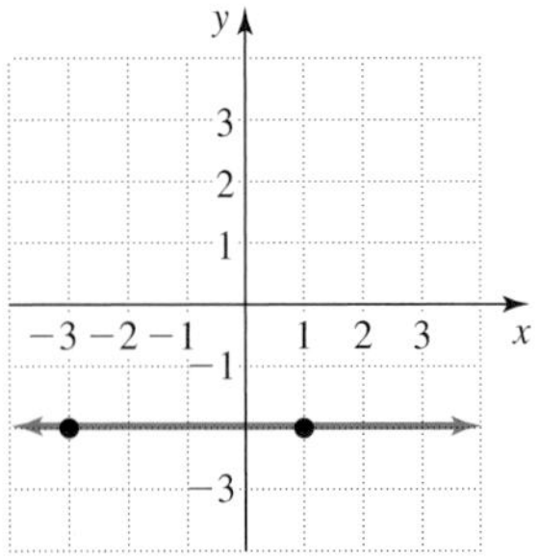

15.

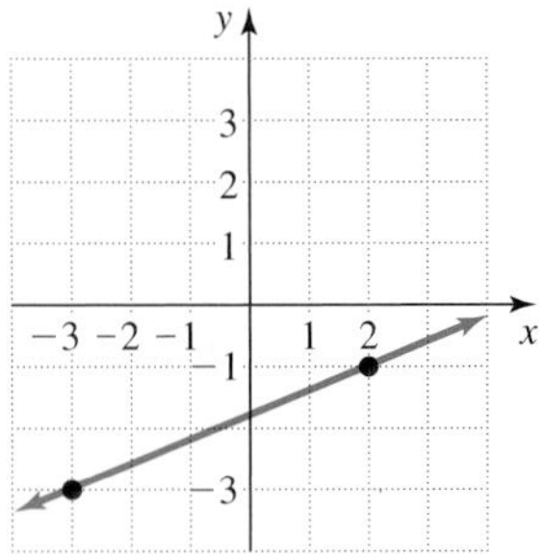

16.

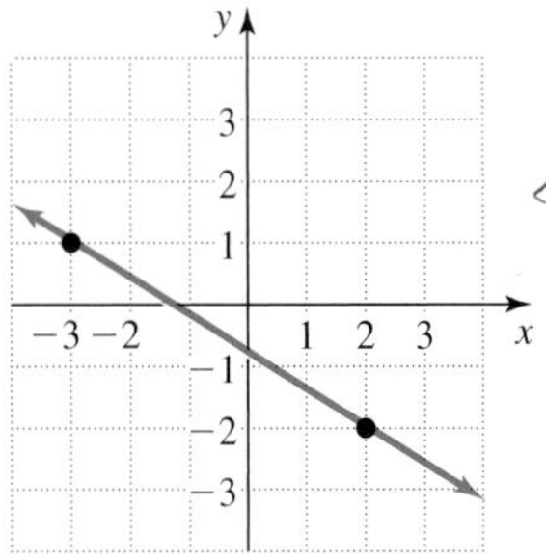

17.

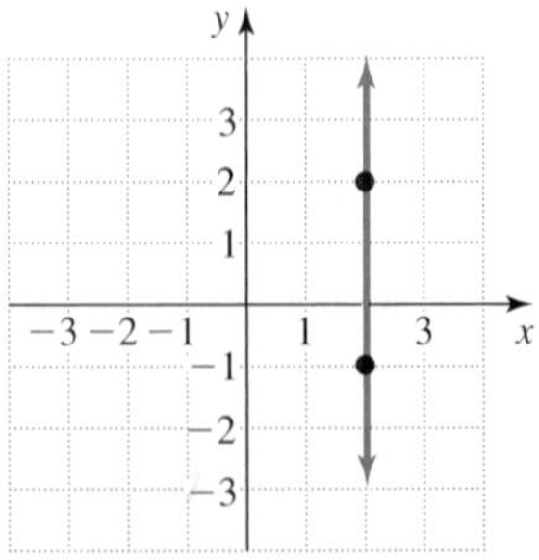

18.

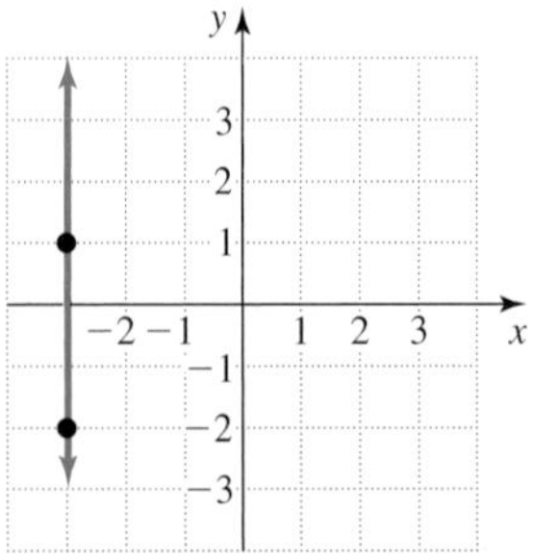

Find the slope of the line that goes through each pair of points. See Examples 3 and 4.

19. $(1, 2), (3, 6)$

20. $(2, 5), (6, 10)$

21. $(2, 4), (5, -1)$

22. $(3, 1), (6, -2)$

23. $(-2, 4), (5, 9)$

24. $(-1, 3), (3, 5)$

25. $(-2, -3), (-5, 1)$

26. $(-6, -3), (-1, 1)$

27. $(-3, 4), (3, -2)$

28. $(-1, 3), (5, -2)$

29. $\left(\frac{1}{2}, 2\right), \left(-1, \frac{1}{2}\right)$

30. $\left(\frac{1}{3}, 2\right), \left(-\frac{1}{3}, 1\right)$

31. $(2, 3), (2, -9)$

32. $(-3, 6), (8, 6)$

33. $(-2, -5), (9, -5)$

34. $(4, -9), (4, 6)$

35. $(0.3, 0.9), (-0.1, -0.3)$

36. $(-0.1, 0.2), (0.5, 0.8)$

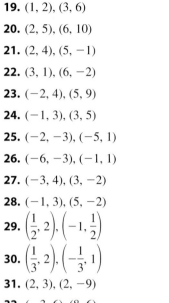

Graph the line with the given point and slope. See Example 5.

37. The line through $(1, 1)$ with slope $\frac{2}{3}$

38. The line through $(2, 3)$ with slope $\frac{1}{2}$

39. The line through $(-2, 3)$ with slope -2

40. The line through $(-2, 5)$ with slope -1

41. The line through $(0, 0)$ with slope $-\frac{2}{5}$

42. The line through $(-1, 4)$ with slope $-\frac{2}{3}$

Solve each problem. See Examples 6 and 7.

43. Draw line l_1 through $(1, -2)$ with slope $\frac{1}{2}$ and line l_2 through $(-1, 1)$ with slope $\frac{1}{2}$.

44. Draw line l_1 through $(0, 3)$ with slope 1 and line l_2 through $(0, 0)$ with slope 1.

45. Draw l_1 through $(1, 2)$ with slope $\frac{1}{2}$, and draw l_2 through $(1, 2)$ with slope -2.

46. Draw l_1 through $(-2, 1)$ with slope $\frac{2}{3}$, and draw l_2 through $(-2, 1)$ with slope $-\frac{3}{2}$.

47. Draw any line l_1 with slope $\frac{3}{4}$. What is the slope of any line perpendicular to l_1? Draw any line l_2 perpendicular to l_1.

48. Draw any line l_1 with slope -1. What is the slope of any line perpendicular to l_1? Draw any line l_2 perpendicular to l_1.

49. Draw l_1 through $(-2, -3)$ and $(4, 0)$. What is the slope of any line parallel to l_1? Draw l_2 through $(1, 2)$ so that it is parallel to l_1.

50. Draw l_1 through $(-4, 0)$ and $(0, 6)$. What is the slope of any line parallel to l_1? Draw l_2 through the origin and parallel to l_1.

51. Draw l_1 through $(-2, 4)$ and $(3, -1)$. What is the slope of any line perpendicular to l_1? Draw l_2 through $(1, 3)$ so that it is perpendicular to l_1.

m = -1?

52. Draw l_1 through $(0, -3)$ and $(3, 0)$. What is the slope of any line perpendicular to l_1? Draw l_2 through the origin so that it is perpendicular to l_1.

In each case, determine whether the lines l_1 and l_2 are parallel, perpendicular, or neither.

53. Line l_1 goes through $(3, 5)$ and $(4, 7)$. Line l_2 goes through $(11, 7)$ and $(12, 9)$.

54. Line l_1 goes through $(-2, -2)$ and $(2, 0)$. Line l_2 goes through $(-2, 5)$ and $(-1, 3)$.

55. Line l_1 goes through $(-1, 4)$ and $(2, 6)$. Line l_2 goes through $(2, -2)$ and $(4, 1)$.

56. Line l_1 goes through $(-2, 5)$ and $(4, 7)$. Line l_2 goes through $(2, 4)$ and $(3, 1)$.

57. Line l_1 goes through $(-1, 4)$ and $(4, 6)$. Line l_2 goes through $(-7, 0)$ and $(3, 4)$.

58. Line l_1 goes through $(1, 2)$ and $(1, -1)$. Line l_2 goes through $(4, 4)$ and $(3, 3)$.

59. Line l_1 goes through $(3, 5)$ and $(3, 6)$. Line l_2 goes through $(-2, 4)$ and $(-3, 4)$.

60. Line l_1 goes through $(-3, 7)$ and $(4, 7)$. Line l_2 goes through $(-5, 1)$ and $(-3, 1)$.

Solve each problem. See Examples 8 and 9.

61. ***Super cost.*** The average cost of 30-second ad during the 1998 Super Bowl was \$1.3 million, and in 2004 it was \$2.4 million (www.adage.com).

a) Find the slope of the line through (1998, 1.3) and (2004, 2.4) and interpret your result.

b) Use the slope to estimate the average cost of an ad in 2002. Is your estimate consistent with the accompanying graph?

c) Use the slope to predict the average cost in 2008?

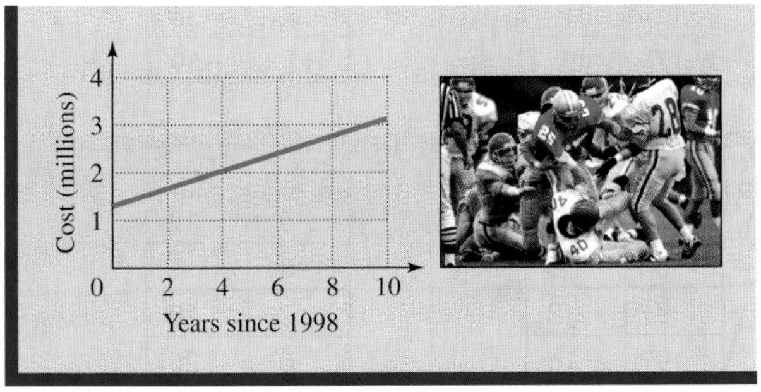

Figure for Exercise 61

62. ***Retirement pay.*** The annual Social Security benefit of a retiree depends on the age at the time of retirement. The accompanying graph gives the annual benefit for persons retiring at ages 62 through 70 in the year 2005 or later (Social Security Administration, www.ssa.gov). What is the annual benefit for a person who retires at age 64? At what retirement age does a person receive an annual benefit of $11,600? Find the slope of each line segment on the graph, and interpret your results. Why do people who postpone retirement until 70 years of age get the highest benefit?

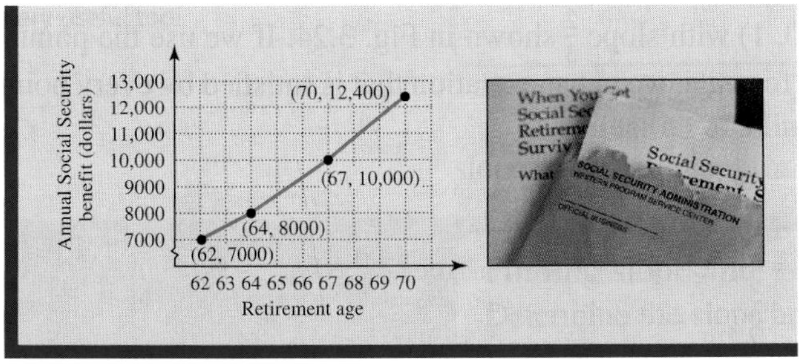

Figure for Exercise 62

63. ***Increasing training.*** The accompanying graph shows the percentage of U.S. workers receiving training by their employers. The percentage went from 5% in 1982 to 25% in 2002 (Department of Labor, www.dol.gov). Find the slope of this line. Interpret your result.

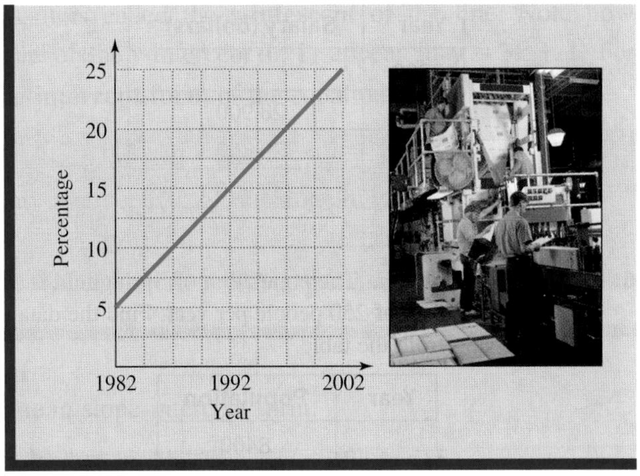

Figure for Exercise 63

64. ***Saving for retirement.*** Financial advisors at Fidelity Investments, Boston, use the accompanying table as a measure of whether a client is on the road to a comfortable retirement.

a) Graph these points and draw a line through them.

b) What is the slope of the line?

c) By what percentage of your salary should you be increasing your savings every year?

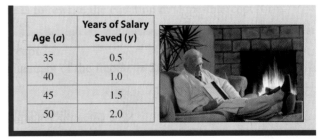

Age (a)	Years of Salary Saved (y)
35	0.5
40	1.0
45	1.5
50	2.0

Figure for Exercise 64

65. ***Increasing salary.*** An elementary school teacher gets a raise of $400 per year. Find the data that is missing from the table on the next page.

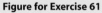

2. How can you determine the slope and y-intercept from the slope-intercept form.

3. What is the standard form for the equation of a line?

4. How can you graph a line when the equation is in slope-intercept form?

5. What form is used in this section to write an equation of a line from a description of the line?

6. What makes lines look perpendicular on a graph?

Write an equation for each line. Use slope-intercept form if possible. See Example 1.

7.

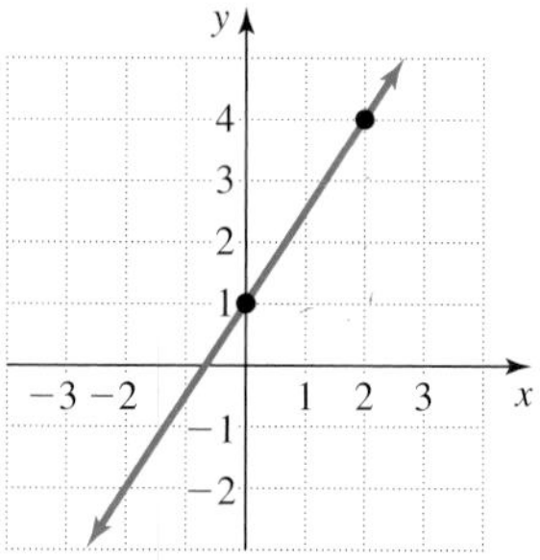

8.

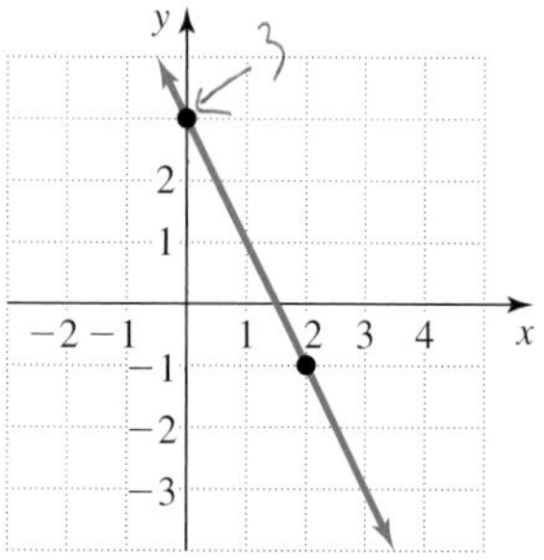

9.

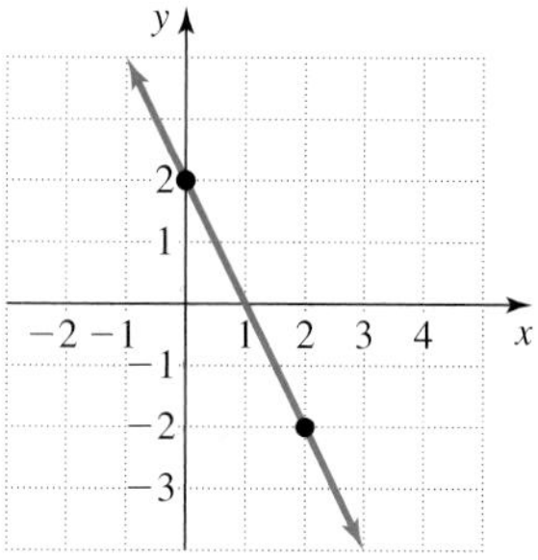

10.

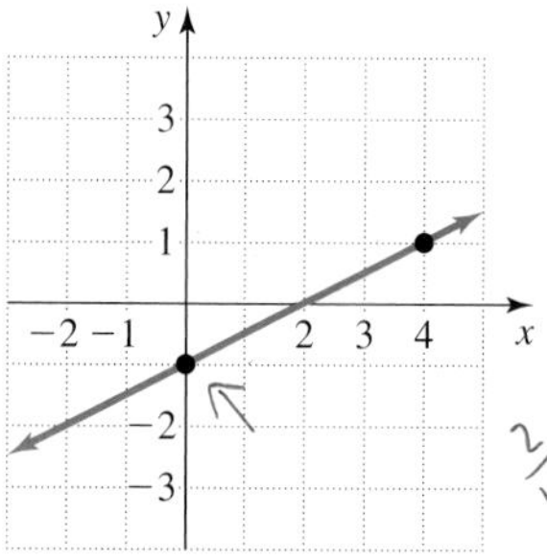

11.

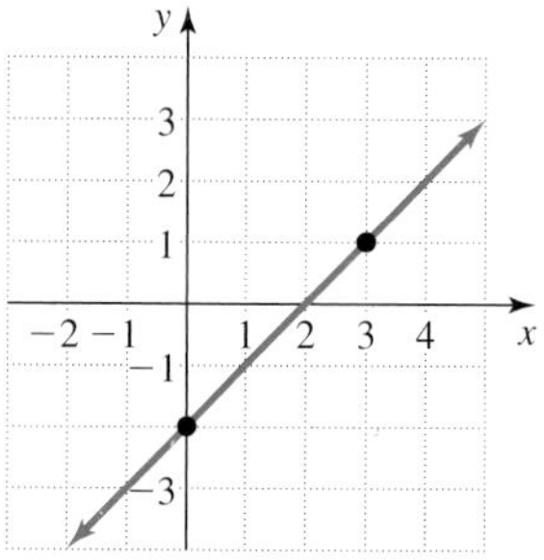

12.

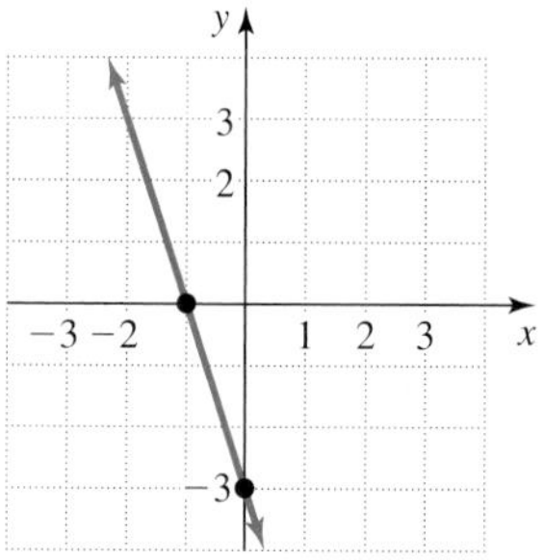

13.

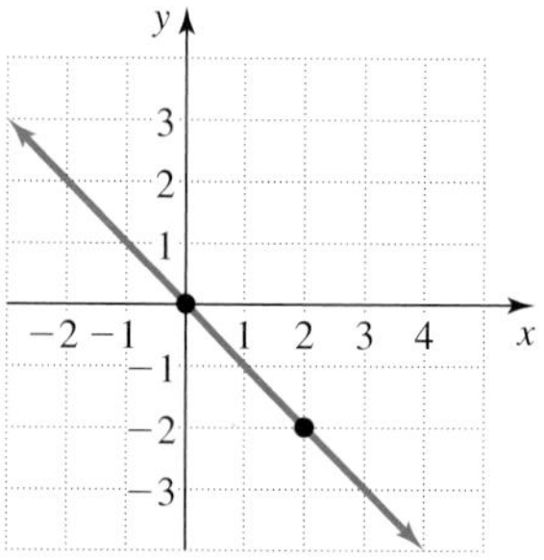

14.

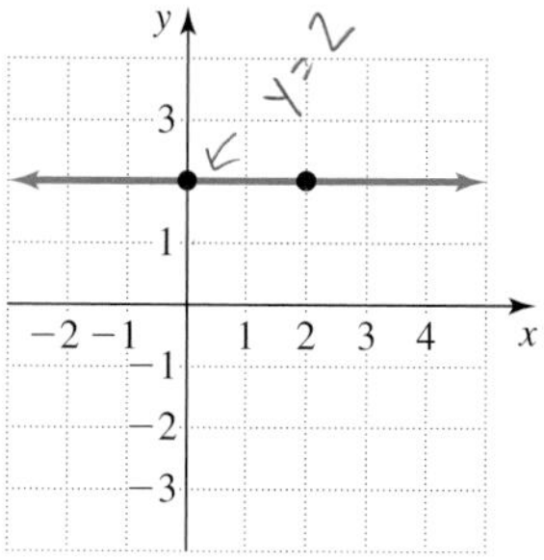

15.

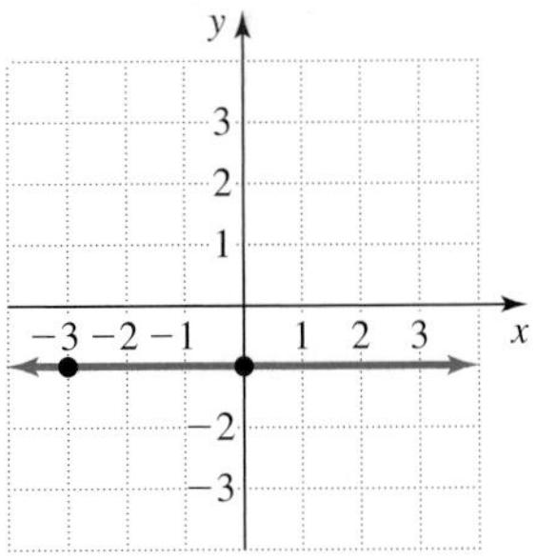

16.

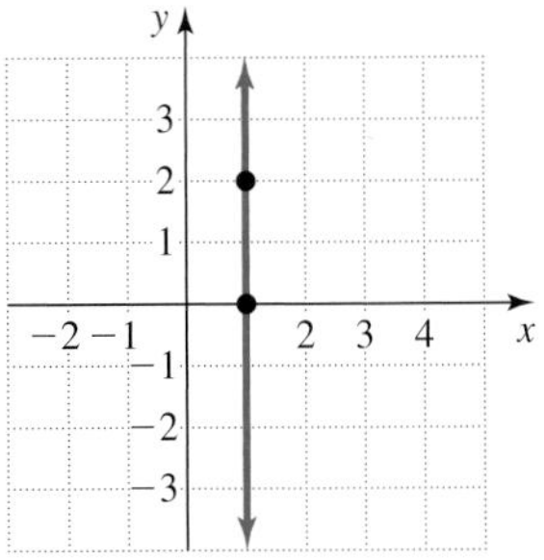

17.

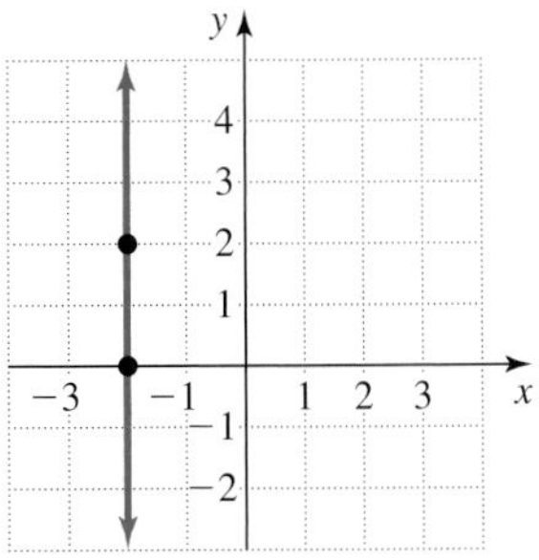

18.

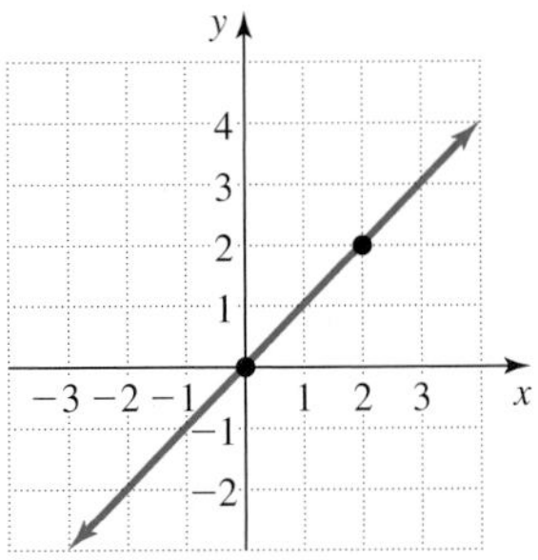

Find the slope and y-intercept for each line that has a slope and y-intercept. See Example 2.

19. $y = 3x - 9$

20. $y = -5x + 4$

21. $y = -\frac{1}{2}x + 3$

22. $y = \frac{1}{4}x + 2$

23. $y = 4$

24. $y = -5$

25. $y = -3x$

26. $y = 2x$

27. $x + y = 5$

28. $x - y = 4$

29. $x - 2y = 4$

30. $x + 2y = 3$

31. $2x - 5y = 10$

32. $2x + 3y = 9$

33. $2x - y + 3 = 0$

34. $3x - 4y - 8 = 0$

35. $x = -3$

36. $\frac{2}{3}x = 4$

Write each equation in standard form using only integers. See Example 3.

37. $y = -x + 2$

38. $y = 3x - 5$

39. $y = \frac{1}{2}x + 3$

40. $y = \frac{2}{3}x - 4$

41. $y = \frac{3}{2}x - \frac{1}{3}$

42. $y = \frac{4}{5}x + \frac{2}{3}$

43. $y = -\frac{3}{5}x + \frac{7}{10}$

44. $y = -\frac{2}{3}x - \frac{5}{6}$

45. $\frac{3}{5}x + 6 = 0$

46. $\frac{1}{2}x - 9 = 0$

47. $\frac{3}{4}y = \frac{5}{2}$

48. $\frac{2}{3}y = \frac{1}{9}$

49. $\frac{x}{2} = \frac{3y}{5}$

50. $\frac{x}{8} = -\frac{4y}{5}$

51. $y = 0.02x + 0.5$

52. $0.2x = 0.03y - 0.1$

Draw the graph of each line using its y-intercept and its slope. See Examples 4 and 5.

53. $y = 2x - 1$

54. $y = 3x - 2$

55. $y = -3x + 5$

56. $y = -4x + 1$

57. $y = \frac{3}{4}x - 2$

58. $y = \frac{3}{2}x - 4$

59. $2y + x = 0$

60. $2x + y = 0$

61. $3x - 2y = 10$

62. $4x + 3y = 9$

63. $4y + x = 8$

64. $y + 4x = 8$

65. $y - 2 = 0$

66. $y + 5 = 0$

In each case determine whether the lines are parallel, perpendicular, or neither.

67. $y = 3x - 4$
$y = 3x - 9$

68. $y = -5x + 7$
$y = \frac{1}{5}x - 6$

69. $y = 2x - 1$
$y = -2x + 1$

70. $y = x + 7$
$y = -x + 2$

71. $y = 3$
$y = -\frac{1}{3}$

72. $y = 3x + 2$
$y = \frac{1}{3}x - 4$

73. $y = -4x + 1$
$y = \frac{1}{4}x - 5$

74. $y = \frac{1}{3}x + \frac{1}{2}$
$y = \frac{1}{3}x - 2$

Write an equation in slope-intercept form, if possible, for each line. See Example 6.

75. The line through $(0, -4)$ with slope $\frac{1}{2}$

76. The line through $(0, 4)$ with slope $-\frac{1}{2}$

77. The line through $(0, 3)$ that is parallel to the line $y = 2x - 1$

78. The line through $(0, -2)$ that is parallel to the line $y = -\frac{1}{3}x + 6$

79. The line through $(0, 6)$ that is perpendicular to the line $y = 3x - 5$

80. The line through $(0, -1)$ that is perpendicular to the line $y = x$

81. The line with y-intercept $(0, 3)$ that is parallel to the line $2x + y = 5$

82. The line through the origin that is parallel to the line $y - 3x = -3$

83. The line through $(2, 3)$ that runs parallel to the x-axis

84. The line through $(-3, 5)$ that runs parallel to the y-axis

85. The line through (0, 4) that is perpendicular to $2x - 3y = 6$

86. The line through (0, −1) that is perpendicular to $2x - 5y = 10$

87. The line through (0, 4) and (5, 0)

88. The line through (0, −3) and (4, 0)

Solve each problem. See Example 7.

89. ***Marginal cost.*** A manufacturer plans to spend \$150,000 on research and development for a new lawn mower and then \$200 to manufacture each mower. The formula $C = 200n + 150{,}000$ gives the cost in dollars of n mowers. What is the cost of 5000 mowers? What is the cost of 5001 mowers? By how much did the one extra lawn mower increase the cost? (The increase in cost is called the *marginal cost* of the 5001st lawn mower.)

90. ***Marginal revenue.*** A defense attorney charges her client \$4000 plus \$120 per hour. The formula $R = 120n + 4000$ gives her revenue in dollars for n hours of work. What is her revenue for 100 hours of work? What is her revenue for 101 hours of work? By how much did the one extra hour of work increase the revenue? (The increase in revenue is called the *marginal revenue* for the 101st hour.)

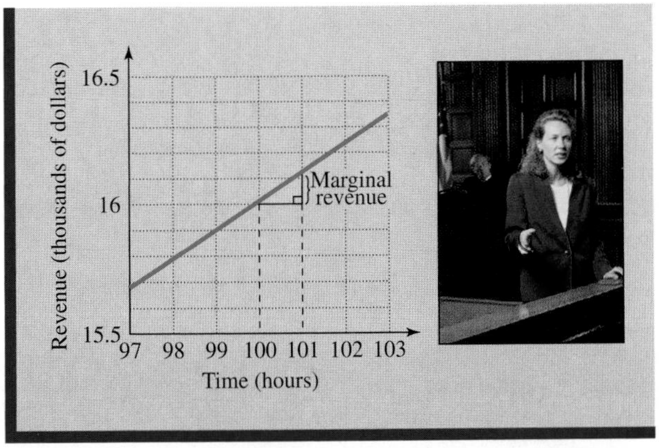

Figure for Exercise 90

91. ***In-house training.*** The accompanying graph shows the percentage of U.S. workers receiving training by their employers (Department of Labor, www.dol.gov). The percentage went from 5% in 1982 to 25% in 2002.

a) Find and interpret the slope of the line.
b) Write the equation of the line in slope-intercept form.
c) What is the meaning of the y-intercept?

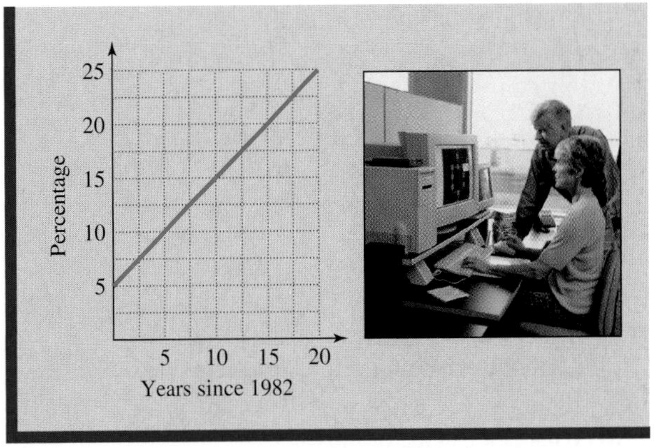

Figure for Exercise 91

d) Use your equation to predict the percentage that will be receiving training in 2010.

92. ***Single women.*** The percentage of women in the 20–24 age group who have never married went from 55% in 1970 to 73% in 2000 (Census Bureau, www.census.gov). Let 1970 be year 0 and 2000 be year 30.

a) Find and interpret the slope of the line through the points (0, 55) and (30, 73).
b) Find the equation of the line in part (a).
c) What is the meaning of the y-intercept?
d) Use the equation to predict the percentage in 2010.
e) If this trend continues, then in what year will the percentage of women in the 20–24 age group who have never married reach 100%?

93. ***Pansies and snapdragons.*** A nursery manager plans to spend \$100 on 6-packs of pansies at 50 cents per pack and snapdragons at 25 cents per pack. The equation $0.50x + 0.25y = 100$ can be used to model this situation.

a) What do x and y represent?

b) Graph the equation.

c) Write the equation in slope-intercept form.

d) What is the slope of the line?

e) What does the slope tell you?

94. ***Pens and pencils.*** A bookstore manager plans to spend \$60 on pens at 30 cents each and pencils at 10 cents each. The equation $0.10x + 0.30y = 60$ can be used to model this situation.

a) What do x and y represent?

b) Graph the equation.

c) Write the equation in slope-intercept form.

d) What is the slope of the line?

e) What does the slope tell you?

Graphing Calculator Exercises

Graph each pair of straight lines on your graphing calculator using a viewing window that makes the lines look perpendicular. Answers may vary.

95. $y = 12x - 100,\ y = -\frac{1}{12}x + 50$

96. $2x - 3y = 300,\ 3x + 2y = -60$

3.4 The Point-Slope Form

In this Section

- **Point-Slope Form**
- **Parallel Lines**
- **Perpendicular Lines**
- **Applications**

In Section 3.3 we wrote the equation of a line given its slope and y-intercept. In this section you will learn to write the equation of a line given the slope and *any* other point on the line.

Point-Slope Form

Consider a line through the point (4, 1) with slope $\frac{2}{3}$ as shown in Fig. 3.28. Because the slope can be found by using any two points on the line, we use (4, 1) and an arbitrary

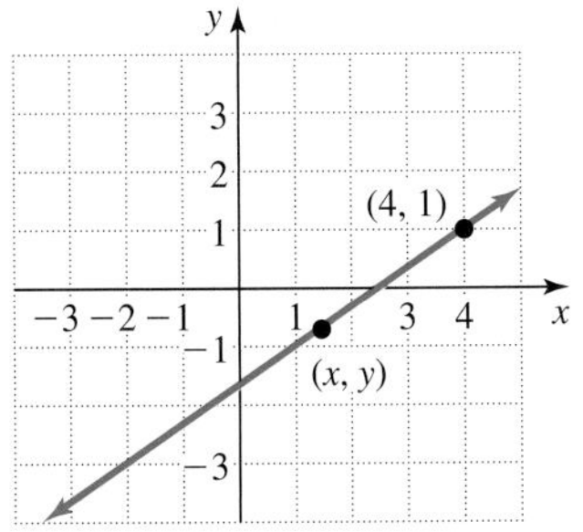

Figure 3.28

point (x, y) in the formula for slope:

$$\frac{y_2 - y_1}{x_2 - x_1} = m \quad \text{Slope formula}$$

$$\frac{y - 1}{x - 4} = \frac{2}{3} \quad \text{Let } m = \tfrac{2}{3}, (x_1, y_1) = (4, 1), \text{ and } (x_2, y_2) = (x, y).$$

$$y - 1 = \frac{2}{3}(x - 4) \quad \text{Multiply each side by } x - 4.$$

Note how the coordinates of the point (4, 1) and the slope $\frac{2}{3}$ appear in the above equation. We can use the same procedure to get the equation of any line given one point on the line and the slope. The resulting equation is called the **point-slope form** of the equation of the line.

Helpful Hint

If a point (x, y) is on a line with slope m through (x_1, y_1), then

$$\frac{y - y_1}{x - x_1} = m.$$

Multiplying each side of this equation by $x - x_1$ gives us the point-slope form.

Point-Slope Form

The equation of the line through the point (x_1, y_1) with slope m is

$$y - y_1 = m(x - x_1).$$

EXAMPLE 1

Writing an equation given a point and a slope

Find the equation of the line through $(-2, 3)$ with slope $\frac{1}{2}$, and write it in slope-intercept form.

Solution

Because we know a point and the slope, we can use the point-slope form:

$$y - y_1 = m(x - x_1) \quad \text{Point-slope form}$$

$$y - 3 = \frac{1}{2}[x - (-2)] \quad \text{Substitute } m = \tfrac{1}{2} \text{ and } (x_1, y_1) = (-2, 3).$$

$$y - 3 = \frac{1}{2}(x + 2) \quad \text{Simplify.}$$

$$y - 3 = \frac{1}{2}x + 1 \quad \text{Distributive property}$$

$$y = \frac{1}{2}x + 4 \quad \text{Slope-intercept form}$$

Alternate Solution

Replace m by $\frac{1}{2}$, x by -2, and y by 3 in the slope-intercept form:

$$y = mx + b \quad \text{Slope-intercept form}$$

$$3 = \frac{1}{2}(-2) + b \quad \text{Substitute } m = \tfrac{1}{2} \text{ and } (x, y) = (-2, 3).$$

$$3 = -1 + b \quad \text{Simplify.}$$

$$4 = b$$

Since $b = 4$, we can write $y = \frac{1}{2}x + 4$.

Now do Exercises 7–22

The alternate solution to Example 1 is shown because many students have seen that method in the past. This does not mean that you should ignore the point-slope form. It is always good to know more than one method to accomplish a task. The good thing about using the point-slope form is that you immediately write down the equation and then you simplify it. In the alternate solution, the last thing you do is to write the equation.

The point-slope form can be used to find the equation of a line for *any* given point and slope. However, if the given point is the y-intercept, then it is simpler to use the slope-intercept form. Note that it is not necessary that the slope be given, because the slope can be found from any two points. So if we know two points on a line, then we can find the slope and use the slope with either one of the points in the point-slope form.

EXAMPLE 2

Writing an equation given two points

Find the equation of the line that contains the points $(-3, -2)$ and $(4, -1)$, and write it in standard form.

Solution

First find the slope using the two given points:

$$m = \frac{-2 - (-1)}{-3 - 4} = \frac{-1}{-7} = \frac{1}{7}$$

Now use one of the points, say $(-3, -2)$, and slope $\frac{1}{7}$ in the point-slope form:

$$y - y_1 = m(x - x_1) \quad \text{Point-slope form}$$

$$y - (-2) = \frac{1}{7}[x - (-3)] \quad \text{Substitute.}$$

$$y + 2 = \frac{1}{7}(x + 3) \quad \text{Simplify.}$$

$$7(y + 2) = 7 \cdot \frac{1}{7}(x + 3) \quad \text{Multiply each side by 7.}$$

$$7y + 14 = x + 3$$

$$7y = x - 11 \quad \text{Subtract 14 from each side.}$$

$$-x + 7y = -11 \quad \text{Subtract } x \text{ from each side.}$$

$$x - 7y = 11 \quad \text{Multiply each side by } -1.$$

The equation in standard form is $x - 7y = 11$. Using the other given point, $(4, -1)$, would give the same final equation in standard form. Try it.

Now do Exercises 23-42

Calculator Close-Up

Graph $y = (x + 3)/7 - 2$ to see that the line goes through $(-3, -2)$ and $(4, -1)$.

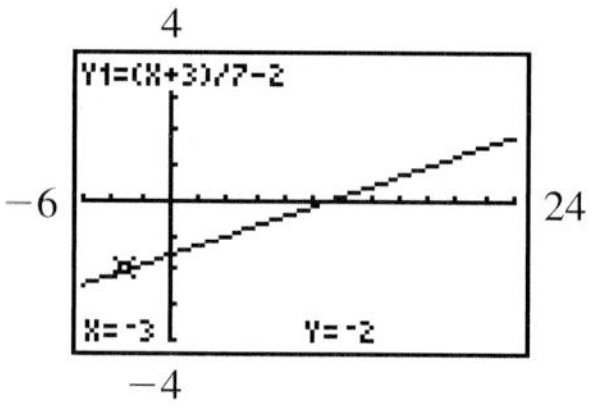

Note that the form of the equation does not matter on the calculator as long as it is solved for y.

Parallel Lines

In Section 3.2 you learned that parallel lines have the same slope. We will use this fact in Example 3.

EXAMPLE 3

Using point-slope form with parallel lines

Find the equation of each line. Write the answer in slope-intercept form.

a) The line through $(2, -1)$ that is parallel to $y = -3x + 9$

b) The line through $(3, 4)$ that is parallel to $2x - 3y = 6$

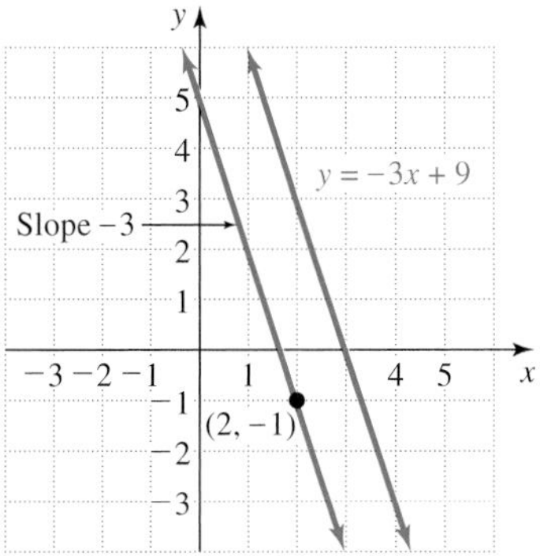

Figure 3.29

Solution

a) The slope of $y = -3x + 9$ and any line parallel to it is -3. See Fig. 3.29. Now use the point $(2, -1)$ and slope -3 in point-slope form:

$$y - y_1 = m(x - x_1) \quad \text{Point-slope form}$$
$$y - (-1) = -3(x - 2) \quad \text{Substitute.}$$
$$y + 1 = -3x + 6 \quad \text{Simplify.}$$
$$y = -3x + 5 \quad \text{Slope-intercept form}$$

Since $-1 = -3(2) + 5$ is correct, the line $y = -3x + 5$ goes through $(2, -1)$. It is certainly parallel to $y = -3x + 9$. So $y = -3x + 5$ is the desired equation.

b) Solve $2x - 3y = 6$ for y to determine its slope:

$$2x - 3y = 6$$
$$-3y = -2x + 6$$
$$y = \frac{2}{3}x - 2$$

So the slope of $2x - 3y = 6$ and any line parallel to it is $\frac{2}{3}$. Now use the point $(3, 4)$ and slope $\frac{2}{3}$ in the point-slope form:

$$y - y_1 = m(x - x_1) \quad \text{Point-slope form}$$
$$y - 4 = \frac{2}{3}(x - 3) \quad \text{Substitute.}$$
$$y - 4 = \frac{2}{3}x - 2 \quad \text{Simplify.}$$
$$y = \frac{2}{3}x + 2 \quad \text{Slope-intercept form}$$

Since $4 = \frac{2}{3}(3) + 2$ is correct, the line $y = \frac{2}{3}x + 2$ contains the point $(3, 4)$. Since $y = \frac{2}{3}x + 2$ and $y = \frac{2}{3}x - 2$ have the same slope, they are parallel. So the equation is $y = \frac{2}{3}x + 2$.

Now do Exercises 49–50

Perpendicular Lines

In Section 3.2 you learned that lines with slopes m and $-\frac{1}{m}$ (for $m \neq 0$) are perpendicular to each other. For example, the lines

$$y = -2x + 7 \quad \text{and} \quad y = \frac{1}{2}x - 8$$

are perpendicular to each other. In the next example we will write the equation of a line that is perpendicular to a given line and contains a given point.

EXAMPLE 4

Writing an equation given a point and a perpendicular line

Write the equation of the line that is perpendicular to $3x + 2y = 8$ and contains the point $(1, -3)$. Write the answer in slope-intercept form.

Solution

First graph $3x + 2y = 8$ and a line through $(1, -3)$ that is perpendicular to $3x + 2y = 8$ as shown in Fig. 3.30. The right angle symbol is used in the figure to indicate that the lines are perpendicular. Now write $3x + 2y = 8$ in slope-intercept form to determine its slope:

$$3x + 2y = 8$$
$$2y = -3x + 8$$
$$y = -\frac{3}{2}x + 4 \quad \text{Slope-intercept form}$$

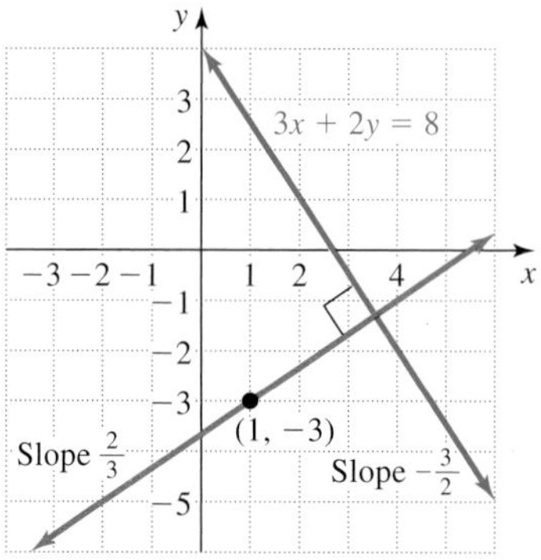

Figure 3.30

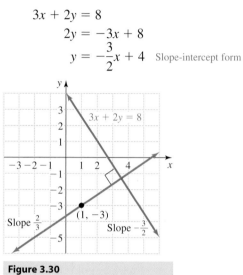

The slope of the given line is $-\frac{3}{2}$. The slope of any line perpendicular to it is $\frac{2}{3}$. Now we use the point-slope form with the point $(1, -3)$ and the slope $\frac{2}{3}$:

$$y - y_1 = m(x - x_1) \quad \text{Point-slope form}$$
$$y - (-3) = \frac{2}{3}(x - 1)$$
$$y + 3 = \frac{2}{3}x - \frac{2}{3}$$
$$y = \frac{2}{3}x - \frac{2}{3} - 3 \quad \text{Subtract 3 from each side.}$$
$$y = \frac{2}{3}x - \frac{11}{3} \quad \text{Slope-intercept form}$$

So $y = \frac{2}{3}x - \frac{11}{3}$ is the equation of the line that contains $(1, -3)$ and is perpendicular to $3x + 2y = 8$. Check that $(1, -3)$ satisfies $y = \frac{2}{3}x - \frac{11}{3}$.

Now do Exercises 47–48

Calculator Close-Up

Graph $y_1 = (2/3)x - 11/3$ and $y_2 = (-3/2)x + 4$ as shown:

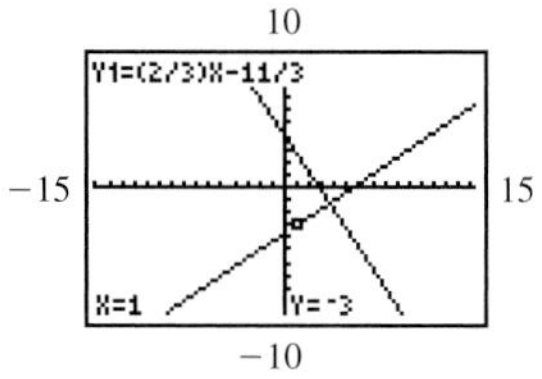

Because the lines look perpendicular and y_1 goes through $(1, -3)$, the graph supports the answer to Example 4.

Applications

We use the point-slope form to find the equation of a line given two points on the line. In Example 5 we use that same procedure to find a linear equation that relates two variables in an applied situation.

EXAMPLE 5

Writing a formula given two points

A contractor charges \$30 for installing 100 feet of pipe and \$120 for installing 500 feet of pipe. To determine the charge he uses a linear equation that gives the charge C in terms of the length L. Find the equation and find the charge for installing 240 feet of pipe.

Solution

Because C is determined from L, we let C take the place of the dependent variable y and let L take the place of the independent variable x. So the ordered pairs are in the form (L, C). We can use the slope formula to find the slope of the line through the two points (100, 30) and (500, 120) shown in Fig. 3.31.

$$m = \frac{120 - 30}{500 - 100} = \frac{90}{400} = \frac{9}{40}$$

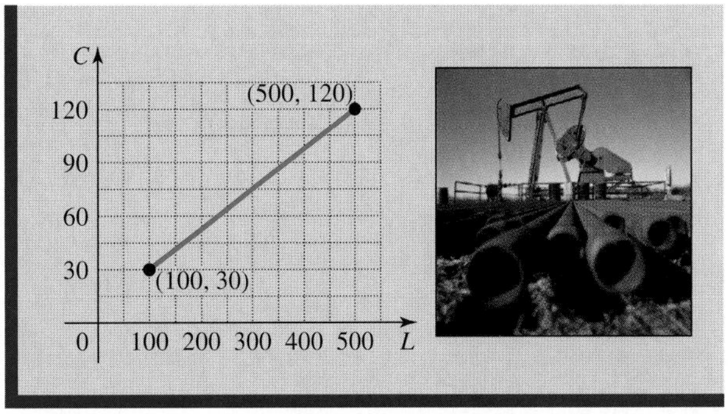

Figure 3.31

Now we use the point-slope form with the point (100, 30) and slope $\frac{9}{40}$:

$$y - y_1 = m(x - x_1)$$

$$C - 30 = \frac{9}{40}(L - 100)$$

$$C - 30 = \frac{9}{40}L - \frac{45}{2}$$

$$C = \frac{9}{40}L - \frac{45}{2} + 30$$

$$C = \frac{9}{40}L + \frac{15}{2}$$

Note that $C = \frac{9}{40}L + \frac{15}{2}$ means that the charge is $\frac{9}{40}$ dollars/foot plus a fixed charge of $\frac{15}{2}$ dollars (or \$7.50). We can now find C when $L = 240$:

$$C = \frac{9}{40} \cdot 240 + \frac{15}{2}$$

$$C = 54 + 7.5$$

$$C = 61.5$$

The charge for installing 240 feet of pipe is \$61.50.

Now do Exercises 77–92

Study Tip

When working a test, scan the problems and pick out the ones that are the easiest for you. Do them first. Save the harder problems till last.

Warm-Ups

True or false? Explain your answer.

1. The formula $y = m(x - x_1)$ is the point-slope form for a line.
2. It is impossible to find the equation of a line through (2, 5) and (−3, 1).
3. The point-slope form will not work for the line through (3, 4) and (3, 6).
4. The equation of the line through the origin with slope 1 is $y = x$.
5. The slope of the line $5x + y = 4$ is 5.
6. The slope of any line perpendicular to the line $y = 4x - 3$ is $-\frac{1}{4}$.
7. The slope of any line parallel to the line $x + y = 1$ is -1.
8. The line $2x - y = -1$ goes through the point $(-2, -3)$.
9. The lines $2x + y = 4$ and $y = -2x + 7$ are parallel.
10. The equation of the line through (0, 0) perpendicular to $y = x$ is $y = -x$.

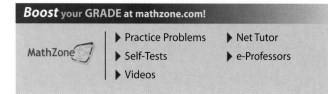

3.4 Exercises

Reading and Writing *After reading this section, write out the answers to these questions. Use complete sentences.*

1. What is the point-slope form for the equation of a line?
2. For what is the point-slope form used?
3. What is the procedure for finding the equation of a line when given two points on the line?
4. How can you find the slope of a line when given the equation of the line?
5. What is the relationship between the slopes of parallel lines?
6. What is the relationship between the slopes of perpendicular lines?

Write each equation in slope-intercept form. See Example 1.

7. $y - 1 = 5(x + 2)$
8. $y + 3 = -3(x - 6)$
9. $3x - 4y = 80$
10. $2x + 3y = 90$
11. $y - \frac{1}{2} = \frac{2}{3}\left(x - \frac{1}{4}\right)$
12. $y + \frac{2}{3} = -\frac{1}{2}\left(x - \frac{2}{5}\right)$

Find the equation of the line that goes through the given point and has the given slope. Write the answer in slope-intercept form. See Example 1.

13. (1, 2), 3 **14.** (2, 5), 4

15. (2, 4), $\frac{1}{2}$ **16.** (4, 6), $\frac{1}{2}$

17. (2, 3), $\frac{1}{3}$ **18.** (1, 4), $\frac{1}{4}$

19. (−2, 5), $-\frac{1}{2}$ **20.** (−3, 1), $-\frac{1}{3}$

21. (−1, −7), −6 **22.** (−1, −5), −8

Write each equation in standard form using only integers. See Example 2.

23. $y - 3 = 2(x - 5)$ **24.** $y + 2 = -3(x - 1)$

25. $y = \frac{1}{2}x - 3$ **26.** $y = \frac{1}{3}x + 5$

27. $y - 2 = \frac{2}{3}(x - 4)$ **28.** $y + 1 = \frac{3}{2}(x + 4)$

Find the equation of the line through each given pair of points. Write the answer in standard form using only integers. See Example 2.

29. (1, 3), (2, 5) **30.** (2, 5), (3, 9)

31. (1, 1), (2, 2) **32.** (−1, 1), (1, −1)

33. (1, 2), (5, 8) **34.** (3, 5), (8, 15)

35. (−2, −1), (3, −4) **36.** (−1, −3), (2, −1)

37. (−2, 0), (0, 2) **38.** (0, 3), (5, 0)

39. (2, 4), (2, 6) **40.** (−3, 5), (−3, −1)

41. (−3, 9), (3, 9) **42.** (2, 5), (4, 5)

The lines in each figure are perpendicular. Find the equation (in slope-intercept form) for the solid line.

43.

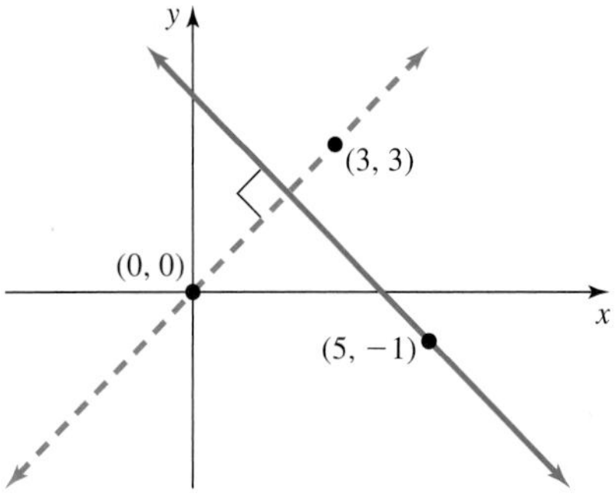

44.

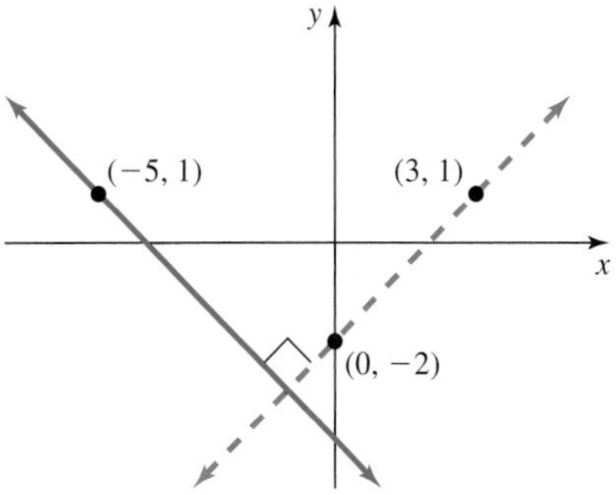

45.

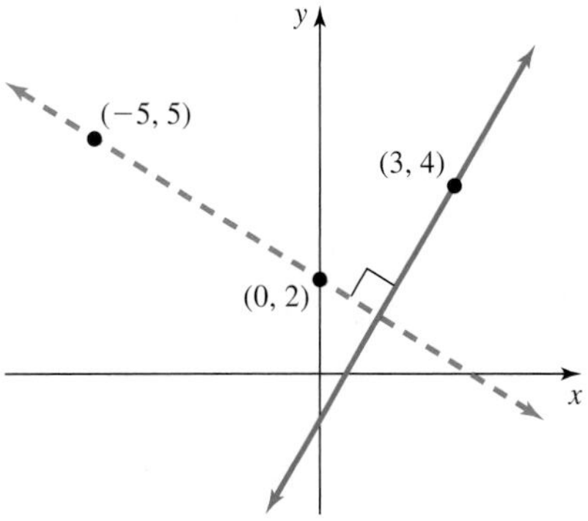

46.

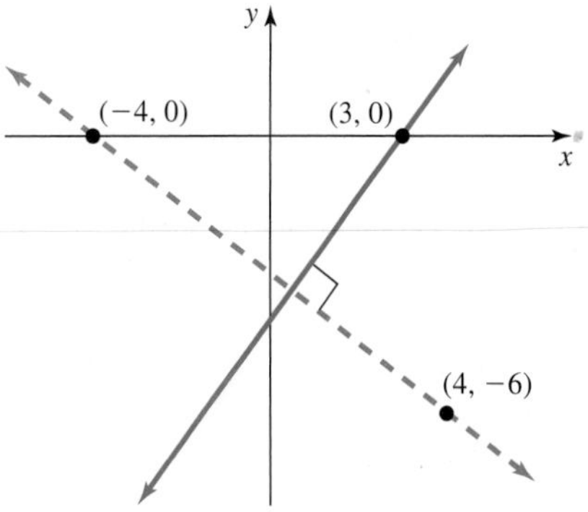

Find the equation of each line. Write each answer in slope-intercept form. See Examples 3 and 4.

47. The line contains the point (3, 4) and is perpendicular to $y = 3x - 1$.

48. The line contains the point $(-2, 3)$ and is perpendicular to $y = 2x + 7$.

49. The line is parallel to $y = x - 9$ and goes through the point (7, 10).

50. The line is parallel to $y = -x + 5$ and goes through the point $(-3, 6)$.

51. The line is perpendicular to $3x - 2y = 10$ and passes through the point (1, 1).

52. The line is perpendicular to $x - 5y = 4$ and passes through the point $(-1, 1)$.

53. The line is parallel to $2x + y = 8$ and contains the point $(-1, -3)$.

54. The line is parallel to $-3x + 2y = 9$ and contains the point $(-2, 1)$.

55. The line goes through $(-1, 2)$ and is perpendicular to $3x + y = 5$.

56. The line goes through (1, 2) and is perpendicular to $y = \frac{1}{2}x - 3$.

57. The line goes through (2, 3) and is parallel to $-2x + y = 6$.

58. The line goes through (1, 4) and is parallel to $x - 2y = 6$.

Find the equation of each line in the form $y = mx + b$ if possible.

59. The line through (3, 2) with slope 0

60. The line through (3, 2) with undefined slope

61. The line through (3, 2) and the origin

62. The line through the origin that is perpendicular to $y = \frac{2}{3}x$

63. The line through the origin that is parallel to the line through (5, 0) and (0, 5)

64. The line through the origin that is perpendicular to the line through $(-3, 0)$ and $(0, -3)$

65. The line through $(-30, 50)$ that is perpendicular to the line $x = 400$

66. The line through $(20, -40)$ that is parallel to the line $y = 6000$

67. The line through $(-5, -1)$ that is perpendicular to the line through (0, 0) and (3, 5)

68. The line through (3, 1) that is parallel to the line through $(-3, -2)$ and (0, 0)

For each line described here choose the correct equation from (a) through (h).

69. The line through (1, 3) and (2, 5)

70. The line through (1, 3) and (5, 2)

71. The line through (1, 3) with no x-intercept

72. The line through (1, 3) with no y-intercept

73. The line through (1, 3) with x-intercept (5, 0)

74. The line through (1, 3) with y-intercept $(0, -5)$

75. The line through (1, 3) with slope -2

76. The line through (1, 3) with slope $\frac{1}{2}$

a) $x + 4y = 13$ **b)** $x = 1$
c) $x - 2y = -5$ **d)** $y = 8x - 5$
e) $y = 2x + 1$ **f)** $y = 3$
g) $2x + y = 5$ **h)** $3x + 4y = 15$

Solve each problem. See Example 5.

77. ***Automated tellers.*** ATM volume reached 10.6 billion transactions in 1996 and 14.2 billion transactions in 2000 as shown in the accompanying graph. If 1996 is year 0 and 2000 is year 4, then the line goes through the points (0, 10.6) and (4, 14.2).

a) Find and interpret the slope of the line.
b) Write the equation of the line in slope-intercept form.
c) Use your equation from part (b) to predict the number of transactions at automated teller machines in 2010.

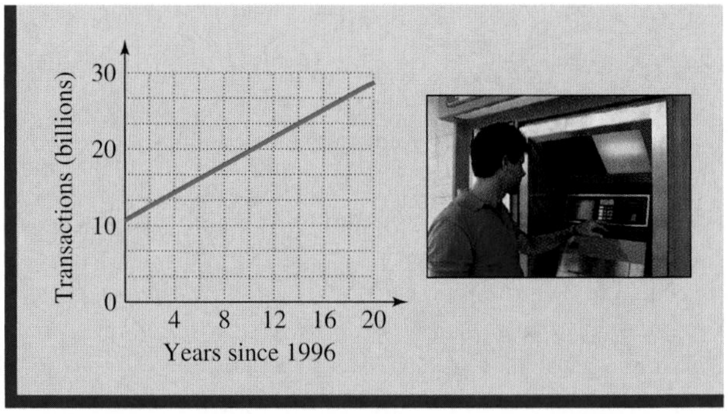

Figure for Exercise 77

78. ***Direct deposit.*** The percentage of workers receiving direct deposit of their paychecks went from 32% in 1994 to 60% in 2004 (www.directdeposit.com). Let 1994 be year 0 and 2004 be year 10.

a) Write the equation of the line through (0, 32) and (10, 60) to model the growth of direct deposit.
b) Use the graph on the next page to predict the year in which 100% of all workers will receive direct deposit of their paychecks.
c) Use the equation from part (a) to predict the year in which 100% of all workers will receive direct deposit.

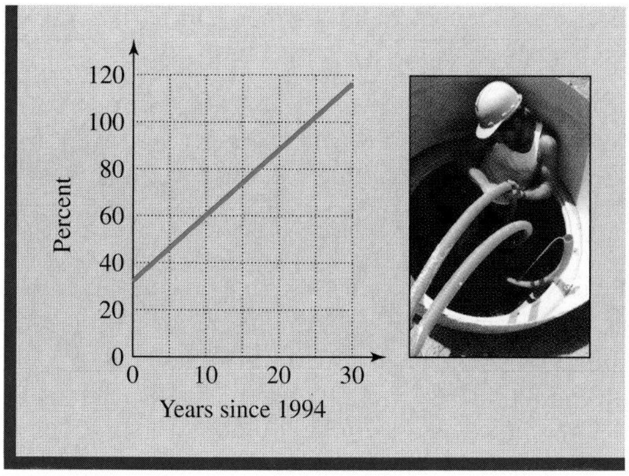

Figure for Exercise 78

79. ***Gross domestic product.*** The U.S. gross domestic product (GDP) per employed person increased from \$62.7 thousand in 1996 to \$71.6 thousand in 2002 (Bureau of Labor Statistics, www.bls.gov). Let 1996 be year 6 and 2002 be year 12.

a) Find the equation of the line through (6, 62.7) and (12, 71.6) to model the gross domestic product.

b) What do x and y represent in your equation?

c) Use the equation to predict the GDP per employed person in 2010.

d) Graph the equation.

80. ***Age at first marriage.*** The median age at first marriage for females increased from 24.5 years in 1995 to 25.1 years in 2000 (U.S. Census Bureau, www.census.gov). Let 1995 be year 5 and 2000 be year 10.

a) Find the equation of the line through (5, 24.5) and (10, 25.1).

b) What do x and y represent in your equation?

c) Interpret the slope of this line.

d) In what year will the median age be 30.

e) Graph the equation.

81. ***Plumbing charges.*** Pete the plumber worked 2 hours at Millie's house and charged her \$70. He then worked 4 hours at Rosalee's house and charged her \$110. To determine the amount he charges Pete uses a linear equation that gives the charge C in terms of the number of hours worked n. Find the equation and find the charge for 7 hours at Fred's house.

82. ***Interior angles.*** The sum of the measures of the interior angles of a triangle is 180°. The sum of the measures of the interior angles of a square is 360°. Let S represent the sum of the measures of the interior angles of a polygon and n represent the number of sides of the polygon. There is a linear equation that gives S in terms of n. Find the equation and find the sum of the measures of the interior angles of the stop sign shown in the accompanying figure.

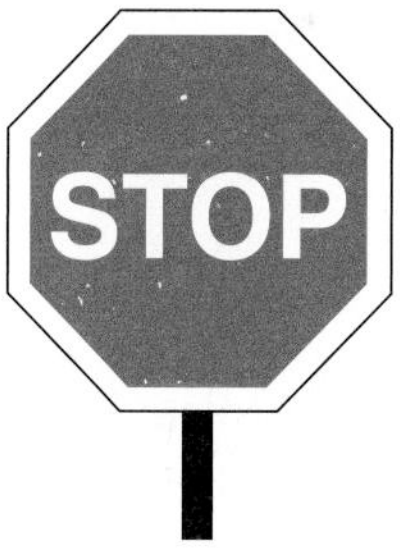

Figure for Exercise 82

83. ***Shoe sizes.*** If a child's foot is 7.75 inches long, then the child wears a size 13 shoe. If a child's foot is 5.75 inches long, then the child wears a size 7 shoe. Let S represent the shoe size and L represent the length of the foot in inches. There is a linear equation that gives S in terms of L. Find the equation and find the shoe size for a child with a 6.25-inch foot. See the figure on the next page.

84. ***Celsius to Fahrenheit.*** Water freezes at 0°C or 32°F and boils at 100°C or 212°F. There is a linear equation that expresses the number of degrees Fahrenheit (F) in terms

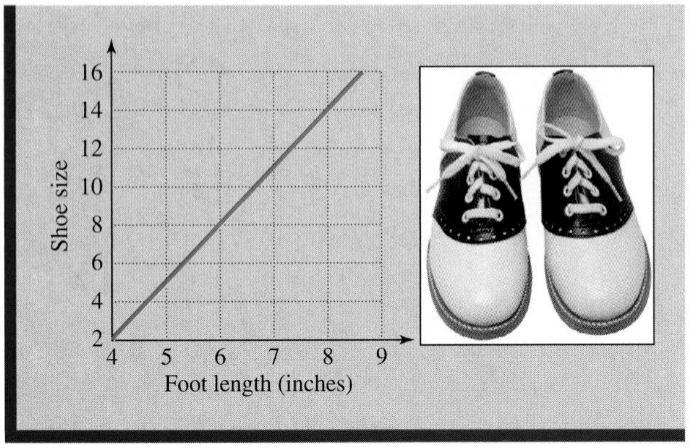

Figure for Exercise 83

of the number of degrees Celsius (C). Find the equation and find the Fahrenheit temperature when the Celsius temperature is 45°.

85. ***Velocity of a projectile.*** A ball is thrown downward from the top of a tall building. Its velocity is 42 feet per second after 1 second and 74 feet per second after 2 seconds. There is a linear equation that expresses the velocity v in terms of the time t. Find the equation and find the velocity after 3.5 seconds.

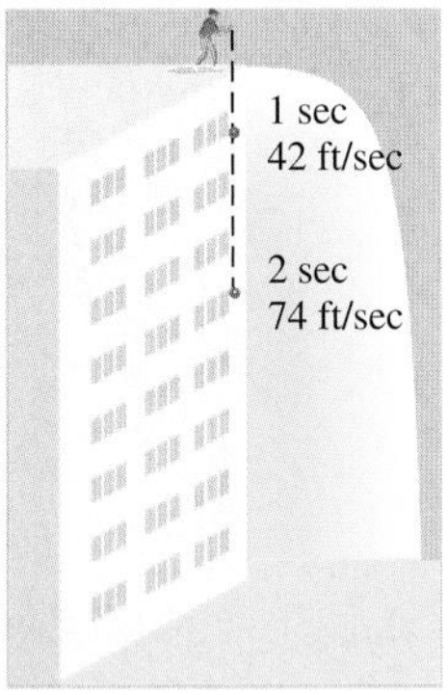

Figure for Exercise 85

86. ***Natural gas.*** The cost of 1000 cubic feet of natural gas is \$39 and the cost of 3000 cubic feet is \$99. There is a linear equation that expresses the cost C in terms of the number of cubic feet n. Find the equation and find the cost of 2400 cubic feet of natural gas.

87. ***Expansion joint.*** When the temperature is 90°F the width of an expansion joint on a bridge is 0.75 inch. When the temperature is 30°F the width is 1.25 inches. There is a linear equation that expresses the width w in terms of the temperature t.

a) Find the equation.
b) What is the width when the temperature is 80°F?
c) What is the temperature when the width is 1 inch?

88. ***Perimeter of a rectangle.*** A rectangle has a fixed width and a variable length. Let P represent the perimeter and L represent the length. $P = 28$ inches when $L = 6.5$ inches and $P = 36$ inches when $L = 10.5$ inches. There is a linear equation that expresses P in terms of L.

a) Find the equation.
b) What is the perimeter when the $L = 40$ inches?
c) What is the length when $P = 215$ inches?
d) What is the width of the rectangle?

89. ***Stretching a spring.*** A weight of 3 pounds stretches a spring 1.8 inches beyond its natural length and weight of 5 pounds stretches the same spring 3 inches beyond its natural length. Let A represent the amount of stretch and w the weight. There is a linear equation that expresses A in terms of w. Find the equation and find the amount that the spring will stretch with a weight of 6 pounds.

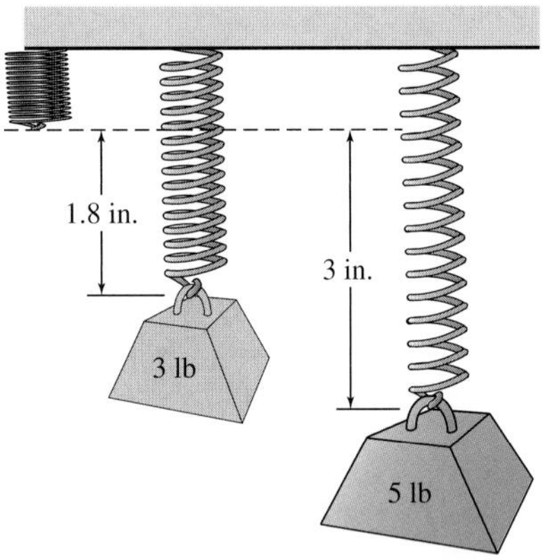

Figure for Exercise 89

90. ***Velocity of a bullet.*** A gun is fired straight upward. The bullet leaves the gun at 100 feet per second (time $t = 0$). After 2 seconds the velocity of the bullet is 36 feet per second. There is a linear equation that gives the velocity v in terms of the time t. Find the equation and find the velocity after 3 seconds.

91. ***Enzyme concentration.*** The amount of light absorbed by a certain liquid depends on the concentration of an enzyme in the liquid. A concentration of 2 milligrams

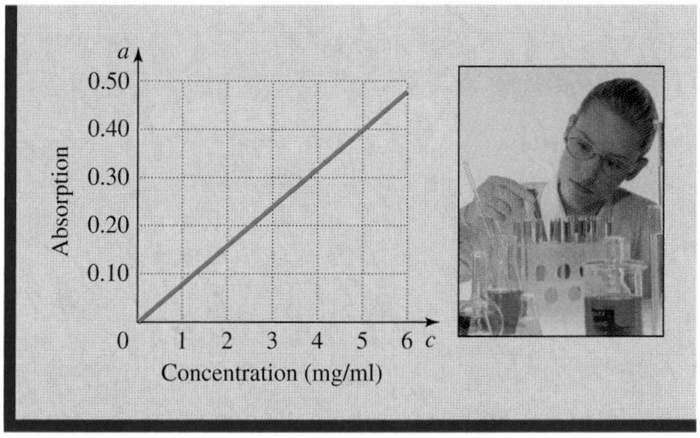

Figure for Exercise 91

per milliliter (mg/ml) produces an absorption of 0.16 and a concentration of 5 mg/ml produces an absorption of 0.40. There is a linear equation that expresses the absorption a in terms of the concentration c.

a) Find the equation.
b) What is the absorption when the concentration is 3 mg/ml?
c) Use the graph above to estimate the concentration when the absorption is 0.50.

92. ***Basal energy requirement.*** The basal energy requirement B is the number of calories that a person needs to maintain the life process. For a 28-year-old female with a height of 160 centimeters and a weight of 45 kilograms (kg), B is 1300 calories. If her weight increases to 50 kg, then B is 1365 calories. There is a linear equation that expresses B in terms of her weight w. Find the equation and find the basal energy requirement if her weight is 53.2 kg.

Getting More Involved

93. ***Exploration***

Each linear equation in the following table is given in standard form $Ax + By = C$. In each case identify A, B, and the slope of the line.

Equation	A	B	Slope
$2x + 3y = 9$			
$4x - 5y = 6$			
$\frac{1}{2}x + 3y = 1$			
$2x - \frac{1}{3}y = 7$			

94. ***Exploration***

Find a pattern in the table of Exercise 93 and write a formula for the slope of $Ax + By = C$, where $B \neq 0$.

Graphing Calculator Exercises

95. Graph each equation on a graphing calculator. Choose a viewing window that includes both the x- and y-intercepts. Use the calculator output to help you draw the graph on paper.

a) $y = 20x - 300$
b) $y = -30x + 500$
c) $2x - 3y = 6000$

96. Graph $y = 2x + 1$ and $y = 1.99x - 1$ on a graphing calculator. Are these lines parallel? Explain your answer.

97. Graph $y = 0.5x + 0.8$ and $y = 0.5x + 0.7$ on a graphing calculator. Find a viewing window in which the two lines are separate.

98. Graph $y = 3x + 1$ and $y = -\frac{1}{3}x + 2$ on a graphing calculator. Do the lines look perpendicular? Explain.

3.5 Variation

In this Section

- **Direct Variation**
- **Finding the Constant**
- **Inverse Variation**
- **Joint Variation**

If $y = 5x$, then the value of y depends on the value of x. As x varies, so does y. Simple relationships like $y = 5x$ are customarily expressed in terms of variation. In this section you will learn the language of variation and learn to write formulas from verbal descriptions.

Direct Variation

Suppose you average 60 miles per hour on the freeway. The distance D that you travel depends on the amount of time T that you travel. Using the formula $D = R \cdot T$, we can write

$$D = 60T.$$

Consider the possible values for T and D given in the following table.

T (hours)	1	2	3	4	5	6
D (miles)	60	120	180	240	300	360

The graph of $D = 60T$ is shown in Fig. 3.32. Note that as T gets larger, so does D. In this situation we say that D *varies directly with* T, or D is *directly proportional* to T. The constant rate of 60 miles per hour is called the **variation constant** or **proportionality constant.** Notice that $D = 60T$ is simply a linear equation. We are just introducing some new terms to express an old idea.

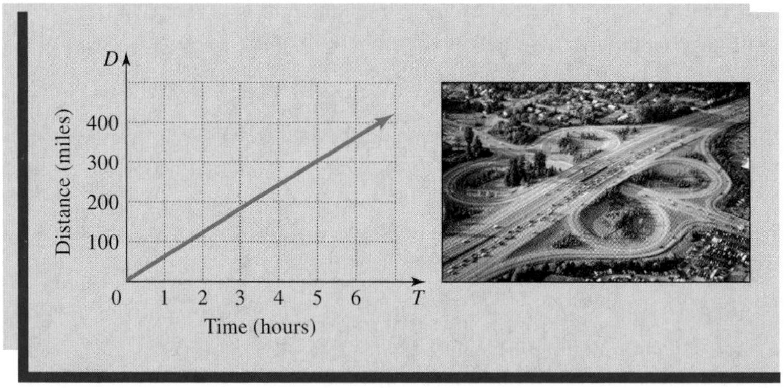

Figure 3.32

Direct Variation

The statement **"y varies directly as x"** or **"y is directly proportional to x"** means that

$$y = kx$$

for some constant k. The constant of variation k is a fixed nonzero real number.

CAUTION Direct variation refers only to equations of the form $y = kx$ (lines through the origin). We do *not* refer to $y = 3x + 5$ as a direct variation.

Finding the Constant

If we know one ordered pair in a direct variation, then we can find the constant of variation.

EXAMPLE 1

Finding a constant of variation

Natasha is traveling by car, and the distance D that she travels varies directly as the rate R at which she drives. At 45 miles per hour, Natasha travels 135 miles. Find the constant of variation, and write a formula for D in terms of R.

Solution

Because D varies directly as R, there is a constant k such that

$$D = kR.$$

Because $D = 135$ when $R = 45$, we can write

$$135 = k \cdot 45$$

or

$$3 = k.$$

So $D = 3R$.

Now do Exercises 15–16

Helpful Hint

In any variation problem you must first determine the general form of the relationship. Because this problem involves direct variation, the general form is $y = kx$.

In Example 2 we find the constant of variation and use it to solve a variation problem.

EXAMPLE 2

A direct variation problem

Your electric bill at Middle States Electric Co-op varies directly with the amount of electricity that you use. If the bill for 2800 kilowatts of electricity is \$196, then what is the bill for 4000 kilowatts of electricity?

Solution

Because the amount A of the electric bill varies directly as the amount E of electricity used, we have

$$A = kE$$

for some constant k. Because 2800 kilowatts cost \$196, we have

$$196 = k \cdot 2800$$

or

$$0.07 = k.$$

So $A = 0.07E$. Now if $E = 4000$ we get

$$A = 0.07(4000) = 280.$$

The bill for 4000 kilowatts would be \$280.

Now do Exercises 25–26

Study Tip

If your grades are not what you would like and you are doing all of your work, then see your teacher for advice. Your grade varies directly with the amount of effort that you put forth.

Inverse Variation

If you plan to make a 400-mile trip by car, the time it will take depends on your rate of speed. Using the formula $D = RT$, we can write

$$T = \frac{400}{R}.$$

Consider the possible values for R and T given in the following table:

R (mph)	10	20	40	50	80	100
T (hours)	40	20	10	8	5	4

The graph of $T = \frac{400}{R}$ is shown in Fig. 3.33. As your rate increases, the time for the trip decreases. In this situation we say that the time is *inversely proportional* to the speed. Note that the graph of $T = \frac{400}{R}$ is not a straight line because $T = \frac{400}{R}$ is not a linear equation.

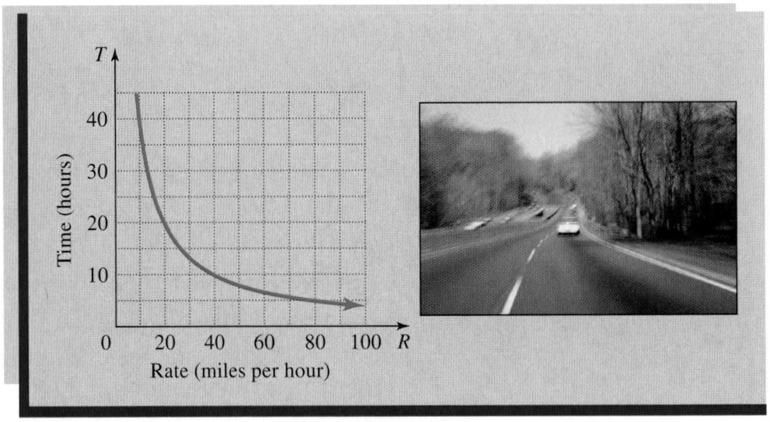

Figure 3.33

Inverse Variation

The statement **"y varies inversely as x"** or **"y is inversely proportional to x"** means that

$$y = \frac{k}{x}$$

for some nonzero constant of variation k.

CAUTION The constant of variation is usually positive because most physical examples involve positive quantities. However, the definitions of direct and inverse variation do not rule out a negative constant.

EXAMPLE 3

An inverse variation problem

The volume of a gas in a cylinder is inversely proportional to the pressure on the gas. If the volume is 12 cubic centimeters when the pressure on the gas is 200 kilograms per square centimeter, then what is the volume when the pressure is 150 kilograms per square centimeter? See Fig. 3.34.

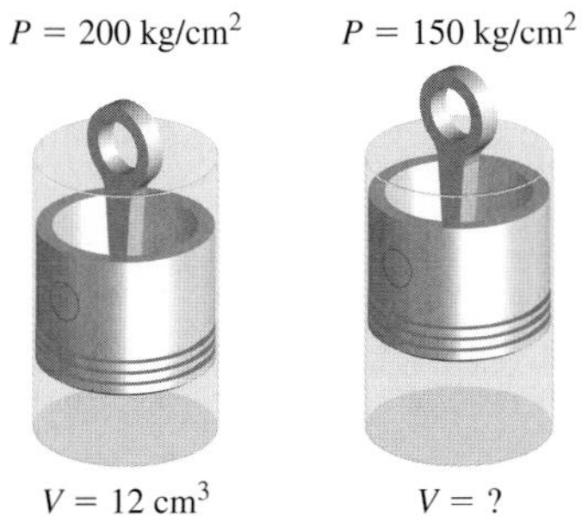

Figure 3.34

Solution

Because the volume V is inversely proportional to the pressure P, we have

$$V = \frac{k}{P}$$

for some constant k. Because $V = 12$ when $P = 200$, we can find k:

$$12 = \frac{k}{200}$$

$$200 \cdot 12 = 200 \cdot \frac{k}{200} \quad \text{Multiply each side by 200.}$$

$$2400 = k$$

Now to find V when $P = 150$, we can use the formula $V = \frac{2400}{P}$:

$$V = \frac{2400}{150} = 16$$

So the volume is 16 cubic centimeters when the pressure is 150 kilograms per square centimeter.

Now do Exercises 27–28

Joint Variation

If the price of carpet is \$30 per square yard, then the cost C of carpeting a rectangular room depends on the width W (in yards) and the length L (in yards). As the width or length of the room increases, so does the cost. We can write the cost in terms of the two variables L and W:

$$C = 30LW$$

We say that C *varies jointly* as L and W.

Joint Variation

The statement **"y varies jointly as x and z"** or **"y is jointly proportional to x and z"** means that

$$y = kxz$$

for some nonzero constant of variation k.

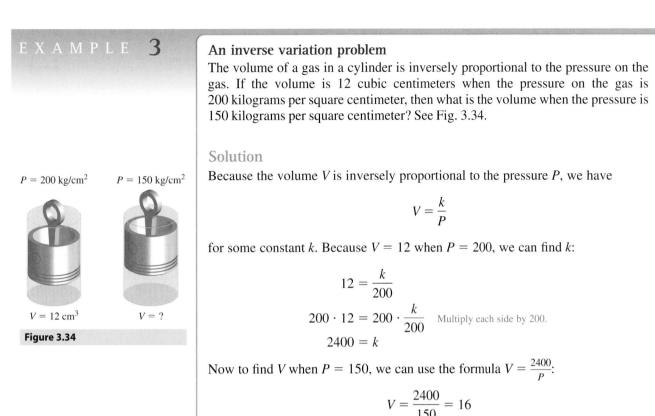

EXAMPLE 4

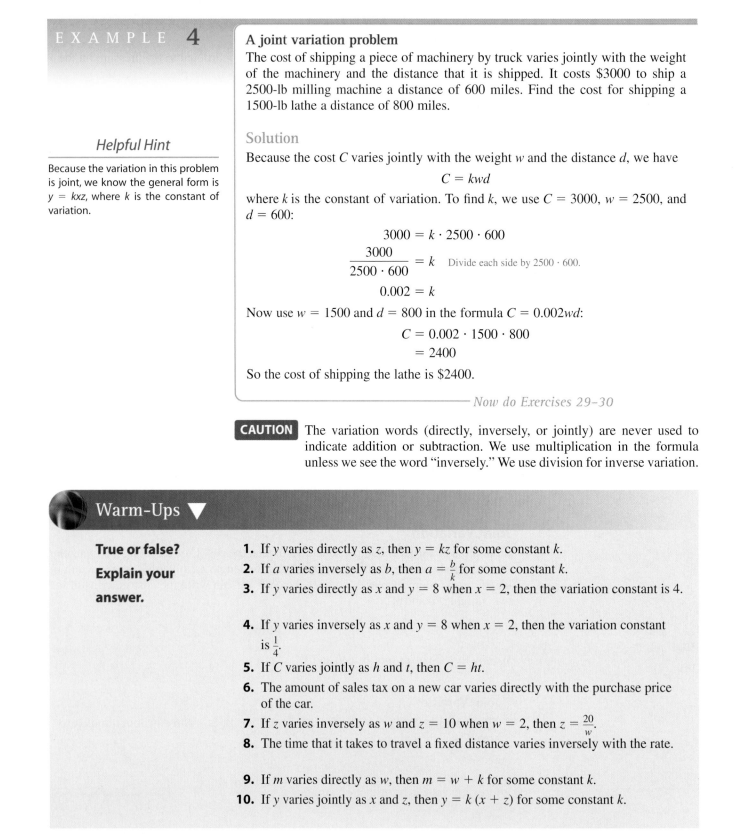

A joint variation problem

The cost of shipping a piece of machinery by truck varies jointly with the weight of the machinery and the distance that it is shipped. It costs $3000 to ship a 2500-lb milling machine a distance of 600 miles. Find the cost for shipping a 1500-lb lathe a distance of 800 miles.

Helpful Hint

Because the variation in this problem is joint, we know the general form is $y = kxz$, where k is the constant of variation.

Solution

Because the cost C varies jointly with the weight w and the distance d, we have

$$C = kwd$$

where k is the constant of variation. To find k, we use $C = 3000$, $w = 2500$, and $d = 600$:

$$3000 = k \cdot 2500 \cdot 600$$

$$\frac{3000}{2500 \cdot 600} = k \quad \text{Divide each side by } 2500 \cdot 600.$$

$$0.002 = k$$

Now use $w = 1500$ and $d = 800$ in the formula $C = 0.002wd$:

$$C = 0.002 \cdot 1500 \cdot 800$$

$$= 2400$$

So the cost of shipping the lathe is $2400.

Now do Exercises 29–30

CAUTION The variation words (directly, inversely, or jointly) are never used to indicate addition or subtraction. We use multiplication in the formula unless we see the word "inversely." We use division for inverse variation.

Warm-Ups ▼

True or false? Explain your answer.

1. If y varies directly as z, then $y = kz$ for some constant k.
2. If a varies inversely as b, then $a = \frac{b}{k}$ for some constant k.
3. If y varies directly as x and $y = 8$ when $x = 2$, then the variation constant is 4.
4. If y varies inversely as x and $y = 8$ when $x = 2$, then the variation constant is $\frac{1}{4}$.
5. If C varies jointly as h and t, then $C = ht$.
6. The amount of sales tax on a new car varies directly with the purchase price of the car.
7. If z varies inversely as w and $z = 10$ when $w = 2$, then $z = \frac{20}{w}$.
8. The time that it takes to travel a fixed distance varies inversely with the rate.
9. If m varies directly as w, then $m = w + k$ for some constant k.
10. If y varies jointly as x and z, then $y = k(x + z)$ for some constant k.

3.5 Exercises

Boost your GRADE at mathzone.com!

MathZone
- Practice Problems
- Self-Tests
- Videos
- Net Tutor
- e-Professors

Reading and Writing *After reading this section, write out the answers to these questions. Use complete sentences.*

1. What does it mean to say that y varies directly as x?

2. What is a variation constant?

3. What does it mean to say that y is inversely proportional to x?

4. What does it mean to say that y varies jointly as x and z?

Write a formula that expresses the relationship described by each statement. Use k for the constant in each case. See Examples 1–4.

5. T varies directly as h.

6. m varies directly as p.

7. y varies inversely as r.

8. u varies inversely as n.

9. R is jointly proportional to t and s.

10. W varies jointly as u and v.

11. i is directly proportional to b.

12. p is directly proportional to x.

13. A is jointly proportional to y and m.

14. t is inversely proportional to e.

Find the variation constant, and write a formula that expresses the indicated variation. See Example 1.

15. y varies directly as x, and $y = 5$ when $x = 3$.

16. m varies directly as w, and $m = \frac{1}{2}$ when $w = \frac{1}{4}$.

17. A varies inversely as B, and $A = 3$ when $B = 2$.

18. c varies inversely as d, and $c = 5$ when $d = 2$.

19. m varies inversely as p, and $m = 22$ when $p = 9$.

20. s varies inversely as v, and $s = 3$ when $v = 4$.

21. A varies jointly as t and u, and $A = 24$ when $t = 6$ and $u = 2$.

22. N varies jointly as p and q, and $N = 720$ when $p = 3$ and $q = 2$.

23. T varies directly as u, and $T = 9$ when $u = 2$.

24. R varies directly as p, and $R = 30$ when $p = 6$.

Solve each variation problem. See Examples 2–4.

25. Y varies directly as x, and $Y = 100$ when $x = 20$. Find Y when $x = 5$.

26. n varies directly as q, and $n = 39$ when $q = 3$. Find n when $q = 8$.

27. a varies inversely as b, and $a = 3$, when $b = 4$. Find a when $b = 12$.

28. y varies inversely as w, and $y = 9$ when $w = 2$. Find y when $w = 6$.

29. P varies jointly as s and t, and $P = 56$ when $s = 2$ and $t = 4$. Find P when $s = 5$ and $t = 3$.

30. B varies jointly as u and v, and $B = 12$ when $u = 4$ and $v = 6$. Find B when $u = 5$ and $v = 8$.

Use the given formula to fill in the missing entries in each table and determine whether b varies directly or inversely as a.

31. $b = \dfrac{300}{a}$

a	b
$\frac{1}{2}$	
1	
	10
900	

32. $b = \dfrac{500}{a}$

a	b
$\frac{1}{5}$	
1	
	10
1500	

33. $b = \dfrac{3}{4}a$

a	b
$\frac{1}{3}$	
8	
	9
20	

34. $b = \dfrac{2}{3}a$

a	b
$\frac{1}{2}$	
3	
	6
21	

For each table, determine whether y varies directly or inversely as x and find a formula for y in terms of x.

35.

x	y
2	7
3	10.5
4	14
5	17.5

36.

x	y
10	5
15	7.5
20	10
25	12.5

37.

x	y
2	10
4	5
10	2
20	1

38.

x	y
5	100
10	50
50	10
250	2

Solve each problem.

39. ***Distance.*** With the cruise control set at 65 mph, the distance traveled varies directly with the time spent traveling. Fill in the missing entries in the following table.

Time (hours)	1	2	3	4
Distance (miles)				

40. ***Cost.*** With gas selling for \$1.60 per gallon, the cost of filling your tank varies directly with the amount of gas that you pump. Fill in the missing entries in the following table.

Amount (gallons)	5	10	15	20
Cost (dollars)				

41. ***Time.*** The time that it takes to complete a 400-mile trip varies inversely with your average speed. Fill in the missing entries in the following table.

Speed (mph)	20	40	50	
Time (hours)				2

42. ***Amount.*** The amount of gasoline that you can buy for \$20 varies inversely with the price per gallon. Fill in the missing entries in the following table.

Price per gallon (dollars)	1	2	4	
Amount (gallons)				2

43. ***Carpeting.*** The cost C of carpeting a rectangular living room with \$20 per square yard carpet varies jointly with the length L and the width W. Fill in the missing entries in the following table.

Length (yd)	Width (yd)	Cost (\$)
8	10	
10		2400
	14	3360

44. ***Waterfront property.*** At \$50 per square foot, the price of a rectangular waterfront lot varies jointly with the length and width. Fill in the missing entries in the following table.

Length (ft)	Width (ft)	Cost (\$)
60	100	
80		360,000
	150	750,000

45. ***Aluminum flatboat.*** The weight of an aluminum flatboat varies directly with the length of the boat. If a 12-foot boat weighs 86 pounds, then what is the weight of a 14-foot boat?

46. ***Christmas tree.*** The price of a Christmas tree varies directly with the height. If a 5-foot tree costs \$20, then what is the price of a 6-foot tree?

47. ***Sharing the work.*** The time it takes to erect the big circus tent varies inversely as the number of elephants working on the job. If it takes four elephants 75 minutes, then how long would it take six elephants?

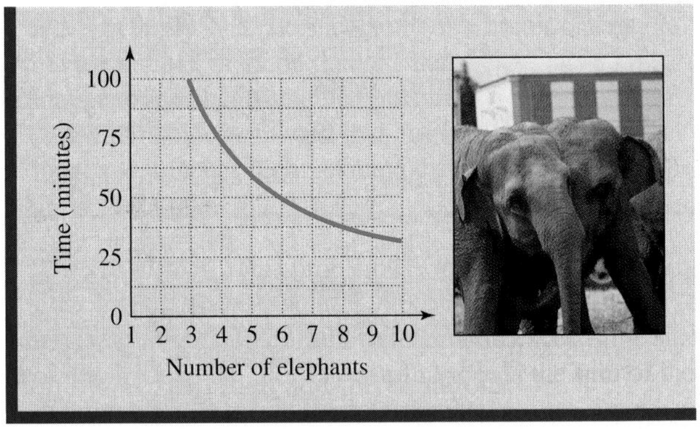

Figure for Exercise 47

48. ***Gas laws.*** The volume of a gas is inversely proportional to the pressure on the gas. If the volume is 6 cubic centimeters when the pressure on the gas is 8 kilograms per square centimeter, then what is the volume when the pressure is 12 kilograms per square centimeter?

49. ***Steel tubing.*** The cost of steel tubing is jointly proportional to its length and diameter. If a 10-foot tube with a 1-inch diameter costs \$5.80, then what is the cost of a 15-foot tube with a 2-inch diameter?

50. ***Sales tax.*** The amount of sales tax varies jointly with the number of Cokes purchased and the price per Coke. If the sales tax on eight Cokes at 65 cents each is 26 cents, then what is the sales tax on six Cokes at 90 cents each?

51. ***Approach speed.*** The approach speed of an airplane is directly proportional to its landing speed. If the approach speed for a Piper Cheyenne is 90 mph with a landing speed of 75 mph, then what is the landing speed for an airplane with an approach speed of 96 mph?

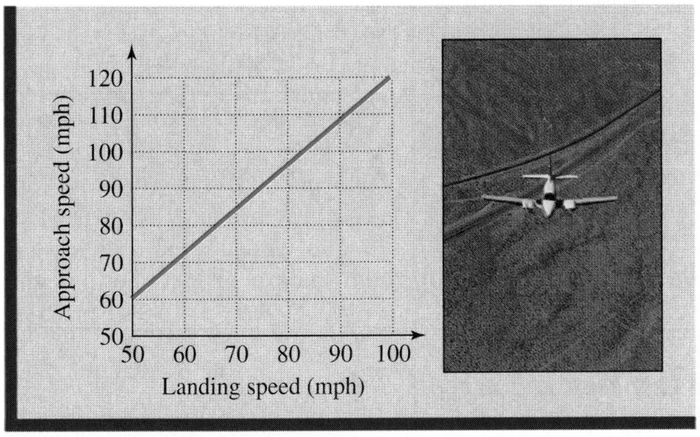

Figure for Exercise 51

52. ***Ideal waist size.*** According to Dr. Aaron R. Folsom of the University of Minnesota School of Public Health, your maximum ideal waist size is directly proportional to your hip size. For a woman with 40-inch hips, the maximum ideal waist size is 32 inches. What is the maximum ideal waist size for a woman with 35-inch hips?

53. ***Sugar Pops.*** The number of days that it takes to eat a large box of Sugar Pops varies inversely with the size of the family. If a family of three eats a box in 7 days, then how many days does it take a family of seven?

54. ***Cost of CDs.*** The cost for manufacturing a CD varies inversely with the number of CDs made. If the cost is \$2.50 per CD when 10,000 are made, then what is the cost per CD when 100,000 are made.

Getting More Involved

55. ***Discussion***

If y varies directly as x, then the graph of the equation is a straight line. What is its slope? What is the y-intercept? If $y = 3x + 2$, then does y vary directly as x? Which straight lines correspond to direct variations?

56. ***Writing***

Write a summary of the three types of variation. Include an example of each type that is not found in this text.

Collaborative Activities

Grouping: Three to four students

Topic: Plotting points, graphing lines

Inches or Centimeters?

In this activity you will generate data by measuring in both inches and centimeters the height of each member of your group. Then you will plot the points on a graph and use any two of your points to find the conversion formula for converting inches to centimeters.

Part I: Measure the height of each person in your group and fill out a table like the one shown here:

Name	Height in Inches	Height in Centimeters

Part II: The numbers for inches and centimeters from the table will give you three or four ordered pairs to graph. Plot these points on a graph. Let inches be the horizontal x-axis and centimeters be the vertical y-axis. Let each mark on the axes represent 10 units. When graphing, you will need to estimate the place to plot fractional values.

Part III: Use any two of your points to find an equation of the line you have graphed. What is the slope of your line? Where does it cross the horizontal axis?

Extension: Look up the conversion formula for converting inches to centimeters. Is it the same as the one you found by measuring? If it is different, what could account for the difference?

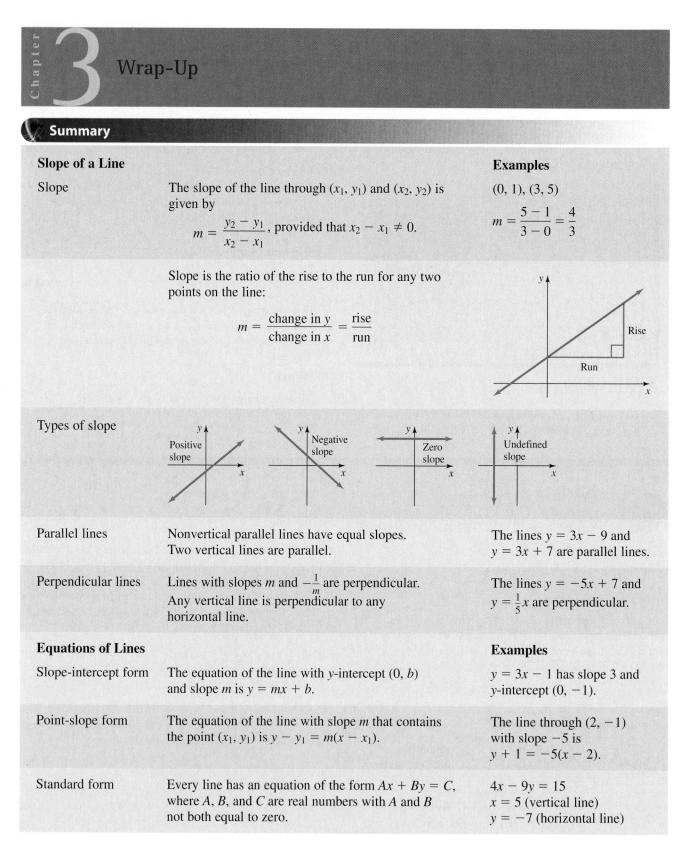

Chapter 3 Wrap-Up

Summary

Slope of a Line

Examples

Slope

The slope of the line through (x_1, y_1) and (x_2, y_2) is given by

$$m = \frac{y_2 - y_1}{x_2 - x_1}, \text{ provided that } x_2 - x_1 \neq 0.$$

Example: $(0, 1), (3, 5)$

$$m = \frac{5 - 1}{3 - 0} = \frac{4}{3}$$

Slope is the ratio of the rise to the run for any two points on the line:

$$m = \frac{\text{change in } y}{\text{change in } x} = \frac{\text{rise}}{\text{run}}$$

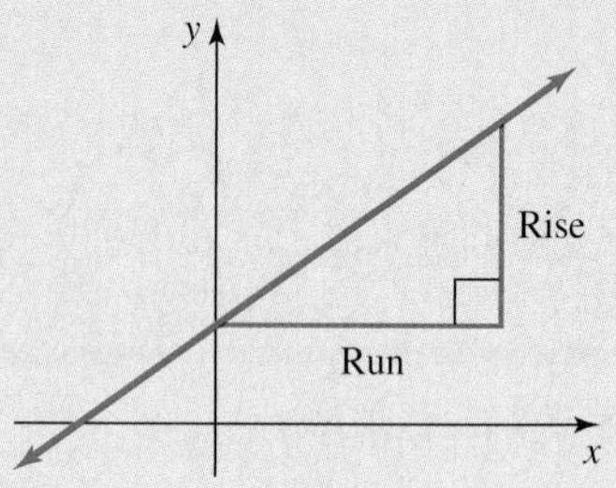

Types of slope

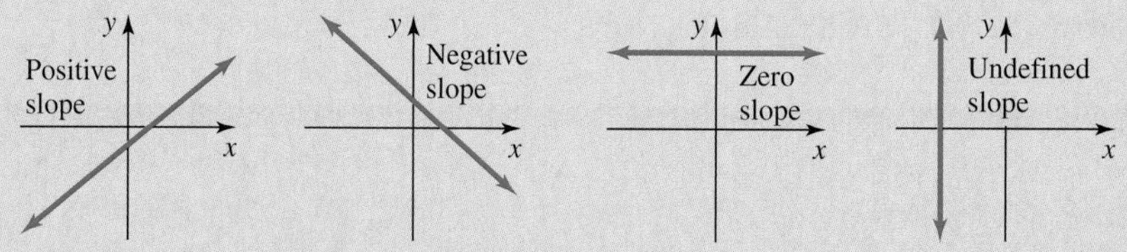

Parallel lines

Nonvertical parallel lines have equal slopes. Two vertical lines are parallel.

Example: The lines $y = 3x - 9$ and $y = 3x + 7$ are parallel lines.

Perpendicular lines

Lines with slopes m and $-\frac{1}{m}$ are perpendicular. Any vertical line is perpendicular to any horizontal line.

Example: The lines $y = -5x + 7$ and $y = \frac{1}{5}x$ are perpendicular.

Equations of Lines

Examples

Slope-intercept form

The equation of the line with y-intercept $(0, b)$ and slope m is $y = mx + b$.

Example: $y = 3x - 1$ has slope 3 and y-intercept $(0, -1)$.

Point-slope form

The equation of the line with slope m that contains the point (x_1, y_1) is $y - y_1 = m(x - x_1)$.

Example: The line through $(2, -1)$ with slope -5 is $y + 1 = -5(x - 2)$.

Standard form

Every line has an equation of the form $Ax + By = C$, where A, B, and C are real numbers with A and B not both equal to zero.

Examples:
$4x - 9y = 15$
$x = 5$ (vertical line)
$y = -7$ (horizontal line)

Graphing a line using y-intercept and slope	1. Write the equation in slope-intercept form. 2. Plot the y-intercept. 3. Use the rise and run to locate a second point. 4. Draw a line through the two points.

Variation		**Examples**
Direct	If $y = kx$, then y varies directly as x.	$D = 50T$
Inverse	If $y = \frac{k}{x}$, then y varies inversely as x.	$R = \dfrac{400}{T}$
Joint	If $y = kxz$, then y varies jointly as x and z.	$V = 6LW$

Enriching Your Mathematical Word Power

For each mathematical term, choose the correct meaning.

1. graph of an equation
a. the Cartesian coordinate system
b. two number lines that intersect at a right angle
c. the x-axis and y-axis
d. an illustration in the coordinate plane that shows all ordered pairs that satisfy an equation

2. x-coordinate
a. the first number in an ordered pair
b. the second number in an ordered pair
c. a point on the x-axis
d. a point where a graph crosses the x-axis

3. y-intercept
a. the second number in an ordered pair
b. a point at which a graph intersects the y-axis
c. any point on the y-axis
d. the point where the y-axis intersects the x-axis

4. coordinate plane
a. a matching plane
b. when the x-axis is coordinated with the y-axis
c. a plane with a rectangular coordinate system
d. a coordinated system for graphs

5. slope
a. the change in x divided by the change in y
b. a measure of the steepness of a line
c. the run divided by the rise
d. the slope of a line

6. slope-intercept form
a. $y = mx + b$
b. rise over run
c. the point at which a line crosses the y-axis
d. $y - y_1 = m(x - x_1)$

7. point-slope form
a. $Ax + By = C$
b. rise over run
c. $y - y_1 = m(x - x_1)$
d. the slope of a line at a single point

8. independent variable
a. a rational constant
b. an irrational constant
c. the first variable of an ordered pair
d. the second variable of an ordered pair

9. dependent variable
a. an irrational variable
b. a rational variable
c. the first variable of an ordered pair
d. the second variable of an ordered pair

10. direct variation
a. $y = \pi$
b. $y = kx$
c. $y = k/x$
d. $y = kxz$

11. inverse variation
a. $y = \pi$
b. $y = kx$
c. $y = k/x$
d. $y = kxz$

12. joint variation
a. $y = \pi$
b. $y = kx$
c. $y = k/x$
d. $y = kxz$

Review Exercises

3.1 *For each point, name the quadrant in which it lies or the axis on which it lies.*

1. $(-2, 5)$

2. $(-3, -5)$

3. $(3, 0)$

4. $(9, 10)$

5. $(0, -6)$

6. $(0, \pi)$

7. $(1.414, -3)$

8. $(-4, 1.732)$

Study Tip

Note how the review exercises are arranged according to the sections in this chapter. If you are having trouble with a certain type of problem, refer back to the appropriate section for examples and explanations.

Complete the given ordered pairs so that each ordered pair satisfies the given equation.

9. $y = 3x - 5$: $(0, \quad)$, $(-3, \quad)$, $(4, \quad)$

10. $y = -2x + 1$: $(9, \quad)$, $(3, \quad)$, $(-1, \quad)$

11. $2x - 3y = 8$: $(0, \quad)$, $(3, \quad)$, $(-6, \quad)$

12. $x + 2y = 1$: $(0, \quad)$, $(-2, \quad)$, $(2, \quad)$

Sketch the graph of each equation by finding three ordered pairs that satisfy each equation.

13. $y = -3x + 4$

14. $y = 2x - 6$

15. $x + y = 7$

16. $x - y = 4$

3.2 *Determine the slope of the line that goes through each pair of points.*

17. $(0, 0)$ and $(1, 1)$

18. $(-1, 1)$ and $(2, -2)$

19. $(-2, -3)$ and $(0, 0)$

20. $(-1, -2)$ and $(4, -1)$

21. $(-4, -2)$ and $(3, 1)$

22. $(0, 4)$ and $(5, 0)$

3.3 *Find the slope and y-intercept for each line.*

23. $y = 3x - 18$

24. $y = -x + 5$

25. $2x - y = 3$

26. $x - 2y = 1$

27. $4x - 2y - 8 = 0$

28. $3x + 5y + 10 = 0$

Sketch the graph of each equation.

29. $y = \frac{2}{3}x - 5$

30. $y = \frac{3}{2}x + 1$

31. $2x + y = -6$

32. $-3x - y = 2$

33. $y = -4$

34. $x = 9$

Determine the equation of each line. Write the answer in standard form using only integers as the coefficients.

35. The line through (0, 4) with slope $\frac{1}{3}$

36. The line through (−2, 0) with slope $-\frac{3}{4}$

37. The line through the origin that is perpendicular to the line $y = 2x - 1$

38. The line through (0, 9) that is parallel to the line $3x + 5y = 15$

39. The line through (3, 5) that is parallel to the x-axis

40. The line through (−2, 4) that is perpendicular to the x-axis

3.4 *Write each equation in slope-intercept form.*

41. $y - 3 = \frac{2}{3}(x + 6)$

42. $y + 2 = -6(x - 1)$

43. $3x - 7y - 14 = 0$

44. $1 - x - y = 0$

45. $y - 5 = -\frac{3}{4}(x + 1)$

46. $y + 8 = -\frac{2}{5}(x - 2)$

Determine the equation of each line. Write the answer in slope-intercept form.

47. The line through (−4, 7) with slope −2

48. The line through (9, 0) with slope $\frac{1}{2}$

49. The line through the two points (−2, 1) and (3, 7)

50. The line through the two points (4, 0) and (−3, −5)

51. The line through $(3, -5)$ that is parallel to the line $y = 3x - 1$

52. The line through $(4, 0)$ that is perpendicular to the line $x + y = 3$

Solve each problem.

53. ***Rental charge.*** The charge for renting an air hammer for two days is \$113 and the charge for five days is \$209. The charge C is determined by the number of days n using a linear equation. Find the equation and find the charge for a four-day rental.

54. ***Time on a treadmill.*** After 2 minutes on a treadmill, Jenny has a heart rate of 82. After 3 minutes she has a heart rate of 86. Assume that there is a linear equation that gives her heart rate h in terms of time on the treadmill t. Find the equation and use it to predict her heart rate after 10 minutes on the treadmill.

55. ***Probability of rain.*** If the probability p of rain is 90%, the probability q that it does not rain is 10%. If the probability of rain is 80%, then the probability that it does not rain is 20%. There is a linear equation that gives q in terms of p.

a) Find the equation.

b) Use the accompanying graph to determine the probability of rain if the probability that it does not rain is 0.

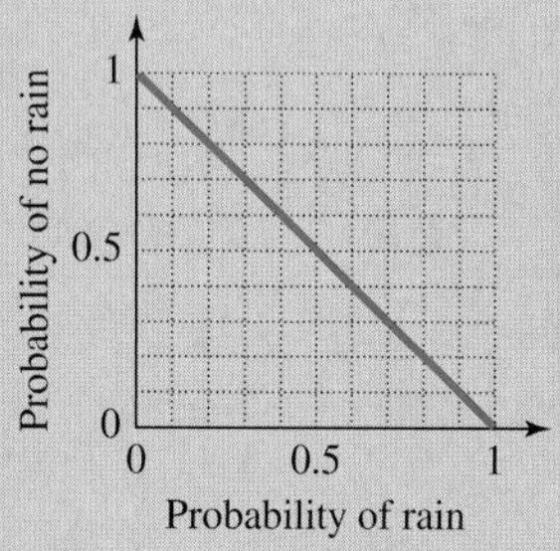

Figure for Exercise 55

56. ***Social Security benefits.*** If you earned an average of \$25,000 over your working life and you retire after 2005 at age 62, 63, or 64, then your annual Social Security benefit will be \$7000, \$7500, or \$8000, respectively (Social Security Administration, www.ssa.gov). There is a linear equation that gives the annual benefit b in terms of age a for these three years. Find the equation.

57. ***Predicting freshman GPA.*** A researcher who is studying the relationship between ACT score and grade point average for freshman gathered the data shown in the accompanying table. Find the equation of the line in slope-intercept form that goes through these points.

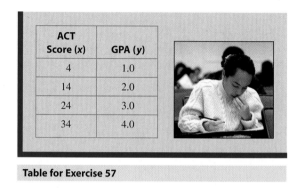

ACT Score (x)	GPA (y)
4	1.0
14	2.0
24	3.0
34	4.0

Table for Exercise 57

58. ***Interest rates.*** A credit manager rates each applicant for a car loan on a scale of 1 through 5 and then determines the interest rate from the accompanying table. Find the equation of the line in slope-intercept form that goes through these points.

Credit Rating	Interest Rate (%)
1	24
2	20
3	16
4	12
5	8

Table for Exercise 58

3.5 *Solve each variation problem.*

59. Suppose y varies directly as w. If $y = 48$ when $w = 4$, then what is y when $w = 11$?

60. Suppose m varies directly as t. If $m = 13$ when $t = 2$, then what is m when $t = 6$?

61. If y varies inversely as v and $y = 8$ when $v = 6$, then what is y when $v = 24$?

62. If y varies inversely as r and $y = 9$ when $r = 3$, then what is y when $r = 9$?

63. Suppose y varies jointly as u and v, and $y = 72$ when $u = 3$ and $v = 4$. Find y when $u = 5$ and $v = 2$.

64. Suppose q varies jointly as s and t, and $q = 10$ when $s = 4$ and $t = 3$. Find q when $s = 25$ and $t = 6$.

65. ***Taxi fare.*** The cost of a taxi ride varies directly with the length of the ride in minutes. A 12-minute ride costs $9.00.

a) Write the cost in terms of the length of the ride.
b) What is the cost of a 20-minute ride?
c) Is the cost increasing or decreasing as the length of the ride increases?

66. ***Applying shingles.*** The number of hours that it takes to apply 296 bundles of shingles varies inversely with the number of roofers working on the job. Three workers can complete the job in 40 hours.

a) Write the number of hours in terms of the number of roofers on the job.
b) How long would it take five roofers to complete the job?
c) Is the time to complete the job increasing or decreasing as the number of workers increases?

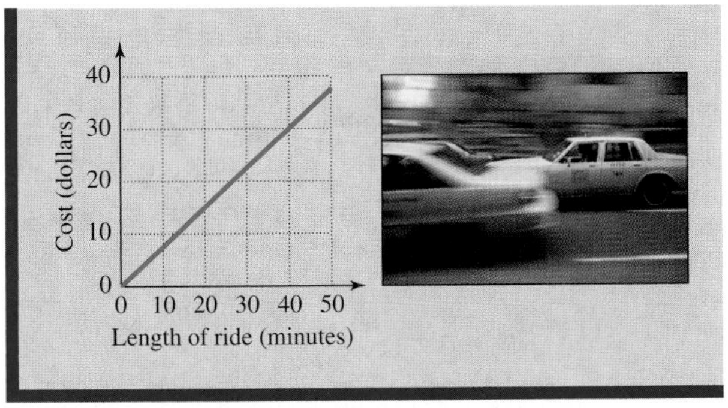

Figure for Exercise 65

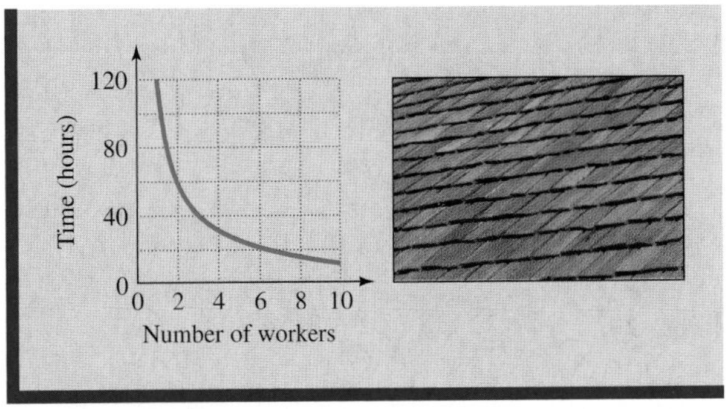

Figure for Exercise 66

Chapter 3 Test

For each point, name the quadrant in which it lies or the axis on which it lies.

1. $(-2, 7)$

2. $(-\pi, 0)$

3. $(3, -6)$

4. $(0, 1785)$

Find the slope of the line through each pair of points.

5. $(3, 3)$ and $(4, 4)$

6. $(-2, -3)$ and $(4, -8)$

Find the slope of each line.

7. The line $y = 3x - 5$

8. The line $y = 3$

9. The line $x = 5$

10. The line $2x - 3y = 4$

Write the equation of each line. Give the answer in slope-intercept form.

11. The line through $(0, 3)$ with slope $-\frac{1}{2}$

12. The line through $(-1, -2)$ with slope $\frac{3}{7}$

Write the equation of each line. Give the answer in standard form using only integers as the coefficients.

13. The line through $(2, -3)$ that is perpendicular to the line $y = -3x + 12$

14. The line through $(3, 4)$ that is parallel to the line $5x + 3y = 9$

Study Tip

Before you take an in-class exam on this chapter, work the sample test given here. Set aside one hour to work this test and use the answers in the back of this book to grade yourself. Even though your instructor might not ask exactly the same questions, you will get a good idea of your test readiness.

Sketch the graph of each equation.

15. $y = \frac{1}{2}x - 3$

16. $2x - 3y = 6$

17. $y = 4$

18. $x = -2$

Solve each problem.

19. Julie's mail-order CD club charges a shipping and handling fee of \$2.50 plus \$0.75 per CD for each order shipped. Write the shipping and handling fee S in terms of the number n of CDs in the order.

20. A 10-ounce soft drink sells for 50 cents, and a 16-ounce soft drink sells for 68 cents. The price P is determined from the volume of the cup v by a linear equation. Find the equation and find the price for a 20-ounce soft drink.

21. The price of a watermelon varies directly with its weight. If the price of a 30-pound watermelon is \$4.20, then what is the price of a 20-pound watermelon?

22. The number of days that Jerry spends on the road is inversely proportional to his sales for the previous month. If Jerry spent 15 days on the road when his previous month's sales were \$75,000, then how many days would he spend on the road when his previous month's sales were \$60,000? Does his road time increase or decrease as his sales increase?

23. The labor cost for installing ceramic floor tile in a rectangular room varies jointly with the length and width. For a room that is 8 feet by 10 feet the cost is \$400. For a room that is 9 feet by 12 feet the cost is \$540. What is the cost for a room that is 11 feet by 14 feet?

Graph Paper

Use these grids for graphing. Make as many copies of this page as you need.

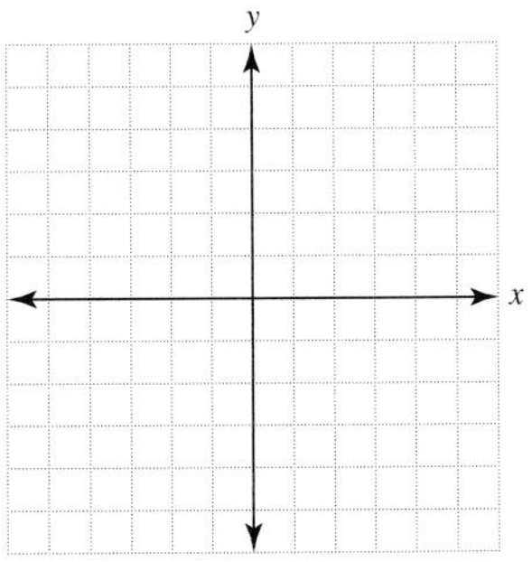

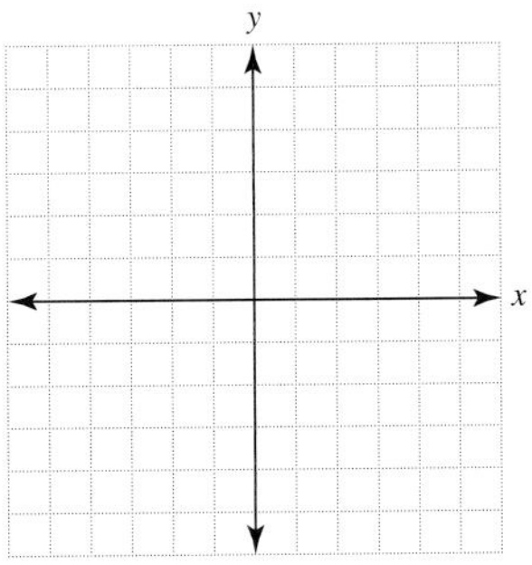

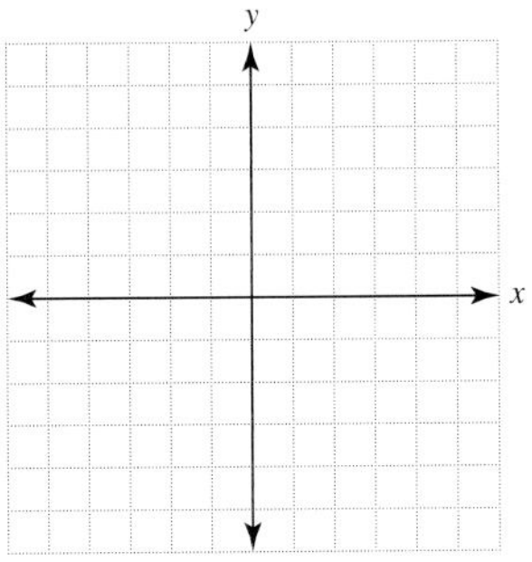

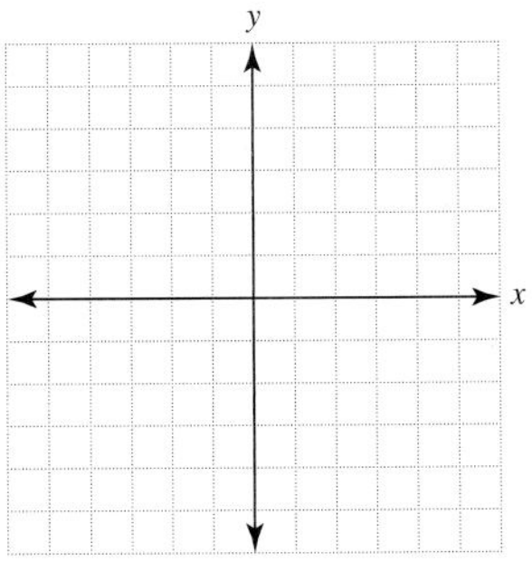

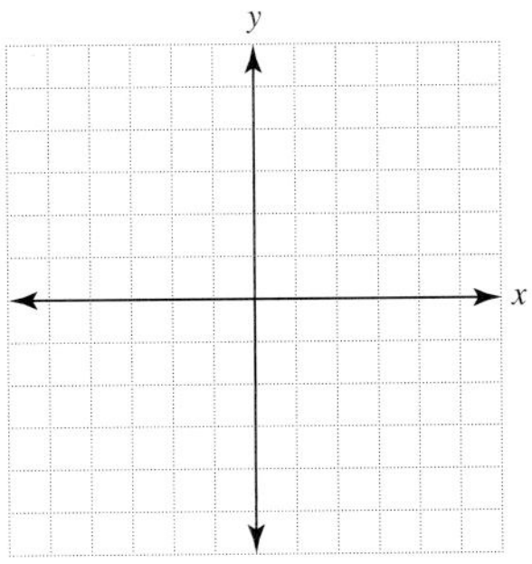

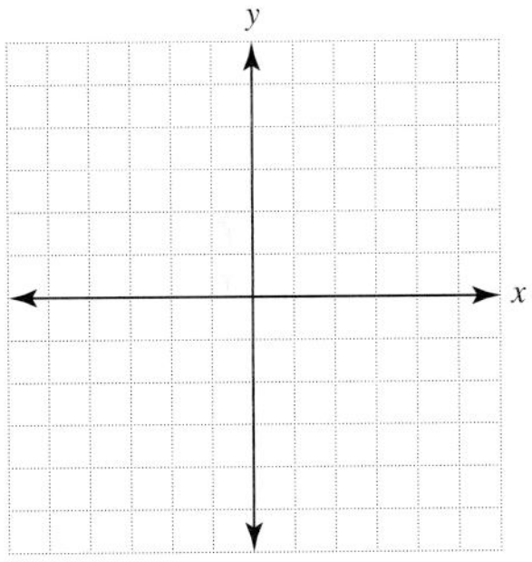

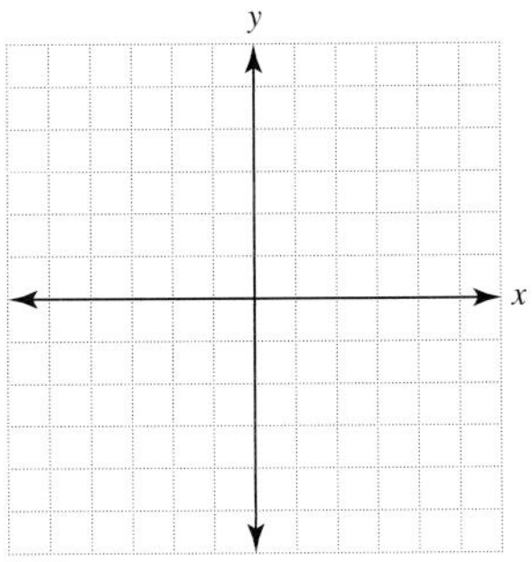

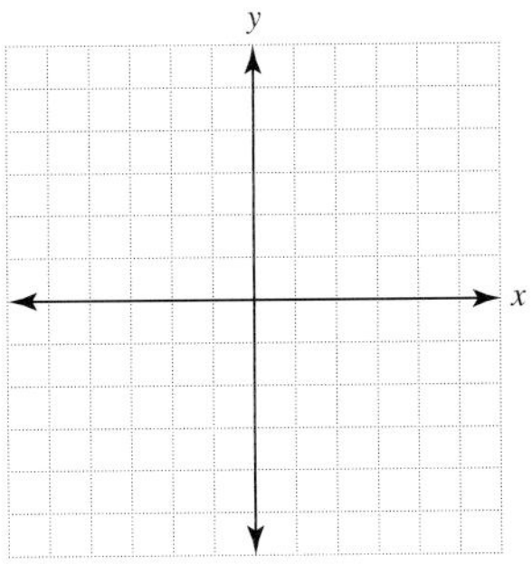

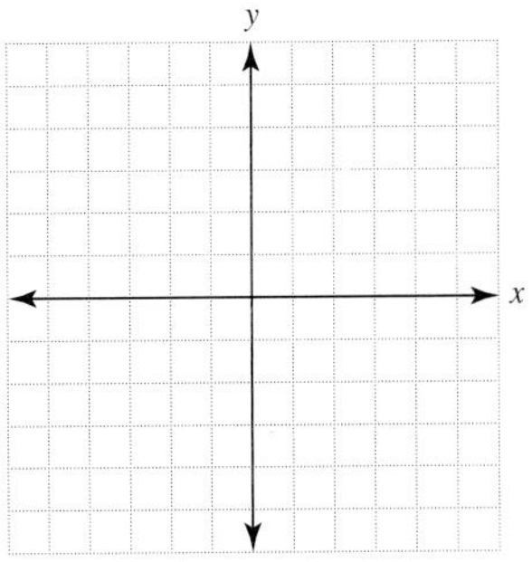

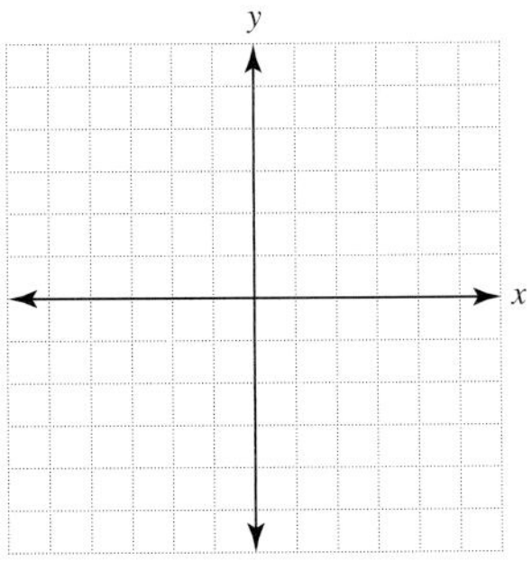

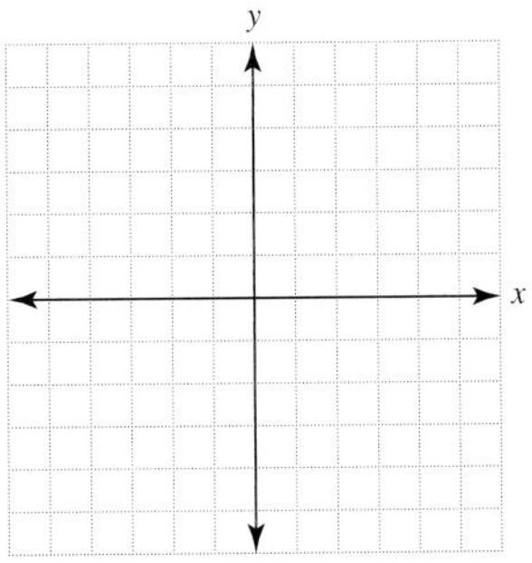

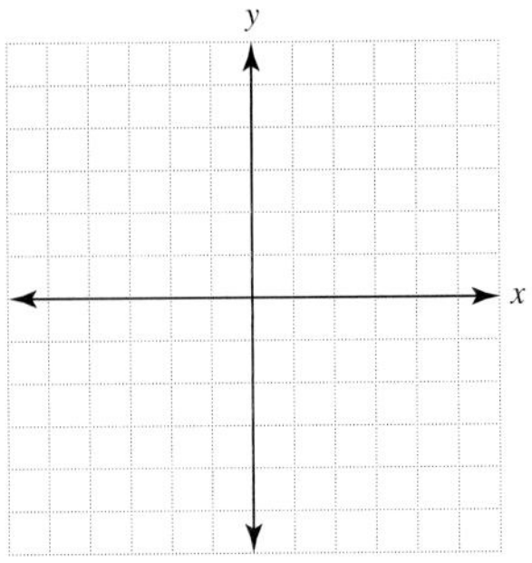

Making Connections | A Review of Chapters 1–3

Simplify each arithmetic expression.

1. $9 - 5 \cdot 2$

2. $-4 \cdot 5 - 7 \cdot 2$

3. $3^2 - 2^3$

4. $3^2 \cdot 2^3$

5. $(-4)^2 - 4(1)(5)$

6. $-4^2 - 4 \cdot 3$

7. $\dfrac{-5 - 9}{2 - (-2)}$

8. $\dfrac{6 - 3.6}{6}$

9. $\dfrac{1 - \frac{1}{2}}{4 - (-1)}$

10. $\dfrac{4 - (-6)}{1 - \frac{1}{3}}$

Simplify the given expression or solve the given equation, whichever is appropriate.

11. $4x - (-9x)$

12. $4(x - 9) - x$

13. $5(x - 3) + x = 0$

14. $5 - 2(x - 1) = x$

15. $\dfrac{1}{2} - \dfrac{1}{3}$

16. $\dfrac{1}{4} + \dfrac{1}{6}$

17. $\dfrac{1}{2}x - \dfrac{1}{3} = \dfrac{1}{4}x + \dfrac{1}{6}$

18. $\dfrac{2}{3}x + \dfrac{1}{5} = \dfrac{3}{5}x - \dfrac{1}{15}$

19. $\dfrac{4x - 8}{2}$

20. $\dfrac{-5x - 10}{-5}$

21. $\dfrac{6 - 2(x - 3)}{2} = 1$

22. $\dfrac{20 - 5(x - 5)}{5} = 6$

23. $-4(x - 9) - 4 = -4x$

24. $4(x - 6) = -4(6 - x)$

Study Tip

Don't wait until the final exam to review material. Do some review on a regular basis. The Making Connections exercises on this page can be used to review, compare, and contrast different concepts that you have studied. A good time to work these exercises is between a test and the start of new material.

Solve each inequality. State the solution set using interval notation.

25. $2x - 3 > 6$

26. $5 - 3x < 7$

27. $51 - 2x \leq 3x + 1$

28. $4x - 80 \geq 60 - 3x$

29. $-1 < 4 - 2x \leq 5$

30. $1 - 2x \leq x + 1 < 3 - 2x$

Solve each equation for y.

31. $3\pi y + 2 = t$

32. $x = \dfrac{y - b}{m}$

33. $3x - 3y - 12 = 0$

34. $2y - 3 = 9$

35. $\dfrac{y}{2} - \dfrac{y}{4} = \dfrac{1}{5}$

36. $0.6y - 0.06y = 108$

Solve.

37. ***Financial planning.*** Financial advisors at Fidelity Investments use the information in the accompanying graph as a guide for retirement investing.

a) What is the slope of the line segment for ages 35 through 50?

b) What is the slope of the line segment for ages 50 through 65?

c) If a 38-year-old man is making $40,000 per year, then what percent of his income should he be saving?

d) If a 58-year-old woman has an annual salary of $60,000, then how much should she have saved and how much should she be saving per year?

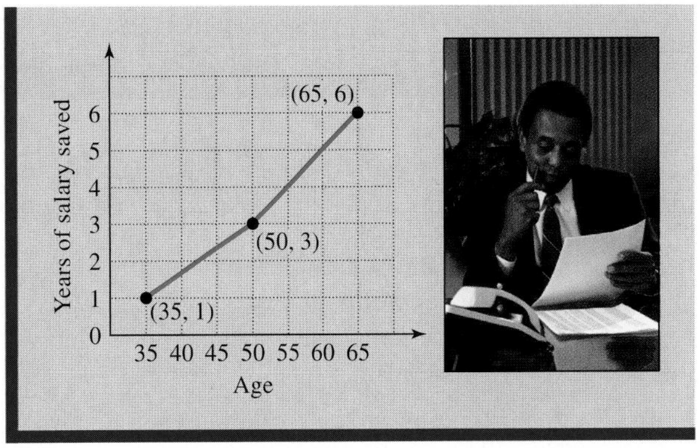

Figure for Exercise 37

Critical Thinking | For Individual or Group Work | Chapter 3

These exercises can be solved by a variety of techniques, which may or may not require algebra. So be creative and think critically. Explain all answers. Answers are in the Instructor's Edition of this text.

1. ***Share and share alike.*** A chocolate bar consists of two rows of small squares with four squares in each row as shown in (a) of the accompanying figure. You want to share it with your friends.

a) How many times must you break it to get it divided into 8 small squares?

b) If the bar has 3 rows of 5 squares in each row as shown in (b) of the accompanying figure, then how many breaks does it take to separate it into 15 small squares?

c) If the bar is divided into m rows with n small squares in each row, then how many breaks does it take to separate it into mn small squares?

(a) (b)

Figure for Exercise 1

2. ***Straight time.*** Starting at 8 A.M. determine the number of times in the next 24 hours for which the hour and minute hands on a clock form a 180° angle.

3. ***Dividing days by months.*** For how many days of the year do you get a whole number when you divide the day number by the month number? For example, for December 24, the result of 24 divided by 12 is 2.

4. ***Crossword fanatic.*** Ms. Smith loves to work the crossword puzzle in her daily newspaper. To keep track of her efforts, she gives herself 2 points for every crossword puzzle that she completes correctly and deducts 3 points for every crossword puzzle that she fails to complete or completes incorrectly. For the month of June her total score was zero. How many puzzles did she solve correctly in June?

Photo for Exercise 4

5. ***Counting ones.*** If you write down the integers between 1 and 100 inclusive, then how many times will you write the number one?

6. ***Smallest sum.*** What is the smallest possible sum that can be obtained by adding five positive integers that have a product of 48?

7. ***Mind control.*** Each student in your class should think of an integer between 2 and 9 inclusive. Multiply your integer by 9. Think of the sum of the digits in your answer. Subtract 5 from your answer. Think of the letter in the alphabet that corresponds to the last answer. Think of a state that begins with that letter. Think of the second letter in the name of the state. Think of a large mammal that begins with that letter. Think of the color of that animal. What is the color that is on everyone's mind? Explain.

8. ***Four-digit numbers.*** How many four-digit whole numbers are there such that the thousands digit is odd, the hundreds digit is even, and all four digits are different? How many four-digit whole numbers are there such that the thousands digit is even, hundreds digit is odd, and all four digits are different?

Chapter

4

Systems of Linear Equations and Inequalities

What determines the prices of the products that you buy? Why do prices of some products go down while the prices of others go up? Economists theorize that prices result from the collective decisions of consumers and producers. Ideally, the demand or quantity purchased by consumers depends only on the price, and price is a function of the supply. Theoretically, if the demand is greater than the supply, then prices rise and manufacturers produce more to meet the demand. As the supply of goods increases, the price comes down. The price at which the supply is equal to the demand is called the equilibrium price.

However, what happens in the real world does not always match the theory. Manufacturers cannot always control the supply, and factors other than price can affect a consumer's decision to buy. For example, droughts in Brazil decreased the supply of coffee and drove coffee prices up. Floods in California did the same to the prices of produce. With one of the most abundant wheat crops ever in 1994, cattle gained weight more quickly, increasing the supply of cattle ready for market. With supply going up, prices went down. Decreased demand for beef in Japan and Mexico drove the price of beef down further. With lower prices, consumers should be buying more beef, but increased competition from chicken and pork products, as well as health concerns, have kept consumer demand low.

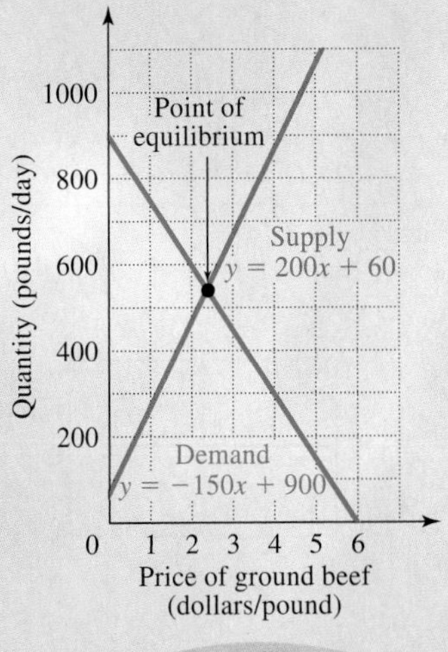

The two functions that govern supply and demand form a system of equations. In this chapter you will learn how to solve systems of equations.

In Exercise 53 of Section 4.2 you will see an example of supply and demand equations for ground beef.

4.1 The Graphing Method

In this Section

- Solving a System of Linear Equations by Graphing
- Independent, Inconsistent, and Dependent Equations
- Applications

You studied linear equations in two variables in Chapter 3. In this section you will learn to solve systems of linear equations in two variables and use systems to solve problems.

Solving a System of Linear Equations by Graphing

Consider the linear equation $y = 2x - 1$. The graph of this equation is a straight line, and every point on the line is a solution to the equation. Now consider a second linear equation, $x + y = 2$. The graph of this equation is also a straight line, and every point on the line is a solution to this equation. The pair of equations

$$y = 2x - 1$$
$$x + y = 2$$

is called a **system of equations.** A point that satisfies both equations is called a **solution to the system.**

EXAMPLE 1

A solution to a system

Determine whether the point $(-1, 3)$ is a solution to each system of equations.

a) $3x - y = -6$
$x + 2y = 5$

b) $y = 2x - 1$
$x + y = 2$

Solution

a) If we let $x = -1$ and $y = 3$ in both equations of the system, we get the following equations:

$$3(-1) - 3 = -6 \quad \text{Correct}$$
$$-1 + 2(3) = 5 \quad \text{Correct}$$

Because both of these equations are correct, $(-1, 3)$ is a solution to the system.

b) If we let $x = -1$ and $y = 3$ in both equations of the system, we get the following equations:

$$3 = 2(-1) - 1 \quad \text{Incorrect}$$
$$-1 + 3 = 2 \quad \text{Correct}$$

Because the first equation is not satisfied by $(-1, 3)$, the point $(-1, 3)$ is not a solution to the system.

Now do Exercises 7–12

Calculator Close-Up

Graph $y_1 = 3x + 6$ and $y_2 = (5 - x)/2$ to see that $(-1, 3)$ is on both lines.

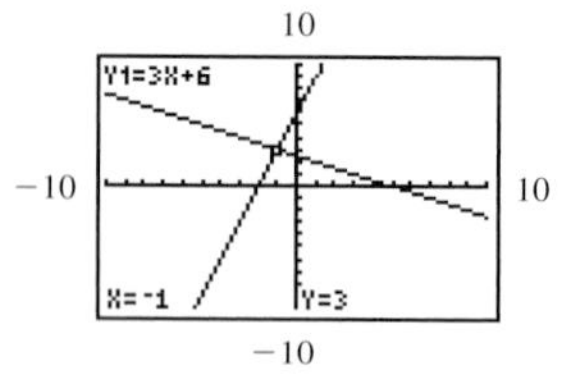

Graph $y_1 = 2x - 1$ and $y_2 = 2 - x$ to see that $(-1, 3)$ is on one line but not the other.

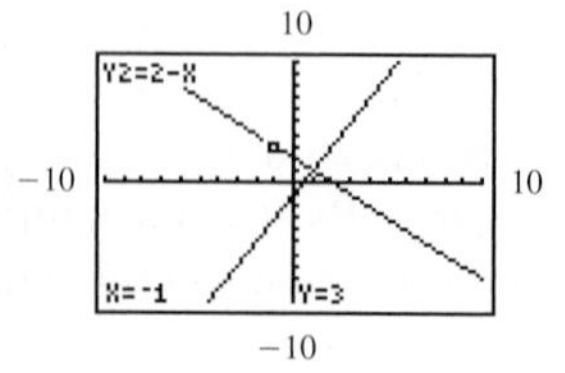

If we graph each equation of a system on the same coordinate plane, then we may be able to see the points that they have in common. Any point that is on both graphs is a solution to the system.

EXAMPLE 2

Using graphing to solve a system

Solve the system by graphing:

$$y = 2x - 1$$
$$x + y = 2$$

Solution

We first write each equation in slope-intercept form:

$$y = 2x - 1$$
$$y = -x + 2$$

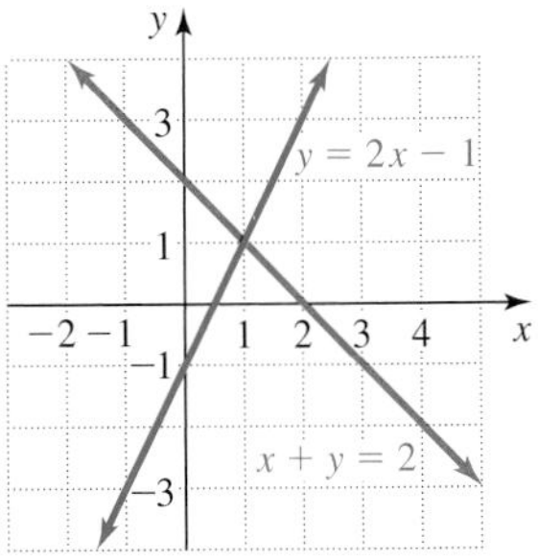

Figure 4.1

Use the y-intercept and the slope to draw the graphs as in Fig. 4.1. From the graph, it appears that these lines intersect at (1, 1). To be certain, we check that (1, 1) satisfies both equations. Let $x = 1$ and $y = 1$ in the original equations:

$$y = 2x - 1 \qquad x + y = 2$$
$$1 = 2(1) - 1 \qquad 1 + 1 = 2$$

Because these equations are both true, (1, 1) is the solution to the system.

Now do Exercises 15–26

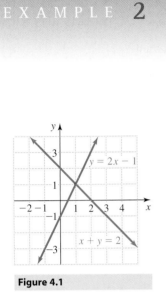

EXAMPLE 3

A system of parallel lines

Solve the system by graphing:

$$3y = 2x - 6$$
$$2x - 3y = 3$$

Solution

Write each equation in slope-intercept form to get the following system:

$$y = \frac{2}{3}x - 2$$
$$y = \frac{2}{3}x - 1$$

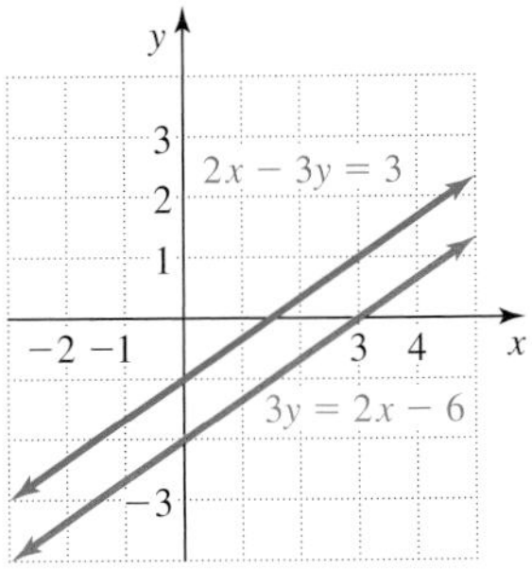

Figure 4.2

Each line has slope $\frac{2}{3}$, but they have different y-intercepts. Their graphs are shown in Fig. 4.2. Because these two lines have the same slope, they are parallel and there is no point on both lines. The system has no solution.

Now do Exercises 27–30

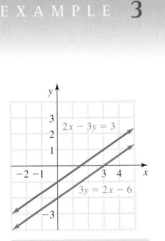

Independent, Inconsistent, and Dependent Equations

When two lines are positioned in a coordinate plane, they might intersect at a single point, they might have no intersection, or they might coincide. There are no other possibilities. If they intersect at a single point, then there is exactly one solution to the corresponding system of linear equations. If the lines are parallel, then there is no solution to the system. If the lines are coincident, then any point on the line satisfies both equations of the system.

Independent, Inconsistent, and Dependent

The equations of a system of two linear equations are

1. **independent** if the lines intersect in exactly one point,
2. **inconsistent** if the lines are parallel, and
3. **dependent** if the lines coincide.

Figure 4.3 shows each of the cases.

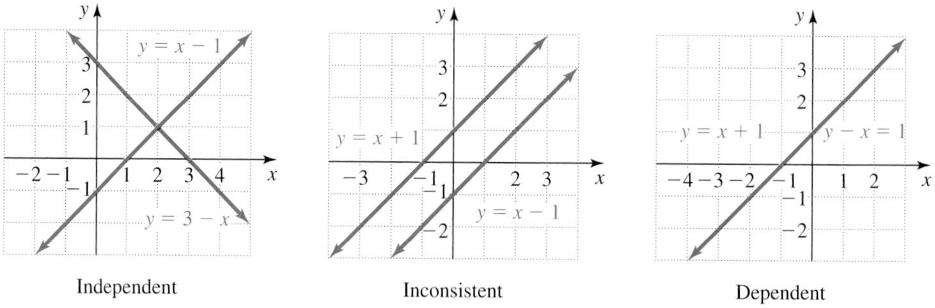

Figure 4.3

Because the system of Example 2 had a single point for the solution, the equations are independent (or the system is independent). Because the graphs of the equations in Example 3 were parallel, the equations are inconsistent and there is no solution to that system. Example 4 illustrates a system of dependent equations.

EXAMPLE 4

A dependent system of equations

Solve the system by graphing:

$$4x - 2y = 6$$
$$y - 2x = -3$$

Solution

Rewrite both equations in slope-intercept form for easy graphing:

$$\begin{aligned} 4x - 2y &= 6 \\ -2y &= -4x + 6 \\ y &= 2x - 3 \end{aligned} \qquad \begin{aligned} y - 2x &= -3 \\ y &= 2x - 3 \end{aligned}$$

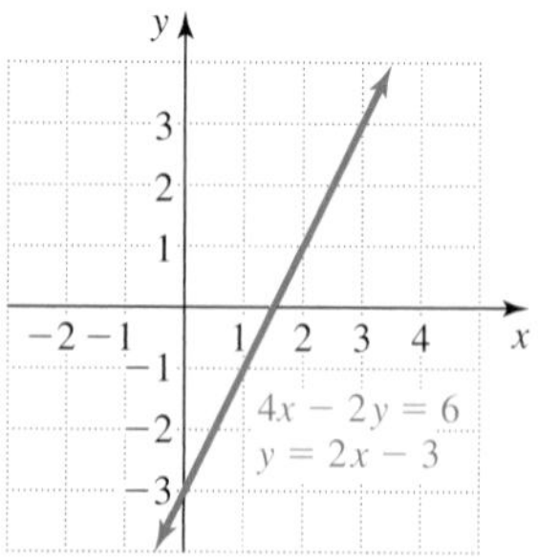

Figure 4.4

By writing the equations in slope-intercept form, we discover that they are identical. The graphs of the system are shown in Fig. 4.4. Because the graphs of the two equations are identical, any point on the line satisfies both equations. The set of points on that line is written as

$$\{(x, y) \mid y = 2x - 3\}.$$

The system is dependent.

Now do Exercises 31–38

Calculator Close-Up

With a graphing calculator, you can graph both equations of a system in a single viewing window. The TRACE feature can then be used to estimate the solution to an independent system. You could also use ZOOM to "blow up" the intersection and get more accuracy. Many calculators have an intersect feature, which can find a point of intersection. First graph $y_1 = 2x - 1$ and $y_2 = 2 - x$.

From the CALC menu choose intersect.

Verify the curves (or lines) that you want to intersect by pressing ENTER. After you make a guess as to the intersection by positioning the cursor or entering a number, the calculator will find the intersection.

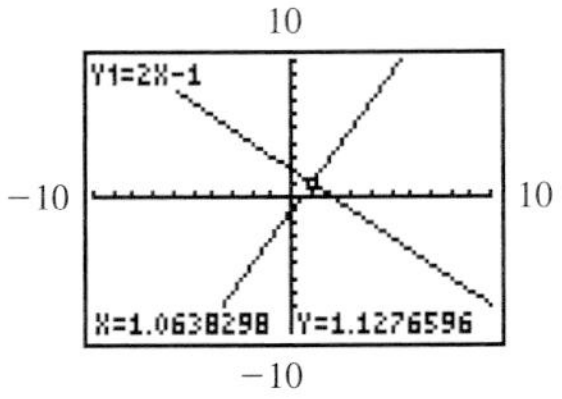

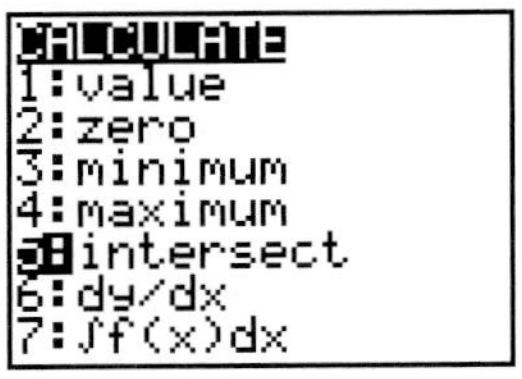

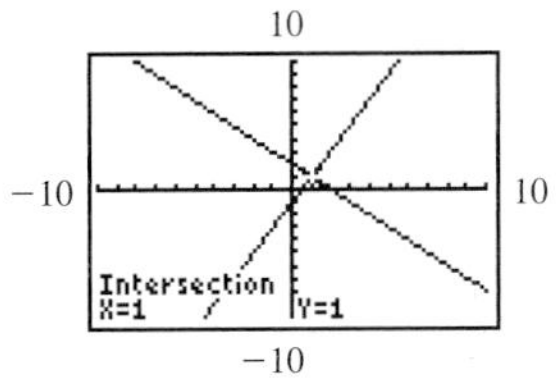

Applications

In Example 5 we see the ideas of supply, demand, and equilibrium price. Supply is the quantity producers are willing to make or supply. Demand is the quantity consumers will purchase. Both supply and demand depend on the price. As the price increases, producers increase the supply to take advantage of the rising prices. However, as the price increases, consumer demand decreases. The equilibrium price is the price at which supply equals demand.

EXAMPLE 5

Supply and demand

Monthly demand for Greeny Babies (small toy frogs) is given by the equation $y = 8000 - 400x$, while monthly supply is given by the equation $y = 400x$, where x is the price in dollars. Graph the two equations and find the equilibrium price and the demand at the equilibrium price.

Solution

The graph of $y = 8000 - 400x$ goes through (0, 8000) and (20, 0). The graph of $y = 400x$ goes through (0, 0) and (20, 8000). The two lines cross at (10, 4000) as shown in Fig. 4.5. So the equilibrium price is \$10 and the monthly demand is 4000 Greeny Babies.

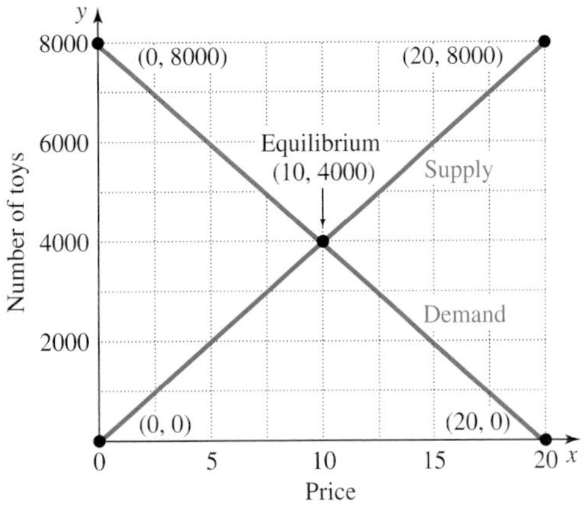

Figure 4.5

Now do Exercises 53–56

Warm-Ups ▼

True or false? Explain your answer.

The statements refer to the following systems:

a) $y = 2x - 5$, $y = -2x - 5$ **b)** $y = 3x - 4$, $y = 3x + 5$ **c)** $x + y = 9$, $y = 9 - x$

1. The ordered pair $(1, -3)$ satisfies the equation $y = 2x - 5$.
2. The ordered pair $(1, -3)$ satisfies the equation $y = -2x - 5$.
3. The ordered pair $(1, -3)$ is a solution to system (a).
4. System (a) is inconsistent.
5. System (b) has no solution.
6. The equations of system (b) are inconsistent.
7. System (c) is dependent.
8. The set of ordered pairs that satisfy system (c) is $\{(x, y) \mid y = 9 - x\}$.
9. Two distinct straight lines in the coordinate plane either are parallel or intersect each other in exactly one point.
10. Any system of two linear equations can be solved by graphing.

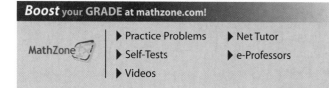

4.1 Exercises

Reading and Writing *After reading this section, write out the answers to these questions. Use complete sentences.*

1. What is a system of equations?
2. What is a solution to a system of equations?
3. What method was used to solve a system of equations?
4. What is an independent system?
5. What is a dependent system?
6. What is an inconsistent system?

Which of the given points is a solution to the given system? See Example 1.

7. $2x + y = 4$, $x - y = 5$ (6, 1), (3, −2), (2, 4)
8. $2x - 3y = -5$, $y = x + 1$ (−1, 1), (3, 4), (2, 3)
9. $6x - 2y = 4$, $y = 3x - 2$ (0, −2), (2, 4), (3, 7)
10. $y = -2x + 5$, $4x + 2y = 10$ (9, −13), (−1, 7), (0, 5)
11. $2x - y = 3$, $2x - y = 2$ (3, 3), (5, 7), (7, 11)
12. $y = x + 5$, $y = x - 3$ (1, −2), (3, 0), (6, 3)

Use the given graph to find an ordered pair that satisfies each system of equations. Check that your answer satisfies both equations of each system.

13. $y = 3x + 9$
$2x + 3y = 5$

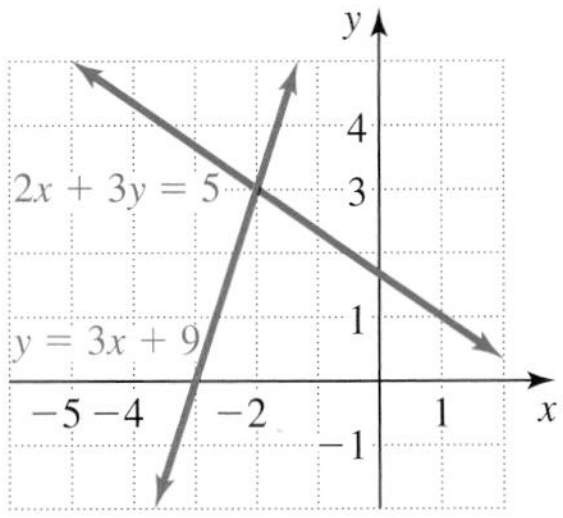

14. $x - 2y = 5$
$y = -\frac{2}{3}x + 1$

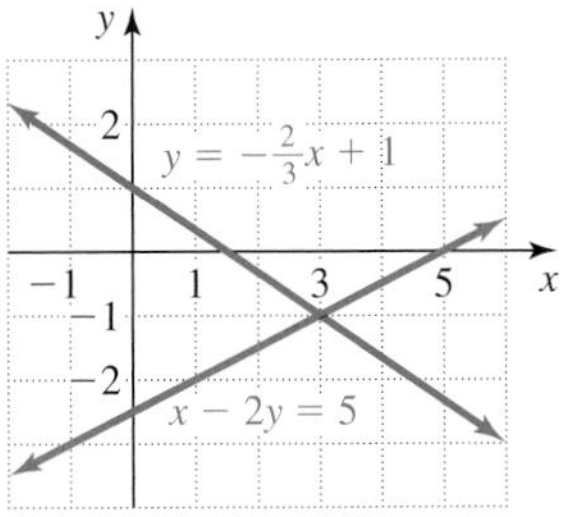

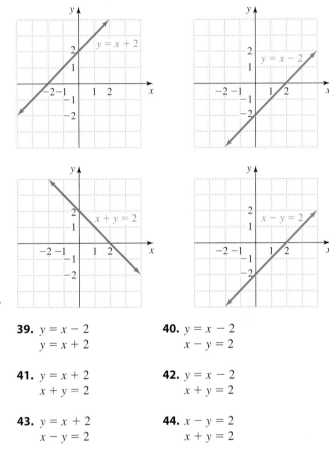

Solve each system by graphing. See Example 2.

15. $y = 2x$
$y = -x + 6$

16. $y = 3x$
$y = -x + 4$

17. $3x - y = 1$
$2y - 3x = 1$

18. $2x + y = 3$
$x + y = 1$

19. $x - y = 5$
$x + y = -5$

20. $y + 4x = 10$
$2x - y = 2$

21. $2y + x = 4$
$2x - y = -7$

22. $2x + y = -1$
$x + y = -2$

23. $y = x$
$x + y = 0$

24. $x = 2y$
$0 = 9x - y$

25. $y = 2x - 1$
$x - 2y = -4$

26. $y = x - 1$
$2x - y = 0$

Solve each system by graphing and indicate whether the system is independent, inconsistent, or dependent. See Examples 3 and 4.

27. $x - y = 3$
$3x = 3y + 12$

28. $x - y = 3$
$3x = 3y + 9$

29. $x - y = 3$
$3x = y + 5$

30. $3x + 2y = 6$
$2x - y = 4$

31. $2x + y = 3$
$6x - 9 = -3y$

32. $4y - 2x = -16$
$x - 2y = 8$

33. $x - y = 0$
$5x = 5y$

34. $y = -3x + 1$
$2 - 2y = 6x$

35. $x - y = -1$
$y = \frac{1}{2}x - 1$

36. $y = \frac{1}{3}x + 2$
$y = -\frac{1}{3}x$

37. $y - 4x = 4$
$y + 4x = -4$

38. $2y = -3x + 6$
$2y = -3x - 2$

Use the following graphs to determine whether the systems in Exercises 39–44 are independent, inconsistent, or dependent.

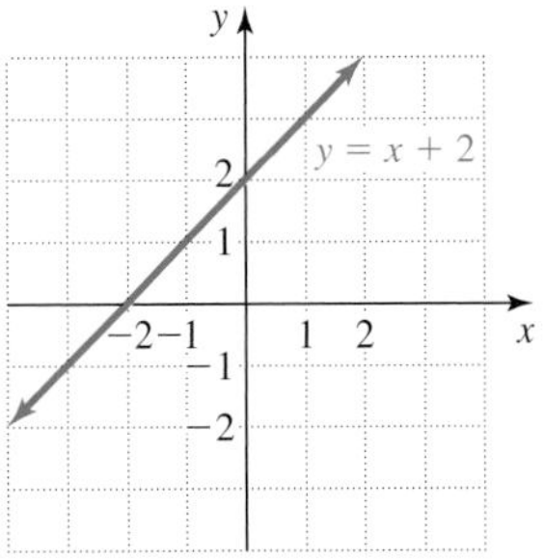

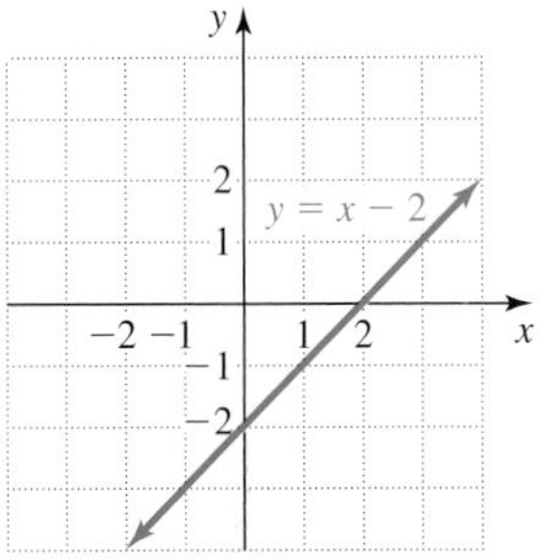

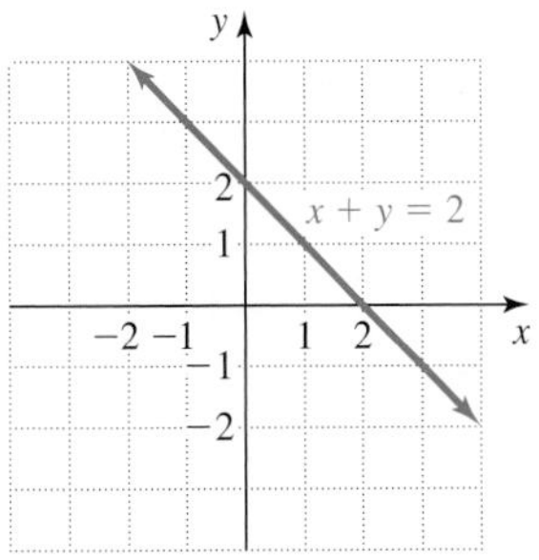

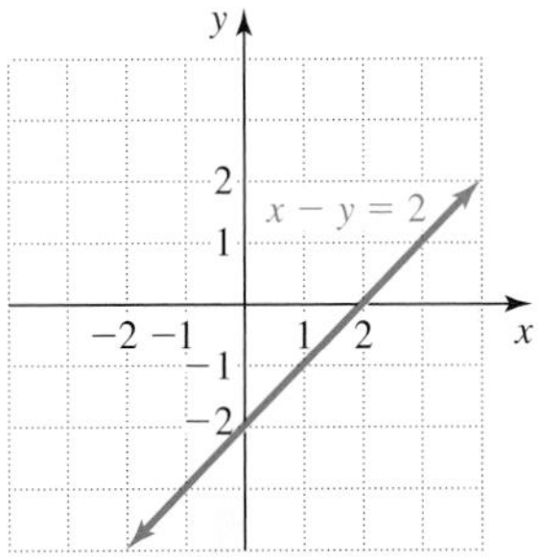

39. $y = x - 2$
$y = x + 2$

40. $y = x - 2$
$x - y = 2$

41. $y = x + 2$
$x + y = 2$

42. $y = x - 2$
$x + y = 2$

43. $y = x + 2$
$x - y = 2$

44. $x - y = 2$
$x + y = 2$

Determine whether each system is independent, inconsistent, or dependent.

45. $y = \frac{1}{2}x + 3$
$y = \frac{1}{2}x - 5$

46. $y = -3x - 60$
$y = \frac{1}{3}x - 60$

47. $y = 4x + 3$
$y = 3 + 4x$

48. $y = 5x - 4$
$y = 4 + 5x$

49. $y = \frac{1}{2}x + 3$
$y = -3x - 1$

50. $y = -x - 1$
$y = -1 - x$

51. $2x - 3y = 5$
$2x - 3y = 7$

52. $x + y = 1$
$2x + 2y = 2$

Solve each problem by using the graphing method. See Example 5.

53. ***Competing pizzas.*** Mamma's Pizza charges \$10 plus \$2 per topping for a deep dish pizza. Papa's Pizza charges \$5 plus \$3 per topping for a similar pizza. The equations $C = 2n + 10$ and $C = 3n + 5$ express the cost C at each restaurant in terms of the number of toppings n.

a) Solve this system of equations by examining the accompanying graph.
b) Interpret the solution.

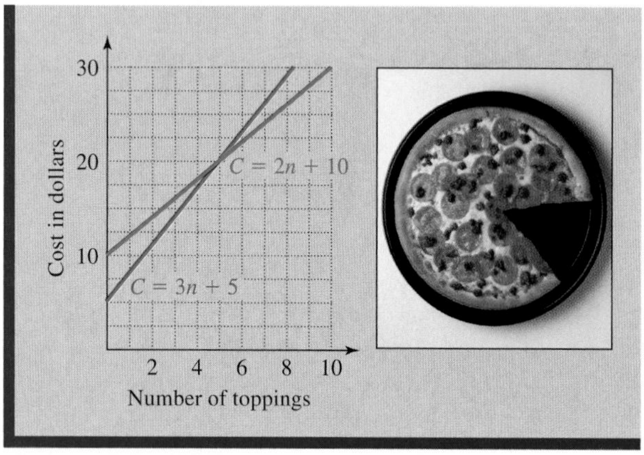

Figure for Exercise 53

54. ***Equilibrium price.*** A manufacturer plans to supply y units of its model 1020P CD player per month when the retail price is p dollars per player, where $y = 6p + 100$. Consumer studies show that consumer demand for the model 1020P is y units per month, where $y = -3p + 910$.

a) Fill in the missing entries in the following table.

Price	Supply	Demand
\$ 0		
50		
100		
300		

b) Use the data in part (a) to graph both linear equations on the same coordinate system.
c) What is the price at which the supply is equal to the demand, the *equilibrium price?*

55. ***Cost of two copiers.*** An office manager figures the total cost in dollars for a certain used Xerox copier is given by $C = 800 + 0.05x$, where x is the number of copies made. She is also considering a used Panasonic copier for which the total cost is $C = 500 + 0.07x$.

a) Fill in the missing entries in the following table.

Number of Copies	Cost Xerox	Cost Panasonic
0		
5000		
10,000		
20,000		

b) Use the data from part (a) to graph both equations on the same coordinate system.
c) For what number of copies is the total cost the same for either copier.
d) If she plans to buy another copier before 10,000 copies are made, then which copier is cheaper?

56. ***Flat tax proposals.*** Representative Schneider has proposed a flat income tax of 15% on earnings in excess of \$10,000.

Under his proposal the tax T for a person earning E dollars is given by $T = 0.15(E - 10{,}000)$. Representative Humphries has proposed that the income tax should be 20% on earnings in excess of \$20,000, or $T = 0.20(E - 20{,}000)$. Graph both linear equations on the same coordinate system. For what earnings would you pay the same amount of income tax under either plan? Under which plan does a rich person pay less income tax?

Getting More Involved

57. *Discussion*

If both $(-1, 3)$ and $(2, 7)$ satisfy a system of two linear equations, then what can you say about the system?

58. *Cooperative learning*

Working in groups, write an independent system of two linear equations whose solution is $(3, 5)$. Each group should then give its system to another group to solve.

59. *Cooperative learning*

Working in groups, write an inconsistent system of linear equations such that $(-2, 3)$ satisfies one equation and $(1, 4)$ satisfies the other. Each group should then give its system to another group to solve.

60. *Cooperative learning*

Suppose that $2x + 3y = 6$ is one equation of a system. Find the second equation given that $(4, 8)$ satisfies the second equation and the system is inconsistent.

Graphing Calculator Exercises

Solve each system by graphing each pair of equations on a graphing calculator and using the calculator to estimate the point of intersection. Give the coordinates of the intersection to the nearest tenth.

61. $y = 2.5x - 6.2$
$y = -1.3x + 8.1$

62. $y = 305x + 200$
$y = -201x - 999$

63. $2.2x - 3.1y = 3.4$
$5.4x + 6.2y = 7.3$

64. $34x - 277y = 1$
$402x + 306y = 12{,}000$

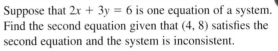

4.2 The Substitution Method

In this Section

- Solving a System of Linear Equations by Substitution
- Inconsistent and Dependent Systems
- Applications

Solving a system by graphing is certainly limited by the accuracy of the graph. If the lines intersect at a point whose coordinates are not integers, then it is difficult to identify the solution from a graph. In this section we introduce a method for solving systems of linear equations in two variables that does not depend on a graph and is totally accurate.

Solving a System of Linear Equations by Substitution

The next example shows how to solve a system without graphing. The method is called **substitution.**

EXAMPLE 1

Solving a system by substitution

Solve:

$$2x - 3y = 9$$
$$y - 4x = -8$$

Solution

First solve the second equation for y:

$$y - 4x = -8$$
$$y = 4x - 8$$

Now substitute $4x - 8$ for y in the first equation:

$$2x - 3y = 9$$
$$2x - 3(4x - 8) = 9 \quad \text{Substitute } 4x - 8 \text{ for } y.$$
$$2x - 12x + 24 = 9 \quad \text{Simplify.}$$
$$-10x + 24 = 9$$
$$-10x = -15$$
$$x = \frac{-15}{-10}$$
$$= \frac{3}{2}$$

Use the value $x = \frac{3}{2}$ in $y = 4x - 8$ to find y:

$$y = 4 \cdot \frac{3}{2} - 8$$
$$= -2$$

Check that $\left(\frac{3}{2}, -2\right)$ satisfies both of the original equations. The solution to the system is $\left(\frac{3}{2}, -2\right)$.

Now do Exercises 7–14

EXAMPLE 2

Solving a system by substitution

Solve:

$$3x + 4y = 5$$
$$x = y - 1$$

Solution

Because the second equation is already solved for x in terms of y, we can substitute $y - 1$ for x in the first equation:

$$3x + 4y = 5$$
$$3(y - 1) + 4y = 5 \quad \text{Replace } x \text{ with } y - 1.$$
$$3y - 3 + 4y = 5 \quad \text{Simplify.}$$
$$7y - 3 = 5$$
$$7y = 8$$
$$y = \frac{8}{7}$$

Calculator Close-Up

To check Example 2, graph

$$y_1 = (5 - 3x)/4$$

and

$$y_2 = x + 1.$$

Use the intersect feature of your calculator to find the point of intersection.

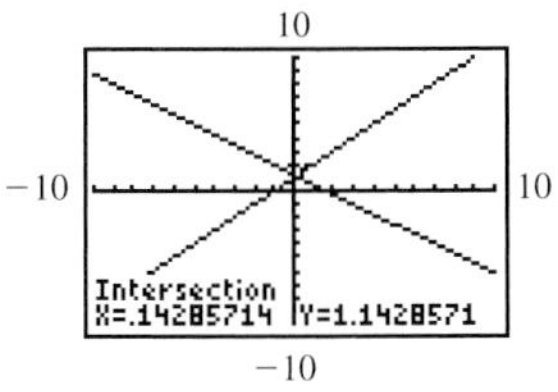

Now use the value $y = \frac{8}{7}$ in one of the original equations to find x. The simplest one to use is $x = y - 1$:

$$x = \frac{8}{7} - 1$$

$$x = \frac{1}{7}$$

Check that $\left(\frac{1}{7}, \frac{8}{7}\right)$ satisfies both equations. The solution to the system is $\left(\frac{1}{7}, \frac{8}{7}\right)$.

Now do Exercises 15–18

Use the following strategy for solving by substitution.

Strategy for Solving a System by Substitution

1. Solve one of the equations for one variable in terms of the other. Choose the equation that is easiest to solve for x or y.
2. Substitute this value into the other equation to eliminate one of the variables.
3. Solve for the remaining variable.
4. Insert this value into one of the original equations to find the value of the other variable.
5. Check your solution in both equations.

Inconsistent and Dependent Systems

Examples 3 and 4 illustrate how the inconsistent and dependent cases appear when we use substitution to solve the system.

EXAMPLE 3

An inconsistent system

Solve by substitution:

$$3x - 6y = 9$$
$$x = 2y + 5$$

Solution

Use $x = 2y + 5$ to replace x in the first equation:

$$
\begin{aligned}
3x - 6y &= 9 \\
3(2y + 5) - 6y &= 9 \quad \text{Replace } x \text{ by } 2y + 5. \\
6y + 15 - 6y &= 9 \quad \text{Simplify.} \\
15 &= 9
\end{aligned}
$$

No values for x and y will make 15 equal to 9. So there is no ordered pair that satisfies both equations. This system is inconsistent. It has no solution. The equations are the equations of parallel lines.

Now do Exercises 19–22

Calculator Close-Up

To check Example 3, graph $y_1 = (3x - 9)/6$ and $y_2 = (x - 5)/2$. Since the lines appear to be parallel, there is no solution to the system.

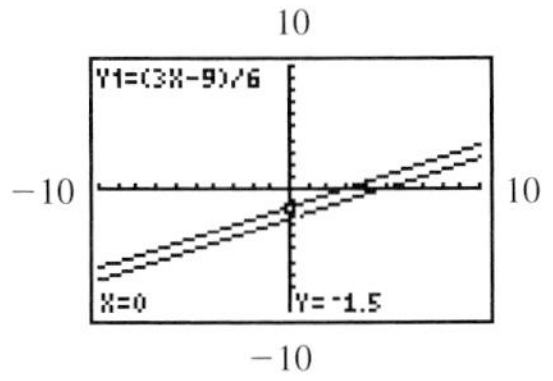

Let $x = 10{,}500$ in the equation $y = 25{,}000 - x$ to find y:

$$y = 25{,}000 - 10{,}500$$
$$= 14{,}500$$

Check these values for x and y in the original problem. Mrs. Robinson invested \$10,500 at 6% and \$14,500 at 8%.

Now do Exercises 43–54

Warm-Ups ▼

True or false? Explain your answer.

For Exercises 1–7, use the following systems:

a) $y = x - 7$
$2x + 3y = 4$

b) $x + 2y = 1$
$2x - 4y = 0$

1. If we substitute $x - 7$ for y in system (a), we get $2x + 3(x - 7) = 4$.
2. The x-coordinate of the solution to system (a) is 5.
3. The solution to system (a) is $(5, -2)$.
4. The point $\left(\frac{1}{2}, \frac{1}{4}\right)$ satisfies system (b).
5. It would be difficult to solve system (b) by graphing.
6. Either x or y could be eliminated by substitution in system (b).
7. System (b) is a dependent system.
8. Solving an inconsistent system by substitution will result in a false statement.
9. Solving a dependent system by substitution results in an equation that is always true.
10. Any system of two linear equations can be solved by substitution.

4.2 Exercises

Reading and Writing *After reading this section, write out the answers to these questions. Use complete sentences.*

1. What method is used in this section to solve systems of equations?

2. What is wrong with the graphing method for solving systems?

3. What is a dependent system?

4. What is an inconsistent system?

5. What happens when you try to solve a dependent system by substitution?

6. What happens when you try to solve an inconsistent system by substitution?

Solve each system by the substitution method. See Examples 1 and 2.

7. $y = x + 3$
$2x - 3y = -11$

8. $y = x - 5$
$x + 2y = 8$

9. $x = 2y - 4$
$2x + y = 7$

10. $x = y - 2$
$-2x + y = -1$

11. $2x + y = 5$
$5x + 2y = 8$

12. $5y - x = 0$
$6x - y = 2$

13. $x + y = 0$
$3x + 2y = -5$

14. $x - y = 6$
$3x + 4y = -3$

15. $x + y = 1$
$4x - 8y = -4$

16. $x - y = 2$
$3x - 6y = 8$

17. $2x + 3y = 2$
$4x - 9y = -1$

18. $x - 2y = 1$
$3x + 10y = -1$

Solve each system by substitution, and identify each system as independent, dependent, or inconsistent. See Examples 3 and 4.

19. $x - 2y = -2$
$x + 2y = 8$

20. $y = -3x + 1$
$y = 2x + 4$

21. $x = 4 - 2y$
$4y + 2x = -8$

22. $y - 3 = 2(x - 1)$
$y = 2x + 3$

23. $21x - 35 = 7y$
$3x - y = 5$

24. $2x + y = 3x$
$3x - y = 2y$

25. $y + 1 = 5(x + 1)$
$y = 5x - 1$

26. $3x - 2y = 7$
$3x + 2y = 7$

27. $2x + 5y = 5$
$3x - 5y = 6$

28. $x + 5y = 4$
$x + 5y = 4y$

Solve each system by the graphing method shown in Section 4.1 and by substitution.

29. $x + y = 5$
$x - y = 1$

30. $x + y = 6$
$2x - y = 3$

31. $y = x - 2$
$y = 4 - x$

32. $y = 2x - 3$
$y = -x + 3$

33. $y = 3x - 2$
$y - 3x = 1$

34. $x + y = 5$
$y = 2 - x$

Determine whether each system is independent, inconsistent, or dependent.

35. $y = -4x + 3$
$y = -4x - 6$

36. $y = -3x - 6$
$y = 3x - 6$

37. $y = x$
$x = y$

38. $y = x$
$y = x + 5$

39. $y = x$
$y = -x$

40. $y = 3x$
$3x - y = 0$

41. $x - y = 4$
$x - y = 5$

42. $y = 1$
$y + 3 = 4$

Write a system of two equations in two unknowns for each problem. Solve each system by substitution. See Example 5.

43. ***Rectangular patio.*** The length of a rectangular patio is twice the width. If the perimeter is 84 feet, then what are the length and width?

44. ***Rectangular lot.*** The width of a rectangular lot is 50 feet less than the length. If the perimeter is 900 feet, then what are the length and width?

45. ***Investing in the future.*** Mrs. Miller invested \$20,000 and received a total of \$1600 in interest. If she invested part of the money at 10% and the remainder at 5%, then how much did she invest at each rate?

46. ***Stocks and bonds.*** Mr. Walker invested \$30,000 in stocks and bonds and had a total return of \$2880 in one year. If his stock investment returned 10% and his bond investment returned 9%, then how much did he invest in each?

47. ***Gross receipts.*** Two of the highest grossing movies of all time were *Titanic* and *Star Wars* with total receipts of \$1062 million (www.movieweb.com). If the gross receipts for *Titanic* exceeded the gross receipts for *Star Wars* by \$140 million, then what were the gross receipts for each movie?

48. ***Tennis court dimensions.*** The singles court in tennis is four yards longer than it is wide. If its perimeter is 44 yards, then what are the length and width?

49. ***Mowing and shoveling.*** When Mr. Wilson came back from his vacation, he paid Frank \$50 for mowing his lawn three times and shoveling his sidewalk two times. During Mr. Wilson's vacation last year, Frank earned \$45 for mowing the lawn two times and shoveling the sidewalk three times. How much does Frank make for mowing the lawn once? How much does Frank make for shoveling the sidewalk once?

50. ***Burgers and fries.*** Donna ordered four burgers and one order of fries at the Hamburger Palace. However, the waiter put three burgers and two orders of fries in the bag and charged Donna the correct price for three burgers and two orders of fries, \$3.15. When Donna discovered the mistake, she went back to complain. She found out that the price for four burgers and one order of fries is \$3.45 and decided to keep what she had. What is the price of one burger, and what is the price of one order of fries?

51. ***Racing rules.*** According to NASCAR rules, no more than 52% of a car's total weight can be on any pair of tires. For optimal performance a driver of a 1150-pound car wants to have 50% of its weight on the left rear and left front tires and 48% of its weight on the left rear and right front tires. If the right front weight is determined to be 264 pounds, then what amount of weight should be on the left rear and left front? Are the NASCAR rules satisfied with this weight distribution?

52. ***Weight distribution.*** A driver of a 1200-pound car wants to have 50% of the car's weight on the left front and left rear tires, 48% on the left rear and right front tires, and 51% on the left rear and right rear tires. How much weight should be on each of these tires?

53. ***Price of hamburger.*** A grocer will supply y pounds of ground beef per day when the retail price is x dollars per pound, where $y = 200x + 60$. Consumer studies show that consumer demand for ground beef is y pounds per day, where $y = -150x + 900$. What is the price at which the supply is equal to the demand, the equilibrium price? See the accompanying figure.

54. ***Tweedle Dum and Dee.*** Tweedle Dum said to Tweedle Dee "The sum of my weight and twice yours is 361 pounds." Tweedle Dee said to Tweedle Dum "Contrariwise the sum of my weight and twice yours is 362 pounds." Find the weight of each.

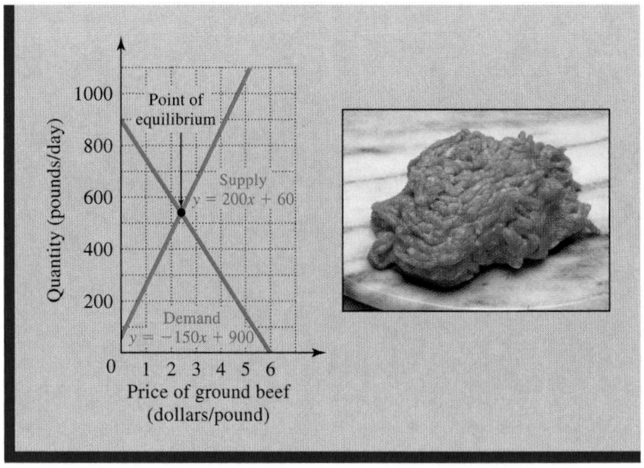

Figure for Exercise 53

Graphing Calculator Exercise

55. ***Life expectancy.*** Since 1950, the life expectancy of a U.S. male born in year x is modeled by the formula

$$y = 0.165x - 256.7,$$

and the life expectancy of a U.S. female born in year x is modeled by

$$y = 0.186x - 290.6$$

(National Center for Health Statistics, www.cdc.gov).

a) Find the life expectancy of a U.S. male born in 1975 and a U.S. female born in 1975.

b) Graph both equations on your graphing calculator for $1950 < x < 2050$.

c) Will U.S. males ever catch up with U.S. females in life expectancy?

d) Assuming that these equations were valid before 1950, solve the system to find the year of birth for which U.S. males and females had the same life expectancy.

4.3 The Addition Method

In this Section

- **Solving a System of Linear Equations by Addition**
- **Inconsistent and Dependent Systems**
- **Applications**

In Section 4.2 we solved systems of equations by using substitution. We substituted one equation into the other to eliminate a variable. The addition method of this section is another method for eliminating a variable to solve a system of equations.

Solving a System of Linear Equations by Addition

In the substitution method we solve for one variable in terms of the other variable. When doing this, we may get an expression involving fractions, which must be substituted into the other equation. The addition method avoids fractions and is easier to use on certain systems.

EXAMPLE 1

Solving a system by addition

Solve: $3x - y = 5$

$2x + y = 10$

Calculator Close-Up

To check Example 1, graph $y_1 = 3x - 5$ and $y_2 = 10 - 2x$. The lines appear to intersect at (3, 4).

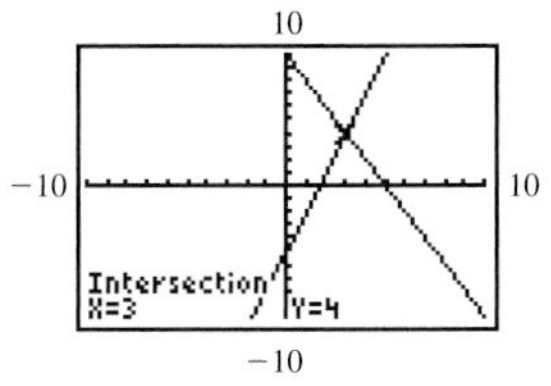

Solution

The addition property of equality enables us to add the same number to each side of an equation. If we assume that x and y are numbers that satisfy $3x - y = 5$, then adding these equations is equivalent to adding 5 to each side of $2x + y = 10$:

$$\begin{aligned} 3x - y &= 5 \\ 2x + y &= 10 \quad \text{Add.} \\ \hline 5x \quad &= 15 \quad -y + y = 0 \\ x &= 3 \end{aligned}$$

Note that the y-term was eliminated when we added the equations because the coefficients of y in the two equations were opposites. Now use $x = 3$ in either one of the original equations to find y:

$$\begin{aligned} 2x + y &= 10 \\ 2(3) + y &= 10 \quad \text{Let } x = 3. \\ y &= 4 \end{aligned}$$

Check that (3, 4) satisfies both equations. The solution to the system is (3, 4).

Now do Exercises 7–10

The addition method is based on the addition property of equality. We are adding equal quantities to each side of an equation. The form of the equations does not matter as long as the equal signs and the like terms are in line.

In Example 1, y was eliminated by the addition because the coefficients of y in the two equations were opposites. If no variable will be eliminated by addition, we can use the multiplication property of equality to change the coefficients of the variables. In Example 2 the coefficient of x in one equation is a multiple of the coefficient of x in the other equation. We use the multiplication property of equality to get opposite coefficients for x.

EXAMPLE 2

Solving a system by addition

Solve: $-x + 4y = -14$
$2x - 3y = 18$

Solution

If we add these equations as they are written, we will not eliminate any variables. However, if we multiply each side of the first equation by 2, then we will be adding $-2x$ and $2x$, and x will be eliminated:

$$2(-x + 4y) = 2(-14) \quad \text{Multiply each side by 2.}$$
$$2x - 3y = 18$$

$$-2x + 8y = -28$$
$$2x - 3y = 18 \quad \text{Add.}$$
$$5y = -10$$
$$y = -2$$

Now replace y by -2 in one of the original equations:

$$-x + 4(-2) = -14$$
$$-x - 8 = -14$$
$$-x = -6$$
$$x = 6$$

Check $x = 6$ and $y = -2$ in the original equations.

$$-6 + 4(-2) = -14$$
$$2(6) - 3(-2) = 18$$

The solution to the system is $(6, -2)$.

Now do Exercises 11–20

In Example 3 we need to use a multiple of each equation to eliminate a variable by addition.

EXAMPLE 3

Study Tip

Read carefully. Ask yourself questions and look for the answers. Every sentence says something pertaining to the subject. You cannot read a mathematics textbook like you read a novel. You can read a novel passively, but a textbook requires more concentration and retention.

Solving a system by addition

Solve: $2x + 3y = 7$
$3x + 4y = 10$

Solution

To eliminate x by addition, the coefficients of x in the two equations must be opposites. So we multiply the first equation by -3 and the second by 2:

$$-3(2x + 3y) = -3(7)$$
$$2(3x + 4y) = 2(10)$$

$$-6x - 9y = -21$$
$$6x + 8y = 20 \quad \text{Add.}$$
$$-y = -1$$
$$y = 1$$

Replace y with 1 in one of the original equations:

$$\begin{aligned} 2x + 3y &= 7 \\ 2x + 3(1) &= 7 \\ 2x + 3 &= 7 \\ 2x &= 4 \\ x &= 2 \end{aligned}$$

Check $x = 2$ and $y = 1$ in the original equations.

$$\begin{aligned} 2(2) + 3(1) &= 7 \\ 3(2) + 4(1) &= 10 \end{aligned}$$

The solution to the system is (2, 1).

Now do Exercises 21–26

If the equations have fractions, you can multiply each equation by the LCD to eliminate the fractions. Once the fractions are cleared, it is easier to see how to eliminate a variable by addition.

EXAMPLE 4

A system involving fractions

Solve: $\frac{1}{2}x - \frac{2}{3}y = 2$

$\frac{1}{4}x + \frac{1}{2}y = 6$

Solution

Multiply the first equation by 6 and the second by 4 to eliminate the fractions:

$$\begin{aligned} 6\left(\frac{1}{2}x - \frac{2}{3}y\right) &= 6 \cdot 2 \\ 4\left(\frac{1}{4}x + \frac{1}{2}y\right) &= 4 \cdot 6 \\ 3x - 4y &= 12 \\ x + 2y &= 24 \end{aligned}$$

Now multiply $x + 2y = 24$ by 2 to get $2x + 4y = 48$, and then add:

$$\begin{aligned} 3x - 4y &= 12 \\ 2x + 4y &= 48 \\ \hline 5x \quad &= 60 \\ x &= 12 \end{aligned}$$

Let $x = 12$ in $x + 2y = 24$:

$$\begin{aligned} 12 + 2y &= 24 \\ 2y &= 12 \\ y &= 6 \end{aligned}$$

Check $x = 12$ and $y = 6$ in the original equations. The solution is (12, 6).

Now do Exercises 27–30

Use the following strategy to solve a system by addition.

Strategy for Solving a System by Addition

1. Write both equations in standard form.
2. If a variable will be eliminated by adding, then add the equations.
3. If necessary, obtain multiples of one or both equations so that a variable will be eliminated by adding the equations.
4. After one variable is eliminated, solve for the remaining variable.
5. Use the value of the remaining variable to find the value of the eliminated variable.
6. Check the solution in the original system.

Inconsistent and Dependent Systems

When the addition method is used, an inconsistent system will be indicated by a false statement. A dependent system will be indicated by an equation that is always true.

EXAMPLE 5

Inconsistent and dependent systems

Use the addition method to solve each system.

a) $-2x + 3y = 9$
$2x - 3y = 18$

b) $2x - y = 1$
$4x - 2y = 2$

Solution

a) Add the equations:

$$\begin{array}{r} -2x + 3y = 9 \\ 2x - 3y = 18 \\ \hline 0 = 27 \end{array} \quad \text{False}$$

There is no solution to the system. The system is inconsistent.

b) Multiply the first equation by -2, and then add the equations:

$$-2(2x - y) = -2(1)$$
$$4x - 2y = 2$$

$$\begin{array}{r} -4x + 2y = -2 \\ 4x - 2y = 2 \\ \hline 0 = 0 \end{array} \quad \text{True}$$

Because the equation $0 = 0$ is correct for any value of x, the system is dependent. The set of points satisfying the system is $\{(x, y) \mid 2x - y = 1\}$.

Now do Exercises 31–46

Applications

In Example 6 we solve a problem using a system of equations and the addition method.

EXAMPLE 6

Milk and bread

Lea purchased two gallons of milk and three loaves of bread for \$8.25. Yesterday she purchased five gallons of milk and two loaves of bread for \$13.75. What is the price of a single gallon of milk? What is the price of a single loaf of bread?

Solution

Let x represent the price of one gallon of milk. Let y represent the price of one loaf of bread. We can write two equations about the milk and bread:

$$2x + 3y = 8.25 \quad \text{Today's purchase}$$
$$5x + 2y = 13.75 \quad \text{Yesterday's purchase}$$

To eliminate x, multiply the first equation by -5 and the second by 2:

$$-5(2x + 3y) = -5(8.25)$$
$$2(5x + 2y) = 2(13.75)$$

$$\begin{aligned} -10x - 15y &= -41.25 \\ 10x + 4y &= 27.50 \quad \text{Add.} \\ \hline -11y &= -13.75 \\ y &= 1.25 \end{aligned}$$

Replace y by 1.25 in the first equation:

$$\begin{aligned} 2x + 3(1.25) &= 8.25 \\ 2x + 3.75 &= 8.25 \\ 2x &= 4.50 \\ x &= 2.25 \end{aligned}$$

A gallon of milk costs \$2.25, and a loaf of bread costs \$1.25.

Now do Exercises 55–82

Helpful Hint

You can see from Example 6, that the standard form $Ax + By = C$ occurs naturally in accounting. This form will occur whenever we have the price each and quantity of two items and we want to express the total cost.

Warm-Ups ▼

True or false? Explain your answer.

Use the following systems for these exercises:

a) $3x + 2y = 7$, $4x - 5y = -6$ **b)** $y = -3x + 2$, $2y + 6x - 4 = 0$ **c)** $y = x - 5$, $x = y + 6$

1. To eliminate x by addition in system (a), we multiply the first equation by 4 and the second equation by 3.
2. Either variable in system (a) can be eliminated by the addition method.
3. The ordered pair (1, 2) is a solution to system (a).

4. The addition method can be used to eliminate a variable in system (b).

5. Both (0, 2) and (1, −1) satisfy system (b).

6. The solution to system (c) is $\{(x, y) \mid y = x - 5\}$.

7. System (c) is independent.

8. System (b) is inconsistent.

9. System (a) is dependent.

10. The graphs of the equations in system (c) are parallel lines.

4.3 Exercises

Reading and Writing *After reading this section, write out the answers to these questions. Use complete sentences.*

1. What method is used in this section to solve systems of equations?

2. What three methods have now been presented for solving a system of linear equations?

3. What do the addition method and the substitution method have in common?

4. What do we sometimes do before we add the equations?

5. How do you decide which variable to eliminate when using the addition method?

6. How do you identify inconsistent and dependent systems when using the addition method?

Solve each system by the addition method. See Examples 1–4.

7. $2x + y = 5$
$3x - y = 10$

8. $3x - y = 3$
$4x + y = 11$

9. $x + 2y = 7$
$-x + 3y = 18$

10. $x + 2y = 7$
$-x + 4y = 5$

11. $x + 2y = 2$
$-4x + 3y = 25$

12. $2x - 3y = -7$
$5x + y = -9$

13. $x + 3y = 4$
$2x - y = -1$

14. $x - y = 0$
$x - 2y = 0$

15. $y = 4x - 1$
$y = 3x + 7$

16. $x = 3y + 45$
$x = 2y + 40$

17. $4x = 3y + 1$
$2x = y - 1$

18. $2x = y - 9$
$x = -1 - 3y$

19. $2x - 5y = -22$
$-6x + 3y = 18$

20. $4x - 3y = 7$
$5x + 6y = -1$

21. $2x + 3y = 4$
$-3x + 5y = 13$

22. $-5x + 3y = 1$
$2x - 7y = 17$

23. $2x - 5y = 11$
$3x - 2y = 11$

24. $4x - 3y = 17$
$3x - 5y = 21$

25. $5x + 4y = 13$
$2x + 3y = 8$

26. $4x + 3y = 8$
$6x + 5y = 14$

27. $\frac{1}{2}x + \frac{1}{3}y = 8$
$\frac{1}{3}x - \frac{1}{2}y = 1$

28. $\frac{1}{5}x + \frac{1}{10}y = 5$
$\frac{1}{2}x - \frac{1}{5}y = 8$

29. $\frac{2}{3}x + \frac{3}{4}y = 28$
$\frac{1}{2}x - \frac{3}{8}y = 6$

30. $\frac{2}{5}x - \frac{1}{10}y = 1$
$\frac{3}{10}x + \frac{2}{3}y = 23$

Use either the addition method or substitution to solve each system. State whether the system is independent, inconsistent, or dependent. See Example 5.

31. $x + y = 5$
$x + y = 6$

32. $x + y = 5$
$x + 2y = 6$

33. $x + y = 5$
$2x + 2y = 10$

34. $2x + 3y = 4$
$2x - 3y = 4$

35. $2x = y + 3$
$2y = 4x - 6$

36. $y = 2x - 1$
$2x - y + 5 = 0$

37. $x + 3y = 3$
$5x = 15 - 15y$

38. $y - 3x = 2$
$5y = -15x + 10$

39. $6x - 2y = -2$
$\frac{1}{3}y = x + \frac{4}{3}$

40. $x + y = 8$
$\frac{1}{3}x - \frac{1}{2}y = 1$

41. $\frac{1}{2}x - \frac{2}{3}y = -6$
$-\frac{3}{4}x - \frac{1}{2}y = -18$

42. $\frac{1}{2}x - y = 3$
$\frac{1}{5}x + 2y = 6$

43. $0.04x + 0.09y = 7$
$x + y = 100$

44. $0.08x - 0.05y = 0.2$
$2x + y = 140$

45. $0.1x - 0.2y = -0.01$
$0.3x + 0.5y = 0.08$

46. $0.5y = 0.2x - 0.25$
$0.1y = 0.8x - 1.57$

Use a calculator to assist you in finding the exact solution to each system.

47. $2.33x - 4.58y = 16.319$
$4.98x + 3.44y = -2.162$

48. $234x - 499y = 1337$
$282x + 312y = 51{,}846$

Solve each system by graphing (Section 4.1), by substitution (Section 4.2), and by addition.

49. $x + y = 7$
$x - y = 1$

50. $x + y = 8$
$2x - y = 4$

51. $y = x - 3$
$y = 5 - x$

52. $y = 2x - 5$
$y = -x + 4$

53. $y = 2x - 1$
$y - 2x = 3$

54. $x + y = 3$
$y = 4 - x$

Use two variables and a system of equations to solve each problem. See Example 6.

55. ***Cars and trucks.*** An automobile dealer had 250 vehicles on his lot during the month of June. He must pay a monthly inventory tax of \$3 per car and \$4 per truck. If his tax bill for June was \$850, then how many cars and how many trucks did he have on his lot during June?

Photo for Exercise 55

56. ***Inventory tax.*** A dealer had 120 vehicles on his lot during June. He must pay a monthly inventory tax of \$6 per car and \$5 per truck. If his tax bill for June was \$640, then how many cars and how many trucks did he have on his lot during June?

57. ***The meter maid.*** Rita opened a parking meter and removed 40 coins consisting of dimes and nickels. If the value of these coins is \$3.20, then how many coins of each type does she have?

58. ***Paying the penalty.*** Candy paid her library fine with 30 coins consisting of nickels and dimes. If the fine was \$2.40, then how many coins of each type did she use?

59. ***A great start.*** George paid for his \$4.15 breakfast with 31 coins consisting of dimes and quarters. How many coins of each type did he use?

60. ***Coin collecting.*** Andrew paid \$2000 for 24 rare coins from a dealer. If all of them were nickels and quarters and the total face value was \$4.40, then how many coins of each type did he buy?

61. ***Adults and children.*** The Audubon Zoo charges \$5.50 for each adult admission and \$2.75 for each child. The total bill for the 30 people on the Spring Creek Elementary School kindergarten field trip was \$99. How many adults and how many children went on the field trip?

62. ***Concert revenue.*** A total of 1000 tickets were sold for *The Grinch Who Stole Christmas.* Children's tickets were \$6 each and adult tickets were \$10 each. If the total revenue

Photo for Exercise 57

was $8400, then how many children's tickets and how many adult tickets were sold?

63. ***Coffee and doughnuts.*** Jorge has worked at Dandy Doughnuts so long that he has memorized the amounts for many of the common orders. For example, six doughnuts and five coffees cost $4.35, while four doughnuts and three coffees cost $2.75. What are the prices of one cup of coffee and one doughnut?

64. ***Time for a change.*** Jimmy has worked at the Donut Shop so long that he has memorized the amounts of the common orders. For example, four doughnuts and two coffees cost $4.80, while five doughnuts and three coffees cost $6.45. What are the prices for one doughnut and for one coffee?

65. ***Marketing research.*** American Marketing Corporation found 147 smokers among 660 adults surveyed. If one-fourth of the men and one-fifth of the women were smokers, then how many men and how many women were surveyed?

66. ***Marketing research.*** The Independent Marketing Research Corporation found 130 smokers among 300 adults surveyed. If one-half of the men and one-third of the women were smokers, then how many men and how many women were in the survey?

67. ***Time and a half.*** In one month, Shelly earned $1800 for 210 hours of work. If she earns $8 per hour for regular time and $12 per hour for overtime, then how many hours of each type did she work?

68. ***Good job.*** In one month Hector worked 210 hours and earned $3276. If he makes $14 per hour for regular time and $21 per hour for overtime, then how many hours of each type did he work?

69. ***Investing wisely.*** Janet invested a total of $8000. Part of the money was placed into an account that earned an annual simple interest rate of 7%. The rest of the money was placed into an account that earned an annual simple interest rate of 10%. If the total interest earned in 1 year was $710, then how much was invested at each rate?

70. ***Group investing.*** An investment club split a total of $10,000 into two investments. One of the investments returned 7.5% (simple interest) and the other returned 9% after 1 year. If the total amount of the return for 1 year was $810, then how much was placed in each investment?

71. ***Stocks and bonds.*** Mr. Taylor invested a total of $21,000 in stocks and bonds and realized a total return of $1950 in 1 year. If his stock investment returned 11% and his bond investment returned 7%, then how much did he invest in each?

72. ***Mutual funds.*** Belinda split her total investment of $40,000 between a Dreyfus fund with an annual yield of 15% and a Templeton fund with an annual yield of 12%. If she made $5460 on these investments in one year, then how much did she invest in each fund?

73. ***Equal interest.*** John invested a total of $4800 in two business ventures. One investment returned 15% and the other returned 9% after one year. If the dollar amount of the return for each investment was the same, then how much did he invest in each venture?

74. ***Earning a tenth.*** Latonya invested $5000 in a safe investment that she figured would return 8% after 1 year. She plans to invest in a riskier investment that should earn 15% in 1 year. How much should she put in the riskier investment so that her total return is 10% of her total investment?

75. ***Making cloth.*** A manufacturer uses fibers that are 20% synthetic along with fibers that are 40% synthetic to make a fabric. If 800 pounds of 35% synthetic fabric are made from these fibers, then how many pounds of each type of fiber were used in making the fabric?

76. ***Mixing acid.*** How many gallons of a 14% acid solution must be mixed with 10 gallons of a 30% acid solution to make a 19% acid solution?

77. ***Mixing fertilizer.*** How many ounces of 10% nitrogen fertilizer must be combined with 22% nitrogen fertilizer to make 120 ounces of an 18% nitrogen fertilizer?

78. ***Chocolate sauce.*** How many ounces of pure Swiss chocolate must be added to 100 ounces of chocolate topping that is 20% Swiss chocolate to get a mixture that is 60% Swiss chocolate?

79. ***Mixing metals.*** A metallurgist combined a metal that costs \$4.40 per ounce with a metal that costs \$2.40 per ounce. How many ounces of each were used to make a mixture of 100 ounces costing \$3.68 per ounce?

80. ***Caramel corn.*** A snack food is made by mixing 190 pounds of caramel corn that costs \$0.60 per pound with nuts that cost \$4.00 per pound. How many pounds of nuts are needed to get a mixture that costs \$0.77 per pound?

81. ***Middle grade.*** A butcher combined \$4.20 per pound hamburger with \$3.10 per pound hamburger. How many pounds of each were used to get a 100-pound mixture that is worth \$3.76 per pound?

82. ***Chocolate mix.*** Find the selling price per pound of a mixture made from 25 pounds of chocolate which costs \$6.60 per pound and 75 pounds of fudge that costs \$4.80 per pound.

Getting More Involved

83. ***Discussion***

Compare and contrast the three methods for solving systems of linear equations in two variables that were presented in this chapter. What are the advantages and disadvantages of each method? How do you choose which method to use?

84. ***Exploration***

Consider the following system:

$$a_1x + b_1y = c_1$$
$$a_2x + b_2y = c_2$$

a) Multiply the first equation by a_2 and the second equation by $-a_1$. Add the resulting equations and solve for y to get a formula for y in terms of the a's, b's, and c's.

b) Multiply the first equation by b_2 and the second by $-b_1$. Add the resulting equations and solve for x to get a formula for x in terms of the a's, b's, and c's.

c) Use the formulas that you found in (a) and (b) to find the solution to the following system:

$$2x + 3y = 7$$
$$5x + 4y = 14$$

4.4 Graphing Linear Inequalities in Two Variables

In this Section

- Definition
- Graphing a Linear Inequality
- Using a Test Point to Graph an Inequality
- Applications

You studied linear equations and inequalities in one variable in Chapter 2. In this section we extend the ideas of linear equations in two variables to study linear inequalities in two variables.

Definition

Linear inequalities in two variables have the same form as linear equations in two variables. An inequality symbol is used in place of the equal sign.

Linear Inequality in Two Variables

If A, B, and C are real numbers with A and B not both zero, then

$$Ax + By < C$$

is called a **linear inequality in two variables.** In place of $<$, we can also use $\leq$, $>$, or $\geq$.

The inequalities

$$3x - 4y \leq 8, \qquad y > 2x - 3, \qquad \text{and} \qquad x - y + 9 < 0$$

are linear inequalities. Not all of these are in the form of the definition, but they could all be rewritten in that form.

An ordered pair is a solution to an inequality in two variables if the ordered pair satisfies the inequality.

EXAMPLE 1

Satisfying a linear inequality

Determine whether each point satisfies the inequality $2x - 3y \geq 6$.

a) (4, 1) **b)** (3, 0) **c)** (3, −2)

Study Tip

Write about what you read in the text. Sum things up in your own words. Write out important facts on note cards. When you have a few spare minutes in between classes review your note cards. Try to get the information on the cards into your memory.

Solution

a) To determine whether (4, 1) is a solution to the inequality, we replace x by 4 and y by 1 in the inequality $2x - 3y \geq 6$:

$$2(4) - 3(1) \geq 6$$
$$8 - 3 \geq 6$$
$$5 \geq 6 \quad \text{Incorrect}$$

So (4, 1) does not satisfy the inequality $2x - 3y \geq 6$.

b) Replace x by 3 and y by 0:

$$2(3) - 3(0) \geq 6$$
$$6 \geq 6 \quad \text{Correct}$$

So the point (3, 0) satisfies the inequality $2x - 3y \geq 6$.

c) Replace x by 3 and y by −2:

$$2(3) - 3(-2) \geq 6$$
$$6 + 6 \geq 6$$
$$12 \geq 6 \quad \text{Correct}$$

So the point (3, −2) satisfies the inequality $2x - 3y \geq 6$.

Now do Exercises 7–12

Graphing a Linear Inequality

The graph of a linear inequality in two variables consists of all points in the rectangular coordinate system that satisfy the inequality. For example, the graph of the inequality

$$y > x + 2$$

consists of all points where the y-coordinate is larger than the x-coordinate plus 2. Consider the point (3, 5) on the line

$$y = x + 2.$$

The y-coordinate of (3, 5) is equal to the x-coordinate plus 2. If we choose a point with a larger y-coordinate, such as (3, 6), it satisfies the inequality and it is above the line $y = x + 2$. In fact, any point above the line $y = x + 2$ satisfies $y > x + 2$. Likewise, all points below the line $y = x + 2$ satisfy the inequality $y < x + 2$. See Fig. 4.6.

Helpful Hint

Why do we keep drawing graphs? When we solve $2x + 1 = 7$, we don't bother to draw a graph showing 3, because the solution set is so simple. However, the solution set to a linear inequality is a very large set of ordered pairs. Graphing gives us a way to visualize the solution set.

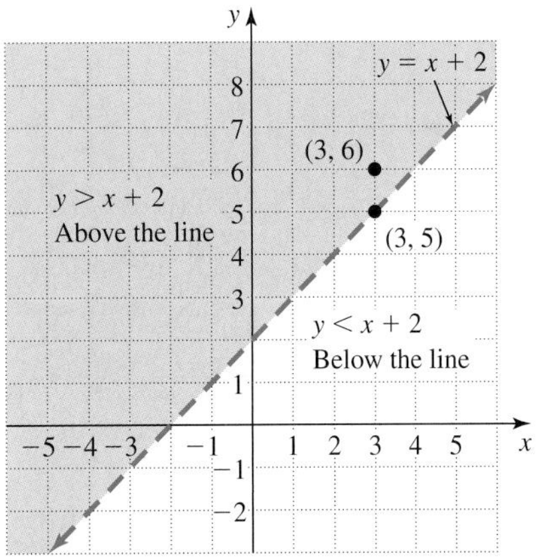

Figure 4.6

To graph the inequality, we shade all points above the line $y = x + 2$. To indicate that the line is not included in the graph of $y > x + 2$, we use a dashed line.

The procedure for graphing linear inequalities is summarized as follows.

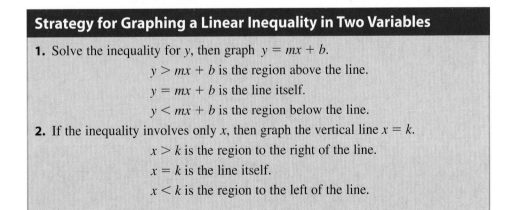

Strategy for Graphing a Linear Inequality in Two Variables

1. Solve the inequality for y, then graph $y = mx + b$.

$y > mx + b$ is the region above the line.

$y = mx + b$ is the line itself.

$y < mx + b$ is the region below the line.

2. If the inequality involves only x, then graph the vertical line $x = k$.

$x > k$ is the region to the right of the line.

$x = k$ is the line itself.

$x < k$ is the region to the left of the line.

EXAMPLE 2

Graphing a linear inequality

Graph each inequality.

a) $y < \frac{1}{3}x + 1$

b) $y \geq -2x + 3$

c) $2x - 3y < 6$

Solution

a) The set of points satisfying this inequality is the region below the line

$$y = \frac{1}{3}x + 1.$$

To show this region, we first graph the boundary line. The slope of the line is $\frac{1}{3}$, and the y-intercept is (0, 1). We draw the line dashed because it is not part of the graph of $y < \frac{1}{3}x + 1$. In Fig. 4.7 the graph is the shaded region.

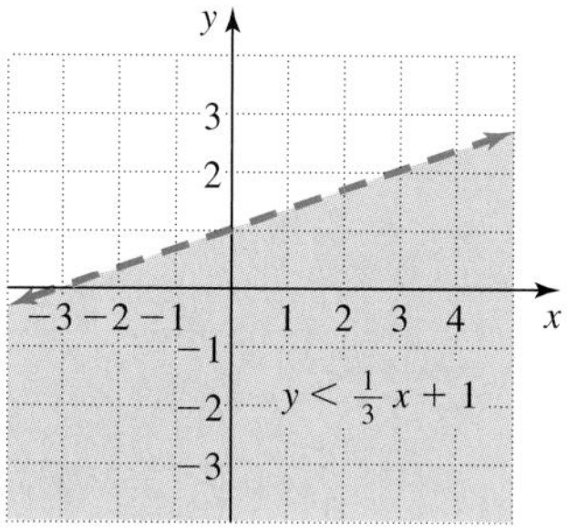

Figure 4.7

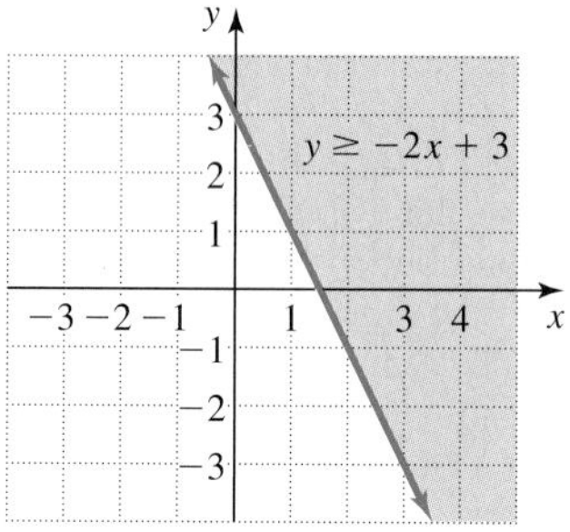

Figure 4.8

b) Because the inequality symbol is $\geq$, every point on or above the line satisfies this inequality. We use the fact that the slope of this line is -2 and the y-intercept is (0, 3) to draw the graph of the line. To show that the line $y = -2x + 3$ is included in the graph, we make it a solid line and shade the region above. See Fig. 4.8.

c) First solve for y:

$$2x - 3y < 6$$
$$-3y < -2x + 6$$
$$y > \frac{2}{3}x - 2 \quad \text{Divide by } -3 \text{ and reverse the inequality.}$$

To graph this inequality, we first graph the line with slope $\frac{2}{3}$ and y-intercept $(0, -2)$. We use a dashed line for the boundary because it is not included, and we shade the region above the line. Remember, "less than" means below the line and "greater than" means above the line only when the inequality is solved for y. See Fig. 4.9 for the graph.

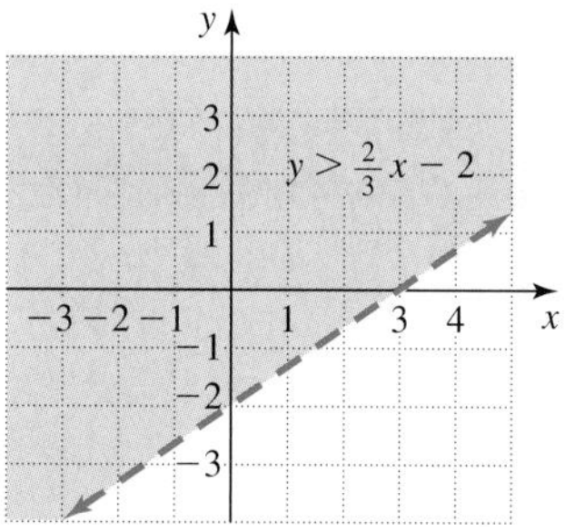

Figure 4.9

Now do Exercises 13–26

EXAMPLE 3

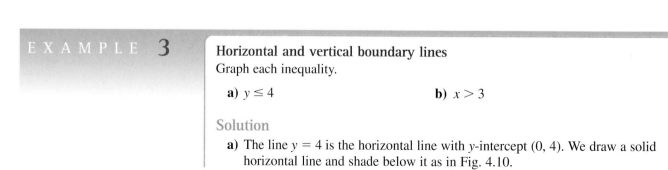

Horizontal and vertical boundary lines

Graph each inequality.

a) $y \leq 4$

b) $x > 3$

Solution

a) The line $y = 4$ is the horizontal line with y-intercept (0, 4). We draw a solid horizontal line and shade below it as in Fig. 4.10.

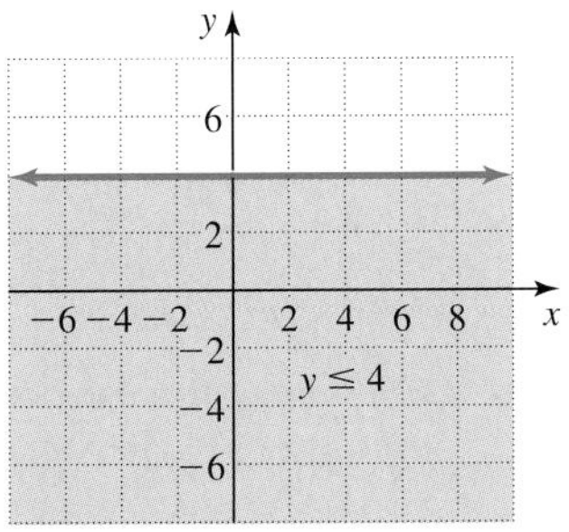

Figure 4.10

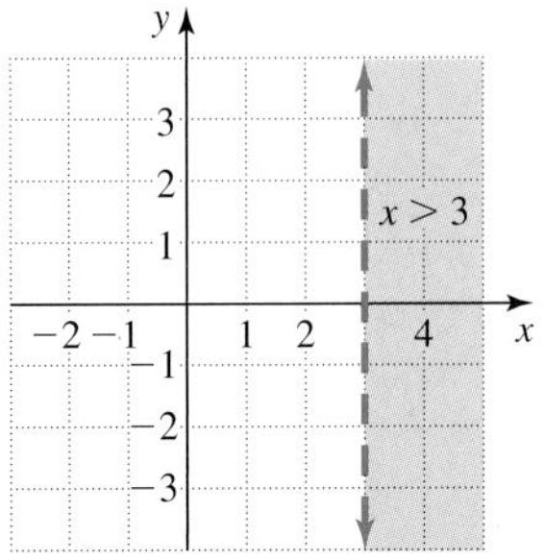

Figure 4.11

b) The line $x = 3$ is a vertical line through (3, 0). Any point to the right of this line has an x-coordinate larger than 3. The graph is shown in Fig. 4.11.

Now do Exercises 27–30

Using a Test Point to Graph an Inequality

The graph of a linear equation such as $2x - 3y = 6$ separates the coordinate plane into two regions. One region satisfies the inequality $2x - 3y > 6$, and the other region satisfies the inequality $2x - 3y < 6$. We can tell which region satisfies which inequality by testing a point in one region. With this method it is not necessary to solve the inequality for y.

EXAMPLE 4

Using a test point

Graph the inequality $2x - 3y > 6$.

Solution

First graph the equation $2x - 3y = 6$ using the x-intercept (3, 0) and the y-intercept (0, −2) as shown in Fig. 4.12. Select a point on one side of the line, say (0, 1), to test in the inequality. Because

$$2(0) - 3(1) > 6$$

is false, the region on the other side of the line satisfies the inequality. The graph of $2x - 3y > 6$ is shown in Fig. 4.13.

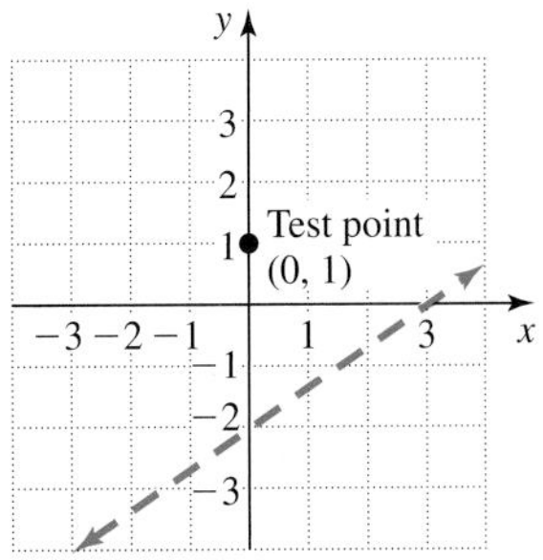

Figure 4.12

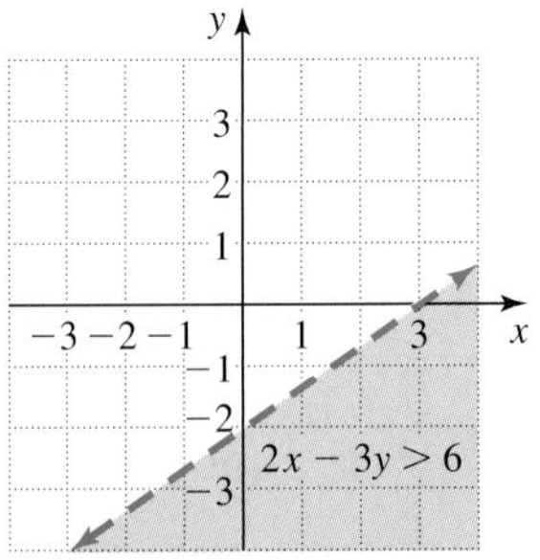

Figure 4.13

Now do Exercises 37–48

Helpful Hint

Some people always like to choose (0, 0) as the test point for lines that do not go through (0, 0). The arithmetic for testing (0, 0) is generally easier than for any other point.

Applications

The values of variables used in applications are often restricted to nonnegative numbers. So solutions to inequalities in these applications are graphed in the first quadrant only.

EXAMPLE 5

Manufacturing tables

The Ozark Furniture Company can obtain at most 8000 board feet of oak lumber for making two types of tables. It takes 50 board feet to make a round table and 80 board feet to make a rectangular table. Write an inequality that limits the possible number of tables of each type that can be made. Draw a graph showing all possibilities for the number of tables that can be made.

Solution

If x is the number of round tables and y is the number of rectangular tables, then x and y satisfy the inequality

$$50x + 80y \le 8000.$$

Now find the intercepts for the line $50x + 80y = 8000$:

$$\begin{aligned} 50 \cdot 0 + 80y &= 8000 \\ 80y &= 8000 \\ y &= 100 \end{aligned} \qquad \begin{aligned} 50x + 80 \cdot 0 &= 8000 \\ 50x &= 8000 \\ x &= 160 \end{aligned}$$

Draw the line through (0, 100) and (160, 0). Because (0, 0) satisfies the inequality, the number of tables must be below the line. Since the number of tables cannot be negative, the number of tables made must be below the line and in the first quadrant as shown in Fig. 4.14. Assuming that Ozark will not make a fraction of a table, only points in Fig. 4.14 with whole-number coordinates are practical.

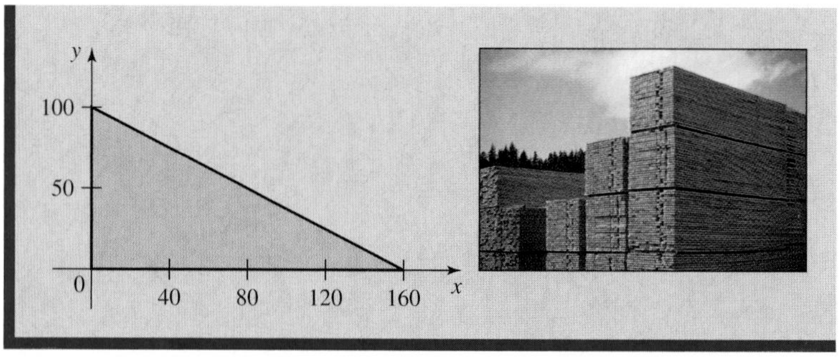

Figure 4.14

Now do Exercises 49–52

Warm-Ups ▼

True or false? Explain your answer.

1. The point $(-1, 4)$ satisfies the inequality $y > 3x + 1$.
2. The point $(2, -3)$ satisfies the inequality $3x - 2y \geq 12$.
3. The graph of the inequality $y > x + 9$ is the region above the line $y = x + 9$.
4. The graph of the inequality $x < y + 2$ is the region below the line $x = y + 2$.
5. The graph of $x = 3$ is a single point on the x-axis.
6. The graph of $y \leq 5$ is the region below the horizontal line $y = 5$.
7. The graph of $x < 3$ is the region to the left of the vertical line $x = 3$.
8. In graphing the inequality $y \geq x$ we use a dashed boundary line.
9. The point $(0, 0)$ is on the graph of the inequality $y \geq x$.
10. The point $(0, 0)$ lies above the line $y = 2x + 1$.

4.4 Exercises

Reading and Writing *After reading this section, write out the answers to these questions. Use complete sentences.*

1. What is a linear inequality in two variables?
2. How can you tell if an ordered pair satisfies a linear inequality in two variables?
3. How do you determine whether to draw the boundary line of the graph of a linear inequality dashed or solid?
4. How do you decide which side of the boundary line to shade?
5. What is the test point method?
6. What is the advantage of the test point method?

Determine which of the points following each inequality satisfy that inequality. See Example 1.

7. $x - y > 5$ $(2, 3), (-3, -9), (8, 3)$
8. $2x + y < 3$ $(-2, 6), (0, 3), (3, 0)$
9. $y \geq -2x + 5$ $(3, 0), (1, 3), (-2, 5)$
10. $y \leq -x + 6$ $(2, 0), (-3, 9), (-4, 12)$
11. $x > -3y + 4$ $(2, 3), (7, -1), (0, 5)$
12. $x < -y - 3$ $(1, 2), (-3, -4), (0, -3)$

Graph each inequality. See Examples 2 and 3.

13. $y < x + 4$
14. $y < 2x + 2$

15. $y > -x + 3$

16. $y < -2x + 1$

17. $y > \frac{2}{3}x - 3$

18. $y < \frac{1}{2}x + 1$

19. $y \le -\frac{2}{5}x + 2$

20. $y \ge -\frac{1}{2}x + 3$

21. $y - x \ge 0$

22. $x - 2y \le 0$

23. $x > y - 5$

24. $2x < 3y + 6$

25. $x - 2y + 4 \le 0$

26. $2x - y + 3 \ge 0$

27. $y \ge 2$

28. $y < 7$

29. $x > 9$

30. $x \le 1$

31. $x + y \le 60$

32. $x - y \le 90$

33. $x \leq 100y$

34. $y \geq 600x$

35. $3x - 4y \leq 8$

36. $2x + 5y \geq 10$

Graph each inequality. Use the test point method of Example 4.

37. $2x - 3y < 6$

38. $x - 4y > 4$

39. $x - 4y \leq 8$

40. $3y - 5x \geq 15$

41. $y - \frac{7}{2}x \leq 7$

42. $\frac{2}{3}x + 3y \leq 12$

43. $x - y < 5$

44. $y - x > -3$

45. $3x - 4y < -12$

46. $4x + 3y > 24$

47. $x < 5y - 100$

48. $-x > 70 - y$

Solve each problem. See Example 5.

49. ***Storing the tables.*** Ozark Furniture Company must store its oak tables before shipping. A round table is packaged in a carton with a volume of 25 cubic feet (ft^3), and a rectangular table is packaged in a carton with a volume of 35 ft^3. The warehouse has at most 3850 ft^3 of space available for these tables. Write an inequality that limits the possible number of tables of each type that can be stored, and graph the inequality in the first quadrant.

Photo for Exercise 50

50. ***Maple rockers.*** Ozark Furniture Company can obtain at most 3000 board feet of maple lumber for making its classic and modern maple rocking chairs. A classic maple rocker requires 15 board feet of maple, and a modern rocker requires 12 board feet of maple. Write an inequality that limits the possible number of maple rockers of each type that can be made, and graph the inequality in the first quadrant.

51. ***Pens and notebooks.*** A student has at most \$4 to spend on pens at \$0.25 each and notebooks at \$0.40 each. Write an inequality that limits the possibilities for the number of pens (x) and the number of notebooks (y) that can be purchased. Graph the inequality in the first quadrant.

52. ***Enzyme concentration.*** A food chemist tests enzymes for their ability to break down pectin in fruit juices (Dennis Callas, *Snapshots of Applications in Mathematics*). Excess pectin makes juice cloudy. In one test, the chemist measures the concentration of the enzyme, c, in milligrams per milliliter and the fraction of light absorbed by the liquid, a. If $a > 0.07c + 0.02$, then the enzyme is working as it should. Graph the inequality in the first quadrant.

Getting More Involved

53. ***Discussion***

When asked to graph the inequality $x + 2y < 12$, a student found that (0, 5) and (8, 0) both satisfied $x + 2y < 12$. The student then drew a dashed line through these two points and shaded the region below the line. What is wrong with this method? Do all of the points graphed by this student satisfy the inequality?

54. ***Writing***

Compare and contrast the two methods presented in this section for graphing linear inequalities. What are the advantages and disadvantages of each method? How do you choose which method to use?

4.5 Graphing Systems of Linear Inequalities

In this Section

- The Solution to a System of Two Inequalities
- Graphing a System of Inequalities
- Systems with No Solution

In Section 4.4 you learned how to solve a linear inequality. In this section you will solve systems of linear inequalities.

The Solution to a System of Two Inequalities

A point is a solution to a system of two equations if it satisfies both equations. Similarly, a point is a solution to a system of two inequalities if it satisfies both inequalities.

EXAMPLE 1

Satisfying a system of inequalities

Determine whether each point is a solution to the system of inequalities:

$$2x + 3y < 6$$
$$y > 2x - 1$$

a) $(-3, 2)$ **b)** $(4, -3)$ **c)** $(5, 1)$

Solution

a) The point $(-3, 2)$ is a solution to the system if it satisfies both inequalities. Let $x = -3$ and $y = 2$ in each inequality:

$2x + 3y < 6$	$y > 2x - 1$
$2(-3) + 3(2) < 6$	$2 > 2(-3) - 1$
$0 < 6$	$2 > -7$

Because both inequalities are satisfied, the point $(-3, 2)$ is a solution to the system.

b) Let $x = 4$ and $y = -3$ in each inequality:

$2x + 3y < 6$	$y > 2x - 1$
$2(4) + 3(-3) < 6$	$-3 > 2(4) - 1$
$-1 < 6$	$-3 > 7$

Because only one inequality is satisfied, the point $(4, -3)$ is not a solution to the system.

c) Let $x = 5$ and $y = 1$ in each inequality:

$2x + 3y < 6$	$y > 2x - 1$
$2(5) + 3(1) < 6$	$1 > 2(5) - 1$
$13 < 6$	$1 > 9$

Because neither inequality is satisfied, the point $(5, 1)$ is not a solution to the system.

Now do Exercises 7–12

Study Tip

Read the text and recite to yourself what you have read. Ask questions and answer them out loud. Listen to your answers to see if they are complete and correct. Would other students understand your answers?

Graphing a System of Inequalities

There are infinitely many points that satisfy a typical system of inequalities. The best way to describe the solution to a system of inequalities is with a graph showing all points that satisfy the system. When we graph the points that satisfy a system, we say that we are graphing the system.

EXAMPLE 2

Graphing a system of inequalities

Graph all ordered pairs that satisfy the following system of inequalities:

$$y > x - 2$$
$$y < -2x + 3$$

Solution

We want a graph showing all points that satisfy both inequalities. The lines $y = x - 2$ and $y = -2x + 3$ divide the coordinate plane into four regions as shown in Fig. 4.15. To determine which of the four regions contains points that satisfy the system, we check one point in each region to see whether it satisfies both inequalities. The points are shown in Fig. 4.15.

Check (0, 0):
$0 > 0 - 2$ Correct
$0 < -2(0) + 3$ Correct

Check (0, 5):
$5 > 0 - 2$ Correct
$5 < -2(0) + 3$ Incorrect

Check (0, −5):
$-5 > 0 - 2$ Incorrect
$-5 < -2(0) + 3$ Correct

Check (4, 0):
$0 > 4 - 2$ Incorrect
$0 < -2(4) + 3$ Incorrect

The only point that satisfies both inequalities of the system is (0, 0). So every point in the region containing (0, 0) also satisfies both inequalities. The points that satisfy the system are graphed in Fig. 4.16.

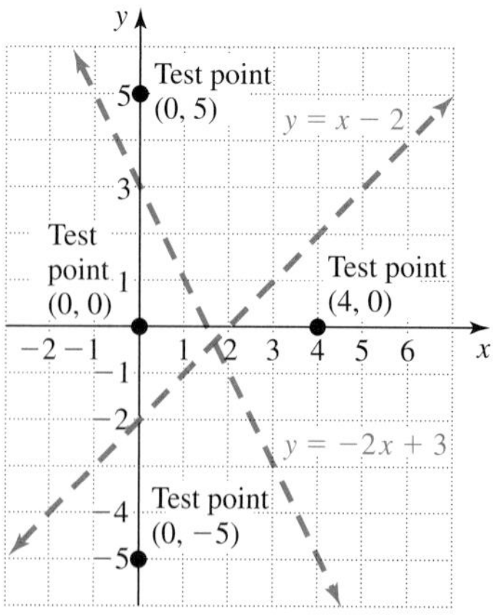

Figure 4.15

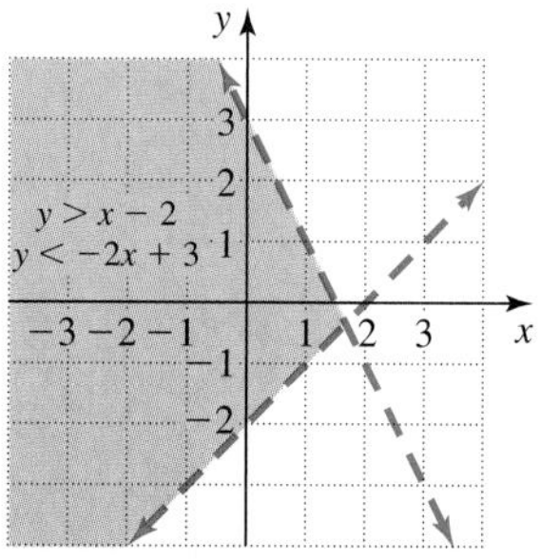

Figure 4.16

Now do Exercises 13–16

EXAMPLE 3

Graphing a system of inequalities

Graph all ordered pairs that satisfy the following system of inequalities:

$$y > -3x + 4$$
$$2y - x > 2$$

Solution

First graph the equations $y = -3x + 4$ and $2y - x = 2$. Now we select the points (0, 0), (0, 2), (0, 6), and (5, 0). We leave it to you to check each point in the system of inequalities. You will find that only (0, 6) satisfies the system. So only the region containing (0, 6) is shaded in Fig. 4.17.

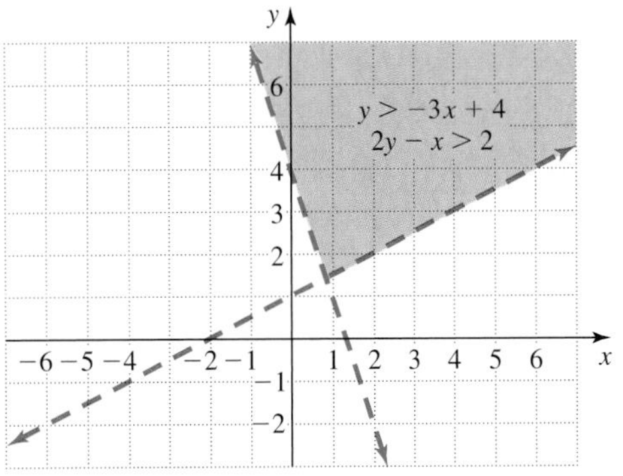

Figure 4.17

Now do Exercises 17–20

EXAMPLE 4

Horizontal and vertical boundary lines

Graph the system of inequalities:

$$x > 4$$
$$y < 3$$

Solution

We first graph the vertical line $x = 4$ and the horizontal line $y = 3$. The points that satisfy both inequalities are those points that lie to the right of the vertical line $x = 4$ and below the horizontal line $y = 3$. See Fig. 4.18 for the graph of the system.

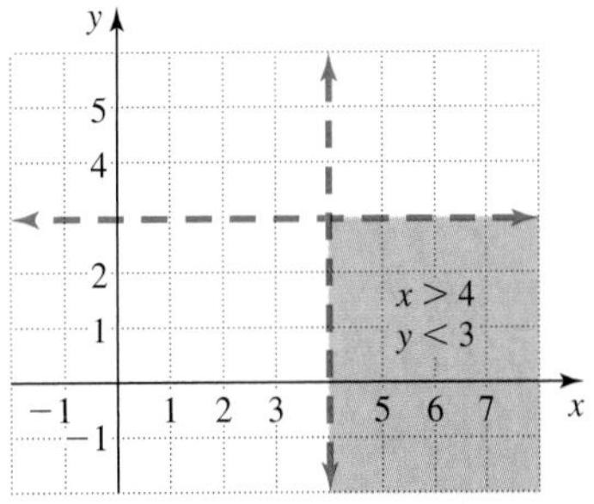

Figure 4.18

Now do Exercises 21–24

EXAMPLE 5

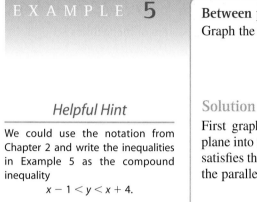

Between parallel lines

Graph the system of inequalities:

$$y < x + 4$$
$$y > x - 1$$

Helpful Hint

We could use the notation from Chapter 2 and write the inequalities in Example 5 as the compound inequality

$$x - 1 < y < x + 4.$$

Solution

First graph the parallel lines $y = x + 4$ and $y = x - 1$. These lines divide the plane into three regions. Check (0, 0), (0, 6), and (0, −4) in the system. Only (0, 0) satisfies the system. So the solution to the system consists of all points in between the parallel lines, as shown in Fig. 4.19.

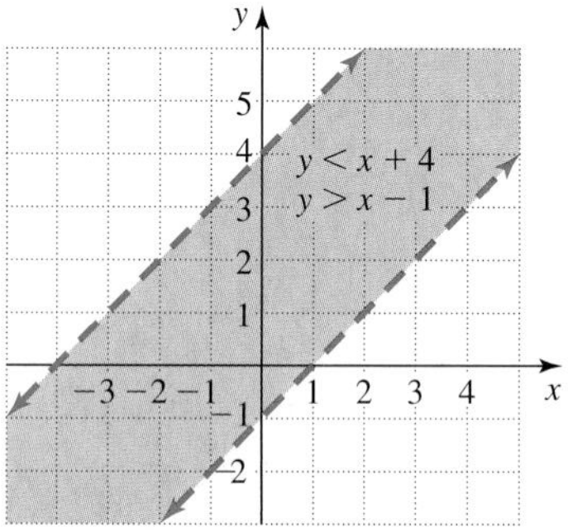

Figure 4.19

Now do Exercises 25–44

Systems with No Solution

The solution set to a system of inequalities can be empty even when the solution sets to the individual inequalities are not empty.

EXAMPLE 6

Systems of inequalities with no solution

Solve each system of inequalities.

a) $y > x + 5$
$y < x - 5$

b) $x \geq 2$
$x \leq -3$

c) $y > 5$
$y < 1$

Solution

a) Note that the lines $y = x + 5$ and $y = x - 5$ are parallel lines and $y = x + 5$ lies above $y = x - 5$. There are no points in the coordinate plane that lie above $y = x + 5$ and below $y = x - 5$. So there are no ordered pairs that satisfy both $y > x + 5$ and $y < x - 5$. The solution set to the system is the empty set, $\emptyset$.

b) Note that $x = 2$ and $x = -3$ are vertical parallel lines and $x = 2$ lies to the right of $x = -3$. There are no points in the coordinate plane that lie on or to the right of $x = 2$ and on or to the left of $x = -3$. So there are no ordered pairs that satisfy both $x \geq 2$ and $x \leq -3$. The solution set to the system is the empty set, $\emptyset$.

c) Note that $y = 5$ and $y = 1$ are horizontal parallel lines and $y = 5$ lies above $y = 1$. There are no points in the coordinate plane that lie above $y = 5$ and below $y = 1$. So there are no ordered pairs that satisfy both $y > 5$ and $y < 1$. The solution set to the system is the empty set, $\varnothing$.

Now do Exercises 45–54

Warm-Ups

True or false? Explain your answer.

Use the following systems for Exercises 1–7.

a) $y > -3x + 5$
$y < 2x - 3$

b) $y > 2x - 3$
$y < 2x + 3$

c) $x + y > 4$
$x - y < 0$

1. The point $(2, -3)$ is a solution to system (a).
2. The point $(5, 0)$ is a solution to system (a).
3. The point $(0, 0)$ is a solution to system (b).
4. The graph of system (b) is the region between two parallel lines.
5. You can use $(0, 0)$ as a test point for system (c).
6. The point $(2, 2)$ satisfies system (c).
7. The point $(4, 5)$ satisfies system (c).
8. The inequality $x + y > 4$ is equivalent to the inequality $y < -x + 4$.
9. The graph of $y < 2x + 3$ is the region below the line $y = 2x + 3$.
10. There is no ordered pair that satisfies $y < 2x - 3$ and $y > 2x + 3$.

4.5 Exercises

Reading and Writing *After reading this section, write out the answers to these questions. Use complete sentences.*

1. What is a system of linear inequalities in two variables?

2. How can you tell if an ordered pair satisfies a system of linear inequalities in two variables?

3. How do we usually describe the solution set to a system of inequalities in two variables?

4. How do you decide whether the boundary lines are solid or dashed?

5. How do you use the test point method for a system of linear inequalities?

6. How do you select test points?

Determine which of the points following each system is a solution to the system. See Example 1.

7. $x - y < 5$ $(4, 3), (8, 2), (-3, 0)$
$2x + y > 3$

8. $x + y < 4$ $(2, -3), (1, 1), (0, -1)$
$2x - y < 3$

9. $y > -2x + 1$ $(-3, 2), (-1, 5), (3, 6)$
$y < 3x + 5$

10. $y < -x + 7$ $(-3, 8), (0, 8), (-5, 15)$
$y < -x + 9$

11. $x > 3$ $(-5, 4), (9, -5), (6, 0)$
$y < -2$

12. $y < -5$ $(-2, 4), (0, -7), (6, -9)$
$x < 1$

Graph each system of inequalities. See Examples 2–5.

13. $y > -x - 1$
$y > x + 1$

14. $y < x + 3$
$y < -2x + 4$

15. $y < 2x - 3$
$y > -x + 2$

16. $y > 2x - 1$
$y < -x - 4$

17. $x + y > 5$
$x - y < 3$

18. $2x + y < 3$
$x - 2y > 2$

19. $2x - 3y < 6$
$x - y > 3$

20. $3x - 2y > 6$
$x + y < 4$

21. $x > 5$
$y > 5$

22. $x < 3$
$y > 2$

23. $y < -1$
$x > -3$

24. $y > -2$
$x < 1$

25. $y > 2x - 4$
$y < 2x + 1$

26. $y < -2x + 3$
$y > -2x$

27. $y > x$
$x > 3$

28. $y < x$
$y < 1$

29. $y > -x$
$x < -1$

30. $y < -x$
$y > -3$

31. $x > 1$
$y - 2x < 3$

32. $y < 2$
$2x + 3y < 6$

33. $2x - 5y < 5$
$x + 2y > 4$

34. $3x + 2y < 2$
$-x - 2y > 4$

35. $x + y > 3$
$x + y > 1$

36. $x - y < 5$
$x - y < 3$

37. $y > 3x + 2$
$y < 3x + 3$

38. $y > x$
$y < -x$

39. $x + y < 5$
$x - y > -1$

40. $2x - y > 4$
$x - 5y < 5$

41. $2x - 3y < 6$
$3x + 4y < 12$

42. $x - 3y > 3$
$x + 2y < 4$

43. $3x - 5y < 15$
$3x + 2y < 12$

44. $x - 4y < 0$
$x + y > 0$

Determine whether or not the solution set to each system of inequalities is the empty set. See Example 6.

45. $y > x$
$y < x - 2$

46. $y < x + 1$
$y > x + 9$

47. $y < 2x$
$y > 3x$

48. $y \geq -9x$
$y \leq 9x$

49. $y \leq 5x - 1$
$y \geq 5x + 2$

50. $y \geq 3x + 8$
$y \leq 3x - 8$

51. $y < 2$
$y > 3$

52. $y < 5$
$y > -5$

53. $x < 4$
$x > -1$

54. $x < -5$
$x > 5$

The graph of each of the following systems is one of the four quadrants in the rectangular coordinate system. Name the quadrant.

55. $y < 0$
$x < 0$

56. $y < 0$
$x > 0$

57. $y > 0$
$x < 0$

58. $y > 0$
$x > 0$

Solve each problem.

59. ***Strawberries and blueberries.*** A manager of a produce stand has at most \$60 to spend on strawberries at \$2 per pint and blueberries at \$3 per pint. If x represents the number of pints of strawberries and y represents the number of pints of blueberries, then x and y must satisfy $2x + 3y \leq 60$. Since she cannot purchase a negative number of pints, we also have $x \geq 0$ and $y \geq 0$. Graph this system of three inequalities.

60. ***Target heart rate.*** For beneficial exercise, experts recommend that your target heart rate y should be between 65% and 75% of the maximum heart rate for your age x. That is,

$$y > 0.65(220 - x) \quad \text{and} \quad y < 0.75(220 - x).$$

Graph this system of inequalities for $20 < x < 70$.

61. ***Making and storing the tables.*** The Ozark Furniture Company can obtain at most 8000 board feet of oak lumber for making round and rectangular tables. The tables must be stored in a warehouse that has at most 3850 ft^3 of space available for the tables. A round table requires 50 board feet of lumber and 25 ft^3 of warehouse space. A rectangular table requires 80 board feet of lumber and 35 ft^3 of warehouse space. Write a system of inequalities that limits the possible number of tables of each type that can be made and stored. Graph the system.

62. ***Allocating resources.*** Wausaukee Enterprises makes yard barns in two sizes. One small barn requires \$250 in materials and 20 hours of labor, and one large barn requires \$400 in materials and 30 hours of labor. Wausaukee has at most \$4000 to spend on materials and at most 300 hours of labor available. Write a system of inequalities that limits the possible number of barns of each type that can be built. Graph the system.

Photo for Exercise 62

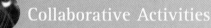

Collaborative Activities

Grouping: Three to four students per group

Topic: Systems of equations

Which Cider?

For the Fall Harvest Festival your math club decides to sell apple cider. You can get a good deal on bulk apple cider by buying 30 gallons for \$120. At the club meeting, one member suggests buying the apple cider from her uncle, who has a nearby organic apple orchard. He will lower his price for the club to \$6 per gallon. You find that you can only get paper cups in batches of 100 at \$3 per 100. To decide which cider to buy, analyze the two options using the profit equation: Profit = Sales − Cost.

1. If you want to sell all 30 gallons in 8-ounce cups, how many cups will you need to buy?

2. What will your total costs be for paper cups and bulk cider? For paper cups and local cider?

3. The club wants to sell the cider for \$1 per cup. Write a profit equation (Profit = Sales − Cost) for each type of cider in terms of number of cups sold. Let c = number of cups sold and P = profit.

4. Graph both equations using a graphing calculator or on graph paper. Let the vertical (y-axis) be profit and the horizontal (x-axis) be the number of cups sold. Decide on a scale for your graph. If using a graphing calculator set your x- and y-max to at least 300.

5. How many cups do you need to sell to make a profit for local cider? For bulk cider?

6. Is there a point at which the number of cups and the profit are the same for both types?

The member who wants the local cider points out that your club could sell it for more, since it will be fresher and of higher quality. She suggests selling it for \$1.75 per cup.

7. Write an equation for local cider at \$1.75 per cup. Graph this equation with the one you had for bulk cider.

8. When are the profits greater for the local cider? Estimate this from your graph. Find this answer algebraically, using elimination or substitution.

9. Decide which cider your club should sell and at which price.

Chapter 4 Wrap-Up

Summary

Systems of Linear Equations in Two Variables		**Examples**
Graphing method	Sketch each graph and identify the points they have in common.	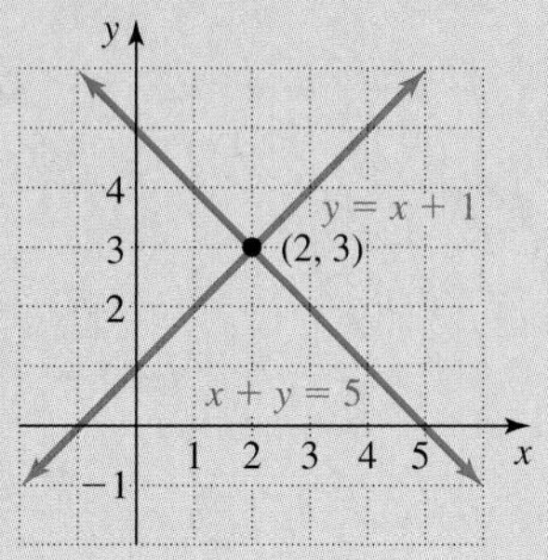
Substitution method	Solve one equation for one variable in terms of the other, then substitute into the other equation.	$y = x - 4$ $x + y = 9$ $x + (x - 4) = 9$
Addition method	Multiply each equation as necessary to eliminate a variable upon addition of the equations.	$5x - 3y = 4$ $x + 3y = 1$ $6x \quad = 5$
Independent	Only one point satisfies both equations. The graphs cross at one point.	$y = x - 4$ $y = 2x + 5$
Inconsistent	No solution The graphs are parallel lines.	$y = 5x - 3$ $y = 5x + 1$
Dependent	Infinitely many solutions One equation is a multiple of the other. The graphs coincide.	$5x + 3y = 2$ $10x + 6y = 4$

Linear Inequalities in Two Variables		**Examples**
Graphing the solution to an inequality in two variables	1. Solve the inequality for y, then graph $y = mx + b$. $y > mx + b$ is the region above the line. $y = mx + b$ is the line itself. $y < mx + b$ is the region below the line. Remember that "less than" means below the line and "greater than" means above the line only when the inequality is solved for y.	$y > x + 3$ $y = x + 3$ $y < x + 3$

	2. If the inequality involves only x, then graph the vertical line $x = k$. $x > k$ is the region to the right of the line. $x = k$ is the line itself. $x < k$ is the region to the left of the line.	$x > 5$ Region to right of vertical line $x = 5$
Test points	A linear inequality may also be graphed by graphing the equation and then testing a point to determine which region satisfies the inequality.	$x + y > 4$ (0, 6) satisfies the inequality.
Graphing a system of inequalities	Graph the equations and use test points to see which regions satisfy both inequalities.	$x + y > 4$ $x - y < 1$ (0, 6) satisfies the system.

Enriching Your Mathematical Word Power

For each mathematical term, choose the correct meaning.

1. system of equations
a. a systematic method for classifying equations
b. a method for solving an equation
c. two or more equations
d. the properties of equality

2. independent linear system
a. a system with exactly one solution
b. an equation that is satisfied by every real number
c. equations that are identical
d. a system of lines

3. inconsistent system
a. a system with no solution
b. a system of inconsistent equations
c. a system that is incorrect
d. a system that we are not sure how to solve

4. dependent system
a. a system that is independent
b. a system that depends on a variable
c. a system that has no solution
d. a system for which the graphs coincide

5. substitution method
a. replacing the variables by the correct answer
b. a method of eliminating a variable by substituting one equation into the other
c. the replacement method
d. any method of solving a system

6. linear inequality in two variables
a. when two lines are not equal
b. line segments that are unequal in length
c. an inequality of the form $Ax + By \geq C$ or with another symbol of inequality
d. an inequality of the form $Ax^2 + By^2 < C^2$

7. rational numbers
a. the numbers 1, 2, 3, and so on
b. the integers
c. numbers that make sense
d. numbers of the form a/b where a and b are integers with $b \neq 0$

8. irrational numbers
a. the cube roots
b. numbers that cannot be expressed as a ratio of integers
c. numbers that do not make sense
d. the integers

9. additive identity
a. the number 0
b. the number 1
c. the opposite of a number
d. when two sums are identical

10. multiplicative identity
a. the number 0
b. the number 1
c. the reciprocal
d. when two products are identical

Review Exercises

4.1 *Solve each system by graphing.*

1. $y = 2x + 1$
$x + y = 4$

2. $y = -x + 1$
$y = -x + 3$

3. $y = 2x + 3$
$y = -2x - 1$

4. $x + y = 6$
$x - y = -10$

4.2 *Solve each system by the substitution method.*

5. $y = 3x$
$2x + 3y = 22$

6. $x + y = 3$
$3x - 2y = -11$

7. $x = y - 5$
$2x - 3y = -7$

8. $2x + y = 5$
$6x - 9 = 3y$

4.3 *In Exercises 9–20, solve each system by the addition method. Indicate whether each system is independent, inconsistent, or dependent.*

9. $x - y = 4$
$2x + y = 5$

10. $x + 2y = -5$
$x - 3y = 10$

11. $2x - 4y = 8$
$x - 2y = 4$

12. $x + 3y = 7$
$2x + 6y = 5$

13. $y = 3x - 5$
$2y = -x - 3$

14. $3x + 4y = 6$
$4x + 3y = 1$

15. $2x + 7y = 0$
$7x + 2y = 0$

16. $3x - 5y = 1$
$10y = 6x - 1$

17. $x - y = 6$
$2x - 12 = 2y$

18. $y = 4x$
$y = 3x$

19. $y = 4x$
$y = 4x + 3$

20. $3x - 5y = 21$
$4x + 7y = -13$

4.4 *Graph each inequality.*

21. $y > \frac{1}{3}x - 5$

22. $y < \frac{1}{2}x + 2$

23. $y \leq -2x + 7$

24. $y \geq x - 6$

25. $y \leq 8$

26. $x \geq -6$

27. $2x + 3y \leq -12$

28. $x - 3y < 9$

4.5 *Graph each system of inequalities.*

29. $x < 5$
$y < 4$

30. $y > -2$
$x < 1$

31. $x + y < 2$
$y > 2x - 3$

32. $x - y > 4$
$2y > x - 4$

33. $y > 5x - 7$
$y < 5x + 1$

34. $y > x - 6$
$y < x - 5$

35. $y < 3x + 5$
$y < 3x$

36. $y > -2x$
$y > -3x$

Miscellaneous

Use a system of equations in two variables to solve each problem. Solve the system by the method of your choice.

37. ***Apples and oranges.*** Two apples and three oranges cost \$1.95, and three apples and two oranges cost \$2.05. What are the costs of one apple and one orange?

Photo for Exercise 37

38. ***Small or medium.*** Three small drinks and one medium drink cost \$2.30, and two small drinks and four medium drinks cost \$3.70. What is the cost of one small drink? What is the cost of one medium drink?

39. ***Gambling fever.*** After a long day at the casinos in Biloxi, Louis returned home and told his wife Lois that he had won \$430 in \$5 bills and \$10 bills. On counting them again, he realized that he had mixed up the number of bills of each denomination, and he had really won only \$380. How many bills of each denomination does Louis have?

40. ***Diversifying investments.*** Diane invested her \$10,000 bonus in a municipal bond fund and an emerging market fund. In one year the amount invested in the bond fund earned 8%, and the amount invested in the emerging market fund earned 10%. If the total income from these two investments for one year was \$880, then how much did she invest in each fund?

41. ***Protein and carbohydrates.*** One serving of green beans contains 1 gram of protein and 4 grams of carbohydrates. One serving of chicken soup contains 3 grams of protein and 9 grams of carbohydrates. The Westdale Diet recommends a lunch of 13 grams of protein and 43 grams of carbohydrates. How many servings of each are necessary to obtain the recommended amounts?

Photo for Exercise 41

42. ***Advertising revenue.*** A television station aired four 30-second commercials and three 60-second commercials during the first hour of the midnight movie. During the second hour, it aired six 30-second commercials and five 60-second commercials. The advertising revenue for the first hour was \$7700, and that for the second hour was \$12,300. What is the cost of each type of commercial?

Chapter 4 Test

Solve the system by graphing.

1. $x + y = 2$
$y = 2x + 5$

Solve each system by substitution.

2. $y = 2x - 3$
$2x + 3y = 7$

3. $x - y = 4$
$3x - 2y = 11$

Solve each system by the addition method.

4. $2x + 5y = 19$
$4x - 3y = -1$

5. $3x - 2y = 10$
$2x + 5y = 13$

Determine whether each system is independent, inconsistent, or dependent.

6. $y = 4x - 9$
$y = 4x + 8$

7. $3x - 3y = 12$
$y = x - 4$

8. $y = 2x$
$y = 5x$

Graph each inequality.

9. $y > 3x - 5$

10. $x - y < 3$

11. $x - 2y \geq 4$

Graph each system of inequalities.

12. $x < 6$
$y > -1$

13. $2x + 3y > 6$
$3x - y < 3$

14. $y > 3x - 4$
$3x - y > 3$

For each problem, write a system of equations in two variables. Use the method of your choice to solve each system.

15. Kathy and Chris studied a total of 54 hours for the CPA exam. If Chris studied only one-half as many hours as Kathy, then how many hours did each of them study?

16. The Rest-Is-Easy Motel just outside Amarillo rented five singles and three doubles on Monday night for a total of \$188. On Tuesday night it rented three singles and four doubles for a total of \$170. On Wednesday night it rented only one single and one double. How much rent did the motel receive on Wednesday night?

*Making*Connections | A Review of Chapters 1–4

Solve each equation.

1. $2(x - 5) + 3x = 25$

2. $3x - 5 = 0$

3. $\frac{x}{3} - \frac{2}{5} = \frac{x}{2} - \frac{12}{5}$

4. $x - 0.05x = 950$

5. $3(x - 5) - 5x = 5 - 2(x - 4)$

6. $7x - 4(5 - x) = 5(2x - 4) + x$

Solve each inequality in one variable. State the solution set using interval notation and sketch the graph on a number line.

7. $3(2 - x) < -6$

8. $-3 \le 2x - 4 \le 6$

9. $4 \ge 5 - x$

Sketch the graph of each equation.

10. $y = 3x - 7$

11. $y = 5 - x$

12. $y = x - 1$

13. $y = x + 1$

14. $y = -2x + 4$

15. $y = -4x - 1$

Graph each inequality in two variables.

16. $y \ge 3x - 7$

17. $x - 2y < 6$

18. $x > 1$

Write the equation of the line going through each pair of points.

19. $(0, 36)$ and $(8, 84)$

20. $(1, 88)$ and $(12, 11)$

Study Tip

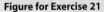

Don't wait until the final exam to review material. Do some review on a regular basis. The Making Connections exercises on this page can be used to review, compare, and contrast different concepts that you have studied. A good time to work these exercises is between a test and the start of new material.

Solve the problem.

21. ***Decreasing market share.*** The market share for Toys "R" Us went from 25% of the toy market in 1990 to 16% in 2000 as shown in the accompanying graph (*Forbes*, www.forbes.com).

a) Write the market share p in terms of x, where x is the number of years since 1990.

b) Wal-Mart's market share of the toy market went from 10% in 1990 to 19% in 2000. Write Wal-Mart's market share p in terms of x, where x is the number of years since 1990.

c) Solve the system of equations that you found in parts (a) and (b) to find the year in which Wal-Mart passed up Toys "R" Us in market share for toys.

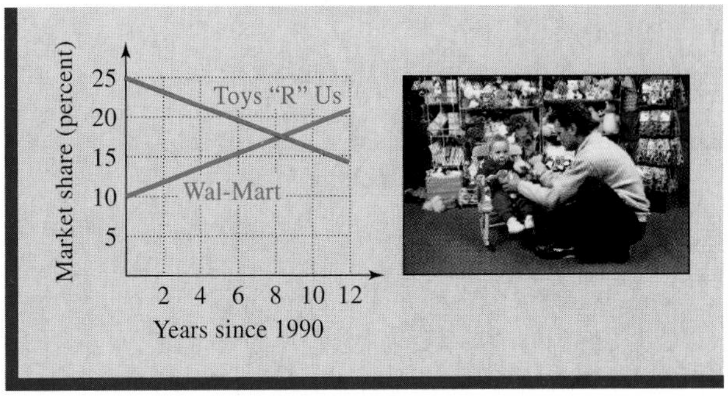

Figure for Exercise 21

Critical **Thinking** | For Individual or Group Work | Chapter 4

These exercises can be solved by a variety of techniques, which may or may not require algebra. So be creative and think critically. Explain all answers. Answers are in the Instructor's Edition of this text.

1. ***Throwing darts.*** A dart board contains a region worth 9 points and a region worth 4 points as shown in the accompanying figure. If you are allowed to throw as many darts as you wish, then what is the largest possible total score that you *cannot* get?

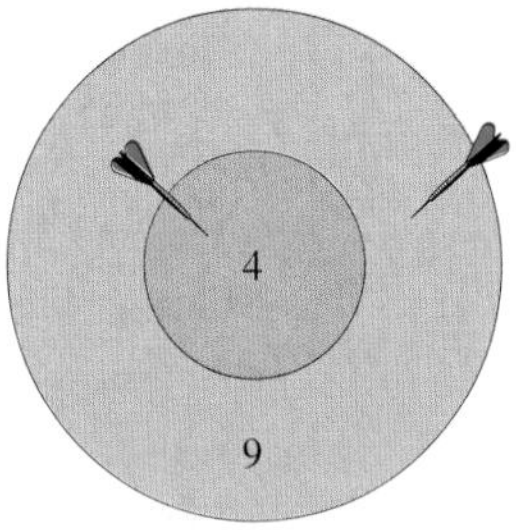

Figure for Exercise 1

2. ***Counting squares.*** A square checkerboard is made up of 36 alternately colored 1 inch by 1 inch squares.

a) What is the total number of squares that are visible on this checkerboard? (*Hint:* Count the 6 by 6 squares, then the 5 by 5 squares, and so on.)

b) How many are visible on a checkerboard that has 64 alternately colored 1 inch by 1 inch squares?

3. ***Four fours.*** Check out these equations:

$$\frac{4+4}{4+4} = 1, \quad \frac{4}{4} + \frac{4}{4} = 2, \quad 4 - 4^{4-4} = 3.$$

a) Using exactly four 4's write arithmetic expressions whose values are 4, 5, 6, and so on. How far can you go?

b) Repeat this exercise using four 5's, three 4's, and three 5's.

4. ***Four coins.*** Place four coins on a table with heads facing downward. On each move you must turn over exactly three coins. Count the number of moves it takes to get all four coins with heads facing upward. What is the minimum number of moves necessary to get all four heads facing upward?

5. ***Snakes and iguanas.*** A woman has a collection of snakes and iguanas. Her young son observed that the reptiles have a total of 50 eyes and 56 feet. How many reptiles of each type does the woman have?

Photo for Exercise 5

6. ***Hungry bugs.*** If it takes a colony of termites one day to devour a block of wood that is 2 inches wide, 2 inches long, and 2 inches high, then how long will it take them to devour a block of wood that is 4 inches wide, 4 inches long, and 4 inches high. Assume that they keep eating at the same rate.

7. ***Ancient history.*** This problem is from the second century. Four numbers have a sum of 9900. The second exceeds the first by one-seventh of the first. The third exceeds the sum of the first two by 300. The fourth exceeds the sum of the first three by 300. Find the four numbers.

8. ***Related digits.*** What is the largest four-digit number such that the second digit is one-fourth of the third digit, the third digit is twice the first digit, and the last digit is the same as the first digit?

Chapter 5

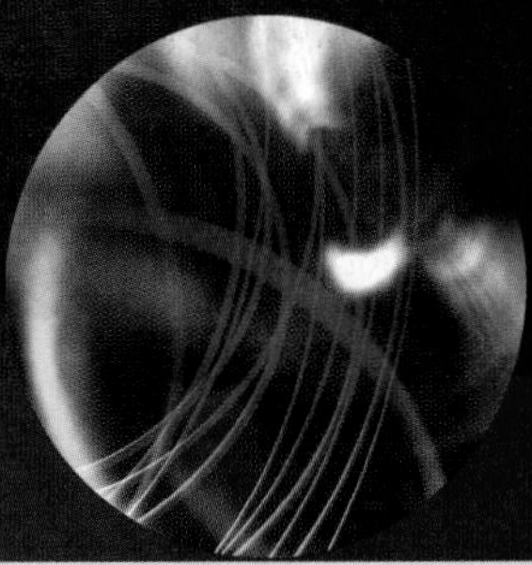

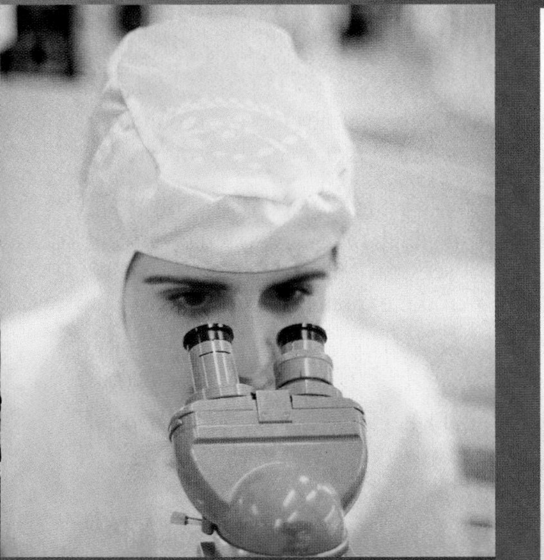

Polynomials and Exponents

The nineteenth-century physician and physicist Jean Louis Marie Poiseuille (1799–1869) is given credit for discovering a formula associated with the circulation of blood through arteries. Poiseuille's law, as it is known, can be used to determine the velocity of blood in an artery at a given distance from the center of the artery. The formula states that the flow of blood in an artery is faster toward the center of the blood vessel and is slower toward the outside. Blood flow can also be affected by a person's blood pressure, the length of the blood vessel, and the viscosity of the blood itself.

In later years, Poiseuille's continued interest in blood circulation led him to experiments to show that blood pressure rises and falls when a person exhales and inhales. In modern medicine, physicians can use Poiseuille's law to determine how much the radius of a blocked blood vessel must be widened to create a healthy flow of blood.

In this chapter you will study polynomials, the fundamental expressions of algebra. Polynomials are to algebra what integers are to arithmetic. We use polynomials to represent quantities in general, such as perimeter, area, revenue, and the volume of blood flowing through an artery.

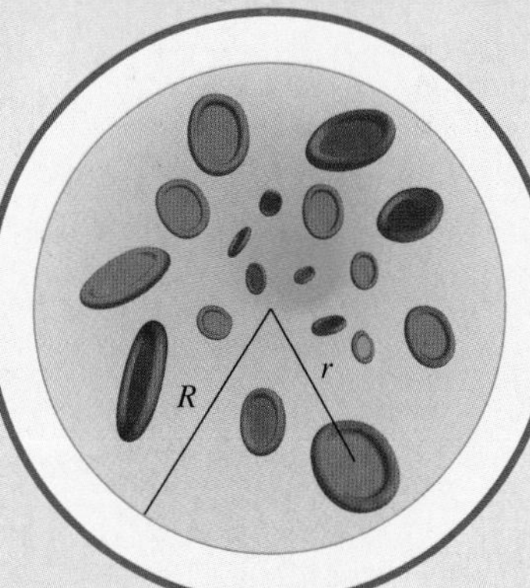

In Exercise 89 of Section 5.4, you will see Poiseuille's law represented by a polynomial.

5.1 Addition and Subtraction of Polynomials

In this Section

- **Polynomials**
- **Using Function Notation with Polynomials**
- **Addition of Polynomials**
- **Subtraction of Polynomials**
- **Applications**

We first used polynomials in Chapter 1 but did not identify them as polynomials. Polynomials also occurred in the equations and inequalities of Chapter 2. In this section we will define polynomials and begin a thorough study of polynomials.

Polynomials

In Chapter 1 we defined a **term** as an expression containing a number or the product of a number and one or more variables raised to powers. Some examples of terms are

$$4x^3,\ -x^2y^3,\ 6ab, \text{ and } -2.$$

A **polynomial** is a single term or a finite sum of terms in which the powers of the variables are positive integers. Later in this text you will see that negative integers and fractions can be used as exponents, but in a polynomial the exponents must be positive integers. For example,

$$4x^3 + (-15x^2) + x^1 + (-2)$$

is a polynomial. Because it is simpler to write addition of a negative as subtraction and $x^1 = x$, this polynomial is usually written as

$$4x^3 - 15x^2 + x - 2.$$

Study Tip

Many commuting students find it difficult to get help. Students are often stuck on a minor point that can be easily resolved over the telephone. So ask your instructor if he or she will answer questions over the telephone and during what hours.

The **degree of a polynomial** in one variable is the highest power of the variable in the polynomial. So $4x^3 - 15x^2 + x - 2$ has degree 3 and $7w - w^2$ has degree 2. The **degree of a term** is the power of the variable in the term. Because the last term has no variable, its degree is 0. The degree of x is 1 because $x = x^1$.

$$4x^3 - 15x^2 + x - 2$$

Third-degree term ($4x^3$), Second-degree term ($15x^2$), First-degree term (x), Zero-degree term (2)

A single number is called a **constant** and so the last term is the **constant term.** The degree of a polynomial consisting of a single number such as 8 is 0.

The number preceding the variable in each term is called the **coefficient** of that variable or the coefficient of that term. In $4x^3 - 15x^2 + x - 2$ the coefficient of x^3 is 4, the coefficient of x^2 is -15, and the coefficient of x is 1 because $x = 1 \cdot x$.

EXAMPLE 1

Identifying coefficients

Determine the coefficients of x^3 and x^2 in each polynomial:

a) $x^3 + 5x^2 - 6$ **b)** $4x^6 - x^3 + x$

Solution

a) Write the polynomial as $1 \cdot x^3 + 5x^2 - 6$ to see that the coefficient of x^3 is 1 and the coefficient of x^2 is 5.

b) The x^2-term is missing in $4x^6 - x^3 + x$. Because $4x^6 - x^3 + x$ can be written as

$$4x^6 - 1 \cdot x^3 + 0 \cdot x^2 + x,$$

the coefficient of x^3 is -1 and the coefficient of x^2 is 0.

Now do Exercises 7–12

For simplicity we generally write polynomials in one variable with the exponents decreasing from left to right and the constant term last. So we write

$$x^3 - 4x^2 + 5x + 1 \qquad \text{rather than} \qquad -4x^2 + 1 + 5x + x^3.$$

When a polynomial is written with decreasing exponents, the coefficient of the first term is called the **leading coefficient.**

Certain polynomials are given special names. A **monomial** is a polynomial that has one term, a **binomial** is a polynomial that has two terms, and a **trinomial** is a polynomial that has three terms. For example, $3x^5$ is a monomial, $2x - 1$ is a binomial, and $4x^6 - 3x + 2$ is a trinomial.

EXAMPLE 2

Types of polynomials

Identify each polynomial as a monomial, binomial, or trinomial and state its degree.

a) $5x^2 - 7x^3 + 2$ **b)** $x^{43} - x^2$ **c)** $5x$ **d)** -12

Solution

a) The polynomial $5x^2 - 7x^3 + 2$ is a third-degree trinomial.

b) The polynomial $x^{43} - x^2$ is a binomial with degree 43.

c) Because $5x = 5x^1$, this polynomial is a monomial with degree 1.

d) The polynomial -12 is a monomial with degree 0.

Now do Exercises 13–24

Study Tip

Be active in class. Don't be embarrassed to ask questions or answer questions. You can often learn more from a wrong answer than a right one. Your instructor knows that you are not yet an expert in algebra. Instructors love active classrooms and they will not think less of you for speaking out.

Using Function Notation with Polynomials

A polynomial such as $x^2 - x + 3$ has a value if x is replaced by a real number. For example, if $x = 2$, then replacing x with 2 yields

$$x^2 - x + 3 = 2^2 - 2 + 3 = 5.$$

So the value of the polynomial is 5 when $x = 2$. Putting 2 into the polynomial gives an output of 5. To make it easier to discuss polynomials and their values, polynomials are often named with letters. For example, if $P = x^2 - x + 3$, then $P = 5$ when $x = 2$.

Another notation that is commonly used in mathematics, computer science, and on graphing calculators is to follow the letter that names the polynomial with the number used for x and write the result as a single equation. So $P = 5$ when $x = 2$ is written as $P(2) = 5$ and read as "P evaluated at 2 is 5" or simply "P of 2 is 5." Think of $P(2) = 5$ as P being applied to the input number 2 and yielding an output of 5. This notation is called **function notation.** (See Section 9.7 for more information on functions.) Using function notation we write the polynomial as $P(x) = x^2 - x + 3$ rather than $P = x^2 - x + 3$. [Read $P(x)$ as "P of x."]

Function notation is very useful when we are evaluating a polynomial at several values of x. For example, if $P(x) = x^2 - x + 3$, then $P(0) = 3$, $P(1) = 3$, and $P(2) = 5$. Note that in function notation $P(x)$ does *not* mean P times x.

49. $(5.76x^2 - 3.14x - 7.09) + (3.9x^2 + 1.21x + 5.6)$

50. $(8.5x^2 + 3.27x - 9.33) + (x^2 - 4.39x - 2.32)$

Perform the indicated operation. See Example 5.

51. $(x - 2) - (5x - 8)$

52. $(x - 7) - (3x - 1)$

53. $(m - 2) - (m + 3)$

54. $(m + 5) - (m + 9)$

55. $(2z^2 - 3z) - (3z^2 - 5z)$

56. $(z^2 - 4z) - (5z^2 - 3z)$

57. $(w^5 - w^3) - (-w^4 + w^2)$

58. $(w^6 - w^3) - (-w^2 + w)$

59. $(t^2 - 3t + 4) - (t^2 - 5t - 9)$

60. $(t^2 - 6t + 7) - (5t^2 - 3t - 2)$

61. $(9 - 3y - y^2) - (2 + 5y - y^2)$

62. $(4 - 5y + y^3) - (2 - 3y + y^2)$

63. $(3.55x - 879) - (26.4x - 455.8)$

64. $(345.56x - 347.4) - (56.6x + 433)$

Add or subtract the polynomials as indicated. See Examples 4 and 5.

65. Add:
$$\begin{array}{r} 3a - 4 \\ \underline{a + 6} \end{array}$$

66. Add:
$$\begin{array}{r} 2w - 8 \\ \underline{w + 3} \end{array}$$

67. Subtract:
$$\begin{array}{r} 3x + 11 \\ \underline{5x + 7} \end{array}$$

68. Subtract:
$$\begin{array}{r} 4x + 3 \\ \underline{2x + 9} \end{array}$$

69. Add:
$$\begin{array}{r} a - b \\ \underline{a + b} \end{array}$$

70. Add:
$$\begin{array}{r} s - 6 \\ \underline{s - 1} \end{array}$$

71. Subtract:
$$\begin{array}{r} -3m + 1 \\ \underline{2m - 6} \end{array}$$

72. Subtract:
$$\begin{array}{r} -5n + 2 \\ \underline{3n - 4} \end{array}$$

73. Add:
$$\begin{array}{r} 2x^2 - x - 3 \\ \underline{2x^2 + x + 4} \end{array}$$

74. Add:
$$\begin{array}{r} -x^2 + 4x - 6 \\ \underline{3x^2 - x - 5} \end{array}$$

75. Subtract:
$$\begin{array}{rrrr} 3a^3 - 5a^2 & & + 7 \\ \underline{2a^3 + 4a^2} & \underline{- 2a} & \underline{} \end{array}$$

76. Subtract:
$$\begin{array}{rrr} -2b^3 + 7b^2 & & - 9 \\ \underline{b^3 } & \underline{- 4b} & \underline{- 2} \end{array}$$

77. Subtract:
$$\begin{array}{l} x^2 - 3x + 6 \\ \underline{x^2 - 3} \end{array}$$

78. Subtract:
$$\begin{array}{r} x^4 - 3x^2 + 2 \\ \underline{3x^4 - 2x^2 } \end{array}$$

79. Add:
$$\begin{array}{r} y^3 + 4y^2 - 6y - 5 \\ \underline{y^3 + 3y^2 + 2y - 9} \end{array}$$

80. Add:
$$\begin{array}{r} q^2 - 4q + 9 \\ \underline{-3q^2 - 7q + 5} \end{array}$$

Perform the indicated operations. See Example 6.

81. $(4m - 2) + (2m + 4) - (9m - 1)$

82. $(-5m - 6) + (8m - 3) - (-5m + 3)$

83. $(6y - 2) - (8y + 3) - (9y - 2)$

84. $(-5y - 1) - (8y - 4) - (y + 3)$

85. $(-x^2 - 5x + 4) + (6x^2 - 8x + 9) - (3x^2 - 7x + 1)$

86. $(-8x^2 + 5x - 12) + (-3x^2 - 9x + 18)$
$- (-3x^2 + 9x - 4)$

87. $(-6z^4 - 3z^3 + 7z^2) - (5z^3 + 3z^2 - 2)$
$+ (z^4 - z^2 + 5)$

88. $(-v^3 - v^2 - 1) - (v^4 - v^2 - v - 1) + (v^3 - 3v^2 + 6)$

Solve each problem. See Example 7.

89. ***Profitable pumps.*** Walter Waterman, of Walter's Water Pumps in Winnipeg has found that when he produces x water pumps per month, his revenue is $R(x) = x^2 + 400x + 300$ dollars. His cost for producing x water pumps per month is $C(x) = x^2 + 300x - 200$ dollars. Write a polynomial that represents his monthly profit $P(x)$ for x water pumps. Evaluate this profit polynomial for $x = 50$.

90. ***Manufacturing costs.*** Ace manufacturing has determined that the cost of labor for producing x transmissions is $L(x) = 0.3x^2 + 400x + 550$ dollars, while the cost of materials is $M(x) = 0.1x^2 + 50x + 800$ dollars.

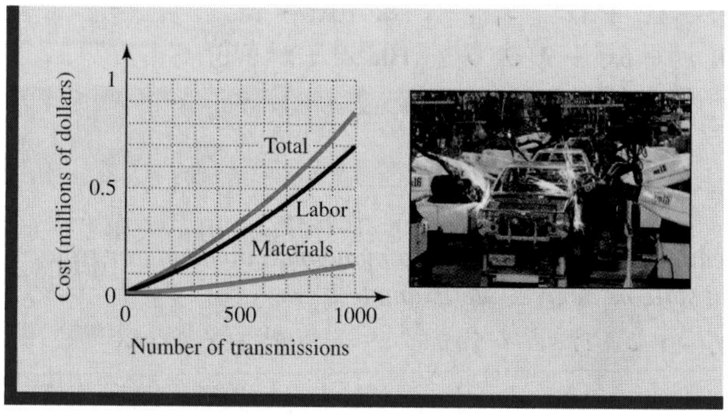

Figure for Exercise 90

a) Write a polynomial $T(x)$ that represents the total cost of materials and labor for producing x transmissions.

b) Evaluate the total cost polynomial for $x = 500$.

c) Find the cost of labor for 500 transmissions and the cost of materials for 500 transmissions.

91. ***Perimeter of a triangle.*** The shortest side of a triangle is x meters, and the other two sides are $3x - 1$ and $2x + 4$ meters. Write a polynomial $P(x)$ that represents the perimeter and then evaluate the perimeter polynomial if x is 4 meters.

92. ***Perimeter of a rectangle.*** The width of a rectangular playground is $2x - 5$ feet, and the length is $3x + 9$ feet. Write a polynomial $P(x)$ that represents the perimeter and then evaluate this perimeter polynomial if x is 4 feet.

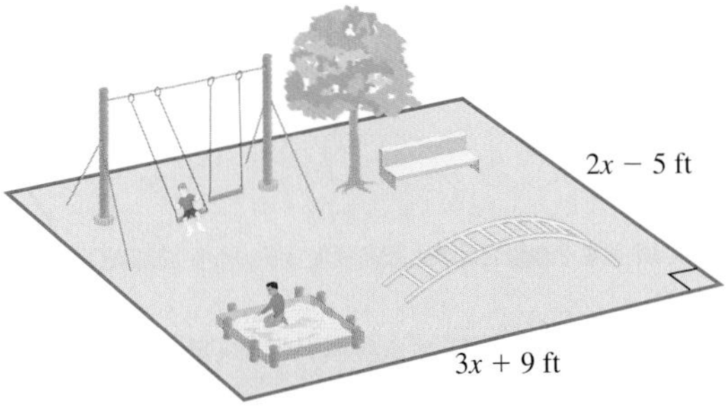

Figure for Exercise 92

93. ***Before and after.*** Jessica traveled $2x + 50$ miles in the morning and $3x - 10$ miles in the afternoon. Write a polynomial that represents the total distance that she traveled. Find the total distance if $x = 20$.

94. ***Total distance.*** Hanson drove his rig at x mph for 3 hours, then increased his speed to $x + 15$ mph and drove for 2 more hours. Write a polynomial that represents the total distance that he traveled. Find the total distance if $x =$ 45 mph.

95. ***Sky divers.*** Bob and Betty simultaneously jump from two airplanes at different altitudes. Bob's altitude t seconds after leaving the plane is $-16t^2 + 6600$ feet. Betty's altitude t seconds after leaving the plane is $-16t^2 + 7400$ feet. Write a polynomial that represents the difference between their altitudes t seconds after leaving the planes. What is the difference between their altitudes 3 seconds after leaving the planes?

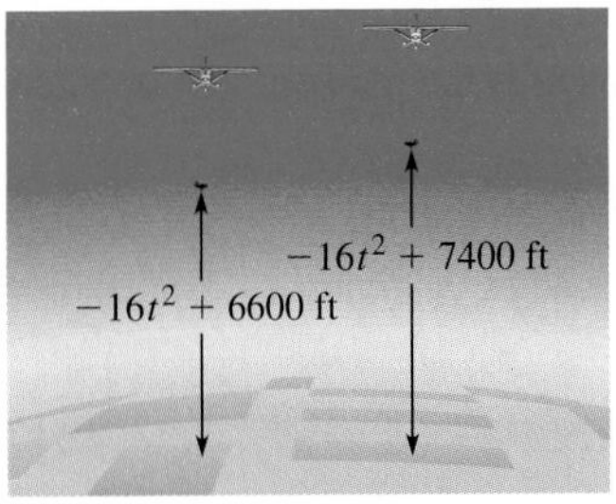

Figure for Exercise 95

96. ***Height difference.*** A red ball and a green ball are simultaneously tossed into the air. The red ball is given an initial velocity of 96 feet per second, and its height t seconds after it is tossed is $-16t^2 + 96t$ feet. The green ball is given an initial velocity of 80 feet per second, and its height t seconds after it is tossed is $-16t^2 + 80t$ feet.

a) Find a polynomial that represents the difference in the heights of the two balls.

b) How much higher is the red ball 2 seconds after the balls are tossed?

c) In reality, when does the difference in the heights stop increasing?

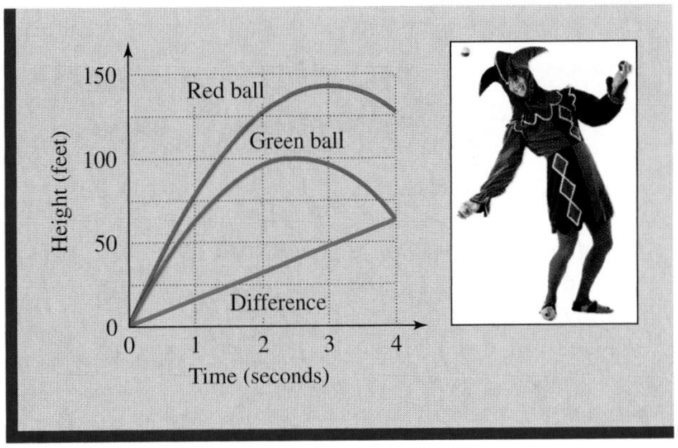

Figure for Exercise 96

97. ***Total interest.*** Donald received $0.08(x + 554)$ dollars interest on one investment and $0.09(x + 335)$ interest on another investment. Write a polynomial that represents the total interest he received. What is the total interest if $x = 1000$?

98. ***Total acid.*** Deborah figured that the amount of acid in one bottle of solution is $0.12x$ milliliters and the amount of acid in another bottle of solution is $0.22(75 - x)$ milliliters.

Find a polynomial that represents the total amount of acid? What is the total amount of acid if $x = 50$?

99. ***Harris-Benedict for females.*** The Harris-Benedict polynomial

$$655.1 + 9.56w + 1.85h - 4.68a$$

represents the number of calories needed to maintain a female at rest for 24 hours, where w is her weight in kilograms, h is her height in centimeters, and a is her age in years. Find the number of calories needed by a 30-year-old 54-kilogram female who is 157 centimeters tall.

100. ***Harris-Benedict for males.*** The Harris-Benedict polynomial

$$66.5 + 13.75w + 5.0h - 6.78a$$

represents the number of calories needed to maintain a male at rest for 24 hours, where w is his weight in kilograms, h is his height in centimeters, and a is his age in years. Find the number of calories needed by a 40-year-old 90-kilogram male who is 185 centimeters tall.

Getting More Involved

101. ***Discussion***

Is the sum of two natural numbers always a natural number? Is the sum of two integers always an integer? Is the sum of two polynomials always a polynomial? Explain.

102. ***Discussion***

Is the difference of two natural numbers always a natural number? Is the difference of two rational numbers always a rational number? Is the difference of two polynomials always a polynomial? Explain.

103. ***Writing***

Explain why the polynomial $2^4 - 7x^3 + 5x^2 - x$ has degree 3 and not degree 4.

104. ***Discussion***

Which of the following polynomials does not have degree 2? Explain.

a) πr^2 **b)** $\pi^2 - 4$
c) $y^2 - 4$ **d)** $x^2 - x^4$
e) $a^2 - 3a + 9$

5.2 Multiplication of Polynomials

In this Section

- **Multiplying Monomials with the Product Rule**
- **Multiplying Polynomials**
- **The Opposite of a Polynomial**
- **Applications**

You learned to multiply some polynomials in Chapter 1. In this section you will learn how to multiply any two polynomials.

Multiplying Monomials with the Product Rule

To multiply two monomials, such as x^3 and x^5, recall that

$$x^3 = x \cdot x \cdot x \qquad \text{and} \qquad x^5 = x \cdot x \cdot x \cdot x \cdot x.$$

So

$$x^3 \cdot x^5 = \underbrace{\overbrace{(x \cdot x \cdot x)}^{\text{3 factors}}\overbrace{(x \cdot x \cdot x \cdot x \cdot x)}^{\text{5 factors}}}_{\text{8 factors}} = x^8.$$

The exponent of the product of x^3 and x^5 is the sum of the exponents 3 and 5. This example illustrates the **product rule** for multiplying exponential expressions.

Product Rule

If a is any real number and m and n are any positive integers, then

$$a^m \cdot a^n = a^{m+n}.$$

EXAMPLE 1

Multiplying monomials

Find the indicated products.

a) $x^2 \cdot x^4 \cdot x$ **b)** $(-2ab)(-3ab)$ **c)** $-4x^2y^2 \cdot 3xy^5$ **d)** $(3a)^2$

Solution

a) $x^2 \cdot x^4 \cdot x = x^2 \cdot x^4 \cdot x^1$

$= x^7$ Product rule

b) $(-2ab)(-3ab) = (-2)(-3) \cdot a \cdot a \cdot b \cdot b$

$= 6a^2b^2$ Product rule

c) $(-4x^2y^2)(3xy^5) = (-4)(3)x^2 \cdot x \cdot y^2 \cdot y^5$

$= -12x^3y^7$ Product rule

d) $(3a)^2 = 3a \cdot 3a$

$= 9a^2$

Now do Exercises 7–22

CAUTION Be sure to distinguish between adding and multiplying monomials. You can add like terms to get $3x^4 + 2x^4 = 5x^4$, but you cannot combine the terms in $3w^5 + 6w^2$. However, you can multiply any two monomials: $3x^4 \cdot 2x^4 = 6x^8$ and $3w^5 \cdot 6w^2 = 18w^7$.

Multiplying Polynomials

To multiply a monomial and a polynomial, we use the distributive property.

EXAMPLE 2

Multiplying monomials and polynomials

Find each product.

a) $3x^2(x^3 - 4x)$ **b)** $(y^2 - 3y + 4)(-2y)$ **c)** $-a(b - c)$

Solution

a) $3x^2(x^3 - 4x) = 3x^2 \cdot x^3 - 3x^2 \cdot 4x$ Distributive property

$= 3x^5 - 12x^3$

b) $(y^2 - 3y + 4)(-2y) = y^2(-2y) - 3y(-2y) + 4(-2y)$ Distributive property

$= -2y^3 - (-6y^2) + (-8y)$

$= -2y^3 + 6y^2 - 8y$

c) $-a(b - c) = (-a)b - (-a)c$ Distributive property

$= -ab + ac$

$= ac - ab$

Note in part (c) that either of the last two binomials is the correct answer. The last one is just a little simpler to read.

Now do Exercises 23–36

Study Tip

When doing homework or taking notes, use a pencil with an eraser. Everyone makes mistakes. If you get a problem wrong, don't start over. Check your work for errors and use the eraser. It is better to find out where you went wrong than to simply get the right answer.

Just as we use the distributive property to find the product of a monomial and a polynomial, we can use the distributive property to find the product of two binomials and the product of a binomial and a trinomial.

EXAMPLE 3

Multiplying polynomials

Use the distributive property to find each product.

a) $(x + 2)(x + 5)$ **b)** $(x + 3)(x^2 + 2x - 7)$

Solution

a) First multiply each term of $x + 5$ by $x + 2$:

$$\begin{aligned}(x + 2)(x + 5) &= (x + 2)x + (x + 2)5 && \text{Distributive property}\\ &= x^2 + 2x + 5x + 10 && \text{Distributive property}\\ &= x^2 + 7x + 10 && \text{Combine like terms.}\end{aligned}$$

b) First multiply each term of the trinomial by $x + 3$:

$$\begin{aligned}(x + 3)(x^2 + 2x - 7) &= (x + 3)x^2 + (x + 3)2x + (x + 3)(-7) && \text{Distributive property}\\ &= x^3 + 3x^2 + 2x^2 + 6x - 7x - 21 && \text{Distributive property}\\ &= x^3 + 5x^2 - x - 21 && \text{Combine like terms.}\end{aligned}$$

Now do Exercises 37–48

Products of polynomials can also be found by arranging the multiplication vertically like multiplication of whole numbers.

EXAMPLE 4

Multiplying vertically

Find each product.

a) $(x - 2)(3x + 7)$ **b)** $(x + y)(a + 3)$

Solution

a)

$$\begin{array}{rrl} & 3x + 7 & \\ & x - 2 & \\ \hline & -6x - 14 & \leftarrow -2 \text{ times } 3x + 7 \\ 3x^2 + 7x & & \leftarrow x \text{ times } 3x + 7 \\ \hline 3x^2 + & x - 14 & \text{Add.} \end{array}$$

b)

$$\begin{array}{rr} & x + y \\ & a + 3 \\ \hline & 3x + 3y \\ ax + ay & \\ \hline ax + ay & + 3x + 3y \end{array}$$

Now do Exercises 49–64

Helpful Hint

Many students find vertical multiplication easier than applying the distributive property twice horizontally. However, you should learn both methods because horizontal multiplication will help you with factoring by grouping in Section 6.2.

Examples 2 to 4 illustrate the following rule.

Multiplication of Polynomials

To multiply polynomials, multiply each term of one polynomial by every term of the other polynomial, then combine like terms.

The Opposite of a Polynomial

Note the result of multiplying the difference $a - b$ by -1:

$$-1(a - b) = -a + b = b - a$$

Because multiplying by -1 is the same as taking the opposite, we can write

$$-(a - b) = b - a.$$

So $a - b$ and $b - a$ are opposites or additive inverses of each other. If a and b are replaced by numbers, the values of $a - b$ and $b - a$ are additive inverses. For example, $3 - 7 = -4$ and $7 - 3 = 4$.

CAUTION The opposite of $a + b$ is $-a - b$, *not* $a - b$.

EXAMPLE 5

Opposite of a polynomial

Find the opposite of each polynomial.

a) $x - 2$ **b)** $9 - y^2$

c) $a + 4$ **d)** $-x^2 + 6x - 3$

Solution

a) $-(x - 2) = 2 - x$ **b)** $-(9 - y^2) = y^2 - 9$

c) $-(a + 4) = -a - 4$ **d)** $-(-x^2 + 6x - 3) = x^2 - 6x + 3$

Now do Exercises 65–72

Applications

EXAMPLE 6

Multiplying polynomials

A parking lot is 20 yards wide and 30 yards long. If the college increases the length and width by the same amount to handle an increasing number of cars, then what polynomial represents the area of the new lot? What is the new area if the increase is 15 yards?

Solution

If x is the amount of increase, then the new lot will be $x + 20$ yards wide and $x + 30$ yards long as shown in Fig. 5.1. Multiply the length and width to get the area:

$$\begin{aligned}(x + 20)(x + 30) &= (x + 20)x + (x + 20)30\\ &= x^2 + 20x + 30x + 600\\ &= x^2 + 50x + 600\end{aligned}$$

The polynomial $x^2 + 50x + 600$ represents the area of the new lot. If $x = 15$, then

$$x^2 + 50x + 600 = (15)^2 + 50(15) + 600 = 1575.$$

If the increase is 15 yards, then the area of the lot will be 1575 square yards.

Now do Exercises 93–104

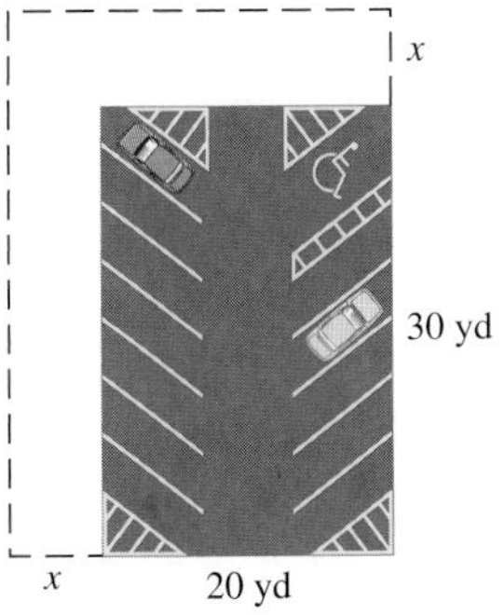

Figure 5.1

Warm-Ups

True or false? Explain your answer.

1. $3x^3 \cdot 5x^4 = 15x^{12}$ for any value of x.
2. $3x^2 \cdot 2x^7 = 5x^9$ for any value of x.
3. $(3y^3)^2 = 9y^6$ for any value of y.
4. $-3x(5x - 7x^2) = -15x^3 + 21x^2$ for any value of x.
5. $2x(x^2 - 3x + 4) = 2x^3 - 6x^2 + 8x$ for any number x.
6. $-2(3 - x) = 2x - 6$ for any number x.
7. $(a + b)(c + d) = ac + ad + bc + bd$ for any values of a, b, c, and d.
8. $-(x - 7) = 7 - x$ for any value of x.
9. $83 - 37 = -(37 - 83)$
10. The opposite of $x + 3$ is $x - 3$ for any number x.

5.2 Exercises

Reading and Writing *After reading this section, write out the answers to these questions. Use complete sentences.*

1. What is the product rule for exponents?
2. Why is the sum of two monomials not necessarily a monomial?
3. What property of the real numbers is used when multiplying a monomial and a polynomial?
4. What property of the real numbers is used when multiplying two binomials?
5. How do we multiply any two polynomials?
6. How do we find the opposite of a polynomial?

Find each product. See Example 1.

7. $3x^2 \cdot 9x^3$
8. $5x^7 \cdot 3x^5$
9. $2a^3 \cdot 7a^8$
10. $3y^{12} \cdot 5y^{15}$
11. $-6x^2 \cdot 5x^2$
12. $-2x^2 \cdot 8x^5$
13. $(-9x^{10})(-3x^7)$
14. $(-2x^2)(-8x^9)$
15. $-6st \cdot 9st$
16. $-12sq \cdot 3s$
17. $3wt \cdot 8w^7t^6$
18. $h^8k^3 \cdot 5h$
19. $(5y)^2$
20. $(6x)^2$
21. $(2x^3)^2$
22. $(3y^5)^2$

Find each product. See Example 2.

23. $x(x + y^2)$
24. $x^2(x - y)$
25. $4y^2(y^5 - 2y)$
26. $6t^3(t^5 + 3t^2)$
27. $-3y(6y - 4)$
28. $-9y(y^2 - 1)$
29. $(y^2 - 5y + 6)(-3y)$

30. $(x^3 - 5x^2 - 1)7x^2$

31. $-x(y^2 - x^2)$

32. $-ab(a^2 - b^2)$

33. $(3ab^3 - a^2b^2 - 2a^3b)5a^3$

34. $(3c^2d - d^3 + 1)8cd^2$

35. $-\frac{1}{2}t^2v(4t^3v^2 - 6tv - 4v)$

36. $-\frac{1}{3}m^2n^3(-6mn^2 + 3mn - 12)$

Use the distributive property to find each product. See Example 3.

37. $(x + 1)(x + 2)$ **38.** $(x + 6)(x + 3)$

39. $(x - 3)(x + 5)$ **40.** $(y - 2)(y + 4)$

41. $(t - 4)(t - 9)$ **42.** $(w - 3)(w - 5)$

43. $(x + 1)(x^2 + 2x + 2)$ **44.** $(x - 1)(x^2 + x + 1)$

45. $(3y + 2)(2y^2 - y + 3)$ **46.** $(4y + 3)(y^2 + 3y + 1)$

47. $(y^2z - 2y^4)(y^2z + 3z^2 - y^4)$

48. $(m^3 - 4mn^2)(6m^4n^2 - 3m^6 + m^2n^4)$

Find each product vertically. See Example 4.

49. $\begin{array}{r} 2a - 3 \\ \underline{a + 5} \end{array}$ **50.** $\begin{array}{r} 2w - 6 \\ \underline{w + 5} \end{array}$

51. $\begin{array}{r} 7x + 30 \\ \underline{2x + 5} \end{array}$ **52.** $\begin{array}{r} 5x + 7 \\ \underline{3x + 6} \end{array}$

53. $\begin{array}{r} 5x + 2 \\ \underline{4x - 3} \end{array}$ **54.** $\begin{array}{r} 4x + 3 \\ \underline{2x - 6} \end{array}$

55. $\begin{array}{r} m - 3n \\ \underline{2a + b} \end{array}$ **56.** $\begin{array}{r} 3x + 7 \\ \underline{a - 2b} \end{array}$

57. $\begin{array}{r} x^2 + 3x - 2 \\ \underline{x + 6} \end{array}$ **58.** $\begin{array}{r} -x^2 + 3x - 5 \\ \underline{x - 7} \end{array}$

59. $\begin{array}{r} 2a^3 - 3a^2 + 4 \\ \underline{-2a - 3} \end{array}$ **60.** $\begin{array}{r} -3x^2 + 5x - 2 \\ \underline{-5x - 6} \end{array}$

61. $\begin{array}{r} x - y \\ \underline{x + y} \end{array}$ **62.** $\begin{array}{r} a^2 + b^2 \\ \underline{a^2 - b^2} \end{array}$

63. $\begin{array}{r} x^2 - xy + y^2 \\ \underline{x + y} \end{array}$ **64.** $\begin{array}{r} 4w^2 + 2wv + v^2 \\ \underline{2w - v} \end{array}$

Find the opposite of each polynomial. See Example 5.

65. $3t - u$ **66.** $-3t - u$

67. $3x + y$ **68.** $x - 3y$

69. $-3a^2 - a + 6$

70. $3b^2 - b - 6$

71. $3v^2 + v - 6$

72. $-3t^2 + t - 6$

Perform the indicated operation.

73. $-3x(2x - 9)$ **74.** $-1(2 - 3x)$

75. $2 - 3x(2x - 9)$ **76.** $6 - 3(4x - 8)$

77. $(2 - 3x) + (2x - 9)$ **78.** $(2 - 3x) - (2x - 9)$

79. $(6x^6)^2$ **80.** $(-3a^3b)^2$

81. $3ab^3(-2a^2b^7)$ **82.** $-4xst \cdot 8xs$

83. $(5x + 6)(5x + 6)$ **84.** $(5x - 6)(5x - 6)$

85. $(5x - 6)(5x + 6)$ **86.** $(2x - 9)(2x + 9)$

87. $2x^2(3x^5 - 4x^2)$ **88.** $4a^3(3ab^3 - 2ab^3)$

89. $(m - 1)(m^2 + m + 1)$ **90.** $(a + b)(a^2 - ab + b^2)$

91. $(3x - 2)(x^2 - x - 9)$

92. $(5 - 6y)(3y^2 - y - 7)$

Solve each problem. See Example 6.

93. ***Office space.*** The length of a professor's office is x feet, and the width is $x + 4$ feet. Write a polynomial that represents the area. Find the area if $x = 10$ ft.

94. ***Swimming space.*** The length of a rectangular swimming pool is $2x - 1$ meters, and the width is $x + 2$ meters. Write a polynomial that represents the area. Find the area if x is 5 meters.

95. ***Area.*** A roof truss is in the shape of a triangle with height of x feet and a base of $2x + 1$ feet. Write a polynomial $A(x)$ that represents the area of the triangle. Find $A(5)$. See the figure on the next page.

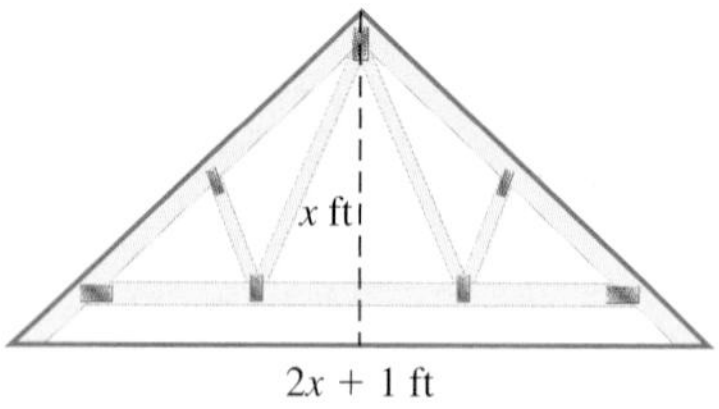

Figure for Exercise 95

96. ***Volume.*** The length, width, and height of a box are x, $2x$, and $3x - 5$ inches, respectively. Write a polynomial $V(x)$ that represents its volume. Find $V(3)$.

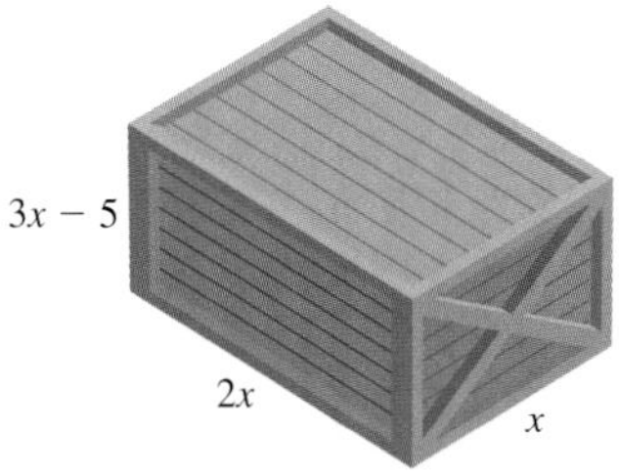

Figure for Exercise 96

97. ***Number pairs.*** If two numbers differ by 5, then what polynomial represents their product?

98. ***Number pairs.*** If two numbers have a sum of 9, then what polynomial represents their product?

99. ***Area of a rectangle.*** The length of a rectangle is $2.3x + 1.2$ meters, and its width is $3.5x + 5.1$ meters. What polynomial represents its area?

100. ***Patchwork.*** A quilt patch cut in the shape of a triangle has a base of $5x$ inches and a height of $1.732x$ inches. What polynomial represents its area?

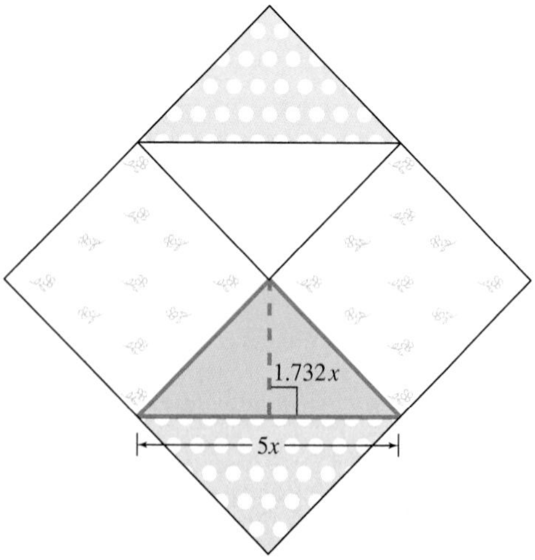

Figure for Exercise 100

101. ***Total revenue.*** If a promoter charges p dollars per ticket for a concert in Tulsa, then she expects to sell $40{,}000 - 1000p$ tickets to the concert. How many tickets will she sell if the tickets are \$10 each? Find the total revenue when the tickets are \$10 each. What polynomial represents the total revenue expected for the concert when the tickets are p dollars each?

102. ***Manufacturing shirts.*** If a manufacturer charges p dollars each for rugby shirts, then he expects to sell $2000 - 100p$ shirts per week. What polynomial represents the total revenue expected for a week? How many shirts will be sold if the manufacturer charges \$20 each for the shirts? Find the total revenue when the shirts are sold for \$20 each. Use the bar graph to determine the price that will give the maximum total revenue.

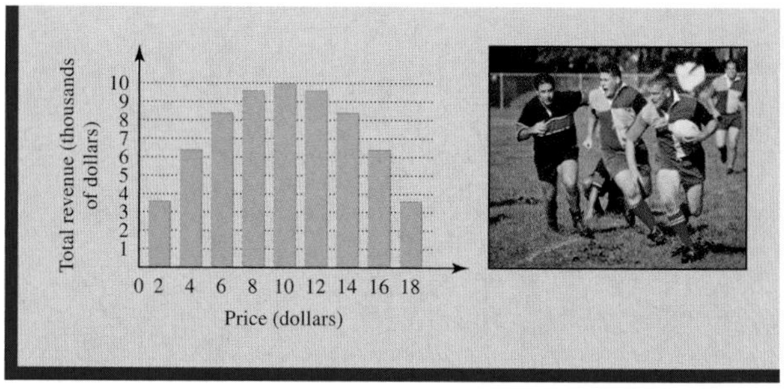

Figure for Exercise 102

103. ***Periodic deposits.*** At the beginning of each year for 5 years, an investor invests \$10 in a mutual fund with an average annual return of r. If we let $x = 1 + r$, then at the end of the first year (just before the next investment) the value is $10x$ dollars. Because \$10 is then added to the $10x$ dollars, the amount at the end of the second year is $(10x + 10)x$ dollars. Find a polynomial that represents the value of the investment at the end of the fifth year. Evaluate this polynomial if $r = 10\%$.

104. ***Increasing deposits.*** At the beginning of each year for 5 years, an investor invests in a mutual fund with an average annual return of r. The first year, she invests \$10; the second year, she invests \$20; the third year, she invests \$30; the fourth year, she invests \$40; the fifth year, she invests \$50. Let $x = 1 + r$ as in Exercise 103 and write a polynomial in x that represents the value of the investment at the end of the fifth year. Evaluate this polynomial for $r = 8\%$.

Getting More Involved

105. *Discussion*

Name all properties of the real numbers that are used in finding the following products:

a) $-2ab^3c^2 \cdot 5a^2bc$ **b)** $(x^2 + 3)(x^2 - 8x - 6)$

106. *Discussion*

Find the product of 27 and 436 without using a calculator. Then use the distributive property to find the product $(20 + 7)(400 + 30 + 6)$ as you would find the product of a binomial and a trinomial. Explain how the two methods are related.

5.3 Multiplication of Binomials

In this Section

- The FOIL Method
- Multiplying Binomials Quickly

In Section 5.2 you learned to multiply polynomials. In this section you will learn a rule that makes multiplication of binomials simpler.

The FOIL Method

We can use the distributive property to find the product of two binomials. For example,

$$
\begin{aligned}
(x + 2)(x + 3) &= (x + 2)x + (x + 2)3 && \text{Distributive property} \\
&= x^2 + 2x + 3x + 6 && \text{Distributive property} \\
&= x^2 + 5x + 6 && \text{Combine like terms.}
\end{aligned}
$$

There are four terms in $x^2 + 2x + 3x + 6$. The term x^2 is the product of the *first* terms of each binomial, x and x. The term $3x$ is the product of the two *outer* terms, 3 and x. The term $2x$ is the product of the two *inner* terms, 2 and x. The term 6 is the product of the last terms of each binomial, 2 and 3. We can connect the terms multiplied by lines as follows:

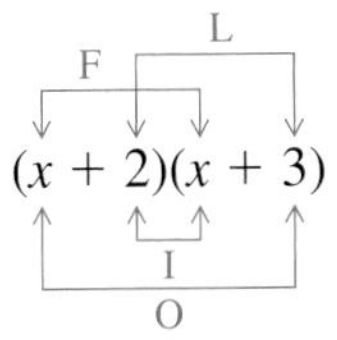

F = First terms
O = Outer terms
I = Inner terms
L = Last terms

If you remember the word FOIL, you can get the product of the two binomials much faster than writing out all of the steps above. This method is called the **FOIL method.** The name should make it easier to remember.

EXAMPLE 1

Using the FOIL method

Find each product.

a) $(x + 2)(x - 4)$ **b)** $(2x + 5)(3x - 4)$

c) $(a - b)(2a - b)$ **d)** $(x + 3)(y + 5)$

Helpful Hint

You may have to practice FOIL a while to get good at it. However, the better you are at FOIL, the easier you will find factoring in Chapter 6.

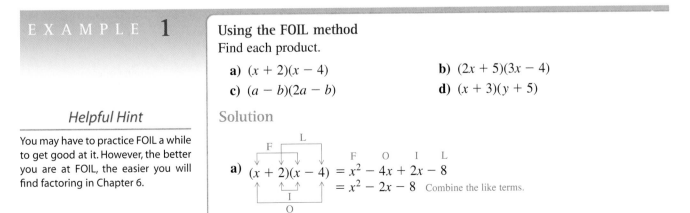

Solution

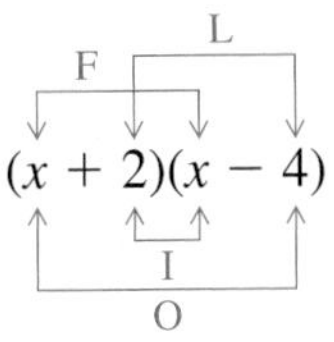

a) $(x + 2)(x - 4) = \overset{F}{x^2} - \overset{O}{4x} + \overset{I}{2x} - \overset{L}{8}$

$= x^2 - 2x - 8$ Combine the like terms.

b) $(2x + 5)(3x - 4) = 6x^2 - 8x + 15x - 20$
$= 6x^2 + 7x - 20$ Combine the like terms.

c) $(a - b)(2a - b) = 2a^2 - ab - 2ab + b^2$
$= 2a^2 - 3ab + b^2$

d) $(x + 3)(y + 5) = xy + 5x + 3y + 15$ There are no like terms to combine.

Now do Exercises 5–28

FOIL can be used to multiply any two binomials. The binomials in Example 2 have higher powers than those of Example 1.

EXAMPLE 2

Using the FOIL method

Find each product.

a) $(x^3 - 3)(x^3 + 6)$ **b)** $(2a^2 + 1)(a^2 + 5)$

Solution

a) $(x^3 - 3)(x^3 + 6) = x^6 + 6x^3 - 3x^3 - 18$
$= x^6 + 3x^3 - 18$

b) $(2a^2 + 1)(a^2 + 5) = 2a^4 + 10a^2 + a^2 + 5$
$= 2a^4 + 11a^2 + 5$

Now do Exercises 29–40

Multiplying Binomials Quickly

The outer and inner products in the FOIL method are often like terms, and we can combine them without writing them down. Once you become proficient at using FOIL, you can find the product of two binomials without writing anything except the answer.

EXAMPLE 3

Using FOIL to find a product quickly

Find each product. Write down only the answer.

a) $(x + 3)(x + 4)$ **b)** $(2x - 1)(x + 5)$ **c)** $(a - 6)(a + 6)$

Solution

a) $(x + 3)(x + 4) = x^2 + 7x + 12$ Combine like terms: $3x + 4x = 7x$.

b) $(2x - 1)(x + 5) = 2x^2 + 9x - 5$ Combine like terms: $10x - x = 9x$.

c) $(a - 6)(a + 6) = a^2 - 36$ Combine like terms: $6a - 6a = 0$.

Now do Exercises 41–64

EXAMPLE 4

Products of three binomials

Find each product.

a) $(b - 1)(b + 2)(b - 3)$

b) $\left(\frac{1}{2}x + 3\right)\left(\frac{1}{2}x - 3\right)(2x + 5)$

Solution

a) Use FOIL to find $(b - 1)(b + 2) = b^2 + b - 2$. Then use the distributive property to multiply $b^2 + b - 2$ and $b - 3$:

$$\begin{aligned}(b - 1)(b + 2)(b - 3) &= (b^2 + b - 2)(b - 3) && \text{FOIL}\\ &= (b^2 + b - 2)b + (b^2 + b - 2)(-3) && \text{Distributive property}\\ &= b^3 + b^2 - 2b - 3b^2 - 3b + 6 && \text{Distributive property}\\ &= b^3 - 2b^2 - 5b + 6 && \text{Combine like terms.}\end{aligned}$$

b)
$$\begin{aligned}\left(\frac{1}{2}x + 3\right)\left(\frac{1}{2}x - 3\right)(2x + 5) &= \left(\frac{1}{4}x^2 - 9\right)(2x + 5) && \text{FOIL}\\ &= \frac{1}{2}x^3 + \frac{5}{4}x^2 - 18x - 45 && \text{FOIL}\end{aligned}$$

Now do Exercises 65–72

EXAMPLE 5

Area of a garden

Sheila has a square garden with sides of length x feet. If she increases the length by 7 feet and decreases the width by 2 feet, then what trinomial represents the area of the new rectangular garden?

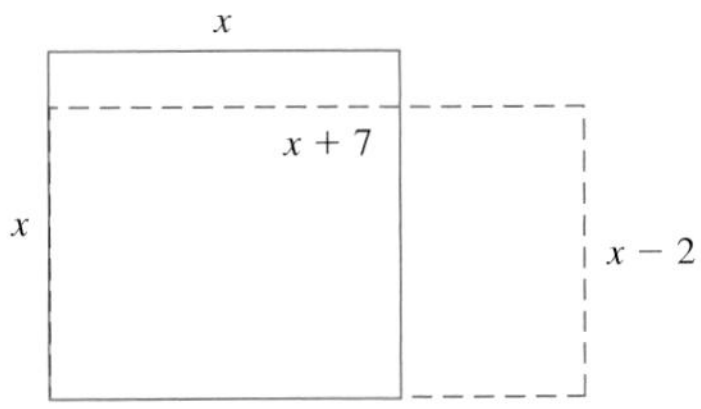

Figure 5.2

Solution

The length of the new garden is $x + 7$ and the width is $x - 2$ as shown in Fig. 5.2. The area is $(x + 7)(x - 2)$ or $x^2 + 5x - 14$ square feet.

Now do Exercises 95–98

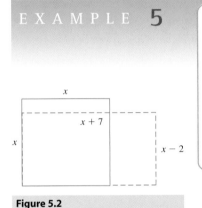

Warm-Ups ▼

True or false? Explain your answer.

1. $(x + 3)(x + 2) = x^2 + 6$
2. $(x + 2)(y + 1) = xy + x + 2y + 2$
3. $(3a - 5)(2a + 1) = 6a^2 + 3a - 10a - 5$
4. $(y + 3)(y - 2) = y^2 + y - 6$
5. $(x^2 + 2)(x^2 + 3) = x^4 + 5x^2 + 6$
6. $(3a^2 - 2)(3a^2 + 2) = 9a^2 - 4$
7. $(t + 3)(t + 5) = t^2 + 8t + 15$
8. $(y - 9)(y - 2) = y^2 - 11y - 18$
9. $(x + 4)(x - 7) = x^2 + 4x - 28$
10. It is not necessary to learn FOIL as long as you can get the answer.

5.3 Exercises

Reading and Writing *After reading this section, write out the answers to these questions. Use complete sentences.*

1. What property of the real numbers do we usually use to find the product of two binomials?

2. What does FOIL stand for?

3. What is the purpose of FOIL?

4. What is the maximum number of terms that can be obtained when two binomials are multiplied?

Use FOIL to find each product. See Example 1.

5. $(x + 2)(x + 4)$

6. $(x + 3)(x + 5)$

7. $(a + 1)(a + 4)$

8. $(w + 3)(w + 6)$

9. $(x + 9)(x + 10)$

10. $(x + 5)(x + 7)$

11. $(2x + 1)(x + 3)$

12. $(3x + 2)(2x + 1)$

13. $(a - 3)(a + 2)$

14. $(b - 1)(b + 2)$

15. $(2x - 1)(x - 2)$

16. $(2y - 5)(y - 2)$

17. $(2a - 3)(a + 1)$

18. $(3x - 5)(x + 4)$

19. $(w - 50)(w - 10)$

20. $(w - 30)(w - 20)$

21. $(y - a)(y + 5)$

22. $(a + t)(3 - y)$

23. $(5 - w)(w + m)$

24. $(a - h)(b + t)$

25. $(2m - 3t)(5m + 3t)$

26. $(2x - 5y)(x + y)$

27. $(5a + 2b)(9a + 7b)$

28. $(11x + 3y)(x + 4y)$

Use FOIL to find each product. See Example 2.

29. $(x^2 - 5)(x^2 + 2)$

30. $(y^2 + 1)(y^2 - 2)$

31. $(h^3 + 5)(h^3 + 5)$

32. $(y^6 + 1)(y^6 - 4)$

33. $(3b^3 + 2)(b^3 + 4)$

34. $(5n^4 - 1)(n^4 + 3)$

35. $(y^2 - 3)(y - 2)$

36. $(x - 1)(x^2 - 1)$

37. $(3m^3 - n^2)(2m^3 + 3n^2)$

38. $(6y^4 - 2z^2)(6y^4 - 3z^2)$

39. $(3u^2v - 2)(4u^2v + 6)$

40. $(5y^3w^2 + z)(2y^3w^2 + 3z)$

Find each product. Try to write only the answer. See Example 3.

41. $(b + 4)(b + 5)$

42. $(y + 8)(y + 4)$

43. $(x - 3)(x + 9)$

44. $(m + 7)(m - 8)$

45. $(a + 5)(a + 5)$

46. $(t - 4)(t - 4)$

47. $(2x - 1)(2x - 1)$

48. $(3y + 4)(3y + 4)$

49. $(z - 10)(z + 10)$

50. $(3h - 5)(3h + 5)$

51. $(a + b)(a + b)$

52. $(x - y)(x - y)$

53. $(a - 1)(a - 2)$

54. $(b - 8)(b - 1)$

55. $(2x - 1)(x + 3)$

56. $(3y + 5)(y - 3)$

57. $(5t - 2)(t - 1)$

58. $(2t - 3)(2t - 1)$

59. $(h - 7)(h - 9)$

60. $(h - 7w)(h - 7w)$

61. $(h + 7w)(h + 7w)$

62. $(h - 7q)(h + 7q)$

63. $(2h^2 - 1)(2h^2 - 1)$

64. $(3h^2 + 1)(3h^2 + 1)$

Find each product. See Example 4.

65. $(a + 1)(a - 2)(a + 5)$

66. $(y - 1)(y + 3)(y - 4)$

67. $(h + 2)(h + 3)(h + 4)$

68. $(m - 1)(m - 3)(m - 5)$

69. $\left(\frac{1}{2}x + 4\right)\left(\frac{1}{2}x - 4\right)(4x - 8)$

70. $\left(\frac{1}{3}w - 3\right)\left(\frac{1}{3}w + 3\right)(w - 6)$

71. $\left(x + \frac{1}{2}\right)\left(x - \frac{1}{2}\right)(x + 8)$

72. $\left(x + \frac{1}{3}\right)\left(x - \frac{1}{3}\right)(x + 9)$

Perform the indicated operations.

73. $(x + 10)(x + 5)$

74. $(x + 4)(x + 8)$

75. $\left(x + \frac{1}{2}\right)\left(x + \frac{1}{2}\right)$

76. $\left(x + \frac{1}{3}\right)\left(x + \frac{1}{6}\right)$

77. $\left(4x + \frac{1}{2}\right)\left(2x + \frac{1}{4}\right)$

78. $\left(3x + \frac{1}{6}\right)\left(6x + \frac{1}{3}\right)$

79. $\left(2a + \frac{1}{2}\right)\left(4a - \frac{1}{2}\right)$

80. $\left(3b + \frac{2}{3}\right)\left(6b - \frac{1}{3}\right)$

81. $\left(\frac{1}{2}x - \frac{1}{3}\right)\left(\frac{1}{4}x + \frac{1}{2}\right)$

82. $\left(\frac{2}{3}t - \frac{1}{4}\right)\left(\frac{1}{2}t - \frac{1}{2}\right)$

83. $a(a + 3)(a + 4)$

84. $w(w + 5)(w + 9)$

85. $x^3(x + 6)(x + 7)$

86. $x^2(x^2 + 1)(x^2 + 8)$

87. $-2x^4(3x - 1)(2x + 5)$

88. $4xy^3(2x - y)(3x + y)$

89. $(x - 1)(x + 1)(x + 3)$

90. $(a - 3)(a + 4)(a - 5)$

91. $(3x - 2)(3x + 2)(x + 5)$

92. $(x - 6)(9x + 4)(9x - 4)$

93. $(x - 1)(x + 2) - (x + 3)(x - 4)$

94. $(k - 4)(k + 9) - (k - 3)(k + 7)$

Solve each problem. See Example 5.

95. ***Area of a rug.*** Find a trinomial that represents the area of a rectangular rug whose sides are $x + 3$ feet and $2x - 1$ feet.

96. ***Area of a parallelogram.*** Find a trinomial that represents the area of a parallelogram whose base is $3x + 2$ meters and whose height is $2x + 3$ meters.

97. ***Area of a sail.*** The sail of a tall ship is triangular in shape with a base of $4.57x + 3$ meters and a height of $2.3x - 1.33$ meters. Find a polynomial that represents the area of the triangle.

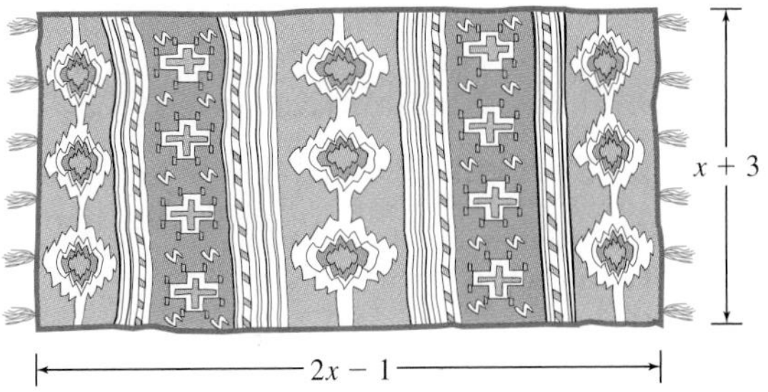

Figure for Exercise 95

98. ***Area of a square.*** A square has a side of length $1.732x + 1.414$ meters. Find a polynomial that represents its area.

Getting More Involved

99. ***Exploration***

Find the area of each of the four regions shown in the figure. What is the total area of the four regions? What does this exercise illustrate?

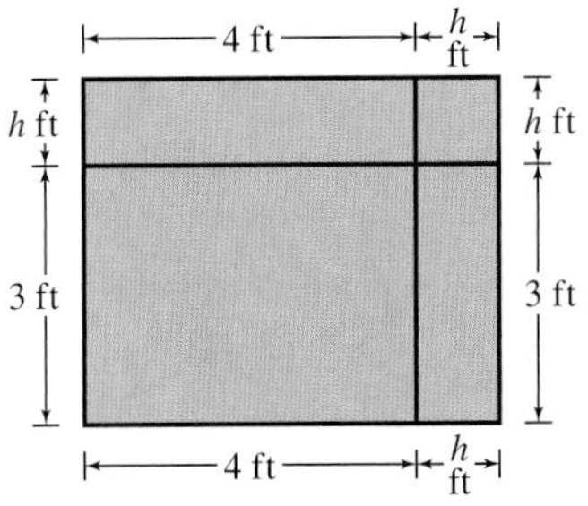

Figure for Exercise 99

100. ***Exploration***

Find the area of each of the four regions shown in the figure. What is the total area of the four regions? What does this exercise illustrate?

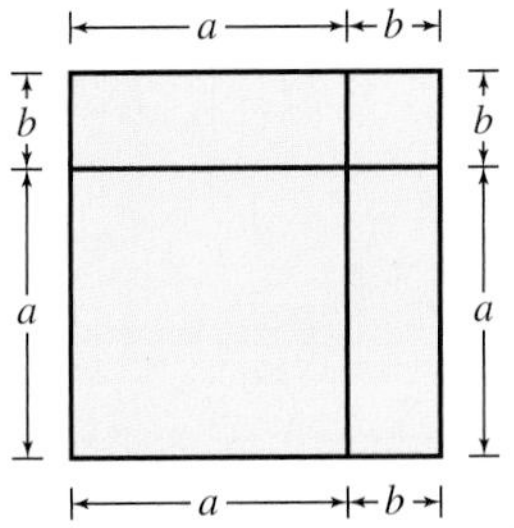

Figure for Exercise 100

5.4 Special Products

In this Section

- **The Square of a Binomial**
- **Product of a Sum and a Difference**
- **Higher Powers of Binomials**
- **Applications to Area**

In Section 5.3 you learned the FOIL method to make multiplying binomials simpler. In this section you will learn rules for squaring binomials and for finding the product of a sum and a difference. These products are called **special products.**

Helpful Hint

To visualize the square of a sum, draw a square with sides of length $a + b$ as shown.

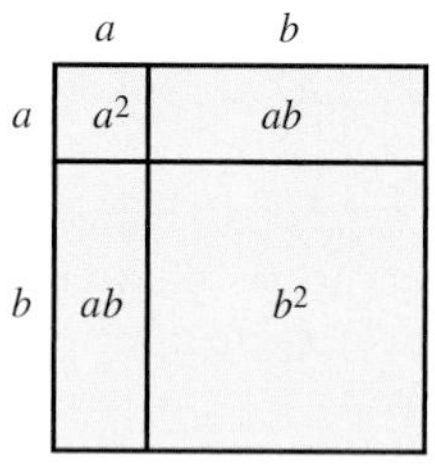

The area of the large square is $(a + b)^2$. It comes from four terms as stated in the rule for the square of a sum.

The Square of a Binomial

To compute $(a + b)^2$, the square of a binomial, we can write it as $(a + b)(a + b)$ and use FOIL:

$$\begin{aligned}(a + b)^2 &= (a + b)(a + b)\\ &= a^2 + ab + ab + b^2\\ &= a^2 + 2ab + b^2\end{aligned}$$

So to square $a + b$, *we square the first term* (a^2), *add twice the product of the two terms* $(2ab)$, *then add the square of the last term* (b^2). The square of a binomial occurs so frequently that it is helpful to learn this new rule to find it. The rule for squaring a sum is given symbolically as follows.

The Square of a Sum

$$(a + b)^2 = a^2 + 2ab + b^2$$

EXAMPLE 1

Using the rule for squaring a sum

Find the square of each sum.

a) $(x + 3)^2$ **b)** $(2a + 5)^2$

Solution

a) $(x + 3)^2 = \underset{\text{Square of first}}{x^2} + \underset{\text{Twice the product}}{\underbrace{2(x)(3)}} + \underset{\text{Square of last}}{3^2} = x^2 + 6x + 9$

b) $$\begin{aligned}(2a + 5)^2 &= (2a)^2 + 2(2a)(5) + 5^2\\ &= 4a^2 + 20a + 25\end{aligned}$$

Now do Exercises 7–20

CAUTION Do not forget the middle term when squaring a sum. The equation $(x + 3)^2 = x^2 + 6x + 9$ is an identity, but $(x + 3)^2 = x^2 + 9$ is not an identity. For example, if $x = 1$ in $(x + 3)^2 = x^2 + 9$, then we get $4^2 = 1^2 + 9$, which is false.

When we use FOIL to find $(a - b)^2$, we see that

$$\begin{aligned}(a - b)^2 &= (a - b)(a - b)\\ &= a^2 - ab - ab + b^2\\ &= a^2 - 2ab + b^2.\end{aligned}$$

So to square $a - b$, *we square the first term* (a^2), *subtract twice the product of the two terms* $(-2ab)$, *and add the square of the last term* (b^2). The rule for squaring a difference is given symbolically as follows.

The Square of a Difference

$$(a - b)^2 = a^2 - 2ab + b^2$$

EXAMPLE 2

Using the rule for squaring a difference

Find the square of each difference.

a) $(x - 4)^2$ **b)** $(4b - 5y)^2$

Solution

a) $(x - 4)^2 = x^2 - 2(x)(4) + 4^2$
$= x^2 - 8x + 16$

b) $(4b - 5y)^2 = (4b)^2 - 2(4b)(5y) + (5y)^2$
$= 16b^2 - 40by + 25y^2$

Now do Exercises 21–34

Helpful Hint

Many students keep using FOIL to find the square of a sum or difference. However, learning the new rules for these special cases will pay off in the future.

Product of a Sum and a Difference

If we multiply the sum $a + b$ and the difference $a - b$ by using FOIL, we get

$$\begin{aligned}(a + b)(a - b) &= a^2 - ab + ab - b^2\\ &= a^2 - b^2.\end{aligned}$$

The inner and outer products have a sum of 0. So *the product of a sum and a difference of the same two terms is equal to the difference of two squares.*

The Product of a Sum and a Difference

$$(a + b)(a - b) = a^2 - b^2$$

EXAMPLE 3

Product of a sum and a difference

Find each product.

a) $(x + 2)(x - 2)$ **b)** $(b + 7)(b - 7)$ **c)** $(3x - 5)(3x + 5)$

Solution

a) $(x + 2)(x - 2) = x^2 - 4$

b) $(b + 7)(b - 7) = b^2 - 49$

c) $(3x - 5)(3x + 5) = 9x^2 - 25$

Now do Exercises 35–46

Helpful Hint

You can use

$$(a + b)(a - b) = a^2 - b^2$$

to perform mental arithmetic tricks like

$$\begin{aligned}19 \cdot 21 &= (20 - 1)(20 + 1)\\ &= 400 - 1\\ &= 399.\end{aligned}$$

What is $29 \cdot 31$? $28 \cdot 32$?

Higher Powers of Binomials

To find a power of a binomial that is higher than 2, we can use the rule for squaring a binomial along with the method of multiplying binomials using the distributive property. Finding the second or higher power of a binomial is called **expanding the binomial** because the result has more terms than the original.

EXAMPLE 4

Higher powers of a binomial

Expand each binomial.

a) $(x + 4)^3$ **b)** $(y - 2)^4$

Study Tip

Correct answers often have more than one form. If your answer to an exercise doesn't agree with the one in the back of this text, try to determine if it is simply a different form of the answer. For example, $\frac{1}{2}x$ and $\frac{x}{2}$ look different but they are equivalent expressions.

Solution

a)
$$\begin{aligned}(x + 4)^3 &= (x + 4)^2(x + 4)\\ &= (x^2 + 8x + 16)(x + 4)\\ &= (x^2 + 8x + 16)x + (x^2 + 8x + 16)4\\ &= x^3 + 8x^2 + 16x + 4x^2 + 32x + 64\\ &= x^3 + 12x^2 + 48x + 64\end{aligned}$$

b)
$$\begin{aligned}(y - 2)^4 &= (y - 2)^2(y - 2)^2\\ &= (y^2 - 4y + 4)(y^2 - 4y + 4)\\ &= (y^2 - 4y + 4)(y^2) + (y^2 - 4y + 4)(-4y) + (y^2 - 4y + 4)(4)\\ &= y^4 - 4y^3 + 4y^2 - 4y^3 + 16y^2 - 16y + 4y^2 - 16y + 16\\ &= y^4 - 8y^3 + 24y^2 - 32y + 16\end{aligned}$$

Now do Exercises 47–54

Applications to Area

EXAMPLE 5

Area of a pizza

A pizza parlor saves money by making all of its round pizzas one inch smaller in radius than advertised. Write a trinomial for the actual area of a pizza with an advertised radius of r inches.

Solution

A pizza advertised as r inches has an actual radius of $r - 1$ inches. The actual area is $\pi(r - 1)^2$:

$$\pi(r - 1)^2 = \pi(r^2 - 2r + 1) = \pi r^2 - 2\pi r + \pi.$$

So $\pi r^2 - 2\pi r + \pi$ is a trinomial representing the actual area.

Now do Exercises 85–94

Warm-Ups

True or false? Explain your answer.

1. $(2 + 3)^2 = 2^2 + 3^2$
2. $(x + 3)^2 = x^2 + 6x + 9$ for any value of x.
3. $(3 + 5)^2 = 9 + 30 + 25$
4. $(2x + 7)^2 = 4x^2 + 28x + 49$ for any value of x.
5. $(y + 8)^2 = y^2 + 64$ for any value of y.
6. The product of a sum and a difference of the same two terms is equal to the difference of two squares.
7. $(40 - 1)(40 + 1) = 1599$
8. $49 \cdot 51 = 2499$
9. $(x - 3)^2 = x^2 - 3x + 9$ for any value of x.
10. The square of a sum is equal to a sum of two squares.

5.4 Exercises

Reading and Writing *After reading this section, write out the answers to these questions. Use complete sentences.*

1. What are the special products?
2. What is the rule for squaring a sum?
3. Why do we need a new rule to find the square of a sum when we already have FOIL?
4. What happens to the inner and outer products in the product of a sum and a difference?
5. What is the rule for finding the product of a sum and a difference?
6. How can you find higher powers of binomials?

Square each binomial. See Example 1.

7. $(x + 1)^2$
8. $(y + 2)^2$
9. $(y + 4)^2$
10. $(z + 3)^2$
11. $(m + 6)^2$
12. $(w + 7)^2$
13. $(3x + 8)^2$
14. $(2m + 7)^2$
15. $(s + t)^2$
16. $(x + z)^2$
17. $(2x + y)^2$
18. $(3t + v)^2$
19. $(2t + 3h)^2$
20. $(3z + 5k)^2$

Square each binomial. See Example 2.

21. $(p - 2)^2$
22. $(b - 5)^2$
23. $(a - 3)^2$
24. $(w - 4)^2$
25. $(t - 1)^2$
26. $(t - 6)^2$
27. $(3t - 2)^2$
28. $(5a - 6)^2$

29. $(s - t)^2$

30. $(r - w)^2$

31. $(3a - b)^2$

32. $(4w - 7)^2$

33. $(3z - 5y)^2$

34. $(2z - 3w)^2$

Find each product. See Example 3.

35. $(a - 5)(a + 5)$

36. $(x - 6)(x + 6)$

37. $(y - 1)(y + 1)$

38. $(p + 2)(p - 2)$

39. $(3x - 8)(3x + 8)$

40. $(6x + 1)(6x - 1)$

41. $(r + s)(r - s)$

42. $(b - y)(b + y)$

43. $(8y - 3a)(8y + 3a)$

44. $(4u - 9v)(4u + 9v)$

45. $(5x^2 - 2)(5x^2 + 2)$

46. $(3y^2 + 1)(3y^2 - 1)$

Expand each binomial. See Example 4.

47. $(x + 1)^3$

48. $(y - 1)^3$

49. $(2a - 3)^3$

50. $(3w - 1)^3$

51. $(a - 3)^4$

52. $(2b + 1)^4$

53. $(a + b)^4$

54. $(2a - 3b)^4$

Find each product.

55. $(a - 20)(a + 20)$

56. $(1 - x)(1 + x)$

57. $(x + 8)(x + 7)$

58. $(x - 9)(x + 5)$

59. $(4x - 1)(4x + 1)$

60. $(9y - 1)(9y + 1)$

61. $(9y - 1)^2$

62. $(4x - 1)^2$

63. $(2t - 5)(3t + 4)$

64. $(2t + 5)(3t - 4)$

65. $(2t - 5)^2$

66. $(2t + 5)^2$

67. $(2t + 5)(2t - 5)$

68. $(3t - 4)(3t + 4)$

69. $(x^2 - 1)(x^2 + 1)$

70. $(y^3 - 1)(y^3 + 1)$

71. $(2y^3 - 9)^2$

72. $(3z^4 - 8)^2$

73. $(2x^3 + 3y^2)^2$

74. $(4y^5 + 2w^3)^2$

75. $\left(\frac{1}{2}x + \frac{1}{3}\right)^2$

76. $\left(\frac{2}{3}y - \frac{1}{2}\right)^2$

77. $(0.2x - 0.1)^2$

78. $(0.1y + 0.5)^2$

79. $(a + b)^3$

80. $(2a - 3b)^3$

81. $(1.5x + 3.8)^2$

82. $(3.45a - 2.3)^2$

83. $(3.5t - 2.5)(3.5t + 2.5)$

84. $(4.5h + 5.7)(4.5h - 5.7)$

Solve each problem. See Example 5.

85. ***Shrinking garden.*** Rose's garden is a square with sides of length x feet. Next spring she plans to make it rectangular by lengthening one side 5 feet and shortening the other side by 5 feet. What polynomial represents the new area? By how much will the area of the new garden differ from that of the old garden?

86. ***Square lot.*** Sam lives on a lot that he thought was a square, 157 feet by 157 feet. When he had it surveyed, he discovered that one side was actually 2 feet longer than he thought and the other was actually 2 feet shorter than he thought. How much less area does he have than he thought he had?

87. ***Area of a circle.*** Find a polynomial that represents the area of a circle whose radius is $b + 1$ meters. Use the value 3.14 for π.

88. ***Comparing dart boards.*** A toy store sells two sizes of circular dartboards. The larger of the two has a radius that is 3 inches greater than that of the other. The radius of the smaller dartboard is t inches. Find a polynomial that represents the difference in area between the two dartboards.

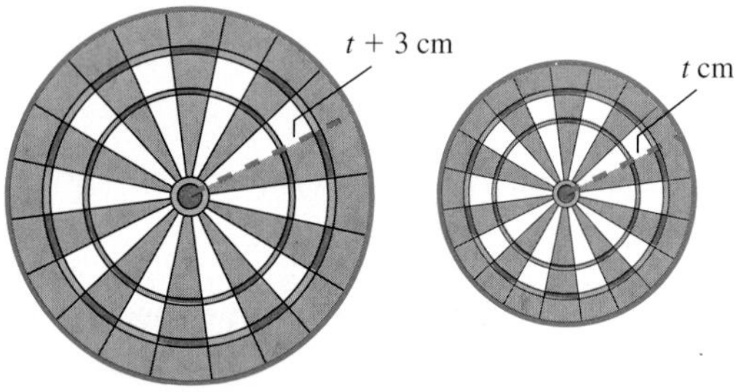

Figure for Exercise 88

89. ***Poiseuille's law.*** According to the nineteenth-century physician Poiseuille, the velocity (in centimeters per second) of blood r centimeters from the center of an artery of radius R centimeters is given by

$$v = k(R - r)(R + r),$$

where k is a constant. Rewrite the formula using a special product rule.

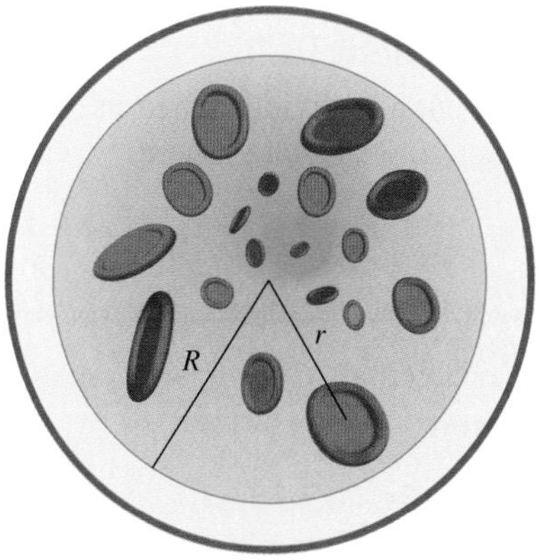

Figure for Exercise 89

90. ***Going in circles.*** A promoter is planning a circular race track with an inside radius of r feet and a width of w feet. The cost in dollars for paving the track is given by the formula

$$C = 1.2\pi[(r + w)^2 - r^2].$$

Use a special product rule to simplify this formula. What is the cost of paving the track if the inside radius is 1000 feet and the width of the track is 40 feet?

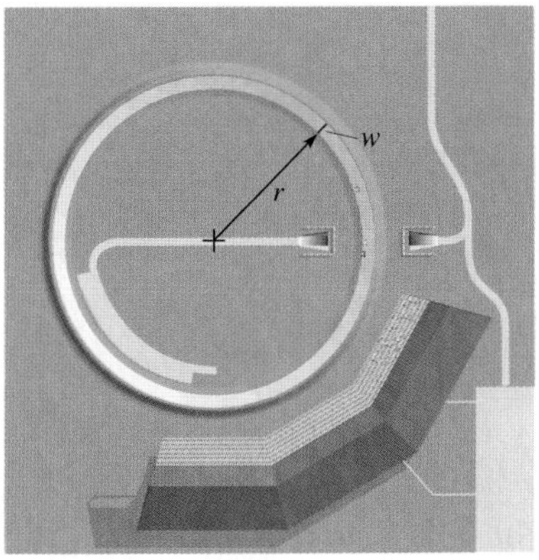

Figure for Exercise 90

91. ***Compounded annually.*** P dollars is invested at annual interest rate r for 2 years. If the interest is compounded annually, then the polynomial $P(1 + r)^2$ represents the value of the investment after 2 years. Rewrite this expression without parentheses. Evaluate the polynomial if $P = \$200$ and $r = 10\%$.

92. ***Compounded semiannually.*** P dollars is invested at annual interest rate r for 1 year. If the interest is compounded semiannually, then the polynomial $P\left(1 + \frac{r}{2}\right)^2$ represents the value of the investment after 1 year. Rewrite this expression without parentheses. Evaluate the polynomial if $P = \$200$ and $r = 10\%$.

93. ***Investing in treasury bills.*** An investment advisor uses the polynomial $P(1 + r)^{10}$ to predict the value in 10 years of a client's investment of P dollars with an average annual return r. The accompanying graph shows historic average annual returns for the last 20 years for various asset classes (T. Rowe Price, www.troweprice.com). Use the historical average return to predict the value in 10 years of an investment of \$10,000 in U.S. treasury bills.

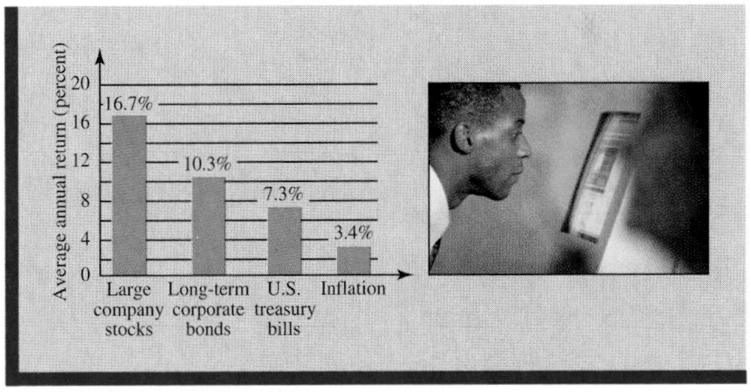

Figure for Exercises 93 and 94

94. ***Comparing investments.*** How much more would the investment in Exercise 93 be worth in 10 years if the client invests in large company stocks rather than U.S. treasury bills?

Getting More Involved

95. *Writing*

What is the difference between the equations $(x + 5)^2 = x^2 + 10x + 25$ and $(x + 5)^2 = x^2 + 25$?

96. *Writing*

Is it possible to square a sum or a difference without using the rules presented in this section? Why should you learn the rules given in this section?

5.5 Division of Polynomials

In this Section

- **Dividing Monomials Using the Quotient Rule**
- **Dividing a Polynomial by a Monomial**
- **Dividing a Polynomial by a Binomial**

You multiplied polynomials in Section 5.2. In this section you will learn to divide polynomials.

Study Tip

Establish a regular routine of eating, sleeping, and exercise. The ability to concentrate depends on adequate sleep, decent nutrition, and the physical well-being that comes with exercise.

Dividing Monomials Using the Quotient Rule

In Chapter 1 we used the definition of division to divide signed numbers. Because the definition of division applies to any division, we restate it here.

Division of Real Numbers

If a, b, and c are any numbers with $b \neq 0$, then

$$a \div b = c \quad \text{provided that} \quad c \cdot b = a.$$

If $a \div b = c$, we call a the **dividend,** b the **divisor,** and c (or $a \div b$) the **quotient.**

You can find the quotient of two monomials by writing the quotient as a fraction and then reducing the fraction. For example,

$$x^5 \div x^2 = \frac{x^5}{x^2} = \frac{x \cdot x \cdot x \cdot \cancel{x} \cdot \cancel{x}}{\cancel{x} \cdot \cancel{x}} = x^3.$$

You can be sure that x^3 is correct by checking that $x^3 \cdot x^2 = x^5$. You can also divide x^2 by x^5, but the result is not a monomial:

$$x^2 \div x^5 = \frac{x^2}{x^5} = \frac{1 \cdot \cancel{x} \cdot \cancel{x}}{x \cdot x \cdot x \cdot \cancel{x} \cdot \cancel{x}} = \frac{1}{x^3}$$

Note that the exponent 3 can be obtained in either case by subtracting 5 and 2. These examples illustrate the quotient rule for exponents.

Quotient Rule

Suppose $a \neq 0$, and m and n are positive integers.

$$\text{If } m \geq n, \text{ then } \frac{a^m}{a^n} = a^{m-n}.$$

$$\text{If } n > m, \text{ then } \frac{a^m}{a^n} = \frac{1}{a^{n-m}}.$$

Note that if you use the quotient rule to subtract the exponents in $x^4 \div x^4$, you get the expression x^{4-4}, or x^0, which has not been defined yet. Because we must have $x^4 \div x^4 = 1$ if $x \neq 0$, we define the zero power of a nonzero real number to be 1. We do not define the expression 0^0.

Zero Exponent

For any nonzero real number a,

$$a^0 = 1.$$

EXAMPLE 1

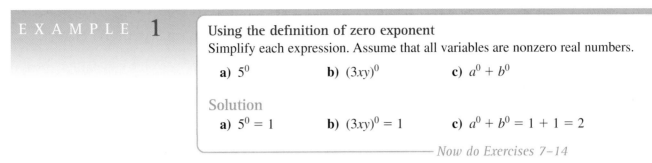

Using the definition of zero exponent

Simplify each expression. Assume that all variables are nonzero real numbers.

a) 5^0 **b)** $(3xy)^0$ **c)** $a^0 + b^0$

Solution

a) $5^0 = 1$ **b)** $(3xy)^0 = 1$ **c)** $a^0 + b^0 = 1 + 1 = 2$

Now do Exercises 7–14

With the definition of zero exponent the quotient rule is valid for all positive integers as stated.

EXAMPLE 2

Using the quotient rule in dividing monomials

Find each quotient.

a) $\dfrac{y^9}{y^5}$ **b)** $\dfrac{12b^2}{3b^7}$

c) $-6x^3 \div (2x^9)$ **d)** $\dfrac{x^8y^2}{x^2y^2}$

Solution

a) $\dfrac{y^9}{y^5} = y^{9-5} = y^4$

Use the definition of division to check that $y^4 \cdot y^5 = y^9$.

b) $\dfrac{12b^2}{3b^7} = \dfrac{12}{3} \cdot \dfrac{b^2}{b^7} = 4 \cdot \dfrac{1}{b^{7-2}} = \dfrac{4}{b^5}$

Use the definition of division to check that

$$\frac{4}{b^5} \cdot 3b^7 = \frac{12b^7}{b^5} = 12b^2.$$

c) $-6x^3 \div (2x^9) = \dfrac{-6x^3}{2x^9} = \dfrac{-3}{x^6}$

Use the definition of division to check that

$$\frac{-3}{x^6} \cdot 2x^9 = \frac{-6x^9}{x^6} = -6x^3.$$

d) $\dfrac{x^8y^2}{x^2y^2} = \dfrac{x^8}{x^2} \cdot \dfrac{y^2}{y^2} = x^6 \cdot y^0 = x^6$

Use the definition of division to check that $x^6 \cdot x^2y^2 = x^8y^2$.

Now do Exercises 15–30

Study Tip

As soon as possible after class, find a quiet place and work on your homework. The longer you wait the harder it is to remember what happened in class.

Note that the parentheses in Example 2(c) are important. According to the order of operations, multiplication and division are performed from left to right in the absence of parentheses. So without parentheses, $-6x^3 \div 2x^9 = \frac{-6x^3}{2} \cdot x^9$.

Dividing a Polynomial by a Monomial

We divided some simple polynomials by monomials in Chapter 1. For example,

$$\frac{6x + 8}{2} = \frac{1}{2}(6x + 8) = 3x + 4.$$

We use the distributive property to take one-half of $6x$ and one-half of 8 to get $3x + 4$. So both $6x$ and 8 are divided by 2. To divide any polynomial by a monomial, we divide each term of the polynomial by the monomial.

EXAMPLE 3

Dividing a polynomial by a monomial

Find the quotient for $(-8x^6 + 12x^4 - 4x^2) \div (4x^2)$.

Solution

$$\frac{-8x^6 + 12x^4 - 4x^2}{4x^2} = \frac{-8x^6}{4x^2} + \frac{12x^4}{4x^2} - \frac{4x^2}{4x^2}$$

$$= -2x^4 + 3x^2 - 1$$

The quotient is $-2x^4 + 3x^2 - 1$. We can check by multiplying.

$$4x^2(-2x^4 + 3x^2 - 1) = -8x^6 + 12x^4 - 4x^2.$$

Now do Exercises 31–38

Study Tip

Play offensive math, not defensive math. A student who says, "Give me a question and I'll see if I can answer it," is playing defensive math. The student is taking a passive approach to learning. A student who takes an active approach and knows the usual questions and answers for each topic is playing offensive math.

Because division by zero is undefined, we will always assume that the divisor is nonzero in any quotient involving variables. For example, the division in Example 3 is valid only if $4x^2 \neq 0$, or $x \neq 0$.

Dividing a Polynomial by a Binomial

Division of whole numbers is often done with a procedure called **long division.** For example, 253 is divided by 7 as follows:

```
               36    ← Quotient
Divisor →    7)253   ← Dividend
              21
               43
               42
                1    ← Remainder
```

Note that $36 \cdot 7 + 1 = 253$. It is always true that

(quotient)(divisor) + (remainder) = dividend.

To divide a polynomial by a binomial, we perform the division like long division of whole numbers. For example, to divide $x^2 - 3x - 10$ by $x + 2$, we get the first term of the quotient by dividing the first term of $x + 2$ into the first term of $x^2 - 3x - 10$.

So divide x^2 by x to get x, then multiply and subtract as follows:

$$\begin{array}{lrl} \text{1 Divide:} & x & x^2 \div x = x \\ \text{2 Multiply:} & x + 2\overline{)x^2 - 3x - 10} & \\ & \underline{x^2 + 2x} & x \cdot (x + 2) = x^2 + 2x \\ \text{3 Subtract:} & -5x & -3x - 2x = -5x \end{array}$$

Now bring down -10 and continue the process. We get the second term of the quotient (below) by dividing the first term of $x + 2$ into the first term of $-5x - 10$. So divide $-5x$ by x to get -5:

$$\begin{array}{lrl} \text{1 Divide:} & x - 5 & -5x \div x = -5 \\ \text{2 Multiply:} & x + 2\overline{)x^2 - 3x - 10} & \\ & \underline{x^2 + 2x} \quad \downarrow & \text{Bring down } -10. \\ & -5x - 10 & \\ & \underline{-5x - 10} & -5(x + 2) = -5x - 10 \\ \text{3 Subtract:} & 0 & -10 - (-10) = 0 \end{array}$$

So the quotient is $x - 5$, and the remainder is 0.

In Example 4 there is a term missing in the dividend. To account for the missing term we insert a term with a zero coefficient.

EXAMPLE 4

Dividing a polynomial by a binomial

Determine the quotient and remainder when $x^3 - 5x - 1$ is divided by $x - 4$.

Solution

Because the x^2-term in the dividend $x^3 - 5x - 1$ is missing, we write $0 \cdot x^2$ for it:

Place x^2 in the quotient because $x^3 \div x = x^2$.

Place $4x$ in the quotient because $4x^2 \div x = 4x$.

Place 11 in the quotient because $11x \div x = 11$.

$$\begin{array}{rl} x^2 + 4x + 11 & \\ x - 4\overline{)x^3 + 0 \cdot x^2 - 5x - 1} & \\ \underline{x^3 - 4x^2} & x^2(x - 4) = x^3 - 4x^2 \\ 4x^2 - 5x & 0 \cdot x^2 - (-4x^2) = 4x^2 \\ \underline{4x^2 - 16x} & 4x(x - 4) = 4x^2 - 16x \\ 11x - 1 & -5x - (-16x) = 11x \\ \underline{11x - 44} & 11(x - 4) = 11x - 44 \\ 43 & -1 - (-44) = 43 \end{array}$$

The quotient is $x^2 + 4x + 11$ and the remainder is 43.

Now do Exercises 39–42

In Example 5 the terms of the dividend are not in order of decreasing exponents and there is a missing term.

EXAMPLE 5

Helpful Hint

Students usually have the most difficulty with the subtraction part of long division. So pay particular attention to that step and double check your work.

Dividing a polynomial by a binomial

Divide $2x^3 - 4 - 7x^2$ by $2x - 3$, and identify the quotient and the remainder.

Solution

Rearrange the dividend as $2x^3 - 7x^2 - 4$. Because the x-term in the dividend is missing, we write $0 \cdot x$ for it:

$$
\begin{array}{r l l}
 & x^2 - 2x - 3 & 2x^3 \div (2x) = x^2 \\
2x - 3 \,\big) & 2x^3 - 7x^2 + 0 \cdot x - 4 & \\
 & 2x^3 - 3x^2 & x^2(2x - 3) = 2x^3 - 3x^2 \\
 & \quad -4x^2 + 0 \cdot x & -7x^2 - (-3x^2) = -4x^2 \\
 & \quad -4x^2 + 6x & -2x(2x - 3) = -4x^2 + 6x \\
 & \qquad -6x - 4 & 0 \cdot x - 6x = -6x \\
 & \qquad -6x + 9 & -3(2x - 3) = -6x + 9 \\
 & \qquad\quad -13 & -4 - (9) = -13
\end{array}
$$

The quotient is $x^2 - 2x - 3$, and the remainder is -13. Note that the degree of the remainder is 0 and the degree of the divisor is 1. To check, we must verify that

$$(2x - 3)(x^2 - 2x - 3) - 13 = 2x^3 - 7x^2 - 4.$$

Now do Exercises 43–56

CAUTION To avoid errors, always write the terms of the divisor and the dividend in descending order of the exponents and insert a zero for any term that is missing.

If we divide both sides of the equation

$$\text{dividend} = (\text{quotient})(\text{divisor}) + (\text{remainder})$$

by the divisor, we get the equation

$$\frac{\text{dividend}}{\text{divisor}} = \text{quotient} + \frac{\text{remainder}}{\text{divisor}}.$$

This fact is used in expressing improper fractions as mixed numbers. For example, if 19 is divided by 5, the quotient is 3 and the remainder is 4. So

$$\frac{19}{5} = 3 + \frac{4}{5} = 3\frac{4}{5}.$$

We can also use this form to rewrite algebraic fractions.

EXAMPLE 6

Rewriting algebraic fractions

Express $\frac{-3x}{x-2}$ in the form

$$\text{quotient} + \frac{\text{remainder}}{\text{divisor}}.$$

Solution

Use long division to get the quotient and remainder:

$$\begin{array}{r|l} & -3 \\ x-2 & -3x + 0 \\ & -3x + 6 \\ \hline & -6 \end{array}$$

Because the quotient is -3 and the remainder is -6, we can write

$$\frac{-3x}{x-2} = -3 + \frac{-6}{x-2}.$$

To check, we must verify that $-3(x - 2) - 6 = -3x$.

Now do Exercises 57–72

CAUTION When dividing polynomials by long division, we do not stop until the remainder is 0 or the degree of the remainder is smaller than the degree of the divisor. For example, we stop dividing in Example 6 because the degree of the remainder -6 is 0 and the degree of the divisor $x - 2$ is 1.

Warm-Ups

True or false? Explain your answer.

1. $y^{10} \div y^2 = y^5$ for any nonzero value of y.
2. $\frac{7x + 2}{7} = x + 2$ for any value of x.
3. $\frac{7x^2}{7} = x^2$ for any value of x.
4. If $3x^2 + 6$ is divided by 3, the quotient is $x^2 + 6$.
5. If $4y^2 - 6y$ is divided by $2y$, the quotient is $2y - 3$.
6. The quotient times the remainder plus the dividend equals the divisor.
7. $(x + 2)(x + 1) + 3 = x^2 + 3x + 5$ for any value of x.
8. If $x^2 + 3x + 5$ is divided by $x + 2$, then the quotient is $x + 1$.
9. If $x^2 + 3x + 5$ is divided by $x + 2$, the remainder is 3.
10. If the remainder is zero, then (divisor)(quotient) = dividend.

5.5 Exercises

Reading and Writing *After reading this section, write out the answers to these questions. Use complete sentences.*

1. What rule is important for dividing monomials?

2. What is the meaning of a zero exponent?

3. How many terms should you get when dividing a polynomial by a monomial?

4. How should the terms of the polynomials be written when dividing with long division?

5. How do you know when to stop the process in long division of polynomials?

6. How do you handle missing terms in the dividend polynomial when doing long division?

Simplify each expression. See Example 1.

7. 9^0 **8.** m^0 **9.** $(-2x^3)^0$

10. $(5a^3b)^0$ **11.** $2 \cdot 5^0 - 3^0$ **12.** $-4^0 - 8^0$

13. $(2x - y)^0$ **14.** $(a^2 + b^2)^0$

Find each quotient. Try to write only the answer. See Example 2.

15. $\dfrac{x^8}{x^2}$ **16.** $\dfrac{y^9}{y^3}$

17. $\dfrac{a^5}{a^{14}}$ **18.** $\dfrac{b^{12}}{b^{19}}$

19. $\dfrac{6a^7}{2a^{12}}$ **20.** $\dfrac{30b^2}{3b^6}$

21. $a^9 \div a^3$ **22.** $b^{12} \div b^4$

23. $-12x^5 \div (3x^9)$ **24.** $-6y^5 \div (-3y^{10})$

25. $-6y^2 \div (6y)$ **26.** $-3a^2b \div (3ab)$

27. $\dfrac{-6x^3y^2}{2x^2y^2}$ **28.** $\dfrac{-4h^2k^4}{-2hk^3}$

29. $\dfrac{-9x^2y^2}{3x^5y^2}$ **30.** $\dfrac{-12z^4y^2}{-2z^{10}y^2}$

Find the quotients. See Example 3.

31. $\dfrac{3x - 6}{3}$

32. $\dfrac{5y - 10}{-5}$

33. $\dfrac{x^5 + 3x^4 - x^3}{x^2}$

34. $\dfrac{6y^6 - 9y^4 + 12y^2}{3y^2}$

35. $\dfrac{-8x^2y^2 + 4x^2y - 2xy^2}{-2xy}$

36. $\dfrac{-9ab^2 - 6a^3b^3}{-3ab^2}$

37. $(x^2y^3 - 3x^3y^2) \div (x^2y)$

38. $(4h^5k - 6h^2k^2) \div (-2h^2k)$

Complete each division and identify the quotient and remainder. See Example 4.

39. $\begin{array}{r} 2 \\ x - 1\overline{)2x - 3} \\ \underline{2x - 2} \end{array}$ **40.** $\begin{array}{r} -3 \\ x + 2\overline{)-3x + 4} \\ \underline{-3x - 6} \end{array}$

41. $\begin{array}{r} x \\ x - 3\overline{)x^2 + 2x + 1} \\ \underline{x^2 - 3x} \end{array}$ **42.** $\begin{array}{r} x \\ x + 4\overline{)x^2 - 3x + 2} \\ \underline{x^2 + 4x} \end{array}$

Find the quotient and remainder for each division. Check by using the fact that dividend = (divisor)(quotient) + remainder. See Example 5.

43. $(x^2 + 5x + 13) \div (x + 3)$

44. $(x^2 + 3x + 6) \div (x + 3)$

45. $(2x) \div (x + 5)$

46. $(5x) \div (x - 1)$

47. $(a^3 + 4a - 3) \div (a - 2)$

48. $(w^3 + 2w^2 - 3) \div (w - 2)$

49. $(x^2 - 3x) \div (x + 1)$

50. $(3x^2) \div (x + 1)$

51. $(h^3 - 27) \div (h - 3)$

52. $(w^3 + 1) \div (w + 1)$

53. $(6x^2 - 13x + 7) \div (3x - 2)$

54. $(4b^2 + 25b - 3) \div (4b + 1)$

55. $(x^3 - x^2 + x - 2) \div (x - 1)$

56. $(a^3 - 3a^2 + 4a - 4) \div (a - 2)$

Write each expression in the form

$$\text{quotient} + \frac{\text{remainder}}{\text{divisor}}.$$

See Example 6.

57. $\frac{3x}{x - 5}$

58. $\frac{2x}{x - 1}$

59. $\frac{-x}{x + 3}$

60. $\frac{-3x}{x + 1}$

61. $\frac{x - 1}{x}$

62. $\frac{a - 5}{a}$

63. $\frac{3x + 1}{x}$

64. $\frac{2y + 1}{y}$

65. $\frac{x^2}{x + 1}$

66. $\frac{x^2}{x - 1}$

67. $\frac{x^2 + 4}{x + 2}$

68. $\frac{x^2 + 1}{x - 1}$

69. $\frac{x^3}{x - 2}$

70. $\frac{x^3 - 1}{x + 1}$

71. $\frac{x^3 + 3}{x}$

72. $\frac{2x^2 + 4}{2x}$

Find each quotient.

73. $-6a^3b \div (2a^2b)$

74. $-14x^7 \div (-7x^2)$

75. $-8w^4t^7 \div (-2w^9t^3)$

76. $-9y^7z^4 \div (3y^3z^{11})$

77. $(3a - 12) \div (-3)$

78. $(-6z + 3z^2) \div (-3z)$

79. $(3x^2 - 9x) \div (3x)$

80. $(5x^3 + 15x^2 - 25x) \div (5x)$

81. $(12x^4 - 4x^3 + 6x^2) \div (-2x^2)$

82. $(-9x^3 + 3x^2 - 15x) \div (-3x)$

83. $(t^2 - 5t - 36) \div (t - 9)$

84. $(b^2 + 2b - 35) \div (b - 5)$

85. $(6w^2 - 7w - 5) \div (3w - 5)$

86. $(4z^2 + 23z - 6) \div (4z - 1)$

87. $(8x^3 + 27) \div (2x + 3)$

88. $(8y^3 - 1) \div (2y - 1)$

89. $(t^3 - 3t^2 + 5t - 6) \div (t - 2)$

90. $(2u^3 - 13u^2 - 8u + 7) \div (u - 7)$

91. $(-6v^2 - 4 + 9v + v^3) \div (v - 4)$

92. $(14y + 8y^2 + y^3 + 12) \div (6 + y)$

Solve each problem.

93. ***Area of a rectangle.*** The area of a rectangular billboard is $x^2 + x - 30$ square meters. If the length is $x + 6$ meters, find a binomial that represents the width.

Figure for Exercise 93

94. ***Perimeter of a rectangle.*** The perimeter of a rectangular backyard is $6x + 6$ yards. If the width is x yards, find a binomial that represents the length.

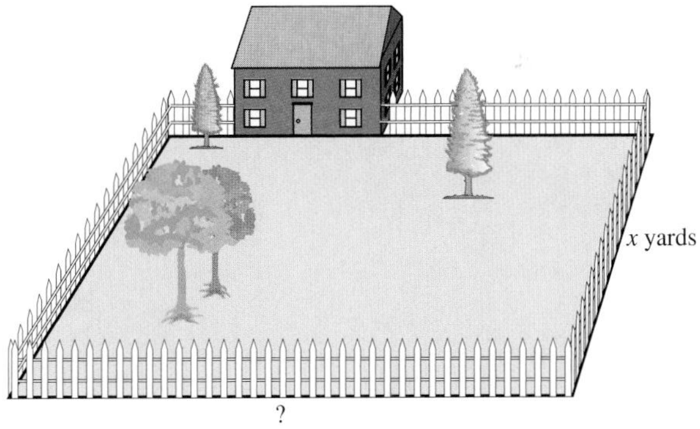

Figure for Exercise 94

Getting More Involved

95. ***Exploration***

Divide $x^3 - 1$ by $x - 1$, $x^4 - 1$ by $x - 1$, and $x^5 - 1$ by $x - 1$. What is the quotient when $x^9 - 1$ is divided by $x - 1$?

96. ***Exploration***

Divide $a^3 - b^3$ by $a - b$ and $a^4 - b^4$ by $a - b$. What is the quotient when $a^8 - b^8$ is divided by $a - b$?

97. ***Discussion***

Are the expressions $\frac{10x}{5x}$, $10x \div 5x$, and $(10x) \div (5x)$ equivalent? Before you answer, review the order of operations in Section 1.5 and evaluate each expression for $x = 3$.

5.6 Nonnegative Integral Exponents

In this Section

- The Product and Quotient Rules
- Raising an Exponential Expression to a Power
- Power of a Product
- Power of a Quotient
- Summary of Rules

The product rule for positive integral exponents was presented in Section 5.2, and the quotient rule was presented in Section 5.5. In this section we review those rules and then further investigate the properties of exponents.

The Product and Quotient Rules

The rules that we have already discussed are summarized below.

The following rules hold for nonnegative integers m and n and $a \neq 0$.

$$a^m \cdot a^n = a^{m+n} \quad \text{Product rule}$$

$$\frac{a^m}{a^n} = a^{m-n} \quad \text{if } m \geq n \quad \text{Quotient rule}$$

$$\frac{a^m}{a^n} = \frac{1}{a^{n-m}} \quad \text{if } n > m$$

$$a^0 = 1 \quad \text{Zero exponent}$$

CAUTION The product and quotient rules apply only if the bases of the expressions are identical. For example, $3^2 \cdot 3^4 = 3^6$, but the product rule cannot be applied to $5^2 \cdot 3^4$. Note also that the bases are not multiplied: $3^2 \cdot 3^4 \neq 9^6$.

Note that in the quotient rule the exponents are always subtracted, as in

$$\frac{x^7}{x^3} = x^4 \quad \text{and} \quad \frac{y^5}{y^8} = \frac{1}{y^3}.$$

If the larger exponent is in the denominator, then the result is placed in the denominator.

EXAMPLE 1 Using the product and quotient rules

Use the rules of exponents to simplify each expression. Assume that all variables represent nonzero real numbers.

a) $2^3 \cdot 2^2$ **b)** $(3x)^0(5x^2)(4x)$ **c)** $\dfrac{8x^2}{-2x^5}$ **d)** $\dfrac{(3a^2b)b^9}{(6a^5)a^3b^2}$

Solution

a) Because the bases are both 2, we can use the product rule:

$$2^3 \cdot 2^2 = 2^5 \quad \text{Product rule}$$
$$= 32 \quad \text{Simplify.}$$

b) $(3x)^0(5x^2)(4x) = 1 \cdot 5x^2 \cdot 4x$ Definition of zero exponent

$= 20x^3$ Product rule

Study Tip

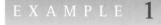

Keep track of your time for one entire week. Account for how you spend every half hour. Add up your totals for sleep, study, work, and recreation. You should be sleeping 50–60 hours per week and studying 1–2 hours for every hour you spend in the classroom.

c) $\dfrac{8x^2}{-2x^5} = -\dfrac{4}{x^3}$ Quotient rule

d) First use the product rule to simplify the numerator and denominator:

$$\frac{(3a^2b)b^9}{(6a^5)a^3b^2} = \frac{3a^2b^{10}}{6a^8b^2} \quad \text{Product rule}$$

$$= \frac{b^8}{2a^6} \quad \text{Quotient rule}$$

Now do Exercises 7–18

Raising an Exponential Expression to a Power

When we raise an exponential expression to a power, we can use the product rule to find the result, as shown in the following example:

$$(w^4)^3 = w^4 \cdot w^4 \cdot w^4 \quad \text{Three factors of } w^4 \text{ because of the exponent 3}$$

$$= w^{12} \quad \text{Product rule}$$

By the product rule we add the three 4's to get 12, but 12 is also the product of 4 and 3. This example illustrates the **power rule** for exponents.

Power Rule

If m and n are nonnegative integers and $a \neq 0$, then

$$(a^m)^n = a^{mn}.$$

In Example 2 we use the new rule along with the other rules.

EXAMPLE 2

Using the power rule

Use the rules of exponents to simplify each expression. Assume that all variables represent nonzero real numbers.

a) $3x^2(x^3)^5$ **b)** $\dfrac{(2^3)^4 \cdot 2^7}{2^5 \cdot 2^9}$ **c)** $\dfrac{3(x^5)^4}{15x^{22}}$

Solution

a) $3x^2(x^3)^5 = 3x^2x^{15}$ Power rule

$= 3x^{17}$ Product rule

b) $\dfrac{(2^3)^4 \cdot 2^7}{2^5 \cdot 2^9} = \dfrac{2^{12} \cdot 2^7}{2^{14}}$ Power rule and product rule

$= \dfrac{2^{19}}{2^{14}}$ Product rule

$= 2^5$ Quotient rule

$= 32$ Evaluate 2^5.

c) $\dfrac{3(x^5)^4}{15x^{22}} = \dfrac{3x^{20}}{15x^{22}} = \dfrac{1}{5x^2}$

Now do Exercises 19–26

Warm-Ups

True or false? Explain your answer.

1. $-3^0 = 1$
2. $2^5 \cdot 2^8 = 4^{13}$
3. $2^3 \cdot 3^2 = 6^5$
4. $(2x)^4 = 2x^4$
5. $(q^3)^5 = q^8$
6. $(-3x^2)^3 = 27x^6$
7. $(ab^3)^4 = a^4b^{12}$
8. $\frac{a^{12}}{a^4} = a^3$
9. $\frac{6w^4}{3w^9} = 2w^5$
10. $\left(\frac{2y^3}{9}\right)^2 = \frac{4y^6}{81}$

5.6 Exercises

Reading and Writing *After reading this section, write out the answers to these questions. Use complete sentences.*

1. What is the product rule for exponents?
2. What is the quotient rule for exponents?
3. Why must the bases be the same in these rules?
4. What is the power rule for exponents?
5. What is the power of a product rule?
6. What is the power of a quotient rule?

For all exercises in this section, assume that the variables represent nonzero real numbers.

Simplify the exponential expressions. See Example 1.

7. $2^2 \cdot 2^5$
8. $x^6 \cdot x^7$
9. $(-3u^8)(-2u^2)$
10. $(3r^4)(-6r^2)$
11. $a^3b^4 \cdot ab^6(ab)^0$
12. $x^2y \cdot x^3y^6(x + y)^0$
13. $\frac{-2a^3}{4a^7}$
14. $\frac{-3t^9}{6t^{18}}$
15. $\frac{2a^5b \cdot 3a^7b^3}{15a^6b^8}$
16. $\frac{3xy^8 \cdot 5xy^9}{20x^3y^{14}}$
17. $2^3 \cdot 5^2$
18. $2^2 \cdot 10^3$

Simplify. See Example 2.

19. $(x^2)^3$
20. $(y^2)^4$
21. $2x^2 \cdot (x^2)^5$
22. $(y^2)^6 \cdot 3y^5$
23. $\frac{(t^2)^5}{(t^3)^4}$
24. $\frac{(r^4)^2}{(r^5)^3}$
25. $\frac{3x(x^5)^2}{6x^3(x^2)^4}$
26. $\frac{5y^3(y^5)^2}{10y^5(y^2)^6}$

Simplify. See Example 3.

27. $(xy^2)^3$
28. $(wy^2)^6$
29. $(-2t^5)^3$
30. $(-3r^3)^3$
31. $(-2x^2y^5)^3$
32. $(-3y^2z^3)^3$
33. $\frac{(a^4b^2c^5)^3}{a^3b^4c}$
34. $\frac{(2ab^2c^3)^5}{(2a^3bc)^4}$

Simplify. See Example 4.

35. $\left(\frac{x^4}{4}\right)^3$
36. $\left(\frac{y^2}{2}\right)^3$
37. $\left(\frac{-2a^2}{b^3}\right)^4$
38. $\left(\frac{-9r^3}{t^5}\right)^2$
39. $\left(\frac{2x^2y}{-4y^2}\right)^3$
40. $\left(\frac{3y^8}{2zy^2}\right)^4$
41. $\left(\frac{-6x^2y^4z^9}{3x^6y^4z^3}\right)^2$
42. $\left(\frac{-10rs^9t^4}{2rs^2t^7}\right)^3$

Simplify each expression. Your answer should be an integer or a fraction. Do not use a calculator.

43. $3^2 + 6^2$

44. $(5 - 3)^2$

45. $(3 + 6)^2$

46. $5^2 - 3^2$

47. $2^3 - 3^3$

48. $3^3 + 4^3$

49. $(2 - 3)^3$

50. $(3 + 4)^3$

51. $\left(\frac{2}{5}\right)^3$

52. $\left(\frac{3}{4}\right)^3$

53. $5^2 \cdot 2^3$

54. $10^3 \cdot 3^3$

55. $2^3 \cdot 2^4$

56. $10^2 \cdot 10^4$

57. $\left(\frac{2^3}{2^5}\right)^2$

58. $\left(\frac{3}{3^3}\right)^2$

Simplify each expression.

59. $x^4 \cdot x^3$

60. $x^5 \cdot x^8$

61. $(a^8)^4$

62. $(b^5)^8$

63. $(a^4b^2)^3$

64. $(x^2t^4)^6$

65. $\frac{x^4}{x^7}$

66. $\frac{m^8}{m^{10}}$

67. $\frac{a^{13}}{a^9}$

68. $\frac{t^9}{t^6}$

69. $\left(\frac{a^3}{b^4}\right)^3$

70. $\left(\frac{t}{m^2}\right)^4$

71. $\left(\frac{x^3}{x^4}\right)^5$

72. $\left(\frac{w^5}{w^2}\right)^8$

73. $3x^4 \cdot 5x^7$

74. $-2y^3(3y)$

75. $(-5x^4)^3$

76. $(4z^3)^3$

77. $-3y^5z^{12} \cdot 9yz^7$

78. $2a^4b^5 \cdot 2a^9b^2$

79. $\frac{-9u^4v^9}{-3u^5v^8}$

80. $\frac{-20a^5b^{13}}{5a^4b^{13}}$

81. $(-xt^2)(-2x^2t)^4$

82. $(-ab)^3(-3ba^2)^4$

83. $\left(\frac{2x^2}{x^4}\right)^3$

84. $\left(\frac{3y^8}{y^5}\right)^2$

85. $\left(\frac{-8a^3b^4}{4c^5}\right)^5$

86. $\left(\frac{-10a^5c}{5a^5b^4}\right)^5$

87. $\left(\frac{-8x^4y^7}{-16x^5y^6}\right)^5$

88. $\left(\frac{-5x^2yz^3}{-5x^2yz}\right)^5$

Solve each problem.

89. ***Long-term investing.*** Sheila invested P dollars at annual rate r for 10 years. At the end of 10 years her investment was worth $P(1 + r)^{10}$ dollars. She then reinvested this money for another 5 years at annual rate r. At the end of the second time period her investment was worth $P(1 + r)^{10}(1 + r)^5$ dollars. Which law of exponents can be used to simplify the last expression? Simplify it.

90. ***CD rollover.*** Ronnie invested P dollars in a 2-year CD with an annual rate of return of r. After the CD rolled over three times, its value was $P((1 + r)^2)^3$. Which law of exponents can be used to simplify the expression? Simplify it.

Getting More Involved

91. ***Writing***

When we square a product, we square each factor in the product. For example, $(3b)^2 = 9b^2$. Explain why we cannot square a sum by simply squaring each term of the sum.

92. ***Writing***

Explain why we define 2^0 to be 1. Explain why $-2^0 \neq 1$.

5.7 Negative Exponents and Scientific Notation

In this Section

- **Negative Integral Exponents**
- **Rules for Integral Exponents**
- **Converting from Scientific Notation**
- **Converting to Scientific Notation**
- **Computations with Scientific Notation**

We defined exponential expressions with positive integral exponents in Chapter 1 and learned the rules for positive integral exponents in Section 5.6. In this section you will first study negative exponents and then see how positive and negative integral exponents are used in scientific notation.

Negative Integral Exponents

If x is nonzero, the reciprocal of x is written as $\frac{1}{x}$. For example, the reciprocal of 2^3 is written as $\frac{1}{2^3}$. To write the reciprocal of an exponential expression in a simpler way, we use a negative exponent. So $2^{-3} = \frac{1}{2^3}$. In general we have the following definition.

Negative Integral Exponents

If a is a nonzero real number and n is a positive integer, then

$$a^{-n} = \frac{1}{a^n}. \quad \text{(If } n \text{ is positive, } -n \text{ is negative.)}$$

EXAMPLE 1

Simplifying expressions with negative exponents

Simplify.

a) 2^{-5} **b)** $(-2)^{-5}$ **c)** $\frac{2^{-3}}{3^{-2}}$

Solution

a) $2^{-5} = \frac{1}{2^5} = \frac{1}{32}$

b) $(-2)^{-5} = \frac{1}{(-2)^5}$ Definition of negative exponent

$= \frac{1}{-32} = -\frac{1}{32}$

c) $\frac{2^{-3}}{3^{-2}} = 2^{-3} \div 3^{-2}$

$= \frac{1}{2^3} \div \frac{1}{3^2}$

$= \frac{1}{8} \div \frac{1}{9} = \frac{1}{8} \cdot \frac{9}{1} = \frac{9}{8}$

Now do Exercises 7–14

Calculator Close-Up

You can evaluate expressions with negative exponents on a calculator as shown here.

```
2^-5▸Frac
            1/32
(-2)^-5▸Frac
           -1/32
2^-3/3^-2▸Frac
             9/8
```

CAUTION In simplifying -5^{-2}, the negative sign preceding the 5 is used after 5 is squared and the reciprocal is found. So $-5^{-2} = -(5^{-2}) = -\frac{1}{25}$.

To evaluate a^{-n}, you can first find the nth power of a and then find the reciprocal. However, the result is the same if you first find the reciprocal of a and then find the nth power of the reciprocal. For example,

$$3^{-2} = \frac{1}{3^2} = \frac{1}{9} \quad \text{or} \quad 3^{-2} = \left(\frac{1}{3}\right)^2 = \frac{1}{3} \cdot \frac{1}{3} = \frac{1}{9}.$$

So the power and the reciprocal can be found in either order. If the exponent is -1, we simply find the reciprocal. For example,

$$5^{-1} = \frac{1}{5}, \quad \left(\frac{1}{4}\right)^{-1} = 4, \quad \text{and} \quad \left(-\frac{3}{5}\right)^{-1} = -\frac{5}{3}.$$

Because $3^{-2} \cdot 3^2 = 1$, the reciprocal of 3^{-2} is 3^2, and we have

$$\frac{1}{3^{-2}} = 3^2.$$

These examples illustrate the following rules.

Helpful Hint

Just because the exponent is negative, it doesn't mean the expression is negative. Note that $(-2)^{-3} = -\frac{1}{8}$ while $(-2)^{-4} = \frac{1}{16}$.

Rules for Negative Exponents

If a is a nonzero real number and n is a positive integer, then

$$a^{-n} = \left(\frac{1}{a}\right)^n, \quad a^{-1} = \frac{1}{a}, \quad \frac{1}{a^{-n}} = a^n, \quad \text{and} \quad \left(\frac{a}{b}\right)^{-n} = \left(\frac{b}{a}\right)^n.$$

EXAMPLE 2

Using the rules for negative exponents

Simplify

a) $\left(\frac{3}{4}\right)^{-3}$ **b)** $10^{-1} + 10^{-1}$ **c)** $\frac{2}{10^{-3}}$

Solution

a) We can find the third power and the reciprocal in either order:

$$\left(\frac{3}{4}\right)^{-3} = \left(\frac{4}{3}\right)^{3} = \frac{64}{27} \qquad \left(\frac{3}{4}\right)^{-3} = \left(\frac{27}{64}\right)^{-1} = \frac{64}{27}$$

b) $10^{-1} + 10^{-1} = \frac{1}{10} + \frac{1}{10} = \frac{2}{10} = \frac{1}{5}$

c) $\frac{2}{10^{-3}} = 2 \cdot \frac{1}{10^{-3}} = 2 \cdot 10^3 = 2 \cdot 1000 = 2000$

Now do Exercises 15–22

Calculator Close-Up

You can use a calculator to demonstrate that the product rule for exponents holds when the exponents are negative numbers.

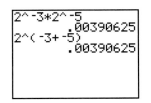

```
2^-3*2^-5
        .00390625
2^(-3+-5)
        .00390625
```

Rules for Integral Exponents

Negative exponents are used to make expressions involving reciprocals simpler looking and easier to write. Negative exponents have the added benefit of working in conjunction with all of the rules of exponents that you learned in Section 5.6. For example, we can use the product rule to get

$$x^{-2} \cdot x^{-3} = x^{-2+(-3)} = x^{-5}$$

and the quotient rule to get

$$\frac{y^3}{y^5} = y^{3-5} = y^{-2}.$$

With negative exponents there is no need to state the quotient rule in two parts as we did in Section 5.6. It can be stated simply as

$$\frac{a^m}{a^n} = a^{m-n}$$

for any integers m and n. We list the rules of exponents here for easy reference.

Helpful Hint

The definitions of the different types of exponents are a really clever mathematical invention. The fact that we have rules for performing arithmetic with those exponents makes the notation of exponents even more amazing.

Rules for Integral Exponents

The following rules hold for nonzero real numbers a and b and any integers m and n.

1. $a^0 = 1$ — Definition of zero exponent
2. $a^m \cdot a^n = a^{m+n}$ — Product rule
3. $\frac{a^m}{a^n} = a^{m-n}$ — Quotient rule
4. $(a^m)^n = a^{mn}$ — Power rule
5. $(ab)^n = a^n \cdot b^n$ — Power of a product rule
6. $\left(\frac{a}{b}\right)^n = \frac{a^n}{b^n}$ — Power of a quotient rule

EXAMPLE 3

The product and quotient rules for integral exponents

Simplify. Write your answers without negative exponents. Assume that the variables represent nonzero real numbers.

a) $b^{-3}b^5$ **b)** $-3x^{-3} \cdot 5x^2$ **c)** $\dfrac{m^{-6}}{m^{-2}}$ **d)** $\dfrac{4y^5}{-12y^{-3}}$

Solution

a) $b^{-3}b^5 = b^{-3+5}$ Product rule

$= b^2$ Simplify.

b) $-3x^{-3} \cdot 5x^2 = -15x^{-1}$ Product rule

$= -\dfrac{15}{x}$ Definition of negative exponent

c) $\dfrac{m^{-6}}{m^{-2}} = m^{-6-(-2)}$ Quotient rule

$= m^{-4}$ Simplify.

$= \dfrac{1}{m^4}$ Definition of negative exponent

Note that we could use the rules for negative exponents and the old quotient rule:

$$\frac{m^{-6}}{m^{-2}} = \frac{m^2}{m^6} = \frac{1}{m^4}$$

d) $\dfrac{4y^5}{-12y^{-3}} = \dfrac{y^{5-(-3)}}{-3} = \dfrac{-y^8}{3}$

Now do Exercises 23–34

In Example 4 we use the power rules with negative exponents.

EXAMPLE 4

The power rules for integral exponents

Simplify each expression. Write your answers with positive exponents only. Assume that all variables represent nonzero real numbers.

a) $(a^{-3})^2$ **b)** $(10x^{-3})^{-2}$ **c)** $\left(\dfrac{4x^{-5}}{y^2}\right)^{-2}$

Solution

a) $(a^{-3})^2 = a^{-3 \cdot 2}$ Power rule

$= a^{-6}$

$= \dfrac{1}{a^6}$ Definition of negative exponent

b) $(10x^{-3})^{-2} = 10^{-2}(x^{-3})^{-2}$ Power of a product rule

$= 10^{-2}x^{(-3)(-2)}$ Power rule

$= \dfrac{x^6}{10^2}$ Definition of negative exponent

$= \dfrac{x^6}{100}$

Calculator Close-Up

You can use a calculator to demonstrate that the power rule for exponents holds when the exponents are negative integers.

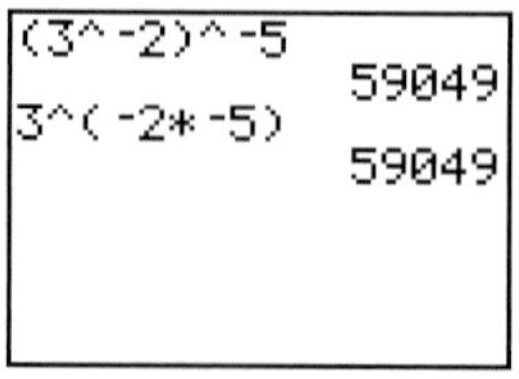

Helpful Hint

The exponent rules in this section apply to expressions that involve only multiplication and division. This is not too surprising since exponents, multiplication, and division are closely related. Recall that $a^3 = a \cdot a \cdot a$ and $a \div b = a \cdot b^{-1}$.

c) $\left(\dfrac{4x^{-5}}{y^2}\right)^{-2} = \dfrac{(4x^{-5})^{-2}}{(y^2)^{-2}}$ Power of a quotient rule

$= \dfrac{4^{-2}x^{10}}{y^{-4}}$ Power of a product rule and power rule

$= 4^{-2} \cdot x^{10} \cdot \dfrac{1}{y^{-4}}$ Because $\frac{a}{b} = a \cdot \frac{1}{b}$.

$= \dfrac{1}{4^2} \cdot x^{10} \cdot y^4$ Definition of negative exponent

$= \dfrac{x^{10}y^4}{16}$ Simplify.

Now do Exercises 35–46

Converting from Scientific Notation

Many of the numbers occurring in science are either very large or very small. The speed of light is 983,569,000 feet per second. One millimeter is equal to 0.000001 kilometer. In scientific notation, numbers larger than 10 or smaller than 1 are written by using positive or negative exponents.

Scientific notation is based on multiplication by integral powers of 10. Multiplying a number by a positive power of 10 moves the decimal point to the right:

$$10(5.32) = 53.2$$
$$10^2(5.32) = 100(5.32) = 532$$
$$10^3(5.32) = 1000(5.32) = 5320$$

Multiplying by a negative power of 10 moves the decimal point to the left:

$$10^{-1}(5.32) = \frac{1}{10}(5.32) = 0.532$$
$$10^{-2}(5.32) = \frac{1}{100}(5.32) = 0.0532$$
$$10^{-3}(5.32) = \frac{1}{1000}(5.32) = 0.00532$$

So if n is a positive integer, multiplying by 10^n moves the decimal point n places to the right and multiplying by 10^{-n} moves it n places to the left.

A number in scientific notation is written as a product of a number between 1 and 10 and a power of 10. The times symbol $\times$ indicates multiplication. For example, 3.27×10^9 and 2.5×10^{-4} are numbers in scientific notation. In scientific notation, there is one digit to the left of the decimal point.

To convert 3.27×10^9 to standard notation, move the decimal point nine places to the right:

$$3.27 \times 10^9 = 3{,}270{,}000{,}000$$

9 places to the right

Of course, it is not necessary to put the decimal point in when writing a whole number.

To convert 2.5×10^{-4} to standard notation, the decimal point is moved four places to the left:

$$2.5 \times 10^{-4} = 0.00025$$

4 places to the left

Calculator Close-Up

On a graphing calculator you can write scientific notation by actually using the power of 10 or press EE to get the letter E, which indicates that the following number is the power of 10.

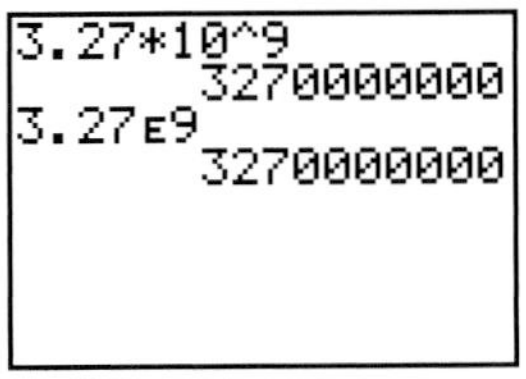

Note that if the exponent is not too large, scientific notation is converted to standard notation when you press ENTER.

In general, we use the following strategy to convert from scientific notation to standard notation.

Strategy for Converting from Scientific Notation to Standard Notation

1. Determine the number of places to move the decimal point by examining the exponent on the 10.
2. Move to the right for a positive exponent and to the left for a negative exponent.

EXAMPLE 5

Converting scientific notation to standard notation

Write in standard notation.

a) 7.02×10^6 **b)** 8.13×10^{-5}

Solution

a) Because the exponent is positive, move the decimal point six places to the right:

$$7.02 \times 10^6 = 7020000. = 7{,}020{,}000$$

b) Because the exponent is negative, move the decimal point five places to the left:

$$8.13 \times 10^{-5} = 0.0000813$$

Now do Exercises 67–74

Study Tip

Remember that everything we do in solving problems is based on principles (which are also called rules, theorems, and definitions). These principles justify the steps we take. Be sure that you understand the reasons. If you just memorize procedures without understanding, you will soon forget the procedures.

Converting to Scientific Notation

To convert a positive number to scientific notation, we just reverse the strategy for converting from scientific notation.

Strategy for Converting to Scientific Notation

1. Count the number of places (n) that the decimal must be moved so that it will follow the first nonzero digit of the number.
2. If the original number was larger than 10, use 10^n.
3. If the original number was smaller than 1, use 10^{-n}.

Remember that the scientific notation for a number larger than 10 will have a positive power of 10 and the scientific notation for a number between 0 and 1 will have a negative power of 10.

Math *at Work* Aerospace Engineering

Aircraft design is a delicate balance between weight and strength. Saving 1 pound of weight could save the plane's operators \$5000 over 20 years. Mathematics is used to calculate the strength of each of a plane's parts and to predict when the material making up a part will fail. If calculations show that one kind of metal isn't strong enough, designers usually have to choose another material or change the design.

As an example, consider an aluminum stringer with a circular cross section. The stringer is used inside the wing of an airplane as shown in the accompanying figure. The aluminum rod has a diameter of 20 mm and will support a load of 5×10^4 Newtons (N). The maximum stress on aluminum is 1×10^8 Pascals (Pa), where 1 Pa = 1 N/m^2. To calculate the stress S on the rod we use S = (load)/(cross sectional area). Note that we must divide the diameter by 2 to get the radius and convert square millimeters to square meters:

$$S = \frac{L}{\pi r^2} = \frac{5 \times 10^4 \text{ N}}{\pi(10 \text{ mm})^2} \cdot \left(\frac{1000 \text{ mm}}{1 \text{ m}}\right)^2 \approx 1.6 \times 10^8 \text{ Pa}$$

Since the stress is 1.6×10^8 Pa and the maximum stress on aluminum is 1×10^8 Pa, the aluminum rod is not strong enough. The design must be changed. The diameter of the aluminum rod could be increased or stronger/lighter metal such as titanium could be used.

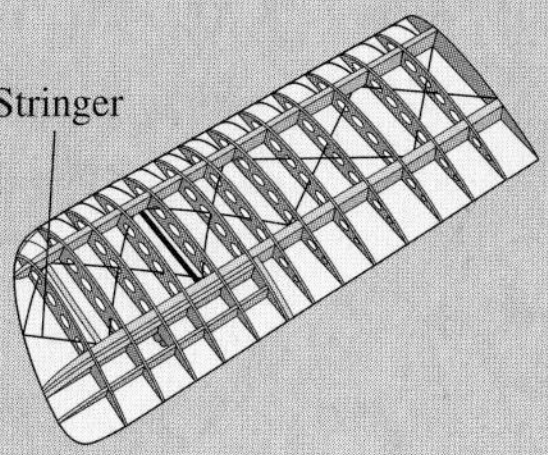

EXAMPLE 6 **Converting numbers to scientific notation**

Write in scientific notation.

a) 7,346,200 **b)** 0.0000348 **c)** 135×10^{-12}

Calculator Close-Up

To convert to scientific notation, set the mode to scientific. In scientific mode all results are given in scientific notation.

```
7346200
           7.3462E6
.0000348
            3.48E-5
135E-12
           1.35E-10
```

Solution

a) Because 7,346,200 is larger than 10, the exponent on the 10 will be positive:

$$7{,}346{,}200 = 7.3462 \times 10^6$$

b) Because 0.0000348 is smaller than 1, the exponent on the 10 will be negative:

$$0.0000348 = 3.48 \times 10^{-5}$$

c) There should be only one nonzero digit to the left of the decimal point:

$$135 \times 10^{-12} = 1.35 \times 10^2 \times 10^{-12} \quad \text{Convert 135 to scientific notation.}$$
$$= 1.35 \times 10^{-10} \quad \text{Product rule}$$

Now do Exercises 75–82

Computations with Scientific Notation

An important feature of scientific notation is its use in computations. Numbers in scientific notation are nothing more than exponential expressions, and you have already studied operations with exponential expressions in this section. We use the same rules of exponents on numbers in scientific notation that we use on any other exponential expressions.

EXAMPLE 7

Using the rules of exponents with scientific notation

Perform the indicated computations. Write the answers in scientific notation.

a) $(3 \times 10^6)(2 \times 10^8)$ **b)** $\dfrac{4 \times 10^5}{8 \times 10^{-2}}$ **c)** $(5 \times 10^{-7})^3$

Solution

a) $(3 \times 10^6)(2 \times 10^8) = 3 \cdot 2 \cdot 10^6 \cdot 10^8 = 6 \times 10^{14}$

b) $\dfrac{4 \times 10^5}{8 \times 10^{-2}} = \dfrac{4}{8} \cdot \dfrac{10^5}{10^{-2}} = \dfrac{1}{2} \cdot 10^{5-(-2)}$ Quotient rule

$= (0.5)10^7$ $\frac{1}{2} = 0.5$

$= 5 \times 10^{-1} \cdot 10^7$ Write 0.5 in scientific notation.

$= 5 \times 10^6$ Product rule

c) $(5 \times 10^{-7})^3 = 5^3(10^{-7})^3$ Power of a product rule

$= 125 \cdot 10^{-21}$ Power rule

$= 1.25 \times 10^2 \times 10^{-21}$ $125 = 1.25 \times 10^2$

$= 1.25 \times 10^{-19}$ Product rule

Now do Exercises 83–94

Calculator Close-Up

With a calculator's built-in scientific notation, some parentheses can be omitted as shown below. Writing out the powers of 10 can lead to errors.

```
4E5/8E-2
                5E6
4*10^5/8*10^-2
                5E2
```

Try these computations with your calculator.

EXAMPLE 8

Converting to scientific notation for computations

Perform these computations by first converting each number into scientific notation. Give your answer in scientific notation.

a) (3,000,000)(0.0002) **b)** $(20{,}000{,}000)^3(0.0000003)$

Solution

a) $(3{,}000{,}000)(0.0002) = 3 \times 10^6 \cdot 2 \times 10^{-4}$ Scientific notation

$= 6 \times 10^2$ Product rule

b) $(20{,}000{,}000)^3(0.0000003) = (2 \times 10^7)^3(3 \times 10^{-7})$ Scientific notation

$= 8 \times 10^{21} \cdot 3 \times 10^{-7}$ Power of a product rule

$= 24 \times 10^{14}$

$= 2.4 \times 10^1 \times 10^{14}$ $24 = 2.4 \times 10^1$

$= 2.4 \times 10^{15}$ Product rule

Now do Exercises 95–102

Warm-Ups

True or false? Explain your answer.

1. $10^{-2} = \frac{1}{100}$
2. $\left(-\frac{1}{5}\right)^{-1} = 5$
3. $3^{-2} \cdot 2^{-1} = 6^{-3}$
4. $\frac{3^{-2}}{3^{-1}} = \frac{1}{3}$
5. $23.7 = 2.37 \times 10^{-1}$
6. $0.000036 = 3.6 \times 10^{-5}$
7. $25 \cdot 10^{7} = 2.5 \times 10^{8}$
8. $0.442 \times 10^{-3} = 4.42 \times 10^{-4}$
9. $(3 \times 10^{-9})^2 = 9 \times 10^{-18}$
10. $(2 \times 10^{-5})(4 \times 10^{4}) = 8 \times 10^{-20}$

5.7 Exercises

Boost your GRADE at mathzone.com!

MathZone
- Practice Problems
- Self-Tests
- Videos
- Net Tutor
- e-Professors

Reading and Writing *After reading this section, write out the answers to these questions. Use complete sentences.*

1. What does a negative exponent mean?
2. What is the correct order for evaluating the operations indicated by a negative exponent?
3. What is the new quotient rule for exponents?
4. How do you convert a number from scientific notation to standard notation?
5. How do you convert a number from standard notation to scientific notation?
6. Which numbers are not usually written in scientific notation?

Variables in all exercises represent positive real numbers. Evaluate each expression. See Example 1.

7. 3^{-1}
8. 3^{-3}
9. $(-2)^{-4}$
10. $(-3)^{-4}$
11. -4^{-2}
12. -2^{-4}
13. $\frac{5^{-2}}{10^{-2}}$
14. $\frac{3^{-4}}{6^{-2}}$

Simplify. See Example 2.

15. $\left(\frac{5}{2}\right)^{-3}$
16. $\left(\frac{4}{3}\right)^{-2}$
17. $6^{-1} + 6^{-1}$
18. $2^{-1} + 4^{-1}$
19. $\frac{10}{5^{-3}}$
20. $\frac{1}{25 \cdot 10^{-4}}$
21. $\frac{1}{4^{-3}} + \frac{3^2}{2^{-1}}$
22. $\frac{2^3}{10^{-2}} - \frac{2}{7^{-2}}$

Simplify. Write answers without negative exponents. See Example 3.

23. $x^{-1}x^{2}$
24. $y^{-3}y^{5}$
25. $-2x^{2} \cdot 8x^{-6}$
26. $5y^{5}(-6y^{-7})$
27. $-3a^{-2}(-2a^{-3})$
28. $(-b^{-3})(-b^{-5})$
29. $\frac{u^{-5}}{u^{3}}$
30. $\frac{w^{-4}}{w^{6}}$

31. $\frac{8t^{-3}}{-2t^{-5}}$

32. $\frac{-22w^{-4}}{-11w^{-3}}$

33. $\frac{-6x^5}{-3x^{-6}}$

34. $\frac{-51y^6}{17y^{-9}}$

Simplify each expression. Write answers without negative exponents. See Example 4.

35. $(x^2)^{-5}$

36. $(y^{-2})^4$

37. $(a^{-3})^{-3}$

38. $(b^{-5})^{-2}$

39. $(2x^{-3})^{-4}$

40. $(3y^{-1})^{-2}$

41. $(4x^2y^{-3})^{-2}$

42. $(6s^{-2}t^4)^{-1}$

43. $\left(\frac{2x^{-1}}{y^{-3}}\right)^{-2}$

44. $\left(\frac{a^{-2}}{3b^3}\right)^{-3}$

45. $\left(\frac{2a^{-3}}{ac^{-2}}\right)^{-4}$

46. $\left(\frac{3w^2}{w^4x^3}\right)^{-2}$

Simplify. Write answers without negative exponents.

47. $2 \cdot 3w^{-5}$

48. $4 \cdot 3m^{-6}$

49. $(2h)^{-3}$

50. $(3t)^{-4}$

51. $(x^{-4})^{-3}(x^{-5})^6$

52. $(y^{-5})^{-6}(y^{-6})^7$

53. $\frac{(b^3)^{-5}}{(b^{-7})^4}$

54. $\frac{(a^9)^{-3}}{(a^{-4})^7}$

55. $\frac{(v^{-3})^6(v^{-5})^{-4}}{(v^{-7})^3}$

56. $\frac{(k^{-3})^4(k^5)^{-5}}{(k^{-5})^4}$

57. $\frac{(c^{-1})^{-12}(c^{-5})^6}{(c^{-4})^0(c^3)^{-3}}$

58. $\frac{(p^{-5})^{-9}(p^{-6})^4}{(p^{-8})^0(p^4)^{-5}}$

59. $2^{-1} \cdot 3^{-1}$

60. $2^{-1} + 3^{-1}$

61. $(2 \cdot 3^{-1})^{-1}$

62. $(2^{-1} + 3)^{-1}$

63. $(x^{-2})^{-3} + 3x^7(-5x^{-1})$

64. $(ab^{-1})^2 - ab(-ab^{-3})$

65. $\frac{a^3b^{-2}}{a^{-1}} + \left(\frac{b^6a^{-2}}{b^5}\right)^{-2}$

66. $\left(\frac{x^{-3}y^{-1}}{2x}\right)^{-3} + \frac{6x^9y^3}{-3x^{-3}}$

Write each number in standard notation. See Example 5.

67. 9.86×10^9

68. 4.007×10^4

69. 1.37×10^{-3}

70. 9.3×10^{-5}

71. 1×10^{-6}

72. 3×10^{-1}

73. 6×10^5

74. 8×10^6

Write each number in scientific notation. See Example 6.

75. 9000

76. 5,298,000

77. 0.00078

78. 0.000214

79. 0.0000085

80. 5,670,000,000

81. 525×10^9

82. 0.0034×10^{-8}

Perform the computations. Write answers in scientific notation. See Example 7.

83. $(3 \times 10^5)(2 \times 10^{-15})$

84. $(2 \times 10^{-9})(4 \times 10^{23})$

85. $\frac{4 \times 10^{-8}}{2 \times 10^{30}}$

86. $\frac{9 \times 10^{-4}}{3 \times 10^{-6}}$

87. $\frac{3 \times 10^{20}}{6 \times 10^{-8}}$

88. $\frac{1 \times 10^{-8}}{4 \times 10^7}$

89. $(3 \times 10^{12})^2$

90. $(2 \times 10^{-5})^3$

91. $(5 \times 10^{4})^3$

92. $(5 \times 10^{14})^{-1}$

93. $(4 \times 10^{32})^{-1}$

94. $(6 \times 10^{11})^2$

Perform the following computations by first converting each number into scientific notation. Write answers in scientific notation. See Example 8.

95. $(4300)(2{,}000{,}000)$

96. $(40{,}000)(4{,}000{,}000{,}000)$

97. $(4{,}200{,}000)(0.00005)$

98. $(0.00075)(4{,}000{,}000)$

99. $(300)^3(0.000001)^5$

100. $(200)^4(0.0005)^3$

101. $\dfrac{(4000)(90{,}000)}{0.00000012}$

102. $\dfrac{(30{,}000)(80{,}000)}{(0.000006)(0.002)}$

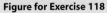

Perform the following computations with the aid of a calculator. Write answers in scientific notation. Round to three decimal places.

103. $(6.3 \times 10^{6})(1.45 \times 10^{-4})$

104. $(8.35 \times 10^{9})(4.5 \times 10^{3})$

105. $(5.36 \times 10^{-4}) + (3.55 \times 10^{-5})$

106. $(8.79 \times 10^{8}) + (6.48 \times 10^{9})$

107. $\dfrac{(3.5 \times 10^{5})(4.3 \times 10^{-6})}{3.4 \times 10^{-8}}$

108. $\dfrac{(3.5 \times 10^{-8})(4.4 \times 10^{-4})}{2.43 \times 10^{45}}$

109. $(3.56 \times 10^{85})(4.43 \times 10^{96})$

110. $(8 \times 10^{99}) + (3 \times 10^{99})$

Solve each problem.

111. ***Distance to the sun.*** The distance from the earth to the sun is 93 million miles. Express this distance in feet. (1 mile = 5280 feet.)

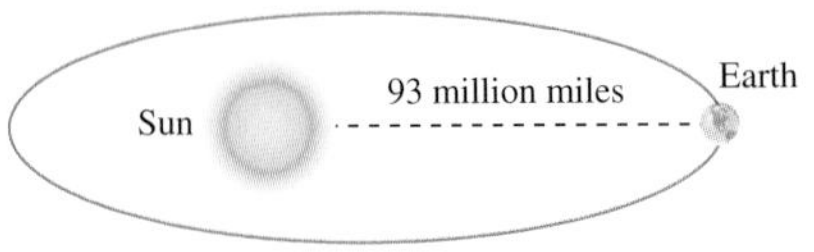

Figure for Exercise 111

112. ***Speed of light.*** The speed of light is 9.83569×10^{8} feet per second. How long does it take light to travel from the sun to the earth? See Exercise 111.

113. ***Warp drive, Scotty.*** How long does it take a spacecraft traveling at 2×10^{35} miles per hour (warp factor 4) to travel 93 million miles?

114. ***Area of a dot.*** If the radius of a very small circle is 2.35×10^{-8} centimeters, then what is the circle's area?

115. ***Circumference of a circle.*** If the circumference of a circle is 5.68×10^{9} feet, then what is its radius?

116. ***Diameter of a circle.*** If the diameter of a circle is 1.3×10^{-12} meters, then what is its radius?

117. ***Present value.*** The present value P that will amount to A dollars in n years with interest compounded annually at annual interest rate r, is given by

$$P = A(1 + r)^{-n}.$$

Find the present value that will amount to \$50,000 in 20 years at 8% compounded annually.

118. ***Investing in stocks.*** U.S. small company stocks have returned an average of 14.9% annually for the last 50 years (T. Rowe Price, www.troweprice.com). Use the present value formula from the previous exercise to find the amount invested today in small company stocks that would be worth \$1 million in 50 years, assuming that small company stocks continue to return 14.9% annually for the next 50 years.

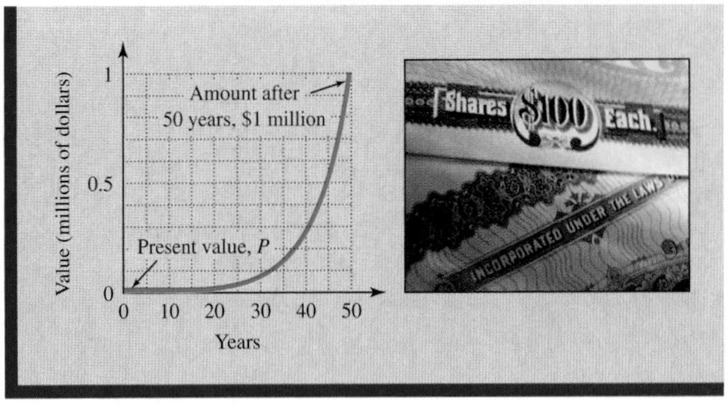

Figure for Exercise 118

Getting More Involved

119. *Exploration*

a) If $w^{-3} < 0$, then what can you say about w?

b) If $(-5)^m < 0$, then what can you say about m?

c) What restriction must be placed on w and m so that $w^m < 0$?

120. *Discussion*

Which of the following expressions is not equal to -1? Explain your answer.

a) -1^{-1}

b) -1^{-2}

c) $(-1^{-1})^{-1}$

d) $(-1)^{-1}$

e) $(-1)^{-2}$

Collaborative Activities

Grouping: Two students per group

Topic: Multiplying Polynomials

Area as a Model of Binomial Multiplication

Drawings and diagrams are often used to illustrate mathematical ideas. In this activity we use areas of rectangles to illustrate multiplication of binomials.

Example. The product $15 \cdot 13$ is the area of a 15 by 13 rectangle. Rewrite the product as $(10 + 5)(10 + 3)$ and make the following drawing.

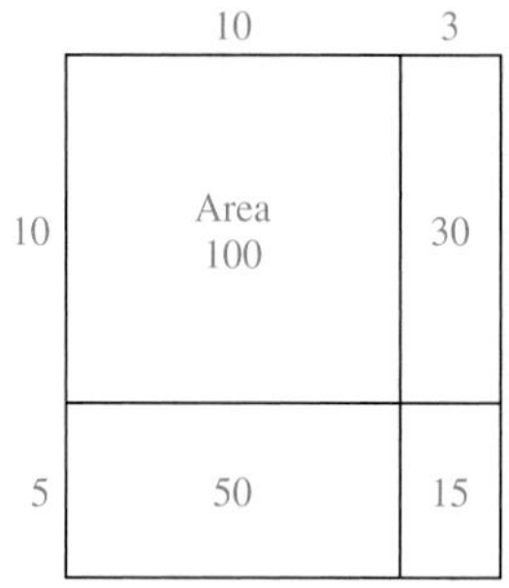

Note that the areas of the four regions are the four parts of FOIL:

$$\begin{aligned}(10 + 5)(10 + 3) &= \overset{\text{First}}{10^2} + \overset{\text{Outer}}{10 \cdot 3} + \overset{\text{Inner}}{5 \cdot 10} + \overset{\text{Last}}{5 \cdot 3} \\ &= 100 + 30 + 50 + 15 \\ &= 195\end{aligned}$$

Exercises

1. Make a drawing using graph paper to illustrate the product $12 \cdot 13$ as the product of two binomials $(10 + 2)(10 + 3)$. Have each square of your graph paper represent one unit. One partner should draw the rectangle and find the total of the areas of the four regions that correspond to FOIL while the other should find $12 \cdot 13$ in the "usual" way. Check your answers.

2. We have been using 10 for the first term of our binomials, but now we will use x. Choose any positive integer for x and draw an $x + 2$ by $x + 4$ rectangle on your graph paper. Have one partner draw the rectangle and find the total of the areas of the four regions that correspond to FOIL while the other will find $(x + 2)(x + 4)$ in the "usual" way. Check your answers.

3. Suppose that the area of the $x + 2$ by $x + 4$ rectangle drawn in the last exercise is actually 120. Does your rectangle have an area of 120? If not, then draw an $x + 2$ by $x + 4$ rectangle with an area of 120. What is x? Is x an integer?

4. Is there an $x + 3$ by $x + 6$ rectangle for which the area is 120 and x is a positive integer?

5. Is there an $x + 7$ by $x + 14$ rectangle for which the area is 120 and x is a positive integer?

6. How many ways are there to partition a 10 by 12 rectangle into four regions using sides $x + a$ and $x + b$, where x, a, and b are positive integers?

Chapter 5 Wrap-Up

Summary

Polynomials		Examples
Term	A number or the product of a number and one or more variables raised to powers	$5x^3, -4x, 7$
Polynomial	A single term or a finite sum of terms	$2x^5 - 9x^2 + 11$
Degree of a polynomial	The highest degree of any of the terms	Degree of $2x - 9$ is 1. Degree of $5x^3 - x^2$ is 3.
Naming a Polynomial	A polynomial can be named with a letter such as P or $P(x)$ (function notation).	$P = x^2 - 1$ $P(x) = x^2 - 1$
Evaluating a polynomial	The value of a polynomial is the real number that is obtained when the variable (x) is replaced with a real number.	If $x = 3$ then $P = 8$, or $P(3) = 8$.

Adding, Subtracting, and Multiplying Polynomials		Examples
Add or subtract polynomials	Add or subtract the like terms.	$(x + 1) + (x - 4) = 2x - 3$ $(x^2 - 3x) - (4x^2 - x)$ $= -3x^2 - 2x$
Multiply monomials	Use the product rule for exponents.	$-2x^5 \cdot 6x^8 = -12x^{13}$
Multiply polynomials	Multiply each term of one polynomial by every term of the other polynomial, then combine like terms.	$\begin{array}{r} x^2 + 2x + 5 \\ x - 1 \\ \hline -x^2 - 2x - 5 \\ x^3 + 2x^2 + 5x \quad\; \\ \hline x^3 + x^2 + 3x - 5 \end{array}$

Binomials		Examples
FOIL	A method for multiplying two binomials quickly	$(x - 2)(x + 3) = x^2 + x - 6$
Square of a sum	$(a + b)^2 = a^2 + 2ab + b^2$	$(x + 3)^2 = x^2 + 6x + 9$
Square of a difference	$(a - b)^2 = a^2 - 2ab + b^2$	$(m - 5)^2 = m^2 - 10m + 25$
Product of a sum and a difference	$(a - b)(a + b) = a^2 - b^2$	$(x + 2)(x - 2) = x^2 - 4$

Dividing Polynomials		**Examples**
Dividing monomials	Use the quotient rule for exponents	$8x^5 \div (2x^2) = 4x^3$
Divide a polynomial by a monomial	Divide each term of the polynomial by the monomial.	$\dfrac{3x^5 + 9x}{3x} = x^4 + 3$
Divide a polynomial by a binomial	If the divisor is a binomial, use long division. (divisor)(quotient) + (remainder) = dividend	$\begin{array}{rl} x - 7 & \leftarrow \text{Quotient} \\ \text{Divisor} \rightarrow x + 2\overline{)x^2 - 5x - 4} & \leftarrow \text{Dividend} \\ \underline{x^2 + 2x} & \\ -7x - 4 & \\ \underline{-7x - 14} & \\ 10 & \leftarrow \text{Remainder} \end{array}$

Rules of Exponents		**Examples**
The following rules hold for any integers *m* and *n*, and nonzero real numbers *a* and *b*.		
Zero exponent	$a^0 = 1$	$2^0 = 1, (-34)^0 = 1$
Product rule	$a^m \cdot a^n = a^{m+n}$	$a^2 \cdot a^3 = a^5$ $3x^6 \cdot 4x^9 = 12x^{15}$
Quotient rule	$\dfrac{a^m}{a^n} = a^{m-n}$	$x^8 \div x^2 = x^6, \dfrac{y^3}{y^3} = y^0 = 1$ $\dfrac{c^7}{c^9} = c^{-2} = \dfrac{1}{c^2}$
Power rule	$(a^m)^n = a^{mn}$	$(2^2)^3 = 2^6, (w^5)^3 = w^{15}$
Power of a product rule	$(ab)^n = a^n b^n$	$(2t)^3 = 8t^3$
Power of a quotient rule	$\left(\dfrac{a}{b}\right)^n = \dfrac{a^n}{b^n}$	$\left(\dfrac{x}{3}\right)^3 = \dfrac{x^3}{27}$

Negative Exponents		**Examples**
Negative integral exponents	If *n* is a positive integer and *a* is a nonzero real number, then $a^{-n} = \dfrac{1}{a^n}$	$3^{-2} = \dfrac{1}{3^2}, x^{-5} = \dfrac{1}{x^5}$
Rules for negative exponents	If *a* is a nonzero real number and *n* is a positive integer, then $a^{-n} = \left(\dfrac{1}{a}\right)^n$, $a^{-1} = \dfrac{1}{a}$, and $\dfrac{1}{a^{-n}} = a^n$.	$\left(\dfrac{2}{3}\right)^{-3} = \left(\dfrac{3}{2}\right)^3, 5^{-1} = \dfrac{1}{5}$ $\dfrac{1}{w^{-8}} = w^8$

Scientific Notation		Examples
Converting from scientific notation	1. Find the number of places to move the decimal point by examining the exponent on the 10. 2. Move to the right for a positive exponent and to the left for a negative exponent.	$5.6 \times 10^3 = 5600$ $9 \times 10^{-4} = 0.0009$
Converting into scientific notation (positive numbers)	1. Count the number of places (n) that the decimal point must be moved so that it will follow the first nonzero digit of the number. 2. If the original number was larger than 10, use 10^n. 3. If the original number was smaller than 1, use 10^{-n}.	$304.6 = 3.046 \times 10^2$ $0.0035 = 3.5 \times 10^{-3}$

Enriching Your Mathematical Word Power

For each mathematical term, choose the correct meaning.

1. term
a. an expression containing a number or the product of a number and one or more variables
b. the amount of time spent in this course
c. a word that describes a number
d. a variable

2. polynomial
a. four or more terms
b. many numbers
c. a sum of four or more numbers
d. a single term or a finite sum of terms

3. degree of a polynomial
a. the number of terms in a polynomial
b. the highest degree of any of the terms of a polynomial
c. the value of a polynomial when $x = 0$
d. the largest coefficient of any of the terms of a polynomial

4. leading coefficient
a. the first coefficient
b. the largest coefficient
c. the coefficient of the first term when a polynomial is written with decreasing exponents
d. the most important coefficient

5. monomial
a. a single polynomial
b. one number
c. an equation that has only one solution
d. a polynomial that has one term

6. FOIL
a. a method for adding polynomials
b. first, outer, inner, last
c. an equation with no solution
d. a polynomial with five terms

7. dividend
a. a in a/b
b. b in a/b
c. the result of a/b
d. what a bank pays on deposits

8. divisor
a. a in a/b
b. b in a/b
c. the result of a/b
d. two visors

9. quotient
a. a in a/b
b. b in a/b
c. a/b
d. the divisor plus the remainder

10. binomial
a. a polynomial with two terms
b. any two numbers
c. the two coordinates in an ordered pair
d. an equation with two variables

11. integral exponent
a. an exponent that is an integer
b. a positive exponent
c. a rational exponent
d. a fractional exponent

12. scientific notation
a. the notation of rational exponents
b. the notation of algebra
c. a notation for expressing large or small numbers with powers of 10
d. radical notation

Review Exercises

5.1 *Perform the indicated operations.*

1. $(2w - 6) + (3w + 4)$

2. $(1 - 3y) + (4y - 6)$

3. $(x^2 - 2x - 5) - (x^2 + 4x - 9)$

4. $(3 - 5x - x^2) - (x^2 - 7x + 8)$

5. $(5 - 3w + w^2) + (w^2 - 4w - 9)$

6. $(-2t^2 + 3t - 4) + (t^2 - 7t + 2)$

7. $(4 - 3m - m^2) - (m^2 - 6m + 5)$

8. $(n^3 - n^2 + 9) - (n^4 - n^3 + 5)$

5.2 *Perform the indicated operations.*

9. $5x^2 \cdot (-10x^9)$

10. $3h^3t^2 \cdot 2h^2t^5$

11. $(-11a^7)^2$

12. $(12b^3)^2$

13. $x - 5(x - 3)$

14. $x - 4(x - 9)$

15. $5x + 3(x^2 - 5x + 4)$

16. $5 + 4x^2(x - 5)$

17. $3m^2(5m^3 - m + 2)$

18. $-4a^4(a^2 + 2a + 4)$

19. $(x - 5)(x^2 - 2x + 10)$

20. $(x + 2)(x^2 - 2x + 4)$

21. $(x^2 - 2x + 4)(3x - 2)$

22. $(5x + 3)(x^2 - 5x + 4)$

5.3 *Perform the indicated operations.*

23. $(q - 6)(q + 8)$

24. $(w + 5)(w + 12)$

25. $(2t - 3)(t - 9)$

26. $(5r + 1)(5r + 2)$

27. $(4y - 3)(5y + 2)$

28. $(11y + 1)(y + 2)$

29. $(3x^2 + 5)(2x^2 + 1)$

30. $(x^3 - 7)(2x^3 + 7)$

5.4 *Perform the indicated operations. Try to write only the answers.*

31. $(z - 7)(z + 7)$

32. $(a - 4)(a + 4)$

33. $(y + 7)^2$

34. $(a + 5)^2$

35. $(w - 3)^2$

36. $(a - 6)^2$

37. $(x^2 - 3)(x^2 + 3)$

38. $(2b^2 - 1)(2b^2 + 1)$

39. $(3a + 1)^2$

40. $(1 - 3c)^2$

41. $(4 - y)^2$

42. $(9 - t)^2$

Study Tip

Note how the review exercises are arranged according to the sections in this chapter. If you are having trouble with a certain type of problem, refer back to the appropriate section for examples and explanations.

5.5 *Find each quotient.*

43. $-10x^5 \div (2x^3)$

44. $-6x^4y^2 \div (-2x^2y^2)$

45. $\dfrac{6a^5b^7c^6}{-3a^3b^9c^6}$

46. $\dfrac{-9h^5t^9r^2}{3h^7t^6r^2}$

47. $\dfrac{3x - 9}{-3}$

48. $\dfrac{7 - y}{-1}$

49. $\dfrac{9x^3 - 6x^2 + 3x}{-3x}$

50. $\dfrac{-8x^3y^5 + 4x^2y^4 - 2xy^3}{2xy^2}$

51. $(a - 1) \div (1 - a)$

52. $(t - 3) \div (3 - t)$

53. $(m^4 - 16) \div (m - 2)$

54. $(x^4 - 1) \div (x - 1)$

Find the quotient and remainder.

55. $(3m^3 - 9m^2 + 18m) \div (3m)$

56. $(8x^3 - 4x^2 - 18x) \div (2x)$

57. $(b^2 - 3b + 5) \div (b + 2)$

58. $(r^2 - 5r + 9) \div (r - 3)$

59. $(4x^2 - 9) \div (2x + 1)$

60. $(9y^3 + 2y) \div (3y + 2)$

61. $(x^3 + x^2 - 11x + 10) \div (x - 1)$

62. $(y^3 - 9y^2 + 3y - 6) \div (y + 1)$

Write each expression in the form

$$\textit{quotient} + \frac{\textit{remainder}}{\textit{divisor}}.$$

63. $\dfrac{2x}{x - 3}$

64. $\dfrac{3x}{x - 4}$

65. $\dfrac{2x}{1 - x}$

66. $\dfrac{3x}{5 - x}$

67. $\dfrac{x^2 - 3}{x + 1}$

68. $\dfrac{x^2 + 3x + 1}{x - 3}$

69. $\dfrac{x^2}{x + 1}$

70. $\dfrac{-2x^2}{x - 3}$

5.6 *Simplify each expression.*

71. $2y^{10} \cdot 3y^{20}$

72. $(-3a^5)(5a^3)$

73. $\frac{-10b^5c^3}{2b^5c^9}$

74. $\frac{-30k^3y^9}{15k^3y^2}$

75. $(b^5)^6$

76. $(y^5)^8$

77. $(-2x^3y^2)^3$

78. $(-3a^4b^6)^4$

79. $\left(\frac{2a}{b}\right)^3$

80. $\left(\frac{3y}{2}\right)^3$

81. $\left(\frac{-6x^2y^5}{-3z^6}\right)^3$

82. $\left(\frac{-3a^4b^8}{6a^3b^{12}}\right)^4$

For the following exercises, assume that all of the variables represent positive real numbers.

5.7 *Simplify each expression. Use only positive exponents in answers.*

83. 2^{-5}

84. -2^{-4}

85. 10^{-3}

86. $5^{-1} \cdot 5^0$

87. x^5x^{-8}

88. $a^{-3}a^{-9}$

89. $\frac{a^{-8}}{a^{-12}}$

90. $\frac{a^{10}}{a^{-4}}$

91. $\frac{a^3}{a^{-7}}$

92. $\frac{b^{-2}}{b^{-6}}$

93. $(x^{-3})^4$

94. $(x^5)^{-10}$

95. $(2x^{-3})^{-3}$

96. $(3y^{-5})^2$

97. $\left(\frac{a}{3b^{-3}}\right)^{-2}$

98. $\left(\frac{a^{-2}}{5b}\right)^{-3}$

Convert each number in scientific notation to a number in standard notation, and convert each number in standard notation to a number in scientific notation.

99. 5000

100. 0.00009

101. 3.4×10^5

102. 5.7×10^{-8}

103. 0.0000461

104. 44,000

105. 5.69×10^{-6}

106. 5.5×10^9

Perform each computation without using a calculator. Write answers in scientific notation.

107. $(3.5 \times 10^8)(2.0 \times 10^{-12})$

108. $(9 \times 10^{12})(2 \times 10^{17})$

109. $(2 \times 10^{-4})^4$

110. $(-3 \times 10^5)^3$

111. (0.00000004)(2,000,000,000)

112. $(3{,}000{,}000{,}000) \div (0.000002)$

113. $(0.0000002)^5$

114. $(50{,}000{,}000{,}000)^3$

Miscellaneous

Perform the indicated operations.

115. $(x + 3)(x + 7)$

116. $(k + 5)(k + 4)$

117. $(t - 3y)(t - 4y)$

118. $(t + 7z)(t + 6z)$

119. $(2x^3)^0 + (2y)^0$

120. $(4y^2 - 9)^0$

121. $(-3ht^6)^3$

122. $(-9y^3c^4)^2$

123. $(2w + 3)(w - 6)$

124. $(3x + 5)(2x - 6)$

125. $(3u - 5v)(3u + 5v)$

126. $(9x^2 - 2)(9x^2 + 2)$

127. $(3h + 5)^2$

128. $(4v - 3)^2$

129. $(x + 3)^3$

130. $(k - 10)^3$

131. $(-7s^2t)(-2s^3t^5)$

132. $-5w^3r^2 \cdot 2w^4r^8$

133. $\left(\frac{k^4m^2}{2k^2m^2}\right)^4$

134. $\left(\frac{-6h^3y^5}{2h^7y^2}\right)^4$

135. $(5x^2 - 8x - 8) - (4x^2 + x - 3)$

136. $(4x^2 - 6x - 8) - (9x^2 - 5x + 7)$

137. $(2x^2 - 2x - 3) + (3x^2 + x - 9)$

138. $(x^2 - 3x - 1) + (x^2 - 2x + 1)$

139. $(x + 4)(x^2 - 5x + 1)$

140. $(2x^2 - 7x + 4)(x + 3)$

141. $(x^2 + 4x - 12) \div (x - 2)$

142. $(a^2 - 3a - 10) \div (a - 5)$

Solve each problem.

143. ***Roundball court.*** The length of a basketball court is 44 feet more than its width w. Find polynomials $P(w)$ and $A(w)$ that represent its perimeter and area. Find $P(50)$ and $A(50)$.

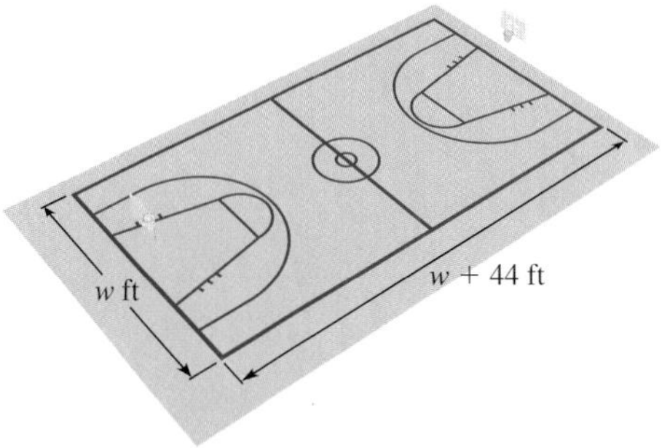

Figure for Exercise 143

144. ***Badminton court.*** The width of a badminton court is 24 feet less than its length x. Find polynomials $P(x)$ and $A(x)$ that represent its perimeter and area. Find $P(44)$ and $A(44)$.

145. ***Smoke alert.*** A retailer of smoke alarms knows that at a price of p dollars each, she can sell $600 - 15p$ smoke alarms per week. Find a polynomial that represents the weekly revenue for the smoke alarms. Find the revenue for a week in which the price is \$12 per smoke alarm. Use the bar graph to find the price per smoke alarm that gives the maximum weekly revenue.

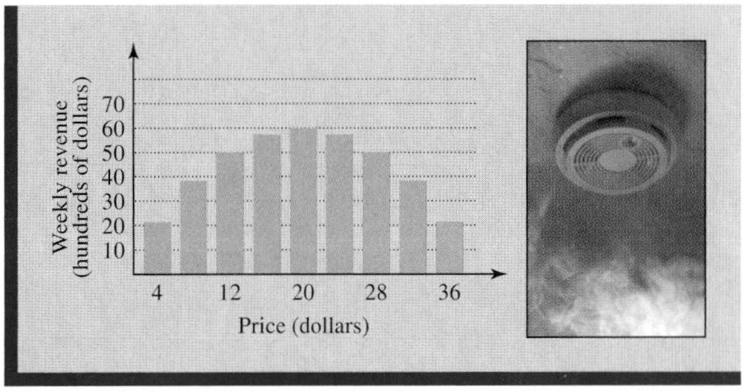

Figure for Exercise 145

146. ***Boom box sales.*** A retailer of boom boxes knows that at a price of q dollars each, he can sell $900 - 3q$ boom boxes per month. Find a polynomial that represents the monthly revenue for the boom boxes. How many boom boxes will he sell if the price is \$300 each?

Chapter 5 Test

Perform the indicated operations.

1. $(7x^3 - x^2 - 6) + (5x^2 + 2x - 5)$

2. $(x^2 - 3x - 5) - (2x^2 + 6x - 7)$

3. $\dfrac{6y^3 - 9y^2}{-3y}$

4. $(x - 2) \div (2 - x)$

5. $(x^3 - 2x^2 - 4x + 3) \div (x - 3)$

6. $3x^2(5x^3 - 7x^2 + 4x - 1)$

Find the products.

7. $(x + 5)(x - 2)$

8. $(3a - 7)(2a + 5)$

9. $(a - 7)^2$

10. $(4x + 3y)^2$

11. $(b - 3)(b + 3)$

12. $(3t^2 - 7)(3t^2 + 7)$

13. $(4x^2 - 3)(x^2 + 2)$

14. $(x - 2)(x + 3)(x - 4)$

Write each expression in the form

$$\textit{quotient} + \frac{\textit{remainder}}{\textit{divisor}}.$$

15. $\dfrac{2x}{x - 3}$

16. $\dfrac{x^2 - 3x + 5}{x + 2}$

Use the rules of exponents to simplify each expression. Write answers without negative exponents.

17. $-5x^3 \cdot 7x^5$

18. $3x^3y \cdot (2xy^4)^2$

19. $-4a^6b^5 \div (2a^5b)$

20. $3x^{-2} \cdot 5x^7$

21. $\left(\dfrac{-2a}{b^2}\right)^5$

22. $\dfrac{-6a^7b^6c^2}{-2a^3b^8c^2}$

23. $\dfrac{6t^{-7}}{2t^9}$

24. $\dfrac{w^{-6}}{w^{-4}}$

25. $(-3s^{-3}t^2)^{-2}$

26. $(-2x^{-6}y)^3$

Study Tip

Before you take an in-class exam on this chapter, work the sample test given here. Set aside one hour to work this test and use the answers in the back of this book to grade yourself. Even though your instructor might not ask exactly the same questions, you will get a good idea of your test readiness.

Convert to scientific notation.

27. 5,433,000

28. 0.0000065

Perform each computation by converting to scientific notation. Give answers in scientific notation.

29. (80,000)(0.000006)

30. $(0.0000003)^4$

Solve each problem.

31. Find the quotient and remainder when $x^2 - 5x + 9$ is divided by $x - 3$.

32. Subtract $3x^2 - 4x - 9$ from $x^2 - 3x + 6$.

33. The width of a pool table is x feet, and the length is 4 feet longer than the width. Find polynomials $A(x)$ and $P(x)$ that represent the area and perimeter of the pool table. Find $A(4)$ and $P(4)$.

34. If a manufacturer charges q dollars each for footballs, then he can sell $3000 - 150q$ footballs per week. Find a polynomial that represents the revenue for one week. Find the weekly revenue if the price is \$8 for each football.

Making Connections | A Review of Chapters 1–5

Evaluate each arithmetic expression.

1. $-16 \div (-2)$

2. $-16 \div \left(-\frac{1}{2}\right)$

3. $(-5)^2 - 3(-5) + 1$

4. $-5^2 - 4(-5) + 3$

5. $2^{15} \div 2^{10}$

6. $2^6 - 2^5$

7. $-3^2 \cdot 4^2$

8. $(-3 \cdot 4)^2$

9. $\left(\frac{1}{2}\right)^3 + \frac{1}{2}$

10. $\left(\frac{2}{3}\right)^2 - \frac{1}{3}$

11. $(5 + 3)^2$

12. $5^2 + 3^2$

13. $3^{-1} + 2^{-1}$

14. $2^{-2} - 3^{-2}$

15. $(30 - 1)(30 + 1)$

16. $(30 - 1) \div (1 - 30)$

Perform the indicated operations.

17. $(x + 3)(x + 5)$

18. $x + 3(x + 5)$

19. $-5t^3v \cdot 3t^2v^6$

20. $(-10t^3v^2) \div (-2t^2v)$

21. $(x^2 + 8x + 15) + (x + 5)$

22. $(x^2 + 8x + 15) - (x + 5)$

23. $(x^2 + 8x + 15) \div (x + 5)$

24. $(x^2 + 8x + 15)(x + 5)$

25. $(-6y^3 + 8y^2) \div (-2y^2)$

26. $(18y^4 - 12y^3 + 3y^2) \div (3y^2)$

Solve each equation.

27. $2x + 1 = 0$

28. $x - 7 = 0$

29. $\frac{3}{4}x - 3 = \frac{1}{2}$

30. $\frac{x}{2} - \frac{3}{4} = \frac{1}{8}$

31. $2(x - 3) = 3(x - 2)$

32. $2(3x - 3) = 3(2x - 2)$

Solve.

33. Find the x-intercept for the line $y = 2x + 1$.

34. Find the y-intercept for the line $y = x - 7$.

35. Find the slope of the line $y = 2x + 1$.

36. Find the slope of the line that goes through $(0, 0)$ and $\left(\frac{1}{2}, \frac{1}{3}\right)$.

37. If $y = \frac{3}{4}x - 3$ and y is $\frac{1}{2}$, then what is x?

38. Find y if $y = \frac{x}{2} - \frac{3}{4}$ and x is $\frac{1}{2}$.

Solve the problem.

39. ***Average cost.*** Pineapple Recording plans to spend \$100,000 to record a new CD by the Woozies and \$2.25 per CD to manufacture the disks. The polynomial $2.25n + 100{,}000$ represents the total cost in dollars for recording and manufacturing n disks. Find an expression that represents the average cost per disk by dividing the total cost by n. Find the average cost per disk for $n = 1000$, 100,000, and 1,000,000. What happens to the large initial investment of \$100,000 if the company sells one million CDs?

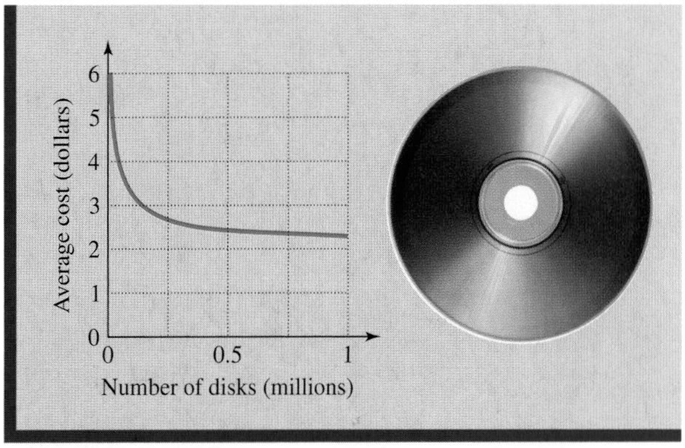

Figure for Exercise 39

Critical **Thinking** | For Individual or Group Work | Chapter 5

These exercises can be solved by a variety of techniques, which may or may not require algebra. So be creative and think critically. Explain all answers. Answers are in the Instructor's Edition of this text.

1. ***Counting cubes.*** What is the total number of cubes that are in each of the following diagrams?

a) 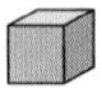

b) 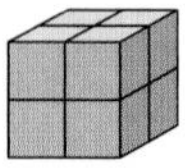

c) 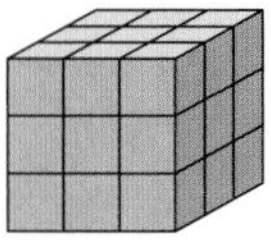

d) 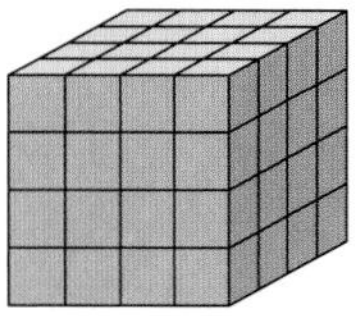

2. ***More cubes.*** Imagine a large cube that is made up of 125 small cubes like those in the previous exercise. What is the total number of cubes that could be found in this arrangement?

3. ***Timely coincidence.*** Starting at 8 A.M. determine the number of times in the next 24 hours for which the hour and minute hands on a clock coincide?

Photo for Exercise 3

4. ***Chess board.*** There are 64 squares on a square chess board. How many squares are neither diagonal squares nor edge squares?

Photo for Exercise 4

5. ***Last digit.*** Find the last digit in 3^{9999}.

6. ***Reconciling remainders.*** Find a positive integer smaller than 500 that has a remainder of 3 when divided by 5, a remainder of 6 when divided by 9, and a remainder of 8 when divided by 11.

7. ***Exact sum.*** Find this sum exactly:

$$\frac{1}{2} + \frac{1}{2^2} + \frac{1}{2^3} + \frac{1}{2^4} + \cdots + \frac{1}{2^{19}}$$

8. ***Ten-digit number.*** Find a 10-digit number whose first digit is the number of 1's in the 10-digit number, whose second digit is the number of 2's in the 10-digit number, whose third digit is the number of 3's in the 10-digit number, and so on. The ninth digit must be the number of nines in the 10-digit number and the tenth digit must be the number of zeros in the 10-digit number.

Chapter

6

Factoring

The sport of skydiving was born in the 1930s soon after the military began using parachutes as a means of deploying troops. Today, skydiving is a popular sport around the world.

With as little as 8 hours of ground instruction, first-time jumpers can be ready to make a solo jump. Without the assistance of oxygen, skydivers can jump from as high as 14,000 feet and reach speeds of more than 100 miles per hour as they fall toward the earth. Jumpers usually open their parachutes between 2000 and 3000 feet and then gradually glide down to their landing area. If the jump and the parachute are handled correctly, the landing can be as gentle as jumping off two steps.

Making a jump and floating to earth are only part of the sport of skydiving. For example, in an activity called "relative work skydiving," a team of as many as 920 free-falling skydivers join together to make geometrically shaped formations. In a related exercise called "canopy relative work," the team members form geometric patterns after their parachutes or canopies have opened. This kind of skydiving takes skill and practice, and teams are not always successful in their attempts.

The amount of time a skydiver has for a free fall depends on the height of the jump and how much the skydiver uses the air to slow the fall.

In Exercises 81 and 82 of Section 6.6 we find the amount of time that it takes a skydiver to fall from a given height.

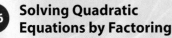

6.1 Factoring Out Common Factors

In this Section

- Prime Factorization of Integers
- Greatest Common Factor
- Finding the Greatest Common Factor for Monomials
- Factoring Out the Greatest Common Factor
- Factoring Out the Opposite of the GCF

In Chapter 5 you learned how to multiply a monomial and a polynomial. In this section you will learn how to reverse that multiplication by finding the greatest common factor for the terms of a polynomial and then factoring the polynomial.

Prime Factorization of Integers

To **factor** an expression means to write the expression as a product. For example, if we start with 12 and write $12 = 4 \cdot 3$, we have factored 12. Both 4 and 3 are **factors** or **divisors** of 12. There are other factorizations of 12:

$$12 = 2 \cdot 6 \qquad 12 = 1 \cdot 12 \qquad 12 = 2 \cdot 2 \cdot 3 = 2^2 \cdot 3$$

The one that is most useful to us is $12 = 2^2 \cdot 3$, because it expresses 12 as a product of *prime numbers.*

Prime Number

A positive integer larger than 1 that has no integral factors other than itself and 1 is called a **prime number.**

The numbers 2, 3, 5, 7, 11, 13, 17, 19, and 23 are the first nine prime numbers. A positive integer larger than 1 that is not a prime is a **composite number.** The numbers 4, 6, 8, 9, 10, and 12 are the first six composite numbers. Every composite number is a product of prime numbers. The **prime factorization** for 12 is $2^2 \cdot 3$.

EXAMPLE 1

Prime factorization

Find the prime factorization for 36.

Solution

We start by writing 36 as a product of two integers:

$$\begin{aligned} 36 &= 2 \cdot 18 && \text{Write 36 as } 2 \cdot 18. \\ &= 2 \cdot 2 \cdot 9 && \text{Replace 18 by } 2 \cdot 9. \\ &= 2 \cdot 2 \cdot 3 \cdot 3 && \text{Replace 9 by } 3 \cdot 3. \\ &= 2^2 \cdot 3^2 && \text{Use exponential notation.} \end{aligned}$$

The prime factorization for 36 is $2^2 \cdot 3^2$.

Now do Exercises 7–12

Helpful Hint

The prime factorization of 36 can be found also with a *factoring tree:*

```
      36
      /\
    2   18
        /\
      2   9
          /\
        3   3
```

So $36 = 2 \cdot 2 \cdot 3 \cdot 3$.

For larger integers, it is better to use the method shown in Example 2 and to recall some divisibility rules. Even numbers are divisible by 2. If the sum of the digits of a number is divisible by 3, then the number is divisible by 3. Numbers that end in 0 or 5 are divisible by 5. Two-digit numbers with repeated digits (11, 22, 33, . . .) are divisible by 11.

EXAMPLE 2

Factoring a large number

Find the prime factorization for 420.

Helpful Hint

The fact that every composite number has a unique prime factorization is known as the fundamental theorem of arithmetic.

Solution

Start by dividing 420 by the smallest prime number that will divide into it evenly (without remainder). The smallest prime divisor of 420 is 2.

$$2\overline{)420}\quad 210$$

Now find the smallest prime that will divide evenly into the quotient, 210. The smallest prime divisor of 210 is 2. Continue this procedure, as follows, until the quotient is a prime number:

	7	$35 \div 5 = 7$
	$5\overline{)35}$	$105 \div 3 = 35$
	$3\overline{)105}$	$210 \div 2 = 105$
	$2\overline{)210}$	
Start here $\rightarrow$	$2\overline{)420}$	

The prime factorization for 420 is $2 \cdot 2 \cdot 3 \cdot 5 \cdot 7$, or $2^2 \cdot 3 \cdot 5 \cdot 7$. Note that it is really not necessary to divide by the smallest prime divisor at each step. We obtain the same factorization if we divide by any prime divisor at each step.

Now do Exercises 13–16

Greatest Common Factor

The largest integer that is a factor of two or more integers is called the **greatest common factor (GCF)** of the integers. For example, 1, 2, 3, and 6 are common factors of 18 and 24. Because 6 is the largest, 6 is the GCF of 18 and 24. We can use prime factorizations to find the GCF. For example, to find the GCF of 8 and 12, we first factor 8 and 12:

$$8 = 2 \cdot 2 \cdot 2 = 2^3 \qquad 12 = 2 \cdot 2 \cdot 3 = 2^2 \cdot 3$$

We see that the factor 2 appears twice in both 8 and 12. So 2^2, or 4, is the GCF of 8 and 12. Notice that 2 is a factor in both 2^3 and $2^2 \cdot 3$ and that 2^2 is the smallest power of 2 in these factorizations. In general, we can use the following strategy to find the GCF.

Strategy for Finding the GCF for Positive Integers

1. Find the prime factorization for each integer.
2. The GCF is the product of the common prime factors using the smallest exponent that appears on each of them.

If two integers have no common prime factors, then their greatest common factor is 1, because 1 is a factor of every integer. For example, 6 and 35 have no common prime factors because $6 = 2 \cdot 3$ and $35 = 5 \cdot 7$. However, because $6 = 1 \cdot 6$ and $35 = 1 \cdot 35$, the GCF for 6 and 35 is 1.

EXAMPLE 3

Greatest common factor

Find the GCF for each group of numbers.

a) 150, 225 **b)** 216, 360, 504 **c)** 55, 168

Solution

a) First find the prime factorization for each number:

$$\begin{array}{r} 5 \\ 5\overline{)25} \\ 3\overline{)75} \\ 2\overline{)150} \end{array} \qquad \begin{array}{r} 5 \\ 5\overline{)25} \\ 3\overline{)75} \\ 3\overline{)225} \end{array}$$

$$150 = 2 \cdot 3 \cdot 5^2 \qquad 225 = 3^2 \cdot 5^2$$

Because 2 is not a factor of 225, it is not a common factor of 150 and 225. Only 3 and 5 appear in both factorizations. Looking at both $2 \cdot 3 \cdot 5^2$ and $3^2 \cdot 5^2$, we see that the smallest power of 5 is 2 and the smallest power of 3 is 1. So the GCF for 150 and 225 is $3 \cdot 5^2$, or 75.

b) First find the prime factorization for each number:

$$216 = 2^3 \cdot 3^3 \qquad 360 = 2^3 \cdot 3^2 \cdot 5 \qquad 504 = 2^3 \cdot 3^2 \cdot 7$$

The only common prime factors are 2 and 3. The smallest power of 2 in the factorizations is 3, and the smallest power of 3 is 2. So the GCF is $2^3 \cdot 3^2$, or 72.

c) First find the prime factorization for each number:

$$55 = 5 \cdot 11 \qquad 168 = 2^3 \cdot 3 \cdot 7$$

Because there are no common factors other than 1, the GCF is 1.

Now do Exercises 17–26

Finding the Greatest Common Factor for Monomials

To find the GCF for a group of monomials, we use the same procedure as that used for integers.

Strategy for Finding the GCF for Monomials

1. Find the GCF for the coefficients of the monomials.
2. Form the product of the GCF for the coefficients and each variable that is common to all of the monomials, where the exponent on each variable is the smallest power of that variable in any of the monomials.

Math *at Work* Kayak Design

Kayaks have been built by the Aleut and Inuit people for the past 4000 years. Today's builders have access to materials and techniques unavailable to the original kayak builders. Modern kayakers incorporate hydrodynamics and materials technology to create designs that are efficient and stable. Builders measure how well their designs work by calculating indicators such as prismatic coefficient, block coefficient, and the midship area coefficient, to name a few.

Even the fitting of a kayak to the paddler is done scientifically. For example, the formula

$$PL = 2 \cdot BL + BS\left(0.38 \cdot EE + 1.2\sqrt{\left(\frac{BW}{2} - \frac{SW}{2}\right)^2 + (SL)^2}\right)$$

can be used to calculate the appropriate paddle length. BL is the length of the paddle's blade. BS is a boating style factor, which is 1.2 for touring, 1.0 for river running, and 0.95 for play boating. EE is the elbow to elbow distance with the paddler's arms straight out to the sides. BW is the boat width and SW is the shoulder width. SL is the spine length, which is the distance measured in a sitting position from the chair seat to the top of the paddler's shoulder. All lengths are in centimeters.

The degree of control a kayaker exerts over the kayak depends largely on the body contact with it. A kayaker wears the kayak. So the choice of a kayak should hinge first on the right body fit and comfort and second on the skill level or intended paddling style. So designing, building, and even fitting a kayak is a blend of art and science.

EXAMPLE 4

Greatest common factor for monomials

Find the greatest common factor for each group of monomials.

a) $15x^2, 9x^3$ **b)** $12x^2y^2, 30x^2yz, 42x^3y$

Solution

a) The GCF for 15 and 9 is 3, and the smallest power of x is 2. So the GCF for the monomials is $3x^2$. If we write these monomials as

$$15x^2 = 5 \cdot 3 \cdot x \cdot x \quad \text{and} \quad 9x^3 = 3 \cdot 3 \cdot x \cdot x \cdot x,$$

we can see that $3x^2$ is the GCF.

b) The GCF for 12, 30, and 42 is 6. For the common variables x and y, 2 is the smallest power of x and 1 is the smallest power of y. So the GCF for the monomials is $6x^2y$.

Now do Exercises 27–38

Study Tip

Success in school depends on effective time management, and effective time management is all about goals. Write down your long-term, short-term, and daily goals. Assess them, develop methods for meeting them, and reward yourself when you do.

Factoring Out the Greatest Common Factor

In Chapter 5 we used the distributive property to multiply monomials and polynomials. For example,

$$6(5x - 3) = 30x - 18.$$

If we start with $30x - 18$ and write

$$30x - 18 = 6(5x - 3),$$

we have factored $30x - 18$. Because multiplication is the last operation to be performed in $6(5x - 3)$, the expression $6(5x - 3)$ is a product. Because 6 is the GCF for 30 and 18, we have **factored out** the GCF.

EXAMPLE 5

Factoring out the greatest common factor

Factor the following polynomials by factoring out the GCF.

a) $25a^2 + 40a$ **b)** $6x^4 - 12x^3 + 3x^2$ **c)** $x^2y^5 + x^6y^3$

Solution

a) The GCF for the coefficients 25 and 40 is 5. Because the smallest power of the common factor a is 1, we can factor $5a$ out of each term:

$$\begin{aligned} 25a^2 + 40a &= 5a \cdot 5a + 5a \cdot 8 \\ &= 5a(5a + 8) \end{aligned}$$

b) The GCF for 6, 12, and 3 is 3. We can factor x^2 out of each term, since the smallest power of x in the three terms is 2. So factor $3x^2$ out of each term as follows:

$$\begin{aligned} 6x^4 - 12x^3 + 3x^2 &= 3x^2 \cdot 2x^2 - 3x^2 \cdot 4x + 3x^2 \cdot 1 \\ &= 3x^2(2x^2 - 4x + 1) \end{aligned}$$

Check by multiplying: $3x^2(2x^2 - 4x + 1) = 6x^4 - 12x^3 + 3x^2$.

c) The GCF for the numerical coefficients is 1. Both x and y are common to each term. Using the lowest powers of x and y, we get

$$\begin{aligned} x^2y^5 + x^6y^3 &= x^2y^3 \cdot y^2 + x^2y^3 \cdot x^4 \\ &= x^2y^3(y^2 + x^4). \end{aligned}$$

Check by multiplying.

Now do Exercises 51–66

Study Tip

The keys to college success are motivation and time management. Anyone who tells you that they are making great grades without studying is probably not telling the truth. Success in college takes effort.

Because of the commutative property of multiplication, the common factor can be placed on either side of the other factor. So in Example 5, the answers could be written as $(5a + 8)5a$, $(2x^2 - 4x + 1)3x^2$, and $(y^2 + x^4)x^2y^3$.

CAUTION If the GCF is one of the terms of the polynomial, then you must remember to leave a 1 in place of that term when the GCF is factored out. For example,

$$ab + b = a \cdot b + 1 \cdot b = b(a + 1).$$

You should always check your answer by multiplying the factors.

In Example 6 the greatest common factor is a binomial. This type of factoring will be used in factoring trinomials in Section 6.2.

EXAMPLE 6

A binomial factor

Factor out the greatest common factor.

a) $(a + b)w + (a + b)6$ **b)** $x(x + 2) + 3(x + 2)$

c) $y(y - 3) - (y - 3)$

Solution

a) The greatest common factor is $a + b$:

$$(a + b)w + (a + b)6 = (a + b)(w + 6)$$

b) The greatest common factor is $x + 2$:

$$x(x + 2) + 3(x + 2) = (x + 3)(x + 2)$$

c) The greatest common factor is $y - 3$:

$$\begin{aligned} y(y - 3) - (y - 3) &= y(y - 3) - 1(y - 3) \\ &= (y - 1)(y - 3) \end{aligned}$$

Now do Exercises 67–74

Factoring Out the Opposite of the GCF

The greatest common factor for $-4x + 2xy$ is $2x$. Note that you can factor out the GCF $(2x)$ or the opposite of the GCF $(-2x)$:

$$-4x + 2xy = 2x(-2 + y) \qquad -4x + 2xy = -2x(2 - y)$$

It is useful to know both of these factorizations. For example, if we wanted to show that $2 - y$ is a factor of $-4x + 2xy$, then the second factorization shows it. Factoring out the opposite of the GCF will be used in factoring by grouping in Section 6.2 and in factoring trinomials with negative leading coefficients in Section 6.4. Remember to check all factoring by multiplying the factors to see if you get the original polynomial.

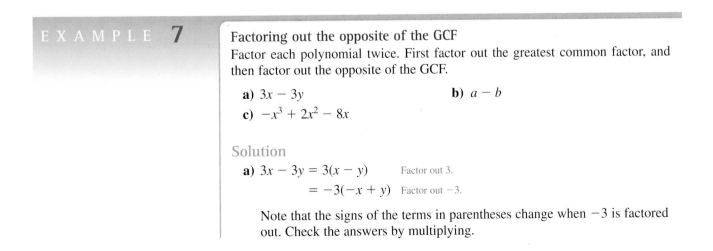

EXAMPLE 7

Factoring out the opposite of the GCF

Factor each polynomial twice. First factor out the greatest common factor, and then factor out the opposite of the GCF.

a) $3x - 3y$ **b)** $a - b$

c) $-x^3 + 2x^2 - 8x$

Solution

a) $3x - 3y = 3(x - y)$ Factor out 3.

$= -3(-x + y)$ Factor out -3.

Note that the signs of the terms in parentheses change when -3 is factored out. Check the answers by multiplying.

b) $a - b = 1(a - b)$ Factor out 1, the GCF of a and b.
$= -1(-a + b)$ Factor out -1.

We can also write $a - b = -1(b - a)$.

c) $-x^3 + 2x^2 - 8x = x(-x^2 + 2x - 8)$ Factor out x.
$= -x(x^2 - 2x + 8)$ Factor out $-x$.

Now do Exercises 75–90

CAUTION Be sure to change the sign of each term in parentheses when you factor out the opposite of the greatest common factor.

Warm-Ups ▼

True or false? Explain your answer.

1. There are only nine prime numbers.
2. The prime factorization of 32 is $2^3 \cdot 3$.
3. The integer 51 is a prime number.
4. The GCF for the integers 12 and 16 is 4.
5. The GCF for the integers 10 and 21 is 1.
6. The GCF for the polynomial $x^5y^3 - x^4y^7$ is x^4y^3.
7. For the polynomial $2x^2y - 6xy^2$ we can factor out either $2xy$ or $-2xy$.
8. The greatest common factor for the polynomial $8a^3b - 12a^2b$ is $4ab$.
9. $x - 7 = 7 - x$ for any real number x.
10. $-3x^2 + 6x = -3x(x - 2)$ for any real number x.

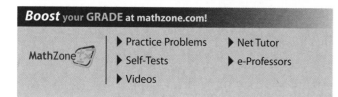

6.1 Exercises

Reading and Writing *After reading this section, write out the answers to these questions. Use complete sentences.*

1. What does it mean to factor an expression?
2. What is a prime number?
3. How do you find the prime factorization for a number?
4. What is the greatest common factor for two numbers?
5. What is the greatest common factor for two monomials?
6. How can you check if you have factored an expression correctly?

Find the prime factorization of each integer. See Examples 1 and 2.

7. 18
8. 20
9. 52
10. 76
11. 98
12. 100
13. 460
14. 345
15. 924
16. 585

Find the greatest common factor (GCF) for each group of integers. See Example 3.

17. 8, 20
18. 18, 42
19. 36, 60
20. 42, 70
21. 40, 48, 88
22. 15, 35, 45
23. 76, 84, 100
24. 66, 72, 120
25. 39, 68, 77
26. 81, 200, 539

Find the greatest common factor (GCF) for each group of monomials. See Example 4.

27. $6x$, $8x^3$
28. $12x^2$, $4x^3$
29. $12x^3$, $4x^2$, $6x$
30. $3y^5$, $9y^4$, $15y^3$
31. $3x^2y$, $2xy^2$
32. $7a^2x^3$, $5a^3x$
33. $24a^2bc$, $60ab^2$
34. $30x^2yz^3$, $75x^3yz^6$
35. $12u^3v^2$, $25s^2t^4$
36. $45m^2n^5$, $56a^4b^8$
37. $18a^3b$, $30a^2b^2$, $54ab^3$
38. $16x^2z$, $40xz^2$, $72z^3$

Complete the factoring of each monomial.

39. $27x = 9(\quad)$
40. $51y = 3y(\quad)$
41. $24t^2 = 8t(\quad)$
42. $18u^2 = 3u(\quad)$
43. $36y^5 = 4y^2(\quad)$
44. $42z^4 = 3z^2(\quad)$
45. $u^4v^3 = uv(\quad)$
46. $x^5y^3 = x^2y(\quad)$
47. $-14m^4n^3 = 2m^4(\quad)$
48. $-8y^3z^4 = 4z^3(\quad)$
49. $-33x^4y^3z^2 = -3x^3yz(\quad)$
50. $-96a^3b^4c^5 = -12ab^3c^3(\quad)$

Factor out the GCF in each expression. See Example 5.

51. $2w + 4t$
52. $6y + 3$
53. $12x - 18y$
54. $24a - 36b$
55. $x^3 - 6x$
56. $10y^4 - 30y^2$
57. $5ax + 5ay$
58. $6wz + 15wa$
59. $h^5 + h^3$
60. $y^6 + y^5$
61. $-2k^7m^4 + 4k^3m^6$
62. $-6h^5t^2 + 3h^3t^6$
63. $2x^3 - 6x^2 + 8x$
64. $6x^3 + 18x^2 + 24x$
65. $12x^4t + 30x^3t - 24x^2t^2$
66. $15x^2y^2 - 9xy^2 + 6x^2y$

Factor out the GCF in each expression. See Example 6.

67. $(x - 3)a + (x - 3)b$
68. $(y + 4)3 + (y + 4)z$
69. $x(x - 1) - 5(x - 1)$
70. $a(a + 1) - 3(a + 1)$
71. $m(m + 9) + (m + 9)$
72. $(x - 2)x - (x - 2)$
73. $a(y + 1)^2 + b(y + 1)^2$
74. $w(w + 2)^2 + 8(w + 2)^2$

First factor out the GCF, and then factor out the opposite of the GCF. See Example 7.

75. $8x - 8y$
76. $2a - 6b$
77. $-4x + 8x^2$
78. $-5x^2 + 10x$
79. $x - 5$
80. $a - 6$
81. $4 - 7a$
82. $7 - 5b$
83. $-24a^3 + 16a^2$
84. $-30b^4 + 75b^3$
85. $-12x^2 - 18x$
86. $-20b^2 - 8b$
87. $-2x^3 - 6x^2 + 14x$
88. $-8x^4 + 6x^3 - 2x^2$
89. $4a^3b - 6a^2b^2 - 4ab^3$
90. $12u^5v^6 + 18u^2v^3 - 15u^4v^5$

Solve each problem by factoring.

91. ***Uniform motion.*** Helen traveled a distance of $20x + 40$ miles at 20 miles per hour on the Yellowhead Highway. Find a binomial that represents the time that she traveled.

92. ***Area of a painting.*** A rectangular painting with a width of x centimeters has an area of $x^2 + 50x$ square centimeters. Find a binomial that represents the length. See the figure on the next page.

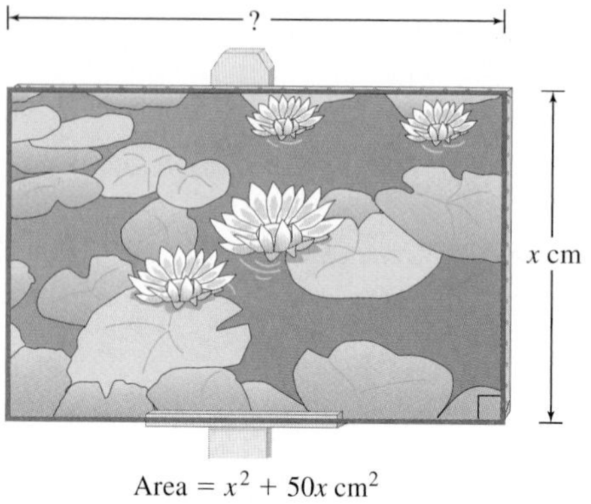

Figure for Exercise 92

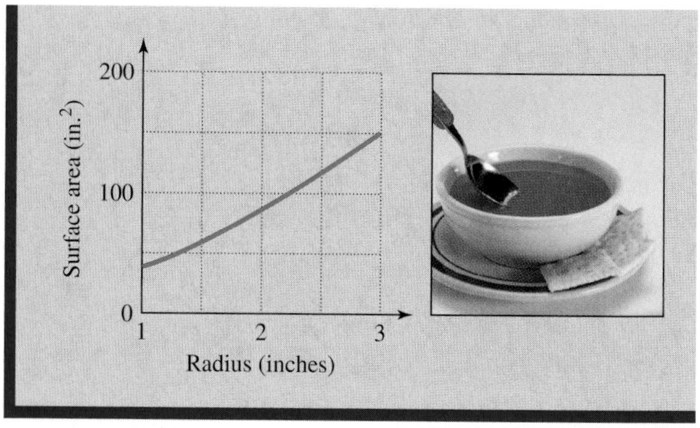

Figure for Exercise 93

93. ***Tomato soup.*** The amount of metal S (in square inches) that it takes to make a can for tomato soup depends on the radius r and height h:

$$S = 2\pi r^2 + 2\pi rh$$

a) Rewrite this formula by factoring out the greatest common factor on the right-hand side.

b) Let $h = 5$ in. and write a formula that expresses S in terms of r.

c) The accompanying graph shows S for r between 1 in. and 3 in. (with $h = 5$ in.). Which of these r-values gives the maximum surface area?

94. ***Amount of an investment.*** The amount of an investment of P dollars for t years at simple interest rate r is given by $A = P + Prt$.

a) Rewrite this formula by factoring out the greatest common factor on the right-hand side.

b) Find A if \$8300 is invested for 3 years at a simple interest rate of 15%.

Getting More Involved

95. ***Discussion***

Is the greatest common factor of $-6x^2 + 3x$ positive or negative? Explain.

96. ***Writing***

Explain in your own words why you use the smallest power of each common prime factor when finding the GCF of two or more integers.

6.2 Factoring the Special Products and Factoring by Grouping

In this Section

- Factoring a Difference of Two Squares
- Factoring a Perfect Square Trinomial
- Factoring Completely
- Factoring by Grouping

In Section 5.4 you learned how to find the special products: the square of a sum, the square of a difference, and the product of a sum and a difference. In this section you will learn how to reverse those operations.

Factoring a Difference of Two Squares

In Section 5.4 you learned that the product of a sum and a difference is a difference of two squares:

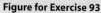

$$(a + b)(a - b) = a^2 - ab + ab - b^2 = a^2 - b^2$$

So a difference of two squares can be factored as a product of a sum and a difference, using the following rule.

Factoring a Difference of Two Squares

For any real numbers a and b,

$$a^2 - b^2 = (a + b)(a - b).$$

Note that the square of an integer is a perfect square. For example, 64 is a perfect square because $64 = 8^2$. The square of a monomial in which the coefficient is an integer is also called a **perfect square** or simply a **square.** For example, $9m^2$ is a perfect square because $9m^2 = (3m)^2$.

EXAMPLE 1

Factoring a difference of two squares

Factor each polynomial.

a) $y^2 - 81$ **b)** $9m^2 - 16$ **c)** $4x^2 - 9y^2$

Solution

a) Because $81 = 9^2$, the binomial $y^2 - 81$ is a difference of two squares:

$$\begin{aligned} y^2 - 81 &= y^2 - 9^2 && \text{Rewrite as a difference of two squares.} \\ &= (y + 9)(y - 9) && \text{Factor.} \end{aligned}$$

Check by multiplying.

b) Because $9m^2 = (3m)^2$ and $16 = 4^2$, the binomial $9m^2 - 16$ is a difference of two squares:

$$\begin{aligned} 9m^2 - 16 &= (3m)^2 - 4^2 && \text{Rewrite as a difference of two squares.} \\ &= (3m + 4)(3m - 4) && \text{Factor.} \end{aligned}$$

Check by multiplying.

c) Because $4x^2 = (2x)^2$ and $9y^2 = (3y)^2$, the binomial $4x^2 - 9y^2$ is a difference of two squares:

$$4x^2 - 9y^2 = (2x + 3y)(2x - 3y)$$

Now do Exercises 7–18

Factoring a Perfect Square Trinomial

In Section 5.4 you learned how to square a binomial using the rule

$$(a + b)^2 = a^2 + 2ab + b^2.$$

You can reverse this rule to factor a trinomial such as $x^2 + 6x + 9$. Notice that

$$x^2 + 6x + 9 = \underset{a^2}{x^2} + \underbrace{2 \cdot x \cdot 3}_{2ab} + \underset{b^2}{3^2}.$$

So if $a = x$ and $b = 3$, then $x^2 + 6x + 9$ fits the form $a^2 + 2ab + b^2$, and

$$x^2 + 6x + 9 = (x + 3)^2.$$

A trinomial that is of the form $a^2 + 2ab + b^2$ or $a^2 - 2ab + b^2$ is called a **perfect square trinomial.** A perfect square trinomial is the square of a binomial. Perfect square trinomials can be identified by using the following strategy.

Strategy for Identifying a Perfect Square Trinomial

A trinomial is a perfect square trinomial if

1. the first and last terms are of the form a^2 and b^2 (perfect squares).

2. the middle term is $2ab$ or $-2ab$.

EXAMPLE 2

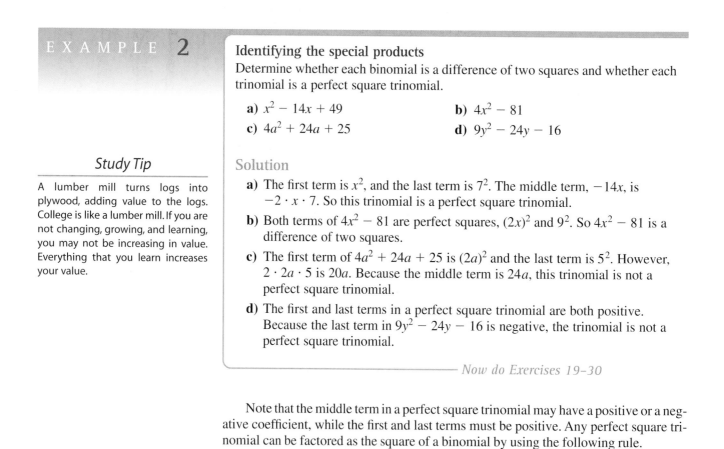

Identifying the special products

Determine whether each binomial is a difference of two squares and whether each trinomial is a perfect square trinomial.

a) $x^2 - 14x + 49$ **b)** $4x^2 - 81$

c) $4a^2 + 24a + 25$ **d)** $9y^2 - 24y - 16$

Study Tip

A lumber mill turns logs into plywood, adding value to the logs. College is like a lumber mill. If you are not changing, growing, and learning, you may not be increasing in value. Everything that you learn increases your value.

Solution

a) The first term is x^2, and the last term is 7^2. The middle term, $-14x$, is $-2 \cdot x \cdot 7$. So this trinomial is a perfect square trinomial.

b) Both terms of $4x^2 - 81$ are perfect squares, $(2x)^2$ and 9^2. So $4x^2 - 81$ is a difference of two squares.

c) The first term of $4a^2 + 24a + 25$ is $(2a)^2$ and the last term is 5^2. However, $2 \cdot 2a \cdot 5$ is $20a$. Because the middle term is $24a$, this trinomial is not a perfect square trinomial.

d) The first and last terms in a perfect square trinomial are both positive. Because the last term in $9y^2 - 24y - 16$ is negative, the trinomial is not a perfect square trinomial.

Now do Exercises 19–30

Note that the middle term in a perfect square trinomial may have a positive or a negative coefficient, while the first and last terms must be positive. Any perfect square trinomial can be factored as the square of a binomial by using the following rule.

Factoring Perfect Square Trinomials

For any real numbers a and b,

$$a^2 + 2ab + b^2 = (a + b)^2$$

$$a^2 - 2ab + b^2 = (a - b)^2.$$

EXAMPLE 3

Factoring perfect square trinomials

Factor.

a) $x^2 - 4x + 4$ **b)** $a^2 + 16a + 64$ **c)** $4x^2 - 12x + 9$

Solution

a) The first term is x^2, and the last term is 2^2. Because the middle term is $-2 \cdot 2 \cdot x$, or $-4x$, this polynomial is a perfect square trinomial:

$$x^2 - 4x + 4 = (x - 2)^2$$

Check by expanding $(x - 2)^2$.

b) $a^2 + 16a + 64 = (a + 8)^2$

Check by expanding $(a + 8)^2$.

c) The first term is $(2x)^2$, and the last term is 3^2. Because $-2 \cdot 2x \cdot 3 = -12x$, the polynomial is a perfect square trinomial. So

$$4x^2 - 12x + 9 = (2x - 3)^2.$$

Check by expanding $(2x - 3)^2$.

Now do Exercises 31–48

Factoring Completely

To factor a polynomial means to write it as a product of simpler polynomials. A polynomial that cannot be factored is called a **prime** or **irreducible polynomial.** The polynomials $3x$, $w + 1$, and $4m - 5$ are prime polynomials. A polynomial is **factored completely** when it is written as a product of prime polynomials. So $(y - 8)(y + 1)$ is a complete factorization. When factoring polynomials, we usually do not factor integers that occur as common factors. So $6x(x - 7)$ is considered to be factored completely even though 6 could be factored.

Some polynomials have a factor common to all terms. To factor such polynomials completely, it is simpler to factor out the greatest common factor (GCF) and then factor the remaining polynomial. Example 4 illustrates factoring completely.

EXAMPLE 4

Factoring completely

Factor each polynomial completely.

a) $2x^3 - 50x$ **b)** $8x^2y - 32xy + 32y$

Solution

a) The greatest common factor of $2x^3$ and $50x$ is $2x$:

$$2x^3 - 50x = 2x(x^2 - 25) \quad \text{Check this step by multiplying.}$$
$$= 2x(x + 5)(x - 5) \quad \text{Difference of two squares}$$

b)
$$8x^2y - 32xy + 32y = 8y(x^2 - 4x + 4) \quad \text{Check this step by multiplying.}$$
$$= 8y(x - 2)^2 \quad \text{Perfect square trinomial}$$

Now do Exercises 49–68

Study Tip

When you take notes, leave space. Go back later and fill in more details, make corrections, or work another problem of the same type.

Remember that factoring reverses multiplication and *every step of factoring can be checked by multiplication.*

Factoring by Grouping

The product of two binomials may be a polynomial with four terms. For example,

$$(x + a)(x + 3) = (x + a)x + (x + a)3$$
$$= x^2 + ax + 3x + 3a.$$

We can factor a polynomial of this type by simply reversing the steps we used to find the product. To reverse these steps, we factor out common factors from the first two terms and from the last two terms. This procedure is called **factoring by grouping.**

EXAMPLE 5

Factoring by grouping

Use grouping to factor each polynomial completely.

a) $xy + 2y + 3x + 6$ **b)** $2x^3 - 3x^2 - 2x + 3$ **c)** $ax + 3y - 3x - ay$

Solution

a) Notice that the first two terms have a common factor of y and the last two terms have a common factor of 3:

$$xy + 2y + 3x + 6 = (xy + 2y) + (3x + 6)$$ Use the associative property to group the terms.

$$= y(x + 2) + 3(x + 2)$$ Factor out the common factors in each group.

$$= (y + 3)(x + 2)$$ Factor out $x + 2$.

b) We can factor x^2 out of the first two terms and 1 out of the last two terms:

$$2x^3 - 3x^2 - 2x + 3 = (2x^3 - 3x^2) + (-2x + 3)$$ Group the terms.

$$= x^2(2x - 3) + 1(-2x + 3)$$

However, we cannot proceed any further because $2x - 3$ and $-2x + 3$ are not the same. To get $2x - 3$ as a common factor, we must factor out -1 from the last two terms:

$$2x^3 - 3x^2 - 2x + 3 = x^2(2x - 3) - 1(2x - 3)$$ Factor out the common factors.

$$= (x^2 - 1)(2x - 3)$$ Factor out $2x - 3$.

$$= (x - 1)(x + 1)(2x - 3)$$ Difference of two squares

c) In $ax + 3y - 3x - ay$ there are no common factors in the first two or the last two terms. However, if we use the commutative property to rewrite the polynomial as $ax - 3x - ay + 3y$, then we can factor by grouping:

$$ax + 3y - 3x - ay = ax - 3x - ay + 3y$$ Rearrange the terms.

$$= x(a - 3) - y(a - 3)$$ Factor out x and $-y$.

$$= (x - y)(a - 3)$$ Factor out $a - 3$.

Now do Exercises 69–84

Warm-Ups

True or false? Explain your answer.

1. The polynomial $x^2 + 16$ is a difference of two squares.
2. The polynomial $x^2 - 8x + 16$ is a perfect square trinomial.
3. The polynomial $9x^2 + 21x + 49$ is a perfect square trinomial.
4. $4x^2 + 4 = (2x + 2)^2$ for any real number x.
5. A difference of two squares is equal to a product of a sum and a difference.
6. The polynomial $16y + 1$ is a prime polynomial.
7. The polynomial $x^2 + 9$ can be factored as $(x + 3)(x + 3)$.
8. The polynomial $4x^2 - 4$ is factored completely as $4(x^2 - 1)$.
9. $y^2 - 2y + 1 = (y - 1)^2$ for any real number y.
10. $2x^2 - 18 = 2(x - 3)(x + 3)$ for any real number x.

6.2 Exercises

Reading and Writing *After reading this section, write out the answers to these questions. Use complete sentences.*

1. What is a perfect square?
2. How do we factor a difference of two squares?
3. How can you recognize if a trinomial is a perfect square?
4. What is a prime polynomial?
5. When is a polynomial factored completely?
6. What should you always look for first when attempting to factor a polynomial completely?

Factor each polynomial. See Example 1.

7. $a^2 - 4$
8. $h^2 - 9$
9. $x^2 - 49$
10. $y^2 - 36$
11. $y^2 - 9x^2$
12. $16x^2 - y^2$
13. $25a^2 - 49b^2$
14. $9a^2 - 64b^2$
15. $121m^2 - 1$
16. $144n^2 - 1$
17. $9w^2 - 25c^2$
18. $144w^2 - 121a^2$

Determine whether each polynomial is a difference of two squares, a perfect square trinomial, or neither of these. See Example 2.

19. $x^2 - 20x + 100$ - perfect square trinomial
20. $x^2 - 10x - 25$
21. $y^2 - 40$
22. $a^2 - 49$
23. $4y^2 + 12y + 9$
24. $9a^2 - 30a - 25$

25. $x^2 - 8x + 64$
26. $x^2 + 4x + 4$
27. $9y^2 - 25c^2$
28. $9x^2 + 4$
29. $9a^2 + 6ab + b^2$
30. $4x^2 - 4xy + y^2$

Factor each perfect square trinomial. See Example 3.

31. $x^2 + 2x + 1$
32. $y^2 + 4y + 4$
33. $a^2 + 6a + 9$
34. $w^2 + 10w + 25$
35. $x^2 + 12x + 36$
36. $y^2 + 14y + 49$
37. $a^2 - 4a + 4$
38. $b^2 - 6b + 9$
39. $4w^2 + 4w + 1$
40. $9m^2 + 6m + 1$
41. $16x^2 - 8x + 1$
42. $25y^2 - 10y + 1$
43. $4t^2 + 20t + 25$
44. $9y^2 - 12y + 4$
45. $9w^2 + 42w + 49$
46. $144x^2 + 24x + 1$
47. $n^2 + 2nt + t^2$
48. $x^2 - 2xy + y^2$

Factor each polynomial completely. See Example 4.

49. $5x^2 - 125$
50. $3y^2 - 27$
51. $-2x^2 + 18$
52. $-5y^2 + 20$
53. $a^3 - ab^2$
54. $x^2y - y$
55. $3x^2 + 6x + 3$
56. $12a^2 + 36a + 27$
57. $-5y^2 + 50y - 125$
58. $-2a^2 - 16a - 32$
59. $x^3 - 2x^2y + xy^2$
60. $x^3y + 2x^2y^2 + xy^3$
61. $-3x^2 + 3y^2$
62. $-8a^2 + 8b^2$
63. $2ax^2 - 98a$
64. $32x^2y - 2y^3$
65. $3ab^2 - 18ab + 27a$
66. $-2a^2b + 8ab - 8b$
67. $-4m^3 + 24m^2n - 36mn^2$
68. $10a^3 - 20a^2b + 10ab^2$

Use grouping to factor each polynomial completely. See Example 5.

69. $bx + by + cx + cy$
70. $3x + 3z + ax + az$
71. $x^3 + x^2 - 4x - 4$
72. $x^3 + x^2 - x - 1$
73. $3a - 3b - xa + xb$
74. $ax - bx - 4a + 4b$
75. $a^3 + 3a^2 + a + 3$
76. $y^3 - 5y^2 + 8y - 40$
77. $xa + ay + 3y + 3x$
78. $x^3 + ax + 3a + 3x^2$
79. $abc - 3 + c - 3ab$
80. $xa + tb + ba + tx$
81. $x^2a - b + bx^2 - a$
82. $a^2m - b^2n + a^2n - b^2m$
83. $y^2 + y + by + b$
84. $ac + mc + aw^2 + mw^2$

Factor each polynomial completely.

85. $6a^3y + 24a^2y^2 + 24ay^3$
86. $8b^5c - 8b^4c^2 + 2b^3c^3$
87. $24a^3y - 6ay^3$
88. $27b^3c - 12bc^3$
89. $2a^3y^2 - 6a^2y$
90. $9x^3y - 18x^2y^2$
91. $ab + 2bw - 4aw - 8w^2$
92. $3am - 6n - an + 18m$

Use factoring to solve each problem.

93. ***Skydiving.*** The height (in feet) above the earth for a skydiver t seconds after jumping from an airplane at 6400 ft is approximated by the formula $h = -16t^2 + 6400$, provided that $t < 5$. Rewrite the formula with the right-hand side factored completely. Use your revised formula to find h when $t = 2$.

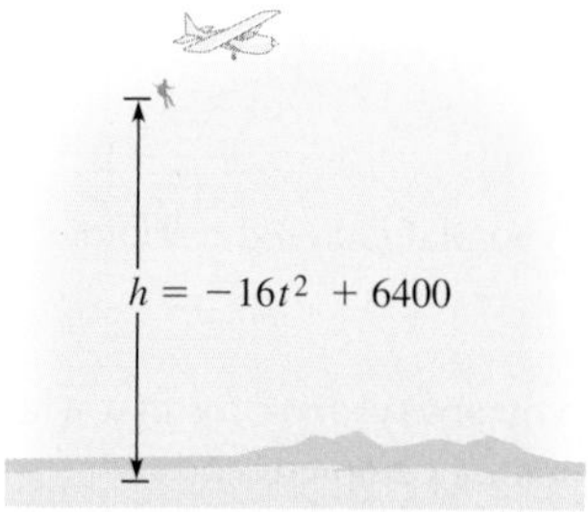

Figure for Exercise 93

94. ***Demand for pools.*** Tropical Pools sells an aboveground model for p dollars each. The monthly revenue from the sale of this model is given by

$$R = -0.08p^2 + 300p.$$

Revenue is the product of the price p and the demand (quantity sold).

a) Factor out the price on the right-hand side of the formula.

b) What is an expression for the monthly demand?

c) What is the monthly demand for this pool when the price is \$3000?

d) Use the graph to estimate the price at which the revenue is maximized. Approximately how many pools will be sold monthly at this price?

e) What is the approximate maximum revenue?

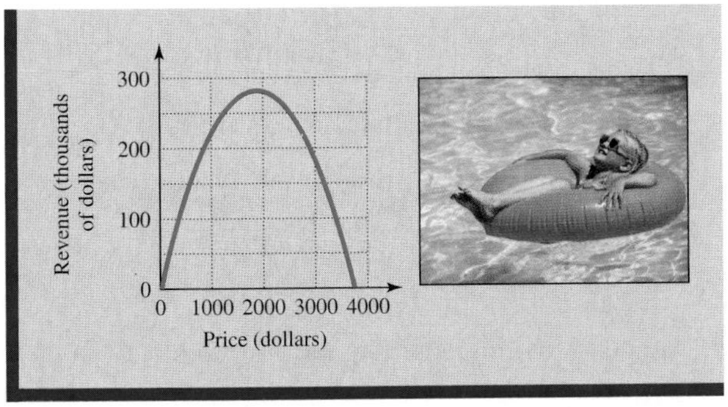

Figure for Exercise 94

f) Use the accompanying graph to estimate the price at which the revenue is zero.

95. ***Volume of a tank.*** The volume of a fish tank with a square base and height y is $y^3 - 6y^2 + 9y$ cubic inches. Find the length of a side of the square base.

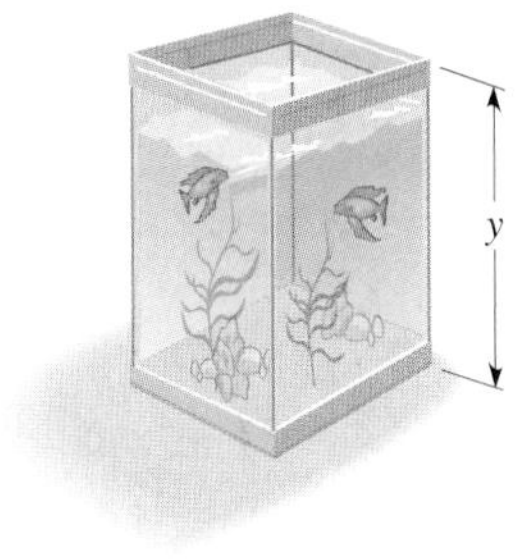

Figure for Exercise 95

Getting More Involved

96. ***Discussion***

For what real number k, does $3x^2 - k$ factor as $3(x - 2)(x + 2)$?

97. ***Writing***

Explain in your own words how to factor a four-term polynomial by grouping.

98. ***Writing***

Explain how you know that $x^2 + 1$ is a prime polynomial.

6.3 Factoring $ax^2 + bx + c$ with $a = 1$

In this Section

- Factoring $ax^2 + bx + c$ with $a = 1$
- Prime Polynomials
- Factoring with Two Variables
- Factoring Completely

In this section we will factor the type of trinomials that result from multiplying two different binomials. We will do this only for trinomials in which the coefficient of x^2, the leading coefficient, is 1. Factoring trinomials with leading coefficient not equal to 1 will be done in Section 6.4.

Factoring $ax^2 + bx + c$ with $a = 1$

Let's look closely at an example of finding the product of two binomials using the distributive property:

$$\begin{aligned}(x + 2)(x + 3) &= (x + 2)x + (x + 2)3 && \text{Distributive property}\\ &= x^2 + 2x + 3x + 6 && \text{Distributive property}\\ &= x^2 + 5x + 6 && \text{Combine like terms.}\end{aligned}$$

To factor $x^2 + 5x + 6$, we reverse these steps as shown in Example 1.

EXAMPLE 1

Factoring a trinomial

Factor.

a) $x^2 + 5x + 6$ **b)** $x^2 + 8x + 12$ **c)** $a^2 - 9a + 20$

Solution

a) The coefficient 5 is the sum of two numbers that have a product of 6. The only integers that have a product of 6 and a sum of 5 are 2 and 3. So write $5x$ as $2x + 3x$, then factor by grouping:

$$\begin{aligned} x^2 + 5x + 6 &= x^2 + 2x + 3x + 6 && \text{Replace } 5x \text{ by } 2x + 3x. \\ &= (x^2 + 2x) + (3x + 6) && \text{Group terms together.} \\ &= x(x + 2) + 3(x + 2) && \text{Factor out common factors.} \\ &= (x + 2)(x + 3) && \text{Factor out } x + 2. \end{aligned}$$

b) To factor $x^2 + 8x + 12$, we must find two integers that have a product of 12 and a sum of 8. The pairs of integers with a product of 12 are 1 and 12, 2 and 6, and 3 and 4. Only 2 and 6 have a sum of 8. So write $8x$ as $2x + 6x$ and factor by grouping:

$$\begin{aligned} x^2 + 8x + 12 &= x^2 + 2x + 6x + 12 \\ &= (x + 2)x + (x + 2)6 && \text{Factor out the common factors.} \\ &= (x + 2)(x + 6) && \text{Factor out } x + 2. \end{aligned}$$

Check by using FOIL: $(x + 2)(x + 6) = x^2 + 8x + 12$.

c) To factor $a^2 - 9a + 20$, we need two integers that have a product of 20 and a sum of -9. The integers are -4 and -5. Now replace $-9a$ by $-4a - 5a$ and factor by grouping:

$$\begin{aligned} a^2 - 9a + 20 &= a^2 - 4a - 5a + 20 && \text{Replace } -9a \text{ by } -4a - 5a. \\ &= a(a - 4) - 5(a - 4) && \text{Factor by grouping.} \\ &= (a - 5)(a - 4) && \text{Factor out } a - 4. \end{aligned}$$

Now do Exercises 7–18

Study Tip

Effective time management will allow adequate time for school, social life, and free time. However, at times you will have to sacrifice to do well.

After sufficient practice factoring trinomials, you may be able to skip most of the steps shown in these examples. For example, to factor $x^2 + x - 6$, simply find a pair of integers with a product of -6 and a sum of 1. The integers are 3 and -2, so we can write

$$x^2 + x - 6 = (x + 3)(x - 2)$$

and check by using FOIL.

EXAMPLE 2

Factoring trinomials

Factor.

a) $x^2 + 5x + 4$

b) $y^2 + 6y - 16$

c) $w^2 - 5w - 24$

Solution

a) To get a product of 4 and a sum of 5, use 1 and 4:

$$x^2 + 5x + 4 = (x + 1)(x + 4)$$

Check by using FOIL on $(x + 1)(x + 4)$.

b) To get a product of -16 we need a positive number and a negative number. To also get a sum of 6, use 8 and -2:

$$y^2 + 6y - 16 = (y + 8)(y - 2)$$

Check by using FOIL on $(y + 8)(y - 2)$.

c) To get a product of -24 and a sum of -5, use -8 and 3:

$$w^2 - 5w - 24 = (w - 8)(w + 3)$$

Check by using FOIL.

Now do Exercises 19–26

Polynomials are easiest to factor when they are in the form $ax^2 + bx + c$. So if a polynomial can be rewritten into that form, rewrite it before attempting to factor it. In Example 3 we factor polynomials that need to be rewritten.

EXAMPLE 3

Factoring trinomials

Factor.

a) $2x - 8 + x^2$ **b)** $-36 + t^2 - 9t$

Solution

a) Before factoring, write the trinomial as $x^2 + 2x - 8$. Now, to get a product of -8 and a sum of 2, use -2 and 4:

$$2x - 8 + x^2 = x^2 + 2x - 8 \quad \text{Write in } ax^2 + bx + c \text{ form.}$$
$$= (x + 4)(x - 2) \quad \text{Factor and check by multiplying.}$$

b) Before factoring, write the trinomial as $t^2 - 9t - 36$. Now, to get a product of -36 and a sum of -9, use -12 and 3:

$$-36 + t^2 - 9t = t^2 - 9t - 36 \quad \text{Write in } ax^2 + bx + c \text{ form.}$$
$$= (t - 12)(t + 3) \quad \text{Factor and check by multiplying.}$$

Now do Exercises 27–28

Study Tip

Stay alert for the entire class period. The first 20 minutes is the easiest and the last 20 minutes the hardest. Some students put down their pencils, fold up their notebooks, and daydream for those last 20 minutes. Don't give in. Recognize when you are losing it and force yourself to stay alert. Think of how much time you will have to spend outside of class figuring out what happened during those last 20 minutes.

Prime Polynomials

To factor $x^2 + bx + c$, we try pairs of integers that have a product of c until we find a pair that has a sum of b. If there is no such pair of integers, then the polynomial cannot be factored and it is a prime polynomial. Before you can conclude that a polynomial is prime, you must try *all* possibilities.

EXAMPLE 4

Prime polynomials

Factor.

a) $x^2 + 7x - 6$ **b)** $x^2 + 9$

Solution

a) Because the last term is -6, we want a positive integer and a negative integer that have a product of -6 and a sum of 7. Check all possible pairs of integers:

Product	**Sum**
$-6 = (-1)(6)$	$-1 + 6 = 5$
$-6 = (1)(-6)$	$1 + (-6) = -5$
$-6 = (2)(-3)$	$2 + (-3) = -1$
$-6 = (-2)(3)$	$-2 + 3 = 1$

None of these possible factors of -6 have a sum of 7, so we can be certain that $x^2 + 7x - 6$ cannot be factored. It is a prime polynomial.

b) Because the x-term is missing in $x^2 + 9$, its coefficient is 0. That is, $x^2 + 9 = x^2 + 0x + 9$. So we seek two positive integers or two negative integers that have a product of 9 and a sum of 0. Check all possibilities:

Product	**Sum**
$9 = (3)(3)$	$3 + 3 = 6$
$9 = (-3)(-3)$	$-3 + (-3) = -6$
$9 = (9)(1)$	$9 + 1 = 10$
$9 = (-9)(-1)$	$-9 + (-1) = -10$

None of these pairs of integers have a sum of 0, so we can conclude that $x^2 + 9$ is a prime polynomial. Note that $x^2 + 9$ does not factor as $(x + 3)^2$ because $(x + 3)^2$ has a middle term: $(x + 3)^2 = x^2 + 6x + 9$.

Now do Exercises 29–56

Helpful Hint

Don't confuse $a^2 + b^2$ with the difference of two squares $a^2 - b^2$ which is not a prime polynomial:

$$a^2 - b^2 = (a + b)(a - b).$$

The prime polynomial $x^2 + 9$ in Example 4(b) is a sum of two squares. It can be shown that any sum of two squares (in which there are no common factors) is a prime polynomial.

Sum of Two Squares

If a sum of two squares, $a^2 + b^2$, has no common factor other than 1, then it is a prime polynomial.

Factoring with Two Variables

In Example 5 we factor polynomials that have two variables using the same technique that we used for one variable.

EXAMPLE 5

Polynomials with two variables

Factor.

a) $x^2 + 2xy - 8y^2$ **b)** $a^2 - 7ab + 10b^2$

Solution

a) To get a product of -8 and a sum of 2, use 4 and -2. To get a product of $-8y^2$ use $4y$ and $-2y$:

$$x^2 + 2xy - 8y^2 = (x + 4y)(x - 2y)$$

Check by multiplying $(x + 4y)(x - 2y)$.

b) To get a product of 10 and a sum of -7, use -5 and -2. To get a product of $10b^2$, we use $-5b$ and $-2b$:

$$a^2 - 7ab + 10b^2 = (a - 5b)(a - 2b)$$

Check by multiplying.

Now do Exercises 57–64

Factoring Completely

In Section 6.2 you learned that binomials such as $3x - 5$ (with no common factor) are prime polynomials. In Example 4 of this section we saw a trinomial that is a prime polynomial. There are infinitely many prime trinomials. When factoring a polynomial completely, we could have a factor that is a prime trinomial.

EXAMPLE 6

Factoring completely

Factor each polynomial completely.

a) $x^3 - 6x^2 - 16x$

b) $4x^3 + 4x^2 + 4x$

Solution

a) $x^3 - 6x^2 - 16x = x(x^2 - 6x - 16)$ Factor out the GCF.

$= x(x - 8)(x + 2)$ Factor $x^2 - 6x - 16$.

b) First factor out $4x$, the greatest common factor:

$$4x^3 + 4x^2 + 4x = 4x(x^2 + x + 1)$$

To factor $x^2 + x + 1$, we would need two integers with a product of 1 and a sum of 1. Because there are no such integers, $x^2 + x + 1$ is prime, and the factorization is complete.

Now do Exercises 65–106

Warm-Ups ▼

True or false? Explain your answer.

1. $x^2 - 6x + 9 = (x - 3)^2$
2. $x^2 + 6x + 9 = (x + 3)^2$
3. $x^2 + 10x + 9 = (x - 9)(x - 1)$
4. $x^2 - 8x - 9 = (x - 8)(x - 9)$
5. $x^2 + 8x - 9 = (x + 9)(x - 1)$
6. $x^2 + 8x + 9 = (x + 3)^2$
7. $x^2 - 10xy + 9y^2 = (x - y)(x - 9y)$
8. $x^2 + x + 1 = (x + 1)(x + 1)$
9. $x^2 + xy + 20y^2 = (x + 5y)(x - 4y)$
10. $x^2 + 1 = (x + 1)(x + 1)$

6.3 Exercises

Reading and Writing *After reading this section, write out the answers to these questions. Use complete sentences.*

1. What types of polynomials did we factor in this section?
2. How can you check if you have factored a trinomial correctly?
3. How can you determine if $x^2 + bx + c$ is prime?
4. How do you factor a sum of two squares?
5. When is a polynomial factored completely?
6. What should you always look for first when attempting to factor a polynomial completely?

Factor each trinomial. Write out all of the steps as shown in Example 1.

7. $x^2 + 4x + 3$
8. $y^2 + 6y + 5$
9. $x^2 + 9x + 18$
10. $w^2 + 6w + 8$
11. $a^2 + 7a + 10$
12. $b^2 + 7b + 12$
13. $a^2 - 7a + 12$
14. $m^2 - 9m + 14$
15. $b^2 - 5b - 6$
16. $a^2 + 5a - 6$
17. $x^2 + 3x - 10$
18. $x^2 - x - 12$

Factor each polynomial. If the polynomial is prime, say so. See Examples 2–4.

19. $y^2 + 7y + 10$
20. $x^2 + 8x + 15$
21. $a^2 - 6a + 8$
22. $b^2 - 8b + 15$
23. $m^2 - 10m + 16$
24. $m^2 - 17m + 16$
25. $w^2 + 9w - 10$
26. $m^2 + 6m - 16$

27. $w^2 - 8 - 2w$
28. $-16 + m^2 - 6m$
29. $a^2 - 2a - 12$
30. $x^2 + 3x + 3$
31. $15m - 16 + m^2$
32. $3y + y^2 - 10$
33. $a^2 - 4a + 12$
34. $y^2 - 6y - 8$
35. $z^2 - 25$
36. $p^2 - 1$
37. $h^2 + 49$
38. $q^2 + 4$
39. $m^2 + 12m + 20$
40. $m^2 + 21m + 20$
41. $t^2 - 3t + 10$
42. $x^2 - 5x - 3$
43. $m^2 - 18 - 17m$
44. $h^2 - 36 + 5h$
45. $m^2 - 23m + 24$
46. $m^2 + 23m + 24$
47. $5t - 24 + t^2$
48. $t^2 - 24 - 10t$
49. $t^2 - 2t - 24$
50. $t^2 + 14t + 24$
51. $t^2 - 10t - 200$
52. $t^2 + 30t + 200$
53. $x^2 - 5x - 150$
54. $x^2 - 25x + 150$
55. $13y + 30 + y^2$
56. $18z + 45 + z^2$

Factor each polynomial. See Example 5.

57. $x^2 + 5ax + 6a^2$
58. $a^2 + 7ab + 10b^2$
59. $x^2 - 4xy - 12y^2$
60. $y^2 + yt - 12t^2$
61. $x^2 - 13xy + 12y^2$
62. $h^2 - 9hs + 9s^2$
63. $x^2 + 4xz - 33z^2$
64. $x^2 - 5xs - 24s^2$

Factor each polynomial completely. Use the methods discussed in Sections 6.1 through 6.3. If the polynomial is prime say so. See Example 6.

65. $5x^3 + 5x$
66. $b^3 + 49b$
67. $w^2 - 8w$
68. $x^4 - x^3$
69. $2w^2 - 162$
70. $6w^4 - 54w^2$
71. $-2b^2 - 98$
72. $-a^3 - 100a$
73. $x^3 - 2x^2 - 9x + 18$
74. $x^3 + 7x^2 - x - 7$
75. $4r^2 + 9$
76. $t^2 + 4z^2$
77. $x^2w^2 + 9x^2$
78. $a^4b + a^2b^3$
79. $w^2 - 18w + 81$
80. $w^2 + 30w + 81$
81. $6w^2 - 12w - 18$
82. $9w - w^3$
83. $3y^2 + 75$
84. $5x^2 + 500$
85. $ax + ay + cx + cy$
86. $y^3 + y^2 - 4y - 4$
87. $-2x^2 - 10x - 12$
88. $-a^3 - 2a^2 - a$
89. $32x^2 - 2x^4$
90. $20w^2 + 100w + 40$
91. $3w^2 + 27w + 54$
92. $w^3 - 3w^2 - 18w$
93. $18w^2 + w^3 + 36w$
94. $18a^2 + 3a^3 + 36a$
95. $9y^2 + 1 + 6y$
96. $2a^2 + 1 + 3a$
97. $8vw^2 + 32vw + 32v$
98. $3h^2t + 6ht + 3t$
99. $6x^3y + 30x^2y^2 + 36xy^3$
100. $3x^3y^2 - 3x^2y^2 + 3xy^2$
101. $5 + 8w + 3w^2$
102. $-3 + 2y + 21y^2$
103. $-3y^3 + 6y^2 - 3y$
104. $-4w^3 - 16w^2 + 20w$
105. $a^3 + ab + 3b + 3a^2$
106. $ac + xc + aw^2 + xw^2$

Use factoring to solve each problem.

107. ***Area of a deck.*** A rectangular deck has an area of $x^2 + 6x + 8$ square feet and a width of $x + 2$ feet. Find the length of the deck. See the figure on the next page.

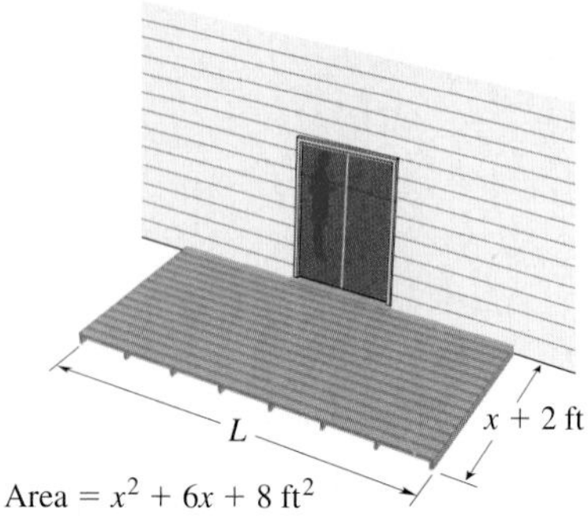

Figure for Exercise 107

108. ***Area of a sail.*** A triangular sail has an area of $x^2 + 5x + 6$ square meters and a height of $x + 3$ meters. Find the length of the sail's base.

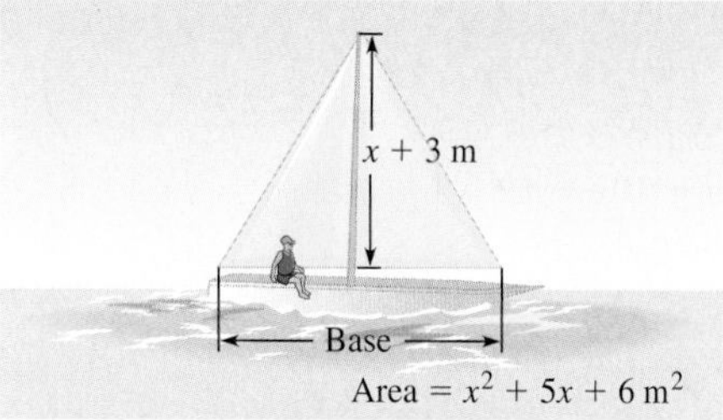

Figure for Exercise 108

109. ***Volume of a cube.*** Hector designed a cubic box with volume x^3 cubic feet. After increasing the dimensions of the bottom, the box has a volume of $x^3 + 8x^2 + 15x$ cubic feet. If each of the dimensions of the bottom was increased by a whole number of feet, then how much was each increase?

110. ***Volume of a container.*** A cubic shipping container had a volume of a^3 cubic meters. The height was decreased by a whole number of meters and the width was increased by a whole number of meters so that the volume of the container is now $a^3 + 2a^2 - 3a$ cubic meters. By how many meters were the height and width changed?

Getting More Involved

111. ***Discussion***

Which of the following products is not equivalent to the others. Explain your answer.

a) $(2x - 4)(x + 3)$ **b)** $(x - 2)(2x + 6)$
c) $2(x - 2)(x + 3)$ **d)** $(2x - 4)(2x + 6)$

112. ***Discussion***

When asked to factor completely a certain polynomial, four students gave the following answers. Only one student gave the correct answer. Which one must it be? Explain your answer.

a) $3(x^2 - 2x - 15)$ **b)** $(3x - 5)(5x - 15)$
c) $3(x - 5)(x - 3)$ **d)** $(3x - 15)(x - 3)$

6.4 Factoring $ax^2 + bx + c$ with $a \neq 1$

In this Section

- **The *ac* Method**
- **Trial and Error**
- **Factoring Completely**

In Section 6.3 we used grouping to factor trinomials with a leading coefficient of 1. In this section we will also use grouping to factor trinomials with a leading coefficient that is not equal to 1.

The *ac* Method

The first step in factoring $ax^2 + bx + c$ with $a = 1$ is to find two numbers with a product of c and a sum of b. If $a \neq 1$, then the first step is to find two numbers with a product of ac and a sum of b. This method is called the ***ac* method.** The strategy for factoring by the ac method follows. Note that this strategy works whether or not the leading coefficient is 1.

Strategy for Factoring $ax^2 + bx + c$ by the *ac* Method

To factor the trinomial $ax^2 + bx + c$:

1. Find two numbers that have a product equal to ac and a sum equal to b.
2. Replace bx by two terms using the two new numbers as coefficients.
3. Factor the resulting four-term polynomial by grouping.

EXAMPLE 1

The *ac* method

Factor each trinomial.

a) $2x^2 + 7x + 6$ **b)** $2x^2 + x - 6$ **c)** $10x^2 + 13x - 3$

Solution

a) In $2x^2 + 7x + 6$ we have $a = 2$, $b = 7$, and $c = 6$. So

$$ac = 2 \cdot 6 = 12.$$

Now we need two integers with a product of 12 and a sum of 7. The pairs of integers with a product of 12 are 1 and 12, 2 and 6, and 3 and 4. Only 3 and 4 have a sum of 7. Replace $7x$ by $3x + 4x$ and factor by grouping:

$$\begin{aligned} 2x^2 + 7x + 6 &= 2x^2 + 3x + 4x + 6 && \text{Replace } 7x \text{ by } 3x + 4x. \\ &= (2x + 3)x + (2x + 3)2 && \text{Factor out the common factors.} \\ &= (2x + 3)(x + 2) && \text{Factor out } 2x + 3. \end{aligned}$$

Check by FOIL.

b) In $2x^2 + x - 6$ we have $a = 2$, $b = 1$, and $c = -6$. So

$$ac = 2(-6) = -12.$$

Now we need two integers with a product of -12 and a sum of 1. We can list the possible pairs of integers with a product of -12 as follows:

1 and -12	2 and -6	3 and -4
-1 and 12	-2 and 6	-3 and 4

Only -3 and 4 have a sum of 1. Replace x by $-3x + 4x$ and factor by grouping:

$$\begin{aligned} 2x^2 + x - 6 &= 2x^2 - 3x + 4x - 6 && \text{Replace } x \text{ by } -3x + 4x. \\ &= (2x - 3)x + (2x - 3)2 && \text{Factor out the common factors.} \\ &= (2x - 3)(x + 2) && \text{Factor out } 2x - 3. \end{aligned}$$

Check by FOIL.

c) Because $ac = 10(-3) = -30$, we need two integers with a product of -30 and a sum of 13. The product is negative, so the integers must have opposite signs. We can list all pairs of factors of -30 as follows:

1 and -30	2 and -15	3 and -10	5 and -6
-1 and 30	-2 and 15	-3 and 10	-5 and 6

The only pair that has a sum of 13 is −2 and 15:

$$\begin{aligned}10x^2 + 13x - 3 &= 10x^2 - 2x + 15x - 3 && \text{Replace } 13x \text{ by } -2x + 15x.\\ &= (5x - 1)2x + (5x - 1)3 && \text{Factor out the common factors.}\\ &= (5x - 1)(2x + 3) && \text{Factor out } 5x - 1.\end{aligned}$$

Check by FOIL.

Now do Exercises 5–40

EXAMPLE 2

Factoring a trinomial in two variables by the *ac* method

Factor $8x^2 - 14xy + 3y^2$

Solution

Since $a = 8$, $b = -14$, and $c = 3$, we have $ac = 24$. Two numbers with a product of 24 and a sum of −14 must both be negative. The possible pairs with a product of 24 follow:

−1 and −24	−3 and −8
−2 and −12	−4 and −6

Only −2 and −12 have a sum of −14. Replace $-14xy$ by $-2xy - 12xy$ and factor by grouping:

$$\begin{aligned}8x^2 - 14xy + 3y^2 &= 8x^2 - 2xy - 12xy + 3y^2\\ &= (4x - y)2x + (4x - y)(-3y)\\ &= (4x - y)(2x - 3y)\end{aligned}$$

Check by FOIL.

Now do Exercises 41–46

Trial and Error

After you have gained some experience at factoring by the *ac* method, you can often find the factors without going through the steps of grouping. For example, consider the polynomial

$$3x^2 + 7x - 6.$$

The factors of $3x^2$ can only be $3x$ and x. The factors of 6 could be 2 and 3 or 1 and 6. We can list all of the possibilities that give the correct first and last terms, without regard to the signs:

$$(3x \quad 3)(x \quad 2) \qquad (3x \quad 2)(x \quad 3) \qquad (3x \quad 6)(x \quad 1) \qquad (3x \quad 1)(x \quad 6)$$

Because the factors of −6 have unlike signs, one binomial factor is a sum and the other binomial is a difference. Now we try some products to see if we get a middle term of $7x$:

$$\begin{aligned}(3x + 3)(x - 2) &= 3x^2 - 3x - 6 && \text{Incorrect}\\ (3x - 3)(x + 2) &= 3x^2 + 3x - 6 && \text{Incorrect}\end{aligned}$$

Helpful Hint

The *ac* method is more systematic than trial and error. However, trial and error can be faster and easier, especially if your first or second trial is correct.

Actually, there is no need to try $(3x \quad 3)(x \quad 2)$ or $(3x \quad 6)(x \quad 1)$ because each contains a binomial with a common factor. A common factor in the binomial causes a common factor in the product. But $3x^2 + 7x - 6$ has no common factor. So the factors must come from either $(3x \quad 2)(x \quad 3)$ or $(3x \quad 1)(x \quad 6)$. So we try again:

$$(3x + 2)(x - 3) = 3x^2 - 7x - 6 \quad \text{Incorrect}$$
$$(3x - 2)(x + 3) = 3x^2 + 7x - 6 \quad \text{Correct}$$

Even though there may be many possibilities in some factoring problems, it is often possible to find the correct factors without writing down every possibility. We can use a bit of guesswork in factoring trinomials. *Try* whichever possibility you think might work. *Check* it by multiplying. If it is not right, then *try again.* That is why this method is called **trial and error.**

EXAMPLE 3

Trial and error

Factor each trinomial using trial and error.

a) $2x^2 + 5x - 3$ **b)** $3x^2 - 11x + 6$

Solution

a) Because $2x^2$ factors only as $2x \cdot x$ and 3 factors only as $1 \cdot 3$, there are only two possible ways to get the correct first and last terms, without regard to the signs:

$$(2x \quad 1)(x \quad 3) \qquad \text{and} \qquad (2x \quad 3)(x \quad 1)$$

Because the last term of the trinomial is negative, one of the missing signs must be $+$, and the other must be $-$. The trinomial is factored correctly as

$$2x^2 + 5x - 3 = (2x - 1)(x + 3).$$

Check by using FOIL.

b) There are four possible ways to factor $3x^2 - 11x + 6$:

$$(3x \quad 1)(x \quad 6) \qquad (3x \quad 2)(x \quad 3)$$
$$(3x \quad 6)(x \quad 1) \qquad (3x \quad 3)(x \quad 2)$$

Because the last term in $3x^2 - 11x + 6$ is positive and the middle term is negative, both signs in the factors must be negative. Because $3x^2 - 11x + 6$ has no common factor, we can rule out $(3x \quad 6)(x \quad 1)$ and $(3x \quad 3)(x \quad 2)$. So the only possibilities left are $(3x - 1)(x - 6)$ and $(3x - 2)(x - 3)$. The trinomial is factored correctly as

$$3x^2 - 11x + 6 = (3x - 2)(x - 3).$$

Check by using FOIL.

Now do Exercises 47–66

Study Tip

Have you ever used excuses to avoid studying? ("Before I can study, I have to do my laundry and go to the bank.") Since the average attention span for one task is approximately 20 minutes, it is better to take breaks from studying to run errands and do laundry than to get everything done before you start studying.

Factoring by trial and error is not just guessing. In fact, if the trinomial has a positive leading coefficient, we can determine in advance whether its factors are sums or differences.

6.4 Exercises

Reading and Writing *After reading this section, write out the answers to these questions. Use complete sentences.*

1. What types of polynomials did we factor in this section?

2. What is the *ac* method of factoring?

3. How can you determine if $ax^2 + bx + c$ is prime?

4. What is the trial-and-error method of factoring?

Find the following. See Example 1.

5. Two integers that have a product of 20 and a sum of 12

6. Two integers that have a product of 36 and a sum of -20

7. Two integers that have a product of -12 and a sum of -4

8. Two integers that have a product of -8 and a sum of 7

Each of the following trinomials is in the form $ax^2 + bx + c$. For each trinomial, find two integers that have a product of ac and a sum of b. Do not factor the trinomials. See Example 1.

9. $6x^2 + 7x + 2$ **10.** $5x^2 + 17x + 6$

11. $6y^2 - 11y + 3$ **12.** $6z^2 - 19z + 10$

13. $12w^2 + w - 1$ **14.** $15t^2 - 17t - 4$

Factor each trinomial using the ac method. See Example 1.

15. $2x^2 + 3x + 1$ **16.** $2x^2 + 11x + 5$

17. $2x^2 + 9x + 4$ **18.** $2h^2 + 7h + 3$

19. $3t^2 + 7t + 2$ **20.** $3t^2 + 8t + 5$

21. $2x^2 + 5x - 3$ **22.** $3x^2 - x - 2$

23. $6x^2 + 7x - 3$ **24.** $21x^2 + 2x - 3$

25. $3x^2 - 5x + 4$ **26.** $6x^2 - 5x + 3$

27. $2x^2 - 7x + 6$ **28.** $3a^2 - 14a + 15$

29. $5b^2 - 13b + 6$ **30.** $7y^2 + 16y - 15$

31. $4y^2 - 11y - 3$ **32.** $35x^2 - 2x - 1$

33. $3x^2 + 2x + 1$ **34.** $6x^2 - 4x - 5$

35. $8x^2 - 2x - 1$ **36.** $8x^2 - 10x - 3$

37. $9t^2 - 9t + 2$ **38.** $9t^2 + 5t - 4$

39. $15x^2 + 13x + 2$ **40.** $15x^2 - 7x - 2$

Use the ac method to factor each trinomial. See Example 2.

41. $4a^2 + 16ab + 15b^2$ **42.** $10x^2 + 17xy + 3y^2$

43. $6m^2 - 7mn - 5n^2$ **44.** $3a^2 + 2ab - 21b^2$

45. $3x^2 - 8xy + 5y^2$ **46.** $3m^2 - 13mn + 12n^2$

Factor each trinomial using trial and error. See Examples 3 and 4.

47. $5a^2 + 6a + 1$ **48.** $7b^2 + 8b + 1$

49. $6x^2 + 5x + 1$ **50.** $15y^2 + 8y + 1$

51. $5a^2 + 11a + 2$ **52.** $3y^2 + 10y + 7$

53. $4w^2 + 8w + 3$ **54.** $6z^2 + 13z + 5$

55. $15x^2 - x - 2$ **56.** $15x^2 + 13x - 2$

57. $8x^2 - 6x + 1$ **58.** $8x^2 - 22x + 5$

59. $15x^2 - 31x + 2$ **60.** $15x^2 + 31x + 2$

61. $4x^2 - 4x + 3$ **62.** $4x^2 + 12x - 5$

63. $2x^2 + 18x - 90$ **64.** $3x^2 + 11x + 10$

65. $3x^2 + x - 10$ **66.** $3x^2 - 17x + 10$

67. $10x^2 - 3xy - y^2$

68. $8x^2 - 2xy - y^2$

69. $42a^2 - 13ab + b^2$

70. $10a^2 - 27ab + 5b^2$

Complete the factoring.

71. $3x^2 + 7x + 2 = (x + 2)(\quad)$

72. $2x^2 - x - 15 = (x - 3)(\quad)$

73. $5x^2 + 11x + 2 = (5x + 1)(\quad)$

74. $4x^2 - 19x - 5 = (4x + 1)(\quad)$

75. $6a^2 - 17a + 5 = (3a - 1)(\quad)$

76. $4b^2 - 16b + 15 = (2b - 5)(\quad)$

Factor each polynomial completely. See Examples 5 and 6.

77. $81w^3 - w$

78. $81w^3 - w^2$

79. $4w^2 + 2w - 30$

80. $2x^2 - 28x + 98$

81. $27 + 12x^2 + 36x$

82. $24y + 12y^2 + 12$

83. $6w^2 - 11w - 35$

84. $8y^2 - 14y - 15$

85. $3x^2z - 3zx - 18z$

86. $a^2b + 2ab - 15b$

87. $9x^3 - 21x^2 + 18x$

88. $-8x^3 + 4x^2 - 2x$

89. $a^2 + 2ab - 15b^2$

90. $a^2b^2 - 2a^2b - 15a^2$

91. $2x^2y^2 + xy^2 + 3y^2$

92. $18x^2 - 6x + 6$

93. $-6t^3 - t^2 + 2t$

94. $-36t^2 - 6t + 12$

95. $12t^4 - 2t^3 - 4t^2$

96. $12t^3 + 14t^2 + 4t$

97. $4x^2y - 8xy^2 + 3y^3$

98. $9x^2 + 24xy - 9y^2$

99. $-4w^2 + 7w - 3$

100. $-30w^2 + w + 1$

101. $-12a^3 + 22a^2b - 6ab^2$

102. $-36a^2b + 21ab^2 - 3b^3$

Solve each problem.

103. ***Height of a ball.*** If a ball is thrown upward at 40 feet per second from a rooftop 24 feet above the ground, then its height above the ground t seconds after it is thrown is given by $h = -16t^2 + 40t + 24$. Rewrite this formula with the polynomial on the right-hand side factored completely. Use the factored version of the formula to find h when $t = 3$.

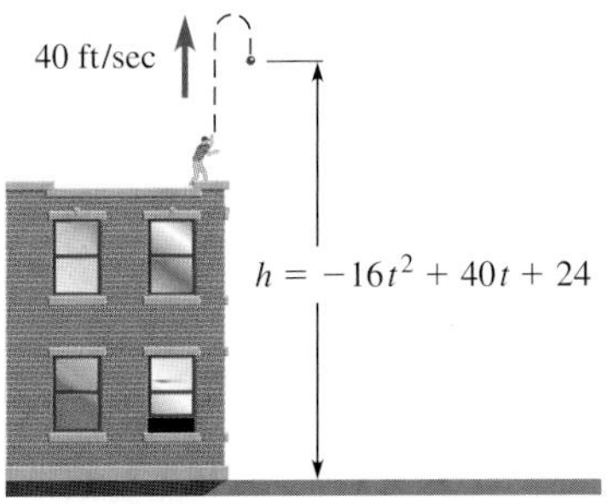

Figure for Exercise 103

104. ***Worker efficiency.*** In a study of worker efficiency at Wong Laboratories it was found that the number of components assembled per hour by the average worker t hours after starting work could be modeled by the formula

$$N(t) = -3t^3 + 23t^2 + 8t.$$

a) Rewrite the formula by factoring the right-hand side completely.

b) Use the factored version of the formula to find $N(3)$.

c) Use the accompanying graph to estimate the time at which the workers are most efficient.

d) Use the accompanying graph to estimate the maximum number of components assembled per hour during an 8-hour shift.

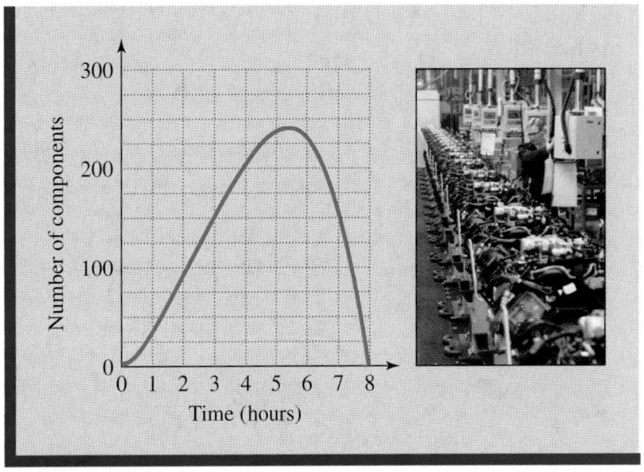

Figure for Exercise 104

Getting More Involved

105. ***Exploration***

Find all positive and negative integers b for which each polynomial can be factored.

a) $x^2 + bx + 3$

b) $3x^2 + bx + 5$

c) $2x^2 + bx - 15$

106. *Exploration*

Find two integers c (positive or negative) for which each polynomial can be factored. Many answers are possible.

a) $x^2 + x + c$

b) $x^2 - 2x + c$

c) $2x^2 - 3x + c$

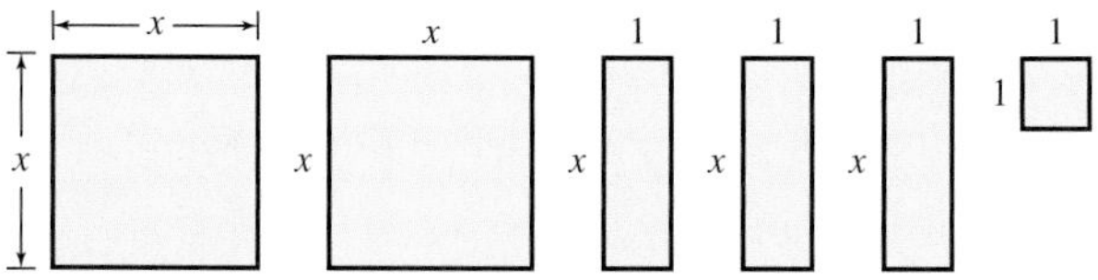

Figure for Exercise 107

107. *Cooperative learning*

Working in groups, cut two large squares, three rectangles, and one small square out of paper that are exactly the same size as shown in the accompanying figure. Then try to place the six figures next to one another so that they form a large rectangle. Do not overlap the pieces or leave any gaps. Explain how factoring $2x^2 + 3x + 1$ can help you solve this puzzle.

108. *Cooperative learning*

Working in groups, cut four squares and eight rectangles out of paper as in the previous exercise to illustrate the trinomial $4x^2 + 7x + 3$. Select one group to demonstrate how to arrange the 12 pieces to form a large rectangle. Have another group explain how factoring the trinomial can help you solve this puzzle.

6.5 Factoring a Difference or Sum of Two Cubes

In this Section

- **Using Division in Factoring**
- **Factoring a Difference or Sum of Two Cubes**
- **Factoring a Difference of Two Fourth Powers**
- **The Factoring Strategy**

In Sections 6.1 to 6.4 we established the general idea of factoring and some special cases. In this section we will see how division relates to factoring and see two more special cases. We will then summarize all of the factoring that we have done with a factoring strategy.

Using Division in Factoring

To find the prime factorization for a large integer such as 1001, you could divide possible factors (prime numbers) into 1001 until you find one that leaves no remainder. If you are told that 13 is a factor (or make a lucky guess), then you could divide 1001 by 13 to get the quotient 77. With this information you can factor 1001:

$$1001 = 77 \cdot 13$$

Now you can factor 77 to get the prime factorization of 1001:

$$1001 = 7 \cdot 11 \cdot 13$$

We can use this same idea with polynomials that are of higher degree than the ones we have been factoring. If we can guess a factor or if we are given a factor, we can use division to find the other factor and then proceed to factor the polynomial completely. Of course, it is harder to guess a factor of a polynomial than it is to guess a factor of an integer. In Example 1 we will factor a third-degree polynomial completely, given one factor.

EXAMPLE 1

Using division in factoring

Factor the polynomial $x^3 + 2x^2 - 5x - 6$ completely, given that the binomial $x + 1$ is a factor of the polynomial.

Solution

Divide the polynomial by the binomial:

$$\begin{array}{r l}
 & x^2 + x - 6 \\
x + 1 \overline{)} & x^3 + 2x^2 - 5x - 6 \\
 & \underline{x^3 + x^2} \\
 & x^2 - 5x \\
 & \underline{x^2 + x} \\
 & -6x - 6 \quad -5x - x = -6x \\
 & \underline{-6x - 6} \\
 & 0 \quad -6 - (-6) = 0
\end{array}$$

Because the remainder is 0, the dividend is the divisor times the quotient:

$$x^3 + 2x^2 - 5x - 6 = (x + 1)(x^2 + x - 6)$$

Now we factor the remaining trinomial to get the complete factorization:

$$x^3 + 2x^2 - 5x - 6 = (x + 1)(x + 3)(x - 2)$$

Now do Exercises 7–16

Study Tip

Everyone has a different attention span. Start by studying 10–15 minutes at a time and then build up to longer periods over time. In your senior year, you should be able to concentrate on one task for 30–45 minutes without a break. Be realistic. When you can't remember what you have read and can no longer concentrate, take a break.

Factoring a Difference or Sum of Two Cubes

We can use division to discover that $a - b$ is a factor of $a^3 - b^3$ (a difference of two cubes) and $a + b$ is a factor of $a^3 + b^3$ (a sum of two cubes):

$$\begin{array}{r l}
 & a^2 + ab + b^2 \\
a - b \overline{)} & a^3 + 0a^2b + 0ab^2 - b^3 \\
 & \underline{a^3 - a^2b} \\
 & a^2b + 0ab^2 \\
 & \underline{a^2b - ab^2} \\
 & ab^2 - b^3 \\
 & \underline{ab^2 - b^3} \\
 & 0
\end{array}
\qquad
\begin{array}{r l}
 & a^2 - ab + b^2 \\
a + b \overline{)} & a^3 + 0a^2b + 0ab^2 + b^3 \\
 & \underline{a^3 + a^2b} \\
 & -a^2b + 0ab^2 \\
 & \underline{-a^2b - ab^2} \\
 & ab^2 + b^3 \\
 & \underline{ab^2 + b^3} \\
 & 0
\end{array}$$

So $a - b$ is a factor of $a^3 - b^3$, and $a + b$ is a factor of $a^3 + b^3$. These results give us two more factoring rules.

Factoring a Difference or Sum of Two Cubes

$$a^3 - b^3 = (a - b)(a^2 + ab + b^2)$$
$$a^3 + b^3 = (a + b)(a^2 - ab + b^2)$$

Note that $a^2 + ab + b^2$ and $a^2 - ab + b^2$ are prime. Do not confuse them with $a^2 + 2ab + b^2$ and $a^2 - 2ab + b^2$, which are not prime because

$$a^2 + 2ab + b^2 = (a + b)^2 \qquad \text{and} \qquad a^2 - 2ab + b^2 = (a - b)^2.$$

These similarities can help you remember the rules for factoring $a^3 - b^3$ and $a^3 + b^3$. Note also how $a^3 - b^3$ compares with $a^2 - b^2$:

$$a^2 - b^2 = (a - b)(a + b)$$
$$a^3 - b^3 = (a - b)(a^2 + ab + b^2)$$

EXAMPLE 2

Factoring a difference or sum of two cubes

Factor each polynomial.

a) $w^3 - 8$ **b)** $x^3 + 1$ **c)** $8y^3 - 27$

Solution

a) Because $8 = 2^3$, $w^3 - 8$ is a difference of two cubes. To factor $w^3 - 8$, let $a = w$ and $b = 2$ in the formula $a^3 - b^3 = (a - b)(a^2 + ab + b^2)$:

$$w^3 - 8 = (w - 2)(w^2 + 2w + 4)$$

b) Because $1 = 1^3$, the binomial $x^3 + 1$ is a sum of two cubes. Let $a = x$ and $b = 1$ in the formula $a^3 + b^3 = (a + b)(a^2 - ab + b^2)$:

$$x^3 + 1 = (x + 1)(x^2 - x + 1)$$

c) $8y^3 - 27 = (2y)^3 - 3^3$ This is a difference of two cubes.

$= (2y - 3)(4y^2 + 6y + 9)$ Let $a = 2y$ and $b = 3$ in the formula.

Now do Exercises 17–32

In Example 2, we used the first three perfect cubes, 1, 8, and 27. You should verify that 1, 8, 27, 64, 125, 216, 343, 512, 729, and 1000 are the first 10 perfect cubes.

CAUTION The polynomial $(a - b)^3$ is not equivalent to $a^3 - b^3$ because if $a = 2$ and $b = 1$, then

$$(a - b)^3 = (2 - 1)^3 = 1^3 = 1$$

and

$$a^3 - b^3 = 2^3 - 1^3 = 8 - 1 = 7.$$

Likewise, $(a + b)^3$ is not equivalent to $a^3 + b^3$.

Factoring a Difference of Two Fourth Powers

A difference of two fourth powers of the form $a^4 - b^4$ is also a difference of two squares, $(a^2)^2 - (b^2)^2$. It can be factored by the rule for factoring a difference of two squares:

$a^4 - b^4 = (a^2)^2 - (b^2)^2$ Write as a difference of two squares.

$= (a^2 - b^2)(a^2 + b^2)$ Difference of two squares

$= (a - b)(a + b)(a^2 + b^2)$ Factor completely.

Note that the sum of two squares $a^2 + b^2$ is prime and cannot be factored.

EXAMPLE 3

Factoring a difference of two fourth powers

Factor each polynomial completely.

a) $x^4 - 16$ **b)** $81m^4 - n^4$

Solution

a) $x^4 - 16 = (x^2)^2 - 4^2$ Write as a difference of two squares.

$= (x^2 - 4)(x^2 + 4)$ Difference of two squares

$= (x - 2)(x + 2)(x^2 + 4)$ Factor completely.

b) $81m^4 - n^4 = (9m^2)^2 - (n^2)^2$ Write as a difference of two squares.

$= (9m^2 - n^2)(9m^2 + n^2)$ Factor.

$= (3m - n)(3m + n)(9m^2 + n^2)$ Factor completely.

Now do Exercises 33–40

The Factoring Strategy

The following is a summary of the ideas that we use to factor a polynomial completely.

Study Tip

Many schools have study skills centers that offer courses, workshops, and individual help on how to study. A search for "study skills" on the World Wide Web will turn up more information than you could possibly read. If you are not having the success in school that you would like, do something about it. What you do now will affect you the rest of your life.

Strategy for Factoring Polynomials Completely

1. If there are any common factors, factor them out first.
2. When factoring a binomial, check to see whether it is a difference of two squares, a difference of two cubes, or a sum of two cubes. *A sum of two squares does not factor.*
3. When factoring a trinomial, check to see whether it is a perfect square trinomial.
4. When factoring a trinomial that is not a perfect square, use the *ac* method or the trial-and-error method.
5. If the polynomial has four terms, try factoring by grouping.
6. Check to see whether any of the factors can be factored again.

We will use the factoring strategy in Example 3.

EXAMPLE 4

Factoring polynomials

Factor each polynomial completely.

a) $2a^2b - 24ab + 72b$

b) $3x^3 + 6x^2 - 75x - 150$

Solution

a) $2a^2b - 24ab + 72b = 2b(a^2 - 12a + 36)$ First factor out the GCF, $2b$.

$= 2b(a - 6)^2$ Factor the perfect square trinomial.

b) $3x^3 + 6x^2 - 75x - 150 = 3[x^3 + 2x^2 - 25x - 50]$ Factor out the GCF, 3.

$= 3[x^2(x + 2) - 25(x + 2)]$ Factor out common factors.

$= 3(x^2 - 25)(x + 2)$ Factor by grouping.

$= 3(x + 5)(x - 5)(x + 2)$ Factor the difference of two squares.

Now do Exercises 41–108

Warm-Ups ▼

True or false? Explain your answer.

1. $x^2 - 4 = (x - 2)^2$ for any real number x.
2. The trinomial $4x^2 + 6x + 9$ is a perfect square trinomial.
3. The polynomial $4y^2 + 25$ is a prime polynomial.
4. $3y + ay + 3x + ax = (x + y)(3 + a)$ for any values of the variables.
5. The polynomial $3x^2 + 51$ cannot be factored.
6. If the GCF is not 1, then you should factor it out first.
7. $x^2 + 9 = (x + 3)^2$ for any real number x.
8. The polynomial $x^2 - 3x - 5$ is a prime polynomial.
9. The polynomial $y^2 - 5y - my + 5m$ can be factored by grouping.
10. The polynomial $x^2 + ax - 3x + 3a$ can be factored by grouping.

6.5 Exercises

Reading and Writing *After reading this section, write out the answers to these questions. Use complete sentences.*

1. What is the relationship between division and factoring?

2. How do we know that $a - b$ is a factor of $a^3 - b^3$?

3. How do we know that $a + b$ is a factor of $a^3 + b^3$?

4. How do you recognize if a polynomial is a sum of two cubes?

5. How do you factor a sum of two cubes?

6. How do you factor a difference of two cubes?

Factor each polynomial completely, given that the binomial following it is a factor of the polynomial. See Example 1.

7. $x^3 + 3x^2 - 10x - 24,\ x + 4$

8. $x^3 - 7x + 6,\ x - 1$

9. $x^3 + 4x^2 + x - 6,\ x - 1$

10. $x^3 - 5x^2 - 2x + 24,\ x + 2$

11. $x^3 - 8,\ x - 2$

12. $x^3 + 27,\ x + 3$

13. $x^3 + 4x^2 - 3x + 10, x + 5$
14. $2x^3 - 5x^2 - x - 6, x - 3$
15. $x^3 + 2x^2 + 2x + 1, x + 1$
16. $x^3 + 2x^2 - 5x - 6, x + 3$

Factor each difference or sum of cubes. See Example 2.

17. $m^3 - 1$
18. $z^3 - 27$
19. $x^3 + 8$
20. $y^3 + 27$
21. $a^3 + 125$
22. $b^3 - 216$
23. $c^3 - 343$
24. $d^3 + 1000$
25. $8w^3 + 1$
26. $125m^3 + 1$
27. $8t^3 - 27$
28. $125n^3 - 8$
29. $x^3 - y^3$
30. $m^3 + n^3$
31. $8t^3 + y^3$
32. $u^3 - 125v^3$

Factor each polynomial completely. See Example 3.

33. $x^4 - y^4$
34. $m^4 - n^4$
35. $x^4 - 1$
36. $a^4 - 81$
37. $16b^4 - 1$
38. $625b^4 - 1$
39. $a^4 - 81b^4$
40. $16a^4 - m^4$

Factor each polynomial completely. If a polynomial is prime, say so. See Example 4.

41. $2x^2 - 18$
42. $3x^3 - 12x$
43. $a^2 + 4$
44. $x^2 + y^2$
45. $4x^2 + 8x - 60$
46. $3x^2 + 18x + 27$
47. $x^3 + 4x^2 + 4x$
48. $a^3 - 5a^2 + 6a$
49. $5max^2 + 20ma$
50. $3bmw^2 - 12bm$
51. $2x^2 - 3x - 1$
52. $3x^2 - 8x - 5$
53. $9x^2 + 6x + 1$
54. $9x^2 + 6x + 3$
55. $9m^2 + 1$
56. $4b^2 + 25$
57. $w^4 - z^4$
58. $y^4 - 1$
59. $6x^2y + xy - 2y$
60. $5x^2y^2 - xy^2 - 6y^2$
61. $y^2 + 10y - 25$
62. $x^2 - 20x + 25$
63. $48a^2 - 24a + 3$
64. $8b^2 + 24b + 18$
65. $16m^2 - 4m - 2$
66. $32a^2 + 4a - 6$
67. $s^4 - 16t^4$
68. $81 - q^4$
69. $9a^2 + 24a + 16$
70. $3x^2 - 18x - 48$
71. $24x^2 - 26x + 6$
72. $4x^2 - 6x - 12$
73. $3m^2 + 27$
74. $5a^2 + 20b^2$
75. $3a^2 - 27a$
76. $a^2 - 25a$
77. $8 - 2x^2$
78. $x^3 + 6x^2 + 9x$
79. $w^2 + 4t^2$
80. $9x^2 + 4y^2$
81. $6x^3 - 5x^2 + 12x$
82. $x^3 + 2x^2 - x - 2$
83. $a^3b - 4ab$
84. $2m^2 - 1800$
85. $x^3 + 2x^2 - 4x - 8$
86. $-2x^3 - 50x$
87. $-7m^3n - 28mn^3$
88. $x^3 - x^2 - x + 1$
89. $2x^3 + 16$
90. $m^2a + 2ma^2 + a^3$
91. $2w^4 - 16w$
92. $m^4n + mn^4$
93. $3a^2w - 18aw + 27w$
94. $8a^3 + 4a$
95. $5x^2 - 500$
96. $25x^2 - 16y^2$
97. $2m + 2n - wm - wn$
98. $aw - 5b - bw + 5a$
99. $3x^4 + 3x$
100. $3a^5 - 81a^2$
101. $4w^2 + 4w - 4$
102. $4w^2 + 8w - 5$
103. $a^4 + 7a^3 - 30a^2$
104. $2y^5 + 3y^4 - 20y^3$
105. $4aw^3 - 12aw^2 + 9aw$
106. $9bn^3 + 15bn^2 - 14bn$
107. $t^2 + 6t + 9$
108. $t^3 + 12t^2 + 36t$

Solve each problem.

109. ***Increasing cube.*** Each of the three dimensions of a cube with a volume of x^3 cubic centimeters is increased by a whole number of centimeters. If the new volume is $x^3 + 10x^2 + 31x + 30$ cubic centimeters and the new height is $x + 2$ centimeters, then what are the new length and width?

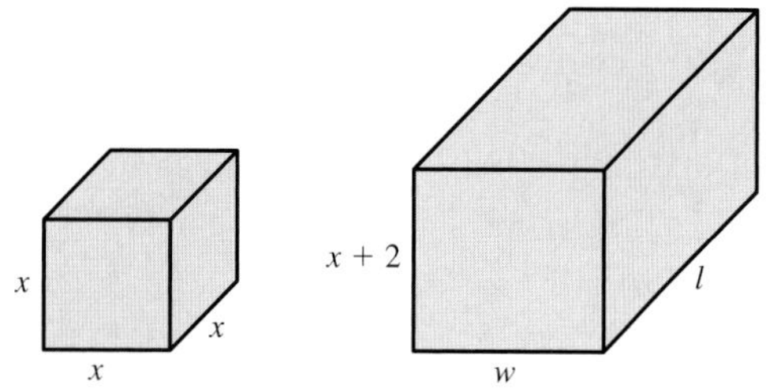

Figure for Exercise 109

110. ***Decreasing cube.*** Each of the three dimensions of a cube with a volume of y^3 cubic centimeters is decreased by a whole number of centimeters. If the new volume is $y^3 - 13y^2 + 54y - 72$ cubic centimeters and the new width is $y - 6$ centimeters, then what are the new length and height?

Getting More Involved

111. *Discussion*

Are there any values for a and b for which $(a + b)^3 = a^3 + b^3$? Find a pair of values for a and b for which $(a + b)^3 \neq a^3 + b^3$. Is $(a + b)^3$ equivalent to $a^3 + b^3$? Explain your answers.

112. *Writing*

Explain why $a^2 + ab + b^2$ and $a^2 - ab + b^2$ are prime polynomials.

6.6 Solving Quadratic Equations by Factoring

In this Section

- **The Zero Factor Property**
- **Applications**

The techniques of factoring can be used to solve equations involving polynomials. These equations cannot be solved by the other methods that you have learned. After you learn to solve equations by factoring, you will use this technique to solve some new types of problems.

The Zero Factor Property

In this chapter you learned to factor polynomials such as $x^2 + x - 6$. The equation $x^2 + x - 6 = 0$ is called a *quadratic equation.*

Quadratic Equation

If a, b, and c are real numbers with $a \neq 0$, then

$$ax^2 + bx + c = 0$$

is called a **quadratic equation.**

A quadratic equation always has a second-degree term because it is specified in the definition that a is not zero. The main idea used to solve quadratic equations, the **zero factor property,** is simply a fact about multiplication by zero.

The Zero Factor Property

The equation $a \cdot b = 0$ is equivalent to

$$a = 0 \quad \text{or} \quad b = 0.$$

So if a product is zero, then one or the other of the factors is zero. In Example 1 we combine the zero factor property with factoring to solve a quadratic equation.

EXAMPLE 1

Helpful Hint

Some students grow up believing that the only way to solve an equation is to "do the same thing to each side." Then along come quadratic equations and the zero factor property. For a quadratic equation, we write an equivalent compound equation that is not obtained by "doing the same thing to each side."

Using the zero factor property

Solve $x^2 + x - 6 = 0$.

Solution

First factor the polynomial on the left-hand side:

$$x^2 + x - 6 = 0$$

$$(x + 3)(x - 2) = 0 \quad \text{Factor the left-hand side.}$$

$$x + 3 = 0 \quad \text{or} \quad x - 2 = 0 \quad \text{Zero factor property}$$

$$x = -3 \quad \text{or} \quad x = 2 \quad \text{Solve each equation.}$$

We now check that -3 and 2 satisfy the original equation.

For $x = -3$:

$$x^2 + x - 6 = (-3)^2 + (-3) - 6 = 9 - 3 - 6 = 0$$

For $x = 2$:

$$x^2 + x - 6 = (2)^2 + (2) - 6 = 4 + 2 - 6 = 0$$

The solutions to $x^2 + x - 6 = 0$ are -3 and 2.

Now do Exercises 7–18

A sentence such as $x = -3$ or $x = 2$, which is made up of two or more equations connected with the word "or" is called a **compound equation.** In Example 2 we again solve a quadratic equation by using the zero factor property to write a compound equation.

EXAMPLE 2

Using the zero factor property

Solve the equation $3x^2 = -3x$.

Solution

First rewrite the equation with 0 on the right-hand side:

$$3x^2 = -3x$$

$$3x^2 + 3x = 0 \quad \text{Add } 3x \text{ to each side.}$$

$$3x(x + 1) = 0 \quad \text{Factor the left-hand side.}$$

$$3x = 0 \quad \text{or} \quad x + 1 = 0 \quad \text{Zero factor property}$$

$$x = 0 \quad \text{or} \quad x = -1 \quad \text{Solve each equation.}$$

Check 0 and -1 in the original equation $3x^2 = -3x$.

For $x = 0$:	For $x = -1$:
$3(0)^2 = -3(0)$	$3(-1)^2 = -3(-1)$
$0 = 0$	$3 = 3$

There are two solutions to the original equation, 0 and -1.

Now do Exercises 19–26

CAUTION If in Example 2 you divide each side of $3x^2 = -3x$ by $3x$, you would get $x = -1$ but not the solution $x = 0$. For this reason we usually do not divide each side of an equation by a variable.

The basic strategy for solving an equation by factoring follows.

Helpful Hint

We have seen quadratic polynomials that cannot be factored. So not all quadratic equations can be solved by factoring. Methods for solving all quadratic equations are presented in Chapter 9.

Strategy for Solving an Equation by Factoring

1. Rewrite the equation with 0 on one side.
2. Factor the other side completely.
3. Use the zero factor property to get simple linear equations.
4. Solve the linear equations.
5. Check the answer in the original equation.
6. State the solution(s) to the original equation.

EXAMPLE 3

Study Tip

Set short-term goals and reward yourself for accomplishing them. When you have solved 10 problems, take a short break and listen to your favorite music.

Using the zero factor property

Solve $(2x + 1)(x - 1) = 14$.

Solution

To write the equation with 0 on the right-hand side, multiply the binomials on the left and then subtract 14 from each side:

$$(2x + 1)(x - 1) = 14 \quad \text{Original equation}$$
$$2x^2 - x - 1 = 14 \quad \text{Multiply the binomials.}$$
$$2x^2 - x - 15 = 0 \quad \text{Subtract 14 from each side.}$$
$$(2x + 5)(x - 3) = 0 \quad \text{Factor.}$$
$$2x + 5 = 0 \quad \text{or} \quad x - 3 = 0 \quad \text{Zero factor property}$$
$$2x = -5 \quad \text{or} \quad x = 3$$
$$x = -\frac{5}{2} \quad \text{or} \quad x = 3$$

Check $-\frac{5}{2}$ and 3 in the original equation:

$$\left(2 \cdot -\frac{5}{2} + 1\right)\left(-\frac{5}{2} - 1\right) = (-5 + 1)\left(-\frac{5}{2} - \frac{2}{2}\right)$$
$$= (-4)\left(-\frac{7}{2}\right)$$
$$= 14$$
$$(2 \cdot 3 + 1)(3 - 1) = (7)(2)$$
$$= 14$$

So the solutions are $-\frac{5}{2}$ and 3.

Now do Exercises 27–32

CAUTION In Example 3 we started with a product of two factors equal to 14. Because there are many pairs of factors that have a product of 14, we *cannot make any conclusion about the factors.* If the product of two factors is 0, then we can conclude that one or the other factor is 0.

If a perfect square trinomial occurs in a quadratic equation with 0 on one side, then there are two identical factors of the trinomial. In this case it is not necessary to set both factors equal to zero. The solution can be found from one factor.

EXAMPLE 4

An equation with a repeated factor

Solve $5x^2 - 30x + 45 = 0$.

Solution

Notice that the trinomial on the left-hand side has a common factor:

$$5x^2 - 30x + 45 = 0$$
$$5(x^2 - 6x + 9) = 0 \quad \text{Factor out the GCF.}$$
$$5(x - 3)^2 = 0 \quad \text{Factor the perfect square trinomial.}$$
$$(x - 3)^2 = 0 \quad \text{Divide each side by 5.}$$
$$x - 3 = 0 \quad \text{Zero factor property}$$
$$x = 3$$

Even though $x - 3$ occurs twice as a factor, it is not necessary to write $x - 3 = 0$ or $x - 3 = 0$. If $x = 3$ in $5x^2 - 30x + 45 = 0$, we get

$$5 \cdot 3^2 - 30 \cdot 3 + 45 = 0,$$

which is correct. So the only solution to the equation is 3.

Now do Exercises 33–36

CAUTION To simplify $5(x-3)^2 = 0$ in Example 4, we divided each side by 5. If we had used the zero factor property, we would have gotten $5 = 0$ or $(x-3)^2 = 0$. Since $5 = 0$ has no solution, we can ignore it and continue to solve $(x-3)^2 = 0$.

If the left-hand side of the equation has more than two factors, we can write an equivalent equation by setting each factor equal to zero.

EXAMPLE 5

An equation with three solutions

Solve $2x^3 - x^2 - 8x + 4 = 0$.

Solution

We can factor the four-term polynomial by grouping:

$$2x^3 - x^2 - 8x + 4 = 0$$

$$x^2(2x-1) - 4(2x-1) = 0 \quad \text{Factor out the common factors.}$$

$$(x^2-4)(2x-1) = 0 \quad \text{Factor out } 2x-1.$$

$$(x-2)(x+2)(2x-1) = 0 \quad \text{Difference of two squares}$$

$$x-2=0 \quad \text{or} \quad x+2=0 \quad \text{or} \quad 2x-1=0 \quad \text{Zero factor property}$$

$$x = 2 \quad \text{or} \quad x = -2 \quad \text{or} \quad x = \frac{1}{2} \quad \text{Solve each equation.}$$

To check let $x = -2, \frac{1}{2}$, and 2 in $2x^3 - x^2 - 8x + 4 = 0$:

$$2(-2)^3 - (-2)^2 - 8(-2) + 4 = 0$$

$$2\left(\frac{1}{2}\right)^3 - \left(\frac{1}{2}\right)^2 - 8\left(\frac{1}{2}\right) + 4 = 0$$

$$2(2)^3 - 2^2 - 8(2) + 4 = 0$$

Since all of these equations are correct, the solutions are $-2, \frac{1}{2}$, and 2.

Now do Exercises 37–44

Helpful Hint

Compare the number of solutions in Examples 1 through 5 to the degree of the polynomial. The number of real solutions to any polynomial equation is less than or equal to the degree of the polynomial. This fact is known as the fundamental theorem of algebra.

Note that all of the equations in this section can be solved by factoring. However, we can write equations involving prime polynomials. Such equations cannot be solved by factoring but can be solved by the methods in Chapter 9.

Applications

There are many problems that can be solved by equations like those we have just discussed.

EXAMPLE 6

Area of a garden

Merida's garden has a rectangular shape with a length that is 1 foot longer than twice the width. If the area of the garden is 55 square feet, then what are the dimensions of the garden?

Solution

If x represents the width of the garden, then $2x + 1$ represents the length. See Fig. 6.1. Because the area of a rectangle is the length times the width, we can write the equation

$$x(2x + 1) = 55.$$

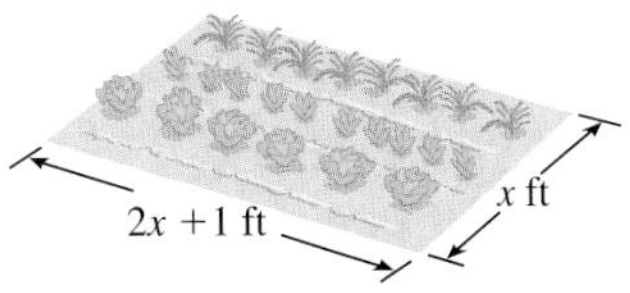

Figure 6.1

We must have zero on the right-hand side of the equation to use the zero factor property. So we rewrite the equation and then factor:

$$2x^2 + x - 55 = 0$$

$$(2x + 11)(x - 5) = 0 \quad \text{Factor.}$$

$$2x + 11 = 0 \quad \text{or} \quad x - 5 = 0 \quad \text{Zero factor property}$$

$$x = -\frac{11}{2} \quad \text{or} \quad x = 5$$

The width is certainly not $-\frac{11}{2}$. So we use $x = 5$ to get the length:

$$2x + 1 = 2(5) + 1 = 11$$

We check by multiplying 11 feet and 5 feet to get the area of 55 square feet. So the width is 5 ft, and the length is 11 ft.

Now do Exercises 63–64

Helpful Hint

To prove the Pythagorean theorem, draw two squares with sides of length $a + b$, and partition them as shown.

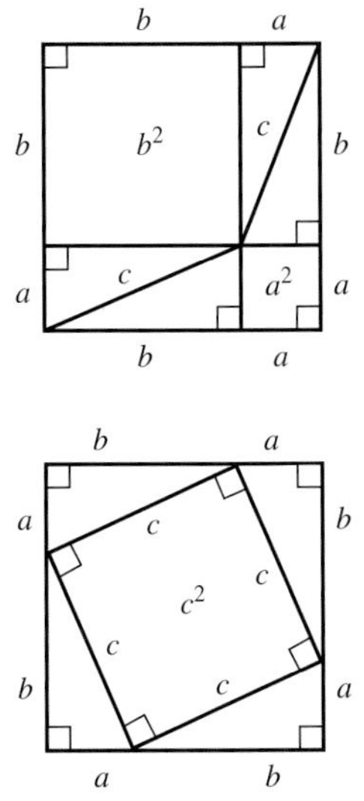

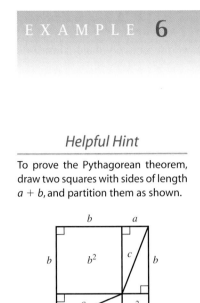

Erasing the four identical triangles from each picture will subtract the same amount of area from each original square. Since we started with equal areas, we will have equal areas after erasing the triangles:

$$a^2 + b^2 = c^2$$

The next application involves a theorem from geometry called the **Pythagorean theorem.** This theorem says that in any right triangle the sum of the squares of the lengths of the legs is equal to the square of the length of the hypotenuse.

The Pythagorean Theorem

The triangle shown in Fig. 6.2 is a right triangle if and only if

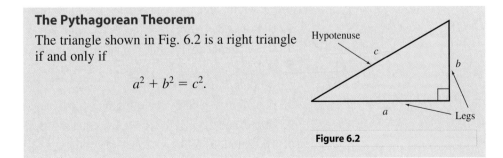

$$a^2 + b^2 = c^2.$$

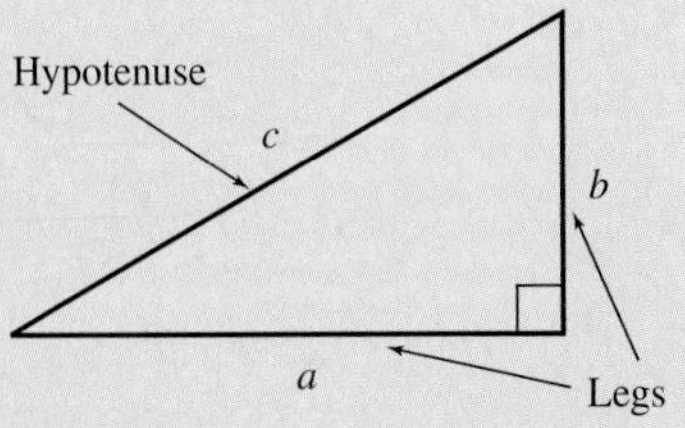

Figure 6.2

EXAMPLE 7

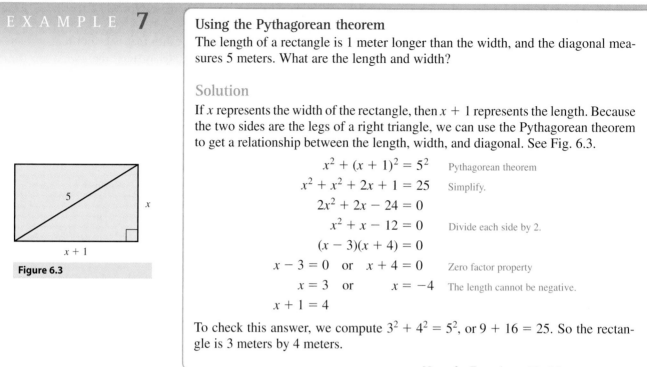

Using the Pythagorean theorem

The length of a rectangle is 1 meter longer than the width, and the diagonal measures 5 meters. What are the length and width?

Solution

If x represents the width of the rectangle, then $x + 1$ represents the length. Because the two sides are the legs of a right triangle, we can use the Pythagorean theorem to get a relationship between the length, width, and diagonal. See Fig. 6.3.

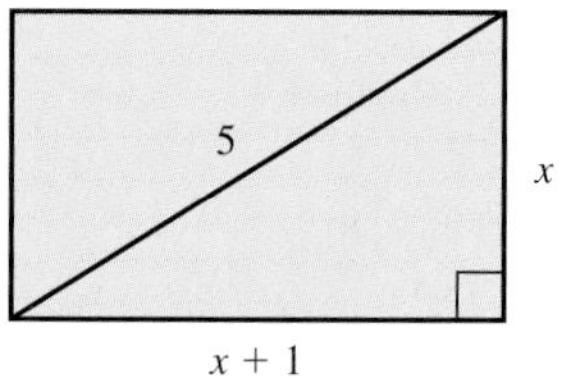

Figure 6.3

$$x^2 + (x + 1)^2 = 5^2 \quad \text{Pythagorean theorem}$$
$$x^2 + x^2 + 2x + 1 = 25 \quad \text{Simplify.}$$
$$2x^2 + 2x - 24 = 0$$
$$x^2 + x - 12 = 0 \quad \text{Divide each side by 2.}$$
$$(x - 3)(x + 4) = 0$$
$$x - 3 = 0 \quad \text{or} \quad x + 4 = 0 \quad \text{Zero factor property}$$
$$x = 3 \quad \text{or} \quad x = -4 \quad \text{The length cannot be negative.}$$
$$x + 1 = 4$$

To check this answer, we compute $3^2 + 4^2 = 5^2$, or $9 + 16 = 25$. So the rectangle is 3 meters by 4 meters.

Now do Exercises 65–66

CAUTION The hypotenuse is the longest side of a right triangle. So if the lengths of the sides of a right triangle are 5 meters, 12 meters, and 13 meters, then the length of the hypotenuse is 13 meters, and $5^2 + 12^2 = 13^2$.

Warm-Ups ▼

True or false? Explain your answer.

1. The equation $x(x + 2) = 3$ is equivalent to $x = 3$ or $x + 2 = 3$.
2. Equations solved by factoring always have two different solutions.
3. The equation $a \cdot d = 0$ is equivalent to $a = 0$ or $d = 0$.
4. If x is the width in feet of a rectangular room and the length is 5 feet longer than the width, then the area is $x^2 + 5x$ square feet.
5. Both 1 and -4 are solutions to the equation $(x - 1)(x + 4) = 0$.
6. If a, b, and c are the sides of any triangle, then $a^2 + b^2 = c^2$.
7. If the perimeter of a rectangular room is 50 feet, then the sum of the length and width is 25 feet.
8. Equations solved by factoring may have more than two solutions.
9. Both 0 and 2 are solutions to the equation $x(x - 2) = 0$.
10. The solutions to $3(x - 2)(x + 5) = 0$ are 3, 2, and -5.

6.6 Exercises

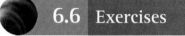

Reading and Writing *After reading this section, write out the answers to these questions. Use complete sentences.*

1. What is a quadratic equation?

2. What is a compound equation?

3. What is the zero factor property?

4. What method is used to solve quadratic equations in this section?

5. Why don't we usually divide each side of an equation by a variable?

6. What is the Pythagorean theorem?

Solve each equation. See Example 1.

7. $(x + 5)(x + 4) = 0$

8. $(a + 6)(a + 5) = 0$

9. $(2x + 5)(3x - 4) = 0$

10. $(3k - 8)(4k + 3) = 0$

11. $x^2 + 3x + 2 = 0$

12. $x^2 + 7x + 12 = 0$

13. $w^2 - 9w + 14 = 0$

14. $t^2 + 6t - 27 = 0$

15. $y^2 - 2y - 24 = 0$

16. $q^2 + 3q - 18 = 0$

17. $2m^2 + m - 1 = 0$

18. $2h^2 - h - 3 = 0$

Solve each equation. See Examples 2 and 3.

19. $x^2 = x$

20. $w^2 = 2w$

21. $m^2 = -7m$

22. $h^2 = -5h$

23. $a^2 + a = 20$

24. $p^2 + p = 42$

25. $2x^2 + 5x = 3$

26. $3x^2 - 10x = -7$

27. $(x + 2)(x + 6) = 12$

28. $(x + 2)(x - 6) = 20$

29. $(a + 3)(2a - 1) = 15$

30. $(b - 3)(3b + 4) = 10$

31. $2(4 - 5h) = 3h^2$

32. $2w(4w + 1) = 1$

Solve each equation. See Examples 4 and 5.

33. $2x^2 + 50 = 20x$

34. $3x^2 + 48 = 24x$

35. $4m^2 - 12m + 9 = 0$

36. $25y^2 + 20y + 4 = 0$

37. $x^3 - 9x = 0$

38. $25x - x^3 = 0$

39. $w^3 + 4w^2 - 4w = 16$

40. $a^3 + 2a^2 - a = 2$

41. $n^3 - 3n^2 + 3 = n$

42. $w^3 + w^2 - 25w = 25$

43. $y^3 - 9y^2 + 20y = 0$

44. $m^3 + 2m^2 - 3m = 0$

Solve each equation.

45. $x^2 - 16 = 0$

46. $x^2 - 36 = 0$

47. $x^2 = 9$

48. $x^2 = 25$

49. $a^3 = a$

50. $x^3 = 4x$

51. $3x^2 + 15x + 18 = 0$

52. $-2x^2 - 2x + 24 = 0$

53. $z^2 + \frac{11}{2}z = -6$

54. $m^2 + \frac{8}{3}m = 1$

55. $(t - 3)(t + 5) = 9$

56. $3x(2x + 1) = 18$

57. $(x - 2)^2 + x^2 = 10$

58. $(x - 3)^2 + (x + 2)^2 = 17$

59. $\frac{1}{16}x^2 + \frac{1}{8}x = \frac{1}{2}$

60. $\frac{1}{18}h^2 - \frac{1}{2}h + 1 = 0$

61. $a^3 + 3a^2 - 25a = 75$

62. $m^4 + m^3 = 100m^2 + 100m$

Solve each problem. See Examples 6 and 7.

63. ***Dimensions of a rectangle.*** The perimeter of a rectangle is 34 feet, and the diagonal is 13 feet long. What are the length and width of the rectangle?

64. ***Address book.*** The perimeter of the cover of an address book is 14 inches, and the diagonal measures 5 inches. What are the length and width of the cover?

Figure for Exercise 64

65. ***Violla's bathroom.*** The length of Violla's bathroom is 2 feet longer than twice the width. If the diagonal measures 13 feet, then what are the length and width?

66. ***Rectangular stage.*** One side of a rectangular stage is 2 meters longer than the other. If the diagonal is 10 meters, then what are the lengths of the sides?

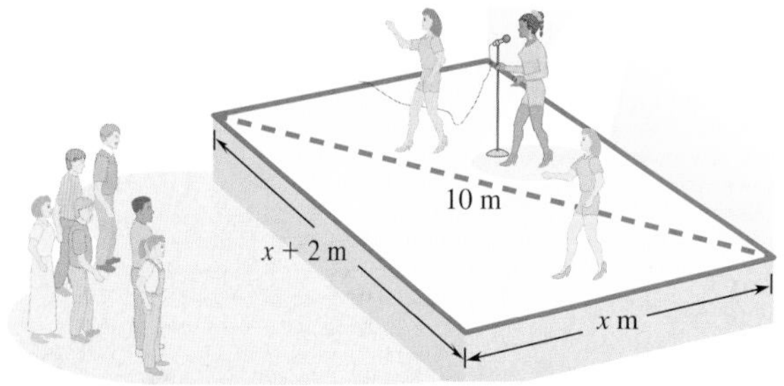

Figure for Exercise 66

67. ***Consecutive integers.*** The sum of the squares of two consecutive integers is 13. Find the integers.

68. ***Consecutive integers.*** The sum of the squares of two consecutive even integers is 52. Find the integers.

69. ***Two numbers.*** The sum of two numbers is 11, and their product is 30. Find the numbers.

70. ***Missing ages.*** Molly's age is twice Anita's. If the sum of the squares of their ages is 80, then what are their ages?

71. ***Three even integers.*** The sum of the squares of three consecutive even integers is 116. Find the integers.

72. ***Two odd integers.*** The product of two consecutive odd integers is 63. Find the integers.

73. ***Consecutive integers.*** The product of two consecutive integers is 5 more than their sum. Find the integers.

74. ***Consecutive even integers.*** If the product of two consecutive even integers is 34 larger than their sum, then what are the integers?

75. ***Two integers.*** Two integers differ by 5. If the sum of their squares is 53, then what are the integers?

76. ***Two negative integers.*** Two negative integers have a sum of -10. If the sum of their squares is 68, then what are the integers?

77. ***Area of a rectangle.*** The area of a rectangle is 72 square feet. If the length is 6 feet longer than the width, then what are the length and the width?

78. ***Area of a triangle.*** The base of a triangle is 4 inches longer than the height. If its area is 70 square inches, then what are the base and the height?

79. ***Legs of a right triangle.*** The hypotenuse of a right triangle is 15 meters. If one leg is 3 meters longer than the other, then what are the lengths of the legs?

80. ***Legs of a right triangle.*** If the longer leg of a right triangle is 1 cm longer than the shorter leg and the hypotenuse is 5 cm, then what are the lengths of the legs?

81. ***Skydiving.*** If there were no air resistance, then the height (in feet) above the earth for a skydiver t seconds after

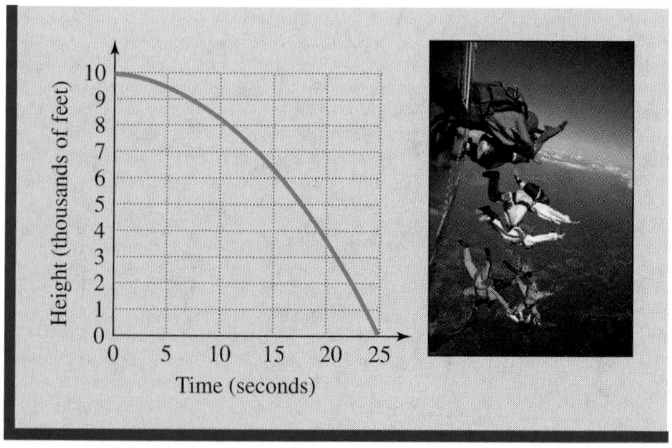

Figure for Exercise 81

jumping from an airplane at 10,000 feet would be given by

$$h(t) = -16t^2 + 10{,}000.$$

a) Find the time that it would take to fall to earth with no air resistance; that is, find t for which $h(t) = 0$. A skydiver actually gets about twice as much free fall time due to air resistance.

b) Use the accompanying graph to determine whether the skydiver (with no air resistance) falls farther in the first 5 seconds or the last 5 seconds of the fall.

c) Is the skydiver's velocity increasing or decreasing as she falls?

82. ***Skydiving.*** If a skydiver jumps from an airplane at a height of 8256 feet, then for the first five seconds, her height above the earth is approximated by the formula $h = -16t^2 + 8256$. How many seconds does it take her to reach 8000 feet?

83. ***Throwing a sandbag.*** If a balloonist throws a sandbag downward at 24 feet per second from an altitude of 720 feet, then its height (in feet) above the ground after t seconds is given by $S = -16t^2 - 24t + 720$. How long does it take for the sandbag to reach the earth? (On the ground, $S = 0$.)

84. ***Throwing a sandbag.*** If the balloonist of Exercise 83 throws his sandbag downward from an altitude of 128 feet with an initial velocity of 32 feet per second, then its altitude after t seconds is given by the formula $S = -16t^2 - 32t + 128$. How long does it take for the sandbag to reach the earth?

85. ***Glass prism.*** One end of a glass prism is in the shape of a triangle with a height that is 1 inch longer than twice the base. If the area of the triangle is 39 square inches, then how long are the base and height?

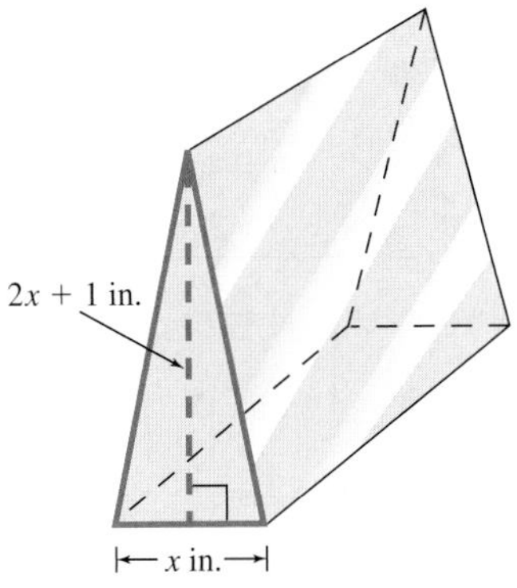

Figure for Exercise 85

86. ***Areas of two circles.*** The radius of a circle is 1 meter longer than the radius of another circle. If their areas differ by 5π square meters, then what is the radius of each?

87. ***Changing area.*** Last year Otto's garden was square. This year he plans to make it smaller by shortening one side 5 feet and the other 8 feet. If the area of the smaller garden will be 180 square feet, then what was the size of Otto's garden last year?

88. ***Dimensions of a box.*** Rosita's Christmas present from Carlos is in a box that has a width that is 3 inches shorter than the height. The length of the base is 5 inches longer than the height. If the area of the base is 84 square inches, then what is the height of the package?

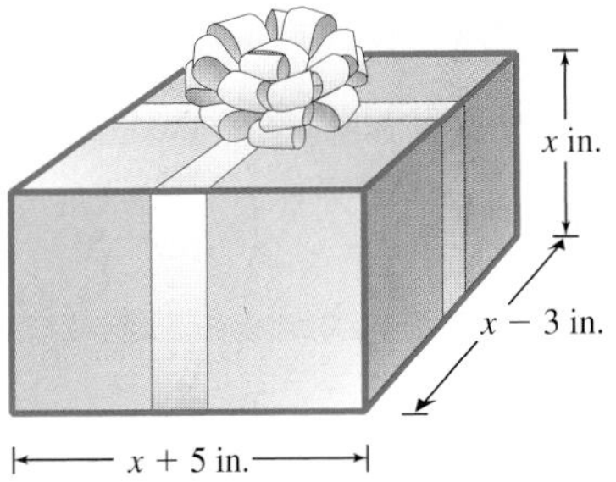

Figure for Exercise 88

89. ***Flying a kite.*** Imelda and Gordon have designed a new kite. While Imelda is flying the kite, Gordon is standing directly below it. The kite is designed so that its altitude is always 20 feet larger than the distance between Imelda and Gordon. What is the altitude of the kite when it is 100 feet from Imelda?

90. ***Avoiding a collision.*** A car is traveling on a road that is perpendicular to a railroad track. When the car is 30 meters from the crossing, the car's new collision detector warns the driver that there is a train 50 meters from the car and heading toward the same crossing. How far is the train from the crossing?

91. ***Carpeting two rooms.*** Virginia is buying carpet for two square rooms. One room is 3 yards wider than the other. If she needs 45 square yards of carpet, then what are the dimensions of each room?

92. ***Winter wheat.*** While finding the amount of seed needed to plant his three square wheat fields, Hank observed that the side of one field was 1 kilometer longer than the side of the smallest field and that the side of the largest field was 3 kilometers longer than the side of the smallest field. If the total area of the three fields is 38 square kilometers, then what is the area of each field?

93. ***Sailing to Miami.*** At point A the captain of a ship determined that the distance to Miami was 13 miles. If she sailed north to point B and then west to Miami, the distance would be 17 miles. If the distance from point A to point B is greater than the distance from point B to Miami, then how far is it from point A to point B?

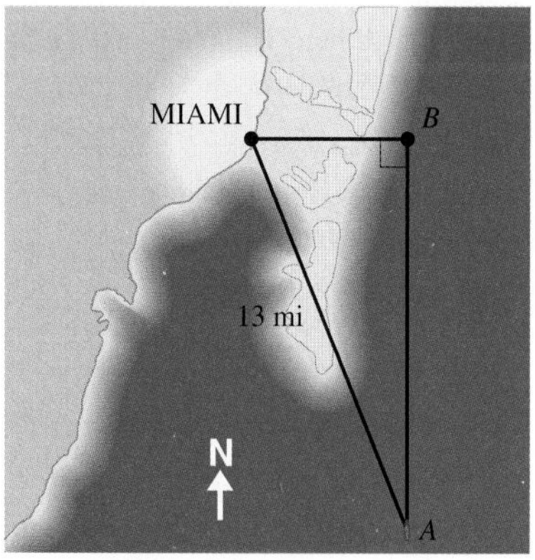

Figure for Exercise 93

94. ***Buried treasure.*** Ahmed has half of a treasure map, which indicates that the treasure is buried in the desert $2x + 6$ paces from Castle Rock. Vanessa has the other half of the map. Her half indicates that to find the treasure, one must get to Castle Rock, walk x paces to the north, and then walk $2x + 4$ paces to the east. If they share their information, then they can find x and save a lot of digging. What is x?

95. ***Emerging markets.*** Catarina's investment of \$16,000 in an emerging market fund grew to \$25,000 in two years. Find the average annual rate of return by solving the equation $16{,}000(1 + r)^2 = 25{,}000$.

96. ***Venture capital.*** Henry invested \$12,000 in a new restaurant. When the restaurant was sold two years later, he received \$27,000. Find his average annual return by solving the equation $12{,}000(1 + r)^2 = 27{,}000$.

Collaborative Activities

Grouping: Three students per group

Topic: Factoring Polynomials

Hannah's Inheritance

Part I: Sally, Kelly, and Hannah inherited property from their father. Sally, who married the neighbor to the east, is given a piece of land adjacent to her husband's property. The land is 5 hectometers wide and the length matches that of her husband's property. Kelly who has married the neighbor to the south is given property that is 4 hectometers wide and its length is the common boundary of her spouse's property. Hannah is to have a square piece that is left after her sister's property is taken out. Hannah wants to find out the dimensions of her land. She knows that her father's land totaled 380 square hectometers. Use the diagram below to find the dimensions of Hannah's land.

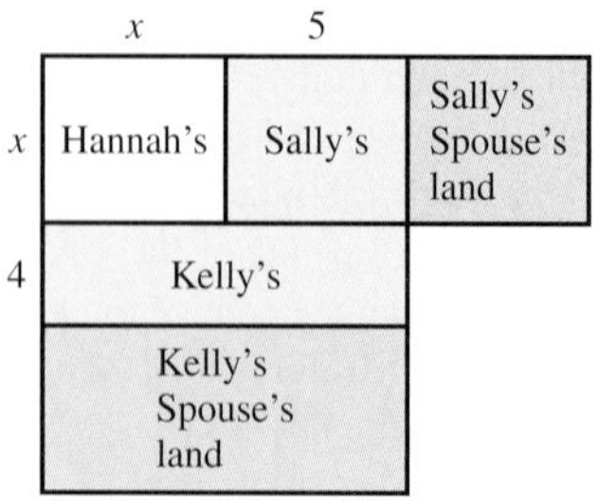

Part II: A few years later Hannah married Harry, a math teacher from town. Now they are getting divorced. As part of the settlement, Harry will get 32% of the land that they now own jointly. However, Harry did not get along with Hannah's sisters and does not want his land to be adjacent to their land. They have agreed that Harry will get a triangular section at the corner as shown in the accompanying figure. Use your answer from Part I to find y and z in the figure.

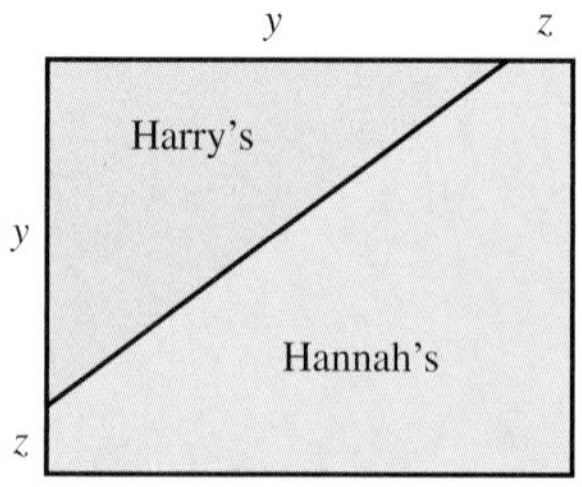

Chapter 6 Wrap-Up

Summary

Factoring		Examples
Prime number	A positive integer larger than 1 that has no integral factors other than 1 and itself	2, 3, 5, 7, 11
Prime polynomial	A polynomial that cannot be factored is prime.	$x^2 + 3$ and $x^2 - x + 5$ are prime.
Strategy for finding the GCF for monomials	1. Find the GCF for the coefficients of the monomials. 2. Form the product of the GCF of the coefficients and each variable that is common to all of the monomials, where the exponent on each variable equals the smallest power of that variable in any of the monomials.	$12x^3yz$, $8x^2y^3$ GCF $= 4x^2y$
Factoring out the GCF	Use the distributive property to factor out the GCF from all terms of a polynomial.	$2x^3 - 4x = 2x(x^2 - 2)$

Special Cases		Examples
Difference of two squares	$a^2 - b^2 = (a + b)(a - b)$	$m^2 - 9 = (m - 3)(m + 3)$
Perfect square trinomial	$a^2 + 2ab + b^2 = (a + b)^2$ $a^2 - 2ab + b^2 = (a - b)^2$	$x^2 + 6x + 9 = (x + 3)^2$ $4h^2 - 12h + 9 = (2h - 3)^2$
Difference or sum of two cubes	$a^3 - b^3 = (a - b)(a^2 + ab + b^2)$ $a^3 + b^3 = (a + b)(a^2 - ab + b^2)$	$t^3 - 8 = (t - 2)(t^2 + 2t + 4)$ $p^3 + 1 = (p + 1)(p^2 - p + 1)$

Factoring Polynomials		Examples
Factoring by grouping	Factor out common factors from groups of terms.	$6x + 6w + ax + aw$ $= 6(x + w) + a(x + w)$ $= (6 + a)(x + w)$
Strategy for factoring $ax^2 + bx + c$ by the ac method	1. Find two numbers that have a product equal to ac and a sum equal to b. 2. Replace bx by two terms using the two new numbers as coefficients. 3. Factor the resulting four-term polynomial by grouping.	$6x^2 + 17x + 12$ $= 6x^2 + 9x + 8x + 12$ $= (2x + 3)3x + (2x + 3)4$ $= (2x + 3)(3x + 4)$

Factoring by trial and error	Try possible factors of the trinomial and check by using FOIL. If incorrect, try again.	$2x^2 + 5x - 12 = (2x - 3)(x + 4)$
Strategy for factoring polynomials completely	1. First factor out any common factors. 2. When factoring a binomial, check to see whether it is a difference of two squares, a difference of two cubes, or a sum of two cubes. Remember that a sum of two squares (with no common factor) is prime. 3. When factoring a trinomial, check to see whether it is a perfect square trinomial. 4. When factoring a trinomial that is not a perfect square, use the *ac* method or trial and error. 5. If the polynomial has four terms, try factoring by grouping. 6. Check to see whether any factors can be factored again.	
Solving Equations		**Examples**
Zero factor property	The equation $a \cdot b = 0$ is equivalent to $a = 0 \quad \text{or} \quad b = 0.$	$x(x - 1) = 0$ $x = 0 \quad \text{or} \quad x - 1 = 0$
Strategy for solving an equation by factoring	1. Rewrite the equation with 0 on the right-hand side. 2. Factor the left-hand side completely. 3. Set each factor equal to zero to get linear equations. 4. Solve the linear equations. 5. Check the answers in the original equation. 6. State the solution(s) to the original equation.	$x^2 + 3x = 18$ $x^2 + 3x - 18 = 0$ $(x + 6)(x - 3) = 0$ $x + 6 = 0 \quad \text{or} \quad x - 3 = 0$ $x = -6 \quad \text{or} \quad x = 3$

Enriching Your Mathematical Word Power

For each mathematical term, choose the correct meaning.

1. factor
- a. to write an expression as a product
- b. to multiply
- c. what two numbers have in common
- d. to FOIL

2. prime number
- a. a polynomial that cannot be factored
- b. a number with no divisors
- c. an integer between 1 and 10
- d. an integer larger than 1 that has no integral factors other than itself and 1

3. greatest common factor
- a. the least common multiple
- b. the least common denominator
- c. the largest integer that is a factor of two or more integers
- d. the largest number in a product

4. prime polynomial
- a. a polynomial that has no factors
- b. a product of prime numbers
- c. a first-degree polynomial
- d. a monomial

5. factor completely
a. to factor by grouping
b. to factor out a prime number
c. to write as a product of primes
d. to factor by trial and error

6. sum of two cubes
a. $(a + b)^3$
b. $a^3 + b^3$
c. $a^3 - b^3$
d. a^3b^3

7. quadratic equation
a. $ax + b = 0$ where $a \neq 0$
b. $ax + b = cx + d$
c. $ax^2 + bx + c = 0$ where $a \neq 0$
d. any equation with four terms

8. zero factor property
a. If $ab = 0$ then $a = 0$ or $b = 0$
b. $a \cdot 0 = 0$ for any a
c. $a = a + 0$ for any real number a
d. $a + (-a) = 0$ for any real number a

9. Pythagorean theorem
a. $a^2 + b^2 = (a + b)^2$
b. a triangle is a right triangle if and only if it has one right angle
c. the legs of a right triangle meet at a 90º angle
d. a theorem that gives a relationship between the two legs and the hypotenuse of a right triangle

10. difference of two squares
a. $a^3 - b^3$
b. $2a - 2b$
c. $a^2 - b^2$
d. $(a - b)^2$

Review Exercises

6.1 *Find the prime factorization for each integer.*

1. 144

2. 121

3. 58

4. 76

5. 150

6. 200

Find the greatest common factor for each group.

7. 36, 90

8. 30, 42, 78

9. $8x$, $12x^2$

10. $6a^2b$, $9ab^2$, $15a^2b^2$

Complete the factorization of each binomial.

11. $3x + 6 = 3(\quad)$

12. $7x^2 + x = x(\quad)$

13. $2a - 20 = -2(\quad)$

14. $a^2 - a = -a(\quad)$

Factor each polynomial by factoring out the GCF.

15. $2a - a^2$

16. $9 - 3b$

17. $6x^2y^2 - 9x^5y$

18. $a^3b^5 + a^3b^2$

19. $3x^2y - 12xy - 9y^2$

20. $2a^2 - 4ab^2 - ab$

6.2 *Factor each polynomial completely.*

21. $y^2 - 400$

22. $4m^2 - 9$

23. $w^2 - 8w + 16$

24. $t^2 + 20t + 100$

25. $4y^2 + 20y + 25$

26. $2a^2 - 4a - 2$

27. $r^2 - 4r + 4$

28. $3m^2 - 75$

29. $8t^3 - 24t^2 + 18t$

30. $t^2 - 9w^2$

31. $x^2 + 12xy + 36y^2$

32. $9y^2 - 12xy + 4x^2$

33. $x^2 + 5x - xy - 5y$

34. $x^2 + xy + ax + ay$

6.3 *Factor each polynomial.*

35. $b^2 + 5b - 24$

36. $a^2 - 2a - 35$

37. $r^2 - 4r - 60$

38. $x^2 + 13x + 40$

39. $y^2 - 6y - 55$

40. $a^2 + 6a - 40$

41. $u^2 + 26u + 120$

42. $v^2 - 22v - 75$

Factor completely.

43. $3t^3 + 12t^2$

44. $-4m^4 - 36m^2$

45. $5w^3 + 25w^2 + 25w$

46. $-3t^3 + 3t^2 - 6t$

47. $2a^3b + 3a^2b^2 + ab^3$

48. $6x^2y^2 - xy^3 - y^4$

49. $9x^3 - xy^2$

50. $h^4 - 100h^2$

6.4 *Factor each polynomial completely.*

51. $14t^2 + t - 3$

52. $15x^2 - 22x - 5$

53. $6x^2 - 19x - 7$

54. $2x^2 - x - 10$

55. $6p^2 + 5p - 4$

56. $3p^2 + 2p - 5$

57. $-30p^3 + 8p^2 + 8p$

58. $-6q^2 - 40q - 50$

59. $6x^2 - 29xy - 5y^2$

60. $10a^2 + ab - 2b^2$

61. $32x^2 + 16xy + 2y^2$

62. $8a^2 + 40ab + 50b^2$

6.5 *Factor completely.*

63. $5x^3 + 40x$

64. $w^2 + 6w + 9$

65. $9x^2 + 3x - 2$

66. $ax^3 + ax$

67. $n^2 + 64$

68. $4t^2 + h^2$

69. $x^3 + 2x^2 - x - 2$

70. $16x^2 - 2x - 3$

71. $x^2y - 16xy^2$

72. $-3x^2 + 27$

73. $w^2 + 4w + 5$

74. $2n^2 + 3n - 1$

75. $a^2 + 2a + 1$

76. $-2w^2 - 12w - 18$

77. $x^3 - x^2 + x - 1$

78. $9x^2y^2 - 9y^2$

79. $a^2 + ab + 2a + 2b$

80. $4m^2 + 20m + 25$

81. $-2x^2 + 16x - 24$

82. $6x^2 + 21x - 45$

83. $m^3 - 1000$

84. $8p^3 + 1$

85. $p^4 - q^4$

86. $z^4 - 81$

Factor each polynomial completely, given that the binomial following it is a factor of the polynomial.

87. $x^3 + x + 10, x + 2$

88. $x^3 - 5x - 12, x - 3$

89. $x^3 + 6x^2 - 7x - 60, x + 4$

90. $x^3 - 4x^2 - 3x - 10, x - 5$

6.6 *Solve each equation.*

91. $x^3 = 5x^2$

92. $2m^2 + 10m = -12$

93. $(a - 2)(a - 3) = 6$

94. $(w - 2)(w + 3) = 50$

95. $2m^2 - 9m - 5 = 0$

96. $12x^2 + 5x - 3 = 0$

97. $m^3 + 4m^2 - 9m = 36$

98. $w^3 + 5w^2 - w = 5$

99. $(x + 3)^2 + x^2 = 5$

100. $(h - 2)^2 + (h + 1)^2 = 9$

101. $p^2 + \frac{1}{4}p - \frac{1}{8} = 0$

102. $t^2 + 1 = \frac{13}{6}t$

Solve each problem.

103. ***Positive numbers.*** Two positive numbers differ by 6, and their squares differ by 96. Find the numbers.

104. ***Consecutive integers.*** Find three consecutive integers such that the sum of their squares is 77.

105. ***Dimensions of a notebook.*** The perimeter of a notebook is 28 inches, and the diagonal measures 10 inches. What are the length and width of the notebook?

106. ***Two numbers.*** The sum of two numbers is 8.5, and their product is 18. Find the numbers.

107. ***Poiseuille's law.*** According to the nineteenth-century physician Poiseuille, the velocity (in centimeters per second) of blood r centimeters from the center of an artery of radius R centimeters is given by $v = kR^2 - kr^2$, where k is a constant. Rewrite the formula by factoring the right-hand side completely.

108. ***Racquetball.*** The volume of rubber (in cubic centimeters) in a hollow rubber ball used in racquetball is given by

$$V = \frac{4}{3}\pi R^3 - \frac{4}{3}\pi r^3,$$

where the inside radius is r centimeters and the outside radius is R centimeters.

a) Rewrite the formula by factoring the right-hand side completely.

b) The accompanying graph shows the relationship between r and V when $R = 3$. Use the graph to estimate the value of r for which $V = 100\text{ cm}^3$.

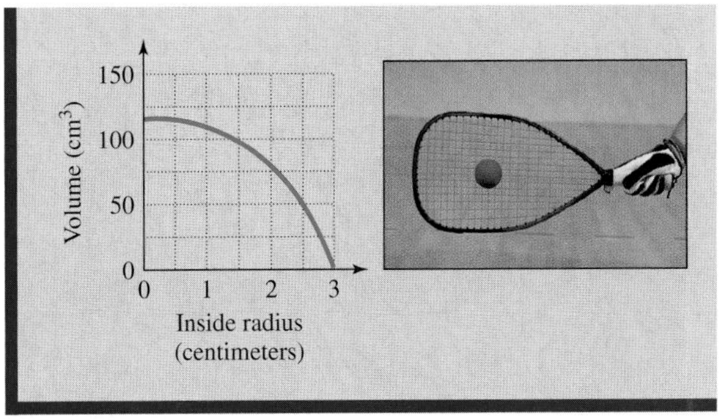

Figure for Exercise 108

109. ***Leaning ladder.*** A 10-foot ladder is placed against a building so that the distance from the bottom of the ladder to the building is 2 feet less than the distance from the top of the ladder to the ground. What is the distance from the bottom of the ladder to the building?

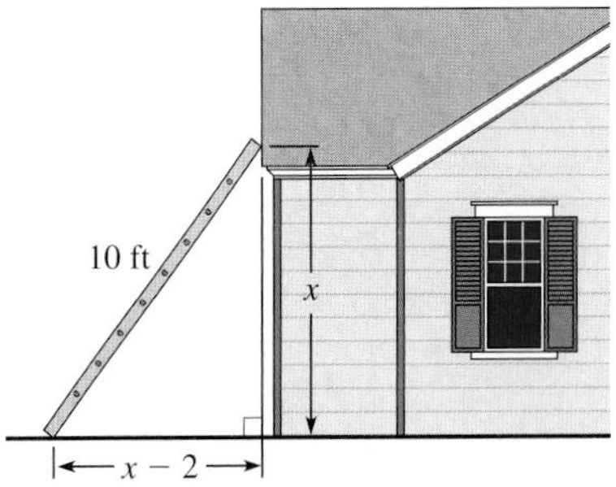

Figure for Exercise 109

110. ***Towering antenna.*** A guy wire of length 50 feet is attached to the ground and to the top of an antenna. The height of the antenna is 10 feet larger than the distance from the base of the antenna to the point where the guy wire is attached to the ground. What is the height of the antenna?

Chapter 6 Test

Give the prime factorization for each integer.

1. 66

2. 336

Find the greatest common factor (GCF) for each group.

3. 48, 80

4. 42, 66, 78

5. $6y^2$, $15y^3$

6. $12a^2b$, $18ab^2$, $24a^3b^3$

Factor each polynomial completely.

7. $5x^2 - 10x$

8. $6x^2y^2 + 12xy^2 + 12y^2$

9. $3a^3b - 3ab^3$

10. $a^2 + 2a - 24$

11. $4b^2 - 28b + 49$

12. $3m^3 + 27m$

13. $ax - ay + bx - by$

14. $ax - 2a - 5x + 10$

15. $6b^2 - 7b - 5$

16. $m^2 + 4mn + 4n^2$

17. $2a^2 - 13a + 15$

18. $z^3 + 9z^2 + 18z$

19. $x^3 + 125$

20. $a^4 - ab^3$

Factor the polynomial completely, given that $x - 1$ is a factor.

21. $x^3 - 6x^2 + 11x - 6$

Solve each equation.

22. $x^2 + 6x + 9 = 0$

23. $2x^2 + 5x - 12 = 0$

24. $3x^3 = 12x$

25. $(2x - 1)(3x + 5) = 5$

Write a complete solution to each problem.

26. If the length of a rectangle is 3 feet longer than the width and the diagonal is 15 feet, then what are the length and width?

27. The sum of two numbers is 4, and their product is -32. Find the numbers.

*Making*Connections | A Review of Chapters 1–6

Simplify each expression.

1. $\dfrac{91 - 17}{17 - 91}$

2. $\dfrac{4 - 18}{-6 - 1}$

3. $5 - 2(7 - 3)$

4. $3^2 - 4(6)(-2)$

5. $2^5 - 2^4$

6. $0.07(37) + 0.07(63)$

Perform the indicated operations.

7. $x \cdot 2x$

8. $x + 2x$

9. $\dfrac{6 + 2x}{2}$

10. $\dfrac{6 \cdot 2x}{2}$

11. $2 \cdot 3y \cdot 4z$

12. $2(3y + 4z)$

13. $2 - (3 - 4z)$

14. $t^8 \div t^2$

15. $t^8 \cdot t^2$

16. $\dfrac{8t^8}{2t^2}$

Solve each inequality. State the solution set in interval notation and sketch its graph.

17. $2x - 5 > 3x + 4$

18. $4 - 5x \le -11$

19. $-\frac{2}{3}x + 3 < -5$

20. $0.05(x - 120) - 24 < 0$

Find the solution set to each equation.

21. $2x - 3 = 0$

22. $2x + 1 = 0$

23. $(x - 3)(x + 5) = 0$

24. $(2x - 3)(2x + 1) = 0$

25. $3x(x - 3) = 0$

26. $x^2 = x$

27. $3x - 3x = 0$

28. $3x - 3x = 1$

29. $0.01x - x + 14.9 = 0.5x$

30. $0.05x + 0.04(x - 40) = 2$

31. $2x^2 = 18$

32. $2x^2 + 7x - 15 = 0$

Solve the problem.

33. ***Another ace.*** Professional tennis players can serve a tennis ball at speeds over 120 mph into a rectangular region that has a perimeter of 69 feet and an area of 283.5 square feet. Find the length and width of the service region.

Photo for Exercise 33

Critical **Thinking** | For Individual or Group Work | Chapter 6

These exercises can be solved by a variety of techniques, which may or may not require algebra. So be creative and think critically. Explain all answers. Answers are in the Instructor's Edition of this text.

1. ***Equilateral triangles.*** Consider the sequence of three equilateral triangles shown in the accompanying figure.

   a) How many equilateral triangles are there in (a) of the accompanying figure?

   b) How many equilateral triangles congruent to the one in (a) can be found in (b) of the accompanying figure? How many are found in (c)?

   c) Suppose the sequence of equilateral triangles shown in (a), (b), and (c) is continued. How many equilateral triangles [congruent to the one in (a)] could be found in the nth such figure?

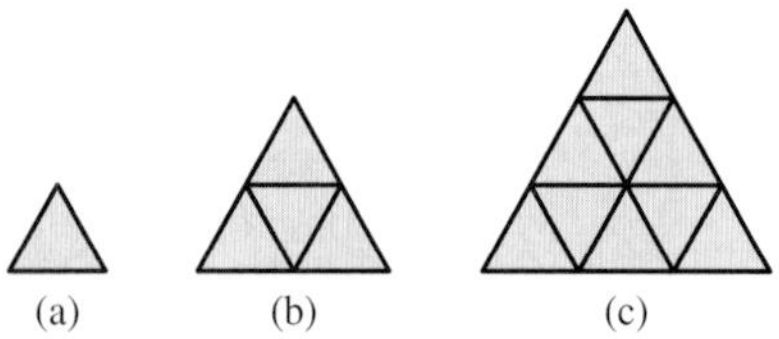

Figure for Exercise 1

2. ***The amazing Amber.*** Amber has been amazing her friends with a math trick. Amber has a friend select a three-digit number and reverse the digits. The friend then finds the difference of the two numbers and reads the first two digits of the difference (from left to right). Amber can always tell the last digit of the difference. Explain how Amber does this.

3. ***Missing proceeds.*** Ruth and Betty sell apples at a farmers market. Ruth's apples sell at 2 for \$1 while Betty's slightly smaller apples sell at 3 for \$1. When Betty leaves to pick up her kids they each have 30 apples and Ruth takes charge of both businesses. To simplify things, Ruth puts all 60 of the apples together and sells them at 5 for \$2. When Betty returns, all of the apples have been sold, but they begin arguing over how to divide up the proceeds. What is the problem? Explain what went wrong.

4. ***Eyes and feet.*** A rancher has some sheep and ostriches. His young daughter observed that the animals have a total of 60 eyes and 86 feet. How many animals of each type does the rancher have?

Photo for Exercise 4

5. ***Evaluation nightmare.*** Evaluate:

$$\frac{9876543210}{9876543211^2 - 9876543210 \cdot 9876543212}$$

6. ***Perfect squares.*** Find a positive integer such that the integer increased by 1 is a perfect square and one-half of the integer increased by 1 is a perfect square. Also find the next two larger positive integers that have this same property.

7. ***Multiplying primes.*** Find the units digit of the product of the first 500 prime numbers.

8. ***Ones and zeros.*** Find the sum of all seven-digit numbers that can be written using only ones or zeros.

Chapter

7

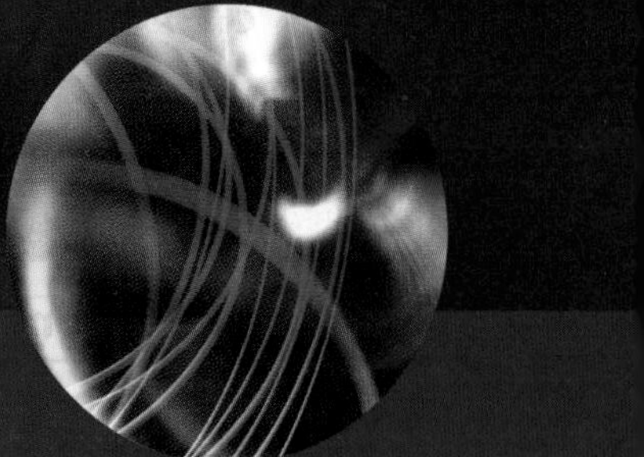

Rational Expressions

Advanced technical developments have made sports equipment faster, lighter, and more responsive to the human body. Behind the more flexible skis, lighter bats, and comfortable athletic shoes lies the science of biomechanics, which is the study of human movement and the factors that influence it.

Designing and testing an athletic shoe go hand in hand. While a shoe is being designed, it is tested in a multitude of ways, including long-term wear, rear foot stability, and strength of materials. Testing basketball shoes usually includes an evaluation of the force applied to the ground by the foot during running, jumping, and landing. Many biomechanics laboratories have a special platform that can measure the force exerted when a player cuts from side to side as well as the force against the bottom of the shoe. Force exerted in landing from a lay-up shot can be as high as 14 times the weight of the body. Side-to-side force is usually about 1 to 2 body weights in a cutting movement.

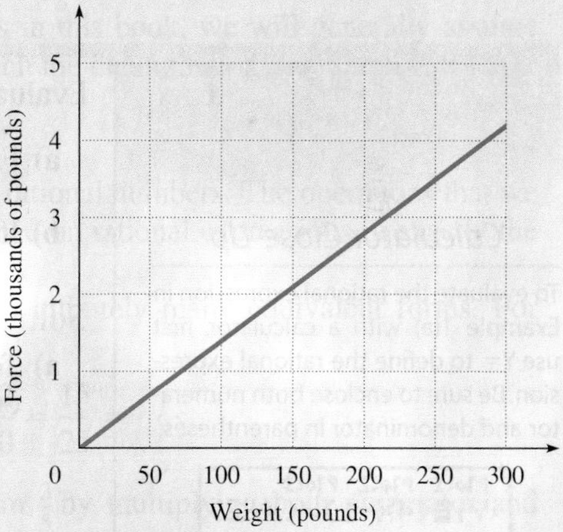

In Exercises 57 and 58 of Section 7.7 you will see how designers of athletic shoes use proportions to find the amount of force on the foot and soles of shoes for activities such as running and jumping.

b) $\dfrac{6x^4y^2}{4xy^5} = \dfrac{\cancel{2} \cdot 3x^4y^2}{\cancel{2} \cdot 2xy^5}$ Factor.

$= \dfrac{3x^{4-1}}{2y^{5-2}}$ Quotient rule

$= \dfrac{3x^3}{2y^3}$

Now do Exercises 41–52

The essential part of reducing is getting a complete factorization for the numerator and denominator. To get a complete factorization, you must use the techniques for factoring from Chapter 6. If there are large integers in the numerator and denominator, you can use the technique shown in Section 6.1 to get a prime factorization of each integer.

EXAMPLE 5

Reducing expressions involving large integers

Reduce $\frac{420}{616}$ to lowest terms.

Solution

Use the method of Section 6.1 to get a prime factorization of 420 and 616:

$$\begin{array}{rr} 7 & 11 \\ 5\overline{)35} & 7\overline{)77} \\ 3\overline{)105} & 2\overline{)154} \\ 2\overline{)210} & 2\overline{)308} \\ \text{Start here} \rightarrow 2\overline{)420} & 2\overline{)616} \end{array}$$

The complete factorization for 420 is $2^2 \cdot 3 \cdot 5 \cdot 7$, and the complete factorization for 616 is $2^3 \cdot 7 \cdot 11$. To reduce the fraction, we divide out the common factors:

$$\frac{420}{616} = \frac{2^2 \cdot 3 \cdot 5 \cdot 7}{2^3 \cdot 7 \cdot 11}$$

$$= \frac{3 \cdot 5}{2 \cdot 11}$$

$$= \frac{15}{22}$$

Now do Exercises 53–60

Dividing $a - b$ by $b - a$

In Section 5.2 you learned that $a - b = -(b - a) = -1(b - a)$. So if $a - b$ is divided by $b - a$, the quotient is -1:

$$\frac{a - b}{b - a} = \frac{-1(b - a)}{b - a}$$

$$= -1$$

We will use this fact in Example 6.

EXAMPLE 6

Expressions with $a - b$ and $b - a$

Reduce to lowest terms.

a) $\dfrac{5x - 5y}{4y - 4x}$ **b)** $\dfrac{m^2 - n^2}{n - m}$

Solution

a) $\dfrac{5x - 5y}{4y - 4x} = \dfrac{5(x - y)}{4(y - x)}$ Factor.

$= \dfrac{5}{4} \cdot (-1)$ $\quad \dfrac{x - y}{y - x} = -1$

$= -\dfrac{5}{4}$

b) $\dfrac{m^2 - n^2}{n - m} = \dfrac{\overset{-1}{\cancel{(m - n)}}(m + n)}{\cancel{n - m}}$ Factor.

$= -1(m + n)$ $\quad \dfrac{m - n}{n - m} = -1$

$= -m - n$

Now do Exercises 61–68

CAUTION We can reduce $\frac{a - b}{b - a}$ to -1, but we cannot reduce $\frac{a - b}{a + b}$. There is no factor that is common to the numerator and denominator of $\frac{a - b}{a + b}$ or $\frac{a + b}{a - b}$.

Factoring Out the Opposite of a Common Factor

If we can factor out a common factor, we can also factor out the opposite of that common factor. For example, from $-3x - 6y$ we can factor out the common factor 3 or the common factor -3:

$$-3x - 6y = 3(-x - 2y) \qquad \text{or} \qquad -3x - 6y = -3(x + 2y)$$

To reduce an expression, it is sometimes necessary to factor out the opposite of a common factor.

EXAMPLE 7

Factoring out the opposite of a common factor

Reduce $\frac{-3w - 3w^2}{w^2 - 1}$ to lowest terms.

Solution

We can factor $3w$ or $-3w$ from the numerator. If we factor out $-3w$, we get a common factor in the numerator and denominator:

$\dfrac{-3w - 3w^2}{w^2 - 1} = \dfrac{-3w\cancel{(1 + w)}}{(w - 1)\cancel{(w + 1)}}$ Factor.

$= \dfrac{-3w}{w - 1}$ Since $1 + w = w + 1$, we divide out $w + 1$.

$= \dfrac{3w}{1 - w}$ Multiply numerator and denominator by -1.

The last step in this reduction is not absolutely necessary, but we usually perform it to make the answer look a little simpler.

Now do Exercises 69–76

The main points to remember for reducing rational expressions are summarized in the following reducing strategy.

Strategy for Reducing Rational Expressions

1. Reducing is done by dividing out all common factors.
2. Factor the numerator and denominator completely to see the common factors.
3. Use the quotient rule to reduce a ratio of two monomials.
4. You may have to factor out a common factor with a negative sign to get identical factors in the numerator and denominator.
5. The quotient of $a - b$ and $b - a$ is -1.

Writing Rational Expressions

Rational expressions occur naturally in applications involving rates.

EXAMPLE 8

Writing rational expressions

Answer each question with a rational expression.

a) If a trucker drives 500 miles in $x + 1$ hours, then what is his average speed?

b) If a wholesaler buys 100 pounds of shrimp for x dollars, then what is the price per pound?

c) If a painter completes an entire house in $2x$ hours, then at what rate is she painting?

Solution

a) Because $R = \frac{D}{T}$, he is averaging $\frac{500}{x + 1}$ mph.

b) At x dollars for 100 pounds, the wholesaler is paying $\frac{x}{100}$ dollars per pound or $\frac{x}{100}$ dollars/pound.

c) By completing 1 house in $2x$ hours, her rate is $\frac{1}{2x}$ house/hour.

Now do Exercises 97–102

Warm-Ups ▼

True or false? Explain your answer.

1. A complete factorization of 3003 is $2 \cdot 3 \cdot 7 \cdot 11 \cdot 13$.
2. A complete factorization of 120 is $2^3 \cdot 3 \cdot 5$.
3. Any number can be used in place of x in the expression $\frac{x - 2}{5}$.
4. We cannot replace x by -1 or 3 in the expression $\frac{x + 1}{x - 3}$.
5. The rational expression $\frac{x + 2}{2}$ reduces to x.
6. $\frac{2x}{2} = x$ for any real number x.

7. $\frac{x^{13}}{x^{20}} = \frac{1}{x^7}$ for any nonzero value of x.
8. $\frac{a^2 + b^2}{a + b}$ reduced to lowest terms is $a + b$.
9. If $a \neq b$, then $\frac{a - b}{b - a} = 1$.
10. The expression $\frac{-3x - 6}{x + 2}$ reduces to -3.

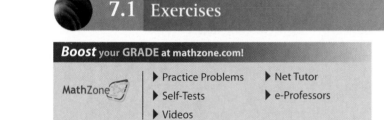

7.1 Exercises

Reading and Writing *After reading this section, write out the answers to these questions. Use complete sentences.*

1. What is a rational number?
2. What is a rational expression?
3. How do you reduce a rational number to lowest terms?
4. How do you reduce a rational expression to lowest terms?
5. How is the quotient rule used in reducing rational expressions?
6. What is the relationship between $a - b$ and $b - a$?

Evaluate each rational expression. See Example 1.

7. Evaluate $\frac{3x - 3}{x + 5}$ for $x = -2$.
8. Evaluate $\frac{3x + 1}{4x - 4}$ for $x = 5$.
9. If $R(x) = \frac{2x + 9}{x}$, find $R(3)$.
10. If $R(x) = \frac{-20x - 2}{x - 8}$, find $R(-1)$.
11. If $R(x) = \frac{x - 5}{x + 3}$, find $R(2)$, $R(-4)$, $R(-3.02)$, and $R(-2.96)$.
12. If $R(x) = \frac{x^2 - 2x - 3}{x - 2}$, find $R(3)$, $R(5)$, $R(2.05)$, and $R(1.999)$.

Which numbers cannot be used in place of the variable in each rational expression? See Example 2.

13. $\frac{x}{x + 1}$
14. $\frac{3x}{x - 7}$
15. $\frac{7a}{3a - 5}$
16. $\frac{84}{3 - 2a}$
17. $\frac{2x + 3}{x^2 - 16}$
18. $\frac{2y + 1}{y^2 - y - 6}$
19. $\frac{p - 1}{2}$
20. $\frac{m + 31}{5}$

Reduce each rational expression to lowest terms. Assume that the variables represent only numbers for which the denominators are nonzero. See Example 3.

21. $\frac{6}{27}$
22. $\frac{14}{21}$
23. $\frac{42}{90}$
24. $\frac{42}{54}$
25. $\frac{36a}{90}$
26. $\frac{56y}{40}$
27. $\frac{78}{30w}$
28. $\frac{68}{44y}$

29. $\frac{6x + 2}{6}$

30. $\frac{2w + 2}{2w}$

31. $\frac{2x + 4y}{6y + 3x}$

32. $\frac{3m + 9w}{3m - 6w}$

33. $\frac{w^2 - 49}{w + 7}$

34. $\frac{a^2 - b^2}{a - b}$

35. $\frac{a^2 - 1}{a^2 + 2a + 1}$

36. $\frac{x^2 - y^2}{x^2 + 2xy + y^2}$

37. $\frac{2x^2 + 4x + 2}{4x^2 - 4}$

38. $\frac{2x^2 + 10x + 12}{3x^2 - 27}$

39. $\frac{3x^2 + 18x + 27}{21x + 63}$

40. $\frac{x^3 - 3x^2 - 4x}{x^2 - 4x}$

Reduce each expression to lowest terms. Assume that all denominators are nonzero. See Example 4.

41. $\frac{x^{10}}{x^7}$

42. $\frac{y^8}{y^5}$

43. $\frac{z^3}{z^8}$

44. $\frac{w^9}{w^{12}}$

45. $\frac{4x^7}{-2x^5}$

46. $\frac{-6y^3}{3y^9}$

47. $\frac{-12m^9n^{18}}{8m^6n^{16}}$

48. $\frac{-9u^9v^{19}}{6u^9v^{14}}$

49. $\frac{6b^{10}c^4}{-8b^{10}c^7}$

50. $\frac{9x^{20}y}{-6x^{25}y^3}$

51. $\frac{30a^3bc}{18a^7b^{17}}$

52. $\frac{15m^{10}n^3}{24m^{12}np}$

Reduce each expression to lowest terms. See Example 5.

53. $\frac{210}{264}$

54. $\frac{616}{660}$

55. $\frac{231}{168}$

56. $\frac{936}{624}$

57. $\frac{630x^5}{300x^9}$

58. $\frac{96y^2}{108y^5}$

59. $\frac{924a^{23}}{448a^{19}}$

60. $\frac{270b^{75}}{165b^{12}}$

Reduce each expression to lowest terms. See Example 6.

61. $\frac{3a - 2b}{2b - 3a}$

62. $\frac{5m - 6n}{6n - 5m}$

63. $\frac{h^2 - t^2}{t - h}$

64. $\frac{r^2 - s^2}{s - r}$

65. $\frac{2g - 6h}{9h^2 - g^2}$

66. $\frac{5a - 10b}{4b^2 - a^2}$

67. $\frac{x^2 - x - 6}{9 - x^2}$

68. $\frac{1 - a^2}{a^2 + a - 2}$

Reduce each expression to lowest terms. See Example 7.

69. $\frac{-x - 6}{x + 6}$

70. $\frac{-5x - 20}{3x + 12}$

71. $\frac{-2y - 6y^2}{3 + 9y}$

72. $\frac{y^2 - 16}{-8 - 2y}$

73. $\frac{-3x - 6}{3x - 6}$

74. $\frac{8 - 4x}{-8x - 16}$

75. $\frac{-12a - 6}{2a^2 + 7a + 3}$

76. $\frac{-2b^2 - 6b - 4}{b^2 - 1}$

Reduce each expression to lowest terms.

77. $\frac{2x^{12}}{4x^8}$

78. $\frac{4x^2}{2x^9}$

79. $\frac{2x + 4}{4x}$

80. $\frac{2x + 4x^2}{4x}$

81. $\frac{a - 4}{4 - a}$

82. $\frac{2b - 4}{2b + 4}$

83. $\frac{2c - 4}{4 - c^2}$

84. $\frac{-2t - 4}{4 - t^2}$

85. $\frac{x^2 + 4x + 4}{x^2 - 4}$

86. $\frac{3x - 6}{x^2 - 4x + 4}$

87. $\frac{-2x - 4}{x^2 + 5x + 6}$

88. $\frac{-2x - 8}{x^2 + 2x - 8}$

89. $\frac{2q^8 + q^7}{2q^6 + q^5}$

90. $\frac{8s^{12}}{12s^6 - 16s^5}$

91. $\frac{u^2 - 6u - 16}{u^2 - 16u + 64}$

92. $\frac{v^2 + 3v - 18}{v^2 + 12v + 36}$

93. $\frac{a^3 - 8}{2a - 4}$

94. $\frac{4w^2 - 12w + 36}{2w^3 + 54}$

95. $\frac{y^3 - 2y^2 - 4y + 8}{y^2 - 4y + 4}$

96. $\frac{mx + 3x + my + 3y}{m^2 - 3m - 18}$

Answer each question with a rational expression. Be sure to include the units. See Example 8.

97. If Sergio drove 300 miles at $x + 10$ miles per hour, then how many hours did he drive?

98. If Carrie walked 40 miles in x hours, then how fast did she walk?

99. If $x + 4$ pounds of peaches cost \$4.50, then what is the cost per pound?

100. If nine pounds of pears cost x dollars, then what is the price per pound?

101. If Ayesha can clean the entire swimming pool in x hours, then how much of the pool does she clean per hour?

102. If Ramon can mow the entire lawn in $x - 3$ hours, then how much of the lawn does he mow per hour?

Solve each problem.

103. ***Annual reports.*** The Crest Meat Company found that the cost per report for printing x annual reports at Peppy Printing is given by the formula

$$C(x) = \frac{150 + 0.60x}{x},$$

where $C(x)$ is in dollars.

a) Use the accompanying graph to estimate the cost per report for printing 1000 reports.

b) Use the formula to find the cost per report for printing 1000, 5000, and 10,000 reports.

c) What happens to the cost per report as the number of reports gets very large?

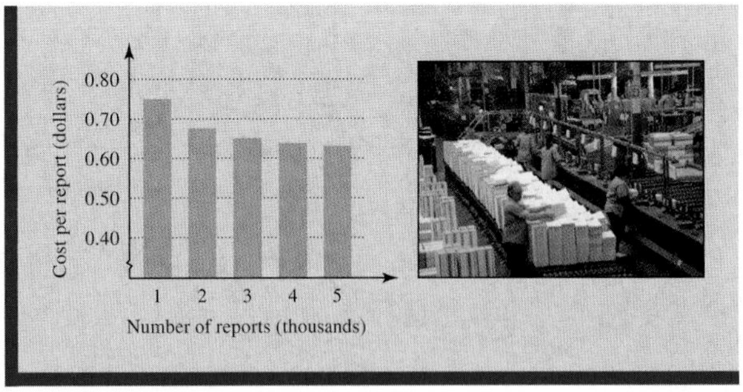

Figure for Exercise 103

104. ***Toxic pollutants.*** The annual cost in dollars for removing $p\%$ of the toxic chemicals from a town's water supply is given by the formula

$$C(p) = \frac{500{,}000}{100 - p}.$$

a) Use the accompanying graph to estimate the cost for removing 90% and 95% of the toxic chemicals.

b) Use the formula to determine the cost for removing 99.5% of the toxic chemicals.

c) What happens to the cost as the percentage of pollutants removed approaches 100%?

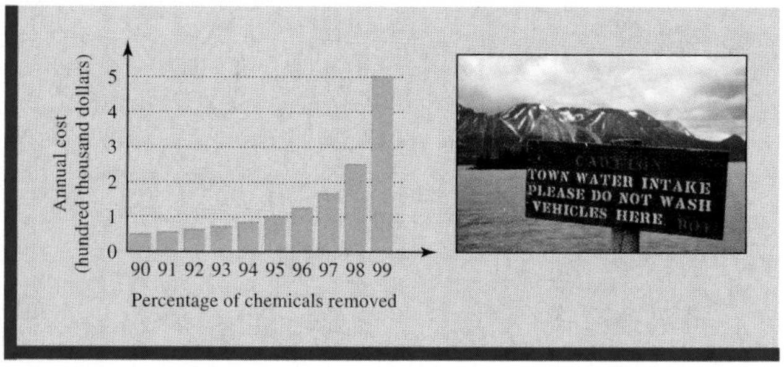

Figure for Exercise 104

7.2 Multiplication and Division

In this Section

- Multiplication of Rational Numbers
- Multiplication of Rational Expressions
- Division of Rational Numbers
- Division of Rational Expressions
- Applications

In Section 7.1 you learned to reduce rational expressions in the same way that we reduce rational numbers. In this section we will multiply and divide rational expressions using the same procedures that we use for rational numbers.

Multiplication of Rational Numbers

Two rational numbers are multiplied by multiplying their numerators and multiplying their denominators.

Multiplication of Rational Numbers

If $b \neq 0$ and $d \neq 0$, then

$$\frac{a}{b} \cdot \frac{c}{d} = \frac{ac}{bd}.$$

EXAMPLE 1

Multiplying rational numbers

Find the product $\frac{6}{7} \cdot \frac{14}{15}$.

Solution

The product is found by multiplying the numerators and multiplying the denominators:

$$\frac{6}{7} \cdot \frac{14}{15} = \frac{84}{105}$$

$$= \frac{21 \cdot 4}{21 \cdot 5} \quad \text{Factor the numerator and denominator}$$

$$= \frac{4}{5} \quad \text{Divide out the GCF 21.}$$

Helpful Hint

Did you know that the line separating the numerator and denominator in a fraction is called the *vinculum?*

The reducing that we did after multiplying is easier to do before multiplying. First factor all terms, reduce, and then multiply:

$$\frac{6}{7} \cdot \frac{14}{15} = \frac{2 \cdot \cancel{3}}{\cancel{7}} \cdot \frac{2 \cdot \cancel{7}}{\cancel{3} \cdot 5}$$
$$= \frac{4}{5}$$

Now do Exercises 5–12

Multiplication of Rational Expressions

We multiply rational expressions in the same way we multiply rational numbers. As with rational numbers, we can factor, reduce, and then multiply.

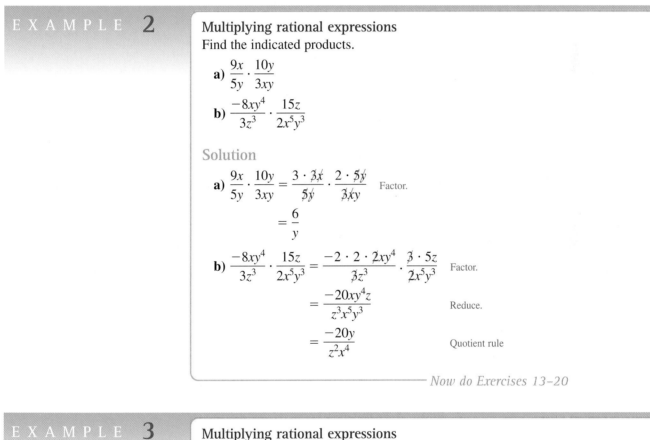

EXAMPLE 2

Multiplying rational expressions

Find the indicated products.

a) $\frac{9x}{5y} \cdot \frac{10y}{3xy}$

b) $\frac{-8xy^4}{3z^3} \cdot \frac{15z}{2x^5y^3}$

Solution

a) $\frac{9x}{5y} \cdot \frac{10y}{3xy} = \frac{3 \cdot \cancel{3}\cancel{x}}{\cancel{5}\cancel{y}} \cdot \frac{2 \cdot \cancel{5}\cancel{y}}{\cancel{3}\cancel{x}y}$ Factor.

$$= \frac{6}{y}$$

b) $\frac{-8xy^4}{3z^3} \cdot \frac{15z}{2x^5y^3} = \frac{-2 \cdot 2 \cdot \cancel{2}xy^4}{\cancel{3}z^3} \cdot \frac{\cancel{3} \cdot 5z}{\cancel{2}x^5y^3}$ Factor.

$= \frac{-20xy^4z}{z^3x^5y^3}$ Reduce.

$= \frac{-20y}{z^2x^4}$ Quotient rule

Now do Exercises 13–20

EXAMPLE 3

Multiplying rational expressions

Find the indicated products.

a) $\frac{2x - 2y}{4} \cdot \frac{2x}{x^2 - y^2}$

b) $\frac{x^2 + 7x + 12}{2x + 6} \cdot \frac{x}{x^2 - 16}$

c) $\frac{a + b}{6a} \cdot \frac{8a^2}{a^2 + 2ab + b^2}$

Study Tip

We are all creatures of habit. When you find a place in which you study successfully, stick with it. Using the same place for studying will help you to concentrate and associate the place with good studying.

Solution

a) $\frac{2x - 2y}{4} \cdot \frac{2x}{x^2 - y^2} = \frac{\cancel{2}(\cancel{x - y})}{\cancel{2} \cdot \cancel{2}} \cdot \frac{\cancel{2} \cdot x}{(\cancel{x - y})(x + y)}$ Factor.

$= \frac{x}{x + y}$ Reduce.

b) $\frac{x^2 + 7x + 12}{2x + 6} \cdot \frac{x}{x^2 - 16} = \frac{(\cancel{x + 3})(\cancel{x + 4})}{2(\cancel{x + 3})} \cdot \frac{x}{(x - 4)(\cancel{x + 4})}$ Factor.

$= \frac{x}{2(x - 4)}$ Reduce.

c) $\frac{a + b}{6a} \cdot \frac{8a^2}{a^2 + 2ab + b^2} = \frac{a + b}{2 \cdot 3a} \cdot \frac{2 \cdot 4a^2}{(a + b)^2}$ Factor.

$= \frac{4a}{3(a + b)}$ Reduce.

Now do Exercises 21–28

Division of Rational Numbers

Division of rational numbers can be accomplished by multiplying by the reciprocal of the divisor.

Division of Rational Numbers

If $b \neq 0$, $c \neq 0$, and $d \neq 0$, then

$$\frac{a}{b} \div \frac{c}{d} = \frac{a}{b} \cdot \frac{d}{c}.$$

EXAMPLE 4

Dividing rational numbers

Find each quotient.

a) $5 \div \frac{1}{2}$ **b)** $\frac{6}{7} \div \frac{3}{14}$

Solution

a) $5 \div \frac{1}{2} = 5 \cdot 2 = 10$

b) $\frac{6}{7} \div \frac{3}{14} = \frac{6}{7} \cdot \frac{14}{3} = \frac{2 \cdot \cancel{3}}{\cancel{7}} \cdot \frac{2 \cdot \cancel{7}}{\cancel{3}} = 4$

Now do Exercises 29–36

Division of Rational Expressions

We divide rational expressions in the same way we divide rational numbers: Invert the divisor and multiply.

EXAMPLE 5

Dividing rational expressions

Find each quotient.

a) $\dfrac{5}{3x} \div \dfrac{5}{6x}$

b) $\dfrac{x^7}{2} \div (2x^2)$

c) $\dfrac{4 - x^2}{x^2 + x} \div \dfrac{x - 2}{x^2 - 1}$

Solution

a) $\dfrac{5}{3x} \div \dfrac{5}{6x} = \dfrac{5}{3x} \cdot \dfrac{6x}{5}$ Invert the divisor and multiply.

$= \dfrac{\cancel{5}}{\cancel{3x}} \cdot \dfrac{2 \cdot \cancel{3x}}{\cancel{5}}$ Factor.

$= 2$ Divide out the common factors.

b) $\dfrac{x^7}{2} \div (2x^2) = \dfrac{x^7}{2} \cdot \dfrac{1}{2x^2}$ Invert and multiply.

$= \dfrac{x^5}{4}$ Quotient rule

c) $\dfrac{4 - x^2}{x^2 + x} \div \dfrac{x - 2}{x^2 - 1} = \dfrac{4 - x^2}{x^2 + x} \cdot \dfrac{x^2 - 1}{x - 2}$ Invert and multiply.

$= \dfrac{\overset{-1}{\cancel{(2 - x)}}(2 + x)}{x\cancel{(x + 1)}} \cdot \dfrac{\cancel{(x + 1)}(x - 1)}{\cancel{x - 2}}$ Factor.

$= \dfrac{-1(2 + x)(x - 1)}{x}$ $\dfrac{2 - x}{x - 2} = -1$

$= \dfrac{-1(x^2 + x - 2)}{x}$ Simplify.

$= \dfrac{-x^2 - x + 2}{x}$

Now do Exercises 37–50

Helpful Hint

A doctor told a nurse to give a patient half of the usual dose of a certain medicine. The nurse figured, "dividing in half means dividing by 1/2 which means multiply by 2." So the patient got four times the prescribed amount and died (true story). There is a big difference between dividing a quantity in half and dividing by one-half.

We sometimes write division of rational expressions using the fraction bar. For example, we can write

$$\frac{a + b}{3} \div \frac{1}{6} \quad \text{as} \quad \frac{\frac{a + b}{3}}{\frac{1}{6}}.$$

No matter how division is expressed, we invert the divisor and multiply.

EXAMPLE 6

Division expressed with a fraction bar

Find each quotient.

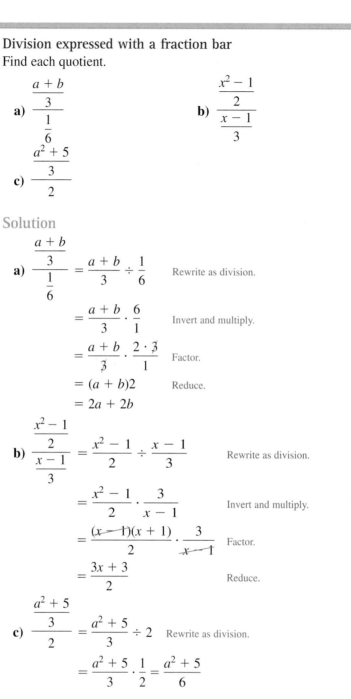

a) $\dfrac{\frac{a+b}{3}}{\frac{1}{6}}$

b) $\dfrac{\frac{x^2-1}{2}}{\frac{x-1}{3}}$

c) $\dfrac{\frac{a^2+5}{3}}{2}$

Solution

a) $\dfrac{\frac{a+b}{3}}{\frac{1}{6}} = \dfrac{a+b}{3} \div \dfrac{1}{6}$ Rewrite as division.

$= \dfrac{a+b}{3} \cdot \dfrac{6}{1}$ Invert and multiply.

$= \dfrac{a+b}{\not{3}} \cdot \dfrac{2 \cdot \not{3}}{1}$ Factor.

$= (a+b)2$ Reduce.

$= 2a + 2b$

b) $\dfrac{\frac{x^2-1}{2}}{\frac{x-1}{3}} = \dfrac{x^2-1}{2} \div \dfrac{x-1}{3}$ Rewrite as division.

$= \dfrac{x^2-1}{2} \cdot \dfrac{3}{x-1}$ Invert and multiply.

$= \dfrac{\cancel{(x-1)}(x+1)}{2} \cdot \dfrac{3}{\cancel{x-1}}$ Factor.

$= \dfrac{3x+3}{2}$ Reduce.

c) $\dfrac{\frac{a^2+5}{3}}{2} = \dfrac{a^2+5}{3} \div 2$ Rewrite as division.

$= \dfrac{a^2+5}{3} \cdot \dfrac{1}{2} = \dfrac{a^2+5}{6}$

Now do Exercises 51–58

Applications

We saw in Section 7.1 that rational expressions can be used to represent rates. Note that there are several ways to write rates. For example, miles per hour is written mph, mi/hr, or $\frac{\text{mi}}{\text{hr}}$. The last way is best when doing operations with rates because it helps us reconcile our answers. Notice how hours "cancels" when we multiply miles per hour and hours in Example 7, giving an answer in miles, as it should be.

EXAMPLE 7

Writing rational expressions

Answer each question with a rational expression.

a) Shasta averaged $\frac{200}{x}$ mph for x hours before she had lunch. How many miles did she drive in the first 3 hours after lunch assuming that she continued to average $\frac{200}{x}$ mph?

b) If a bathtub can be filled in x minutes, then the rate at which it is filling is $\frac{1}{x}$ tub/min. How much of the tub is filled in 10 minutes?

Solution

a) Because $R \cdot T = D$, the distance she traveled after lunch is the product of the rate and time:

$$\frac{200}{x}\frac{\text{mi}}{\cancel{\text{hr}}} \cdot 3\,\cancel{\text{hr}} = \frac{600}{x}\text{ mi}$$

b) Because $R \cdot T = W$, the work completed is the product of the rate and time:

$$\frac{1}{x}\frac{\text{tub}}{\cancel{\text{min}}} \cdot 10\,\cancel{\text{min}} = \frac{10}{x}\text{ tub}$$

Now do Exercises 79–82

Warm-Ups

True or false? Explain your answer.

1. $\frac{2}{3} \cdot \frac{5}{3} = \frac{10}{9}$.
2. The product of $\frac{x-7}{3}$ and $\frac{6}{7-x}$ is -2.
3. Dividing by 2 is equivalent to multiplying by $\frac{1}{2}$.
4. $3 \div x = \frac{1}{3} \cdot x$ for any nonzero number x.
5. Factoring polynomials is essential in multiplying rational expressions.
6. One-half of one-fourth is one-sixth.
7. One-half divided by three is three-halves.
8. The quotient of $(839 - 487)$ and $(487 - 839)$ is -1.
9. $\frac{a}{3} \div 3 = \frac{a}{9}$ for any value of a.
10. $\frac{a}{b} \cdot \frac{b}{a} = 1$ for any nonzero values of a and b.

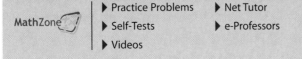

7.2 Exercises

Reading and Writing *After reading this section, write out the answers to these questions. Use complete sentences.*

1. How do you multiply rational numbers?

2. How do you multiply rational expressions?

3. What can be done to simplify the process of multiplying rational numbers or rational expressions?

4. How do you divide rational numbers or rational expressions?

Perform the indicated operation. See Example 1.

5. $\frac{2}{3} \cdot \frac{5}{6}$ **6.** $\frac{3}{4} \cdot \frac{2}{5}$ **7.** $\frac{8}{15} \cdot \frac{35}{24}$

8. $\frac{3}{4} \cdot \frac{8}{21}$ **9.** $\frac{12}{17} \cdot \frac{51}{10}$ **10.** $\frac{25}{48} \cdot \frac{56}{35}$

11. $24 \cdot \frac{7}{20}$ **12.** $\frac{3}{10} \cdot 35$

Perform the indicated operation. See Example 2.

13. $\frac{2x}{3} \cdot \frac{5}{4x}$ **14.** $\frac{3y}{7} \cdot \frac{21}{2y}$

15. $\frac{5a}{12b} \cdot \frac{3ab}{55a}$ **16.** $\frac{3m}{7p} \cdot \frac{35p}{6mp}$

17. $\frac{-2x^6}{7a^5} \cdot \frac{21a^2}{6x}$ **18.** $\frac{5z^3w}{-9y^3} \cdot \frac{-6y^5}{20z^9}$

19. $\frac{15t^3y^5}{20w^7} \cdot 24t^5w^3y^2$ **20.** $22x^2y^3z \cdot \frac{6x^5}{33y^3z^4}$

Perform the indicated operation. See Example 3.

21. $\frac{2x + 2y}{7} \cdot \frac{15}{6x + 6y}$

22. $\frac{3}{a^2 + a} \cdot \frac{2a + 2}{6}$

23. $\frac{3a + 3b}{15} \cdot \frac{10a}{a^2 - b^2}$

24. $\frac{b^3 + b}{5} \cdot \frac{10}{b^2 + b}$

25. $(x^2 - 6x + 9) \cdot \frac{3}{x - 3}$

26. $\frac{12}{4x + 10} \cdot (4x^2 + 20x + 25)$

27. $\frac{16a + 8}{5a^2 + 5} \cdot \frac{2a^2 + a - 1}{4a^2 - 1}$

28. $\frac{6x - 18}{2x^2 - 5x - 3} \cdot \frac{4x^2 + 4x + 1}{6x + 3}$

Perform the indicated operation. See Example 4.

29. $\frac{1}{4} \div \frac{1}{2}$ **30.** $\frac{1}{6} \div \frac{1}{2}$ **31.** $12 \div \frac{2}{5}$

32. $32 \div \frac{1}{4}$ **33.** $\frac{5}{7} \div \frac{15}{14}$ **34.** $\frac{3}{4} \div \frac{15}{2}$

35. $\frac{40}{3} \div 12$ **36.** $\frac{22}{9} \div 9$

Perform the indicated operation. See Example 5.

37. $\frac{x^2}{4} \div \frac{x}{2}$ **38.** $\frac{3}{2a^2} \div \frac{6}{2a}$

39. $\frac{5x^2}{3} \div \frac{10x}{21}$ **40.** $\frac{4u^2}{3v} \div \frac{14u}{15v^6}$

41. $\frac{8m^3}{n^4} \div (12mn^2)$ **42.** $\frac{2p^4}{3q^3} \div (4pq^5)$

43. $\frac{y - 6}{2} \div \frac{6 - y}{6}$

44. $\frac{4 - a}{5} \div \frac{a^2 - 16}{3}$

45. $\frac{x^2 + 4x + 4}{8} \div \frac{(x + 2)^3}{16}$

46. $\frac{a^2 + 2a + 1}{3} \div \frac{a^2 - 1}{a}$

47. $\frac{t^2 + 3t - 10}{t^2 - 25} \div (4t - 8)$

48. $\frac{w^2 - 7w + 12}{w^2 - 4w} \div (w^2 - 9)$

49. $(2x^2 - 3x - 5) \div \frac{2x - 5}{x - 1}$

50. $(6y^2 - y - 2) \div \frac{2y + 1}{3y - 2}$

Perform the indicated operation. See Example 6.

51. $\frac{\frac{x - 2y}{5}}{\frac{1}{10}}$ **52.** $\frac{\frac{3m + 6n}{8}}{\frac{3}{4}}$ **53.** $\frac{\frac{x^2 - 4}{12}}{\frac{x - 2}{6}}$

54. $\frac{\frac{6a^2 + 6}{5}}{\frac{6a + 6}{5}}$ **55.** $\frac{\frac{x^2 + 9}{3}}{5}$ **56.** $\frac{\frac{1}{a - 3}}{4}$

57. $\dfrac{\dfrac{x^2 - y^2}{x - y}}{9}$

58. $\dfrac{\dfrac{x^2 + 6x + 8}{x + 2}}{x + 1}$

Perform the indicated operation.

59. $\dfrac{x - 1}{3} \cdot \dfrac{9}{1 - x}$

60. $\dfrac{2x - 2y}{3} \cdot \dfrac{1}{y - x}$

61. $\dfrac{3a + 3b}{a} \cdot \dfrac{1}{3}$

62. $\dfrac{a - b}{2b - 2a} \cdot \dfrac{2}{5}$

63. $\dfrac{\frac{b}{a}}{\frac{1}{2}}$

64. $\dfrac{\frac{2g}{3h}}{\frac{1}{h}}$

65. $\dfrac{6y}{3} \div (2x)$

66. $\dfrac{8x}{9} \div (18x)$

67. $\dfrac{a^3b^4}{-2ab^2} \cdot \dfrac{a^5b^7}{ab}$

68. $\dfrac{-2a^2}{3a^2} \cdot \dfrac{20a}{15a^3}$

69. $\dfrac{2mn^4}{6mn^2} \div \dfrac{3m^5n^7}{m^2n^4}$

70. $\dfrac{rt^2}{rt^2} \div \dfrac{rt^2}{r^3t^2}$

71. $\dfrac{3x^2 + 16x + 5}{x} \cdot \dfrac{x^2}{9x^2 - 1}$

72. $\dfrac{x^2 + 6x + 5}{x} \cdot \dfrac{x^4}{3x + 3}$

73. $\dfrac{a^2 - 2a + 4}{a^2 - 4} \cdot \dfrac{(a + 2)^3}{2a + 4}$

74. $\dfrac{w^2 - 1}{(w - 1)^2} \cdot \dfrac{w - 1}{w^2 + 2w + 1}$

75. $\dfrac{2x^2 + 19x - 10}{x^2 - 100} \div \dfrac{4x^2 - 1}{2x^2 - 19x - 10}$

76. $\dfrac{x^3 - 1}{x^2 + 1} \div \dfrac{9x^2 + 9x + 9}{x^2 - x}$

77. $\dfrac{9 + 6m + m^2}{9 - 6m + m^2} \cdot \dfrac{m^2 - 9}{m^2 + mk + 3m + 3k}$

78. $\dfrac{3x + 3w + bx + bw}{x^2 - w^2} \cdot \dfrac{6 - 2b}{9 - b^2}$

Solve each problem. Answers could be rational expressions. Be sure to give your answer with appropriate units. See Example 7.

79. ***Distance.*** Florence averaged $\frac{26.2}{x}$ mph for the x hours in which she ran the Boston Marathon. If she ran at that same rate for $\frac{1}{2}$ hour in the Manchac Fun Run, then how many miles did she run at Manchac?

80. ***Work.*** Henry sold 120 magazine subscriptions in $x + 2$ days. If he sold at the same rate for another week, then how many magazines did he sell in the extra week?

81. ***Area of a rectangle.*** If the length of a rectangular flag is x meters and its width is $\frac{5}{x}$ meters, then what is the area of the rectangle?

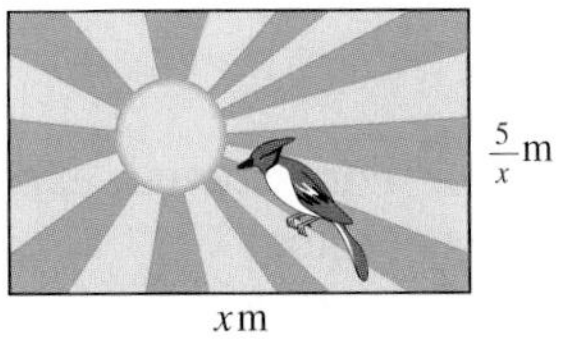

Figure for Exercise 81

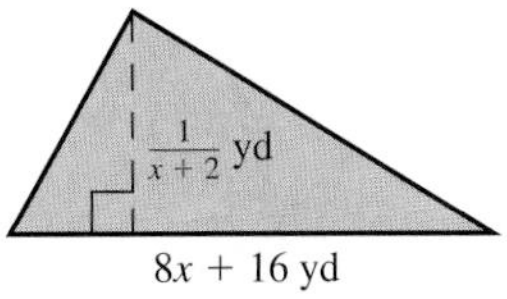

Figure for Exercise 82

82. ***Area of a triangle.*** If the base of a triangle is $8x + 16$ yards and its height is $\frac{1}{x + 2}$ yards, then what is the area of the triangle?

Getting More Involved

83. ***Discussion***

Evaluate each expression.

a) One-half of $\dfrac{1}{4}$ **b)** One-third of 4

c) One-half of $\dfrac{4x}{3}$ **d)** One-half of $\dfrac{3x}{2}$

84. ***Exploration***

Let $R = \dfrac{6x^2 + 23x + 20}{24x^2 + 29x - 4}$ and $H = \dfrac{2x + 5}{8x - 1}$.

a) Find R when $x = 2$ and $x = 3$. Find H when $x = 2$ and $x = 3$.

b) How are these values of R and H related and why?

7.3 Finding the Least Common Denominator

In this Section

- Building Up the Denominator
- Finding the Least Common Denominator
- Converting to the LCD

Every rational expression can be written in infinitely many equivalent forms. Because we can add or subtract only fractions with identical denominators, we must be able to change the denominator of a fraction. You have already learned how to change the denominator of a fraction by reducing. In this section you will learn the opposite of reducing, which is called **building up the denominator.**

Building Up the Denominator

To convert the fraction $\frac{2}{3}$ into an equivalent fraction with a denominator of 21, we factor 21 as $21 = 3 \cdot 7$. Because $\frac{2}{3}$ already has a 3 in the denominator, multiply the numerator and denominator of $\frac{2}{3}$ by the missing factor 7 to get a denominator of 21:

$$\frac{2}{3} = \frac{2}{3} \cdot \frac{7}{7} = \frac{14}{21}$$

For rational expressions the process is the same. To convert the rational expression

$$\frac{5}{x + 3}$$

into an equivalent rational expression with a denominator of $x^2 - x - 12$, first factor $x^2 - x - 12$:

$$x^2 - x - 12 = (x + 3)(x - 4)$$

From the factorization we can see that the denominator $x + 3$ needs only a factor of $x - 4$ to have the required denominator. So multiply the numerator and denominator by the missing factor $x - 4$:

$$\frac{5}{x + 3} = \frac{5(x - 4)}{(x + 3)(x - 4)} = \frac{5x - 20}{x^2 - x - 12}$$

EXAMPLE 1

Building up the denominator

Build each rational expression into an equivalent rational expression with the indicated denominator.

a) $3 = \frac{?}{12}$ **b)** $\frac{3}{w} = \frac{?}{wx}$ **c)** $\frac{2}{3y^3} = \frac{?}{12y^8}$

Solution

a) Because $3 = \frac{3}{1}$, we get a denominator of 12 by multiplying the numerator and denominator by 12:

$$3 = \frac{3}{1} = \frac{3 \cdot 12}{1 \cdot 12} = \frac{36}{12}$$

b) Multiply the numerator and denominator by x:

$$\frac{3}{w} = \frac{3 \cdot x}{w \cdot x} = \frac{3x}{wx}$$

c) To build the denominator $3y^3$ up to $12y^8$, multiply by $4y^5$:

$$\frac{2}{3y^3} = \frac{2 \cdot 4y^5}{3y^3 \cdot 4y^5} = \frac{8y^5}{12y^8}$$

Now do Exercises 5–22

In Example 2 we must factor the original denominator before building up the denominator.

EXAMPLE 2

Building up the denominator

Build each rational expression into an equivalent rational expression with the indicated denominator.

a) $\dfrac{7}{3x - 3y} = \dfrac{?}{6y - 6x}$ **b)** $\dfrac{x - 2}{x + 2} = \dfrac{?}{x^2 + 8x + 12}$

Helpful Hint

Notice that reducing and building up are exactly the opposite of each other. In reducing you remove a factor that is common to the numerator and denominator, and in building up you put a common factor into the numerator and denominator.

Solution

a) Because $3x - 3y = 3(x - y)$, we factor -6 out of $6y - 6x$. This will give a factor of $x - y$ in each denominator:

$$3x - 3y = 3(x - y)$$
$$6y - 6x = -6(x - y) = -2 \cdot 3(x - y)$$

To get the required denominator, we multiply the numerator and denominator by -2 only:

$$\frac{7}{3x - 3y} = \frac{7(-2)}{(3x - 3y)(-2)} = \frac{-14}{6y - 6x}$$

b) Because $x^2 + 8x + 12 = (x + 2)(x + 6)$, we multiply the numerator and denominator by $x + 6$, the missing factor:

$$\frac{x - 2}{x + 2} = \frac{(x - 2)(x + 6)}{(x + 2)(x + 6)} = \frac{x^2 + 4x - 12}{x^2 + 8x + 12}$$

Now do Exercises 23–34

CAUTION When building up a denominator, *both* the numerator and the denominator must be multiplied by the appropriate expression, because that is how we build up fractions.

Finding the Least Common Denominator

We can use the idea of building up the denominator to convert two fractions with different denominators into fractions with identical denominators. For example,

$$\frac{5}{6} \quad \text{and} \quad \frac{1}{4}$$

can both be converted into fractions with a denominator of 12, since $12 = 2 \cdot 6$ and $12 = 3 \cdot 4$:

$$\frac{5}{6} = \frac{5 \cdot 2}{6 \cdot 2} = \frac{10}{12} \qquad \frac{1}{4} = \frac{1 \cdot 3}{4 \cdot 3} = \frac{3}{12}$$

The smallest number that is a multiple of all of the denominators is called the **least common denominator (LCD).** The LCD for the denominators 6 and 4 is 12.

To find the LCD in a systematic way, we look at a complete factorization of each denominator. Consider the denominators 24 and 30:

$$24 = 2 \cdot 2 \cdot 2 \cdot 3 = 2^3 \cdot 3$$
$$30 = 2 \cdot 3 \cdot 5$$

Any multiple of 24 must have three 2's in its factorization, and any multiple of 30 must have one 2 as a factor. So a number with three 2's in its factorization will have enough to be a multiple of both 24 and 30. The LCD must also have one 3 and one 5 in its factorization. *We use each factor the maximum number of times it appears in either factorization.* So the LCD is $2^3 \cdot 3 \cdot 5$:

$$2^3 \cdot 3 \cdot 5 = \overbrace{2 \cdot 2 \cdot 2 \cdot 3}^{24} \cdot 5 = 120, \quad \text{where } \underbrace{2 \cdot 3 \cdot 5}_{30}$$

If we omitted any one of the factors in $2 \cdot 2 \cdot 2 \cdot 3 \cdot 5$, we would not have a multiple of both 24 and 30. That is what makes 120 the *least* common denominator. To find the LCD for two polynomials, we use the same strategy.

Study Tip

Studying in a quiet place is better than studying in a noisy place. There are very few people who can listen to music or a conversation and study at the same time.

Strategy for Finding the LCD for Polynomials

1. Factor each denominator completely. Use exponent notation for repeated factors.
2. Write the product of all of the different factors that appear in the denominators.
3. On each factor, use the highest power that appears on that factor in any of the denominators.

EXAMPLE 3

Finding the LCD

If the given expressions were used as denominators of rational expressions, then what would be the LCD for each group of denominators?

a) 20, 50 **b)** x^3yz^2, x^5y^2z, xyz^5

c) $a^2 + 5a + 6, a^2 + 4a + 4$

Solution

a) First factor each number completely:

$$20 = 2^2 \cdot 5 \qquad 50 = 2 \cdot 5^2$$

The highest power of 2 is 2, and the highest power of 5 is 2. So the LCD of 20 and 50 is $2^2 \cdot 5^2$, or 100.

b) The expressions x^3yz^2, x^5y^2z, and xyz^5 are already factored. For the LCD, use the highest power of each variable. So the LCD is $x^5y^2z^5$.

c) First factor each polynomial.

$$a^2 + 5a + 6 = (a + 2)(a + 3) \qquad a^2 + 4a + 4 = (a + 2)^2$$

The highest power of $(a + 3)$ is 1, and the highest power of $(a + 2)$ is 2. So the LCD is $(a + 3)(a + 2)^2$.

Now do Exercises 35–48

Converting to the LCD

When adding or subtracting rational expressions, we must convert the expressions into expressions with identical denominators. To keep the computations as simple as possible, we use the least common denominator.

EXAMPLE 4

Converting to the LCD

Find the LCD for the rational expressions, and convert each expression into an equivalent rational expression with the LCD as the denominator.

a) $\dfrac{4}{9xy}, \dfrac{2}{15xz}$ **b)** $\dfrac{5}{6x^2}, \dfrac{1}{8x^3y}, \dfrac{3}{4y^2}$

Helpful Hint

What is the difference between LCD, GCF, CBS, and NBC? The LCD for the denominators 4 and 6 is 12. The *least* common denominator is *greater than* or equal to both numbers. The GCF for 4 and 6 is 2. The *greatest* common factor is *less than* or equal to both numbers. CBS and NBC are TV networks.

Solution

a) Factor each denominator completely:

$$9xy = 3^2xy \qquad 15xz = 3 \cdot 5xz$$

The LCD is $3^2 \cdot 5xyz$. Now convert each expression into an expression with this denominator. We must multiply the numerator and denominator of the first rational expression by $5z$ and the second by $3y$:

$$\left.\begin{aligned} \frac{4}{9xy} &= \frac{4 \cdot 5z}{9xy \cdot 5z} = \frac{20z}{45xyz} \\ \frac{2}{15xz} &= \frac{2 \cdot 3y}{15xz \cdot 3y} = \frac{6y}{45xyz} \end{aligned}\right\} \text{Same denominator}$$

b) Factor each denominator completely:

$$6x^2 = 2 \cdot 3x^2 \qquad 8x^3y = 2^3x^3y \qquad 4y^2 = 2^2y^2$$

The LCD is $2^3 \cdot 3 \cdot x^3y^2$ or $24x^3y^2$. Now convert each expression into an expression with this denominator:

$$\frac{5}{6x^2} = \frac{5 \cdot 4xy^2}{6x^2 \cdot 4xy^2} = \frac{20xy^2}{24x^3y^2}$$

$$\frac{1}{8x^3y} = \frac{1 \cdot 3y}{8x^3y \cdot 3y} = \frac{3y}{24x^3y^2}$$

$$\frac{3}{4y^2} = \frac{3 \cdot 6x^3}{4y^2 \cdot 6x^3} = \frac{18x^3}{24x^3y^2}$$

Now do Exercises 49–60

EXAMPLE 5

Converting to the LCD

Find the LCD for the rational expressions

$$\frac{5x}{x^2 - 4} \quad \text{and} \quad \frac{3}{x^2 + x - 6}$$

and convert each into an equivalent rational expression with that denominator.

Solution

First factor the denominators:

$$x^2 - 4 = (x - 2)(x + 2)$$
$$x^2 + x - 6 = (x - 2)(x + 3)$$

The LCD is $(x - 2)(x + 2)(x + 3)$. Now we multiply the numerator and denominator of the first rational expression by $(x + 3)$ and those of the second rational expression by $(x + 2)$. Because each denominator already has one factor of $(x - 2)$, there is no reason to multiply by $(x - 2)$. We multiply each denominator by the factors in the LCD that are missing from that denominator:

$$\frac{5x}{x^2 - 4} = \frac{5x(x + 3)}{(x - 2)(x + 2)(x + 3)} = \frac{5x^2 + 15x}{(x - 2)(x + 2)(x + 3)}$$

$$\frac{3}{x^2 + x - 6} = \frac{3(x + 2)}{(x - 2)(x + 3)(x + 2)} = \frac{3x + 6}{(x - 2)(x + 2)(x + 3)}$$

Same denominator

Now do Exercises 61–72

Note that in Example 5 we multiplied the expressions in the numerators but left the denominators in factored form. The numerators are simplified because it is the numerators that must be added when we add rational expressions in Section 7.4. Because we can add rational expressions with identical denominators, there is no need to multiply the denominators.

Warm-Ups ▼

True or false? Explain your answer.

1. To convert $\frac{2}{3}$ into an equivalent fraction with a denominator of 18, we would multiply only the denominator of $\frac{2}{3}$ by 6.
2. Factoring has nothing to do with finding the least common denominator.
3. $\frac{3}{2ab^2} = \frac{15a^2b^2}{10a^3b^4}$ for any nonzero values of a and b.
4. The LCD for the denominators $2^5 \cdot 3$ and $2^4 \cdot 3^2$ is $2^5 \cdot 3^2$.
5. The LCD for the fractions $\frac{1}{6}$ and $\frac{1}{10}$ is 60.
6. The LCD for the denominators $6a^2b$ and $4ab^3$ is $2ab$.
7. The LCD for the denominators $a^2 + 1$ and $a + 1$ is $a^2 + 1$.
8. $\frac{x}{2} = \frac{x+7}{2+7}$ for any real number x.
9. The LCD for the rational expressions $\frac{1}{x-2}$ and $\frac{3}{x+2}$ is $x^2 - 4$.
10. $x = \frac{3x}{3}$ for any real number x.

7.3 Exercises

Reading and Writing *After reading this section, write out the answers to these questions. Use complete sentences.*

1. What is building up the denominator?

2. How do we build up the denominator of a rational expression?

3. What is the least common denominator for fractions?

4. How do you find the LCD for two polynomial denominators?

Build each rational expression into an equivalent rational expression with the indicated denominator. See Example 1.

5. $\frac{1}{3} = \frac{?}{27}$

6. $\frac{2}{5} = \frac{?}{35}$

7. $\frac{3}{4} = \frac{?}{16}$

8. $\frac{3}{7} = \frac{?}{28}$

9. $2 = \frac{?}{6}$

10. $5 = \frac{?}{12}$

11. $\frac{5}{x} = \frac{?}{ax}$

12. $\frac{x}{3} = \frac{?}{3x}$

13. $7 = \frac{?}{2x}$

14. $6 = \frac{?}{4y}$

15. $\frac{5}{b} = \frac{?}{3bt}$

16. $\frac{7}{2ay} = \frac{?}{2ayz}$

17. $\frac{-9z}{2aw} = \frac{?}{8awz}$

18. $\frac{-7yt}{3x} = \frac{?}{18xyt}$

19. $\frac{2}{3a} = \frac{?}{15a^3}$

20. $\frac{7b}{12c^5} = \frac{?}{36c^8}$

21. $\frac{4}{5xy^2} = \frac{?}{10x^2y^5}$

22. $\frac{5y^2}{8x^3z} = \frac{?}{24x^5z^3}$

Build each rational expression into an equivalent rational expression with the indicated denominator. See Example 2.

23. $\dfrac{5}{x+3} = \dfrac{?}{2x+6}$

24. $\dfrac{4}{a-5} = \dfrac{?}{3a-15}$

25. $\dfrac{5}{2x+2} = \dfrac{?}{-8x-8}$

26. $\dfrac{3}{m-n} = \dfrac{?}{2n-2m}$

27. $\dfrac{8a}{5b^2-5b} = \dfrac{?}{20b^2-20b^3}$

28. $\dfrac{5x}{-6x-9} = \dfrac{?}{18x^2+27x}$

29. $\dfrac{3}{x+2} = \dfrac{?}{x^2-4}$

30. $\dfrac{a}{a+3} = \dfrac{?}{a^2-9}$

31. $\dfrac{3x}{x+1} = \dfrac{?}{x^2+2x+1}$

32. $\dfrac{-7x}{2x-3} = \dfrac{?}{4x^2-12x+9}$

33. $\dfrac{y-6}{y-4} = \dfrac{?}{y^2+y-20}$

34. $\dfrac{z-6}{z+3} = \dfrac{?}{z^2-2z-15}$

If the given expressions were used as denominators of rational expressions, then what would be the LCD for each group of denominators? See Example 3.

35. 12, 16

36. 28, 42

37. 12, 18, 20

38. 24, 40, 48

39. $6a^2$, $15a$

40. $18x^2$, $20xy$

41. $2a^4b$, $3ab^6$, $4a^3b^2$

42. $4m^3nw$, $6mn^5w^8$, $9m^6nw$

43. x^2-16, $x^2+8x+16$

44. x^2-9, x^2+6x+9

45. x, $x+2$, $x-2$

46. y, $y-5$, $y+2$

47. x^2-4x, x^2-16, $2x$

48. y, y^2-3y, $3y$

Find the LCD for the given rational expressions, and convert each rational expression into an equivalent rational expression with the LCD as the denominator. See Example 4.

49. $\dfrac{1}{6}, \dfrac{3}{8}$

50. $\dfrac{5}{12}, \dfrac{3}{20}$

51. $\dfrac{1}{2x}, \dfrac{5}{6x}$

52. $\dfrac{3}{5x}, \dfrac{1}{10x}$

53. $\dfrac{2}{3a}, \dfrac{1}{2b}$

54. $\dfrac{y}{4x}, \dfrac{x}{6y}$

55. $\dfrac{3}{84a}, \dfrac{5}{63b}$

56. $\dfrac{4b}{75a}, \dfrac{6}{105ab}$

57. $\dfrac{1}{3x^2}, \dfrac{3}{2x^5}$

58. $\dfrac{3}{8a^3b^9}, \dfrac{5}{6a^2c}$

59. $\dfrac{x}{9y^5z}, \dfrac{y}{12x^3}, \dfrac{1}{6x^2y}$

60. $\dfrac{5}{12a^6b}, \dfrac{3b}{14a^3}, \dfrac{1}{2ab^3}$

Find the LCD for the given rational expressions, and convert each rational expression into an equivalent rational expression with the LCD as the denominator. See Example 5.

61. $\dfrac{2x}{x-3}, \dfrac{5x}{x+2}$

62. $\dfrac{2a}{a-5}, \dfrac{3a}{a+2}$

63. $\dfrac{4}{a-6}, \dfrac{5}{6-a}$

64. $\dfrac{4}{x-y}, \dfrac{5x}{2y-2x}$

65. $\dfrac{x}{x^2-9}, \dfrac{5x}{x^2-6x+9}$

66. $\dfrac{5x}{x^2-1}, \dfrac{-4}{x^2-2x+1}$

67. $\dfrac{w+2}{w^2-2w-15}, \dfrac{-2w}{w^2-4w-5}$

68. $\dfrac{z-1}{z^2+6z+8}, \dfrac{z+1}{z^2+5z+6}$

69. $\dfrac{-5}{6x-12}, \dfrac{x}{x^2-4}, \dfrac{3}{2x+4}$

70. $\dfrac{3}{4b^2-9}, \dfrac{2b}{2b+3}, \dfrac{-5}{2b^2-3b}$

71. $\dfrac{2}{2q^2-5q-3}, \dfrac{3}{2q^2+9q+4}, \dfrac{4}{q^2+q-12}$

72. $\dfrac{-3}{2p^2+7p-15}, \dfrac{p}{2p^2-11p+12}, \dfrac{2}{p^2+p-20}$

Getting More Involved

73. ***Discussion***

Why do we learn how to convert two rational expressions into equivalent rational expressions with the same denominator?

74. ***Discussion***

Which expression is the LCD for

$$\frac{3x-1}{2^2 \cdot 3 \cdot x^2(x+2)} \quad \text{and} \quad \frac{2x+7}{2 \cdot 3^2 \cdot x(x+2)^2}?$$

a) $2 \cdot 3 \cdot x(x+2)$
b) $36x(x+2)$
c) $36x^2(x+2)^2$
d) $2^3 \cdot 3^3 x^3(x+2)^2$

7.4 Addition and Subtraction

In this Section

- **Addition and Subtraction of Rational Numbers**
- **Addition and Subtraction of Rational Expressions**
- **Applications**

In Section 7.3 you learned how to find the LCD and build up the denominators of rational expressions. In this section we will use that knowledge to add and subtract rational expressions with different denominators.

Addition and Subtraction of Rational Numbers

We can add or subtract rational numbers (or fractions) only with identical denominators according to the following definition.

Addition and Subtraction of Rational Numbers

If $b \neq 0$, then

$$\frac{a}{b} + \frac{c}{b} = \frac{a+c}{b} \quad \text{and} \quad \frac{a}{b} - \frac{c}{b} = \frac{a-c}{b}.$$

EXAMPLE 1

Adding or subtracting fractions with the same denominator

Perform the indicated operations. Reduce answers to lowest terms.

a) $\frac{1}{12} + \frac{7}{12}$ **b)** $\frac{1}{4} - \frac{3}{4}$

Solution

a) $\frac{1}{12} + \frac{7}{12} = \frac{8}{12} = \frac{\not{4} \cdot 2}{\not{4} \cdot 3} = \frac{2}{3}$

b) $\frac{1}{4} - \frac{3}{4} = \frac{-2}{4} = -\frac{1}{2}$

Now do Exercises 5–12

If the rational numbers have different denominators, we must convert them to equivalent rational numbers that have identical denominators and then add or subtract. Of course, it is most efficient to use the least common denominator (LCD), as in the following example.

EXAMPLE 2

Adding or subtracting fractions with different denominators

Find each sum or difference.

a) $\frac{3}{20} + \frac{7}{12}$ **b)** $\frac{1}{6} - \frac{4}{15}$

Helpful Hint

Note how all of the operations with rational expressions are performed according to the rules for fractions. So keep thinking of how you perform operations with fractions and you will improve your skills with fractions and with rational expressions.

Solution

a) Because $20 = 2^2 \cdot 5$ and $12 = 2^2 \cdot 3$, the LCD is $2^2 \cdot 3 \cdot 5$, or 60. Convert each fraction to an equivalent fraction with a denominator of 60:

$$\frac{3}{20} + \frac{7}{12} = \frac{3 \cdot 3}{20 \cdot 3} + \frac{7 \cdot 5}{12 \cdot 5} \quad \text{Build up the denominators.}$$

$$= \frac{9}{60} + \frac{35}{60} \quad \text{Simplify numerators and denominators.}$$

$$= \frac{44}{60} \quad \text{Add the fractions.}$$

$$= \frac{4 \cdot 11}{4 \cdot 15} \quad \text{Factor.}$$

$$= \frac{11}{15} \quad \text{Reduce.}$$

b) Because $6 = 2 \cdot 3$ and $15 = 3 \cdot 5$, the LCD is $2 \cdot 3 \cdot 5$ or 30:

$$\frac{1}{6} - \frac{4}{15} = \frac{1}{2 \cdot 3} - \frac{4}{3 \cdot 5} \quad \text{Factor the denominators.}$$

$$= \frac{1 \cdot 5}{2 \cdot 3 \cdot 5} - \frac{4 \cdot 2}{3 \cdot 5 \cdot 2} \quad \text{Build up the denominators.}$$

$$= \frac{5}{30} - \frac{8}{30} \quad \text{Simplify the numerators and denominators.}$$

$$= \frac{-3}{30} \quad \text{Subtract.}$$

$$= \frac{-1 \cdot 3}{10 \cdot 3} \quad \text{Factor.}$$

$$= -\frac{1}{10} \quad \text{Reduce.}$$

Now do Exercises 13–20

Addition and Subtraction of Rational Expressions

Rational expressions are added or subtracted just like rational numbers. We can add or subtract only rational expressions that have identical denominators.

EXAMPLE 3

Rational expressions with the same denominator

Perform the indicated operations and reduce answers to lowest terms.

a) $\frac{2}{3y} + \frac{4}{3y}$

b) $\frac{2x}{x + 2} + \frac{4}{x + 2}$

c) $\frac{x^2 + 2x}{(x - 1)(x + 3)} - \frac{2x + 1}{(x - 1)(x + 3)}$

Solution

a) $$\frac{2}{3y} + \frac{4}{3y} = \frac{6}{3y} \quad \text{Add the fractions.}$$

$$= \frac{2}{y} \quad \text{Reduce.}$$

b) $$\frac{2x}{x + 2} + \frac{4}{x + 2} = \frac{2x + 4}{x + 2} \quad \text{Add the fractions.}$$

$$= \frac{2(x + 2)}{x + 2} \quad \text{Factor the numerator.}$$

$$= 2 \quad \text{Reduce.}$$

Study Tip

Eliminate the obvious distractions when you study. Disconnect the telephone and put away newspapers, magazines, and unfinished projects. Even the sight of a textbook from another class might keep reminding you of how far behind you are in that class.

c) $\dfrac{x^2 + 2x}{(x-1)(x+3)} - \dfrac{2x+1}{(x-1)(x+3)} = \dfrac{x^2 + 2x - (2x+1)}{(x-1)(x+3)}$ Subtract the fractions.

$= \dfrac{x^2 + 2x - 2x - 1}{(x-1)(x+3)}$ Remove parentheses.

$= \dfrac{x^2 - 1}{(x-1)(x+3)}$ Combine like terms.

$= \dfrac{(x-1)(x+1)}{(x-1)(x+3)}$ Factor.

$= \dfrac{x+1}{x+3}$ Reduce.

Now do Exercises 21–32

CAUTION When subtracting a numerator containing more than one term, be sure to enclose it in parentheses, as in Example 3(c). Because that numerator is a binomial, the sign of each of its terms must be changed for the subtraction.

In Example 4 the rational expressions have different denominators.

EXAMPLE 4

Rational expressions with different denominators

Perform the indicated operations.

a) $\dfrac{5}{2x} + \dfrac{2}{3}$

b) $\dfrac{4}{x^3y} + \dfrac{2}{xy^3}$

c) $\dfrac{a+1}{6} - \dfrac{a-2}{8}$

Helpful Hint

You can remind yourself of the difference between addition and multiplication of fractions with a simple example: If you and your spouse each own 1/7 of Microsoft, then together you own 2/7 of Microsoft. If you own 1/7 of Microsoft, and give 1/7 of your stock to your child, then your child owns 1/49 of Microsoft.

Solution

a) The LCD for $2x$ and 3 is $6x$:

$\dfrac{5}{2x} + \dfrac{2}{3} = \dfrac{5 \cdot 3}{2x \cdot 3} + \dfrac{2 \cdot 2x}{3 \cdot 2x}$ Build up both denominators to $6x$.

$= \dfrac{15}{6x} + \dfrac{4x}{6x}$ Simplify numerators and denominators.

$= \dfrac{15 + 4x}{6x}$ Add the rational expressions.

b) The LCD is x^3y^3.

$\dfrac{4}{x^3y} + \dfrac{2}{xy^3} = \dfrac{4 \cdot y^2}{x^3y \cdot y^2} + \dfrac{2 \cdot x^2}{xy^3 \cdot x^2}$ Build up both denominators to the LCD.

$= \dfrac{4y^2}{x^3y^3} + \dfrac{2x^2}{x^3y^3}$ Simplify numerators and denominators.

$= \dfrac{4y^2 + 2x^2}{x^3y^3}$ Add the rational expressions.

c) Because $6 = 2 \cdot 3$ and $8 = 2^3$, the LCD is $2^3 \cdot 3$, or 24:

$$\frac{a+1}{6} - \frac{a-2}{8} = \frac{(a+1)4}{6 \cdot 4} - \frac{(a-2)3}{8 \cdot 3}$$ Build up both denominators to the LCD 24.

$$= \frac{4a+4}{24} - \frac{3a-6}{24}$$ Simplify numerators and denominators.

$$= \frac{4a+4-(3a-6)}{24}$$ Subtract the rational expressions.

$$= \frac{4a+4-3a+6}{24}$$ Remove the parentheses.

$$= \frac{a+10}{24}$$ Combine like terms.

Now do Exercises 33–48

EXAMPLE 5

Rational expressions with different denominators

Perform the indicated operations:

a) $\dfrac{1}{x^2-9} + \dfrac{2}{x^2+3x}$ **b)** $\dfrac{4}{5-a} - \dfrac{2}{a-5}$

Helpful Hint

Once the denominators are factored as in Example 5(a), you can simply look at each denominator and ask, "What factor does the other denominator(s) have that is missing from this one?" Then use the missing factor to build up the denominator. Repeat until all denominators are identical and you will have the LCD.

Solution

a) $$\frac{1}{x^2-9} + \frac{2}{x^2+3x} = \underbrace{\frac{1}{(x-3)(x+3)}}_{\text{Needs } x} + \underbrace{\frac{2}{x(x+3)}}_{\text{Needs } x-3}$$ The LCD is $x(x-3)(x+3)$.

$$= \frac{1 \cdot x}{(x-3)(x+3)x} + \frac{2(x-3)}{x(x+3)(x-3)}$$

$$= \frac{x}{x(x-3)(x+3)} + \frac{2x-6}{x(x-3)(x+3)}$$

$$= \frac{3x-6}{x(x-3)(x+3)}$$ We usually leave the denominator in factored form.

b) Because $-1(5-a) = a-5$, we can get identical denominators by multiplying only the first expression by -1 in the numerator and denominator:

$$\frac{4}{5-a} - \frac{2}{a-5} = \frac{4(-1)}{(5-a)(-1)} - \frac{2}{a-5}$$

$$= \frac{-4}{a-5} - \frac{2}{a-5}$$

$$= \frac{-6}{a-5}$$ $-4 - 2 = -6$

$$= -\frac{6}{a-5}$$

Now do Exercises 49–66

In Example 6 we combine three rational expressions by addition and subtraction.

EXAMPLE 6

Rational expressions with different denominators

Perform the indicated operations.

$$\frac{x+1}{x^2+2x}+\frac{2x+1}{6x+12}-\frac{1}{6}$$

Solution

The LCD for $x(x+2)$, $6(x+2)$, and 6 is $6x(x+2)$.

$$\frac{x+1}{x^2+2x}+\frac{2x+1}{6x+12}-\frac{1}{6}=\frac{x+1}{x(x+2)}+\frac{2x+1}{6(x+2)}-\frac{1}{6}$$ Factor denominators.

$$=\frac{6(x+1)}{6x(x+2)}+\frac{x(2x+1)}{6x(x+2)}-\frac{1x(x+2)}{6x(x+2)}$$ Build up to the LCD.

$$=\frac{6x+6}{6x(x+2)}+\frac{2x^2+x}{6x(x+2)}-\frac{x^2+2x}{6x(x+2)}$$ Simplify numerators.

$$=\frac{6x+6+2x^2+x-x^2-2x}{6x(x+2)}$$ Combine the numerators.

$$=\frac{x^2+5x+6}{6x(x+2)}$$ Combine like terms.

$$=\frac{(x+3)(x+2)}{6x(x+2)}$$ Factor.

$$=\frac{x+3}{6x}$$ Reduce.

Now do Exercises 67–72

Applications

We have seen how rational expressions can occur in problems involving rates. In Example 7 we see an applied situation in which we add rational expressions.

EXAMPLE 7

Adding work

Harry takes twice as long as Lucy to proofread a manuscript. Write a rational expression for the amount of work they do in 3 hours working together on a manuscript.

Solution

Let x = the number of hours it would take Lucy to complete the manuscript alone and $2x$ = the number of hours it would take Harry to complete the manuscript alone. Make a table showing rate, time, and work completed:

	Rate	Time	Work
Lucy	$\frac{1}{x}\,\frac{\text{msp}}{\text{hr}}$	3 hr	$\frac{3}{x}$ msp
Harry	$\frac{1}{2x}\,\frac{\text{msp}}{\text{hr}}$	3 hr	$\frac{3}{2x}$ msp

Math *at Work* Gravity on the Moon

Hundreds of years before humans even considered traveling beyond the earth, Isaac Newton established the laws of gravity. So when Neil Armstrong made the first human step onto the moon in 1969 he knew what amount of gravitational force to expect. Let's see how he knew.

Newton's equation for the force of gravity between two objects is $F = G\frac{m_1m_2}{d^2}$, where m_1 and m_2 are the masses of the objects (in kilograms), d is the distance (in meters) between the centers of the two objects, and G is the gravitational constant 6.67×10^{-11}. To find the force of gravity for Armstrong on earth use 5.98×10^{24} kg for the mass of the earth, 6.378×10^6 m for the radius of the earth, 80 kg for Armstrong's mass. We get

$$F = 6.67 \times 10^{-11} \cdot \frac{5.98 \times 10^{24}\text{ kg} \cdot 80\text{ kg}}{(6.378 \times 10^6\text{ m})^2} \approx 784\text{ Newtons.}$$

To find the force of gravity for Armstrong on the moon use 7.34×10^{22} kg for the mass of the moon and 1.737×10^6 m for the radius of the moon. We get

$$F = 6.67 \times 10^{-11} \cdot \frac{7.34 \times 10^{22}\text{ kg} \cdot 80\text{ kg}}{(1.737 \times 10^6\text{ m})^2} \approx 130\text{ Newtons.}$$

So the force of gravity for Armstrong on the moon was about one-sixth of the force of gravity for Armstrong on earth. Fortunately, the moon is smaller than the earth. Walking on a planet much larger than the earth would present a real problem in terms of gravitational force.

Now find the sum of each person's work.

$$\frac{3}{x} + \frac{3}{2x} = \frac{2 \cdot 3}{2 \cdot x} + \frac{3}{2x}$$

$$= \frac{6}{2x} + \frac{3}{2x}$$

$$= \frac{9}{2x}$$

So in 3 hours working together they will complete $\frac{9}{2x}$ of the manuscript.

Now do Exercises 83-88

Warm-Ups ▼

True or false? Explain your answer.

1. $\frac{1}{2} + \frac{1}{3} = \frac{2}{5}$

2. $\frac{7}{12} - \frac{1}{12} = \frac{1}{2}$

3. $\frac{3}{5} + \frac{4}{3} = \frac{29}{15}$

4. $\frac{4}{5} - \frac{5}{7} = \frac{3}{35}$

5. $\frac{5}{20} + \frac{3}{4} = 1$

6. $\frac{2}{x} + 1 = \frac{3}{x}$ for any nonzero value of x.

7. $1 + \frac{1}{a} = \frac{a+1}{a}$ for any nonzero value of a.

8. $a - \frac{1}{4} = \frac{3}{4}a$ for any value of a.

9. $\frac{a}{2} + \frac{b}{3} = \frac{3a + 2b}{6}$ for any values of a and b.

10. The LCD for the rational expressions $\frac{1}{x}$ and $\frac{3x}{x-1}$ is $x^2 - 1$.

7.4 Exercises

Reading and Writing *After reading this section, write out the answers to these questions. Use complete sentences.*

1. How do you add or subtract rational numbers?

2. How do you add or subtract rational expressions?

3. What is the least common denominator?

4. Why do we use the *least* common denominator when adding rational expressions?

Perform the indicated operation. Reduce each answer to lowest terms. See Example 1.

5. $\frac{1}{10} + \frac{1}{10}$

6. $\frac{1}{8} + \frac{3}{8}$

7. $\frac{7}{8} - \frac{1}{8}$

8. $\frac{4}{9} - \frac{1}{9}$

9. $\frac{1}{6} - \frac{5}{6}$

10. $-\frac{3}{8} - \frac{7}{8}$

11. $-\frac{7}{8} + \frac{1}{8}$

12. $-\frac{9}{20} + \left(-\frac{3}{20}\right)$

Perform the indicated operation. Reduce each answer to lowest terms. See Example 2.

13. $\frac{1}{3} + \frac{2}{9}$

14. $\frac{1}{4} + \frac{5}{6}$

15. $\frac{7}{16} + \frac{5}{18}$

16. $\frac{7}{6} + \frac{4}{15}$

17. $\frac{1}{8} - \frac{9}{10}$

18. $\frac{2}{15} - \frac{5}{12}$

19. $-\frac{1}{6} - \left(-\frac{3}{8}\right)$

20. $-\frac{1}{5} - \left(-\frac{1}{7}\right)$

Perform the indicated operation. Reduce each answer to lowest terms. See Example 3.

21. $\frac{1}{2x} + \frac{1}{2x}$

22. $\frac{1}{3y} + \frac{2}{3y}$

23. $\frac{3}{2w} + \frac{7}{2w}$

24. $\frac{5x}{3y} + \frac{7x}{3y}$

25. $\frac{3a}{a + 5} + \frac{15}{a + 5}$

26. $\frac{a + 7}{a - 4} + \frac{9 - 5a}{a - 4}$

27. $\frac{q - 1}{q - 4} - \frac{3q - 9}{q - 4}$

28. $\frac{3 - a}{3} - \frac{a - 5}{3}$

29. $\frac{4h - 3}{h(h + 1)} - \frac{h - 6}{h(h + 1)}$

30. $\frac{2t - 9}{t(t - 3)} - \frac{t - 9}{t(t - 3)}$

31. $\frac{x^2 - x - 5}{(x + 1)(x + 2)} + \frac{1 - 2x}{(x + 1)(x + 2)}$

32. $\frac{2x - 5}{(x - 2)(x + 6)} + \frac{x^2 - 2x + 1}{(x - 2)(x + 6)}$

Perform the indicated operation. Reduce each answer to lowest terms. See Example 4.

33. $\frac{1}{a} + \frac{1}{2a}$

34. $\frac{1}{3w} + \frac{2}{w}$

35. $\frac{x}{3} + \frac{x}{2}$

36. $\frac{y}{4} + \frac{y}{2}$

37. $\frac{m}{5} + m$

38. $\frac{y}{4} + 2y$

39. $\frac{1}{x} + \frac{2}{y}$

40. $\frac{2}{a} + \frac{3}{b}$

41. $\frac{3}{2a} + \frac{1}{5a}$

42. $\frac{5}{6y} - \frac{3}{8y}$

43. $\frac{w - 3}{9} - \frac{w - 4}{12}$

44. $\frac{y + 4}{10} - \frac{y - 2}{14}$

45. $\frac{b^2}{4a} - c$

46. $y + \frac{3}{7b}$

47. $\frac{2}{wz^2} + \frac{3}{w^2z}$

48. $\frac{1}{a^5b} - \frac{5}{ab^3}$

Perform the indicated operation. Reduce each answer to lowest terms. See Examples 5 and 6.

49. $\frac{1}{x} + \frac{1}{x + 2}$

50. $\frac{1}{y} + \frac{2}{y + 1}$

51. $\frac{2}{x + 1} - \frac{3}{x}$

52. $\frac{1}{a - 1} - \frac{2}{a}$

53. $\frac{2}{a - b} + \frac{1}{a + b}$

54. $\frac{3}{x + 1} + \frac{2}{x - 1}$

55. $\frac{3}{x^2 + x} - \frac{4}{5x + 5}$

56. $\frac{3}{a^2 + 3a} - \frac{2}{5a + 15}$

57. $\frac{2a}{a^2 - 9} + \frac{a}{a - 3}$

58. $\frac{x}{x^2 - 1} + \frac{3}{x - 1}$

59. $\frac{4}{a - b} + \frac{4}{b - a}$

60. $\frac{2}{x - 3} + \frac{3}{3 - x}$

61. $\frac{3}{2a - 2} - \frac{2}{1 - a}$

62. $\frac{5}{2x - 4} - \frac{3}{2 - x}$

63. $\frac{1}{x^2 - 4} - \frac{3}{x^2 - 3x - 10}$

64. $\frac{2x}{x^2 - 9} + \frac{3x}{x^2 + 4x + 3}$

65. $\frac{3}{x^2 + x - 2} + \frac{4}{x^2 + 2x - 3}$

66. $\frac{x - 1}{x^2 - x - 12} + \frac{x + 4}{x^2 + 5x + 6}$

67. $\frac{1}{a} + \frac{1}{b} + \frac{1}{c}$

68. $\frac{1}{x} + \frac{1}{x^2} + \frac{1}{x^3}$

69. $\frac{2}{x} - \frac{1}{x-1} + \frac{1}{x+2}$

70. $\frac{1}{a} - \frac{2}{a+1} + \frac{3}{a-1}$

71. $\frac{5}{3a-9} - \frac{3}{2a} + \frac{4}{a^2-3a}$

72. $\frac{3}{4c+2} - \frac{c-4}{2c^2+c} - \frac{5}{6c}$

Match each expression in (a)–(f) with the equivalent expression in (A)–(F).

73. a) $\frac{1}{y} + 2$ b) $\frac{1}{y} + \frac{2}{y}$ c) $\frac{1}{y} + \frac{1}{2}$
d) $\frac{1}{y} + \frac{1}{2y}$ e) $\frac{2}{y} + 1$ f) $\frac{y}{2} + 1$
A) $\frac{3}{y}$ B) $\frac{3}{2y}$ C) $\frac{y+2}{2}$
D) $\frac{y+2}{y}$ E) $\frac{y+2}{2y}$ F) $\frac{2y+1}{y}$

74. a) $\frac{1}{x} - x$ b) $\frac{1}{x} - \frac{1}{x^2}$ c) $\frac{1}{x} - 1$
d) $\frac{1}{x^2} - x$ e) $x - \frac{1}{x}$ f) $\frac{1}{x^2} - \frac{1}{x}$
A) $\frac{1-x^3}{x^2}$ B) $\frac{1-x}{x}$ C) $\frac{1-x^2}{x}$
D) $\frac{1-x}{x^2}$ E) $\frac{x^2-1}{x}$ F) $\frac{x-1}{x^2}$

Perform the indicated operation. Reduce each answer to lowest terms.

75. $\frac{3}{2p} - \frac{1}{2p+8}$

76. $\frac{3}{2y} - \frac{3}{2y+4}$

77. $\frac{3}{a^2+3a+2} + \frac{3}{a^2+5a+6}$

78. $\frac{4}{w^2+w} + \frac{12}{w^2-3w}$

79. $\frac{2}{b^2+4b+3} - \frac{1}{b^2+5b+6}$

80. $\frac{9}{m^2-m-2} - \frac{6}{m^2-1}$

81. $\frac{3}{2t} - \frac{2}{t+2} - \frac{3}{t^2+2t}$

82. $\frac{4}{3n} + \frac{2}{n+1} + \frac{2}{n^2+n}$

Solve each problem. See Example 7.

83. ***Perimeter of a rectangle.*** Suppose that the length of a rectangle is $\frac{3}{x}$ feet and its width is $\frac{5}{2x}$ feet. Find a rational expression for the perimeter of the rectangle.

84. ***Perimeter of a triangle.*** The lengths of the sides of a triangle are $\frac{1}{x}$, $\frac{1}{2x}$, and $\frac{2}{3x}$ meters. Find a rational expression for the perimeter of the triangle.

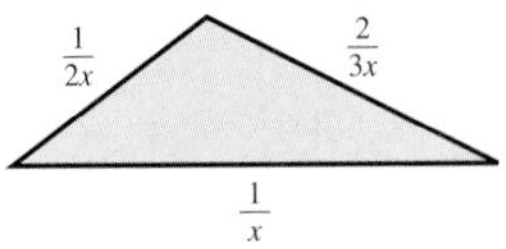

Figure for Exercise 84

85. ***Traveling time.*** Janet drove 120 miles at x mph before 6:00 A.M. After 6:00 A.M., she increased her speed by 5 mph and drove 195 additional miles. Use the fact that $T = \frac{D}{R}$ to complete the following table.

	Rate	Time	Distance
Before	$x \frac{\text{mi}}{\text{hr}}$		120 mi
After	$x + 5 \frac{\text{mi}}{\text{hr}}$		195 mi

Write a rational expression for her total traveling time. Evaluate the expression for $x = 60$.

86. ***Traveling time.*** After leaving Moose Jaw, Hanson drove 200 kilometers at x km/hr and then decreased his speed by 20 km/hr and drove 240 additional kilometers. Make a table like the one in Exercise 85. Write a rational expression for his total traveling time. Evaluate the expression for $x = 100$.

87. ***House painting.*** Kent can paint a certain house by himself in x days. His helper Keith can paint the same house by himself in $x + 3$ days. Suppose that they work together on the job for 2 days. To complete the table, use the fact that the work completed is the product of the rate and the time.

	Rate	Time	Work
Kent	$\frac{1}{x}\ \frac{\text{job}}{\text{day}}$	2 days	
Keith	$\frac{1}{x+3}\ \frac{\text{job}}{\text{day}}$	2 days	

Write a rational expression for the fraction of the house that they complete by working together for 2 days. Evaluate the expression for $x = 6$.

88. ***Barn painting.*** Melanie can paint a certain barn by herself in x days. Her helper Melissa can paint the same barn by herself in $2x$ days. Write a rational expression for the fraction of the barn that they complete in one day by working together. Evaluate the expression for $x = 5$.

Photo for Exercise 88

Getting More Involved

89. *Writing*

Write a step-by-step procedure for adding rational expressions.

90. *Writing*

Explain why fractions must have the same denominator to be added. Use real-life examples.

7.5 Complex Fractions

In this Section

- Complex Fractions
- Using the LCD to Simplify Complex Fractions
- Applications

In this section we will use the idea of least common denominator to simplify complex fractions. Also we will see how complex fractions can arise in applications.

Complex Fractions

A **complex fraction** is a fraction having rational expressions in the numerator, denominator, or both. Consider the following complex fraction:

$$\frac{\frac{1}{2}+\frac{2}{3}}{\frac{1}{4}-\frac{5}{8}} \quad \begin{array}{l}\leftarrow \text{Numerator of complex fraction}\\ \leftarrow \text{Denominator of complex fraction}\end{array}$$

To simplify it, we can combine the fractions in the numerator as follows:

$$\frac{1}{2}+\frac{2}{3}=\frac{1\cdot 3}{2\cdot 3}+\frac{2\cdot 2}{3\cdot 2}=\frac{3}{6}+\frac{4}{6}=\frac{7}{6}$$

We can combine the fractions in the denominator as follows:

$$\frac{1}{4}-\frac{5}{8}=\frac{1\cdot 2}{4\cdot 2}-\frac{5}{8}=\frac{2}{8}-\frac{5}{8}=-\frac{3}{8}$$

Now divide the numerator by the denominator:

$$\frac{\frac{1}{2}+\frac{2}{3}}{\frac{1}{4}-\frac{5}{8}} = \frac{\frac{7}{6}}{-\frac{3}{8}} = \frac{7}{6} \div \left(-\frac{3}{8}\right)$$

$$= \frac{7}{6} \cdot \left(-\frac{8}{3}\right)$$

$$= -\frac{56}{18} = -\frac{28}{9}$$

Using the LCD to Simplify Complex Fractions

A complex fraction can be simplified by writing the numerator and denominator as single fractions and then dividing, as we just did. However, there is a better method. Example 1 shows how to simplify a complex fraction by using the LCD of all of the single fractions in the complex fraction.

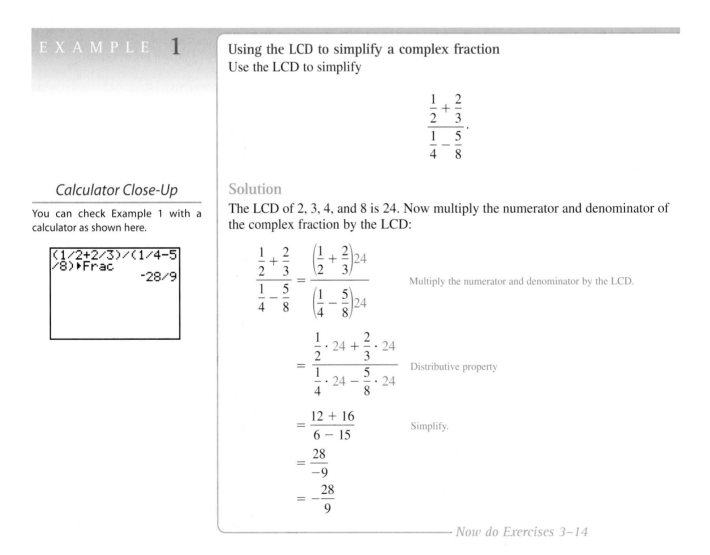

EXAMPLE 1

Using the LCD to simplify a complex fraction

Use the LCD to simplify

$$\frac{\frac{1}{2}+\frac{2}{3}}{\frac{1}{4}-\frac{5}{8}}.$$

Solution

The LCD of 2, 3, 4, and 8 is 24. Now multiply the numerator and denominator of the complex fraction by the LCD:

$$\frac{\frac{1}{2}+\frac{2}{3}}{\frac{1}{4}-\frac{5}{8}} = \frac{\left(\frac{1}{2}+\frac{2}{3}\right)24}{\left(\frac{1}{4}-\frac{5}{8}\right)24} \quad \text{Multiply the numerator and denominator by the LCD.}$$

$$= \frac{\frac{1}{2}\cdot 24 + \frac{2}{3}\cdot 24}{\frac{1}{4}\cdot 24 - \frac{5}{8}\cdot 24} \quad \text{Distributive property}$$

$$= \frac{12+16}{6-15} \quad \text{Simplify.}$$

$$= \frac{28}{-9}$$

$$= -\frac{28}{9}$$

Now do Exercises 3–14

Calculator Close-Up

You can check Example 1 with a calculator as shown here.

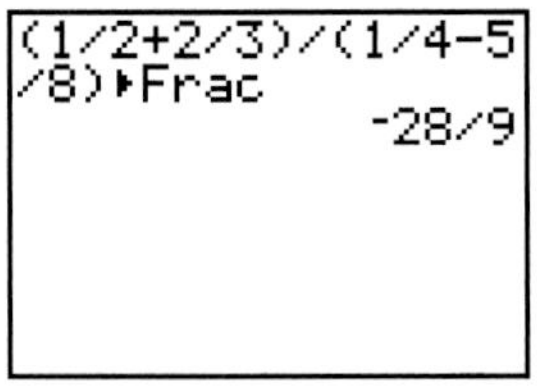

CAUTION We simplify a complex fraction by multiplying the numerator and denominator of the *complex fraction* by the LCD. Do not multiply the numerator and denominator of each fraction in the complex fraction by the LCD.

In Example 2 we simplify a complex fraction involving variables.

EXAMPLE 2

A complex fraction with variables

Simplify

$$\frac{2 - \frac{1}{x}}{\frac{1}{x^2} - \frac{1}{2}}.$$

Solution

The LCD of the denominators x, x^2, and 2 is $2x^2$:

$$\frac{2 - \frac{1}{x}}{\frac{1}{x^2} - \frac{1}{2}} = \frac{\left(2 - \frac{1}{x}\right)(2x^2)}{\left(\frac{1}{x^2} - \frac{1}{2}\right)(2x^2)} \quad \text{Multiply the numerator and denominator by } 2x^2.$$

$$= \frac{2 \cdot 2x^2 - \frac{1}{x} \cdot 2x^2}{\frac{1}{x^2} \cdot 2x^2 - \frac{1}{2} \cdot 2x^2} \quad \text{Distributive property}$$

$$= \frac{4x^2 - 2x}{2 - x^2} \quad \text{Simplify.}$$

The numerator of this answer can be factored, but the rational expression cannot be reduced.

Now do Exercises 15–24

Helpful Hint

When students see addition or subtraction in a complex fraction, they often convert all fractions to the same denominator. This is not wrong, but it is not necessary. Simply multiplying every fraction by the LCD eliminates the denominators of the original fractions.

The general strategy for simplifying a complex fraction is stated as follows.

Strategy for Simplifying a Complex Fraction

1. Find the LCD for all the denominators in the complex fraction.
2. Multiply both the numerator and the denominator of the complex fraction by the LCD. Use the distributive property if necessary.
3. Combine like terms if possible.
4. Reduce to lowest terms when possible.

EXAMPLE 3

Simplifying a complex fraction

Simplify

$$\frac{\frac{1}{x-2}-\frac{2}{x+2}}{\frac{3}{2-x}+\frac{4}{x+2}}.$$

Solution

Because $x - 2$ and $2 - x$ are opposites, we can use $(x - 2)(x + 2)$ as the LCD. Multiply the numerator and denominator by $(x - 2)(x + 2)$:

$$\frac{\frac{1}{x-2}-\frac{2}{x+2}}{\frac{3}{2-x}+\frac{4}{x+2}} = \frac{\frac{1}{x-2}(x-2)(x+2)-\frac{2}{x+2}(x-2)(x+2)}{\frac{3}{2-x}(x-2)(x+2)+\frac{4}{x+2}(x-2)(x+2)}$$

$$= \frac{x+2-2(x-2)}{3(-1)(x+2)+4(x-2)} \qquad \frac{x-2}{2-x}=-1$$

$$= \frac{x+2-2x+4}{-3x-6+4x-8} \qquad \text{Distributive property}$$

$$= \frac{-x+6}{x-14} \qquad \text{Combine like terms.}$$

Now do Exercises 25–40

Applications

As their name suggests, complex fractions arise in some fairly complex situations.

EXAMPLE 4

Fast-food workers

A survey of college students found that $\frac{1}{2}$ of the female students had jobs and $\frac{2}{3}$ of the male students had jobs. It was also found that $\frac{1}{4}$ of the female students worked in fast-food restaurants and $\frac{1}{6}$ of the male students worked in fast-food restaurants. If equal numbers of male and female students were surveyed, then what fraction of the working students worked in fast-food restaurants?

Solution

Let x represent the number of males surveyed. The number of females surveyed is also x. The total number of students working in fast-food restaurants is

$$\frac{1}{4}x + \frac{1}{6}x.$$

The total number of working students in the survey is

$$\frac{1}{2}x + \frac{2}{3}x.$$

So the fraction of working students who work in fast-food restaurants is

$$\frac{\frac{1}{4}x + \frac{1}{6}x}{\frac{1}{2}x + \frac{2}{3}x}.$$

The LCD of the denominators 2, 3, 4, and 6 is 12. Multiply the numerator and denominator by 12 to eliminate the fractions as follows:

$$\frac{\frac{1}{4}x + \frac{1}{6}x}{\frac{1}{2}x + \frac{2}{3}x} = \frac{\left(\frac{1}{4}x + \frac{1}{6}x\right)12}{\left(\frac{1}{2}x + \frac{2}{3}x\right)12} \quad \text{Multiply numerator and denominator by 12.}$$

$$= \frac{3x + 2x}{6x + 8x} \quad \text{Distributive property}$$

$$= \frac{5x}{14x} \quad \text{Combine like terms.}$$

$$= \frac{5}{14} \quad \text{Reduce.}$$

So $\frac{5}{14}$ (or about 36%) of the working students work in fast-food restaurants.

Now do Exercises 55–56

Warm-Ups ▼

True or false? Explain your answer.

1. The LCD for the denominators 4, x, 6, and x^2 is $12x^3$.
2. The LCD for the denominators $a - b$, $2b - 2a$, and 6 is $6a - 6b$.
3. The fraction $\frac{4117}{7983}$ is a complex fraction.
4. The LCD for the denominators $a - 3$ and $3 - a$ is $a^2 - 9$.
5. The largest common denominator for the fractions $\frac{1}{2}, \frac{1}{3}$, and $\frac{1}{4}$ is 24.

Questions 6–10 refer to the following complex fractions:

a) $\dfrac{\frac{1}{2} + \frac{x}{3}}{\frac{1}{4} + \frac{1}{5}}$ **b)** $\dfrac{1 + \frac{2}{b}}{\frac{2}{a} + 5}$ **c)** $\dfrac{x - \frac{1}{2}}{x + \frac{3}{2}}$ **d)** $\dfrac{\frac{1}{2} + \frac{1}{3}}{1 + \frac{1}{2}}$

6. To simplify (a), we multiply the numerator and denominator by $60x$.
7. To simplify (b), we multiply the numerator and denominator by $\frac{ab}{ab}$.

8. The complex fraction (c) is equivalent to $\frac{2x-1}{2x+3}$.

9. If $x \neq -\frac{3}{2}$, then (c) represents a real number.

10. The complex fraction (d) can be written as $\frac{5}{6} \div \frac{3}{2}$.

7.5 Exercises

Reading and Writing *After reading this section, write out the answers to these questions. Use complete sentences.*

1. What is a complex fraction?

2. What are the two ways to simplify a complex fraction?

Simplify each complex fraction. See Example 1.

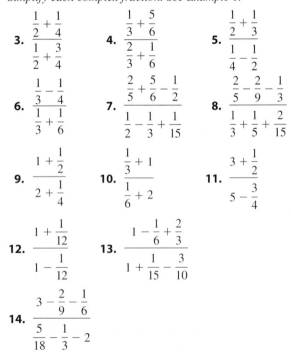

3. $\dfrac{\frac{1}{2}+\frac{1}{4}}{\frac{1}{2}+\frac{3}{4}}$ **4.** $\dfrac{\frac{1}{3}+\frac{5}{6}}{\frac{2}{3}+\frac{1}{6}}$ **5.** $\dfrac{\frac{1}{2}+\frac{1}{3}}{\frac{1}{4}-\frac{1}{2}}$

6. $\dfrac{\frac{1}{3}-\frac{1}{4}}{\frac{1}{3}+\frac{1}{6}}$ **7.** $\dfrac{\frac{2}{5}+\frac{5}{6}-\frac{1}{2}}{\frac{1}{2}-\frac{1}{3}+\frac{1}{15}}$ **8.** $\dfrac{\frac{2}{5}-\frac{2}{9}-\frac{1}{3}}{\frac{1}{3}+\frac{1}{5}+\frac{2}{15}}$

9. $\dfrac{1+\frac{1}{2}}{2+\frac{1}{4}}$ **10.** $\dfrac{\frac{1}{3}+1}{\frac{1}{6}+2}$ **11.** $\dfrac{3+\frac{1}{2}}{5-\frac{3}{4}}$

12. $\dfrac{1+\frac{1}{12}}{1-\frac{1}{12}}$ **13.** $\dfrac{1-\frac{1}{6}+\frac{2}{3}}{1+\frac{1}{15}-\frac{3}{10}}$

14. $\dfrac{3-\frac{2}{9}-\frac{1}{6}}{\frac{5}{18}-\frac{1}{3}-2}$

Simplify each complex fraction. See Example 2.

15. $\dfrac{\frac{1}{a}+\frac{1}{b}}{\frac{2}{a}+\frac{2}{b}}$ **16.** $\dfrac{\frac{1}{x}+\frac{1}{y}}{\frac{3}{x}+\frac{3}{y}}$ **17.** $\dfrac{\frac{1}{a}+\frac{3}{b}}{\frac{1}{b}-\frac{3}{a}}$

18. $\dfrac{\frac{1}{x}-\frac{3}{2}}{\frac{3}{4}+\frac{1}{x}}$ **19.** $\dfrac{5-\frac{3}{a}}{3+\frac{1}{a}}$ **20.** $\dfrac{4+\frac{3}{y}}{1-\frac{2}{y}}$

21. $\dfrac{\frac{1}{2}-\frac{2}{x}}{3-\frac{1}{x^2}}$ **22.** $\dfrac{\frac{2}{a}+\frac{5}{3}}{\frac{3}{a}-\frac{3}{a^2}}$

23. $\dfrac{\frac{3}{2b}+\frac{1}{b}}{\frac{3}{4}-\frac{1}{b^2}}$ **24.** $\dfrac{\frac{3}{2w}+\frac{4}{3w}}{\frac{1}{4w}-\frac{5}{9w}}$

Simplify each complex fraction. See Example 3.

25. $\dfrac{\frac{1}{x+1}+1}{\frac{3}{x+1}+3}$ **26.** $\dfrac{\frac{2}{x+3}+1}{\frac{4}{x+3}+2}$

27. $\dfrac{1-\frac{3}{y+1}}{3+\frac{1}{y+1}}$ **28.** $\dfrac{2-\frac{1}{a-3}}{3-\frac{1}{a-3}}$

29. $\dfrac{x+\frac{4}{x-2}}{x-\frac{x+1}{x-2}}$ **30.** $\dfrac{x-\frac{x-6}{x-1}}{x-\frac{x+15}{x-1}}$

31. $\dfrac{\dfrac{1}{3-x}-5}{\dfrac{1}{x-3}-2}$

32. $\dfrac{\dfrac{2}{x-5}-x}{\dfrac{3x}{5-x}-1}$

33. $\dfrac{1-\dfrac{5}{a-1}}{3-\dfrac{2}{1-a}}$

34. $\dfrac{\dfrac{1}{3}-\dfrac{2}{9-x}}{\dfrac{1}{6}-\dfrac{1}{x-9}}$

35. $\dfrac{\dfrac{1}{m-3}-\dfrac{4}{m}}{\dfrac{3}{m-3}+\dfrac{1}{m}}$

36. $\dfrac{\dfrac{1}{y+3}-\dfrac{4}{y}}{\dfrac{1}{y}-\dfrac{2}{y+3}}$

37. $\dfrac{\dfrac{2}{w-1}-\dfrac{3}{w+1}}{\dfrac{4}{w+1}+\dfrac{5}{w-1}}$

38. $\dfrac{\dfrac{1}{x+2}-\dfrac{3}{x+3}}{\dfrac{2}{x+3}+\dfrac{3}{x+2}}$

39. $\dfrac{\dfrac{1}{a-b}-\dfrac{1}{a+b}}{\dfrac{1}{b-a}+\dfrac{1}{b+a}}$

40. $\dfrac{\dfrac{1}{2+x}-\dfrac{1}{2-x}}{\dfrac{1}{x+2}-\dfrac{1}{x-2}}$

Simplify each complex fraction. Reduce each answer to lowest terms.

41. $\dfrac{1-\dfrac{4}{a^2}}{1+\dfrac{2}{a}-\dfrac{8}{a^2}}$

42. $\dfrac{\dfrac{1}{3}+\dfrac{1}{y}}{\dfrac{y}{3}-\dfrac{3}{y}}$

43. $\dfrac{\dfrac{1}{2}+\dfrac{1}{4x}}{\dfrac{x}{3}-\dfrac{1}{12x}}$

44. $\dfrac{\dfrac{1}{9}+\dfrac{1}{3x}}{\dfrac{x}{9}-\dfrac{1}{x}}$

45. $\dfrac{\dfrac{1}{3}-\dfrac{5}{3x}+\dfrac{2}{x^2}}{\dfrac{1}{3}-\dfrac{3}{x^2}}$

46. $\dfrac{\dfrac{1}{2}-\dfrac{3}{2x}+\dfrac{1}{x^2}}{\dfrac{1}{2}-\dfrac{1}{2x^2}}$

47. $\dfrac{\dfrac{2x-9}{6}}{\dfrac{2x-3}{9}}$

48. $\dfrac{\dfrac{a-5}{12}}{\dfrac{a+2}{15}}$

49. $\dfrac{\dfrac{2x-4y}{xy^2}}{\dfrac{3x-6y}{x^3y}}$

50. $\dfrac{\dfrac{ab+b^2}{4ab^5}}{\dfrac{a+b}{6a^2b^4}}$

51. $\dfrac{\dfrac{a^2+2a-24}{a+1}}{\dfrac{a^2-a-12}{(a+1)^2}}$

52. $\dfrac{\dfrac{y^2-3y-18}{y^2-4}}{\dfrac{y^2+5y+6}{y-2}}$

53. $\dfrac{\dfrac{x}{x+1}}{\dfrac{1}{x^2-1}-\dfrac{1}{x-1}}$

54. $\dfrac{\dfrac{a}{a^2-b^2}}{\dfrac{1}{a+b}+\dfrac{1}{a-b}}$

Solve each problem. See Example 4.

55. ***Sophomore math.*** A survey of college sophomores showed that $\frac{5}{6}$ of the males were taking a mathematics class and $\frac{3}{4}$ of the females were taking a mathematics class. One-third of the males were enrolled in calculus, and $\frac{1}{5}$ of the females were enrolled in calculus. If just as many males as females were surveyed, then what fraction of the surveyed students taking mathematics were enrolled in calculus? Rework this problem assuming that the number of females in the survey was twice the number of males.

56. ***Commuting students.*** At a well-known university, $\frac{1}{4}$ of the undergraduate students commute, and $\frac{1}{3}$ of the graduate students commute. One-tenth of the undergraduate students drive more than 40 miles daily, and $\frac{1}{6}$ of the graduate students drive more than 40 miles daily. If there are twice as many undergraduate students as there are graduate students, then what fraction of the commuters drive more than 40 miles daily?

Photo for Exercise 56

Getting More Involved

57. ***Exploration***

Simplify

$$\frac{1}{1+\frac{1}{2}},\quad \frac{1}{1+\dfrac{1}{1+\frac{1}{2}}},\quad \text{and}\quad \frac{1}{1+\dfrac{1}{1+\dfrac{1}{1+\frac{1}{2}}}}.$$

a) Are these fractions getting larger or smaller as the fractions become more complex?

b) Continuing the pattern, find the next two complex fractions and simplify them.

c) Now what can you say about the values of all five complex fractions?

58. ***Discussion***

A complex fraction can be simplified by writing the numerator and denominator as single fractions and then dividing them or by multiplying the numerator and denominator by the LCD. Simplify the complex fraction

$$\frac{\frac{4}{xy^2} - \frac{6}{xy}}{\frac{2}{x^2} + \frac{4}{x^2y}}$$

by using each of these methods. Compare the number of steps used in each method, and determine which method requires fewer steps.

7.6 Solving Equations with Rational Expressions

In this Section

- **Equations with Rational Expressions**
- **Extraneous Solutions**

Many problems in algebra can be solved by using equations involving rational expressions. In this section you will learn how to solve equations that involve rational expressions, and in Sections 7.7 and 7.8 you will solve problems using these equations.

Equations with Rational Expressions

We solved some equations involving fractions in Section 2.3. In that section the equations had only integers in the denominators. Our first step in solving those equations was to multiply by the LCD to eliminate all of the denominators.

EXAMPLE 1

Integers in the denominators

Solve $\frac{1}{2} - \frac{x-2}{3} = \frac{1}{6}$.

Solution

The LCD for 2, 3, and 6 is 6. Multiply each side of the equation by 6:

$$\frac{1}{2} - \frac{x-2}{3} = \frac{1}{6} \quad \text{Original equation}$$

$$6\left(\frac{1}{2} - \frac{x-2}{3}\right) = 6 \cdot \frac{1}{6} \quad \text{Multiply each side by 6.}$$

$$6 \cdot \frac{1}{2} - \overset{2}{\cancel{6}} \cdot \frac{x-2}{\cancel{3}} = \cancel{6} \cdot \frac{1}{\cancel{6}} \quad \text{Distributive property}$$

$$3 - 2(x-2) = 1 \quad \text{Simplify.}$$

$$3 - 2x + 4 = 1 \quad \text{Distributive property}$$

$$-2x = -6 \quad \text{Subtract 7 from each side.}$$

$$x = 3 \quad \text{Divide each side by } -2.$$

Helpful Hint

Note that it is not necessary to convert each fraction into an equivalent fraction with a common denominator here. Since we can multiply both sides of an equation by any expression we choose, we choose to multiply by the LCD. This tactic eliminates the fractions in one step.

Check $x = 3$ in the original equation:

$$\frac{1}{2} - \frac{3-2}{3} = \frac{1}{2} - \frac{1}{3} = \frac{3}{6} - \frac{2}{6} = \frac{1}{6}$$

The solution to the equation is 3.

Now do Exercises 5–14

CAUTION When a numerator contains a binomial, as in Example 1, the numerator must be enclosed in parentheses when the denominator is eliminated.

To solve an equation involving rational expressions, we usually multiply each side of the equation by the LCD for all the denominators involved, just as we do for an equation with fractions.

EXAMPLE 2

Variables in the denominators

Solve $\frac{1}{x} + \frac{1}{6} = \frac{1}{4}$.

Solution

We multiply each side of the equation by $12x$, the LCD for 4, 6, and x:

$$\frac{1}{x} + \frac{1}{6} = \frac{1}{4} \quad \text{Original equation}$$

$$12x\left(\frac{1}{x} + \frac{1}{6}\right) = 12x\left(\frac{1}{4}\right) \quad \text{Multiply each side by } 12x.$$

$$12\cancel{x} \cdot \frac{1}{\cancel{x}} + \overset{2}{\cancel{12}}x \cdot \frac{1}{\cancel{6}} = \overset{3}{\cancel{12}}x \cdot \frac{1}{\cancel{4}} \quad \text{Distributive property}$$

$$12 + 2x = 3x \quad \text{Simplify.}$$

$$12 = x \quad \text{Subtract } 2x \text{ from each side.}$$

Check that 12 satisfies the original equation:

$$\frac{1}{12} + \frac{1}{6} = \frac{1}{12} + \frac{2}{12} = \frac{3}{12} = \frac{1}{4}$$

The solution to the equation is 12.

Now do Exercises 15–26

EXAMPLE 3

An equation with two solutions

Solve the equation $\frac{100}{x} + \frac{100}{x+5} = 9$.

Solution

The LCD for the denominators x and $x + 5$ is $x(x + 5)$:

$$\frac{100}{x} + \frac{100}{x+5} = 9 \quad \text{Original equation}$$

$$x(x+5)\frac{100}{x} + x(x+5)\frac{100}{x+5} = x(x+5)9 \quad \text{Multiply each side by } x(x+5).$$

$$(x+5)100 + x(100) = (x^2 + 5x)9 \quad \text{All denominators are eliminated.}$$

$$100x + 500 + 100x = 9x^2 + 45x \quad \text{Simplify.}$$

$$500 + 200x = 9x^2 + 45x$$

$$0 = 9x^2 - 155x - 500 \quad \text{Get 0 on one side.}$$

$$0 = (9x + 25)(x - 20) \quad \text{Factor.}$$

$$9x + 25 = 0 \quad \text{or} \quad x - 20 = 0 \quad \text{Zero factor property}$$

$$x = -\frac{25}{9} \quad \text{or} \quad x = 20$$

A check will show that both $-\frac{25}{9}$ and 20 satisfy the original equation.

Now do Exercises 27–34

Study Tip

Your mood for studying should match the mood in which you are tested. Being too relaxed during studying will not match the increased level of activation you attain during a test. Likewise, if you get too tensed-up during a test, you will not do well because your test-taking mood will not match your studying mood.

Extraneous Solutions

In a rational expression we can replace the variable only by real numbers that do not cause the denominator to be 0. When solving equations involving rational expressions, we must check every solution to see whether it causes 0 to appear in a denominator. If a number causes the denominator to be 0, then it cannot be a solution to the equation. A number that appears to be a solution but causes 0 in a denominator is called an **extraneous solution.**

EXAMPLE 4

An equation with an extraneous solution

Solve the equation $\frac{1}{x-2} = \frac{x}{2x-4} + 1$.

Solution

Because the denominator $2x - 4$ factors as $2(x - 2)$, the LCD is $2(x - 2)$.

$$2(x-2)\frac{1}{x-2} = 2(x-2)\frac{x}{2(x-2)} + 2(x-2)\cdot 1 \quad \text{Multiply each side of the original equation by } 2(x-2).$$

$$2 = x + 2x - 4 \quad \text{Simplify.}$$

$$2 = 3x - 4$$

$$6 = 3x$$

$$2 = x$$

Check 2 in the original equation:

$$\frac{1}{2-2} = \frac{2}{2\cdot 2-4} + 1$$

The denominator $2 - 2$ is 0. So 2 does not satisfy the equation, and it is an extraneous solution. The equation has no solutions.

Now do Exercises 35–38

EXAMPLE 5

Another extraneous solution

Solve the equation $\frac{1}{x} + \frac{1}{x-3} = \frac{x-2}{x-3}$.

Solution

The LCD for the denominators x and $x - 3$ is $x(x - 3)$:

$$\frac{1}{x} + \frac{1}{x-3} = \frac{x-2}{x-3} \quad \text{Original equation}$$

$$x(x-3)\cdot\frac{1}{x} + x(x-3)\cdot\frac{1}{x-3} = x(x-3)\cdot\frac{x-2}{x-3} \quad \text{Multiply each side by } x(x-3).$$

$$x - 3 + x = x(x-2)$$

$$2x - 3 = x^2 - 2x$$

$$0 = x^2 - 4x + 3$$

$$0 = (x-3)(x-1)$$

$$x - 3 = 0 \quad \text{or} \quad x - 1 = 0$$

$$x = 3 \quad \text{or} \quad x = 1$$

If $x = 3$, then the denominator $x - 3$ has a value of 0. If $x = 1$, the original equation is satisfied. The only solution to the equation is 1.

Now do Exercises 39–42

Study Tip

Studying in an environment similar to the one in which you will be tested can increase your chances of recalling information. When possible, review for a test in the classroom in which you will take the test.

CAUTION Always be sure to check your answers in the original equation to determine whether they are extraneous solutions.

Warm-Ups ▼

True or false? Explain your answers.

1. The LCD is not used in solving equations with rational expressions.
2. To solve the equation $x^2 = 8x$, we divide each side by x.
3. An extraneous solution is an irrational number.

Use the following equations for Questions 4–10.

a) $\frac{3}{x} + \frac{5}{x-2} = \frac{2}{3}$ **b)** $\frac{1}{x} + \frac{1}{2} = \frac{3}{4}$ **c)** $\frac{1}{x-1} + 2 = \frac{1}{x+1}$

4. To solve Eq. (a), we must add the expressions on the left-hand side.
5. Both 0 and 2 satisfy Eq. (a).

6. To solve Eq. (a), we multiply each side by $3x^2 - 6x$.
7. The only solution to Eq. (b) is 4.
8. Equation (b) is equivalent to $4 + 2x = 3x$.
9. To solve Eq. (c), we multiply each side by $x^2 - 1$.
10. The numbers 1 and -1 do not satisfy Eq. (c).

7.6 Exercises

Boost your GRADE at mathzone.com!

MathZone
- Practice Problems
- Self-Tests
- Videos
- Net Tutor
- e-Professors

Reading and Writing *After reading this section, write out the answers to these questions. Use complete sentences.*

1. What is the typical first step for solving an equation involving rational expressions?

2. What is the difference in procedure for solving an equation involving rational expressions and adding rational expressions?

3. What is an extraneous solution?

4. Why do extraneous solutions sometimes occur for equations with rational expressions?

Solve each equation. See Example 1.

5. $\frac{x}{2} + 1 = \frac{x}{4}$

6. $\frac{x}{3} + 2 = \frac{x}{6}$

7. $\frac{x}{3} - 5 = \frac{x}{2} - 7$

8. $\frac{x}{3} - \frac{x}{2} = \frac{x}{5} - 11$

9. $\frac{y}{5} - \frac{2}{3} = \frac{y}{6} + \frac{1}{3}$

10. $\frac{z}{6} + \frac{5}{4} = \frac{z}{2} - \frac{3}{4}$

11. $\frac{3}{4} - \frac{t - 4}{3} = \frac{t}{12}$

12. $\frac{4}{5} - \frac{v - 1}{10} = \frac{v - 5}{30}$

13. $\frac{1}{5} - \frac{w + 10}{15} = \frac{1}{10} - \frac{w + 1}{6}$

14. $\frac{q}{5} - \frac{q - 1}{2} = \frac{13}{20} - \frac{q + 1}{4}$

Solve each equation. See Example 2.

15. $\frac{1}{x} + \frac{1}{2} = 3$

16. $\frac{2}{x} + \frac{3}{4} = 5$

17. $\frac{1}{x} + \frac{2}{x} = 7$

18. $\frac{5}{x} + \frac{6}{x} = 12$

19. $\frac{1}{x} + \frac{1}{2} = \frac{3}{4}$

20. $\frac{3}{x} + \frac{1}{4} = \frac{5}{8}$

21. $\frac{2}{3x} + \frac{1}{2x} = \frac{7}{24}$

22. $\frac{1}{6x} - \frac{1}{8x} = \frac{1}{72}$

23. $\frac{1}{2} + \frac{a - 2}{a} = \frac{a + 2}{2a}$

24. $\frac{1}{b} + \frac{1}{5} = \frac{b - 1}{5b} + \frac{3}{10}$

25. $\frac{1}{3} - \frac{k + 3}{6k} = \frac{1}{3k} - \frac{k - 1}{2k}$

26. $\frac{3}{p} - \frac{p + 3}{3p} = \frac{2p - 1}{2p} - \frac{5}{6}$

Solve each equation. See Example 3.

27. $\frac{x}{2} = \frac{5}{x + 3}$

28. $\frac{x}{3} = \frac{4}{x + 1}$

29. $\frac{x}{x + 1} = \frac{6}{x + 7}$

30. $\frac{x}{x + 3} = \frac{2}{x - 3}$

31. $\frac{2}{x + 1} = \frac{1}{x} + \frac{1}{6}$

32. $\frac{1}{w + 1} - \frac{1}{2w} = \frac{3}{40}$

33. $\frac{a - 1}{a^2 - 4} + \frac{1}{a - 2} = \frac{a + 4}{a + 2}$

34. $\frac{b + 17}{b^2 - 1} - \frac{1}{b + 1} = \frac{b - 2}{b - 1}$

Solve each equation. Watch for extraneous solutions. See Examples 4 and 5.

35. $\frac{1}{x-1}+\frac{2}{x}=\frac{x}{x-1}$

36. $\frac{4}{x}+\frac{3}{x-3}=\frac{x}{x-3}-\frac{1}{3}$

37. $\frac{5}{x+2}+\frac{2}{x-3}=\frac{x-1}{x-3}$

38. $\frac{6}{y-2}+\frac{7}{y-8}=\frac{y-1}{y-8}$

39. $1+\frac{3y}{y-2}=\frac{6}{y-2}$

40. $\frac{5}{y-3}=\frac{y+7}{2y-6}+1$

41. $\frac{z}{z+1}-\frac{1}{z+2}=\frac{2z+5}{z^2+3z+2}$

42. $\frac{z}{z-2}-\frac{1}{z+5}=\frac{7}{z^2+3z-10}$

Solve each equation.

43. $\frac{a}{4}=\frac{5}{2}$

44. $\frac{y}{3}=\frac{6}{5}$

45. $\frac{w}{6}=\frac{3w}{11}$

46. $\frac{2m}{3}=\frac{3m}{2}$

47. $\frac{5}{x}=\frac{x}{5}$

48. $\frac{-3}{x}=\frac{x}{-3}$

49. $\frac{x-3}{5}=\frac{x-3}{x}$

50. $\frac{a+4}{2}=\frac{a+4}{a}$

51. $\frac{1}{x+2}=\frac{x}{x+2}$

52. $\frac{-3}{w+2}=\frac{w}{w+2}$

53. $\frac{1}{2x-4}+\frac{1}{x-2}=\frac{3}{2}$

54. $\frac{7}{3x-9}-\frac{1}{x-3}=\frac{4}{3}$

55. $\frac{3}{a^2-a-6}=\frac{2}{a^2-4}$

56. $\frac{8}{a^2+a-6}=\frac{6}{a^2-9}$

57. $\frac{4}{c-2}-\frac{1}{2-c}=\frac{25}{c+6}$

58. $\frac{3}{x+1}-\frac{1}{1-x}=\frac{10}{x^2-1}$

59. $\frac{1}{x^2-9}+\frac{3}{x+3}=\frac{4}{x-3}$

60. $\frac{3}{x-2}-\frac{5}{x+3}=\frac{1}{x^2+x-6}$

61. $\frac{3}{2x+4}-\frac{1}{x+2}=\frac{1}{3x+1}$

62. $\frac{5}{2m+6}-\frac{1}{m+1}=\frac{1}{m+3}$

63. $\frac{2t-1}{3t+3}+\frac{3t-1}{6t+6}=\frac{t}{t+1}$

64. $\frac{4w-1}{3w+6}-\frac{w-1}{3}=\frac{w-1}{w+2}$

Solve each problem.

65. ***Lens equation.*** The focal length f for a camera lens is related to the object distance o and the image distance i by the formula

$$\frac{1}{f}=\frac{1}{o}+\frac{1}{i}.$$

See the accompanying figure. The image is in focus at distance i from the lens. For an object that is 600 mm from a 50-mm lens, use $f = 50$ mm and $o = 600$ mm to find i.

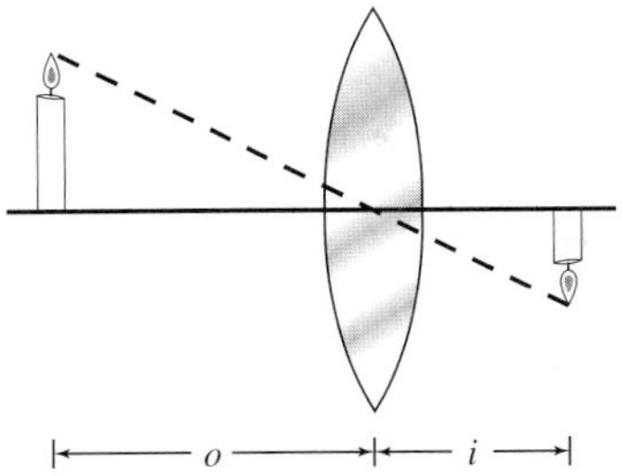

Figure for Exercise 65

66. ***Telephoto lens.*** Use the formula from Exercise 65 to find the image distance i for an object that is 2,000,000 mm from a 250-mm telephoto lens.

Photo for Exercise 66

7.7 Applications of Ratios and Proportions

In this Section

- Ratios
- Proportions

In this section we will use the ideas of rational expressions in ratio and proportion problems. We will solve proportions in the same way we solved equations in Section 7.6.

Ratios

In Chapter 1 we defined a rational number as the *ratio of two integers*. We will now give a more general definition of ratio. If a and b are any real numbers (not just integers), with $b \neq 0$, then the expression $\frac{a}{b}$ is called the **ratio of a and b** or the **ratio of a to b.** The ratio of a to b is also written as $a:b$. A ratio is a comparison of two numbers. Some examples of ratios are

$$\frac{3}{4}, \quad \frac{4.2}{2.1}, \quad \frac{\frac{1}{4}}{\frac{1}{2}}, \quad \frac{3.6}{5}, \quad \text{and} \quad \frac{100}{1}.$$

Ratios are treated just like fractions. We can reduce ratios, and we can build them up. We generally express ratios as ratios of integers. When possible, we will convert a ratio into an equivalent ratio of integers in lowest terms.

EXAMPLE 1

Finding equivalent ratios

Find an equivalent ratio of integers in lowest terms for each ratio.

a) $\frac{4.2}{2.1}$ **b)** $\frac{\frac{1}{4}}{\frac{1}{2}}$ **c)** $\frac{3.6}{5}$

Solution

a) Because both the numerator and the denominator have one decimal place, we will multiply the numerator and denominator by 10 to eliminate the decimals:

$$\frac{4.2}{2.1} = \frac{4.2(10)}{2.1(10)} = \frac{42}{21} = \frac{21 \cdot 2}{21 \cdot 1} = \frac{2}{1}$$ Do not omit the 1 in a ratio.

So the ratio of 4.2 to 2.1 is equivalent to the ratio 2 to 1.

b) This ratio is a complex fraction. We can simplify this expression using the LCD method as shown in Section 7.5. Multiply the numerator and denominator of this ratio by 4:

$$\frac{\frac{1}{4}}{\frac{1}{2}} = \frac{\frac{1}{4} \cdot 4}{\frac{1}{2} \cdot 4} = \frac{1}{2}$$

Study Tip

To get the "big picture," survey the chapter that you are studying. Read the headings to get the general idea of the chapter content. Read the chapter summary to see what is important in the chapter. Repeat this survey procedure several times while you are working in a chapter.

c) We can get a ratio of integers if we multiply the numerator and denominator by 10.

$$\frac{3.6}{5} = \frac{3.6(10)}{5(10)} = \frac{36}{50}$$

$$= \frac{18}{25} \qquad \text{Reduce to lowest terms.}$$

Now do Exercises 7–22

In Example 2 a ratio is used to compare quantities.

EXAMPLE 2

Nitrogen to potash

In a 50-pound bag of lawn fertilizer there are 8 pounds of nitrogen and 12 pounds of potash. What is the ratio of nitrogen to potash?

Solution

The nitrogen and potash occur in this fertilizer in the ratio of 8 pounds to 12 pounds:

$$\frac{8}{12} = \frac{2 \cdot \cancel{4}}{3 \cdot \cancel{4}} = \frac{2}{3}$$

So the ratio of nitrogen to potash is 2 to 3.

Now do Exercises 23–24

EXAMPLE 3

Males to females

In a class of 50 students, there were exactly 20 male students. What was the ratio of males to females in this class?

Solution

Because there were 20 males in the class of 50, there were 30 females. The ratio of males to females was 20 to 30, or 2 to 3.

Now do Exercises 25–26

Ratios give us a means of comparing the size of two quantities. For this reason *the numbers compared in a ratio should be expressed in the same units.* For example, if one dog is 24 inches high and another is 1 foot high, then the ratio of their heights is 2 to 1, not 24 to 1.

EXAMPLE 4

Quantities with different units

What is the ratio of length to width for a poster with a length of 30 inches and a width of 2 feet?

Solution

Because the width is 2 feet, or 24 inches, the ratio of length to width is 30 to 24. Reduce as follows:

$$\frac{30}{24} = \frac{5 \cdot 6}{4 \cdot 6} = \frac{5}{4}$$

So the ratio of length to width is 5 to 4.

Now do Exercises 27–30

Proportions

A **proportion** is any statement expressing the equality of two ratios. The statement

$$\frac{a}{b} = \frac{c}{d} \quad \text{or} \quad a{:}b = c{:}d$$

is a proportion. In any proportion the numbers in the positions of a and d shown here are called the **extremes.** The numbers in the positions of b and c as shown are called the **means.** In the proportion

$$\frac{30}{24} = \frac{5}{4},$$

the means are 24 and 5, and the extremes are 30 and 4.

If we multiply each side of the proportion

$$\frac{a}{b} = \frac{c}{d}$$

by the LCD, bd, we get

$$\frac{a}{b} \cdot bd = \frac{c}{d} \cdot bd$$

or

$$a \cdot d = b \cdot c.$$

We can express this result by saying *that the product of the extremes is equal to the product of the means.* We call this fact the **extremes-means property** or **cross-multiplying.**

Helpful Hint

The extremes-means property or cross-multiplying is nothing new. You can accomplish the same thing by multiplying each side of the equation by the LCD.

Extremes-Means Property (Cross-Multiplying)

Suppose a, b, c, and d are real numbers with $b \neq 0$ and $d \neq 0$. If

$$\frac{a}{b} = \frac{c}{d}, \quad \text{then} \quad ad = bc.$$

We use the extremes-means property to solve proportions.

EXAMPLE 5

Using the extremes-means property

Solve the proportion $\frac{3}{x} = \frac{5}{x + 5}$ for x.

Solution

Instead of multiplying each side by the LCD, we use the extremes-means property:

$$\frac{3}{x} = \frac{5}{x + 5} \quad \text{Original proportion}$$
$$3(x + 5) = 5x \quad \text{Extremes-means property}$$
$$3x + 15 = 5x \quad \text{Distributive property}$$
$$15 = 2x$$
$$\frac{15}{2} = x$$

Check:

$$\frac{3}{\frac{15}{2}} = 3 \cdot \frac{2}{15} = \frac{2}{5}$$

$$\frac{5}{\frac{15}{2} + 5} = \frac{5}{\frac{25}{2}} = 5 \cdot \frac{2}{25} = \frac{2}{5}$$

So $\frac{15}{2}$ is the solution to the equation or the solution to the proportion.

Now do Exercises 31–42

EXAMPLE 6

Solving a proportion

The ratio of men to women at Brighton City College is 2 to 3. If there are 894 men, then how many women are there?

Solution

Because the ratio of men to women is 2 to 3, we have

$$\frac{\text{Number of men}}{\text{Number of women}} = \frac{2}{3}.$$

If x represents the number of women, then we have the following proportion:

$$\frac{894}{x} = \frac{2}{3}$$
$$2x = 2682 \quad \text{Extremes-means property}$$
$$x = 1341$$

The number of women is 1341.

Now do Exercises 43–46

10. $\frac{1000}{200}$ **11.** $\frac{2.5}{3.5}$ **12.** $\frac{4.8}{1.2}$

13. $\frac{0.32}{0.6}$ **14.** $\frac{0.05}{0.8}$ **15.** $\frac{35}{10}$

16. $\frac{88}{33}$ **17.** $\frac{4.5}{7}$ **18.** $\frac{3}{2.5}$

19. $\frac{\frac{1}{2}}{\frac{1}{5}}$ **20.** $\frac{\frac{2}{3}}{\frac{3}{4}}$ **21.** $\frac{5}{\frac{1}{3}}$

22. $\frac{4}{\frac{1}{4}}$

Find a ratio for each of the following, and write it as a ratio of integers in lowest terms. See Examples 2–4.

23. ***Men and women.*** Find the ratio of men to women in a bowling league containing 12 men and 8 women.

24. ***Coffee drinkers.*** Among 100 coffee drinkers, 36 said that they preferred their coffee black and the rest did not prefer their coffee black. Find the ratio of those who prefer black coffee to those who prefer nonblack coffee.

Photo for Exercise 24

25. ***Smokers.*** A life insurance company found that among its last 200 claims, there were six dozen smokers. What is the ratio of smokers to nonsmokers in this group of claimants?

26. ***Hits and misses.*** A woman threw 60 darts and hit the target a dozen times. What is her ratio of hits to misses?

27. ***Violence and kindness.*** While watching television for one week, a consumer group counted 1240 acts of violence and 40 acts of kindness. What is the violence to kindness ratio for television, according to this group?

28. ***Length to width.*** What is the ratio of length to width for the rectangle shown?

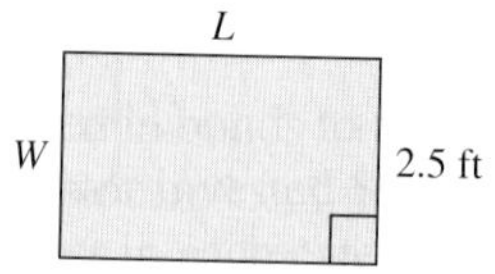

Figure for Exercise 28

29. ***Rise to run.*** What is the ratio of rise to run for the stairway shown in the accompanying figure?

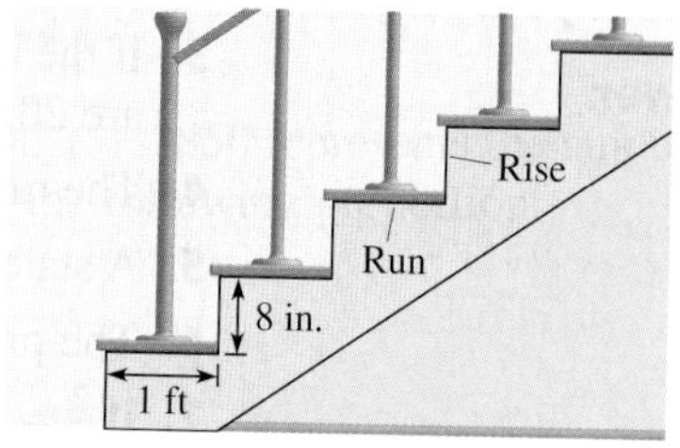

Figure for Exercise 29

30. ***Rise and run.*** If the rise is $\frac{3}{2}$ and the run is 5, then what is the ratio of the rise to the run?

Solve each proportion. See Example 5.

31. $\frac{4}{x} = \frac{2}{3}$ **32.** $\frac{9}{x} = \frac{3}{2}$

33. $\frac{a}{2} = \frac{-1}{5}$ **34.** $\frac{b}{3} = \frac{-3}{4}$

35. $-\frac{5}{9} = \frac{3}{x}$ **36.** $-\frac{3}{4} = \frac{5}{x}$

37. $\frac{10}{x} = \frac{34}{x + 12}$ **38.** $\frac{x}{3} = \frac{x + 1}{2}$

39. $\frac{a}{a + 1} = \frac{a + 3}{a}$ **40.** $\frac{c + 3}{c - 1} = \frac{c + 2}{c - 3}$

41. $\frac{m - 1}{m - 2} = \frac{m - 3}{m + 4}$ **42.** $\frac{h}{h - 3} = \frac{h}{h - 9}$

Use a proportion to solve each problem. See Examples 6–8.

43. ***New shows and reruns.*** The ratio of new shows to reruns on cable TV is 2 to 27. If Frank counted only eight new shows one evening, then how many reruns were there?

44. ***Fast food.*** If four out of five doctors prefer fast food, then at a convention of 445 doctors, how many prefer fast food?

45. ***Voting.*** If 220 out of 500 voters surveyed said that they would vote for the incumbent, then how many votes could the incumbent expect out of the 400,000 voters in the state?

Photo for Exercise 45

46. ***New product.*** A taste test with 200 randomly selected people found that only three of them said that they would buy a box of new Sweet Wheats cereal. How many boxes could the manufacturer expect to sell in a country of 280 million people?

47. ***Basketball blowout.*** As the final buzzer signaled the end of the basketball game, the Lions were 34 points ahead of the Tigers. If the Lions scored 5 points for every 3 scored by the Tigers, then what was the final score?

48. ***The golden ratio.*** The ancient Greeks thought that the most pleasing shape for a rectangle was one for which the ratio of the length to the width was 8 to 5, the golden ratio. If the length of a rectangular painting is 2 ft longer than its width, then for what dimensions would the length and width have the golden ratio?

49. ***Automobile sales.*** The ratio of sports cars to luxury cars sold in Wentworth one month was 3 to 2. If 20 more sports cars were sold than luxury cars, then how many of each were sold that month?

50. ***Foxes and rabbits.*** The ratio of foxes to rabbits in the Deerfield Forest Preserve is 2 to 9. If there are 35 fewer foxes than rabbits, then how many of each are there?

51. ***Inches and feet.*** If there are 12 inches in 1 foot, then how many inches are there in 7 feet?

52. ***Feet and yards.*** If there are 3 feet in 1 yard, then how many yards are there in 28 feet?

53. ***Minutes and hours.*** If there are 60 minutes in 1 hour, then how many minutes are there in 0.25 hour?

54. ***Meters and kilometers.*** If there are 1000 meters in 1 kilometer, then how many meters are there in 2.33 kilometers?

55. ***Miles and hours.*** If Alonzo travels 230 miles in 3 hours, then how many miles does he travel in 7 hours?

56. ***Hiking time.*** If Evangelica can hike 19 miles in 2 days on the Appalachian Trail, then how many days will it take her to hike 63 miles?

57. ***Force on basketball shoes.*** The force exerted on shoe soles in a jump shot is proportional to the weight of the person jumping. If a 70-pound boy exerts a force of 980 pounds on his shoe soles when he returns to the court after a jump, then what force does a 6 ft 8 in. professional ball player weighing 280 pounds exert on the soles of his shoes when he returns to the court after a jump? Use the accompanying graph to estimate the force for a 150-pound player.

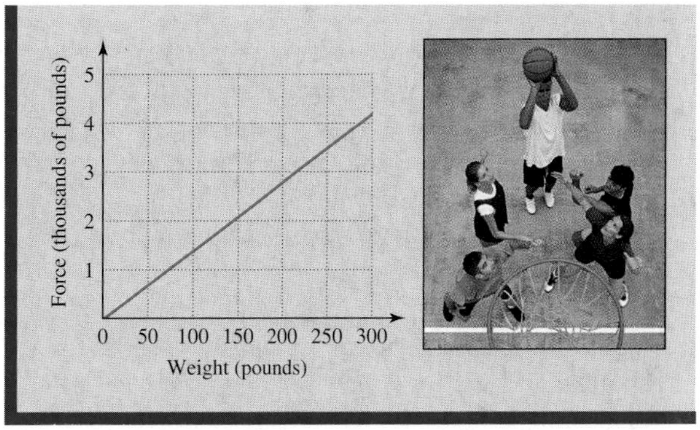

Figure for Exercise 57

58. ***Force on running shoes.*** The ratio of the force on the shoe soles to the weight of a runner is 3 to 1. What force does a 130-pound jogger exert on the soles of her shoes?

59. ***Capture-recapture.*** To estimate the number of trout in Trout Lake, rangers used the capture-recapture method. They caught, tagged, and released 200 trout. One week later, they caught a sample of 150 trout and found that 5 of them were tagged. Assuming that the ratio of tagged trout to the total number of trout in the lake is the same as the ratio of tagged trout in the sample to the number of trout in the sample, find the number of trout in the lake.

60. ***Bear population.*** To estimate the size of the bear population on the Keweenaw Peninsula, conservationists captured, tagged, and released 50 bears. One year later, a random sample of 100 bears included only 2 tagged bears. What is the conservationist's estimate of the size of the bear population?

61. ***Fast-food waste.*** The accompanying figure shows the typical distribution of waste at a fast-food restaurant (U.S. Environmental Protection Agency, www.epa.gov).

a) What is the ratio of customer waste to food waste?

b) If a typical McDonald's generates 67 more pounds of food waste than customer waste per day, then how many pounds of customer waste does it generate?

WASTE GENERATION AT A FAST-FOOD RESTAURANT

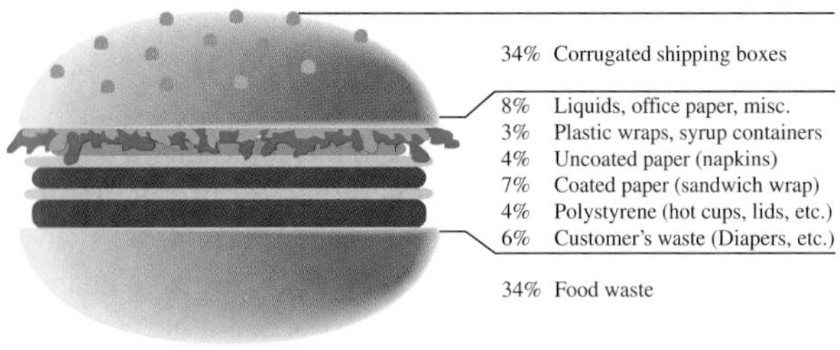

Figure for Exercises 61 and 62

62. ***Corrugated waste.*** Use the accompanying figure to find the ratio of waste from corrugated shipping boxes to waste not from corrugated shipping boxes. If a typical McDonald's generates 81 pounds of waste per day from corrugated shipping boxes, then how many pounds of waste per day does it generate that is not from corrugated shipping boxes?

63. ***Mascara needs.*** In determining warehouse needs for a particular mascara for a chain of 2000 stores, Mike Pittman first determines a need B based on sales figures for the past 52 weeks. He then determines the actual need A from the equation $\frac{A}{B} = k$, where

$$k = 1 + V + C + X - D.$$

He uses $V = 0.22$ if there is a national TV ad and $V = 0$ if not, $C = 0.26$ if there is a national coupon and $C = 0$ if not, $X = 0.36$ if there is a chain-specific ad and $X = 0$ if not, and $D = 0.29$ if there is a special display in the chain and $D = 0$ if not. (D is subtracted because less product is needed in the warehouse when more is on display in the store.) If $B = 4200$ units and there is a special display and a national coupon but no national TV ad and no chain-specific ad, then what is the value of A?

Getting More Involved

64. ***Discussion***

Which of the following equations is not a proportion? Explain.

a) $\frac{1}{2} = \frac{1}{2}$ **b)** $\frac{x}{x + 2} = \frac{4}{5}$

c) $\frac{x}{4} = \frac{9}{x}$ **d)** $\frac{8}{x + 2} - 1 = \frac{5}{x + 2}$

65. ***Discussion***

Find all of the errors in the following solution to an equation.

$$\frac{7}{x} = \frac{8}{x + 3} + 1$$
$$7(x + 3) = 8x + 1$$
$$7x + 3 = 8x$$
$$-x = -3$$
$$x = 3$$

7.8 Applications of Rational Expressions

In this Section

- **Formulas**
- **Uniform Motion Problems**
- **Work Problems**
- **Purchasing Problems**

In this section we will study additional applications of rational expressions.

Formulas

Many formulas involve rational expressions. When solving a formula of this type for a certain variable, we usually multiply each side by the LCD to eliminate the denominators.

EXAMPLE 1

An equation of a line

The equation for the line through $(-2, 4)$ with slope $\frac{3}{2}$ can be written as

$$\frac{y-4}{x+2}=\frac{3}{2}.$$

We studied equations of this type in Chapter 3. Solve this equation for y.

Solution

To isolate y on the left-hand side of the equation, we multiply each side by $x + 2$:

$$\frac{y-4}{x+2}=\frac{3}{2} \quad \text{Original equation}$$

$$(x+2)\cdot\frac{y-4}{x+2}=(x+2)\cdot\frac{3}{2} \quad \text{Multiply by } x+2.$$

$$y-4=\frac{3}{2}x+3 \quad \text{Simplify.}$$

$$y=\frac{3}{2}x+7 \quad \text{Add 4 to each side.}$$

Because the original equation is a proportion, we could have used the extremes-means property to solve it for y.

Now do Exercises 1–8

EXAMPLE 2

Distance, rate, and time

Solve the formula $\frac{D}{T} = R$ for T.

Solution

Because the only denominator is T, we multiply each side by T:

$$\frac{D}{T}=R \quad \text{Original formula}$$

$$T\cdot\frac{D}{T}=T\cdot R \quad \text{Multiply each side by } T.$$

$$D=TR$$

$$\frac{D}{R}=\frac{TR}{R} \quad \text{Divide each side by } R.$$

$$\frac{D}{R}=T \quad \text{Simplify.}$$

The formula solved for T is $T = \frac{D}{R}$.

Now do Exercises 9–14

Study Tip

As you study from the text, think about the material. Ask yourself questions. If you were the professor, what questions would you ask on the test?

In Example 3, different subscripts are used on a variable to indicate that they are different variables. Think of R_1 as the first resistance, R_2 as the second resistance, and R as a combined resistance.

EXAMPLE 3

Total resistance

The formula

$$\frac{1}{R} = \frac{1}{R_1} + \frac{1}{R_2}$$

(from physics) expresses the relationship between different amounts of resistance in a parallel circuit. Solve it for R_2.

Solution

The LCD for R, R_1, and R_2 is RR_1R_2:

$$\frac{1}{R} = \frac{1}{R_1} + \frac{1}{R_2} \quad \text{Original formula}$$

$$RR_1R_2 \cdot \frac{1}{R} = RR_1R_2 \cdot \frac{1}{R_1} + RR_1R_2 \cdot \frac{1}{R_2} \quad \text{Multiply each side by the LCD, } RR_1R_2.$$

$$R_1R_2 = RR_2 + RR_1 \quad \text{All denominators are eliminated.}$$

$$R_1R_2 - RR_2 = RR_1 \quad \text{Get all terms involving } R_2 \text{ onto the left side.}$$

$$R_2(R_1 - R) = RR_1 \quad \text{Factor out } R_2.$$

$$R_2 = \frac{RR_1}{R_1 - R} \quad \text{Divide each side by } R_1 - R.$$

Now do Exercises 15–22

EXAMPLE 4

Finding the value of a variable

In the formula of Example 1, find x if $y = -3$.

Solution

Substitute $y = -3$ into the formula, then solve for x:

$$\frac{y - 4}{x + 2} = \frac{3}{2} \quad \text{Original formula}$$

$$\frac{-3 - 4}{x + 2} = \frac{3}{2} \quad \text{Replace } y \text{ by } -3.$$

$$\frac{-7}{x + 2} = \frac{3}{2} \quad \text{Simplify.}$$

$$3x + 6 = -14 \quad \text{Extremes-means property}$$

$$3x = -20$$

$$x = -\frac{20}{3}$$

Now do Exercises 23–32

Uniform Motion Problems

In uniform motion problems we use the formula $D = RT$. In some problems in which the time is unknown, we can use the formula $T = \frac{D}{R}$ to get an equation involving rational expressions.

EXAMPLE 5

Driving to Florida

Susan drove 1500 miles to Daytona Beach for spring break. On the way back she averaged 10 miles per hour less, and the drive back took her 5 hours longer. Find Susan's average speed on the way to Daytona Beach.

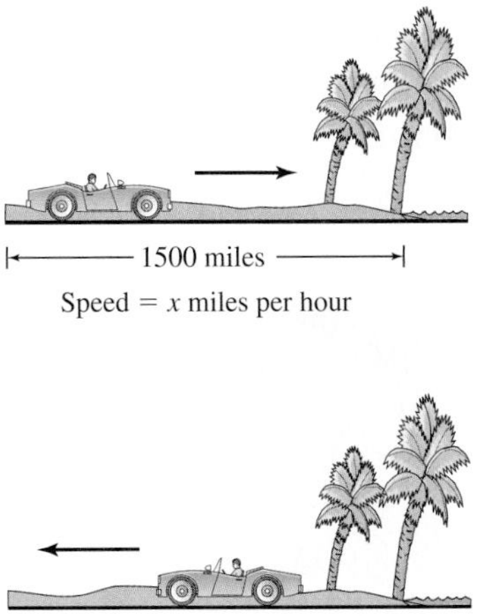

Figure 7.1

Solution

If x represents her average speed going there, then $x - 10$ is her average speed for the return trip. See Fig. 7.1. We use the formula $T = \frac{D}{R}$ to make the following table.

	D	R	T	
Going	1500	x	$\frac{1500}{x}$	← Shorter time
Returning	1500	$x - 10$	$\frac{1500}{x-10}$	← Longer time

Because the difference between the two times is 5 hours, we have

$$\text{longer time} - \text{shorter time} = 5.$$

Using the time expressions from the table, we get the following equation:

$$\frac{1500}{x-10} - \frac{1500}{x} = 5$$

$$x(x-10)\frac{1500}{x-10} - x(x-10)\frac{1500}{x} = x(x-10)5 \quad \text{Multiply by } x(x-10).$$

$$1500x - 1500(x-10) = 5x^2 - 50x$$

$$15{,}000 = 5x^2 - 50x \quad \text{Simplify.}$$

$$3000 = x^2 - 10x \quad \text{Divide each side by 5.}$$

$$0 = x^2 - 10x - 3000$$

$$(x+50)(x-60) = 0 \quad \text{Factor.}$$

$$x + 50 = 0 \quad \text{or} \quad x - 60 = 0$$

$$x = -50 \quad \text{or} \quad x = 60$$

The answer $x = -50$ is a solution to the equation, but it cannot indicate the average speed of the car. Her average speed going to Daytona Beach was 60 mph.

Now do Exercises 33–38

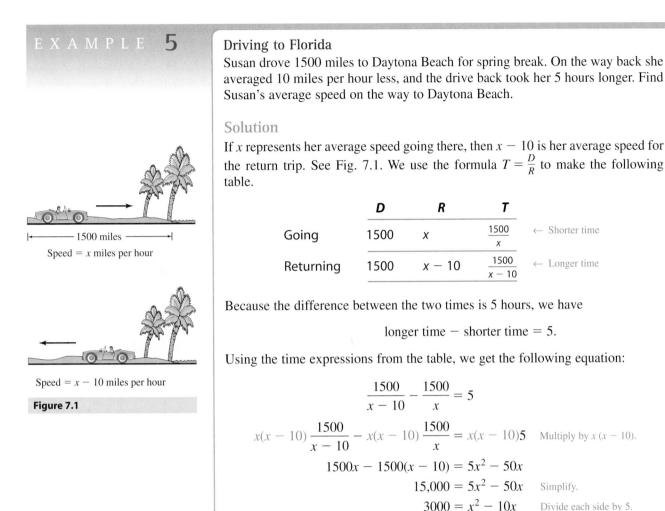

Work Problems

If you can complete a job in 3 hours, then you are working at the rate of $\frac{1}{3}$ of the job per hour. If you work for 2 hours at the rate of $\frac{1}{3}$ of the job per hour, then you will complete $\frac{2}{3}$ of the job. The product of the rate and time is the amount of work completed. For problems involving work, we will always assume that the work is done at a constant rate. So if a job takes x hours to complete, then the rate is $\frac{1}{x}$ of the job per hour.

Helpful Hint

Notice that a work rate is the same as a slope from Chapter 3. The only difference is that the work rates here can contain a variable.

EXAMPLE 6

Shoveling snow

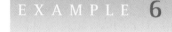

After a heavy snowfall, Brian can shovel all of the driveway in 30 minutes. If his younger brother Allen helps, the job takes only 20 minutes. How long would it take Allen to do the job by himself?

Helpful Hint

The secret to work problems is remembering that the individual rates or the amounts of work can be added when people work together. If your painting rate is 1/10 of the house per day and your helper's rate is 1/5 of the house per day, then your rate together will be 3/10 of the house per day. In 2 days you will paint 2/10 of the house and your helper will paint 2/5 of the house for a total of 3/5 of the house completed.

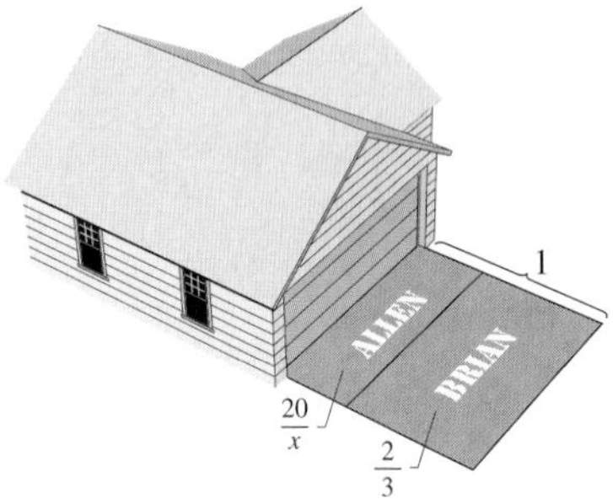

Figure 7.2

Solution

Let x represent the number of minutes it would take Allen to do the job by himself. Brian's rate for shoveling is $\frac{1}{30}$ of the driveway per minute, and Allen's rate for shoveling is $\frac{1}{x}$ of the driveway per minute. We organize all of the information in a table like the table in Example 5.

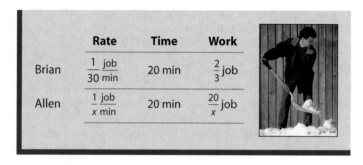

	Rate	Time	Work
Brian	$\frac{1 \text{ job}}{30 \text{ min}}$	20 min	$\frac{2}{3}$ job
Allen	$\frac{1 \text{ job}}{x \text{ min}}$	20 min	$\frac{20}{x}$ job

If Brian works for 20 min at the rate $\frac{1}{30}$ of the job per minute, then he does $\frac{20}{30}$ or $\frac{2}{3}$ of the job, as shown in Fig. 7.2. The amount of work that each boy does is a fraction of the whole job. So the expressions for work in the last column of the table have a sum of 1:

$$\frac{2}{3} + \frac{20}{x} = 1$$

$$3x \cdot \frac{2}{3} + 3x \cdot \frac{20}{x} = 3x \cdot 1 \quad \text{Multiply each side by } 3x.$$

$$2x + 60 = 3x$$

$$60 = x$$

If it takes Allen 60 min to do the job by himself, then he works at the rate of $\frac{1}{60}$ of the job per minute. In 20 minutes he does $\frac{1}{3}$ of the job while Brian does $\frac{2}{3}$. So it would take Allen 60 minutes to shovel the driveway by himself.

Now do Exercises 39–44

Notice the similarities between the uniform motion problem in Example 5 and the work problem in Example 6. In both cases, it is beneficial to make a table. We use $D = R \cdot T$ in uniform motion problems and $W = R \cdot T$ in work problems. The main points to remember when solving work problems are summarized in the following strategy.

Strategy for Solving Work Problems

1. If a job is completed in x hours, then the rate is $\frac{1}{x}$ job/hr.
2. Make a table showing rate, time, and work completed ($W = R \cdot T$) for each person or machine.
3. The total work completed is the sum of the individual amounts of work completed.
4. If the job is completed, then the total work done is 1 job.

Purchasing Problems

Rates are used in uniform motion and work problems. But rates also occur in purchasing problems. If gasoline is 149.9 cents/gallon, then that is the rate at which your bill is increasing as you pump the gallons into your tank. In purchasing problems the product of the rate and the quantity purchased is the total cost.

EXAMPLE 7

Oranges and grapefruit

Tamara bought 50 pounds of fruit consisting of Florida oranges and Texas grapefruit. She paid twice as much per pound for the grapefruit as she did for the oranges. If Tamara bought \$12 worth of oranges and \$16 worth of grapefruit, then how many pounds of each did she buy?

Solution

Let x represent the number of pounds of oranges and $50 - x$ represent the number of pounds of grapefruit. See Fig. 7.3. Make a table.

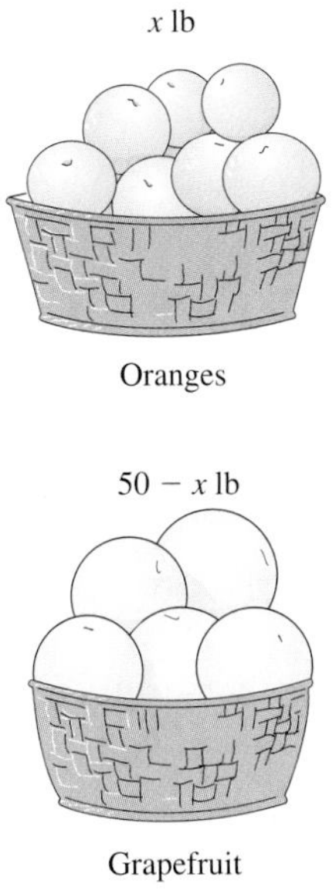

Figure 7.3

	Rate	Quantity	Total Cost
Oranges	$\frac{12}{x}$ dollars/pound	x pounds	12 dollars
Grapefruit	$\frac{16}{50 - x}$ dollars/pound	$50 - x$ pounds	16 dollars

Since the price per pound for the grapefruit is twice that for the oranges, we have:

$$2(\text{price per pound for oranges}) = \text{price per pound for grapefruit}$$

$$2\left(\frac{12}{x}\right) = \frac{16}{50 - x}$$

$$\frac{24}{x} = \frac{16}{50 - x}$$

$$16x = 1200 - 24x \quad \text{Extremes-means property}$$

$$40x = 1200$$

$$x = 30$$

$$50 - x = 20$$

If Tamara purchased 20 pounds of grapefruit for \$16, then she paid \$0.80 per pound. If she purchased 30 pounds of oranges for \$12, then she paid \$0.40 per pound. Because \$0.80 is twice \$0.40, we can be sure that she purchased 20 pounds of grapefruit and 30 pounds of oranges.

Now do Exercises 45–48

Warm-Ups ▼

True or false? Explain your answer.

1. The formula $t = \frac{1-t}{m}$, solved for m, is $m = \frac{1-t}{t}$.
2. To solve $\frac{1}{m} + \frac{1}{n} = \frac{1}{2}$ for m, we multiply each side by $2mn$.
3. If Fiona drives 300 miles in x hours, then her average speed is $\frac{x}{300}$ mph.
4. If Miguel drives 20 hard bargains in x hours, then he is driving $\frac{20}{x}$ hard bargains per hour.
5. If Fred can paint a house in y days, then he paints $\frac{1}{y}$ of the house per day.
6. If $\frac{1}{x}$ is 1 less than $\frac{2}{x+3}$, then $\frac{1}{x} - 1 = \frac{2}{x+3}$.
7. If a and b are nonzero and $a = \frac{m}{b}$, then $b = am$.
8. If $D = RT$, then $T = \frac{D}{R}$.
9. Solving $P + Prt = I$ for P gives $P = I - Prt$.
10. To solve $3R + yR = m$ for R, we must first factor the left-hand side.

7.8 Exercises

Boost your GRADE at mathzone.com!

MathZone
- Practice Problems
- Self-Tests
- Videos
- Net Tutor
- e-Professors

Solve each equation for y. See Example 1.

1. $\frac{y-1}{x-3} = 2$

2. $\frac{y-2}{x-4} = -2$

3. $\frac{y-1}{x+6} = -\frac{1}{2}$

4. $\frac{y+5}{x-2} = -\frac{1}{2}$

5. $\frac{y+a}{x-b} = m$

6. $\frac{y-h}{x+k} = a$

7. $\frac{y-1}{x+4} = -\frac{1}{3}$

8. $\frac{y-1}{x+3} = -\frac{3}{4}$

Solve each formula for the indicated variable. See Examples 2 and 3.

9. $A = \frac{B}{C}$ for C

10. $P = \frac{A}{C+D}$ for A

11. $\frac{1}{a} + m = \frac{1}{p}$ for p

12. $\frac{2}{f} + t = \frac{3}{m}$ for m

13. $F = k\dfrac{m_1 m_2}{r^2}$ for m_1

14. $F = \dfrac{mv^2}{r}$ for r

15. $\dfrac{1}{a} + \dfrac{1}{b} = \dfrac{1}{f}$ for a

16. $\dfrac{1}{R} = \dfrac{1}{R_1} + \dfrac{1}{R_2}$ for R

17. $S = \dfrac{a}{1 - r}$ for r

18. $I = \dfrac{E}{R + r}$ for R

19. $\dfrac{P_1V_1}{T_1} = \dfrac{P_2V_2}{T_2}$ for P_2

20. $\dfrac{P_1V_1}{T_1} = \dfrac{P_2V_2}{T_2}$ for T_1

21. $V = \dfrac{4}{3}\pi r^2 h$ for h

22. $h = \dfrac{S - 2\pi r^2}{2\pi r}$ for S

Find the value of the indicated variable. See Example 4.

23. In the formula of Exercise 9, if $A = 12$ and $B = 5$, find C.

24. In the formula of Exercise 10, if $A = 500$, $P = 100$, and $C = 2$, find D.

25. In the formula of Exercise 11, if $p = 6$ and $m = 4$, find a.

26. In the formula of Exercise 12, if $m = 4$ and $t = 3$, find f.

27. In the formula of Exercise 13, if $F = 32$, $r = 4$, $m_1 = 2$, and $m_2 = 6$, find k.

28. In the formula of Exercise 14, if $F = 10$, $v = 8$, and $r = 6$, find m.

29. In the formula of Exercise 15, if $f = 3$ and $a = 2$, find b.

30. In the formula of Exercise 16, if $R = 3$ and $R_1 = 5$, find R_2.

31. In the formula of Exercise 17, if $S = \dfrac{3}{2}$ and $r = \dfrac{1}{5}$, find a.

32. In the formula of Exercise 18, if $I = 15$, $E = 3$, and $R = 2$, find r.

Show a complete solution to each problem. See Example 5.

33. ***Fast walking.*** Marcie can walk 8 miles in the same time as Frank walks 6 miles. If Marcie walks 1 mile per hour faster than Frank, then how fast does each person walk?

34. ***Upstream, downstream.*** Junior's boat will go 15 miles per hour in still water. If he can go 12 miles downstream in the same amount of time as it takes to go 9 miles upstream, then what is the speed of the current?

35. ***Delivery routes.*** Pat travels 70 miles on her milk route, and Bob travels 75 miles on his route. Pat travels 5 miles per hour slower than Bob, and her route takes her one-half hour longer than Bob's. How fast is each one traveling?

36. ***Ride the peaks.*** Smith bicycled 45 miles going east from Durango, and Jones bicycled 70 miles. Jones averaged 5 miles per hour more than Smith, and his trip took one-half hour longer than Smith's. How fast was each one traveling?

Photo for Exercise 36

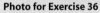

37. ***Walking and running.*** Raffaele ran 8 miles and then walked 6 miles. If he ran 5 miles per hour faster than he walked and the total time was 2 hours, then how fast did he walk?

38. ***Triathlon.*** Luisa participated in a triathlon in which she swam 3 miles, ran 5 miles, and then bicycled 10 miles. Luisa ran twice as fast as she swam, and she cycled three times as fast as she swam. If her total time for the triathlon was 1 hour and 46 minutes, then how fast did she swim?

Show a complete solution to each problem. See Example 6.

39. ***Fence painting.*** Kiyoshi can paint a certain fence in 3 hours by himself. If Red helps, the job takes only 2 hours. How long would it take Red to paint the fence by himself?

40. ***Envelope stuffing.*** Every week, Linda must stuff 1000 envelopes. She can do the job by herself in 6 hours. If

Laura helps, they get the job done in $5\frac{1}{2}$ hours. How long would it take Laura to do the job by herself?

41. ***Garden destroying.*** Mr. McGregor has discovered that a large dog can destroy his entire garden in 2 hours and that a small boy can do the same job in 1 hour. How long would it take the large dog and the small boy working together to destroy Mr. McGregor's garden?

42. ***Draining the vat.*** With only the small valve open, all of the liquid can be drained from a large vat in 4 hours. With only the large valve open, all of the liquid can be drained from the same vat in 2 hours. How long would it take to drain the vat with both valves open?

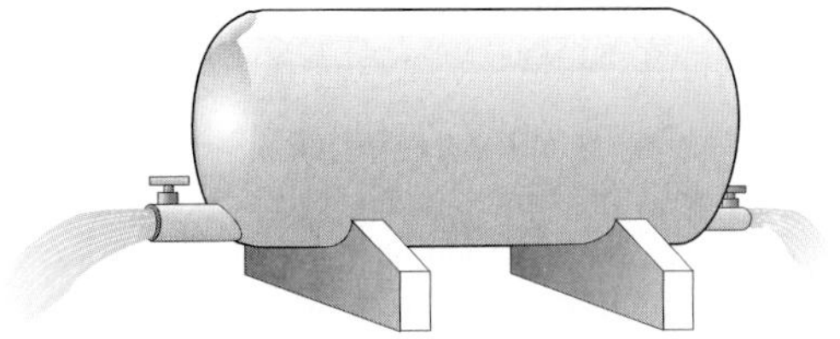

Figure for Exercise 42

43. ***Cleaning sidewalks.*** Edgar can blow the leaves off the sidewalks around the capitol building in 2 hours using a gasoline-powered blower. Ellen can do the same job in 8 hours using a broom. How long would it take them working together?

44. ***Computer time.*** It takes a computer 8 days to print all of the personalized letters for a national sweepstakes. A new computer is purchased that can do the same job in 5 days. How long would it take to do the job with both computers working on it?

Show a complete solution to each problem. See Example 7.

45. ***Apples and bananas.*** Bertha bought 18 pounds of fruit consisting of apples and bananas. She paid \$9 for the apples and \$2.40 for the bananas. If the price per pound of the apples was 3 times that of the bananas, then how many pounds of each type of fruit did she buy?

46. ***Running backs.*** In the playoff game the ball was carried by either Anderson or Brown on 21 plays. Anderson gained 36 yards, and Brown gained 54 yards. If Brown averaged twice as many yards per carry as Anderson, then on how many plays did Anderson carry the ball?

47. ***Fuel efficiency.*** Last week, Joe's Electric Service used 110 gallons of gasoline in its two trucks. The large truck was driven 800 miles, and the small truck was driven 600 miles. If the small truck gets twice as many miles per gallon as the large truck, then how many gallons of gasoline did the large truck use?

Photo for Exercise 46

48. ***Repair work.*** Sally received a bill for a total of 8 hours labor on the repair of her bulldozer. She paid \$50 to the master mechanic and \$90 to his apprentice. If the master mechanic gets \$10 more per hour than his apprentice, then how many hours did each work on the bulldozer?

Show a complete solution to each problem.

49. ***Small plane.*** It took a small plane 1 hour longer to fly 480 miles against the wind than it took the plane to fly the same distance with the wind. If the wind speed was 20 mph, then what is the speed of the plane in calm air?

50. ***Fast boat.*** A motorboat at full throttle takes two hours longer to travel 75 miles against the current than it takes to travel the same distance with the current. If the rate of the current is 5 mph, then what is the speed of the boat at full throttle in still water?

51. ***Light plane.*** At full throttle a light plane flies 275 miles against the wind in the same time as it flies 325 miles with the wind. If the plane flies at 120 mph at full throttle in still air, then what is the wind speed?

52. ***Big plane.*** A six-passenger plane cruises at 180 mph in calm air. If the plane flies 7 miles with the wind in the same amount of time as it flies 5 miles against the wind, then what is the wind speed?

53. ***Two cyclists.*** Ben and Jerry start from the same point and ride their bicycles in opposite directions. If Ben rides twice as fast as Jerry and they are 90 miles apart after four hours, then what is the speed of each rider?

54. ***Catching up.*** A sailboat leaves port and travels due south at an average speed of 9 mph. Four hours later a motorboat leaves the same port and travels due south at an average speed of 21 mph. How long will it take the motorboat to catch the sailboat?

55. ***Road trip.*** The Griswalds averaged 45 mph on their way to Las Vegas and 60 mph on the way back home using the same route. Find the distance from their home to Las Vegas if the total driving time was 70 hours.

56. ***Meeting cyclists.*** Tanya and Lebron start at the same time from opposite ends of a bicycle trail that is 81 miles long. Tanya averages 12 mph and Lebron averages 15 mph. How long does it take for them to meet?

57. ***Filling a fountain.*** Pete's fountain can be filled using a pipe or a hose. The fountain can be filled using the pipe in 6 hours or the hose in 12 hours. How long will it take to fill the fountain using both the pipe and the hose?

58. ***Mowing a lawn.*** Albert can mow a lawn in 40 minutes, while his cousin Vinnie can mow the same lawn in one hour. How long would it take to mow the lawn if Albert and Vinnie work together?

59. ***Printing a report.*** Debra plans to use two computers to print all of the copies of the annual report that are needed for the year-end meeting. The new computer can do the whole job in 2 hours while the old computer can do the whole job in 3 hours. How long will it take to get the job done using both computers simultaneously?

60. ***Installing a dishwasher.*** A plumber can install a dishwasher in 50 min. If the plumber brings his apprentice to help, the job takes 40 minutes. How long would it take the apprentice working alone to install the dishwasher?

61. ***Filling a tub.*** Using the hot and cold water faucets together, a bathtub fills in 8 minutes. Using the hot water faucet alone, the tub fills in 12 minutes. How long does it take to fill the tub using only the cold water faucet?

62. ***Filling a tank.*** A water tank has an inlet pipe and a drain pipe. A full tank can be emptied in 30 minutes if the drain is opened and an empty tank can be filled in 45 minutes with the inlet pipe opened. If both pipes are accidentally opened when the tank is full, then how long will it take to empty the tank?

Collaborative Activities

Grouping: Three students per group

Topic: Distance formula

How Do I Get There from Here?

Suppose that you want to decide whether to ride your bicycle, drive your car, or take the bus to school this year. The best thing to do is to analyze each of your options. Read all the information given here and then have each person in your group pick one mode of transportation. Working individually, answer the questions using the given information, the accompanying map, and the distance formula $D = RT$. Present a case to your group for your type of transportation. Compute costs for one entire 32-week school year (two 16-week semesters).

If you travel by car, you will need to pay for a parking permit, which costs $150/year. Traffic has increased, so it takes 12 minutes to get to school. What is your average speed? Determine the cost (use current gas prices) of your gasoline for a full 32-week academic year assuming your car will get 30 mpg. Are there any other costs associated with using a car?

If you travel by bicycle, then you will need to buy a new bike lock for $25 and two new tubes at $5.00 apiece. As well as cost, determine how fast you would have to bike to beat the car.

If you travel by bus, then it will cost $7.50 per month for a student bus pass. The bus stops at the end of your block. It leaves at 8:30 A.M. and will get to the college at 8:55 A.M. On Mondays and Wednesdays you have a 9:00 A.M. class, four blocks from the bus stop. As well as cost find the average speed of the bus and figure the cost for the full 32-week school year.

When you present your case, include the time needed to get there, speed, total cost, convenience, and any other expenses. Have at least three reasons why this would be the best way to travel. Consider unique features of your area such as traffic, weather, and terrain.

After each of you has presented your case, decide as a group which type of transportation to choose.

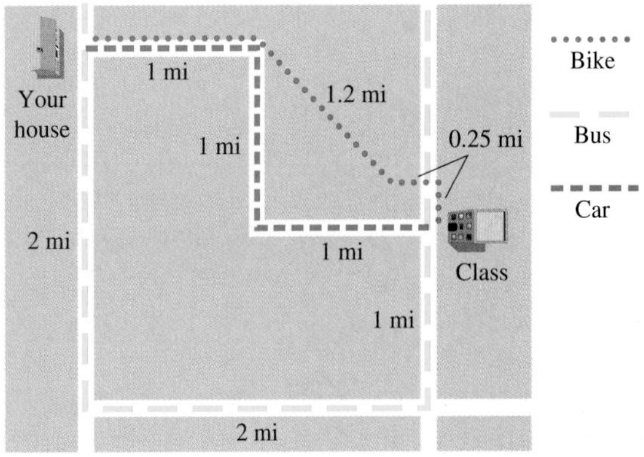

Chapter 7 Wrap-Up

Summary

Rational Expressions		**Examples**
Rational expression	The ratio of two polynomials with the denominator not equal to 0	$\frac{x-1}{x-3} (x \neq 3)$
Rule for reducing rational expressions	If $a \neq 0$ and $c \neq 0$, then $\frac{ab}{ac} = \frac{b}{c}$. (Divide out the common factors.)	$\frac{8x+2}{4x} = \frac{2(4x+1)}{2(2x)} = \frac{4x+1}{2x}$
Quotient rule for exponents	Suppose $a \neq 0$ and m and n are positive integers. If $m \geq n$, then $\frac{a^m}{a^n} = a^{m-n}$. If $m < n$, then $\frac{a^m}{a^n} = \frac{1}{a^{n-m}}$.	$\frac{x^7}{x^5} = x^2$ $\frac{x^2}{x^5} = \frac{1}{x^3}$

Multiplication and Division of Rational Expressions		**Examples**
Multiplication	If $b \neq 0$ and $d \neq 0$, then $\frac{a}{b} \cdot \frac{c}{d} = \frac{ac}{bd}$.	$\frac{3}{x^3} \cdot \frac{6}{x^5} = \frac{18}{x^8}$
Division	If $b \neq 0$, $c \neq 0$, and $d \neq 0$, then $\frac{a}{b} \div \frac{c}{d} = \frac{a}{b} \cdot \frac{d}{c}$. (Invert the divisor and multiply.)	$\frac{a}{x^3} \div \frac{5}{x^9} = \frac{a}{x^3} \cdot \frac{x^9}{5} = \frac{ax^6}{5}$

Addition and Subtraction of Rational Expressions		**Examples**
Least common denominator	The LCD of a group of denominators is the smallest number that is a multiple of all of them.	8, 12 LCD = 24
Finding the least common denominator	1. Factor each denominator completely. Use exponent notation for repeated factors. 2. Write the product of all of the different factors that appear in the denominators. 3. On each factor, use the highest power that appears on that factor in any of the denominators.	$4ab^3$, $6a^2b$ $4ab^3 = 2^2ab^3$ $6a^2b = 2 \cdot 3a^2b$ $\text{LCD} = 2^2 \cdot 3a^2b^3 = 12a^2b^3$

Addition and subtraction of rational expressions	If $b \neq 0$, then $\frac{a}{b} + \frac{c}{b} = \frac{a+c}{b}$ and $\frac{a}{b} - \frac{c}{b} = \frac{a-c}{b}$. If the denominators are not identical, change each fraction to an equivalent fraction so that all denominators are identical.	$\frac{2x}{x-3} + \frac{7x}{x-3} = \frac{9x}{x-3}$ $\frac{2}{x} + \frac{1}{3x} = \frac{6}{3x} + \frac{1}{3x} = \frac{7}{3x}$
Complex fraction	A rational expression that has fractions in the numerator and/or the denominator	$\dfrac{\frac{1}{2} + \frac{1}{3}}{\frac{1}{3} - \frac{3}{4}}$
Simplifying complex fractions	Multiply the numerator and denominator by the LCD.	$\dfrac{\left(\frac{1}{2} + \frac{1}{3}\right)12}{\left(\frac{1}{3} - \frac{3}{4}\right)12} = \frac{6+4}{4-9} = -2$

Equations with Rational Expressions		**Examples**
Solving equations	Multiply each side by the LCD.	$\frac{1}{x} - \frac{1}{3} = \frac{1}{2x} - \frac{1}{6}$ $6x\left(\frac{1}{x} - \frac{1}{3}\right) = 6x\left(\frac{1}{2x} - \frac{1}{6}\right)$ $6 - 2x = 3 - x$
Proportion	An equation expressing the equality of two ratios	$\frac{a}{b} = \frac{c}{d}$
Extremes-means property (Cross-multiply)	If $b \neq 0$ and $d \neq 0$, then $\frac{a}{b} = \frac{c}{d}$ is equivalent to $ad = bc$. Cross-multiplying is a quick way to eliminate the fractions in a proportion.	$\frac{2}{x-3} = \frac{5}{6}$ $2 \cdot 6 = (x-3)5$ $12 = 5x - 15$

Enriching Your Mathematical Word Power

For each mathematical term, choose the correct meaning.

1. **rational expression**
   a. a fraction
   b. a ratio of two polynomials with denominator not equal to 0
   c. an expression involving fractions
   d. a fraction in which the numerator and denominator contain fractions

2. **complex fraction**
   a. a fraction having rational expressions in the numerator, denominator, or both
   b. a fraction with a large denominator
   c. the sum of two fractions
   d. a fraction with a variable in the denominator

3. building up the denominator
a. the opposite of reducing a fraction
b. finding the least common denominator
c. adding the same number to the numerator and denominator
d. writing a fraction larger

4. least common denominator
a. the largest number that is a multiple of all denominators
b. the sum of the denominators
c. the product of the denominators
d. the smallest number that is a multiple of all denominators

5. extraneous solution
a. a number that appears to be a solution to an equation but does not satisfy the equation
b. an extra solution to an equation
c. the second solution
d. a nonreal solution

6. ratio of a to b
a. b/a
b. a/b
c. $a/(a + b)$
d. ab

7. proportion
a. a ratio
b. two ratios
c. the product of the means equals the product of the extremes
d. a statement expressing the equality of two ratios

8. extremes
a. a and d in $a/b = c/d$
b. b and c in $a/b = c/d$
c. the extremes-means property
d. if $a/b = c/d$ then $ad = bc$

9. means
a. the average of a, b, c, and d
b. a and d in $a/b = c/d$
c. b and c in $a/b = c/d$
d. if $a/b = c/d$, then $(a + b)/2 = (c + d)/2$

10. cross-multiplying
a. $ab = ba$ for any real numbers a and b
b. $(a - b)^2 = (b - a)^2$ for any real numbers a and b
c. if $a/b = c/d$, then $ab = cd$
d. if $a/b = c/d$, then $ad = bc$

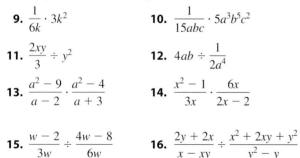

Review Exercises

7.1 *Reduce each rational expression to lowest terms.*

1. $\frac{24}{28}$

2. $\frac{42}{18}$

3. $\frac{2a^3c^3}{8a^5c}$

4. $\frac{39x^6}{15x}$

5. $\frac{6w - 9}{9w - 12}$

6. $\frac{3t - 6}{8 - 4t}$

7. $\frac{x^2 - 1}{3 - 3x}$

8. $\frac{3x^2 - 9x + 6}{10 - 5x}$

7.2 *Perform the indicated operation.*

9. $\frac{1}{6k} \cdot 3k^2$

10. $\frac{1}{15abc} \cdot 5a^3b^5c^2$

11. $\frac{2xy}{3} \div y^2$

12. $4ab \div \frac{1}{2a^4}$

13. $\frac{a^2 - 9}{a - 2} \cdot \frac{a^2 - 4}{a + 3}$

14. $\frac{x^2 - 1}{3x} \cdot \frac{6x}{2x - 2}$

15. $\frac{w - 2}{3w} \div \frac{4w - 8}{6w}$

16. $\frac{2y + 2x}{x - xy} \div \frac{x^2 + 2xy + y^2}{y^2 - y}$

7.3 *Find the least common denominator for each group of denominators.*

17. 36, 54

18. 10, 15, 35

19. $6ab^3$, $8a^7b^2$

20. $20u^4v$, $18uv^5$, $12u^2v^3$

21. $4x$, $6x - 6$

22. $8a$, $6a$, $2a^2 + 2a$

23. $x^2 - 4$, $x^2 - x - 2$

24. $x^2 - 9$, $x^2 + 6x + 9$

Convert each rational expression into an equivalent rational expression with the indicated denominator.

25. $\frac{5}{12} = \frac{?}{36}$

26. $\frac{2a}{15} = \frac{?}{45}$

27. $\frac{2}{3xy} = \frac{?}{15x^2y}$

28. $\frac{3z}{7x^2y} = \frac{?}{42x^3y^8}$

29. $\frac{5}{y - 6} = \frac{?}{12 - 2y}$

30. $\dfrac{-3}{2-t} = \dfrac{?}{2t-4}$

31. $\dfrac{x}{x-1} = \dfrac{?}{x^2-1}$

32. $\dfrac{t}{t-3} = \dfrac{?}{t^2+2t-15}$

7.4 *Perform the indicated operation.*

33. $\dfrac{5}{36} + \dfrac{9}{28}$

34. $\dfrac{7}{30} - \dfrac{11}{42}$

35. $3 - \dfrac{4}{x}$

36. $1 + \dfrac{3a}{2b}$

37. $\dfrac{2}{ab^2} - \dfrac{1}{a^2b}$

38. $\dfrac{3}{4x^3} + \dfrac{5}{6x^2}$

39. $\dfrac{9a}{2a-3} + \dfrac{5}{3a-2}$

40. $\dfrac{3}{x-2} - \dfrac{5}{x+3}$

41. $\dfrac{1}{a-8} - \dfrac{2}{8-a}$

42. $\dfrac{5}{x-14} + \dfrac{4}{14-x}$

43. $\dfrac{3}{2x-4} + \dfrac{1}{x^2-4}$

44. $\dfrac{x}{x^2-2x-3} - \dfrac{3x}{x^2-9}$

7.5 *Simplify each complex fraction.*

45. $\dfrac{\frac{1}{2}-\frac{3}{4}}{\frac{2}{3}+\frac{1}{2}}$

46. $\dfrac{\frac{2}{3}+\frac{5}{8}}{\frac{1}{2}-\frac{3}{8}}$

47. $\dfrac{\frac{1}{a}+\frac{2}{3b}}{\frac{1}{2b}-\frac{3}{a}}$

48. $\dfrac{\frac{3}{xy}-\frac{1}{3y}}{\frac{1}{6x}-\frac{3}{5y}}$

49. $\dfrac{\frac{1}{x-2}-\frac{3}{x+3}}{\frac{2}{x+3}+\frac{1}{x-2}}$

50. $\dfrac{\frac{4}{a+1}+\frac{5}{a^2-1}}{\frac{1}{a^2-1}-\frac{3}{a-1}}$

51. $\dfrac{\frac{x-1}{x-3}}{\frac{1}{x^2-x-6}-\frac{4}{x+2}}$

52. $\dfrac{\frac{6}{a^2+5a+6}-\frac{8}{a+2}}{\frac{2}{a+3}-\frac{4}{a+2}}$

7.6 *Solve each equation.*

53. $\dfrac{-2}{5} = \dfrac{3}{x}$

54. $\dfrac{3}{x} + \dfrac{5}{3x} = 1$

55. $\dfrac{14}{a^2-1} + \dfrac{1}{a-1} = \dfrac{3}{a+1}$

56. $2 + \dfrac{3}{y-5} = \dfrac{2y}{y-5}$

57. $z - \dfrac{3z}{2-z} = \dfrac{6}{z-2}$

58. $\dfrac{1}{x} + \dfrac{1}{3} = \dfrac{1}{2}$

7.7 *Solve each proportion.*

59. $\dfrac{3}{x} = \dfrac{2}{7}$

60. $\dfrac{4}{x} = \dfrac{x}{4}$

61. $\dfrac{2}{w-3} = \dfrac{5}{w}$

62. $\dfrac{3}{t-3} = \dfrac{5}{t+4}$

Solve each problem by using a proportion.

63. ***Taxis in Times Square.*** The ratio of taxis to private automobiles in Times Square at 6:00 P.M. on New Year's Eve was estimated to be 15 to 2. If there were 60 taxis, then how many private automobiles were there?

Photo for Exercise 63

64. ***Student-teacher ratio.*** The student-teacher ratio for Washington High was reported to be 27.5 to 1. If there are 42 teachers, then how many students are there?

65. ***Water and rice.*** At Wong's Chinese Restaurant the secret recipe for white rice calls for a 2 to 1 ratio of water to rice. In one batch the chef used 28 more cups of water than rice. How many cups of each did he use?

Photo for Exercise 65

66. ***Oil and gas.*** An outboard motor calls for a fuel mixture that has a gasoline-to-oil ratio of 50 to 1. How many pints of oil should be added to 6 gallons of gasoline?

7.8 *Solve each formula for the indicated variable.*

67. $\dfrac{y - b}{m} = x$ for y

68. $\dfrac{A}{h} = \dfrac{a + b}{2}$ for a

69. $F = \dfrac{mv + 1}{m}$ for m

70. $m = \dfrac{r}{1 + rt}$ for r

71. $\dfrac{y + 1}{x - 3} = 4$ for y

72. $\dfrac{y - 3}{x + 2} = \dfrac{-1}{3}$ for y

Solve each problem.

73. ***Making a puzzle.*** Tracy, Stacy, and Fred assembled a very large puzzle together in 40 hours. If Stacy worked twice as fast as Fred and Tracy worked just as fast as Stacy, then how long would it have taken Fred to assemble the puzzle alone?

74. ***Going skiing.*** Leon drove 270 miles to the lodge in the same time as Pat drove 330 miles to the lodge. If Pat drove 10 miles per hour faster than Leon, then how fast did each of them drive?

Photo for Exercise 74

75. ***Merging automobiles.*** When Bert and Ernie merged their automobile dealerships, Bert had 10 more cars than Ernie. While 36% of Ernie's stock consisted of new cars, only 25% of Bert's stock consisted of new cars. If they had 33 new cars on the lot after the merger, then how many cars did each one have before the merger?

76. ***Magazine sales.*** A company specializing in magazine sales over the telephone found that in 2500 phone calls, 360 resulted in sales and were made by male callers, and 480 resulted in sales and were made by female callers. If the company gets twice as many sales per call with a woman's voice than with a man's voice, then how many of the 2500 calls were made by females?

77. ***Distribution of waste.*** The accompanying figure shows the distribution of the total municipal solid waste into various categories in 2000 (U.S. Environmental Protection Agency, www.epa.gov). If the paper waste was 59.8 million tons greater than the yard waste, then what was the amount of yard waste generated?

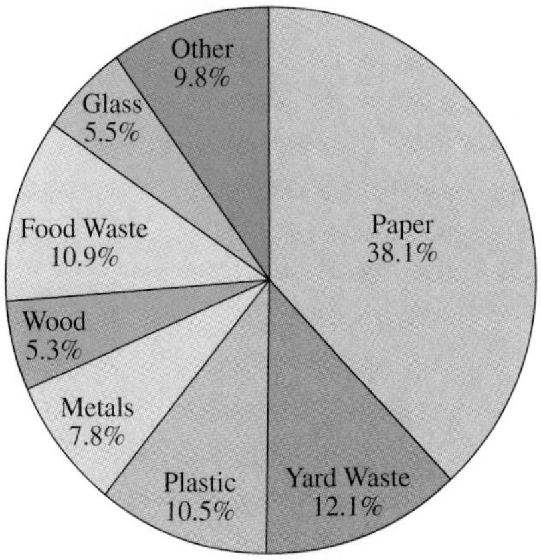

Figure for Exercises 77 and 78

78. ***Total waste.*** Use the information given in Exercise 77 to find the total waste generated in 2000 and the amount of food waste.

Miscellaneous

In place of each question mark, put an expression that makes each equation an identity.

79. $\frac{5}{x} = \frac{?}{2x}$

80. $\frac{?}{a} = \frac{6}{3a}$

81. $\frac{2}{a-5} = \frac{?}{5-a}$

82. $\frac{-1}{a-7} = \frac{1}{?}$

83. $3 = \frac{?}{x}$

84. $2a = \frac{?}{b}$

85. $m \div \frac{1}{2} = ?$

86. $5x \div \frac{1}{x} = ?$

87. $2a \div ? = 12a$

88. $10x \div ? = 20x^2$

89. $\frac{a-1}{a^2-1} = \frac{1}{?}$

90. $\frac{?}{x^2-9} = \frac{1}{x-3}$

91. $\frac{1}{a} - \frac{1}{5} = ?$

92. $\frac{3}{7} - \frac{2}{b} = ?$

93. $\frac{a}{2} - 1 = \frac{?}{2}$

94. $\frac{1}{a} - 1 = \frac{?}{a}$

95. $(a-b) \div (-1) = ?$

96. $(a-7) \div (7-a) = ?$

97. $\frac{\frac{1}{5a}}{2} = ?$

98. $\frac{3a}{\frac{1}{2}} = ?$

For each expression in Exercises 99–118, either perform the indicated operation or solve the equation, whichever is appropriate.

99. $\frac{1}{x} + \frac{1}{2x}$

100. $\frac{1}{y} + \frac{1}{3y} = 2$

101. $\frac{2}{3xy} + \frac{1}{6x}$

102. $\frac{3}{x-1} - \frac{3}{x}$

103. $\frac{5}{a-5} - \frac{3}{5-a}$

104. $\frac{2}{x-2} - \frac{3}{x} = \frac{-1}{x}$

105. $\frac{2}{x-1} - \frac{2}{x} = 1$

106. $\frac{2}{x-2} \cdot \frac{6x-12}{14}$

107. $\frac{-3}{x+2} \cdot \frac{5x+10}{9}$

108. $\frac{3}{10} = \frac{5}{x}$

109. $\frac{1}{-3} = \frac{-2}{x}$

110. $\frac{x^2 - 4}{x} \div \frac{4x - 8}{x}$

111. $\frac{ax + am + 3x + 3m}{a^2 - 9} \div \frac{2x + 2m}{a - 3}$

112. $\frac{-2}{x} = \frac{3}{x + 2}$

113. $\frac{2}{x^2 - 25} + \frac{1}{x^2 - 4x - 5}$

114. $\frac{4}{a^2 - 1} + \frac{1}{2a + 2}$

115. $\frac{-3}{a^2 - 9} - \frac{2}{a^2 + 5a + 6}$

116. $\frac{-5}{a^2 - 4} - \frac{2}{a^2 - 3a + 2}$

117. $\frac{1}{a^2 - 1} + \frac{2}{1 - a} = \frac{3}{a + 1}$

118. $3 + \frac{1}{x - 2} = \frac{2x - 3}{x - 2}$

Chapter 7 Test

What numbers cannot be used for x in each rational expression?

1. $\frac{2x - 1}{x^2 - 1}$

2. $\frac{5}{2 - 3x}$

3. $\frac{1}{x}$

Perform the indicated operation. Write each answer in lowest terms.

4. $\frac{2}{15} - \frac{4}{9}$

5. $\frac{1}{y} + 3$

6. $\frac{3}{a - 2} - \frac{1}{2 - a}$

7. $\frac{2}{x^2 - 4} - \frac{3}{x^2 + x - 2}$

8. $\frac{m^2 - 1}{(m - 1)^2} \cdot \frac{2m - 2}{3m + 3}$

9. $\frac{a - b}{3} \div \frac{b^2 - a^2}{6}$

10. $\frac{5a^2b}{12a} \cdot \frac{2a^3b}{15ab^6}$

Simplify each complex fraction.

11. $\dfrac{\frac{2}{3} + \frac{4}{5}}{\frac{2}{5} - \frac{3}{2}}$

12. $\dfrac{\frac{2}{x} + \frac{1}{x - 2}}{\frac{1}{x - 2} - \frac{3}{x}}$

Solve each equation.

13. $\frac{3}{x} = \frac{7}{5}$

14. $\frac{x}{x - 1} - \frac{3}{x} = \frac{1}{2}$

15. $\frac{1}{x} + \frac{1}{6} = \frac{1}{4}$

Solve each formula for the indicated variable.

16. $\frac{y - 3}{x + 2} = \frac{-1}{5}$ for y

17. $M = \frac{1}{3}b(c + d)$ for c

Solve each problem.

18. If $R(x) = \frac{x + 2}{1 - x}$, then what is $R(0.9)$?

19. When all of the grocery carts escape from the supermarket, it takes Reginald 12 minutes to round them up and bring them back. Because Norman doesn't make as much per hour as Reginald, it takes Norman 18 minutes to do the same job. How long would it take them working together to complete the roundup?

20. Brenda and her husband Randy bicycled cross-country together. One morning, Brenda rode 30 miles. By traveling only 5 miles per hour faster and putting in one more hour, Randy covered twice the distance Brenda covered. What was the speed of each cyclist?

21. For a certain time period the ratio of the dollar value of exports to the dollar value of imports for the United States was 2 to 3. If the value of exports during that time period was 48 billion dollars, then what was the value of imports?

*Making*Connections | A Review of Chapters 1–7

Solve each equation.

1. $3x - 2 = 5$

2. $\frac{3}{5}x = -2$

3. $2(x - 2) = 4x$

4. $2(x - 2) = 2x$

5. $2(x + 3) = 6x + 6$

6. $2(3x + 4) + x^2 = 0$

7. $4x - 4x^3 = 0$

8. $\frac{3}{x} = \frac{-2}{5}$

9. $\frac{3}{x} = \frac{x}{12}$

10. $\frac{x}{2} = \frac{4}{x - 2}$

11. $\frac{w}{18} - \frac{w - 1}{9} = \frac{4 - w}{6}$

12. $\frac{x}{x + 1} + \frac{1}{2x + 2} = \frac{7}{8}$

Solve each equation for y.

13. $2x + 3y = c$

14. $\frac{y - 3}{x - 5} = \frac{1}{2}$

15. $2y = ay + c$

16. $\frac{A}{y} = \frac{C}{B}$

17. $\frac{A}{y} + \frac{1}{3} = \frac{B}{y}$

18. $\frac{A}{y} - \frac{1}{2} = \frac{1}{3}$

19. $3y - 5ay = 8$

20. $y^2 - By = 0$

21. $A = \frac{1}{2}h(b + y)$

22. $2(b + y) = b$

Calculate the value of $b^2 - 4ac$ for each choice of a, b, and c.

23. $a = 1, b = 2, c = -15$

24. $a = 1, b = 8, c = 12$

25. $a = 2, b = 5, c = -3$

26. $a = 6, b = 7, c = -3$

Perform each indicated operation.

27. $(3x - 5) - (5x - 3)$

28. $(2a - 5)(a - 3)$

29. $x^7 \div x^3$

30. $\frac{x - 3}{5} + \frac{x + 4}{5}$

31. $\frac{1}{2} \cdot \frac{1}{x}$

32. $\frac{1}{2} + \frac{1}{x}$

33. $\frac{1}{2} \div \frac{1}{x}$

34. $\frac{1}{2} - \frac{1}{x}$

35. $\frac{x - 3}{5} - \frac{x + 4}{5}$

36. $\frac{3a}{2} \div 2$

37. $(x - 8)(x + 8)$

38. $3x(x^2 - 7)$

39. $2a^5 \cdot 5a^9$

40. $x^2 \cdot x^8$

41. $(k - 6)^2$

42. $(j + 5)^2$

43. $(g - 3) \div (3 - g)$

44. $(6x^3 - 8x^2) \div (2x)$

Solve.

45. ***Present value.*** An investor is interested in the amount or present value that she would have to invest today to receive periodic payments in the future. The present value of \$1 in one year and \$1 in 2 years with interest rate r compounded annually is given by the formula

$$P = \frac{1}{1 + r} + \frac{1}{(1 + r)^2}.$$

a) Rewrite the formula so that the right-hand side is a single rational expression.

b) Find P if $r = 7\%$.

c) The present value of \$1 per year for the next 10 years is given by the formula

$$P = \frac{1}{1 + r} + \frac{1}{(1 + r)^2} + \frac{1}{(1 + r)^3} + \cdots + \frac{1}{(1 + r)^{10}}.$$

Use this formula to find P if $r = 5\%$.

Critical **Thinking** | For Individual or Group Work | Chapter 7

These exercises can be solved by a variety of techniques, which may or may not require algebra. So be creative and think critically. Explain all answers. Answers are in the Instructor's Edition of this text.

1. ***Counting paths.*** How many different paths are there from point A to point B in the accompanying figure? You can only move in a downward direction along the line segments.

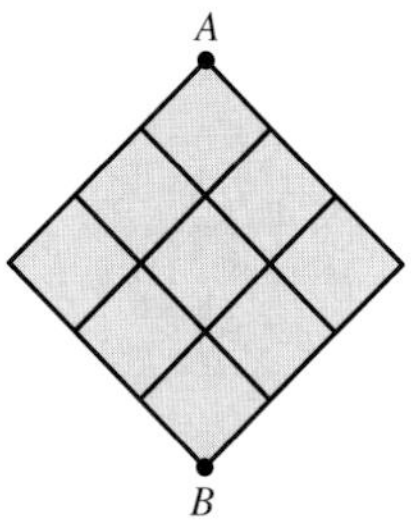

Figure for Exercise 1

2. ***Canadian prime.*** On November 14, 2001, Michael Cameron a 20-year-old Canadian discovered the largest known prime number to date,

$$2^{13,466,917} - 1.$$

This number contains 4,053,946 digits. What is the ones digit? What is the tens digit?

3. ***Good excuse.*** A driver was stopped for doing 70 mph in a 60 mph zone. The driver explained to the officer that her speedometer read 60 mph. However, she recently replaced her stock 24-inch-diameter tires with 25-inch-diameter tires, which caused her to do 10 mph over the speed limit. If her speedometer reads 60 mph, then what is her actual speed with the new tires?

4. ***Palindromic dates.*** October 2, 2001, can be written in *abbreviated* form as 10-2-01, or in *complete* form as 10-02-2001, both of which are palindromic dates (they read the same forward or backward). What was the last palindromic date in abbreviated form prior to 10-1-01? What was the last palindromic date in complete form prior to 10-1-01?

5. ***Integral solutions.*** Find all integral solutions to

$$x^2y - y^3 = 105.$$

6. ***Jogging observation.*** While jogging on a circular track, Heather observes that one-fifth of the joggers in front of her plus five-sixths of the joggers behind her is equal to the total number of joggers. How many joggers are on the track?

Photo for Exercise 6

7. ***Volume of a box.*** The area of the top of a box is 30 in.^{2} and the area of the front is 12 in.^{2} If the surface area of the box is 164 in.^{2}, then what is the volume of the box?

8. ***Making triangles.*** Brenda has a 12-in. loop necklace that she plays with on her desk by forming triangles with it.

a) Find all possible triangles with integer sides that she can form with her necklace and classify each as a right, obtuse, scalene, equilateral, or isosceles triangle.

b) Which triangle has the largest area?

Chapter 8

Powers and Roots

Sailing—the very word conjures up images of warm summer breezes, sparkling blue water, and white billowing sails. But to boat builders, sailing is a serious business. Yacht designers know that the ocean is a dangerous and unforgiving place. It is their job to build boats that are not only fast, comfortable, and fun, but capable of withstanding the punishment inflicted by the wind and waves.

Designing sailboats is a technical balancing act. A good boat has just the right combination of length, width (or beam), sail area, and displacement. A boat can always be made faster by increasing the sail area, but too much sail area increases the chance of capsize. Making the boat wider decreases the chance of capsize, but causes more resistance from the water and slows down the boat.

There are four ratios commonly used to measure performance and safety for a yacht: ballast-displacement ratio, displacement-length ratio, sail area-displacement ratio, and capsize screening value. The formulas for these ratios involve powers and roots, which is the subject of this chapter.

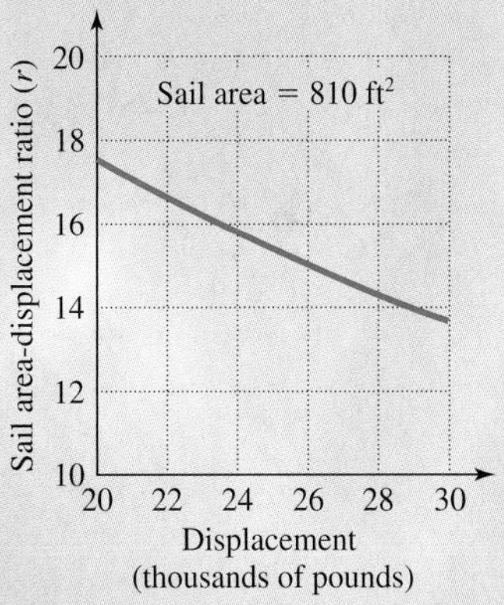

In Exercise 89 of Section 8.5 you will find the sail area-displacement ratio for a sailboat.

8.1 Roots, Radicals, and Rules

In this Section

- **Fundamentals**
- **Roots and Variables**
- **Product Rule for Radicals**
- **Quotient Rule for Radicals**

In Section 5.6 you learned the basic facts about powers. In this section you will study roots and see how powers and roots are related.

Fundamentals

We use the idea of roots to reverse powers. Because $3^2 = 9$ and $(-3)^2 = 9$, both 3 and -3 are square roots of 9. Because $2^4 = 16$ and $(-2)^4 = 16$, both 2 and -2 are fourth roots of 16. Because $2^3 = 8$ and $(-2)^3 = -8$, there is only one real cube root of 8 and only one real cube root of -8. The cube root of 8 is 2 and the cube root of -8 is -2.

_n_th Roots

If $a = b^n$ for a positive integer n, then b is an ***n*th root of *a*.** If $a = b^2$, then b is a **square root** of a. If $a = b^3$, then b is the **cube root** of a.

The following box contains three more definitions concerning roots.

Even Roots, Principal Roots, and Odd Roots

If n is a positive *even* integer and a is positive, then there are two real nth roots of a. We call these roots **even roots.**

The positive even root of a positive number is called the **principal root.**

If n is a positive *odd* integer and a is any real number, then there is only one real nth root of a. We call that root an **odd root.**

So the principal square root of 9 is 3 and the principal fourth root of 16 is 2 and these roots are even roots. Because $2^5 = 32$, the fifth root of 32 is 2 and 2 is an odd root.

We use the **radical symbol** $\sqrt{}$ to signify roots.

$\sqrt[n]{a}$

If n is a positive *even* integer and a is positive, then $\sqrt[n]{a}$ denotes the *principal nth root of a.*

If n is a positive *odd* integer, then $\sqrt[n]{a}$ denotes the nth root of a.

If n is any positive integer, then $\sqrt[n]{0} = 0$.

We read $\sqrt[n]{a}$ as "the nth root of a." In the notation $\sqrt[n]{a}$, n is the **index of the radical** and a is the **radicand.** For square roots the index is omitted, and we simply write $\sqrt{a}$.

EXAMPLE 1

Evaluating radical expressions

Find the following roots:

a) $\sqrt{25}$ **b)** $\sqrt[3]{-27}$ **c)** $\sqrt[6]{64}$ **d)** $-\sqrt{4}$

Solution

a) Because $5^2 = 25$, $\sqrt{25} = 5$.

b) Because $(-3)^3 = -27$, $\sqrt[3]{-27} = -3$.

c) Because $2^6 = 64$, $\sqrt[6]{64} = 2$.

d) Because $\sqrt{4} = 2$, $-\sqrt{4} = -(\sqrt{4}) = -2$.

Now do Exercises 7–30

Calculator Close-Up

We can use the radical symbol to find a square root on a graphing calculator, but for other roots we use the *x*th root symbol as shown. The *x*th root symbol is in the MATH menu.

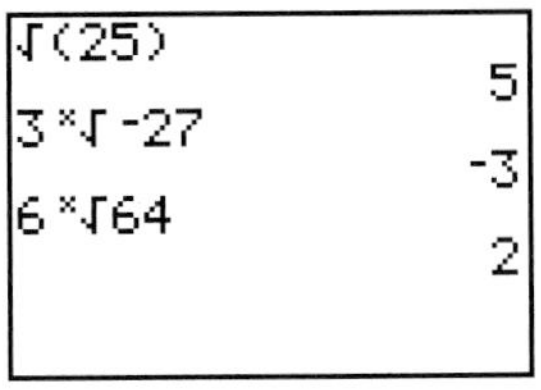

CAUTION In radical notation, $\sqrt{4}$ represents the *principal square root of* 4, so $\sqrt{4} = 2$. Note that -2 is also a square root of 4, but $\sqrt{4} \neq -2$.

Note that even roots of negative numbers are omitted from the definition of *n*th roots because even powers of real numbers are never negative. So no real number can be an even root of a negative number. Expressions such as

$$\sqrt{-9}, \quad \sqrt[4]{-81}, \quad \text{and} \quad \sqrt[6]{-64}$$

are not real numbers. Square roots of negative numbers will be discussed in Section 9.5 when we discuss the imaginary numbers.

Roots and Variables

Consider the result of squaring a power of x:

$$(x^1)^2 = x^2, \quad (x^2)^2 = x^4, \quad (x^3)^2 = x^6, \quad \text{and} \quad (x^4)^2 = x^8.$$

When a power of x is squared, the exponent is multiplied by 2. So any even power of x is a perfect square.

Perfect Squares

The following expressions are perfect squares:

$$x^2, \quad x^4, \quad x^6, \quad x^8, \quad x^{10}, \quad x^{12}, \quad \ldots$$

Calculator Close-Up

A calculator can provide numerical support for this discussion of roots. Note that $\sqrt{(-3)^2} = 3$ not -3 because $\sqrt{x^2} \neq x$ when x is negative. Note also that the calculator will not evaluate $\sqrt{-3^2}$ because $\sqrt{-3^2} = \sqrt{-9}$.

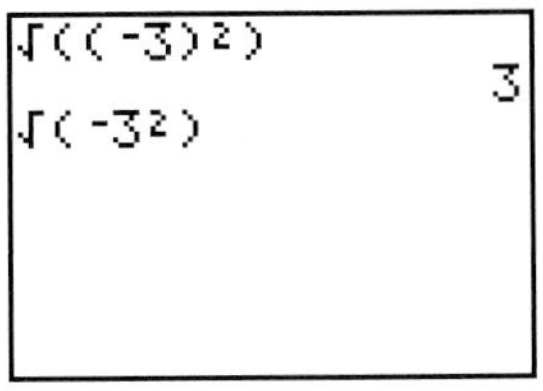

Since taking a square root reverses the operation of squaring, the square root of an even power of x is found by dividing the exponent by 2. Provided x is nonnegative (see the next Caution), we have:

$$\sqrt{x^2} = x^1 = x, \quad \sqrt{x^4} = x^2, \quad \sqrt{x^6} = x^3, \quad \text{and} \quad \sqrt{x^8} = x^4.$$

CAUTION If x is negative, equations like $\sqrt{x^2} = x$ and $\sqrt{x^6} = x^3$ are not correct because the radical represents the nonnegative square root but x and x^3 are negative. That is why we assume x is nonnegative.

If a power of x is cubed, the exponent is multiplied by 3:

$$(x^1)^3 = x^3, \quad (x^2)^3 = x^6, \quad (x^3)^3 = x^9, \quad \text{and} \quad (x^4)^3 = x^{12}.$$

So if the exponent is a multiple of 3, we have a perfect cube.

Perfect Cubes

The following expressions are perfect cubes:

$$x^3,\quad x^6,\quad x^9,\quad x^{12},\quad x^{15},\quad \ldots$$

Since the cube root reverses the operation of cubing, the cube root of any of these perfect cubes is found by dividing the exponent by 3:

$$\sqrt[3]{x^3} = x^1 = x,\quad \sqrt[3]{x^6} = x^2,\quad \sqrt[3]{x^9} = x^3,\quad \text{and}\quad \sqrt[3]{x^{12}} = x^4.$$

If the exponent is divisible by 4, we have a perfect fourth power, and so on.

EXAMPLE 2

Roots of exponential expressions

Find each root. Assume that all variables represent nonnegative real numbers.

a) $\sqrt{x^{22}}$ **b)** $\sqrt[3]{t^{18}}$ **c)** $\sqrt[5]{s^{30}}$

Solution

a) $\sqrt{x^{22}} = x^{11}$ because $(x^{11})^2 = x^{22}$.

b) $\sqrt[3]{t^{18}} = t^6$ because $(t^6)^3 = t^{18}$.

c) $\sqrt[5]{s^{30}} = s^6$ because one-fifth of 30 is 6.

Now do Exercises 41–56

Product Rule for Radicals

Consider the expression $\sqrt{2} \cdot \sqrt{3}$. If we square this product, we get

$$\begin{aligned}(\sqrt{2} \cdot \sqrt{3})^2 &= (\sqrt{2})^2(\sqrt{3})^2 && \text{Power of a product rule}\\ &= 2 \cdot 3 && (\sqrt{2})^2 = 2 \text{ and } (\sqrt{3})^2 = 3\\ &= 6.\end{aligned}$$

The number $\sqrt{6}$ is the unique positive number whose square is 6. Because we squared $\sqrt{2} \cdot \sqrt{3}$ and obtained 6, we must have $\sqrt{6} = \sqrt{2} \cdot \sqrt{3}$. This example illustrates the product rule for radicals.

Product Rule for Radicals

The *n*th root of a product is equal to the product of the *n*th roots. In symbols,

$$\sqrt[n]{ab} = \sqrt[n]{a} \cdot \sqrt[n]{b},$$

provided all of these roots are real numbers.

EXAMPLE 3

Using the product rule for radicals

Simplify each radical. Assume that all variables represent positive real numbers.

a) $\sqrt{4y}$ **b)** $\sqrt{3y^8}$

Solution

a)
$$\begin{aligned}\sqrt{4y} &= \sqrt{4} \cdot \sqrt{y} && \text{Product rule for radicals}\\ &= 2\sqrt{y} && \text{Simplify.}\end{aligned}$$

b) $\sqrt{3y^8} = \sqrt{3} \cdot \sqrt{y^8}$ Product rule for radicals

$= \sqrt{3} \cdot y^4$ $\sqrt{y^8} = y^4$

$= y^4\sqrt{3}$ A radical is usually written last in a product.

Now do Exercises 57–72

Quotient Rule for Radicals

Because $\sqrt{2} \cdot \sqrt{3} = \sqrt{6}$, we have $\sqrt{6} \div \sqrt{3} = \sqrt{2}$, or

$$\sqrt{2} = \sqrt{\frac{6}{3}} = \frac{\sqrt{6}}{\sqrt{3}}.$$

This example illustrates the quotient rule for radicals.

Quotient Rule for Radicals

The nth root of a quotient is equal to the quotient of the nth roots. In symbols,

$$\sqrt[n]{\frac{a}{b}} = \frac{\sqrt[n]{a}}{\sqrt[n]{b}},$$

provided that all of these roots are real numbers and $b \neq 0$.

In Example 4 we use the quotient rule to simplify radical expressions.

EXAMPLE 4

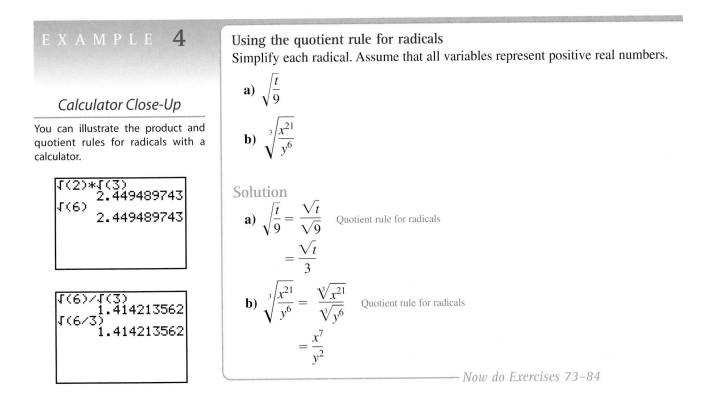

Calculator Close-Up

You can illustrate the product and quotient rules for radicals with a calculator.

```
√(2)*√(3)
         2.449489743
√(6)
         2.449489743
```

```
√(6)/√(3)
         1.414213562
√(6/3)
         1.414213562
```

Using the quotient rule for radicals

Simplify each radical. Assume that all variables represent positive real numbers.

a) $\sqrt{\frac{t}{9}}$

b) $\sqrt[3]{\frac{x^{21}}{y^6}}$

Solution

a) $\sqrt{\frac{t}{9}} = \frac{\sqrt{t}}{\sqrt{9}}$ Quotient rule for radicals

$= \frac{\sqrt{t}}{3}$

b) $\sqrt[3]{\frac{x^{21}}{y^6}} = \frac{\sqrt[3]{x^{21}}}{\sqrt[3]{y^6}}$ Quotient rule for radicals

$= \frac{x^7}{y^2}$

Now do Exercises 73–84

Math *at Work* Just Average

In common usage the word "average" can mean many things. Average might mean "most common." (The average person works for a living.) Average can also mean "middle." When you average four test scores, you add them and divide by 4. In statistics, this average is actually called the *mean*. The mean is a single number that is used to measure or describe the entire set of test scores. It is an estimate of the "middle" of the scores. There are several other precisely defined terms in statistics that also are measures of the middle of a set of scores.

The second most popular measure of the center (after the mean) is the *median*. The median is a number that separates the data into two equal parts. Half of the scores are above the median and half below. The median price of a new house in 2003 was \$195,200, whereas the mean was \$242,500. So half of the new houses sold for less than \$195,200 and half sold for more. The mean of \$242,500 is higher than the median due to some very expensive houses.

Less popular measures of the middle are the *geometric mean*, the *harmonic mean*, and the *quadratic mean*. The geometric mean is the nth root of the product of the n scores. The harmonic mean is n divided by the sum of the reciprocals of the n scores. For the quadratic mean, we find the sum of the squares of the scores, divide by n, and then take the square root. The test scores 45, 64, 78, and 99 have a mean of 71.5, a geometric mean of 68.7, a harmonic mean of 65.8, and a quadratic mean of 74.2.

So what does average really mean?

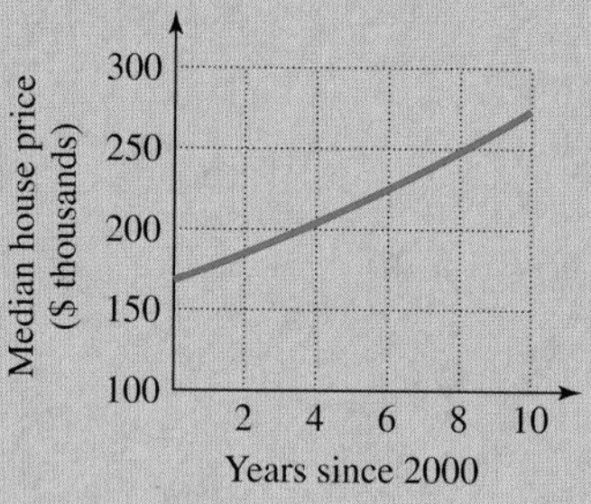

Warm-Ups ▼

True or false? Explain your answer.

1. $\sqrt{2} \cdot \sqrt{2} = 2$
2. $\sqrt[3]{2} \cdot \sqrt[3]{2} = 2$
3. $\sqrt[3]{-27} = -3$
4. $\sqrt{-25} = -5$
5. $\sqrt[4]{16} = 2$
6. $\sqrt{9} = 3$
7. $\sqrt{2^9} = 2^3$
8. $\sqrt{17} \cdot \sqrt{17} = 289$
9. If $w \geq 0$, then $\sqrt{w^2} = w$.
10. If $t \geq 0$, then $\sqrt[4]{t^{12}} = t^3$.

8.1 Exercises

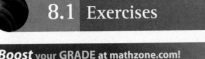

Reading and Writing *After reading this section, write out the answers to these questions. Use complete sentences.*

1. How do you know if b is an nth root of a?

2. What is a principal root?

3. What is the difference between an even root and an odd root?

4. What symbol is used to indicate an nth root?

5. What is the product rule for radicals?

6. What is the quotient rule for radicals?

For all of the exercises in this section assume that all variables represent positive real numbers.

Find each root. See Example 1.

7. $\sqrt{36}$ **8.** $\sqrt{49}$

9. $\sqrt[5]{32}$ **10.** $\sqrt[4]{81}$

11. $\sqrt[3]{1000}$ **12.** $\sqrt[4]{16}$

13. $\sqrt[4]{-16}$

14. $\sqrt{-1}$

15. $\sqrt{0}$ **16.** $\sqrt{1}$

17. $\sqrt[3]{-1}$ **18.** $\sqrt[3]{0}$

19. $\sqrt[3]{1}$ **20.** $\sqrt[4]{1}$

21. $\sqrt[4]{-81}$

22. $\sqrt[6]{-64}$

23. $\sqrt[6]{64}$ **24.** $\sqrt[7]{128}$

25. $\sqrt[5]{-32}$ **26.** $\sqrt[3]{-125}$

27. $-\sqrt{100}$ **28.** $-\sqrt{36}$

29. $\sqrt[4]{-50}$

30. $-\sqrt{-144}$

Simplify each expression.

31. $\sqrt{9+16}$ **32.** $\sqrt{9}+\sqrt{16}$

33. $\sqrt{9}\cdot\sqrt{4}$ **34.** $\sqrt{9\cdot 4}$

35. $\dfrac{\sqrt{36}}{\sqrt{4}}$ **36.** $\sqrt{\dfrac{36}{4}}$

37. $\sqrt{64+36}$ **38.** $\sqrt{64}+\sqrt{36}$

39. $\sqrt{25-16}$ **40.** $\sqrt{25}-\sqrt{16}$

Find each root. See Example 2.

41. $\sqrt{m^2}$ **42.** $\sqrt{m^6}$

43. $\sqrt[5]{y^{15}}$ **44.** $\sqrt[4]{m^8}$

45. $\sqrt[3]{y^{15}}$ **46.** $\sqrt{m^8}$

47. $\sqrt[3]{m^3}$ **48.** $\sqrt[4]{x^4}$

49. $\sqrt{3^6}$ **50.** $\sqrt{4^2}$

51. $\sqrt{2^{10}}$ **52.** $\sqrt[3]{2^{99}}$

53. $\sqrt[3]{5^9}$ **54.** $\sqrt[3]{10^{18}}$

55. $\sqrt{10^{20}}$ **56.** $\sqrt{10^{18}}$

Use the product rule for radicals to simplify each expression. See Example 3.

57. $\sqrt{9y}$ **58.** $\sqrt{16n}$

59. $\sqrt{4a^2}$ **60.** $\sqrt{36n^2}$

61. $\sqrt{x^4y^2}$ **62.** $\sqrt{w^6t^2}$

63. $\sqrt{5m^{12}}$ **64.** $\sqrt{7z^{16}}$

65. $\sqrt[3]{8y}$ **66.** $\sqrt[3]{27z^2}$

67. $\sqrt[3]{-27w^3}$ **68.** $\sqrt[3]{-125m^6}$

69. $\sqrt[4]{16s}$ **70.** $\sqrt[4]{81w}$

71. $\sqrt[3]{-125a^9y^6}$ **72.** $\sqrt[3]{-27z^3w^{15}}$

Simplify each radical. See Example 4.

73. $\sqrt{\dfrac{t}{4}}$ **74.** $\sqrt{\dfrac{w}{36}}$

75. $\sqrt{\dfrac{625}{16}}$ **76.** $\sqrt{\dfrac{9}{144}}$

77. $\sqrt[3]{\dfrac{t}{8}}$ **78.** $\sqrt[3]{\dfrac{a}{27}}$

79. $\sqrt[3]{\dfrac{-8x^6}{y^3}}$ **80.** $\sqrt[3]{\dfrac{-27y^{36}}{1000}}$

81. $\sqrt{\dfrac{4a^6}{9}}$ **82.** $\sqrt{\dfrac{9a^2}{49b^4}}$

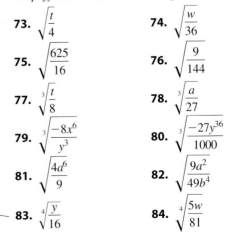

83. $\sqrt[4]{\dfrac{y}{16}}$ **84.** $\sqrt[4]{\dfrac{5w}{81}}$

Use a calculator to find the approximate value of each expression to three decimal places.

85. $\sqrt{3} + \sqrt{5}$

86. $\sqrt{7} - \sqrt{3}$

87. $\dfrac{\sqrt{5} + \sqrt{2}}{\sqrt{3} - 4}$

88. $\dfrac{\sqrt{2} - \sqrt{3}}{1 - \sqrt{5}}$

89. $\sqrt{7.1^2 - 4(1.2)(3)}$

90. $\sqrt{3^2 - 4(-2)(0.2)}$

91. $\dfrac{-3 + \sqrt{3^2 - 4(1)(-2.9)}}{2}$

92. $\dfrac{8 + \sqrt{(-8)^2 - 4(1.3)(-6.2)}}{2(1.3)}$

Solve each problem.

93. ***Economic order quantity.*** When a part is needed for a space shuttle external fuel tank, Joseph Bursavich at Martin Marietta determines the most economic order quantity E by using the formula $E = \sqrt{\frac{2AS}{I}}$, where A is the quantity that the plant will use in one year, S is the cost of setup for making the part, and I is the cost of holding one unit in stock for one year. Find the most economic order quantity if $S = \$5290$, $A = 20$, and $I = \$100$.

94. ***Diagonal of a box.*** The length of the diagonal D of the box shown in the figure can be found from the formula

$$D = \sqrt{L^2 + W^2 + H^2},$$

where L, W, and H represent the length, width, and height of the box, respectively. If the box has length 6 inches, width 4 inches, and height 3 inches, then what is the length of the diagonal to the nearest tenth of an inch?

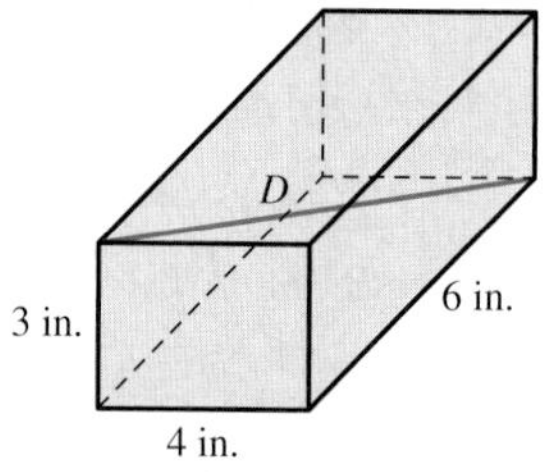

Figure for Exercise 94

95. ***Buena vista.*** The formula $V = 1.22\sqrt{A}$ gives the view in miles from horizon to horizon at an altitude of A feet (Delta Airlines brochure).

a) Use the formula to find the view to the nearest mile from an altitude of 35,000 feet.

b) Use the accompanying graph to estimate the altitude of an airplane from which the view is 100 miles.

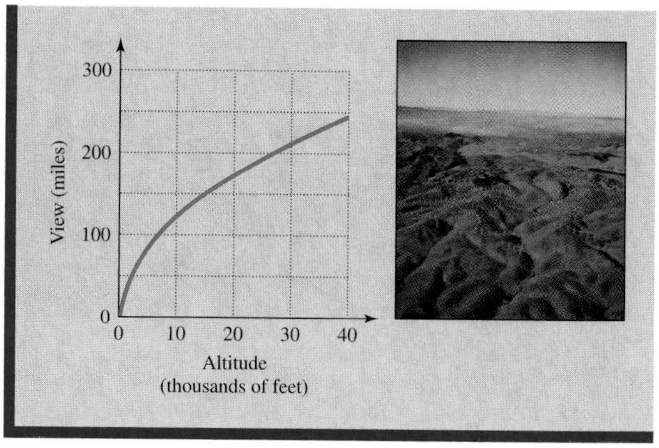

Figure for Exercise 95

96. ***Sailing speed.*** To find the maximum speed in knots (nautical miles per hour) for a sailboat, sailors use the formula $M = 1.3\sqrt{w}$, where w is the length of the waterline in feet.

a) If the length of the waterline for the sloop *John B.* is 38 feet, then what is the maximum speed for the *John B.?*

b) Use the accompanying graph to estimate the length of the waterline for a boat for which the maximum speed is 6 knots.

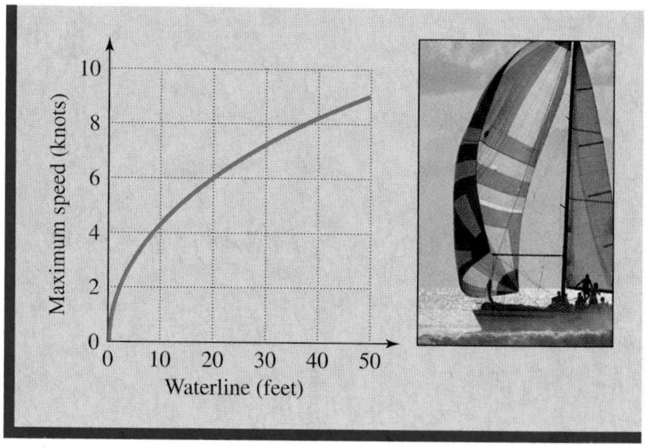

Figure for Exercise 96

Getting More Involved

97. ***Discussion***

Determine whether each equation is correct.

a) $\sqrt{(-5)^2} = -5$

b) $\sqrt[3]{(-2)^3} = -2$

c) $\sqrt[4]{(-3)^4} = -3$

d) $\sqrt[5]{(-7)^5} = -7$

98. ***Writing***

If x is a negative number and $\sqrt[n]{x^n} = x$, then what can you say about n? Explain your answer.

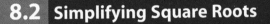

8.2 Simplifying Square Roots

In this Section

- **Using the Product Rule**
- **Rationalizing the Denominator**
- **Simplified Form of a Square Root**

In Section 8.1 you learned to simplify some radical expressions using the product rule. In this section you will learn three basic rules to follow for writing expressions involving square roots in simplest form. These rules can be extended to radicals with index greater than 2, but we will not do that in this text.

Using the Product Rule

We can use the product rule to simplify square roots of certain numbers. For example,

$$\begin{aligned} \sqrt{45} &= \sqrt{9 \cdot 5} && \text{Factor 45 as } 9 \cdot 5. \\ &= \sqrt{9} \cdot \sqrt{5} && \text{Product rule for radicals} \\ &= 3\sqrt{5} && \sqrt{9} = 3 \end{aligned}$$

Because 45 is not a perfect square, we cannot write $\sqrt{45}$ without the radical symbol. However, $3\sqrt{5}$ is considered a simpler expression that represents the exact value of $\sqrt{45}$. When simplifying square roots, we can factor the perfect squares out of the radical and replace them with their square roots. Look for the factors

$$4, \quad 9, \quad 16, \quad 25, \quad 36, \quad 49, \quad \text{and so on.}$$

EXAMPLE 1

Simplifying radicals using the product rule

Simplify.

a) $\sqrt{12}$ **b)** $\sqrt{50}$ **c)** $\sqrt{72}$

Solution

a) Because $12 = 4 \cdot 3$, we can use the product rule to write

$$\sqrt{12} = \sqrt{4} \cdot \sqrt{3} = 2\sqrt{3}.$$

b) $\sqrt{50} = \sqrt{25} \cdot \sqrt{2} = 5\sqrt{2}$

c) Note that 4, 9, and 36 are perfect squares and are factors of 72. In factoring out a perfect square, it is most efficient to use the largest perfect square:

$$\sqrt{72} = \sqrt{36} \cdot \sqrt{2} = 6\sqrt{2}$$

If we had factored out 9, we could still get the correct answer as follows:

$$\sqrt{72} = \sqrt{9} \cdot \sqrt{8} = 3 \cdot \sqrt{8} = 3 \cdot \sqrt{4} \cdot \sqrt{2} = 3 \cdot 2\sqrt{2} = 6\sqrt{2}$$

Now do Exercises 7–18

Calculator Close-Up

You can use a calculator to see that $\sqrt{12}$ and $2\sqrt{3}$ agree for the first 10 digits (out of infinitely many). Having the same first 10 digits does not make $\sqrt{12} = 2\sqrt{3}$. The product rule for radicals guarantees that they are equal.

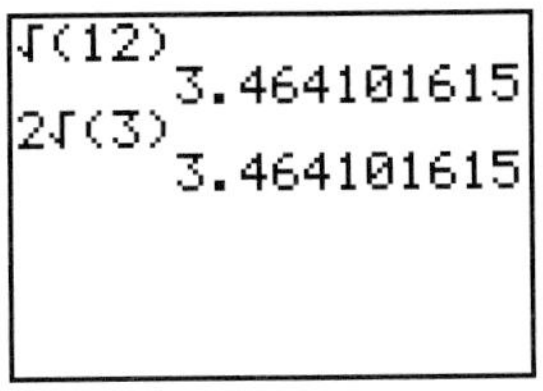

Rationalizing the Denominator

Radicals such as $\sqrt{2}$, $\sqrt{3}$, and $\sqrt{5}$ are irrational numbers. So a fraction such as $\frac{3}{\sqrt{5}}$ has an irrational denominator. Because fractions with rational denominators are considered simpler than fractions with irrational denominators, we usually convert fractions with irrational denominators to equivalent ones with rational denominators. That is, we **rationalize the denominator.**

EXAMPLE 2

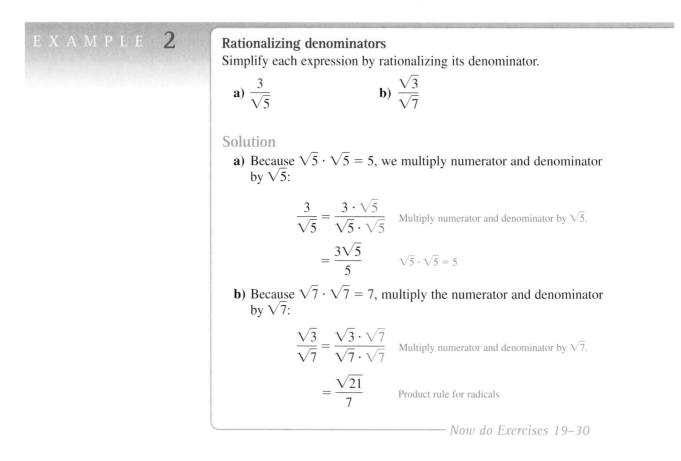

Rationalizing denominators

Simplify each expression by rationalizing its denominator.

a) $\dfrac{3}{\sqrt{5}}$ **b)** $\dfrac{\sqrt{3}}{\sqrt{7}}$

Solution

a) Because $\sqrt{5} \cdot \sqrt{5} = 5$, we multiply numerator and denominator by $\sqrt{5}$:

$$\frac{3}{\sqrt{5}} = \frac{3 \cdot \sqrt{5}}{\sqrt{5} \cdot \sqrt{5}} \quad \text{Multiply numerator and denominator by } \sqrt{5}.$$

$$= \frac{3\sqrt{5}}{5} \quad \sqrt{5} \cdot \sqrt{5} = 5$$

b) Because $\sqrt{7} \cdot \sqrt{7} = 7$, multiply the numerator and denominator by $\sqrt{7}$:

$$\frac{\sqrt{3}}{\sqrt{7}} = \frac{\sqrt{3} \cdot \sqrt{7}}{\sqrt{7} \cdot \sqrt{7}} \quad \text{Multiply numerator and denominator by } \sqrt{7}.$$

$$= \frac{\sqrt{21}}{7} \quad \text{Product rule for radicals}$$

Now do Exercises 19–30

Simplified Form of a Square Root

When we simplify any expression, we try to write a "simpler" expression that is equivalent to the original. However, one person's idea of simpler is sometimes different from another person's. For a square root the expression must satisfy three conditions to be in simplified form. These three conditions provide specific rules to follow for simplifying square roots.

Simplified Form for Square Roots

An expression involving a square root is in **simplified form** if it has

1. *no* perfect-square factors inside the radical,
2. *no* fractions inside the radical, and
3. *no* radicals in the denominator.

Because a decimal is a form of a fraction, a simplified square root should not contain any decimal numbers. Also, a simplified expression should use the fewest number of radicals possible. So we write $\sqrt{6}$ rather than $\sqrt{2} \cdot \sqrt{3}$ even though both $\sqrt{2}$ and $\sqrt{3}$ are both in simplified form.

EXAMPLE 3

Simplified form for square roots

Write each radical expression in simplified form.

a) $\sqrt{300}$ **b)** $\sqrt{\frac{2}{5}}$ **c)** $\frac{\sqrt{10}}{\sqrt{6}}$

Solution

a) We must remove the perfect square factor of 100 from inside the radical:

$$\sqrt{300} = \sqrt{100 \cdot 3} = \sqrt{100} \cdot \sqrt{3} = 10\sqrt{3}$$

b) We first use the quotient rule to remove the fraction $\frac{2}{5}$ from inside the radical:

$$\sqrt{\frac{2}{5}} = \frac{\sqrt{2}}{\sqrt{5}} \quad \text{Quotient rule for radicals}$$

$$= \frac{\sqrt{2} \cdot \sqrt{5}}{\sqrt{5} \cdot \sqrt{5}} \quad \text{Rationalize the denominator.}$$

$$= \frac{\sqrt{10}}{5} \quad \text{Product rule for radicals}$$

c) The numerator and denominator have a common factor of $\sqrt{2}$:

$$\frac{\sqrt{10}}{\sqrt{6}} = \frac{\sqrt{2} \cdot \sqrt{5}}{\sqrt{2} \cdot \sqrt{3}} \quad \text{Product rule for radicals}$$

$$= \frac{\sqrt{5}}{\sqrt{3}} \quad \text{Reduce.}$$

$$= \frac{\sqrt{5} \cdot \sqrt{3}}{\sqrt{3} \cdot \sqrt{3}} \quad \text{Rationalize the denominator.}$$

$$= \frac{\sqrt{15}}{3} \quad \text{Product rule for radicals}$$

Note that we could have simplified $\frac{\sqrt{10}}{\sqrt{6}}$ by first using the quotient rule to get $\frac{\sqrt{10}}{\sqrt{6}} = \sqrt{\frac{10}{6}}$ and then reducing $\frac{10}{6}$. Another way to simplify $\frac{\sqrt{10}}{\sqrt{6}}$ is to first multiply the numerator and denominator by $\sqrt{6}$. You should try these alternatives. Of course, the simplified form is $\frac{\sqrt{15}}{3}$ by any method.

Now do Exercises 31–42

Calculator Close-Up

Using a calculator to check simplification problems will help you to understand the concepts.

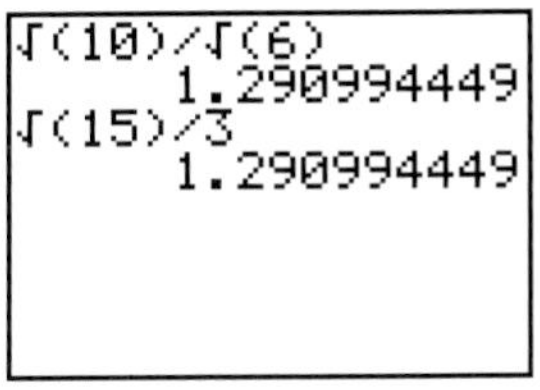

In Example 4 we simplify some expressions involving variables. Remember that *any exponential expression with an even exponent is a perfect square.*

EXAMPLE 4

Radicals containing variables

Simplify each expression. All variables represent nonnegative real numbers.

a) $\sqrt{x^3}$ **b)** $\sqrt{8a^9}$ **c)** $\sqrt{18a^4b^7}$

Solution

a) $\sqrt{x^3} = \sqrt{x^2 \cdot x}$ The largest perfect square factor of x^3 is x^2.

$= \sqrt{x^2} \cdot \sqrt{x}$ Product rule for radicals

$= x\sqrt{x}$ For any nonnegative x, $\sqrt{x^2} = x$.

b) $\sqrt{8a^9} = \sqrt{4a^8} \cdot \sqrt{2a}$ The largest perfect square factor of $8a^9$ is $4a^8$.

$= 2a^4\sqrt{2a}$ $\sqrt{4a^8} = 2a^4$

c) $\sqrt{18a^4b^7} = \sqrt{9a^4b^6} \cdot \sqrt{2b}$ Factor out the perfect squares.

$= 3a^2b^3\sqrt{2b}$ $\sqrt{9a^4b^6} = 3a^2b^3$

Now do Exercises 43–54

If square roots of variables appear in the denominator, then we rationalize the denominator.

EXAMPLE 5

Radicals containing variables

Simplify each expression. All variables represent positive real numbers.

a) $\frac{5}{\sqrt{a}}$ **b)** $\sqrt{\frac{a}{b}}$ **c)** $\frac{\sqrt{2}}{\sqrt{6a}}$

Helpful Hint

If you are going to compute the value of a radical expression with a calculator, it doesn't matter if the denominator is rational. However, rationalizing the denominator provides another opportunity to practice building up the denominator of a fraction and multiplying radicals.

Solution

a) $\frac{5}{\sqrt{a}} = \frac{5 \cdot \sqrt{a}}{\sqrt{a} \cdot \sqrt{a}}$ Multiply numerator and denominator by $\sqrt{a}$.

$= \frac{5\sqrt{a}}{a}$ $\sqrt{a} \cdot \sqrt{a} = a$

b) $\sqrt{\frac{a}{b}} = \frac{\sqrt{a}}{\sqrt{b}}$ Quotient rule for radicals

$= \frac{\sqrt{a} \cdot \sqrt{b}}{\sqrt{b} \cdot \sqrt{b}}$ Rationalize the denominator.

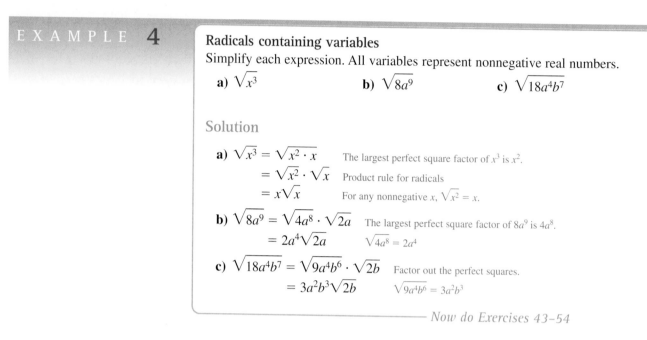

$= \frac{\sqrt{ab}}{b}$ Product rule for radicals

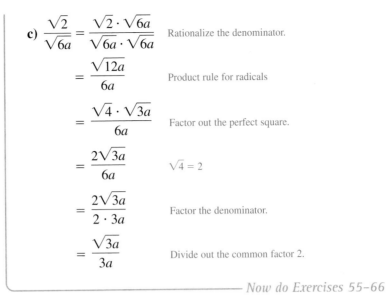

c) $\dfrac{\sqrt{2}}{\sqrt{6a}} = \dfrac{\sqrt{2} \cdot \sqrt{6a}}{\sqrt{6a} \cdot \sqrt{6a}}$ Rationalize the denominator.

$= \dfrac{\sqrt{12a}}{6a}$ Product rule for radicals

$= \dfrac{\sqrt{4} \cdot \sqrt{3a}}{6a}$ Factor out the perfect square.

$= \dfrac{2\sqrt{3a}}{6a}$ $\sqrt{4} = 2$

$= \dfrac{2\sqrt{3a}}{2 \cdot 3a}$ Factor the denominator.

$= \dfrac{\sqrt{3a}}{3a}$ Divide out the common factor 2.

Now do Exercises 55–66

CAUTION Do not attempt to reduce an expression like the one in Example 5(c):

$$\frac{\sqrt{3a}}{3a}$$

You cannot divide out common factors when one is inside a radical.

Warm-Ups ▼

True or false? Explain your answer.

1. $\sqrt{20} = 2\sqrt{5}$

2. $\sqrt{18} = 9\sqrt{2}$

3. $\dfrac{1}{\sqrt{3}} = \dfrac{\sqrt{3}}{3}$

4. $\dfrac{9}{4} = \dfrac{3}{2}$

5. $\sqrt{a^3} = a\sqrt{a}$ for any positive value of a.

6. $\sqrt{a^9} = a^3$ for any positive value of a.

7. $\sqrt{y^{17}} = y^8\sqrt{y}$ for any positive value of y.

8. $\dfrac{\sqrt{6}}{2} = \sqrt{3}$

9. $\sqrt{4} = \sqrt{2}$

10. $\sqrt{283} = 17$

8.2 Exercises

Reading and Writing *After reading this section, write out the answers to these questions. Use complete sentences.*

1. How do we simplify a radical with the product rule?

2. Which integers are perfect squares?

3. What does it mean to rationalize a denominator?

4. What is simplified form for a square root?

5. How do you simplify a square root that contains a variable?

6. How can you tell if an exponential expression is a perfect square?

Assume that all variables in the exercises represent positive real numbers.

Simplify each radical. See Example 1.

7. $\sqrt{8}$ **8.** $\sqrt{20}$ **9.** $\sqrt{24}$

10. $\sqrt{75}$ **11.** $\sqrt{28}$ **12.** $\sqrt{40}$

13. $\sqrt{90}$ **14.** $\sqrt{200}$ **15.** $\sqrt{500}$

16. $\sqrt{98}$ **17.** $\sqrt{150}$ **18.** $\sqrt{120}$

Simplify each expression by rationalizing the denominator. See Example 2.

19. $\frac{1}{\sqrt{5}}$ **20.** $\frac{1}{\sqrt{6}}$ **21.** $\frac{3}{\sqrt{2}}$

22. $\frac{4}{\sqrt{3}}$ **23.** $\frac{\sqrt{3}}{\sqrt{2}}$ **24.** $\frac{\sqrt{7}}{\sqrt{6}}$

25. $\frac{-3}{\sqrt{10}}$ **26.** $\frac{-4}{\sqrt{5}}$ **27.** $\frac{-10}{\sqrt{17}}$

28. $\frac{-3}{\sqrt{19}}$ **29.** $\frac{\sqrt{11}}{\sqrt{7}}$ **30.** $\frac{\sqrt{10}}{\sqrt{3}}$

Write each radical expression in simplified form. See Example 3.

31. $\sqrt{63}$ **32.** $\sqrt{48}$ **33.** $\sqrt{\frac{3}{2}}$

34. $\sqrt{\frac{3}{5}}$ **35.** $\sqrt{\frac{5}{8}}$ **36.** $\sqrt{\frac{5}{18}}$

37. $\frac{\sqrt{6}}{\sqrt{10}}$ **38.** $\frac{\sqrt{12}}{\sqrt{20}}$ **39.** $\frac{\sqrt{75}}{\sqrt{3}}$

40. $\frac{\sqrt{45}}{\sqrt{5}}$ **41.** $\frac{\sqrt{15}}{\sqrt{10}}$ **42.** $\frac{\sqrt{30}}{\sqrt{21}}$

Simplify each expression. See Example 4.

43. $\sqrt{a^8}$ **44.** $\sqrt{y^{10}}$ **45.** $\sqrt{a^9}$

46. $\sqrt{t^{11}}$ **47.** $\sqrt{8a^6}$ **48.** $\sqrt{18w^9}$

49. $\sqrt{20a^4b^9}$ **50.** $\sqrt{12x^2y^3}$ **51.** $\sqrt{27x^3y^3}$

52. $\sqrt{45x^5y^3}$ **53.** $\sqrt{27a^3b^8c^2}$ **54.** $\sqrt{125x^3y^9z^4}$

Simplify each expression. See Example 5.

55. $\frac{1}{\sqrt{x}}$ **56.** $\frac{1}{\sqrt{2x}}$ **57.** $\frac{\sqrt{2}}{\sqrt{3a}}$

58. $\frac{\sqrt{5}}{\sqrt{2b}}$ **59.** $\frac{\sqrt{3}}{\sqrt{15y}}$ **60.** $\frac{\sqrt{5}}{\sqrt{10x}}$

61. $\sqrt{\frac{3x}{2y}}$ **62.** $\sqrt{\frac{6}{5w}}$ **63.** $\sqrt{\frac{10y}{15x}}$

64. $\sqrt{\dfrac{6x}{4y}}$ **65.** $\sqrt{\dfrac{8x^3}{y}}$ **66.** $\sqrt{\dfrac{8s^5}{t}}$

Simplify each expression.

67. $\sqrt{8a}$ **68.** $\sqrt{12b}$ **69.** $\sqrt{w^4}$

70. $\sqrt{m^6}$ **71.** $\sqrt{z^5}$ **72.** $\sqrt{p^9}$

73. $\sqrt{\dfrac{p}{3}}$ **74.** $\sqrt{\dfrac{a}{7}}$ **75.** $\sqrt{\dfrac{b}{a}}$

76. $\sqrt{\dfrac{w}{2z}}$ **77.** $\dfrac{ab}{\sqrt{a}}$ **78.** $\dfrac{wx^2}{\sqrt{wx}}$

79. $\sqrt{80x^3}$ **80.** $\sqrt{90y^{80}}$ **81.** $\sqrt{9y^9x^{15}}$

82. $\sqrt{48x^2y^7}$ **83.** $\dfrac{20x^6}{\sqrt{5x^5}}$ **84.** $\dfrac{7x^7y}{\sqrt{7x^9}}$

85. $\dfrac{-22p^2}{p\sqrt{6pq}}$ **86.** $\dfrac{-30t^5}{t^2\sqrt{3t}}$

87. $\dfrac{a^3b^7\sqrt{a^2b^3c^4}}{\sqrt{abc}}$ **88.** $\dfrac{3n^4b^5\sqrt{n^2b^2c^7}}{\sqrt{nbc}}$

89. $\dfrac{\sqrt{4xy^2}}{x^9y^3\sqrt{6xy^3}}$ **90.** $\dfrac{\sqrt{8m^3n^2}}{m^3n^2\sqrt{6mn^3}}$

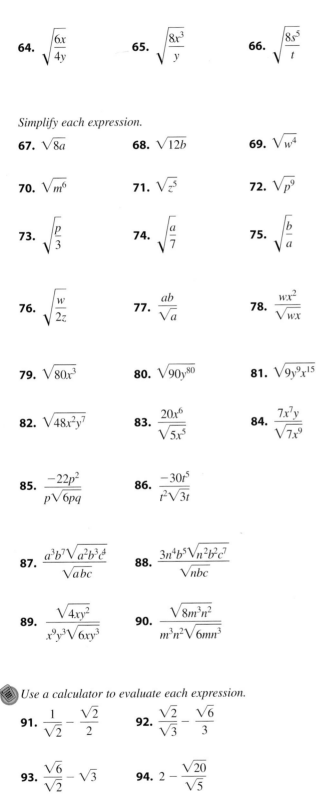

Use a calculator to evaluate each expression.

91. $\dfrac{1}{\sqrt{2}} - \dfrac{\sqrt{2}}{2}$ **92.** $\dfrac{\sqrt{2}}{\sqrt{3}} - \dfrac{\sqrt{6}}{3}$

93. $\dfrac{\sqrt{6}}{\sqrt{2}} - \sqrt{3}$ **94.** $2 - \dfrac{\sqrt{20}}{\sqrt{5}}$

Solve each problem.

95. ***Economic order quantity.*** The formula for economic order quantity

$$E = \sqrt{\frac{2AS}{I}}$$

was used in Exercise 93 of Section 8.1.

a) Express the right-hand side in simplified form.

b) Find E when $A = 23$, $S = \$4566$, and $I = \$80$.

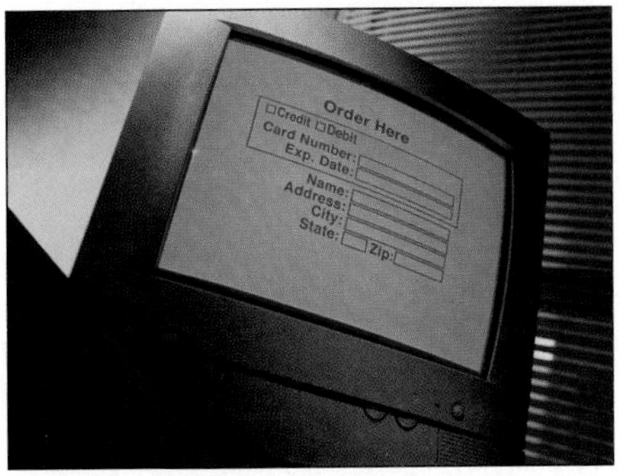

Photo for Exercise 95

96. ***Landing speed.*** Aircraft design engineers determine the proper landing speed V (in ft/sec) by using the formula

$$V = \sqrt{\frac{841L}{CS}},$$

where L is the gross weight of the aircraft in pounds, C is the coefficient of lift, and S is the wing surface area in square feet.

a) Express the right-hand side in simplified form.

b) Find V when $L = 8600$ pounds, $C = 2.81$, and $S = 200$ square feet.

Photo for Exercise 96

8.3 Operations with Radicals

In this Section

- **Adding and Subtracting Radicals**
- **Multiplying Radicals**
- **Dividing Radicals**

In this section you will learn how to perform the basic operations of arithmetic with radical expressions.

Adding and Subtracting Radicals

Consider the sum $2\sqrt{3} + 5\sqrt{3}$. When you studied like terms, you learned that

$$2x + 5x = 7x$$

is true for any value of x. If $x = \sqrt{3}$, then we get

$$2\sqrt{3} + 5\sqrt{3} = 7\sqrt{3}.$$

The expressions $2\sqrt{3}$ and $5\sqrt{3}$ are called **like radicals** because they can be combined just as we combine like terms. We can also add or subtract other radicals as long as they have *the same index and the same radicand.* For example,

$$8\sqrt[3]{w} - 6\sqrt[3]{w} = 2\sqrt[3]{w}.$$

EXAMPLE 1

Combining like radicals

Simplify the following expressions by combining like radicals. Assume that the variables represent nonnegative numbers.

a) $2\sqrt{5} + 7\sqrt{5}$

b) $3\sqrt[3]{2} - 9\sqrt[3]{2}$

c) $\sqrt{2} - 5\sqrt{a} + 4\sqrt{2} - 3\sqrt{a}$

Solution

a) $2\sqrt{5} + 7\sqrt{5} = 9\sqrt{5}$

b) $3\sqrt[3]{2} - 9\sqrt[3]{2} = -6\sqrt[3]{2}$

c) $\sqrt{2} - 5\sqrt{a} + 4\sqrt{2} - 3\sqrt{a} = \sqrt{2} + 4\sqrt{2} - 5\sqrt{a} - 3\sqrt{a}$

$= 5\sqrt{2} - 8\sqrt{a}$ Combine like radicals only.

Now do Exercises 7–18

Calculator Close-Up

A calculator can show you when a rule is applied incorrectly. For example, we cannot say that $\sqrt{2} + \sqrt{5}$ is equal to $\sqrt{7}$.

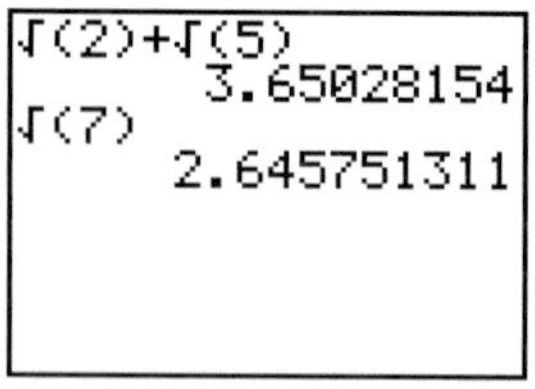

CAUTION We cannot combine the terms in the expressions

$$\sqrt{2} + \sqrt{5}, \quad \sqrt[3]{x} - \sqrt{x}, \quad \text{or} \quad 3\sqrt{2} + \sqrt[4]{6}$$

because the radicands are unequal in the first expression, the indices are unequal in the second, and both the radicands and the indices are unequal in the third.

In Example 2 we simplify the radicals before adding or subtracting.

EXAMPLE 2

Simplifying radicals before combining like terms

Simplify. Assume that all variables represent nonnegative real numbers.

a) $\sqrt{12} + \sqrt{75}$ **b)** $\sqrt{8x^3} + x\sqrt{18x}$ **c)** $\frac{4}{\sqrt{2}} - \frac{\sqrt{3}}{\sqrt{6}}$

Solution

a) $\sqrt{12} + \sqrt{75} = \sqrt{4} \cdot \sqrt{3} + \sqrt{25} \cdot \sqrt{3}$ Product rule for radicals

$= 2\sqrt{3} + 5\sqrt{3}$ Simplify.

$= 7\sqrt{3}$ Combine like radicals.

b) $\sqrt{8x^3} + x\sqrt{18x} = \sqrt{4x^2} \cdot \sqrt{2x} + x\sqrt{9} \cdot \sqrt{2x}$ Product rule for radicals

$= 2x\sqrt{2x} + 3x\sqrt{2x}$ Simplify.

$= 5x\sqrt{2x}$ Combine like radicals.

c) $\frac{4}{\sqrt{2}} - \frac{\sqrt{3}}{\sqrt{6}} = \frac{4 \cdot \sqrt{2}}{\sqrt{2} \cdot \sqrt{2}} - \frac{\sqrt{3} \cdot \sqrt{6}}{\sqrt{6} \cdot \sqrt{6}}$ Rationalize the denominators.

$= \frac{4\sqrt{2}}{2} - \frac{\sqrt{18}}{6}$ Simplify.

$= \frac{4\sqrt{2}}{2} - \frac{3\sqrt{2}}{6}$ $\sqrt{18} = \sqrt{9} \cdot \sqrt{2} = 3\sqrt{2}$

$= \frac{4\sqrt{2}}{2} - \frac{\sqrt{2}}{2}$ Reduce $\frac{3\sqrt{2}}{6}$ to $\frac{\sqrt{2}}{2}$.

$= \frac{3\sqrt{2}}{2}$ $4\sqrt{2} - \sqrt{2} = 3\sqrt{2}$

Now do Exercises 19–30

Multiplying Radicals

We have been using the product rule for radicals

$$\sqrt[n]{ab} = \sqrt[n]{a} \cdot \sqrt[n]{b}$$

to express a root of a product as a product of the roots of the factors. When we rationalized denominators in Section 8.2, we used the product rule to multiply radicals. We will now study multiplication of radicals in more detail.

EXAMPLE 3

Multiplying radical expressions

Multiply and simplify. Assume that variables represent positive numbers.

a) $\sqrt{2} \cdot \sqrt{5}$

b) $2\sqrt{5} \cdot 3\sqrt{6}$

c) $\sqrt{2a^2} \cdot \sqrt{6a}$

d) $\sqrt[3]{4} \cdot \sqrt[3]{2}$

Helpful Hint

Students often write

$$\sqrt{15} \cdot \sqrt{15} = \sqrt{225} = 15.$$

Although this is correct, you should get used to the idea that

$$\sqrt{15} \cdot \sqrt{15} = 15.$$

Because the definition of square root, $\sqrt{a} \cdot \sqrt{a} = a$ for any positive number a.

Solution

a) $\sqrt{2} \cdot \sqrt{5} = \sqrt{2 \cdot 5}$ Product rule for radicals
$= \sqrt{10}$

b) $2\sqrt{5} \cdot 3\sqrt{6} = 2 \cdot 3 \cdot \sqrt{5} \cdot \sqrt{6}$
$= 6\sqrt{30}$ Product rule for radicals

c) $\sqrt{2a^2} \cdot \sqrt{6a} = \sqrt{12a^3}$ Product rule for radicals
$= \sqrt{4a^2} \cdot \sqrt{3a}$ Factor out the perfect square.
$= 2a\sqrt{3a}$ Simplify.

d) $\sqrt[3]{4} \cdot \sqrt[3]{2} = \sqrt[3]{8}$ Product rule for radicals
$= 2$

Now do Exercises 31–42

A sum such as $\sqrt{6} + \sqrt{2}$ is in its simplest form, and so it is treated like a binomial when it occurs in a product.

EXAMPLE 4

Using the distributive property with radicals

Find the product: $3\sqrt{3}(\sqrt{6} + \sqrt{2})$

Solution

$$3\sqrt{3}(\sqrt{6} + \sqrt{2}) = 3\sqrt{3} \cdot \sqrt{6} + 3\sqrt{3} \cdot \sqrt{2}$$ Distributive property
$$= 3\sqrt{18} + 3\sqrt{6}$$ Product rule for radicals
$$= 3 \cdot 3\sqrt{2} + 3\sqrt{6}$$ $\sqrt{18} = \sqrt{9} \cdot \sqrt{2} = 3\sqrt{2}$
$$= 9\sqrt{2} + 3\sqrt{6}$$

Now do Exercises 43–48

In Example 5 we use the FOIL method to find products of expressions involving radicals.

EXAMPLE 5

Using FOIL to multiply radicals

Multiply and simplify.

a) $(\sqrt{3} + 5)(\sqrt{3} - 2)$ **b)** $(\sqrt{5} - 2)(\sqrt{5} + 2)$
c) $(2\sqrt{3} + \sqrt{5})(\sqrt{3} - \sqrt{5})$

Solution

a) $(\sqrt{3} + 5)(\sqrt{3} - 2) = \overset{F}{3} - \overset{O}{2\sqrt{3}} + \overset{I}{5\sqrt{3}} - \overset{L}{10}$
$= 3\sqrt{3} - 7$ Add the like terms.

b) The product $(\sqrt{5} - 2)(\sqrt{5} + 2)$ is the product of a sum and a difference. Recall that $(a - b)(a + b) = a^2 - b^2$.

$$\begin{aligned}(\sqrt{5} - 2)(\sqrt{5} + 2) &= (\sqrt{5})^2 - 2^2 \\ &= 5 - 4 \\ &= 1\end{aligned}$$

c)
$$\begin{aligned}(2\sqrt{3} + \sqrt{5})(\sqrt{3} - \sqrt{5}) &= \overset{\text{F}}{2\sqrt{3}\sqrt{3}} - \overset{\text{O}}{2\sqrt{3}\sqrt{5}} + \overset{\text{I}}{\sqrt{5}\sqrt{3}} - \overset{\text{L}}{\sqrt{5}\sqrt{5}} \\ &= 6 - 2\sqrt{15} + \sqrt{15} - 5 \\ &= 1 - \sqrt{15}\end{aligned}$$

Now do Exercises 49–60

Study Tip

Review, review, review! Don't wait until the end of a chapter to review. Do a little review every time you study for this course.

Dividing Radicals

In Section 8.1 we used the quotient rule for radicals to write a square root of a quotient as a quotient of square roots. We can also use the quotient rule for radicals to divide radicals of the same index. For example,

$$\frac{\sqrt{10}}{\sqrt{2}} = \sqrt{\frac{10}{2}} = \sqrt{5}.$$

Division of radicals is simplest when the quotient of the radicands is a whole number, as it was in the example $\sqrt{10} \div \sqrt{2} = \sqrt{5}$. If the quotient of the radicands is not a whole number, then we divide by rationalizing the denominator, as shown in Example 6.

EXAMPLE 6

Dividing radicals

Divide and simplify.

a) $\sqrt{30} \div \sqrt{3}$ **b)** $(5\sqrt{2}) \div (2\sqrt{5})$ **c)** $(15\sqrt{6}) \div (3\sqrt{2})$

Solution

a) $\sqrt{30} \div \sqrt{3} = \dfrac{\sqrt{30}}{\sqrt{3}} = \sqrt{10}$

b)
$$\begin{aligned}(5\sqrt{2}) \div (2\sqrt{5}) &= \frac{5\sqrt{2}}{2\sqrt{5}} = \frac{5\sqrt{2} \cdot \sqrt{5}}{2\sqrt{5} \cdot \sqrt{5}} && \text{Rationalize the denominator.} \\ &= \frac{5\sqrt{10}}{2 \cdot 5} && \text{Product rule for radicals} \\ &= \frac{\sqrt{10}}{2} && \text{Reduce.}\end{aligned}$$

Note that $\sqrt{10} \div 2 \neq \sqrt{5}$.

c) $(15\sqrt{6}) \div (3\sqrt{2}) = \dfrac{15\sqrt{6}}{3\sqrt{2}} = 5\sqrt{3}$ $\sqrt{6} \div \sqrt{2} = \sqrt{3}$

Now do Exercises 61–68

CAUTION You can use the quotient rule to divide roots of the same index only. For example,

$$\frac{\sqrt{14}}{\sqrt{2}} = \sqrt{7} \quad \text{but} \quad \frac{\sqrt{14}}{2} \neq \sqrt{7}.$$

In Example 7 we simplify expressions with radicals in the numerator and whole numbers in the denominator.

EXAMPLE 7

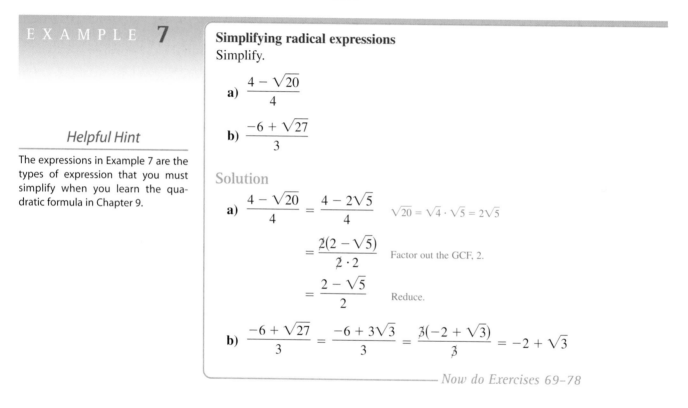

Simplifying radical expressions

Simplify.

a) $\dfrac{4 - \sqrt{20}}{4}$

b) $\dfrac{-6 + \sqrt{27}}{3}$

Helpful Hint

The expressions in Example 7 are the types of expression that you must simplify when you learn the quadratic formula in Chapter 9.

Solution

a) $\dfrac{4 - \sqrt{20}}{4} = \dfrac{4 - 2\sqrt{5}}{4}$ $\quad \sqrt{20} = \sqrt{4} \cdot \sqrt{5} = 2\sqrt{5}$

$= \dfrac{\cancel{2}(2 - \sqrt{5})}{\cancel{2} \cdot 2}$ $\quad$ Factor out the GCF, 2.

$= \dfrac{2 - \sqrt{5}}{2}$ $\quad$ Reduce.

b) $\dfrac{-6 + \sqrt{27}}{3} = \dfrac{-6 + 3\sqrt{3}}{3} = \dfrac{\cancel{3}(-2 + \sqrt{3})}{\cancel{3}} = -2 + \sqrt{3}$

Now do Exercises 69–78

CAUTION In the expression $\frac{2 - \sqrt{5}}{2}$ you cannot divide out the remaining 2's because 2 is not a *factor* of the numerator.

In Example 5(b) we used the rule for the product of a sum and a difference to get $(\sqrt{5} - 2)(\sqrt{5} + 2) = 1$. If we apply the same rule to other products of this type, we also get a rational number as the result. For example,

$$(\sqrt{7} + \sqrt{2})(\sqrt{7} - \sqrt{2}) = 7 - 2 = 5.$$

Expressions such as $\sqrt{5} + 2$ and $\sqrt{5} - 2$ are called **conjugates** of each other. The conjugate of $\sqrt{7} + \sqrt{2}$ is $\sqrt{7} - \sqrt{2}$. We can use conjugates to simplify a radical expression that has a sum or a difference in its denominator.

EXAMPLE 8

Rationalizing the denominator using conjugates

Simplify each expression.

a) $\dfrac{\sqrt{3}}{\sqrt{7} - \sqrt{2}}$

b) $\dfrac{4}{6 + \sqrt{2}}$

Helpful Hint

The word conjugate is used in many contexts in mathematics. According to the dictionary, conjugate means joined together, especially as in a pair.

Solution

a) $\dfrac{\sqrt{3}}{\sqrt{7} - \sqrt{2}} = \dfrac{\sqrt{3}(\sqrt{7} + \sqrt{2})}{(\sqrt{7} - \sqrt{2})(\sqrt{7} + \sqrt{2})}$ Multiply by $\sqrt{7} + \sqrt{2}$, the conjugate of $\sqrt{7} - \sqrt{2}$.

$= \dfrac{\sqrt{21} + \sqrt{6}}{7 - 2}$

$= \dfrac{\sqrt{21} + \sqrt{6}}{5}$

b) $\dfrac{4}{6 + \sqrt{2}} = \dfrac{4(6 - \sqrt{2})}{(6 + \sqrt{2})(6 - \sqrt{2})}$ Multiply by $6 - \sqrt{2}$, the conjugate of $6 + \sqrt{2}$.

$= \dfrac{24 - 4\sqrt{2}}{36 - 2}$

$= \dfrac{24 - 4\sqrt{2}}{34}$

$= \dfrac{\cancel{2}(12 - 2\sqrt{2})}{\cancel{2} \cdot 17} = \dfrac{12 - 2\sqrt{2}}{17}$

Now do Exercises 79–86

Warm-Ups ▼

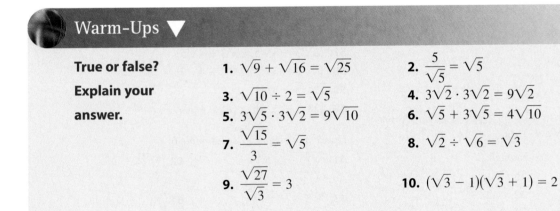

True or false? Explain your answer.

1. $\sqrt{9} + \sqrt{16} = \sqrt{25}$
2. $\dfrac{5}{\sqrt{5}} = \sqrt{5}$
3. $\sqrt{10} \div 2 = \sqrt{5}$
4. $3\sqrt{2} \cdot 3\sqrt{2} = 9\sqrt{2}$
5. $3\sqrt{5} \cdot 3\sqrt{2} = 9\sqrt{10}$
6. $\sqrt{5} + 3\sqrt{5} = 4\sqrt{10}$
7. $\dfrac{\sqrt{15}}{3} = \sqrt{5}$
8. $\sqrt{2} \div \sqrt{6} = \sqrt{3}$
9. $\dfrac{\sqrt{27}}{\sqrt{3}} = 3$
10. $(\sqrt{3} - 1)(\sqrt{3} + 1) = 2$

8.3 Exercises

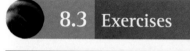

Reading and Writing *After reading this section, write out the answers to these questions. Use complete sentences.*

1. What are like radicals?
2. How do we combine like radicals?
3. What operations can be performed with radicals?
4. What method can we use to multiply a sum of two square roots by a sum of two square roots?
5. What radical expressions are conjugates of each other?
6. How do you rationalize a denominator that contains a sum of two radicals?

Assume that all variables in these exercises represent only positive real numbers.

Simplify each expression by combining like radicals. See Example 1.

7. $4\sqrt{5} + 3\sqrt{5}$ **8.** $\sqrt{2} + \sqrt{2}$

9. $\sqrt[3]{2} + \sqrt[3]{2}$ **10.** $4\sqrt[3]{6} - 7\sqrt[3]{6}$

11. $3u\sqrt{11} + 5u\sqrt{11}$

12. $9m\sqrt{5} - 12m\sqrt{5}$

13. $\sqrt{2} + \sqrt{3} - 5\sqrt{2} + 3\sqrt{3}$

14. $8\sqrt{6} - \sqrt{2} - 3\sqrt{6} + 5\sqrt{2}$

15. $3\sqrt{y} - \sqrt{x} - 4\sqrt{y} - 3\sqrt{x}$

16. $5\sqrt{7} - \sqrt{a} + 3\sqrt{7} - 5\sqrt{a}$

17. $3x\sqrt{y} - \sqrt{a} + 2x\sqrt{y} + 3\sqrt{a}$

18. $a\sqrt{b} + 5a\sqrt{b} - 2\sqrt{a} + 3\sqrt{a}$

Simplify each expression. See Example 2.

19. $\sqrt{24} + \sqrt{54}$ **20.** $\sqrt{12} + \sqrt{27}$

21. $2\sqrt{27} - 4\sqrt{75}$ **22.** $\sqrt{2} - \sqrt{18}$

23. $\sqrt{3a} - \sqrt{12a}$ **24.** $\sqrt{5w} - \sqrt{45w}$

25. $\sqrt{x^3} + x\sqrt{4x}$ **26.** $\sqrt{27x^3} + 5x\sqrt{12x}$

27. $\dfrac{1}{\sqrt{3}} + \dfrac{\sqrt{2}}{\sqrt{6}}$ **28.** $\dfrac{3}{\sqrt{5}} + \dfrac{\sqrt{2}}{\sqrt{10}}$

29. $\dfrac{1}{\sqrt{3}} + \sqrt{12}$ **30.** $\dfrac{1}{\sqrt{2}} + 3\sqrt{8}$

Multiply and simplify. See Example 3.

31. $\sqrt{7} \cdot \sqrt{11}$ **32.** $\sqrt{3} \cdot \sqrt{13}$

33. $2\sqrt{6} \cdot 3\sqrt{6}$ **34.** $4\sqrt{2} \cdot 3\sqrt{2}$

35. $-3\sqrt{5} \cdot 4\sqrt{2}$ **36.** $-8\sqrt{3} \cdot 3\sqrt{2}$

37. $\sqrt{2a^3} \cdot \sqrt{6a^5}$ **38.** $\sqrt{3w^7} \cdot \sqrt{w^9}$

39. $\sqrt[3]{9} \cdot \sqrt[3]{3}$ **40.** $\sqrt[3]{-25} \cdot \sqrt[3]{5}$

41. $\sqrt[3]{-4m^2} \cdot \sqrt[3]{2m}$ **42.** $\sqrt[3]{100m^4} \cdot \sqrt[3]{10m^2}$

Multiply and simplify. See Example 4.

43. $\sqrt{2}(\sqrt{2} + \sqrt{3})$ **44.** $\sqrt{3}(\sqrt{3} - \sqrt{2})$

45. $3\sqrt{2}(2\sqrt{6} + \sqrt{10})$ **46.** $2\sqrt{3}(\sqrt{6} + 2\sqrt{15})$

47. $2\sqrt{5}(\sqrt{5} - 3\sqrt{10})$ **48.** $\sqrt{6}(\sqrt{24} - 6)$

Multiply and simplify. See Example 5.

49. $(\sqrt{5} - 4)(\sqrt{5} + 3)$

50. $(\sqrt{6} - 2)(\sqrt{6} - 3)$

51. $(\sqrt{3} - 1)(\sqrt{3} + 1)$

52. $(\sqrt{6} + 2)(\sqrt{6} - 2)$

53. $(\sqrt{5} - \sqrt{2})(\sqrt{5} + \sqrt{2})$

54. $(\sqrt{3} - \sqrt{6})(\sqrt{3} + \sqrt{6})$

55. $(2\sqrt{5} + 1)(3\sqrt{5} - 2)$

56. $(2\sqrt{2} + 3)(4\sqrt{2} + 4)$

57. $(2\sqrt{3} - 3\sqrt{5})(3\sqrt{3} + 4\sqrt{5})$

58. $(4\sqrt{3} + 3\sqrt{7})(2\sqrt{3} + 4\sqrt{7})$

59. $(2\sqrt{3} + 5)^2$

60. $(3\sqrt{2} + \sqrt{6})^2$

Divide and simplify. See Example 6.

61. $\sqrt{10} \div \sqrt{5}$ **62.** $\sqrt{14} \div \sqrt{2}$

63. $\sqrt{5} \div \sqrt{3}$ **64.** $\sqrt{3} \div \sqrt{2}$

65. $(4\sqrt{5}) \div (3\sqrt{6})$

66. $(3\sqrt{7}) \div (4\sqrt{3})$

67. $(5\sqrt{14}) \div (3\sqrt{2})$

68. $(4\sqrt{15}) \div (5\sqrt{2})$

Simplify each expression. See Example 7.

69. $\dfrac{2 + \sqrt{8}}{2}$

70. $\dfrac{3 + \sqrt{18}}{3}$

71. $\dfrac{-4 + \sqrt{20}}{2}$

72. $\dfrac{-6 + \sqrt{45}}{3}$

73. $\dfrac{4 - \sqrt{20}}{6}$

74. $\dfrac{-6 - \sqrt{27}}{6}$

75. $\dfrac{-4 - \sqrt{24}}{-6}$

76. $\dfrac{-3 - \sqrt{27}}{-3}$

77. $\dfrac{3 + \sqrt{12}}{6}$

78. $\dfrac{3 + \sqrt{8}}{3}$

Simplify each expression. See Example 8.

79. $\dfrac{5}{\sqrt{3} - \sqrt{2}}$

80. $\dfrac{3}{\sqrt{6} + \sqrt{2}}$

81. $\dfrac{\sqrt{3}}{\sqrt{5} - \sqrt{3}}$

82. $\dfrac{\sqrt{2}}{\sqrt{2} + \sqrt{5}}$

83. $\dfrac{2 + \sqrt{3}}{5 - \sqrt{3}}$

84. $\dfrac{\sqrt{2} - \sqrt{3}}{\sqrt{3} - 1}$

85. $\dfrac{\sqrt{7} - 5}{2\sqrt{7} + 1}$

86. $\dfrac{\sqrt{5} + 4}{3\sqrt{2} - \sqrt{5}}$

Simplify.

87. $\sqrt{5a} + \sqrt{20a}$

88. $a\sqrt{6} \cdot a\sqrt{12}$

89. $\sqrt{75} \div \sqrt{6}$

90. $\sqrt{24} - \sqrt{150}$

91. $(5 + 3\sqrt{5})^2$

92. $(\sqrt{6} - \sqrt{5})(\sqrt{6} + \sqrt{5})$

93. $\sqrt{5} + \dfrac{\sqrt{20}}{3}$

94. $\dfrac{5}{\sqrt{8} - \sqrt{3}}$

Use a calculator to find the approximate value of each expression to three decimal places.

95. $\dfrac{2 + \sqrt{3}}{2}$

96. $\dfrac{-2 + \sqrt{3}}{-6}$

97. $\dfrac{-4 - \sqrt{6}}{5 - \sqrt{3}}$

98. $\dfrac{-5 - \sqrt{2}}{\sqrt{3} + \sqrt{7}}$

Solve each problem.

99. Find the exact area and perimeter of the given rectangle.

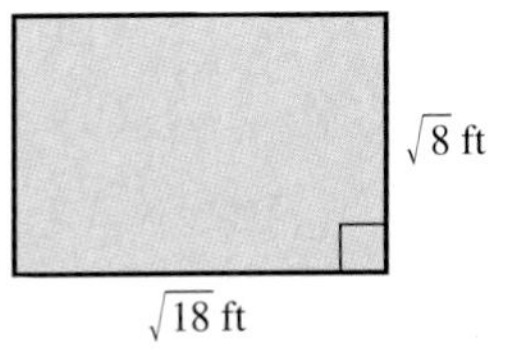

Figure for Exercise 99

100. Find the exact area and perimeter of the given triangle.

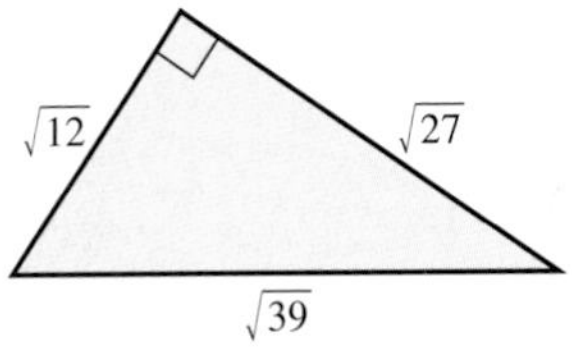

Figure for Exercise 100

101. Find the exact volume in cubic meters of the given rectangular box.

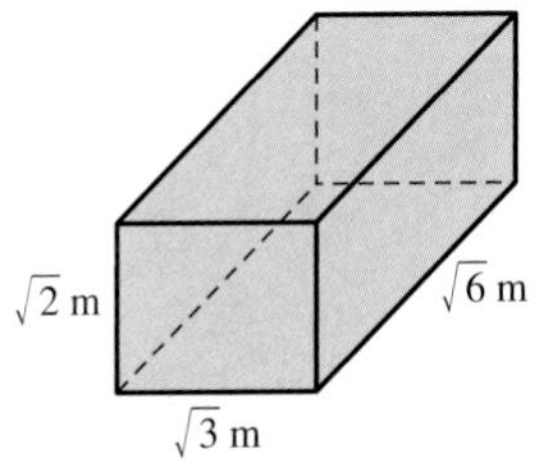

Figure for Exercise 101

102. Find the exact surface area of the box in Exercise 101.

8.4 Solving Equations with Radicals and Exponents

In this Section

- The Square Root Property
- Obtaining Equivalent Equations
- Squaring Each Side of an Equation
- Solving for the Indicated Variable
- Applications

Equations involving radicals and exponents occur in many applications. In this section you will learn to solve equations of this type, and you will see how these equations occur in applications.

The Square Root Property

An equation of the form $x^2 = k$ can have two solutions, one solution, or no solutions, depending on the value of k. For example,

$$x^2 = 4$$

has two solutions because $(2)^2 = 4$ and $(-2)^2 = 4$. So $x^2 = 4$ is equivalent to the compound equation

$$x = 2 \quad \text{or} \quad x = -2,$$

which is also written as $x = \pm 2$ and is read as "x equals positive or negative 2." The only solution to $x^2 = 0$ is 0. The equation

$$x^2 = -4$$

has no real solution because the square of every real number is nonnegative.

These examples illustrate the **square root property.**

Square Root Property (Solving $x^2 = k$)

For $k > 0$ the equation $x^2 = k$ is equivalent to the compound equation

$$x = \sqrt{k} \quad \text{or} \quad x = -\sqrt{k}. \qquad (\text{Also written } x = \pm\sqrt{k}.)$$

For $k = 0$ the equation $x^2 = k$ is equivalent to $x = 0$.

For $k < 0$ the equation $x^2 = k$ has no real solution.

CAUTION The expression $\sqrt{9}$ has a value of 3 only, but the equation $x^2 = 9$ has two solutions: 3 and -3.

EXAMPLE 1

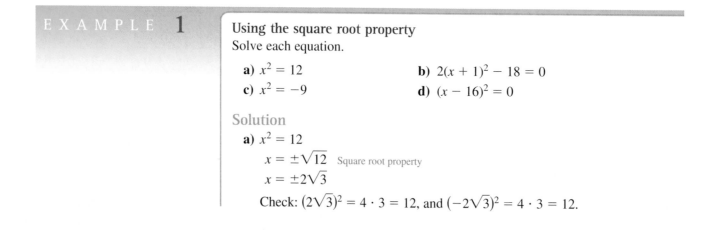

Using the square root property

Solve each equation.

a) $x^2 = 12$ **b)** $2(x + 1)^2 - 18 = 0$

c) $x^2 = -9$ **d)** $(x - 16)^2 = 0$

Solution

a) $x^2 = 12$

$x = \pm\sqrt{12}$ Square root property

$x = \pm 2\sqrt{3}$

Check: $(2\sqrt{3})^2 = 4 \cdot 3 = 12$, and $(-2\sqrt{3})^2 = 4 \cdot 3 = 12$.

Helpful Hint

We do not say "take the square root of each side." We are not doing the same thing to each side of $x^2 = 12$ when we write $x = \pm\sqrt{12}$. This is the second time that we have seen a rule for obtaining an equivalent equation without "doing the same thing to each side." (What was the first?)

b) $2(x + 1)^2 - 18 = 0$

$$2(x + 1)^2 = 18 \quad \text{Add 18 to each side.}$$

$$(x + 1)^2 = 9 \quad \text{Divide each side by 2.}$$

$$x + 1 = \pm\sqrt{9} \quad \text{Square root property}$$

$$x + 1 = 3 \quad \text{or} \quad x + 1 = -3$$

$$x = 2 \quad \text{or} \quad x = -4$$

Check 2 and -4 in the original equation. Both -4 and 2 are solutions to the equation.

c) The equation $x^2 = -9$ has no real solution because no real number has a square that is negative.

d) $(x - 16)^2 = 0$

$$x - 16 = 0 \quad \text{Square root property}$$

$$x = 16$$

Check: $(16 - 16)^2 = 0$. The equation has only one solution, 16.

Now do Exercises 7–20

Obtaining Equivalent Equations

When solving equations, we use a sequence of equivalent equations with each one simpler than the last. To get an equivalent equation, we can

1. add the same number to each side,
2. subtract the same number from each side,
3. multiply each side by the same nonzero number, or
4. divide each side by the same nonzero number.

However, "doing the same thing to each side" is not the only way to obtain an equivalent equation. In Chapter 6 we used the zero factor property to obtain equivalent equations. For example, by the zero factor property the equation

$$(x - 3)(x + 2) = 0$$

is equivalent to the compound equation

$$x - 3 = 0 \quad \text{or} \quad x + 2 = 0.$$

Study Tip

Make sure that you know what your instructor expects from you. You can determine what your instructor feels is important by looking at the examples that your instructor works in class and the homework assigned. When in doubt, ask your instructor what you will be responsible for and write down the answer.

In this section you just learned how to obtain equivalent equations by the square root property. This property tells us how to write an equation that is equivalent to the equation $x^2 = k$. Note that the square root property does not tell us to "take the square root of each side." Because a real number might have two, one, or no square roots, taking *the* square root of each side can lead to errors. To become proficient at solving equations, we must understand these methods. One of our main goals in algebra is to keep expanding our skills for solving equations.

Squaring Each Side of an Equation

Some equations involving radicals can be solved by squaring each side:

$$\sqrt{x} = 5$$

$$(\sqrt{x})^2 = 5^2 \quad \text{Square each side.}$$

$$x = 25$$

All three of these equations are equivalent. Because $\sqrt{25} = 5$ is correct, 25 satisfies the original equation.

However, squaring each side does not necessarily produce an equivalent equation. For example, consider the equation

$$x = 3.$$

Squaring each side, we get

$$x^2 = 9.$$

Both 3 and -3 satisfy $x^2 = 9$, but only 3 satisfies the original equation $x = 3$. So $x^2 = 9$ is not equivalent to $x = 3$. The extra solution to $x^2 = 9$ is called an **extraneous solution.**

These two examples illustrate the **squaring property of equality.**

Squaring Property of Equality

When we square each side of an equation, the solutions to the new equation include all of the solutions to the original equation. However, the new equation might have extraneous solutions.

This property means that *we may square each side of an equation, but we must check all of our answers to eliminate extraneous solutions.*

EXAMPLE 2

Using the squaring property of equality

Solve each equation.

a) $\sqrt{x^2 - 16} = 3$ **b)** $x = \sqrt{2x + 3}$

Solution

a)

$$\begin{aligned} \sqrt{x^2 - 16} &= 3 \\ (\sqrt{x^2 - 16})^2 &= 3^2 && \text{Square each side.} \\ x^2 - 16 &= 9 \\ x^2 &= 25 \\ x &= \pm 5 && \text{Square root property} \end{aligned}$$

Check each solution:

$$\sqrt{5^2 - 16} = \sqrt{25 - 16} = \sqrt{9} = 3$$

$$\sqrt{(-5)^2 - 16} = \sqrt{25 - 16} = \sqrt{9} = 3$$

So both 5 and -5 are solutions to the equation.

b)

$$\begin{aligned} x &= \sqrt{2x + 3} \\ x^2 &= (\sqrt{2x + 3})^2 && \text{Square each side.} \\ x^2 &= 2x + 3 \\ x^2 - 2x - 3 &= 0 && \text{Solve by factoring.} \\ (x - 3)(x + 1) &= 0 && \text{Factor.} \end{aligned}$$

$$x - 3 = 0 \quad \text{or} \quad x + 1 = 0 \qquad \text{Zero factor property}$$

$$x = 3 \quad \text{or} \quad x = -1$$

Study Tip

Most instructors believe that what they do in class is important. If you miss class, you miss what is important to your instructor. You miss what is most likely to appear on the test.

Check in the original equation:

Check $x = 3$:	Check $x = -1$:
$3 = \sqrt{2 \cdot 3 + 3}$	$-1 = \sqrt{2(-1) + 3}$
$3 = \sqrt{9}$ Correct	$-1 = \sqrt{1}$ Incorrect

Because -1 does not satisfy the original equation, -1 is an extraneous solution. The only solution is 3.

Now do Exercises 21–30

The equations in Example 3 have radicals on both sides of the equation.

EXAMPLE 3

Radicals on both sides

Solve each equation.

a) $\sqrt{x - 3} = \sqrt{2x + 5}$ **b)** $\sqrt{x^2 - 4x} = \sqrt{2 - 3x}$

Solution

a)

$$\sqrt{x - 3} = \sqrt{2x + 5}$$

$$(\sqrt{x - 3})^2 = (\sqrt{2x + 5})^2 \quad \text{Square each side.}$$

$$x - 3 = 2x + 5 \quad \text{Simplify.}$$

$$x - 8 = 2x$$

$$-8 = x$$

Check $x = -8$ in the original equation:

$$\sqrt{-8 - 3} = \sqrt{2(-8) + 5}$$

$$\sqrt{-11} = \sqrt{-11}$$

Because $\sqrt{-11}$ is not a real number -8 does not satisfy the equation and the equation has no solution.

b)

$$\sqrt{x^2 - 4x} = \sqrt{2 - 3x}$$

$$x^2 - 4x = 2 - 3x \quad \text{Square each side.}$$

$$x^2 - x - 2 = 0$$

$$(x - 2)(x + 1) = 0$$

$$x - 2 = 0 \quad \text{or} \quad x + 1 = 0 \quad \text{Zero factor property}$$

$$x = 2 \quad \text{or} \quad x = -1$$

Check each solution in the original equation:

Check $x = 2$:	Check $x = -1$:
$\sqrt{2^2 - 4 \cdot 2} = \sqrt{2 - 3 \cdot 2}$	$\sqrt{(-1)^2 - 4(-1)} = \sqrt{2 - 3(-1)}$
$\sqrt{-4} = \sqrt{-4}$	$\sqrt{5} = \sqrt{5}$

Because $\sqrt{-4}$ is not a real number, 2 is an extraneous solution. The only solution to the original equation is -1.

Now do Exercises 31–36

In Example 4, one of the sides of the equation is a binomial. When we square each side, we must be sure to square the binomial properly.

EXAMPLE 4

Squaring each side of an equation

Solve the equation $x + 2 = \sqrt{-2 - 3x}$.

Solution

$$x + 2 = \sqrt{-2 - 3x}$$

$$(x + 2)^2 = (\sqrt{-2 - 3x})^2 \quad \text{Square each side.}$$

$$x^2 + 4x + 4 = -2 - 3x \quad \text{Square the binomial on the left side.}$$

$$x^2 + 7x + 6 = 0$$

$$(x + 6)(x + 1) = 0 \quad \text{Factor.}$$

$$x + 6 = 0 \quad \text{or} \quad x + 1 = 0$$

$$x = -6 \quad \text{or} \quad x = -1$$

Check these solutions in the original equation:

Check $x = -6$:

$$-6 + 2 = \sqrt{-2 - 3(-6)}$$

$$-4 = \sqrt{16} \quad \text{Incorrect}$$

Check $x = -1$:

$$-1 + 2 = \sqrt{-2 - 3(-1)}$$

$$1 = \sqrt{1} \quad \text{Correct}$$

The solution -6 does not check. The only solution to the original equation is -1.

Now do Exercises 37–42

Solving for the Indicated Variable

In Example 5 we use the square root property to solve a formula for an indicated variable.

EXAMPLE 5

Solving for a variable

Solve the formula $A = \pi r^2$ for r.

Solution

$$A = \pi r^2$$

$$\frac{A}{\pi} = r^2 \quad \text{Divide each side by } \pi.$$

$$\pm\sqrt{\frac{A}{\pi}} = r \quad \text{Square root property}$$

The formula solved for r is

$$r = \pm\sqrt{\frac{A}{\pi}}.$$

If r is the radius of a circle with area A, then r is positive and

$$r = \sqrt{\frac{A}{\pi}}.$$

Now do Exercises 43–50

Applications

Equations involving exponents occur in many applications. If the exact answer to a problem is an irrational number in radical notation, it is usually helpful to find a decimal approximation for the answer.

EXAMPLE 6

Finding the side of a square with a given diagonal

If the diagonal of a square window is 10 feet long, then what are the exact and approximate lengths of a side? Round the approximate answer to two decimal places.

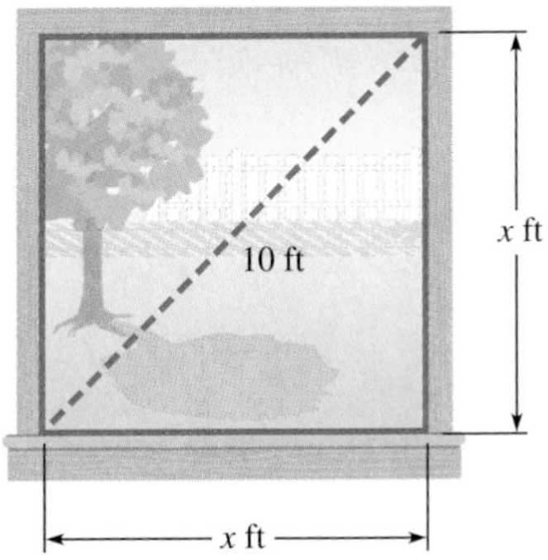

Figure 8.1

Solution

First make a sketch as in Fig. 8.1. Let x be the length of a side. The Pythagorean theorem tells us that the sum of the squares of the sides is equal to the diagonal squared:

$$x^2 + x^2 = 10^2$$
$$2x^2 = 100$$
$$x^2 = 50$$
$$x = \pm\sqrt{50}$$
$$= \pm 5\sqrt{2}$$

Because the length of a side must be positive, we disregard the negative solution. The exact length of a side is $5\sqrt{2}$ feet. Use a calculator to get $5\sqrt{2} \approx 7.07$. The symbol $\approx$ means "is approximately equal to." The approximate length of a side is 7.07 feet.

Now do Exercises 83–94

Warm-Ups

True or false? Explain your answer.

1. The equation $x^2 = 9$ is equivalent to the equation $x = 3$.
2. The equation $x^2 = -16$ has no real solution.
3. The equation $a^2 = 0$ has no solution.
4. Both $-\sqrt{5}$ and $\sqrt{5}$ are solutions to $x^2 + 5 = 0$.
5. The equation $-x^2 = 9$ has no real solution.
6. To solve $\sqrt{x + 4} = \sqrt{2x - 9}$, first take the square root of each side.
7. All extraneous solutions give us a denominator of zero.
8. Squaring both sides of $\sqrt{x} = -1$ will produce an extraneous solution.
9. The equation $x^2 - 3 = 0$ is equivalent to $x = \pm\sqrt{3}$.
10. The equation $-2 = \sqrt{6x^2 - x - 8}$ has no solution.

8.4 Exercises

Reading and Writing *After reading this section, write out the answers to these questions. Use complete sentences.*

1. What is the square root property?

2. When do we take the square root of each side of an equation?

3. What new techniques were introduced in this section for solving equations?

4. Is there any way to obtain an equivalent equation other than doing the same thing to each side?

5. Which property for solving equations can give extraneous roots?

6. Which property of equality does not always give you an equivalent equation?

Solve each equation. See Example 1.

7. $x^2 = 16$ **8.** $x^2 = 49$

9. $x^2 - 40 = 0$ **10.** $x^2 - 24 = 0$

11. $3x^2 = 2$ **12.** $2x^2 = 3$

13. $9x^2 = -4$ **14.** $25x^2 + 1 = 0$

15. $(x - 1)^2 = 4$ **16.** $(x + 3)^2 = 9$

17. $2(x - 5)^2 + 1 = 7$ **18.** $3(x - 6)^2 - 4 = 11$

19. $(x + 19)^2 = 0$ **20.** $5x^2 + 5 = 5$

Solve each equation. See Examples 2 and 3.

21. $\sqrt{x - 9} = 9$ **22.** $\sqrt{x + 3} = 4$

23. $\sqrt{2x - 3} = -4$ **24.** $\sqrt{3x - 5} = -9$

25. $4 = \sqrt{x^2 - 9}$ **26.** $1 = \sqrt{x^2 - 1}$

27. $x = \sqrt{18 - 3x}$ **28.** $x = \sqrt{6x + 27}$

29. $x = \sqrt{x}$ **30.** $x = \sqrt{2x}$

31. $\sqrt{x + 1} = \sqrt{2x - 5}$ **32.** $\sqrt{1 - 3x} = \sqrt{x + 5}$

33. $\sqrt{1 - x} = \sqrt{2x - 14}$ **34.** $\sqrt{2x - 3} = \sqrt{x - 5}$

35. $3\sqrt{2x - 1} + 3 = 5$ **36.** $4\sqrt{x + 5} - 3 = 9$

Solve each equation. See Example 4.

37. $x - 3 = \sqrt{2x - 6}$ **38.** $x - 1 = \sqrt{3x - 5}$

39. $\sqrt{x + 13} = x + 1$ **40.** $x + 1 = \sqrt{22 - 2x}$

41. $\sqrt{10x - 44} = x - 2$ **42.** $\sqrt{8x - 7} = x + 1$

Solve each formula for the indicated variable. See Example 5.

43. $V = \pi r^2 h$ for r **44.** $V = \frac{4}{3}\pi r^2 h$ for r

45. $a^2 + b^2 = c^2$ for b **46.** $y = ax^2 + c$ for x

47. $b^2 - 4ac = 0$ for b **48.** $s = \frac{1}{2}gt^2 + v$ for t

49. $v = \sqrt{2pt}$ for t **50.** $y = \sqrt{2x}$ for x

Solve each equation.

51. $x^2 = \frac{1}{4}$ **52.** $x^2 = \frac{1}{9}$

53. $x^2 + \frac{1}{4} = 0$ **54.** $x^2 + \frac{1}{9} = 0$

55. $\sqrt{x} = \frac{1}{9}$ **56.** $\sqrt{x} = \frac{1}{4}$

57. $\sqrt{x^2 - 15} = 7$ **58.** $\sqrt{2x - 1} = 3$

59. $\sqrt{x} = 2x$ **60.** $\sqrt{2x} = 3x$

61. $(x + 1)^2 = 4$ **62.** $(x + 2)^2 = 16$

63. $3x^2 - 6 = 0$ **64.** $5x^2 + 3 = 0$

65. $\sqrt{2x - 3} = \sqrt{3x + 1}$ **66.** $\sqrt{2x - 4} = \sqrt{x - 9}$

67. $(2x - 1)^2 = 8$ **68.** $(3x - 2)^2 = 18$

69. $\sqrt{2x - 9} = 0$
70. $\sqrt{5 - 3x} = 0$
71. $x + 1 = \sqrt{2x + 10}$
72. $x - 3 = \sqrt{2x + 18}$
73. $3(x + 1)^2 - 27 = 0$
74. $2(x - 3)^2 - 50 = 0$
75. $(2x - 5)^2 = 0$
76. $(3x - 1)^2 = 0$

Use a calculator to find approximate solutions to each equation. Round your answers to three decimal places.

77. $x^2 = 3.25$
78. $(x + 1)^2 = 20.3$
79. $\sqrt{x + 2} = 1.73$
80. $\sqrt{2.3x - 1.4} = 3.3$
81. $1.3(x - 2.4)^2 = 5.4$
82. $-2.4x^2 = -9.55$

Find the exact answer to each problem. If the answer is irrational, then find an approximation to three decimal places. See Example 6.

83. ***Side of a square.*** Find the length of the side of a square whose area is 18 square feet.

84. ***Side of a field.*** Find the length of the side of a square wheat field whose area is 75 square miles.

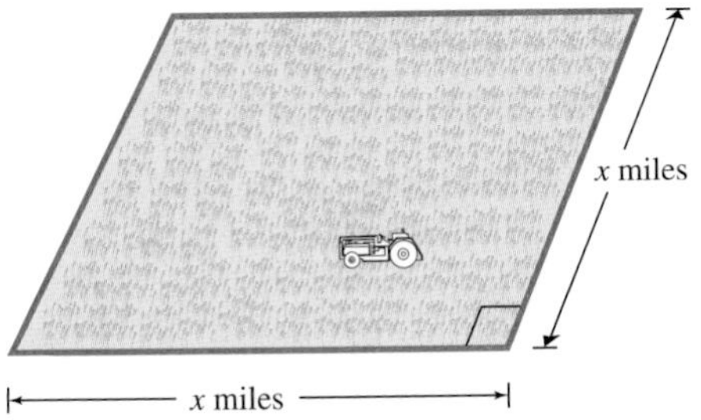

Figure for Exercise 84

85. ***Side of a table.*** Find the length of the side of a square coffee table whose diagonal is 6 feet.

86. ***Side of a square.*** Find the length of the side of a square whose diagonal measures 1 yard.

87. ***Diagonal of a tile.*** Find the length of the diagonal of a square floor tile whose sides measure 1 foot each.

88. ***Diagonal of a sandbox.*** The sandbox at Totland is shaped like a square with an area of 20 square meters. Find the length of the diagonal of the square.

89. ***Diagonal of a tub.*** Find the length of the diagonal of a rectangular bathtub with sides of 3 feet and 4 feet.

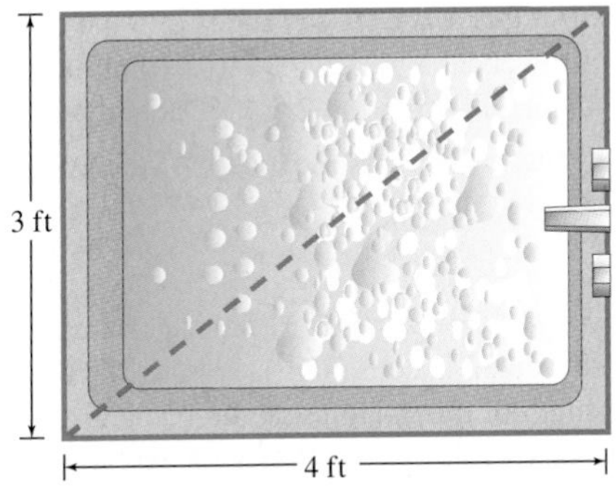

Figure for Exercise 89

90. ***Diagonal of a rectangle.*** What is the length of the diagonal of a rectangular office whose sides are 6 feet and 8 feet?

91. ***Falling bodies.*** If we neglect air resistance, then the number of feet that a body falls from rest during t seconds is given by $s = 16t^2$. How long does it take a pine cone to fall from the top of a 100-foot pine tree?

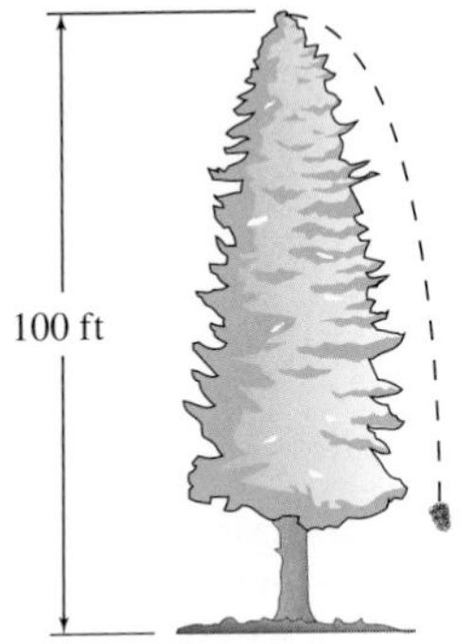

Figure for Exercise 91

92. ***America's favorite pastime.*** A baseball diamond is actually a square, 90 feet on each side. How far is it from home plate to second base?

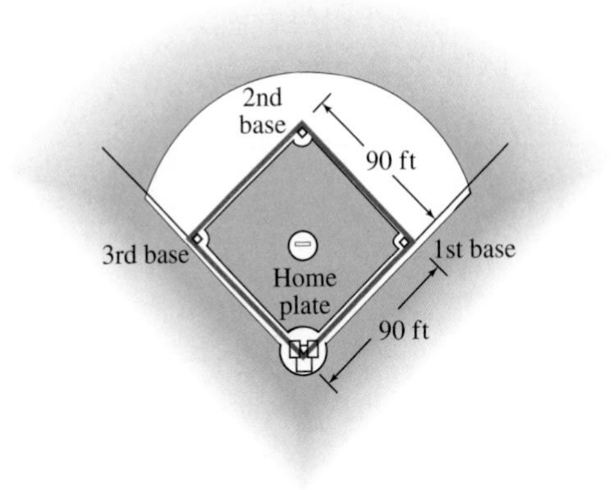

Figure for Exercise 92

93. ***Length of a guy wire.*** A guy wire from the top of a 200-foot tower is to be attached to a point on the ground whose distance from the base of the tower is $\frac{2}{3}$ of the height of the tower. Find the length of the guy wire.

94. ***America's favorite pastime.*** The size of a rectangular television screen is commonly given by the manufacturer as the length of the diagonal of the rectangle. If a television screen measures 10 inches wide and 8 inches high, then what is the exact length of the diagonal of the screen? What is the approximate size of this television screen to the nearest inch?

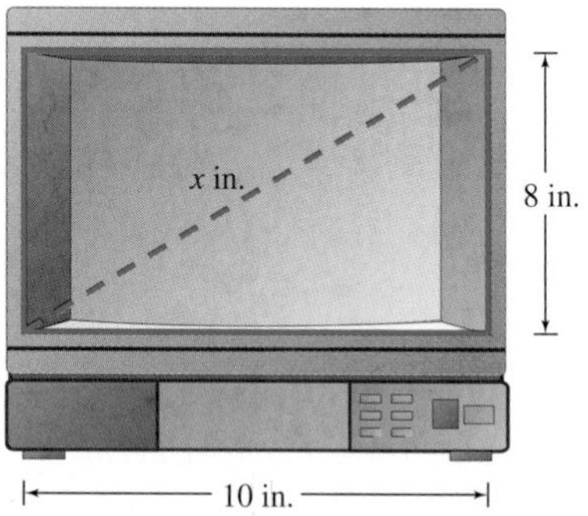

Figure for Exercise 94

8.5 Fractional Exponents

In this Section

- Fractional Exponents
- Using the Rules

You have learned how to use exponents to express powers of numbers and radicals to express roots. In this section you will see that roots can be expressed with exponents also. The advantage of using exponents to express roots is that the rules of exponents can be applied to the expressions.

Calculator Close-Up

You can find the fifth root of 2 using radical notation or exponent notation. Note that the fractional exponent 1/5 must be in parentheses. If you use the decimal equivalent 0.2, then no parentheses are needed

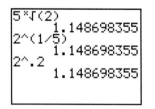

```
5 ×√(2)
        1.148698355
2^(1/5)
        1.148698355
2^.2
        1.148698355
```

Fractional Exponents

The nth root of a number can be expressed by using radical notation or the exponent $1/n$. For example, $8^{1/3}$ and $\sqrt[3]{8}$ both represent the cube root of 8, and we have

$$8^{1/3} = \sqrt[3]{8} = 2.$$

Definition of $a^{1/n}$

If n is any positive integer, then

$$a^{1/n} = \sqrt[n]{a},$$

provided that $\sqrt[n]{a}$ is a real number.

Later in this section we will see that using exponent $1/n$ for nth root is compatible with the rules for integral exponents that we already know.

EXAMPLE 1

Radicals or exponents

Write each radical expression using exponent notation and each exponential expression using radical notation.

a) $\sqrt[3]{35}$ **b)** $\sqrt[4]{xy}$ **c)** $\sqrt{-x}$

d) $5^{1/2}$ **e)** $a^{1/5}$ **f)** $-5^{1/3}$

Solution

a) $\sqrt[3]{35} = 35^{1/3}$ **b)** $\sqrt[4]{xy} = (xy)^{1/4}$ **c)** $\sqrt{-x} = (-x)^{1/2}$

d) $5^{1/2} = \sqrt{5}$ **e)** $a^{1/5} = \sqrt[5]{a}$ **f)** $-5^{1/3} = -\sqrt[3]{5}$

Now do Exercises 7–18

In Example 2 we evaluate some exponential expressions.

EXAMPLE 2

Finding roots

Evaluate each expression.

a) $4^{1/2}$ **b)** $(-8)^{1/3}$ **c)** $81^{1/4}$ **d)** $(-9)^{1/2}$ **e)** $-9^{1/2}$

Solution

a) $4^{1/2} = \sqrt{4} = 2$

b) $(-8)^{1/3} = \sqrt[3]{-8} = -2$

c) $81^{1/4} = \sqrt[4]{81} = 3$

d) Because $(-9)^{1/2}$ or $\sqrt{-9}$ is an even root of a negative number, it is not a real number.

e) Because the exponent $1/2$ is applied to 9 only, $-9^{1/2} = -\sqrt{9} = -3$.

Now do Exercises 19–32

We now extend the definition of exponent $1/n$ to include any fraction or rational number as an exponent. The numerator of the rational number indicates the power, and the denominator indicates the root. For example, the expression

$$8^{2/3}$$

(Power: 2; Root: 3)

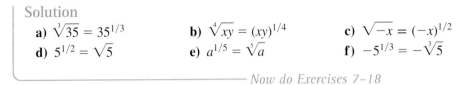

represents the square of the cube root of 8. So we have

$$8^{2/3} = (8^{1/3})^2 = (2)^2 = 4.$$

Helpful Hint

Note that in $a^{m/n}$ we do not require m/n to be reduced. As long as the nth root of a is real, then the value of $a^{m/n}$ is the same whether or not m/n is in lowest terms.

Definition of $a^{m/n}$

If m and n are positive integers, then

$$a^{m/n} = (a^{1/n})^m,$$

provided that $a^{1/n}$ is a real number.

We define negative rational exponents just like negative integral exponents.

Definition of $a^{-m/n}$

If m and n are positive integers and $a \neq 0$, then

$$a^{-m/n} = \frac{1}{a^{m/n}},$$

provided that $a^{1/n}$ is a real number.

EXAMPLE 3

Radicals or exponents

Write each radical expression using exponent notation and each exponential expression using radical notation.

a) $\sqrt[3]{x^2}$ **b)** $\frac{1}{\sqrt[4]{m^3}}$ **c)** $5^{2/3}$ **d)** $a^{-2/5}$

Solution

a) $\sqrt[3]{x^2} = x^{2/3}$ **b)** $\frac{1}{\sqrt[4]{m^3}} = \frac{1}{m^{3/4}} = m^{-3/4}$

c) $5^{2/3} = \sqrt[3]{5^2}$ **d)** $a^{-2/5} = \frac{1}{\sqrt[5]{a^2}}$

Now do Exercises 33–44

Study Tip

If you must miss class, let your instructor know. Be sure to get notes from a reliable classmate. Take good notes yourself, in case a classmate comes to you for notes.

To evaluate an expression with a negative rational exponent, remember that the denominator indicates root, the numerator indicates power, and the negative sign indicates reciprocal:

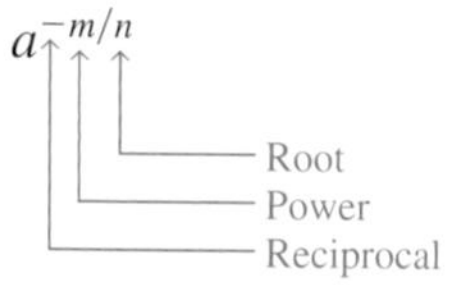

The root, power, and reciprocal can be evaluated in any order. However, to evaluate $a^{-m/n}$ mentally it is usually simplest to use the following strategy.

Strategy for Evaluating $a^{-m/n}$ Mentally

1. Find the nth root of a.
2. Raise your result to the mth power.
3. Find the reciprocal.

For example, to evaluate $8^{-2/3}$ mentally, we find the cube root of 8 (which is 2), square 2 to get 4, then find the reciprocal of 4 to get $\frac{1}{4}$. In print $8^{-2/3}$ could be written for evaluation as $((8^{1/3})^2)^{-1}$ or $\frac{1}{(8^{1/3})^2}$. We prefer the latter because it looks a little simpler.

EXAMPLE 4

Rational exponents

Evaluate each expression.

a) $27^{2/3}$ **b)** $4^{-3/2}$ **c)** $81^{-3/4}$ **d)** $(-8)^{-5/3}$ **e)** $-4^{-3/2}$

Solution

a) Because the exponent is 2/3, we find the cube root of 27, which is 3, and then square it to get 9. In symbols,

$$27^{2/3} = (27^{1/3})^2 = 3^2 = 9.$$

Calculator Close-Up

A negative fractional exponent indicates a reciprocal, a root, and a power. To find $4^{-3/2}$ you can find the reciprocal first, the square root first, or the third power first as shown here.

```
(1/4)^(3/2)
              .125
(√(4))^-3
              .125
(4³)^(-1/2)
              .125
```

b) Because the exponent is $-3/2$, we find the principal square root of 4, which is 2, cube it to get 8, and then find the reciprocal to get $\frac{1}{8}$. In symbols,

$$4^{-3/2} = \frac{1}{(4^{1/2})^3} = \frac{1}{2^3} = \frac{1}{8}.$$

c) Because the exponent is $-3/4$ we find the principal fourth root of 81, which is 3, cube it to get 27, and then find the reciprocal to get $\frac{1}{27}$. In symbols,

$$81^{-3/4} = \frac{1}{(81^{1/4})^3} = \frac{1}{3^3} = \frac{1}{27}.$$

d) $$(-8)^{-5/3} = \frac{1}{((-8)^{1/3})^5} = \frac{1}{(-2)^5} = \frac{1}{-32} = -\frac{1}{32}$$

e) Because the exponent $-3/2$ is applied to 4 only,

$$-4^{-3/2} = -\frac{1}{(\sqrt{4})^3} = -\frac{1}{8}.$$

Now do Exercises 45–60

CAUTION An expression with a negative base and a negative exponent can have a positive or a negative value. For example,

$$(-8)^{-5/3} = -\frac{1}{32} \quad \text{and} \quad (-8)^{-2/3} = \frac{1}{4}.$$

Using the Rules

As we mentioned earlier, the advantage of using exponents to express roots is that the rules for integral exponents can also be used for rational exponents. We will use those rules in Example 5.

EXAMPLE 5

Using the rules of exponents

Simplify each expression. Write answers with positive exponents. Assume that all variables represent positive real numbers.

a) $2^{1/2} \cdot 2^{3/2}$ **b)** $\dfrac{x}{x^{2/3}}$ **c)** $(b^{1/2})^{1/3}$

d) $(x^4y^{-6})^{1/2}$ **e)** $\left(\dfrac{x^6}{y^3}\right)^{-2/3}$

Helpful Hint

Look what happens when we apply legitimate rules to an illegitimate expression:

$$(-1)^{2/2} = (-1)^1 = -1$$

and

$$(-1)^{2/2} = ((-1)^2)^{1/2} = 1^{1/2} = 1$$

So we conclude that $1 = -1$! However, $(-1)^{2/2}$ is not a legal expression because there is no real square root of -1.

Solution

a) $2^{1/2} \cdot 2^{3/2} = 2^{1/2+3/2}$ Product rule

$= 2^2$ $\quad \frac{1}{2} + \frac{3}{2} = \frac{4}{2} = 2$

$= 4$

b) $\dfrac{x}{x^{2/3}} = x^{1-2/3}$ Quotient rule

$= x^{1/3}$ $\quad 1 - \frac{2}{3} = \frac{3}{3} - \frac{2}{3} = \frac{1}{3}$

c) $(b^{1/2})^{1/3} = b^{(1/2)\cdot(1/3)}$ Power rule

$= b^{1/6}$ $\quad \frac{1}{2} \cdot \frac{1}{3} = \frac{1}{6}$

d) $(x^4y^{-6})^{1/2} = (x^4)^{1/2}(y^{-6})^{1/2}$ Power of a product rule

$= x^2y^{-3}$ Power rule

$= \frac{x^2}{y^3}$ Definition of negative exponent

e) $\left(\frac{x^6}{y^3}\right)^{-2/3} = \left(\frac{y^3}{x^6}\right)^{2/3}$ Negative exponent rule

$= \frac{(y^3)^{2/3}}{(x^6)^{2/3}}$ Power of a quotient rule

$= \frac{y^2}{x^4}$ Power rule

Now do Exercises 61–76

Warm-Ups ▼

True or false? Explain your answer.

1. $9^{1/3} = \sqrt[3]{9}$
2. $8^{5/3} = \sqrt[5]{8^3}$
3. $(-16)^{1/2} = -16^{1/2}$
4. $9^{-3/2} = \frac{1}{27}$
5. $6^{-1/2} = \frac{\sqrt{6}}{6}$
6. $\frac{2}{2^{1/2}} = 2^{1/2}$
7. $2^{1/2} \cdot 2^{1/2} = 4^{1/2}$
8. $16^{-1/4} = -2$
9. $6^{1/6} \cdot 6^{1/6} = 6^{1/3}$
10. $(2^8)^{3/4} = 2^6$

8.5 Exercises

Boost your GRADE at mathzone.com!

MathZone
- Practice Problems
- Self-Tests
- Videos
- Net Tutor
- e-Professors

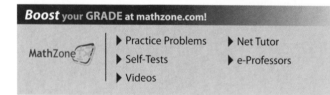

Reading and Writing *After reading this section, write out the answers to these questions. Use complete sentences.*

1. How do we indicate an nth root using exponents?
2. How do we indicate the mth power of the nth root using exponents?
3. What is the meaning of a negative rational exponent?
4. Which rules of exponents hold for rational exponents?
5. In what order must you perform the operations indicated by a negative rational exponent?
6. When is $a^{-m/n}$ a real number?

Write each radical expression using exponent notation and each exponential expression using radical notation. See Example 1.

7. $\sqrt[4]{7}$
8. $\sqrt[3]{cbs}$
9. $9^{1/5}$
10. $3^{1/2}$
11. $\sqrt{5x}$
12. $\sqrt{3y}$

13. $(-a)^{1/2}$

14. $(-b)^{1/5}$

15. $-\sqrt{x}$

16. $-\sqrt[3]{5}$

17. $-6^{1/4}$

18. $-a^{1/2}$

Evaluate each expression. See Example 2.

19. $25^{1/2}$

20. $16^{1/2}$

21. $125^{1/3}$

22. $16^{1/4}$

23. $(-125)^{1/3}$

24. $(-32)^{1/5}$

25. $(-8)^{1/3}$

26. $(-27)^{1/3}$

27. $(-4)^{1/2}$

28. $(-16)^{1/4}$

29. $-4^{1/2}$

30. $-16^{1/4}$

31. $-64^{1/3}$

32. $-36^{1/2}$

Write each radical expression using exponent notation and each exponential expression using radical notation. See Example 3.

33. $\sqrt[3]{w^7}$

34. $\sqrt{a^5}$

35. $\dfrac{1}{\sqrt[3]{2^{10}}}$

36. $\sqrt[3]{\dfrac{1}{a^2}}$

37. $w^{-3/4}$

38. $6^{-5/3}$

39. $(ab)^{3/2}$

40. $(3m)^{-1/5}$

41. $-\sqrt[4]{t^3}$

42. $-\sqrt{p^3}$

43. $-a^{-3/2}$

44. $-w^{-3/4}$

Evaluate each expression. See Example 4.

45. $125^{2/3}$

46. $1000^{2/3}$

47. $25^{3/2}$

48. $16^{3/2}$

49. $27^{-4/3}$

50. $16^{-3/4}$

51. $4^{-3/2}$

52. $25^{-3/2}$

53. $(-27)^{-1/3}$

54. $(-8)^{-4/3}$

55. $(-16)^{-1/4}$

56. $(-100)^{-3/2}$

57. $-16^{3/4}$

58. $-8^{2/3}$

59. $-8^{-2/3}$

60. $-27^{-2/3}$

Simplify each expression. Write answers with positive exponents only. See Example 5.

61. $x^{1/4}x^{1/4}$

62. $y^{1/3}y^{2/3}$

63. $n^{1/2}n^{-1/3}$

64. $w^{-1/4}w^{3/5}$

65. $\dfrac{x^2}{x^{1/2}}$

66. $\dfrac{a^{1/2}}{a^{1/3}}$

67. $\dfrac{8t^{1/2}}{4t^{1/4}}$

68. $\dfrac{6w^{1/4}}{3w^{1/3}}$

69. $(x^6)^{1/3}$

70. $(y^{-4})^{1/2}$

71. $(5^{-1/4})^{-1/2}$

72. $(7^{-3/4})^6$

73. $(x^2y^6)^{1/2}$

74. $(t^3w^6)^{1/3}$

75. $(9x^{-2}y^8)^{-1/2}$

76. $(4w^{-2}t^{-4})^{-1/2}$

Evaluate each expression.

77. $16^{-1/2} + 2^{-1}$

78. $4^{-1/2} - 8^{-2/3}$

79. $27^{-1/6} \cdot 27^{-1/2}$

80. $32^{-1/10} \cdot 32^{-1/10}$

81. $\dfrac{81^{5/6}}{81^{1/12}}$

82. $\dfrac{25^{-3/4}}{25^{3/4}}$

83. $(3^{-4} \cdot 6^8)^{-1/4}$

84. $(-2^{-9} \cdot 3^6)^{-1/3}$

Solve each problem.

85. ***Yacht dimensions.*** Since 1988, a yacht competing for the America's Cup must satisfy the inequality

$$L + 1.25S^{1/2} - 9.8D^{1/3} \leq 16.296,$$

where L is the boat's length in meters, S is the sail area in square meters, and D is the displacement in cubic meters (www.americascupnews.com). Does a boat with a displacement of 21.8 m^3, a sail area of 305.4 m^2, and a length of 21.5 m satisfy the inequality? If the length and displacement are not changed, then what is the maximum number of square meters of sail that could be added and still have the boat satisfy the inequality? See the figure on the next page.

Photo for Exercise 85

86. ***Surface area.*** If A is the surface area of a cube and V is its volume, then

$$A = 6V^{2/3}.$$

Find the surface area of a cube that has a volume of 27 cubic centimeters. What is the surface area of a cube whose sides measure 5 centimeters each?

87. ***Average annual return.*** The average annual return on an investment, r, is given by the formula

$$r = \left(\frac{S}{P}\right)^{1/n} - 1$$

where P is the original investment and S is the value of the investment after n years. An investment of \$10,000 in 1994 in T. Rowe Price's Equity Income Fund amounted to \$30,468 in 2004 (www.troweprice.com). What was the average annual return to the nearest tenth of a percent?

88. ***Population growth.*** The U.S. population grew from 248.7 million in 1990 to 294.4 million in 2004 (U.S. Census Bureau, www.census.gov). Use the formula from Exercise 87 to find the average annual rate of growth to the nearest tenth of a percent.

89. ***Sail area-displacement ratio.*** The sail area-displacement ratio r for a boat with sail area A (in square feet) and displacement d (in pounds) is given by

$$r = A(d/64)^{-2/3}.$$

The Tartan 4100 has a sail area of 810 ft^2 and a displacement of 23,245 pounds.

a) Find r for this boat.

b) Use the accompanying graph to determine whether the ratio is increasing or decreasing as the displacement increases, with sail area fixed at 810 ft^2.

Figure for Exercise 89

c) Estimate the displacement for $r = 14$ using the accompanying graph.

90. ***Piano tuning.*** The note middle C on a piano is tuned so that the string vibrates at 262 cycles per second, or 262 Hz (Hertz). The C note that is one octave higher is tuned to 524 Hz. Tuning for the 11 notes in between using the method of *equal temperament* is $262 \cdot 2^{n/12}$, where n takes the values 1 through 11. Find the tuning rounded to the nearest whole Hertz for those 11 notes.

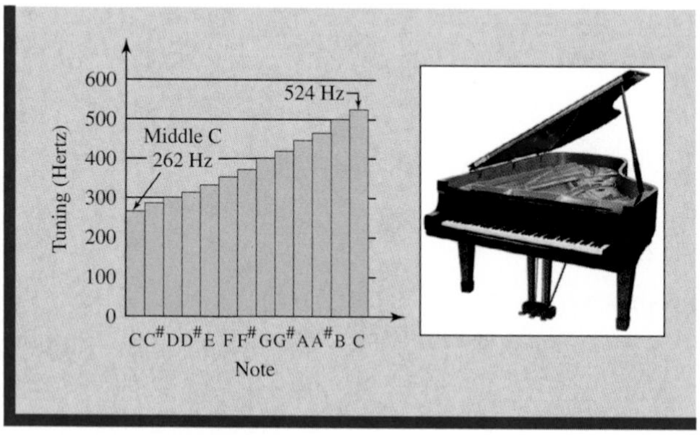

Figure for Exercise 90

Getting More Involved

91. ***Discussion***

If $a^{-m/n} < 0$, then what can you conclude about the values of a, m, and n?

92. ***Discussion***

If $a^{-m/n}$ is not a real number, then what can you conclude about the values of a, m, and n?

93. ***Exploration***

Arrange the following expressions in order from smallest to largest. Use a calculator if you need it.

$$3^0, 3^{-8/9}, 3^{8/9}, 3^{-2/3}, 3^{-1/2}, 3^{2/3}, 3^{1/2}$$

Now arrange the following expressions in order from smallest to largest. Do not use a calculator.

$$7^{-3/4}, 7^0, 7^{1/3}, 7^{-1/3}, 7^{3/4}, 7^{9/10}, 7^{-9/10}$$

Use a calculator to check both arrangements. If $5^m < 5^n$, what can you say about m and n?

Collaborative Activities

Grouping: Four students per group

Topic: Square roots, Pythagorean theorem

Ray's Rafters

Robyn and Ray have started a small business making roof trusses. Each truss has a width w and a height h as shown in the accompanying diagram. The rafter length r includes the part above the building as well as the *lookout,* which is two feet long and hangs over the edge of the building to create the eaves.

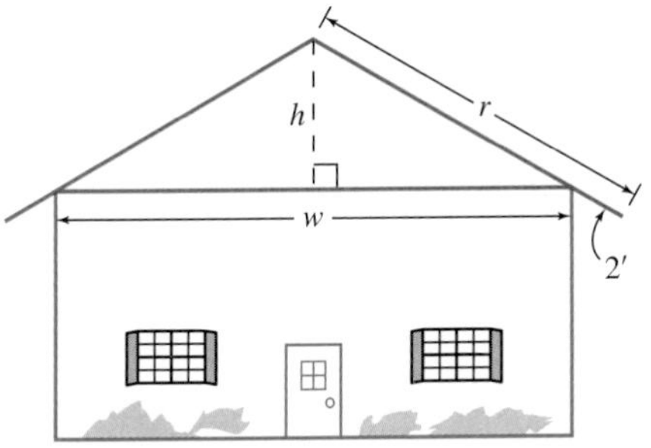

The customer specifies the width and pitch of the roof. Robyn and Ray must determine h and r. For example, a customer orders trusses with a width of 36 feet and a 3/12 pitch. To find the height, Robyn sets up the proportion

$$\frac{3}{12} = \frac{h}{18}$$

and solves it to get $h = \frac{9}{2}$. Why did Robyn use 18 in the proportion? Ray then uses the Pythagorean theorem to find the hypotenuse of the triangle:

$$\sqrt{\left(\frac{9}{2}\right)^2 + 18^2} = \sqrt{\frac{1377}{4}} = \frac{9\sqrt{17}}{2}$$

He adds 2 feet for the lookout:

$$\frac{9\sqrt{17}}{2} + \frac{4}{2} = \frac{9\sqrt{17} + 4}{2}$$

Ray then calculates that $r \approx 20.6$ feet.

To save time Robyn and Ray want to make some tables that they could use to find h and r for various widths and pitches. Your job is to fill in the tables for Robyn and Ray. In each case find r as a simplified radical and find a decimal approximation to the nearest tenth of a foot.

Table for 3/12 pitch

w	**h**	***r*** **(radical)**	***r*** **(decimal)**
36			
48			
60			

Table for 4/12 pitch

w	**h**	***r*** **(radical)**	***r*** **(decimal)**
24			
36			
60			

Chapter 8 Wrap-Up

Summary

Powers and Roots		**Examples**
nth roots	If $a = b^n$ for a positive integer n, then b is an nth root of a.	2 and -2 are fourth roots of 16.
Principal root	The positive even root of a positive number	The principal fourth root of 16 is 2.
Radical notation	If n is a positive even integer and a is positive, then the symbol $\sqrt[n]{a}$ denotes the principal nth root of a. If n is a positive odd integer, then the symbol $\sqrt[n]{a}$ denotes the nth root of a. If n is any positive integer, then $\sqrt[n]{0} = 0$.	$\sqrt[4]{16} = 2$ $\sqrt[4]{16} \neq -2$ $\sqrt[3]{-8} = -2$, $\sqrt[3]{8} = 2$ $\sqrt[5]{0} = 0$, $\sqrt[6]{0} = 0$
Definition of $a^{1/n}$	If n is any positive integer, then $a^{1/n} = \sqrt[n]{a}$ provided that $\sqrt[n]{a}$ is a real number.	$8^{1/3} = \sqrt[3]{8} = 2$ $(-4)^{1/2}$ is not real.
Definition of $a^{m/n}$	If m and n are positive integers, then $a^{m/n} = (a^{1/n})^m$, provided that $a^{1/n}$ is a real number.	$8^{2/3} = (8^{1/3})^2 = 2^2 = 4$ $(-16)^{3/4}$ is not real.
Definition of $a^{-m/n}$	If m and n are positive integers and $a \neq 0$, then $a^{-m/n} = \frac{1}{a^{m/n}}$, provided that $a^{1/n}$ is a real number.	$8^{-2/3} = \dfrac{1}{8^{2/3}} = \dfrac{1}{4}$

Rules of Exponents | **Examples**

The following rules hold for any rational numbers m and n and nonzero real numbers a and b, provided that all expressions represent real numbers.

Zero exponent	$a^0 = 1$	$8^0 = 1$, $(3x - y)^0 = 1$
Product rule	$a^m a^n = a^{m+n}$	$3^3 \cdot 3^4 = 3^7$, $x^5 x^{-2} = x^3$
Quotient rule	$\dfrac{a^m}{a^n} = a^{m-n}$	$\dfrac{3^5}{3^7} = 3^{-2}$, $\dfrac{x}{x^{1/4}} = x^{3/4}$
Power rule	$(a^m)^n = a^{mn}$	$(2^2)^3 = 2^6$ $(w^{3/4})^4 = w^3$
Power of a product rule	$(ab)^n = a^n b^n$	$(2t)^{1/2} = 2^{1/2} t^{1/2}$
Power of a quotient rule	$\left(\dfrac{a}{b}\right)^n = \dfrac{a^n}{b^n}$	$\left(\dfrac{x}{3}\right)^{-3} = \dfrac{x^{-3}}{3^{-3}}$

Rules for Radicals		**Examples**
The following rules hold, provided that all roots are real numbers and n is a positive integer.		
Product rule for radicals	$\sqrt[n]{ab} = \sqrt[n]{a} \cdot \sqrt[n]{b}$	$\sqrt{2} \cdot \sqrt{3} = \sqrt{6}$ $\sqrt{9y} = 3\sqrt{y}$
Quotient rule for radicals	$\sqrt[n]{\frac{a}{b}} = \frac{\sqrt[n]{a}}{\sqrt[n]{b}}$	$\sqrt{\frac{5}{4}} = \frac{\sqrt{5}}{2}$, $\frac{\sqrt{6}}{\sqrt{2}} = \sqrt{\frac{6}{2}} = \sqrt{3}$
Simplified form for square roots	A square root expression is in simplified form if it has 1. *no* perfect square factors inside the radical,	$\sqrt{12} = \sqrt{4 \cdot 3} = 2\sqrt{3}$
	2. *no* fractions inside the radical, and	$\sqrt{\frac{5}{2}} = \frac{\sqrt{5}}{\sqrt{2}}$
	3. *no* radicals in the denominator.	$\frac{\sqrt{5}}{\sqrt{2}} = \frac{\sqrt{5} \cdot \sqrt{2}}{\sqrt{2} \cdot \sqrt{2}} = \frac{\sqrt{10}}{2}$

Solving Equations Involving Squares and Square Roots		**Examples**
Square root property (solving $x^2 = k$)	If $k > 0$, the equation $x^2 = k$ is equivalent to $x = \sqrt{k}$ or $x = -\sqrt{k}$ (also written $x = \pm\sqrt{k}$). If $k = 0$, the equation $x^2 = k$ is equivalent to $x = 0$.	$x^2 = 6$ $x = \pm\sqrt{6}$ $t^2 = 0$ $t = 0$
	If $k < 0$, the equation $x^2 = k$ has no real solution.	$x^2 = -8$, no solution
Squaring property of equality	Squaring each side of an equation may introduce extraneous solutions. We must check all of our answers.	$\sqrt{x} = -3$ $(\sqrt{x})^2 = (-3)^2$ $x = 9$ Extraneous solution

Enriching Your Mathematical Word Power

For each mathematical term, choose the correct meaning.

1. *n*th root of *a*
a. a square root
b. the root of a^n
c. a number b such that $a^n = b$
d. a number b such that $b^n = a$

2. square of *a*
a. a number b such that $b^2 = a$
b. a^2
c. $|a|$
d. $\sqrt{a}$

3. cube root of *a*
a. a^3
b. a number b such that $b^3 = a$
c. $a/3$
d. a number b such that $b = a^3$

4. principal root
a. the main root
b. the positive even root of a positive number
c. the positive odd root of a negative number
d. the negative odd root of a negative number

5. odd root of *a*
a. the number b such that $b^n = a$ where a is an odd number
b. the opposite of the even root of a
c. the nth root of a
d. the number b such that $b^n = a$ where n is an odd number

6. index of a radical
a. the number n in $n\sqrt{a}$
b. the number n in $\sqrt[n]{a}$
c. the number n in a^n
d. the number n in $\sqrt{a^n}$

7. like radicals
a. radicals with the same index
b. radicals with the same radicand
c. radicals with the same radicand and the same index
d. radicals with even indices

8. rational exponent
a. an exponent that produces a rational number
b. an integral exponent
c. an exponent that is a real number
d. an exponent that is a rational number

Review Exercises

8.1 *Find each root. All variables represent positive real numbers.*

1. $\sqrt[5]{32}$ **2.** $\sqrt[3]{-27}$
3. $\sqrt[3]{1000}$ **4.** $\sqrt{100}$
5. $\sqrt{x^{12}}$ **6.** $\sqrt{a^{10}}$
7. $\sqrt[3]{x^6}$ **8.** $\sqrt[3]{a^9}$
9. $\sqrt{4x^2}$ **10.** $\sqrt{9y^4}$
11. $\sqrt[3]{125x^6}$ **12.** $\sqrt[3]{8y^{12}}$
13. $\sqrt{\dfrac{4x^{16}}{y^{14}}}$ **14.** $\sqrt{\dfrac{9y^8}{t^{10}}}$
15. $\sqrt{\dfrac{w^2}{16}}$ **16.** $\sqrt{\dfrac{a^4}{25}}$

8.2 *Write each expression in simplified form. All variables represent positive real numbers.*

17. $\sqrt{72}$ **18.** $\sqrt{48}$
19. $\dfrac{1}{\sqrt{3}}$ **20.** $\dfrac{2}{\sqrt{5}}$
21. $\sqrt{\dfrac{3}{5}}$ **22.** $\sqrt{\dfrac{5}{6}}$
23. $\dfrac{\sqrt{33}}{\sqrt{3}}$ **24.** $\dfrac{\sqrt{50}}{\sqrt{5}}$
25. $\dfrac{\sqrt{3}}{\sqrt{8}}$ **26.** $\dfrac{\sqrt{2}}{\sqrt{18}}$
27. $\sqrt{y^6}$ **28.** $\sqrt{z^{10}}$
29. $\sqrt{24t^9}$ **30.** $\sqrt{8p^7}$
31. $\sqrt{12m^5t^3}$ **32.** $\sqrt{18p^3q^7}$
33. $\dfrac{\sqrt{2}}{\sqrt{x}}$ **34.** $\dfrac{\sqrt{5}}{\sqrt{y}}$
35. $\sqrt{\dfrac{3a^5}{2s}}$ **36.** $\sqrt{\dfrac{5x^7}{3w}}$

8.3 *Perform each computation and simplify.*

37. $2\sqrt{7} + 8\sqrt{7}$ **38.** $3\sqrt{6} - 5\sqrt{6}$
39. $\sqrt{12} - \sqrt{27}$ **40.** $\sqrt{18} + \sqrt{50}$
41. $2\sqrt{3} \cdot 5\sqrt{3}$ **42.** $-3\sqrt{6} \cdot 2\sqrt{6}$
43. $-3\sqrt{6} \cdot 5\sqrt{3}$ **44.** $4\sqrt{12} \cdot 6\sqrt{8}$
45. $-3\sqrt{3}(5 + \sqrt{3})$
46. $4\sqrt{2}(6 + \sqrt{8})$
47. $-\sqrt{3}(\sqrt{6} - \sqrt{15})$
48. $-\sqrt{2}(\sqrt{6} - \sqrt{2})$
49. $(\sqrt{3} - 5)(\sqrt{3} + 5)$
50. $(\sqrt{2} + \sqrt{7})(\sqrt{2} - \sqrt{7})$
51. $(2\sqrt{5} - \sqrt{6})^2$
52. $(3\sqrt{2} + \sqrt{6})^2$
53. $(4 - 3\sqrt{6})(5 - \sqrt{6})$
54. $(\sqrt{3} - 2\sqrt{5})(\sqrt{3} + 4\sqrt{5})$
55. $3\sqrt{5} \div (6\sqrt{2})$ **56.** $6\sqrt{5} \div (4\sqrt{3})$
57. $\dfrac{2 - \sqrt{20}}{10}$ **58.** $\dfrac{6 - \sqrt{12}}{-2}$
59. $\dfrac{3}{1 - \sqrt{5}}$
60. $\dfrac{\sqrt{2}}{\sqrt{6} + \sqrt{3}}$

8.4 *Solve each equation.*

61. $x^2 = 400$ **62.** $x^2 = 121$
63. $7x^2 = 3$
64. $3x^2 - 7 = 0$

65. $(x - 4)^2 - 18 = 0$

66. $2(x + 1)^2 - 40 = 0$

67. $\sqrt{x} = 9$

68. $\sqrt{x} - 20 = 0$

69. $x = \sqrt{36 - 5x}$

70. $x = \sqrt{2 - x}$

71. $x + 2 = \sqrt{52 + 2x}$

72. $x - 4 = \sqrt{x - 4}$

Solve each formula for t.

73. $t^2 - 8sw = 0$

74. $(t + b)^2 = b^2 - 4ac$

75. $3a = \sqrt{bt}$

76. $a - \sqrt{t} = w$

8.5 *Simplify each expression. Answers with exponents should have positive exponents only.*

77. $25^{-3/2}$

78. $9^{-5/2}$

79. $25^{1/2}$

80. $9^{3/2}$

81. $64^{-1/2}$

82. $125^{-2/3}$

83. $-25^{-3/2}$

84. $-9^{-3/2}$

85. $(-25)^{-3/2}$

86. $(-9)^{-3/4}$

87. $(-8)^{-1/3}$

88. $(-27)^{-4/3}$

89. $x^{-3/5}x^{-2/5}$

90. $t^{-1/3}t^{1/2}$

91. $(-8x^{-6})^{-1/3}$

92. $(-27x^{-9})^{-2/3}$

93. $w^{-3/2} \div w^{-7/2}$

94. $m^{1/3} \div m^{-1/4}$

95. $\left(\dfrac{9t^{-6}}{s^{-4}}\right)^{-1/2}$

96. $\left(\dfrac{8y^{-3}}{x^6}\right)^{-2/3}$

97. $\left(\dfrac{8x^{-12}}{y^{30}}\right)^{2/3}$

98. $\left(\dfrac{16y^{-3/4}}{t^{1/2}}\right)^{-2}$

Solve each problem.

99. ***Depreciation of a Lumina.*** If the cost of a piece of equipment was C dollars and it is sold for S dollars after n years, then the annual depreciation rate is given by

$$r = 1 - \left(\frac{S}{C}\right)^{1/n}.$$

A 1994 Chevrolet Lumina that sold new for \$15,446 sells for \$2336 in 2004 (www.edmunds.com).

a) Find the depreciation rate for this car to the nearest tenth of a percent.

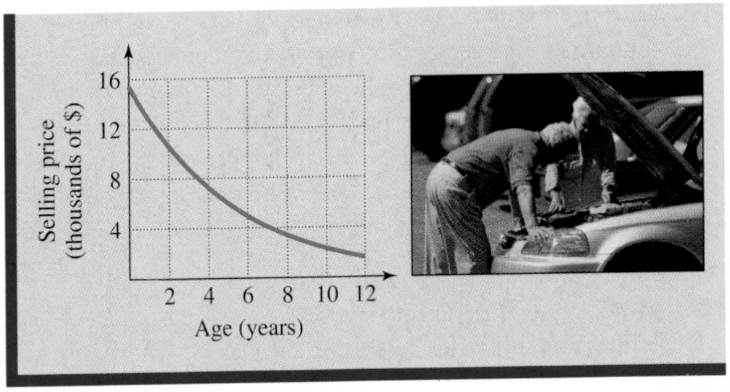

Figure for Exercise 99

b) Use the accompanying graph to determine whether this car depreciated more during the first two years or the last two years shown.

100. ***Depreciation of a Thunderbird.*** A 1996 Ford Thunderbird that sold new for \$16,892 sells for \$3551 in 2004 (www.edmunds.com). Use the formula from the previous exercise to find the depreciation rate to the nearest tenth of a percent.

101. ***Radius of a drop.*** The amount of water in a large raindrop is 0.25 cm^3. Use the formula

$$r = \left(\frac{3V}{4\pi}\right)^{1/3}$$

to find the radius of the spherical drop to the nearest tenth of a centimeter.

102. ***Radius of a circle.*** Solve the formula $A = \pi r^2$ for r.

103. ***Waffle cones.*** A large waffle cone has a height of 6 in. and a radius of 2 in. as shown in the accompanying figure. Find the exact amount of waffle in a cone this size. The formula $A = \pi r\sqrt{r^2 + h^2}$ gives the lateral surface area of a right circular cone with radius r and height h. Be sure to simplify the radical. Use a calculator to find the answer to the nearest square inch.

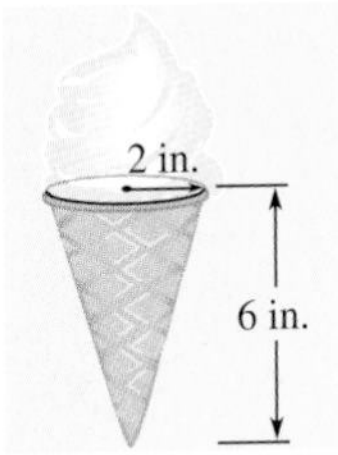

Figure for Exercise 103

104. ***Salting the roads.*** A city manager wants to find the amount of canvas required to cover a conical salt pile that is stored for the winter. The height of the pile is 10 yards, and the diameter of the base is 24 yards. Use the formula in Exercise 103 to find the exact number of square yards of canvas needed. Simplify the radical. Use a calculator to find the answer to the nearest square yard.

105. ***Wide screen TV.*** The screen on a new Panasonic widescreen flat panel television measures 10.8 in. by 19.2 in. Find the diagonal measure of the screen to the nearest inch.

106. ***Wider screen TV.*** The diagonal measure of the screen on a flat panel Samsung television is 40 in. The aspect ratio (ratio of the length to the width) of the screen is 16 to 9. Find the length and width to the nearest tenth of an inch.

Chapter 8 Test

Simplify each expression.

1. $\sqrt{36}$

2. $\sqrt{144}$

3. $\sqrt[3]{-27}$

4. $\sqrt[5]{32}$

5. $16^{1/4}$

6. $\sqrt{24}$

7. $\sqrt{\frac{3}{8}}$

8. $(-4)^{3/2}$

9. $27^{4/3}$

10. $(-27)^{-1/3}$

11. $-8^{4/3}$

12. $\sqrt{8} + \sqrt{2}$

13. $(2 + \sqrt{3})^2$

14. $(3\sqrt{2} - \sqrt{7})(3\sqrt{2} + \sqrt{7})$

15. $\sqrt{21} \div \sqrt{3}$

16. $\sqrt{20} \div \sqrt{3}$

17. $\frac{2 + \sqrt{8}}{2}$

18. $\sqrt{3}(\sqrt{6} - \sqrt{3})$

Simplify. Assume that all variables represent positive real numbers, and write answers with positive exponents only.

19. $y^{1/2} \cdot y^{1/4}$

20. $\frac{6x}{2x^{1/3}}$

21. $(x^3y^9)^{1/3}$

22. $\left(\frac{125w^3}{u^{-12}}\right)^{-1/3}$

23. $\sqrt{\frac{3}{t}}$

24. $\sqrt{4y^6}$

25. $\sqrt[3]{8y^{12}}$

26. $\sqrt{18t^7}$

Solve each equation.

27. $(x + 3)^2 = 36$

28. $\sqrt{x + 7} = 5$

29. $5x^2 = 2$

30. $(3x - 4)^2 = 0$

31. $x - 3 = \sqrt{5x + 9}$

Solve the equation for the specified variable.

32. $S = \pi r^2 h$ for r

33. $a^2 + b^2 = c^2$ for b

Show a complete solution to each problem.

34. Find the exact length of the side of a square whose diagonal is 5 meters.

35. To utilize a center-pivot irrigation system, a farmer planted his crop in a circular field of 100,000 square meters. Find the radius of the circular field to the nearest tenth of a meter.

Making Connections | A Review of Chapters 1–8

Find the solution set to each equation or inequality. For the inequalities, also sketch the graph of the inequality and state the solution set using interval notation.

1. $2x + 3 = 0$

2. $2x = 3$

3. $2x + 3 > 0$

4. $-2x + 3 > 0$

5. $2(x + 3) = 0$

6. $2x^2 = 3$

7. $\dfrac{x}{3} = \dfrac{2}{x}$

8. $\dfrac{x - 1}{x} = \dfrac{x}{x - 2}$

9. $(2x + 3)^2 = 0$

10. $(2x + 3)(x - 3) = 0$

11. $2x^2 + 3 = 0$

12. $(2x + 3)^2 = 1$

13. $(2x + 3)^2 = -1$

14. $\sqrt{2x^2 - 14} = x - 1$

Let $a = 2$, $b = -3$, and $c = -9$. Find the value of each algebraic expression.

15. b^2

16. $-4ac$

17. $b^2 - 4ac$

18. $\sqrt{b^2 - 4ac}$

19. $-b + \sqrt{b^2 - 4ac}$

20. $-b - \sqrt{b^2 - 4ac}$

21. $\dfrac{-b + \sqrt{b^2 - 4ac}}{2a}$

22. $\dfrac{-b - \sqrt{b^2 - 4ac}}{2a}$

Factor each trinomial completely.

23. $x^2 - 6x + 9$

24. $x^2 + 10x + 25$

25. $x^2 + 12x + 36$

26. $x^2 - 20x + 100$

27. $2x^2 - 8x + 8$

28. $3x^2 + 6x + 3$

Perform the indicated operation.

29. $(3 + 2x) - (6 - 5x)$

30. $(5 + 3t)(4 - 5t)$

31. $(8 - 6j)(3 + 4j)$

32. $(1 - u) + (5 + 7u)$

33. $(3 - 4v) - (2 - 5v)$

34. $(2 + t)^2$

35. $(t - 7)(t + 7)$

36. $(3 - 2n)(3 + 2n)$

37. $(1 - m)^2$

38. $(-4 - 6t) - (-3 - 8t)$

39. $(1 + r)(3 - 4r)$

40. $(2 - 6y)(1 + 3y)$

41. $(1 - 2j) + (-6 + 5j)$

42. $(-2 - j) + (4 - 5j)$

43. $\dfrac{4 - 6x}{2}$

44. $\dfrac{-3 - 9p}{3}$

45. $\dfrac{8 - 12q}{-4}$

46. $\dfrac{20 - 5z}{-5}$

Solve the problem.

47. ***Oxygen uptake.*** In studying the oxygen uptake rate for marathon runners, Costill and Fox calculate the power expended P in kilocalories per minute using the formula $P = M(av - b)$, where M is the mass of the runner in kilograms and v is the speed in meters per minute (*Medicine and Science in Sports,* Vol. 1). The constants a and b have values $a = 1.02 \times 10^{-3}$ and $b = 2.62 \times 10^{-2}$.

a) Find P for a 60-kg runner who is running at 300 m/min.

b) Find the velocity of a 55-kg runner who is expending 14 kcal/min.

c) Judging from the accompanying graph of velocity and power expenditure for a 55-kg runner, is power expenditure increasing or decreasing as the velocity increases?

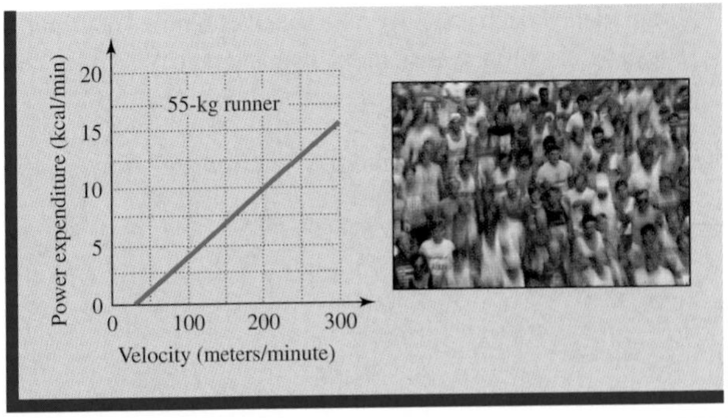

Figure for Exercise 47

Critical Thinking | For Individual or Group Work | Chapter 8

These exercises can be solved by a variety of techniques, which may or may not require algebra. So be creative and think critically. Explain all answers. Answers are in the Instructor's Edition of this text.

1. ***Summing angles.*** Find the sum of the measures of the angles at the points of the irregular five-pointed star shown in the accompanying figure. That is, find

$$m\angle A + m\angle B + m\angle C + m\angle D + m\angle E.$$

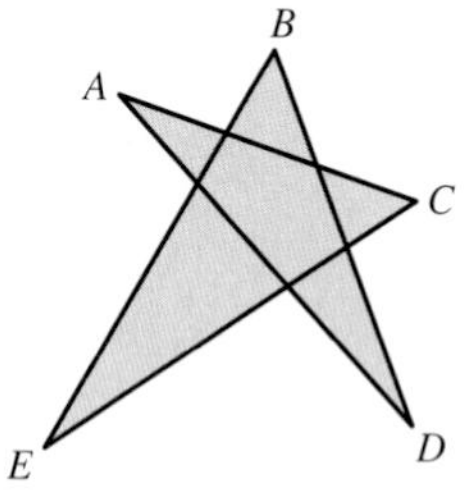

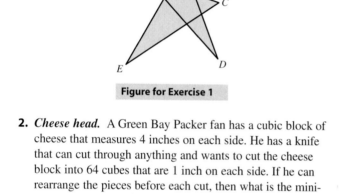

Figure for Exercise 1

2. ***Cheese head.*** A Green Bay Packer fan has a cubic block of cheese that measures 4 inches on each side. He has a knife that can cut through anything and wants to cut the cheese block into 64 cubes that are 1 inch on each side. If he can rearrange the pieces before each cut, then what is the minimum number of cuts that will accomplish this task?

3. ***Wage earners.*** Alice and Beth together have the same hourly wage as Carl. Carl and Don together have the same hourly wage as Eustis. Eustis and Alice together have the same hourly wage as Frank. If Beth's, Don's, and Frank's hourly wages total \$100 and Alice makes \$8 per hour, then what is Frank's hourly wage?

4. ***Factoring fever.*** Express $2^{24} - 1$ as a product of prime numbers without using a calculator.

5. ***Average joggers.*** Two friends jog from their apartment down to the beach at an average speed of 6 miles per hour. They jog back to the apartment at an average of 4 miles per hour. What was the average speed for the entire trip?

6. ***Three pairs.*** Find three integral solutions to

$$180x - y^3 = 0.$$

7. ***Perfect squares.*** For what values of n will the value of

$$\frac{n}{20 - n}$$

be a perfect square?

8. ***Waiting for water.*** A hot water pipe is 1/2 in. in diameter and the shower head is located 70 feet from the hot water tank. If the water runs at 3 gallons per minute, then how long (to the nearest second) does it take for the hot water in the tank to reach the shower head?

Photo for Exercise 8

Chapter 9

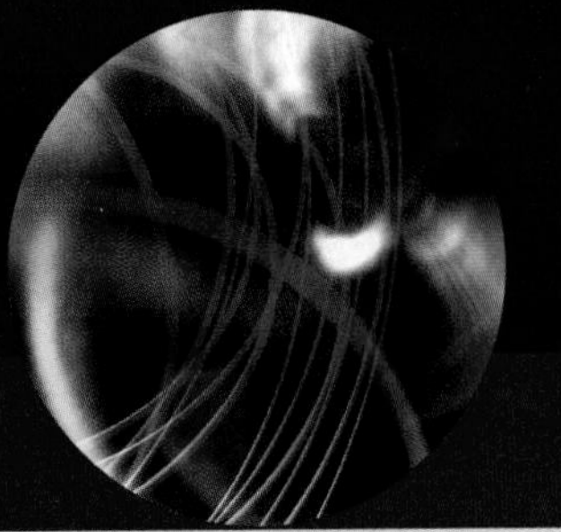

Quadratic Equations, Parabolas, and Functions

Throughout time, humans have been building bridges over waterways. Primitive people threw logs across streams or attached ropes to branches to cross the waters. Later, the Romans built stone structures to span rivers and chasms. Throughout the centuries, bridges have been made of wood and stone and later from cast iron, concrete, and steel. Today's bridges are among the most beautiful and complex creations of modern engineering. Whether the bridge spans a small creek or a four-mile-wide stretch of water, mathematics is a part of its very foundation.

The function of a bridge, the length it must span, and the load it must carry often determine the type of bridge that is built. Some common types designed by civil engineers are cantilevered, arch, cable-stayed, and suspension bridges. The military is known for building trestle bridges and floating or pontoon bridges.

New technology has enabled engineers to build bridges that are stronger, lighter, and less expensive than in the past, as well as being esthetically pleasing. Currently, some engineers are working on making bridges earthquake resistant. Another idea that is being explored is incorporating carbon fibers in cement to warn of small cracks through electronic signals.

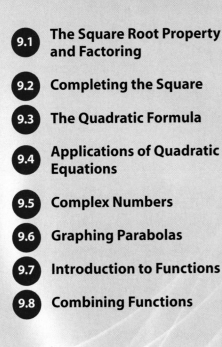

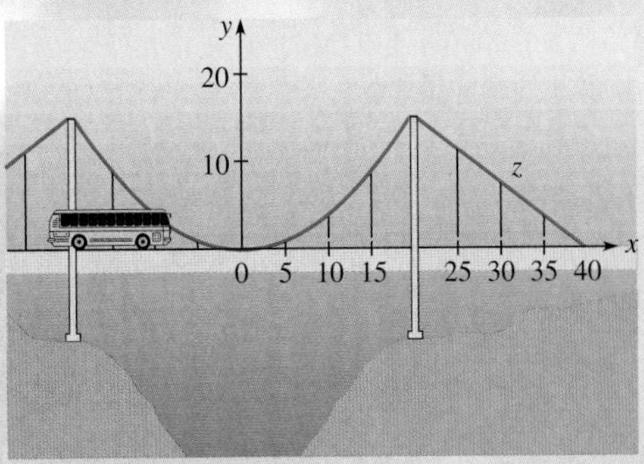

In Exercise 45 of Section 9.6 you will see how quadratic equations are used in designing suspension bridges.

9.1 The Square Root Property and Factoring

In this Section

- Definition
- Using the Square Root Property
- Solving Equations by Factoring

We solved some quadratic equations in Chapters 6 and 7 by factoring. In Chapter 8 we solved some quadratic equations using the square root property. In this section we will review the types that you have already learned to solve. In Section 9.2 you will learn a method by which you can solve any quadratic equation.

Definition

We saw the definition of a quadratic equation in Chapter 6, but we will repeat it here.

Quadratic Equation

A **quadratic equation** is an equation of the form

$$ax^2 + bx + c = 0,$$

where a, b, and c are real numbers with $a \neq 0$.

Equations that can be written in the form of the definition may also be called quadratic equations. In Chapters 6, 7, and 8 we solved quadratic equations such as

$$x^2 = 10, \qquad 5(x - 2)^2 = 20, \qquad \text{and} \qquad x^2 - 5x = -6.$$

Using the Square Root Property

If $b = 0$ in $ax^2 + bx + c = 0$, then the quadratic equation can be solved by using the square root property.

EXAMPLE 1

Using the square root property

Solve the equations.

a) $x^2 - 9 = 0$

b) $2x^2 - 3 = 0$

c) $-3(x + 1)^2 = -6$

Solution

a) Solve the equation for x^2, and then use the square root property:

$$x^2 - 9 = 0$$

$$x^2 = 9 \quad \text{Add 9 to each side.}$$

$$x = \pm 3 \quad \text{Square root property}$$

Check 3 and -3 in the original equation. Both 3 and -3 are solutions to $x^2 - 9 = 0$.

Study Tip

Effective studying involves actively digging into the subject. Make sure that you are making steady progress. At the end of each week, take note of the progress that you have made. What do you know on Friday that you didn't know on Monday?

b) $2x^2 - 3 = 0$

$$2x^2 = 3$$

$$x^2 = \frac{3}{2}$$

$$x = \pm\sqrt{\frac{3}{2}} \quad \text{Square root property}$$

$$x = \pm\frac{\sqrt{3}\cdot\sqrt{2}}{\sqrt{2}\cdot\sqrt{2}} \quad \text{Rationalize the denominator.}$$

$$x = \pm\frac{\sqrt{6}}{2}$$

Check. The solutions to $2x^2 - 3 = 0$ are $\frac{\sqrt{6}}{2}$ and $-\frac{\sqrt{6}}{2}$.

c) This equation is a bit different from the previous two equations. If we actually squared the quantity $(x + 1)$, then we would get a term involving x. Then b would not be equal to zero as it is in the other equations. However, we can solve this equation like the others if we do not square $x + 1$.

$$-3(x + 1)^2 = -6$$

$$(x + 1)^2 = 2 \quad \text{Divide each side by } -3.$$

$$x + 1 = \pm\sqrt{2} \quad \text{Square root property}$$

$$x = -1 \pm \sqrt{2} \quad \text{Subtract 1 from each side.}$$

Check $x = -1 \pm \sqrt{2}$ in the original equation:

$$-3(-1 \pm \sqrt{2} + 1)^2 = -3(\pm\sqrt{2})^2 = -3(2) = -6$$

The solutions are $-1 + \sqrt{2}$ and $-1 - \sqrt{2}$.

Now do Exercises 7–12

EXAMPLE 2

A quadratic equation with no real solution

Solve $x^2 + 12 = 0$.

Solution

The equation $x^2 + 12 = 0$ is equivalent to $x^2 = -12$. Because the square of any real number is nonnegative, this equation has no real solution.

Now do Exercises 13–32

Solving Equations by Factoring

In Chapter 6 you learned to factor trinomials and to use factoring to solve some quadratic equations. Recall that quadratic equations are solved by factoring as follows.

Strategy for Solving Quadratic Equations by Factoring

1. Write the equation with 0 on one side of the equal sign.
2. Factor the other side.
3. Use the zero factor property. (Set each factor equal to 0.)
4. Solve the two linear equations.
5. Check the answers in the original quadratic equation.

EXAMPLE 3

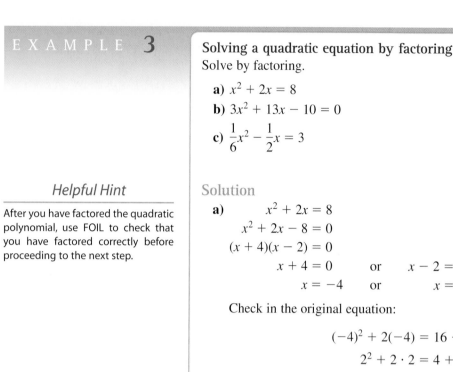

Solving a quadratic equation by factoring

Solve by factoring.

a) $x^2 + 2x = 8$

b) $3x^2 + 13x - 10 = 0$

c) $\frac{1}{6}x^2 - \frac{1}{2}x = 3$

Helpful Hint

After you have factored the quadratic polynomial, use FOIL to check that you have factored correctly before proceeding to the next step.

Solution

a)

$$\begin{aligned} x^2 + 2x &= 8 \\ x^2 + 2x - 8 &= 0 && \text{Get 0 on the right-hand side.} \\ (x + 4)(x - 2) &= 0 && \text{Factor.} \\ x + 4 = 0 \quad &\text{or} \quad x - 2 = 0 && \text{Zero factor property} \\ x = -4 \quad &\text{or} \quad x = 2 && \text{Solve the linear equations.} \end{aligned}$$

Check in the original equation:

$$(-4)^2 + 2(-4) = 16 - 8 = 8$$
$$2^2 + 2 \cdot 2 = 4 + 4 = 8$$

Both -4 and 2 are solutions to the equation.

b)

$$\begin{aligned} 3x^2 + 13x - 10 &= 0 \\ (3x - 2)(x + 5) &= 0 && \text{Factor.} \\ 3x - 2 = 0 \quad &\text{or} \quad x + 5 = 0 && \text{Zero factor property} \\ 3x = 2 \quad &\text{or} \quad x = -5 \\ x = \frac{2}{3} \quad &\text{or} \quad x = -5 \end{aligned}$$

Check in the original equation. Both -5 and $\frac{2}{3}$ are solutions to the equation.

c)

$$\begin{aligned} \frac{1}{6}x^2 - \frac{1}{2}x &= 3 \\ x^2 - 3x &= 18 && \text{Multiply each side by 6.} \\ x^2 - 3x - 18 &= 0 && \text{Get 0 on the right-hand side.} \\ (x - 6)(x + 3) &= 0 && \text{Factor.} \\ x - 6 = 0 \quad &\text{or} \quad x + 3 = 0 && \text{Zero factor property} \\ x = 6 \quad &\text{or} \quad x = -3 \end{aligned}$$

Check in the original equation. The solutions are -3 and 6.

Now do Exercises 33–54

CAUTION You can set each factor equal to zero only when the product of the factors is zero. Note that $x^2 - 3x = 18$ is equivalent to $x(x - 3) = 18$, but you can make no conclusion about two factors that have a product of 18.

Warm-Ups ▼

True or false? Explain your answer.

1. Both −4 and 4 satisfy the equation $x^2 - 16 = 0$.
2. The equation $(x - 3)^2 = 8$ is equivalent to $x - 3 = 2\sqrt{2}$.
3. Every quadratic equation can be solved by factoring.
4. Both −5 and 4 are solutions to $(x - 4)(x + 5) = 0$.
5. The quadratic equation $x^2 = -3$ has no real solutions.
6. The equation $x^2 = 0$ has no real solutions.
7. The equation $(2x + 3)(4x - 5) = 0$ is equivalent to $x = \frac{3}{2}$ or $x = \frac{5}{4}$.
8. The only solution to the equation $(x + 2)^2 = 0$ is −2.
9. $(x - 3)(x - 5) = 4$ is equivalent to $x - 3 = 2$ or $x - 5 = 2$.
10. All quadratic equations have two distinct solutions.

9.1 Exercises

Reading and Writing *After reading this section, write out the answers to these questions. Use complete sentences.*

1. What is a quadratic equation?
2. What property do we use to solve quadratic equations in which $b = 0$?
3. How can a quadratic equation in which $b = 0$ fail to have a real solution?
4. What method is discussed for solving quadratic equations in which $b \neq 0$?
5. When do you need to solve linear equations to find the solutions to a quadratic equation?
6. What new material is presented in this section?

Solve each equation. See Examples 1 and 2.

7. $x^2 = 64$

8. $x^2 = 49$

9. $x^2 = \frac{9}{4}$

10. $x^2 = \frac{25}{81}$

11. $x^2 - 36 = 0$

12. $x^2 - 81 = 0$

13. $x^2 + 10 = 0$

14. $x^2 + 4 = 0$

15. $5x^2 = 50$

16. $7x^2 = 14$

17. $3t^2 - 5 = 0$

18. $5y^2 - 7 = 0$

19. $-3y^2 + 8 = 0$

20. $-5w^2 + 12 = 0$

21. $(x - 3)^2 = 4$

22. $(x + 5)^2 = 9$

23. $(y - 2)^2 = 18$

24. $(m - 5)^2 = 20$

25. $2(x + 1)^2 = \frac{1}{2}$

26. $-3(x - 1)^2 = -\frac{3}{4}$

27. $(x - 1)^2 = \frac{1}{2}$

28. $(y + 2)^2 = \frac{1}{2}$

29. $\left(x + \frac{1}{2}\right)^2 = \frac{1}{2}$

30. $\left(x - \frac{1}{2}\right)^2 = \frac{3}{2}$

31. $(x - 11)^2 = 0$

32. $(x + 45)^2 = 0$

Solve each equation by factoring. See Example 3.

33. $x^2 + 3x + 2 = 0$

34. $x^2 + 6x + 5 = 0$

35. $x^2 - x - 30 = 0$

36. $x^2 + x - 20 = 0$

37. $x^2 - 2x - 15 = 0$

38. $x^2 - x - 12 = 0$

39. $x^2 + 6x + 9 = 0$

40. $x^2 + 10x + 25 = 0$

41. $4x^2 - 4x = 8$

42. $3x^2 + 3x = 90$

43. $3x^2 - 6x = 0$

44. $-5x^2 + 10x = 0$

45. $-4t^2 + 6t = 0$

46. $-6w^2 + 15w = 0$

47. $2x^2 + 11x - 21 = 0$

48. $2x^2 - 5x + 2 = 0$

49. $x^2 - 10x + 25 = 0$

50. $x^2 - 4x + 4 = 0$

51. $x^2 - \frac{7}{2}x = 15$

52. $3x^2 - \frac{2}{5}x = \frac{1}{5}$

53. $\frac{1}{10}a^2 - a + \frac{12}{5} = 0$

54. $\frac{2}{9}w^2 + \frac{5}{3}w - 3 = 0$

Solve each equation.

55. $2x^2 - \frac{1}{2} = 0$

56. $3x^2 - \frac{1}{3} = 0$

57. $(x + 1)^2 = 25$

58. $(x - 3)^2 = 1$

59. $x^2 + 2x - 24 = 0$

60. $x^2 - 5x - 50 = 0$

61. $4x^2 + 36x + 81 = 0$

62. $9x^2 - 30x + 25 = 0$

63. $x^2 - 2x = 2(3 - x)$

64. $x^2 + 2x = \frac{1 + 4x}{2}$

65. $x = \frac{27}{12 - x}$

66. $x = \frac{6}{x + 1}$

67. $\sqrt{3x - 8} = x - 2$

68. $\sqrt{3x - 14} = x - 4$

Solve each problem.

69. ***Side of a square.*** If the diagonal of a square is 5 meters, then what is the length of a side?

70. ***Diagonal of a square.*** If the side of a square is 5 meters, then what is the length of the diagonal?

71. ***Howard's journey.*** Howard walked eight blocks east and then four blocks north to reach the public library. How far was he then from where he started?

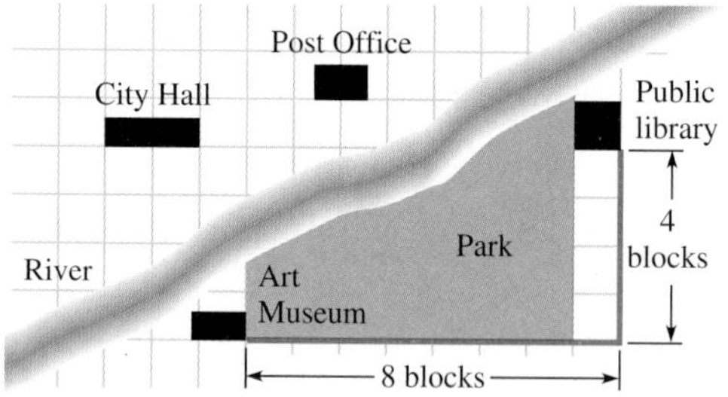

Figure for Exercise 71

72. ***Side and diagonal.*** Each side of a square has length s, and its diagonal has length d. Write a formula for s in terms of d.

73. ***Designing a bridge.*** Find the length d of the diagonal brace shown in the accompanying diagram.

74. ***Designing a bridge.*** Find the length labeled w in the accompanying diagram.

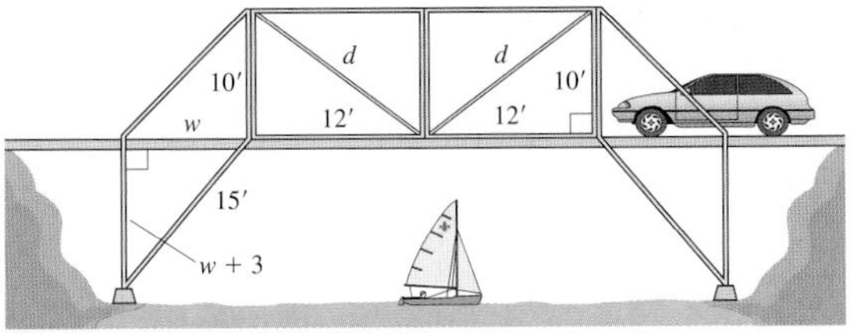

Figure for Exercises 73 and 74

75. ***Two years of interest.*** Tasha deposited \$500 into an account that paid interest compounded annually. At the end of two years she had \$565. Solve the equation $565 = 500(1 + r)^2$ to find the annual rate r to the nearest tenth of a percent.

76. ***Rate of increase.*** The price of a new 2002 Dodge Viper convertible was \$71,725 and a new 2004 Viper convertible was \$80,995 (www.edmunds.com). Find the average annual rate of increase (to the nearest tenth of a percent) for that time period by solving the equation

$$80{,}995 = 71{,}725(1 + r)^2.$$

77. ***Projectile motion.*** If an object is projected upward with initial velocity v_0 ft/sec from an initial height of s_0 feet, then its height s (in feet) t seconds after it is projected is given by the formula $s = -16t^2 + v_0t + s_0$.

a) If a baseball is hit upward at 80 ft/sec from a height of 6 feet, then for what values of t is the baseball 102 feet above the ground?

b) For what value of t is the baseball back at a height of 6 feet?

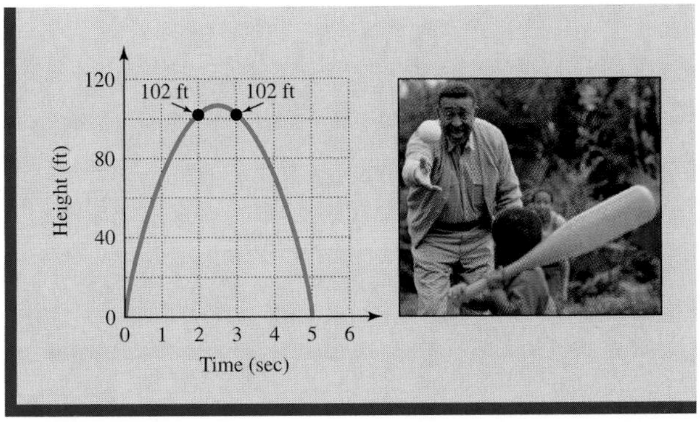

Figure for Exercise 77

78. ***Diving time.*** A springboard diver can perform complicated maneuvers in a short period of time. If a diver springs into the air at 24 ft/sec from a board that is 16 feet above the water, then in how many seconds will she hit the water? Use the formula from Exercise 77.

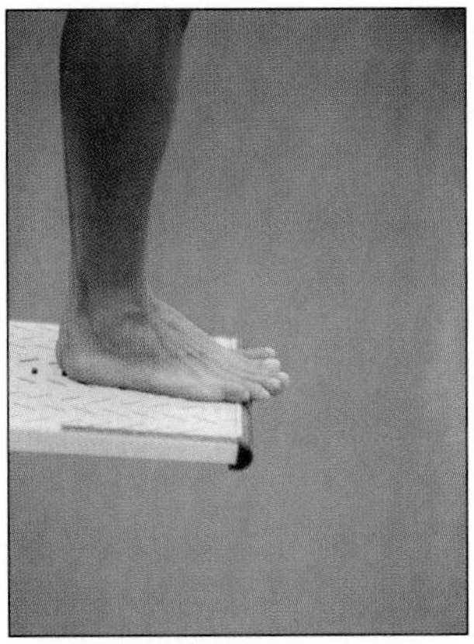

Photo for Exercise 78

79. ***Sum of integers.*** The formula $S = \frac{n^2 + n}{2}$ gives the sum of the first n positive integers. For what value of n is this sum equal to 45?

80. ***Serious reading.*** Kristy's New Year's resolution is to read one page of *Training Your Boa to Squeeze* on January 1, two pages on January 2, three pages on January 3, and so on. On what date will she finish the 136-page book? See Exercise 79.

Getting More Involved

81. ***Writing***

One of the following equations has no real solutions. Find it by inspecting all of the equations (without solving). Explain your answer.

a) $x^2 - 99 = 0$ **b)** $2(v + 77)^2 = 0$
c) $3(y - 22)^2 + 11 = 0$ **d)** $5(w - 8)^2 - 9 = 0$

82. ***Cooperative learning***

For each of three soccer teams A, B, and C to play the other two teams once, it takes three games (AB, AC, and BC). Work in groups to answer the following questions.

a) How many games are required for each team of a four-team league to play every other team once?

b) How many games are required in a five-team soccer league?

c) Find an expression of the form $an^2 + bn + c$ that gives the number of games required in a soccer league of n teams.

d) The Urban Soccer League has fields available for a 120-game season. If the organizers want each team to play every other team once, then how many teams should be in the league?

9.2 Completing the Square

In this Section

- Perfect Square Trinomials
- Solving a Quadratic Equation by Completing the Square
- Applications

The quadratic equations in Section 9.1 were solved by factoring or the square root property, but some quadratic equations cannot be solved by either of those methods. In this section you will learn a method that works on *any* quadratic equation.

Perfect Square Trinomials

The new method for solving any quadratic equation depends on perfect square trinomials. Recall that a perfect square trinomial is the square of a binomial. Just as we recognize the numbers

$$1, \quad 4, \quad 9, \quad 16, \quad 25, \quad 36, \quad \ldots$$

as being the squares of the positive integers, we can recognize a perfect square trinomial. The following is a list of some perfect square trinomials with a leading coefficient of 1:

$$x^2 + 2x + 1 = (x + 1)^2 \qquad x^2 - 2x + 1 = (x - 1)^2$$
$$x^2 + 4x + 4 = (x + 2)^2 \qquad x^2 - 4x + 4 = (x - 2)^2$$
$$x^2 + 6x + 9 = (x + 3)^2 \qquad x^2 - 6x + 9 = (x - 3)^2$$
$$x^2 + 8x + 16 = (x + 4)^2 \qquad x^2 - 8x + 16 = (x - 4)^2$$

To solve quadratic equations using perfect square trinomials, we must be able to determine the last term of a perfect square trinomial when given the first two terms. This process is called **completing the square.** For example, the perfect square trinomial whose first two terms are $x^2 + 6x$ is $x^2 + 6x + 9$.

If the coefficient of x^2 is 1, there is a simple rule for finding the last term in a perfect square trinomial.

Finding the Last Term

The last term of a perfect square trinomial is the square of one-half of the coefficient of the middle term. In symbols, the perfect square trinomial whose first two terms are $x^2 + bx$ is $x^2 + bx + \left(\frac{b}{2}\right)^2$.

EXAMPLE 1

Completing the square

Find the perfect square trinomial whose first two terms are given, and factor the trinomial.

a) $x^2 + 10x$ **b)** $x^2 - 20x$ **c)** $x^2 + 3x$ **d)** $x^2 - x$

Solution

a) One-half of 10 is 5, and 5 squared is 25. So the perfect square trinomial is $x^2 + 10x + 25$. Factor as follows:

$$x^2 + 10x + 25 = (x + 5)^2$$

Helpful Hint

Review the rule for squaring a binomial: square the first term, find twice the product of the two terms, then square the last term. If you are still using FOIL to find the square of a binomial, it is time to learn a more efficient rule.

b) One-half of -20 is -10, and -10 squared is 100. So the perfect square trinomial is $x^2 - 20x + 100$. Factor as follows:

$$x^2 - 20x + 100 = (x - 10)^2$$

c) One-half of 3 is $\frac{3}{2}$, and $\frac{3}{2}$ squared is $\frac{9}{4}$. So the perfect square trinomial is $x^2 + 3x + \frac{9}{4}$. Factor as follows:

$$x^2 + 3x + \frac{9}{4} = \left(x + \frac{3}{2}\right)^2$$

d) One-half of -1 is $-\frac{1}{2}$, and $\left(-\frac{1}{2}\right)^2 = \frac{1}{4}$. So the perfect square is $x^2 - x + \frac{1}{4}$. Factor as follows:

$$x^2 - x + \frac{1}{4} = \left(x - \frac{1}{2}\right)^2$$

Now do Exercises 5–20

Solving a Quadratic Equation by Completing the Square

To complete the squares in Example 1, we simply found the missing last terms. In Examples 2, 3, and 4 we use that process along with the square root property to solve equations of the form $ax^2 + bx + c = 0$. When we use completing the square to solve an equation, we add the appropriate last term to both sides of the equation to obtain an equivalent equation. We first consider an equation in which the coefficient of x^2 is 1.

EXAMPLE 2

Solving by completing the square ($a = 1$)

Solve $x^2 + 6x - 7 = 0$ by completing the square.

Solution

Add 7 to each side of the equation to isolate $x^2 + 6x$:

$$x^2 + 6x = 7$$

Now complete the square for $x^2 + 6x$. One-half of 6 is 3, and $3^2 = 9$.

$x^2 + 6x + 9 = 7 + 9$ Add 9 to each side.

$(x + 3)^2 = 16$ Factor the left side, and simplify the right side.

$x + 3 = \pm 4$ Square root property

$x = -3 \pm 4$

$x = -3 + 4$ or $x = -3 - 4$

$x = 1$ or $x = -7$

Check these answers in the original equation. The solutions are -7 and 1.

Now do Exercises 21–36

Helpful Hint

Always check that you have completed the square correctly by squaring the binomial. In this case $(x + 3)^2 = x^2 + 6x + 9$ is correct.

All of the perfect square trinomials in Examples 1 and 2 have 1 as the leading coefficient. If the leading coefficient is not 1, then we must divide each side of the equation by the leading coefficient to get an equation with a leading coefficient of 1.

The steps to follow in solving a quadratic equation by completing the square are summarized as follows.

Strategy for Solving a Quadratic Equation by Completing the Square

1. The coefficient of x^2 must be 1.
2. Write the equation with only the x^2-terms and the x-terms on the left-hand side.
3. Complete the square on the left-hand side by adding the square of $\frac{1}{2}$ the coefficient of x to both sides of the equation.
4. Factor the perfect square trinomial as the square of a binomial.
5. Apply the square root property.
6. Solve for x and simplify the answer.
7. Check in the original equation.

In Example 3 we solve a quadratic equation in which the coefficient of x^2 is not 1.

EXAMPLE 3

Solving by completing the square ($a \neq 1$)

Solve $2x^2 - 5x - 3 = 0$ by completing the square.

Solution

Our perfect square trinomial must begin with x^2 and not $2x^2$:

$$\frac{2x^2 - 5x - 3}{2} = \frac{0}{2} \quad \text{Divide each side by 2 to get 1 for the coefficient of } x^2.$$

$$x^2 - \frac{5}{2}x - \frac{3}{2} = 0 \quad \text{Simplify.}$$

$$x^2 - \frac{5}{2}x = \frac{3}{2} \quad \text{Write only the } x^2\text{- and } x\text{-terms on the left-hand side.}$$

$$x^2 - \frac{5}{2}x + \frac{25}{16} = \frac{3}{2} + \frac{25}{16} \quad \text{Complete the square: } \frac{1}{2}\left(-\frac{5}{2}\right) = -\frac{5}{4}, \left(-\frac{5}{4}\right)^2 = \frac{25}{16}$$

$$\left(x - \frac{5}{4}\right)^2 = \frac{49}{16} \quad \text{Factor the left-hand side.}$$

$$x - \frac{5}{4} = \pm\frac{7}{4} \quad \text{Square root property}$$

$$x = \frac{5}{4} \pm \frac{7}{4}$$

$$x = \frac{5}{4} + \frac{7}{4} \quad \text{or} \quad x = \frac{5}{4} - \frac{7}{4}$$

$$x = \frac{12}{4} \quad \text{or} \quad x = -\frac{2}{4}$$

$$x = 3 \quad \text{or} \quad x = -\frac{1}{2}$$

Check these answers in the original equation. The solutions to the equation are $-\frac{1}{2}$ and 3.

Now do Exercises 37–42

The equations in Examples 2 and 3 could have been solved by factoring. The quadratic equation in Example 4 cannot be solved by factoring, but it can be solved by completing the square. In fact, every quadratic equation can be solved by completing the square.

EXAMPLE 4

Study Tip

Personal issues can have a tremendous affect on your progress in any course. So do not hesitate to deal with personal problems. If you need help, get it. Most schools have counseling centers that can help you overcome personal issues that are affecting your studies.

Calculator Close-Up

A good way to check an irrational solution with a calculator is to use the answer key (ANS). The value of ANS is the last value calculated by the calculator.

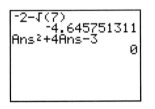

```
-2-√(7)
         -4.645751311
Ans²+4Ans-3
                    0
```

A quadratic equation with irrational solutions

Solve $x^2 + 4x - 3 = 0$ by completing the square.

Solution

$$x^2 + 4x - 3 = 0 \qquad \text{Original equation}$$

$$x^2 + 4x \quad = 3 \qquad \text{Add 3 to each side to isolate the } x^2\text{- and } x\text{-terms.}$$

$$x^2 + 4x + 4 = 3 + 4 \qquad \text{Complete the square by adding 4 to both sides.}$$

$$(x + 2)^2 = 7 \qquad \text{Factor the left-hand side.}$$

$$x + 2 = \pm\sqrt{7} \qquad \text{Square root property}$$

$$x = -2 \pm \sqrt{7}$$

$$x = -2 + \sqrt{7} \quad \text{or} \quad x = -2 - \sqrt{7}$$

Checking answers involving radicals can be done by using the operations with radicals that you learned in Chapter 8. Replace x with $-2 + \sqrt{7}$ in $x^2 + 4x - 3$:

$$(-2 + \sqrt{7})^2 + 4(-2 + \sqrt{7}) - 3 = 4 - 4\sqrt{7} + 7 - 8 + 4\sqrt{7} - 3$$
$$= 0$$

You should check $-2 - \sqrt{7}$. Both $-2 + \sqrt{7}$ and $-2 - \sqrt{7}$ satisfy the equation.

Now do Exercises 43–56

Applications

In Example 5 we use completing the square to solve a geometric problem.

EXAMPLE 5

A geometric problem

The sum of the lengths of the two legs of a right triangle is 8 feet. If the area of the right triangle is 5 square feet, then what are the lengths of the legs?

Solution

If x represents the length of one leg, then $8 - x$ represents the length of the other. See Fig. 9.1 on the next page. The area of a triangle is given by the formula $A = \frac{1}{2}bh$.

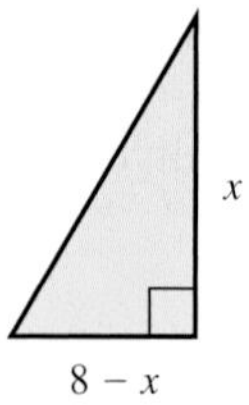

Figure 9.1

Let $A = 5$, $b = x$, and $h = 8 - x$ in this formula:

$$5 = \frac{1}{2}x(8 - x)$$

$$2 \cdot 5 = 2 \cdot \frac{1}{2}x(8 - x) \quad \text{Multiply each side by 2.}$$

$$10 = 8x - x^2$$

$$x^2 - 8x + 10 = 0$$

$$x^2 - 8x = -10 \quad \text{Subtract 10 from each side to isolate the } x^2\text{- and } x\text{-terms.}$$

$$x^2 - 8x + 16 = -10 + 16 \quad \text{Complete the square: } \frac{1}{2}(-8) = -4, (-4)^2 = 16$$

$$(x - 4)^2 = 6 \quad \text{Factor.}$$

$$x - 4 = \pm\sqrt{6}$$

$$x - 4 = \sqrt{6} \quad \text{or} \quad x - 4 = -\sqrt{6}$$

$$x = 4 + \sqrt{6} \quad \text{or} \quad x = 4 - \sqrt{6}$$

If $x = 4 + \sqrt{6}$, then

$$8 - x = 8 - (4 + \sqrt{6}) = 4 - \sqrt{6}.$$

If $x = 4 - \sqrt{6}$, then

$$8 - x = 8 - (4 - \sqrt{6}) = 4 + \sqrt{6}.$$

So there is only one pair of possible lengths for the legs: $4 + \sqrt{6}$ ft and $4 - \sqrt{6}$ ft. Check that the area is 5 square feet:

$$A = \frac{1}{2}(4 + \sqrt{6})(4 - \sqrt{6}) = \frac{1}{2}(16 - 6) = \frac{1}{2}(10) = 5$$

Now do Exercises 77–82

Warm-Ups ▼

True or false? Explain your answer.

1. Completing the square is used for finding the area of a square.
2. The polynomial $x^2 + \frac{2}{3}x + \frac{4}{9}$ is a perfect square trinomial.
3. Every quadratic equation can be solved by factoring.
4. The polynomial $x^2 - x + 1$ is a perfect square trinomial.
5. Every quadratic equation can be solved by completing the square.
6. The solutions to the equation $x - 2 = \pm\sqrt{3}$ are $2 + \sqrt{3}$ and $2 - \sqrt{3}$.
7. There are no real numbers that satisfy $(x + 7)^2 = -5$.
8. To solve $x^2 - 5x = 4$ by completing the square, we can add $\frac{25}{4}$ to each side.
9. One-half of four-fifths is two-fifths.
10. One-half of three-fourths is three-eighths.

9.2 Exercises

Boost your GRADE at mathzone.com!

MathZone

- Practice Problems
- Self-Tests
- Videos
- Net Tutor
- e-Professors

Reading and Writing *After reading this section, write out the answers to these questions. Use complete sentences.*

1. Can every quadratic equation be solved by factoring or the square root property?
2. What method can be used to solve any quadratic equation?
3. How do we find the last term in a perfect square trinomial when we know the first two terms?
4. How do you solve a quadratic equation by completing the square when $a \neq 1$?

Find the perfect square trinomial whose first two terms are given, then factor the trinomial. See Example 1.

5. $x^2 + 6x$
6. $x^2 - 4x$
7. $x^2 + 14x$
8. $x^2 + 16x$
9. $x^2 - 16x$
10. $x^2 - 14x$
11. $t^2 - 18t$
12. $w^2 + 18w$
13. $m^2 + 3m$
14. $n^2 - 5n$
15. $z^2 + z$
16. $v^2 - v$
17. $x^2 - \frac{1}{2}x$
18. $y^2 + \frac{1}{3}y$
19. $y^2 + \frac{1}{4}y$
20. $z^2 - \frac{4}{3}z$

Factor each perfect square trinomial as the square of a binomial.

21. $x^2 + 10x + 25$
22. $x^2 - 6x + 9$
23. $m^2 - 2m + 1$
24. $n^2 + 4n + 4$
25. $x^2 + x + \frac{1}{4}$
26. $y^2 - y + \frac{1}{4}$
27. $t^2 + \frac{1}{3}t + \frac{1}{36}$
28. $v^2 - \frac{2}{3}v + \frac{1}{9}$
29. $x^2 + \frac{2}{5}x + \frac{1}{25}$
30. $y^2 - \frac{1}{4}y + \frac{1}{64}$

Solve each quadratic equation by completing the square. See Examples 2 and 3.

31. $x^2 + 2x - 15 = 0$
32. $x^2 + 2x - 24 = 0$
33. $x^2 - 4x - 21 = 0$
34. $x^2 - 4x - 12 = 0$
35. $x^2 + 6x + 9 = 0$
36. $x^2 - 10x + 25 = 0$
37. $2t^2 - 3t + 1 = 0$
38. $2t^2 - 3t - 2 = 0$
39. $2w^2 - 7w + 6 = 0$
40. $4t^2 + 5t - 6 = 0$
41. $3x^2 + 2x - 1 = 0$
42. $3x^2 - 8x - 3 = 0$

Solve each quadratic equation by completing the square. See Example 4.

43. $x^2 + 2x - 6 = 0$
44. $x^2 + 4x - 4 = 0$
45. $x^2 + 6x + 1 = 0$
46. $x^2 - 6x - 3 = 0$
47. $y^2 - y - 3 = 0$
48. $t^2 + t - 1 = 0$
49. $v^2 + 3v - 3 = 0$
50. $u^2 - 3u + 1 = 0$

51. $2m^2 - m - 4 = 0$

52. $4q^2 + 2q - 1 = 0$

53. $2x^2 + 6x - 3 = 0$

54. $2x^2 - 10x - 1 = 0$

55. $4x^2 - 6x + 1 = 0$

56. $2x^2 - 2x - 5 = 0$

Solve each equation by whichever method is appropriate.

57. $(x - 5)^2 = 7$

58. $x^2 + x = 12$

59. $3n^2 - 5 = 0$

60. $2m^2 + 16 = 0$

61. $4x^2 + 8x - 1 = 0$

62. $2x^2 - 3x - 1 = 0$

63. $3x^2 + 1 = 0$

64. $x^2 + 6x + 7 = 0$

65. $x^2 + 5 = 8x - 3$

66. $2x^2 + 3x = 42 - 2x$

67. $(2x - 7)^2 = 0$

68. $x^2 - 7 = 0$

69. $y^2 + 6y = 11$

70. $y^2 + 6y = 0$

71. $\frac{1}{4}w^2 + \frac{1}{2} = w$

72. $\frac{1}{2}z^2 + \frac{1}{2} = 2z$

73. $t^2 + 0.2t = 0.24$

74. $p^2 - 0.9p + 0.18 = 0$

75. $4x^2 + 4x - 7 = 0$

76. $2x^2 - 8x + 5 = 0$

Use a quadratic equation and completing the square to solve each problem. See Example 5.

77. ***Area of a triangle.*** The sum of the measures of the base and height of a triangle is 10 inches. If the area of the triangle is 11 square inches, then what are the measures of the base and height?

78. ***Dimensions of a rectangle.*** A rectangle has a perimeter of 12 inches and an area of 6 square inches. What are the length and width of the rectangle?

79. ***Missing numbers.*** The sum of two numbers is 12, and their product is 34. What are the numbers?

80. ***More missing numbers.*** The sum of two numbers is 8, and their product is 11. What are the numbers?

81. ***Saving candles.*** Joan has saved the candles from her birthday cake for every year of her life. If Joan has 78 candles, then how old is Joan? (See Exercise 79 of Section 9.1.)

82. ***Raffle tickets.*** A charitable organization is selling chances to win a used Corvette. If the tickets are x dollars each, then the members will sell $5000 - 200x$ tickets. So the total revenue for the tickets is given by $R = x(5000 - 200x)$.

a) What is the revenue if the tickets are sold at \$8 each?

b) For what ticket price is the revenue \$30,000?

c) Use the accompanying graph to estimate the ticket price that will produce the maximum revenue.

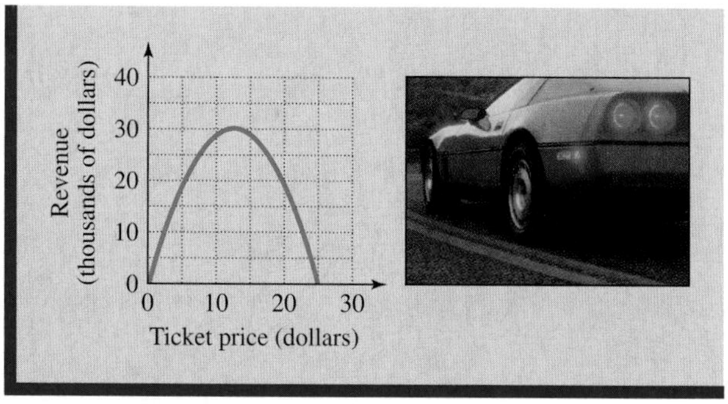

Figure for Exercise 82

Getting More Involved

83. ***Exploration***

a) Find the product $[x - (5 + \sqrt{3})][x - (5 - \sqrt{3})]$.

b) Use completing the square to solve the quadratic equation formed by setting the answer to part (a) equal to zero.

c) Write a quadratic equation (in the form $ax^2 + bx + c = 0$) that has solutions $\frac{3 + \sqrt{2}}{2}$ and $\frac{3 - \sqrt{2}}{2}$.

d) Explain how to find a quadratic equation in the form $ax^2 + bx + c = 0$ for any two given solutions.

9.3 The Quadratic Formula

In this Section

- Developing the Quadratic Formula
- The Discriminant
- Which Method to Use

In Section 9.2 you learned that every quadratic equation can be solved by completing the square. In this section we use completing the square to get a formula, the quadratic formula, for solving any quadratic equation.

Developing the Quadratic Formula

To develop a formula for solving any quadratic equation, we start with the general quadratic equation

$$ax^2 + bx + c = 0$$

and solve it by completing the square. Assume that a is positive for now, and divide each side by a:

$$\frac{ax^2 + bx + c}{a} = \frac{0}{a} \qquad \text{Divide by } a \text{ to get 1 for the coefficient of } x^2.$$

$$x^2 + \frac{b}{a}x + \frac{c}{a} = 0 \qquad \text{Simplify.}$$

$$x^2 + \frac{b}{a}x = -\frac{c}{a} \qquad \text{Isolate the } x^2\text{- and } x\text{-terms.}$$

Now complete the square on the left. One-half of $\frac{b}{a}$ is $\frac{b}{2a}$, and $\left(\frac{b}{2a}\right)^2 = \frac{b^2}{4a^2}$.

$$x^2 + \frac{b}{a}x + \frac{b^2}{4a^2} = \frac{b^2}{4a^2} - \frac{c}{a} \qquad \text{Add } \frac{b^2}{4a^2} \text{ to each side.}$$

$$\left(x + \frac{b}{2a}\right)^2 = \frac{b^2}{4a^2} - \frac{4ac}{4a^2} \qquad \text{Factor on the left-hand side, and get a common denominator on the right-hand side.}$$

$$\left(x + \frac{b}{2a}\right)^2 = \frac{b^2 - 4ac}{4a^2}$$

$$x + \frac{b}{2a} = \pm\sqrt{\frac{b^2 - 4ac}{4a^2}} \qquad \text{Square root property}$$

$$x = -\frac{b}{2a} \pm \frac{\sqrt{b^2 - 4ac}}{2a} \qquad \text{Because } a > 0,\ \sqrt{4a^2} = 2a.$$

$$x = \frac{-b \pm \sqrt{b^2 - 4ac}}{2a} \qquad \text{Combine the two expressions.}$$

We assumed that a was positive so that $\sqrt{4a^2} = 2a$ would be correct. If a is negative, then $\sqrt{4a^2} = -2a$. Either way, the result is the same. It is called the **quadratic formula.** The formula gives x in terms of the coefficients a, b, and c. The quadratic formula is generally used instead of completing the square to solve a quadratic equation that cannot be factored.

Study Tip

Try changing subjects or tasks every hour when you study. The brain does not easily assimilate the same material hour after hour. You will learn and remember more from working on it one hour per day than seven hours on Saturday.

The Quadratic Formula

The solutions to $ax^2 + bx + c = 0$, where $a \neq 0$, are given by

$$x = \frac{-b \pm \sqrt{b^2 - 4ac}}{2a}.$$

EXAMPLE 1

Equations with rational solutions

Use the quadratic formula to solve each equation.

a) $x^2 + 2x - 3 = 0$ **b)** $4x^2 = -9 + 12x$

Solution

a) To use the formula, we first identify a, b, and c. For the equation

$$\underset{\substack{\uparrow \\ a}}{1x^2} + \underset{\substack{\uparrow \\ b}}{2x} - \underset{\substack{\uparrow \\ c}}{3} = 0,$$

$a = 1$, $b = 2$, and $c = -3$. Now use these values in the quadratic formula:

$$x = \frac{-b \pm \sqrt{b^2 - 4ac}}{2a}$$

$$x = \frac{-2 \pm \sqrt{2^2 - 4(1)(-3)}}{2(1)} \qquad 2^2 - 4(1)(-3) = 4 + 12 = 16$$

$$x = \frac{-2 \pm \sqrt{16}}{2}$$

$$x = \frac{-2 \pm 4}{2}$$

$$x = \frac{-2 + 4}{2} \quad \text{or} \quad x = \frac{-2 - 4}{2}$$

$$x = 1 \quad \text{or} \quad x = -3$$

Check these answers in the original equation. The solutions are -3 and 1.

b) Write the equation in the form $ax^2 + bx + c = 0$ to identify a, b, and c:

$$4x^2 = -9 + 12x$$

$$4x^2 - 12x + 9 = 0$$

Now $a = 4$, $b = -12$, and $c = 9$. Use these values in the formula:

$$x = \frac{-(-12) \pm \sqrt{(-12)^2 - 4(4)(9)}}{2(4)}$$

$$x = \frac{12 \pm \sqrt{0}}{8} = \frac{12}{8} = \frac{3}{2}$$

Check. The only solution to the equation is $\frac{3}{2}$.

Now do Exercises 7–14

The equations in Example 1 could have been solved by factoring. (Try it.) The quadratic equation in Example 2 has an irrational solution and cannot be solved by factoring.

EXAMPLE 2

An equation with an irrational solution

Solve $3x^2 - 6x + 1 = 0$.

Solution

For this equation, $a = 3$, $b = -6$, and $c = 1$:

$$x = \frac{-(-6) \pm \sqrt{(-6)^2 - 4(3)(1)}}{2(3)}$$

$$= \frac{6 \pm \sqrt{24}}{6}$$

$$= \frac{6 \pm 2\sqrt{6}}{6} \qquad \sqrt{24} = \sqrt{4}\sqrt{6} = 2\sqrt{6}$$

$$= \frac{2(3 \pm \sqrt{6})}{2(3)} \qquad \text{Numerator and denominator have 2 as a common factor.}$$

$$= \frac{3 \pm \sqrt{6}}{3}$$

The two solutions are the irrational numbers $\frac{3 + \sqrt{6}}{3}$ and $\frac{3 - \sqrt{6}}{3}$.

Now do Exercises 15–20

Calculator Close-Up

Check irrational solutions using the answer key as shown here.

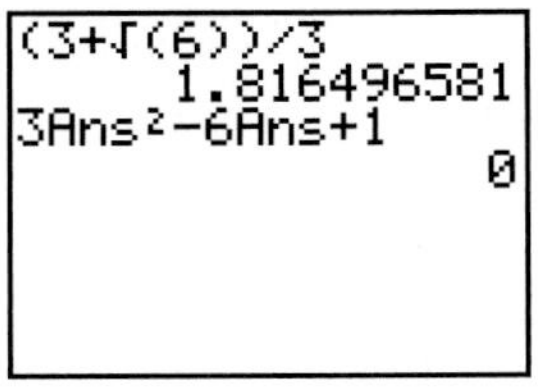

We have seen quadratic equations such as $x^2 = -9$ that do not have any real number solutions. In general, you can conclude that a quadratic equation has no real number solutions if you get a square root of a negative number in the quadratic formula.

EXAMPLE 3

A quadratic equation with no real number solutions

Solve $5x^2 - x + 1 = 0$.

Solution

For this equation we have $a = 5$, $b = -1$, and $c = 1$:

$$x = \frac{1 \pm \sqrt{(-1)^2 - 4(5)(1)}}{2(5)} \qquad b = -1, -b = 1$$

$$x = \frac{1 \pm \sqrt{-19}}{10}$$

The equation has no real solutions because $\sqrt{-19}$ is not real.

Now do Exercises 21–22

The Discriminant

A quadratic equation can have two real solutions, one real solution, or no real number solutions, depending on the value of $b^2 - 4ac$. If $b^2 - 4ac$ is positive, as in Example 1(a) and Example 2, we get two solutions. If $b^2 - 4ac$ is 0, we get only one

Value of $b^2 - 4ac$	**Number of Real Solutions to $ax^2 + bx + c = 0$**
Positive	2
Zero	1
Negative	0

Table 9.1

solution, as in Example 1(b). If $b^2 - 4ac$ is negative, there are no real number solutions, as in Example 3. Table 9.1 summarizes these facts.

The quantity $b^2 - 4ac$ is called the **discriminant** because its value determines the number of real solutions to the quadratic equation.

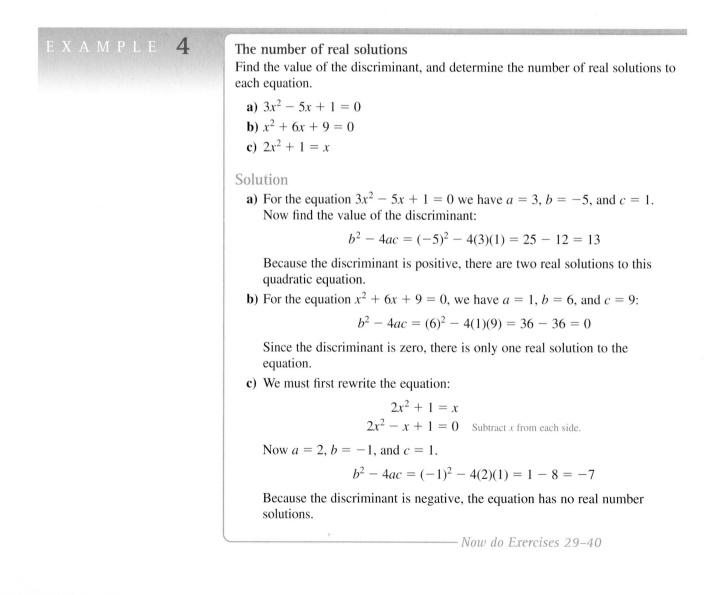

EXAMPLE 4

The number of real solutions

Find the value of the discriminant, and determine the number of real solutions to each equation.

a) $3x^2 - 5x + 1 = 0$

b) $x^2 + 6x + 9 = 0$

c) $2x^2 + 1 = x$

Solution

a) For the equation $3x^2 - 5x + 1 = 0$ we have $a = 3$, $b = -5$, and $c = 1$. Now find the value of the discriminant:

$$b^2 - 4ac = (-5)^2 - 4(3)(1) = 25 - 12 = 13$$

Because the discriminant is positive, there are two real solutions to this quadratic equation.

b) For the equation $x^2 + 6x + 9 = 0$, we have $a = 1$, $b = 6$, and $c = 9$:

$$b^2 - 4ac = (6)^2 - 4(1)(9) = 36 - 36 = 0$$

Since the discriminant is zero, there is only one real solution to the equation.

c) We must first rewrite the equation:

$$2x^2 + 1 = x$$

$$2x^2 - x + 1 = 0 \quad \text{Subtract } x \text{ from each side.}$$

Now $a = 2$, $b = -1$, and $c = 1$.

$$b^2 - 4ac = (-1)^2 - 4(2)(1) = 1 - 8 = -7$$

Because the discriminant is negative, the equation has no real number solutions.

Now do Exercises 29–40

Which Method to Use

If the quadratic equation is simple enough, we can solve it by factoring or by the square root property. These methods should be considered first. *All quadratic equations can be solved by the quadratic formula*. Remember that the quadratic formula is just a shortcut to completing the square and is usually easier to use. However, you should learn completing the square because it is used elsewhere in algebra. The available methods are summarized as follows.

Helpful Hint

If our only intent is to get the answer, then we would probably use a calculator that is programmed with the quadratic formula. However, by learning different methods we gain insight into the problem and get valuable practice with algebra. So be sure you learn all of the methods.

Solving the Quadratic Equation $ax^2 + bx + c = 0$

Method	Comments	Examples
Square root property	Use when $b = 0$.	If $x^2 = 7$, then $x = \pm\sqrt{7}$. If $(x - 2)^2 = 9$, then $x - 2 = \pm 3$.
Factoring	Use when the polynomial can be factored.	$x^2 + 5x + 6 = 0$ $(x + 2)(x + 3) = 0$
Quadratic formula	Use when the first two methods do not apply.	$x^2 + 2x - 6 = 0$ $x = \frac{-2 \pm \sqrt{2^2 - 4 \cdot 1 \cdot (-6)}}{2 \cdot 1}$
Completing the square	Use only when this method is specified.	$x^2 + 4x - 9 = 0$ $x^2 + 4x + 4 = 9 + 4$ $(x + 2)^2 = 13$

Warm-Ups ▼

True or false? Explain your answer.

1. Completing the square is used to develop the quadratic formula.
2. For the equation $x^2 - x + 1 = 0$, we have $a = 1$, $b = -x$, and $c = 1$.
3. For the equation $x^2 - 3 = 5x$, we have $a = 1$, $b = -3$, and $c = 5$.
4. The quadratic formula can be expressed as $x = -b \pm \frac{\sqrt{b^2 - 4ac}}{2a}$.
5. The quadratic equation $2x^2 - 6x = 0$ has two real solutions.
6. All quadratic equations have two distinct real solutions.
7. Some quadratic equations cannot be solved by the quadratic formula.
8. We could solve $2x^2 - 6x = 0$ by factoring, completing the square, or the quadratic formula.
9. The equation $x^2 = x$ is equivalent to $\left(x - \frac{1}{2}\right)^2 = \frac{1}{4}$.
10. The only solution to $x^2 + 6x + 9 = 0$ is -3.

9.3 Exercises

Reading and Writing *After reading this section, write out the answers to these questions. Use complete sentences.*

1. What method presented here can be used to solve any quadratic equation?

2. What is the quadratic formula?

3. What is the quadratic formula used for?

4. What is the discriminant?

5. How can you determine whether there are no real solutions to a quadratic equation?

6. What methods have we studied for solving quadratic equations?

Solve by the quadratic formula. See Examples 1–3.

7. $x^2 + 2x - 15 = 0$

8. $x^2 - 3x - 18 = 0$

9. $x^2 + 10x + 25 = 0$

10. $x^2 - 12x + 36 = 0$

11. $2x^2 + x - 6 = 0$

12. $2x^2 + x - 15 = 0$

13. $4x^2 + 4x - 3 = 0$

14. $4x^2 + 8x + 3 = 0$

15. $x^2 - 6x + 4 = 0$

16. $x^2 - 10x + 19 = 0$

17. $2y^2 - 6y + 3 = 0$

18. $3y^2 + 6y + 2 = 0$

19. $2t^2 + 4t = -1$

20. $w^2 + 2 = 4w$

21. $2x^2 - 2x + 3 = 0$

22. $-2x^2 + 3x - 9 = 0$

23. $8x^2 = 4x$

24. $9y^2 + 3y = -6y$

25. $5w^2 - 3 = 0$

26. $4 - 7z^2 = 0$

27. $\frac{1}{2}h^2 + 7h + \frac{1}{2} = 0$

28. $\frac{1}{4}z^2 - 6z + 3 = 0$

Find the value of the discriminant, and state how many real solutions there are to each quadratic equation. See Example 4.

29. $4x^2 - 4x + 1 = 0$

30. $9x^2 + 6x + 1 = 0$

31. $6x^2 - 7x + 4 = 0$

32. $-3x^2 + 5x - 7 = 0$

33. $-5t^2 - t + 9 = 0$

34. $-2w^2 - 6w + 5 = 0$

35. $4x^2 - 12x + 9 = 0$

36. $9x^2 + 12x + 4 = 0$

37. $x^2 + x + 4 = 0$

38. $y^2 - y + 2 = 0$

39. $x - 5 = 3x^2$

40. $4 - 3x = x^2$

Use the method of your choice to solve each equation.

41. $x^2 + \frac{3}{2}x = 1$

42. $x^2 - \frac{7}{2}x = 2$

43. $(x - 1)^2 + (x - 2)^2 = 5$

44. $x^2 + (x - 3)^2 = 29$

45. $\frac{1}{x} + \frac{1}{x + 2} = \frac{5}{12}$

46. $\frac{1}{x} + \frac{1}{x + 1} = \frac{5}{6}$

47. $x^2 + 6x + 8 = 0$

48. $2x^2 - 5x - 3 = 0$

49. $x^2 - 9x = 0$

50. $x^2 - 9 = 0$

51. $(x + 5)^2 = 9$

52. $(3x - 1)^2 = 0$

53. $x(x - 3) = 2 - 3(x + 4)$

54. $(x - 1)(x + 4) = (2x - 4)^2$

55. $\frac{x}{3} = \frac{x + 2}{x}$

56. $\dfrac{x-2}{x} = \dfrac{5}{x+2}$

57. $2x^2 - 3x = 0$

58. $x^2 = 5$

Use a calculator to find the approximate solutions to each quadratic equation. Round answers to two decimal places.

59. $x^2 - 3x - 3 = 0$

60. $x^2 - 2x - 2 = 0$

61. $x^2 - x - 3.2 = 0$

62. $x^2 - 4.3x + 3 = 0$

63. $5.29x^2 - 3.22x + 0.49 = 0$

64. $2.6x^2 + 3.1x - 5 = 0$

Use a calculator to solve each problem.

65. ***Concert revenues.*** A promoter uses the formula $R = -500x^2 + 20{,}000x$ to predict the concert revenue in dollars when the price of a ticket is x dollars.

a) What is the revenue when the ticket price is \$10?

b) What ticket prices would produce zero revenue?

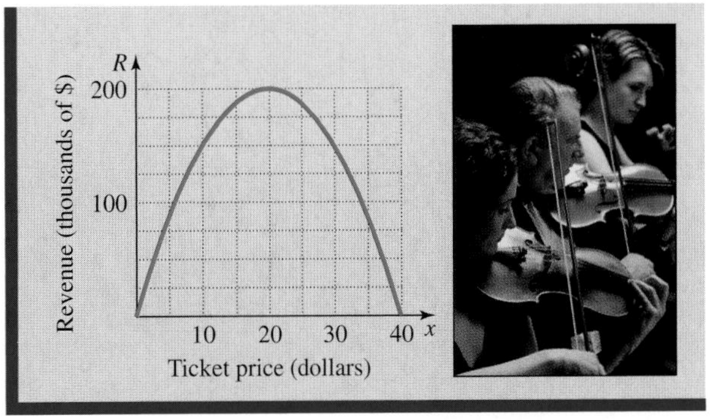

Figure for Exercise 65

c) What is the ticket price if the revenue is \$196,875?

d) Examine the accompanying graph to determine the ticket price that produces the maximum revenue.

66. ***Raffle tickets.*** The formula $R = -200x^2 + 5000x$ was used in Exercise 82 of Section 9.2 to predict the revenue when raffle tickets are sold for x dollars each. For what ticket price is the revenue \$25,000?

9.4 Applications of Quadratic Equations

In this Section

- **Geometric Applications**
- **Work Problems**
- **Vertical Motion**

In this section we will solve problems that involve quadratic equations.

Geometric Applications

Quadratic equations can be used to solve problems involving area.

EXAMPLE 1

Dimensions of a rectangle

The length of a rectangular flower bed is 2 feet longer than the width. If the area is 6 square feet, then what are the exact length and width? Also find the approximate dimensions of the rectangle to the nearest tenth of a foot.

Solution

Let x represent the width, and $x + 2$ represent the length as shown in Fig. 9.2 on the next page. Write an equation using the formula for the area of a rectangle, $A = LW$:

$$x(x + 2) = 6 \quad \text{The area is 6 square feet.}$$

$$x^2 + 2x - 6 = 0$$

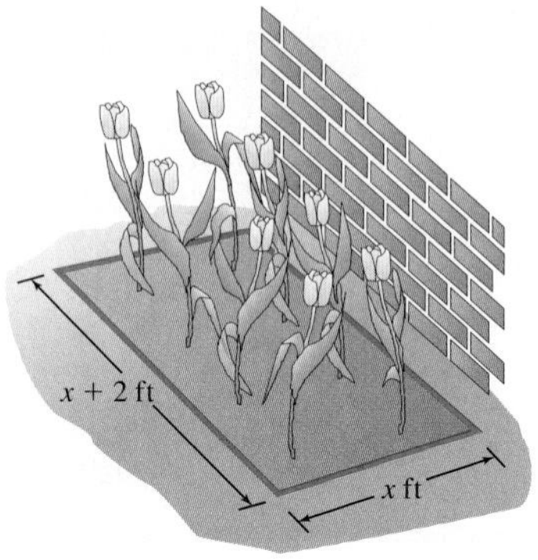

Figure 9.2

We use the quadratic formula to solve the equation:

$$x = \frac{-2 \pm \sqrt{2^2 - 4(1)(-6)}}{2(1)} = \frac{-2 \pm \sqrt{28}}{2}$$

$$= \frac{-2 \pm 2\sqrt{7}}{2} = \frac{2(-1 \pm \sqrt{7})}{2} = -1 \pm \sqrt{7}$$

Because $-1 - \sqrt{7}$ is negative, it cannot be the width of a rectangle. If

$$x = -1 + \sqrt{7},$$

then

$$x + 2 = -1 + \sqrt{7} + 2 = 1 + \sqrt{7}.$$

So the exact width is $-1 + \sqrt{7}$ feet, and the exact length is $1 + \sqrt{7}$ feet. We can check that these dimensions give an area of 6 square feet as follows:

$$LW = (1 + \sqrt{7})(-1 + \sqrt{7}) = -1 - \sqrt{7} + \sqrt{7} + 7 = 6$$

Use a calculator to find the approximate dimensions of 1.6 and 3.6 feet.

Now do Exercises 1–8

Work Problems

The work problems in this section are similar to the work problems that you solved in Chapter 6. However, you will need the quadratic formula to solve the work problems presented in this section.

EXAMPLE 2

Working together

Amy can mow the lawn by herself in 2 hours less time than Bob takes to mow the lawn by himself. When they work together, it takes them only 6 hours to mow the lawn. How long would it take each of them to mow the lawn working alone? Find the exact and approximate answers.

Solution

If x is the number of hours it takes Amy by herself to mow the lawn, then Amy mows at the rate of $\frac{1}{x}$ of the lawn per hour. If $x + 2$ is the number of hours it takes Bob to mow the lawn by himself, then Bob mows at the rate of $\frac{1}{x+2}$ of the lawn per hour. Make a table using the fact that the product of the rate and the time gives the amount of work completed (or the fraction of the lawn mowed).

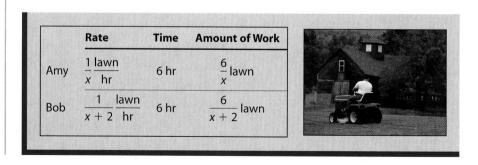

	Rate	Time	Amount of Work
Amy	$\frac{1}{x}\frac{\text{lawn}}{\text{hr}}$	6 hr	$\frac{6}{x}$ lawn
Bob	$\frac{1}{x+2}\frac{\text{lawn}}{\text{hr}}$	6 hr	$\frac{6}{x+2}$ lawn

Helpful Hint

To get familiar with the problem, guess that Amy's time alone is 12 hours and Bob's time alone is 14 hours. In 6 hours of working together, Amy mows 6/12 of the lawn and Bob mows 6/14 of the lawn. Now

$$\frac{6}{12} + \frac{6}{14} = \frac{13}{14}.$$

In 6 hours they would finish only 13/14 of the lawn. So these times are not correct, but they are close.

Because the *total* amount of work done is 1 lawn, we can write the following equation:

$$\frac{6}{x} + \frac{6}{x+2} = 1$$

$$x(x+2)\frac{6}{x} + x(x+2)\frac{6}{x+2} = x(x+2)1 \quad \text{Multiply by the LCD.}$$

$$6x + 12 + 6x = x^2 + 2x$$

$$12x + 12 = x^2 + 2x$$

$$-x^2 + 10x + 12 = 0$$

$$x^2 - 10x - 12 = 0 \quad \text{Multiply each side by } -1.$$

Use the quadratic formula with $a = 1$, $b = -10$, and $c = -12$:

$$x = \frac{10 \pm \sqrt{(-10)^2 - 4(1)(-12)}}{2(1)}$$

$$x = \frac{10 \pm \sqrt{148}}{2} = \frac{10 \pm 2\sqrt{37}}{2} = 5 \pm \sqrt{37}$$

Use a calculator to find that

$$x = 5 - \sqrt{37} \approx -1.08 \quad \text{and} \quad x = 5 + \sqrt{37} \approx 11.08.$$

Because x must be positive, Amy's time alone is $5 + \sqrt{37}$, or approximately 11.1 hours. Because Bob's time alone is 2 hours more than Amy's, Bob's time is $7 + \sqrt{37}$ or approximately 13.1 hours.

Now do Exercises 9–12

Vertical Motion

If an object is projected upward or downward with an initial velocity of v_0 feet per second from an altitude of s_0 feet, then its altitude s in feet after t seconds is given by the formula

$$s = -16t^2 + v_0t + s_0.$$

We use this formula in Example 3.

EXAMPLE 3

Vertical motion

A soccer ball bounces straight up into the air off the head of a soccer player from an altitude of 6 feet with an initial velocity of 40 feet per second. How long does it take the ball to reach the earth? Find the exact answer and an approximate answer.

Solution

The time that it takes the ball to reach the earth is the value of t for which s has a value of 0 in the formula $s = -16t^2 + v_0t + s_0$. To find t, we use $s = 0$, $v_0 = 40$,

Helpful Hint

There is a big difference between Example 3 and Examples 1 and 2. In Example 3 we use a well-known formula that gives the position of a ball at any time and we solve for t. In Examples 1 and 2 we had to decide how two unknown quantities were related and write an equation expressing the relationship.

and $s_0 = 6$:

$$0 = -16t^2 + 40t + 6$$
$$16t^2 - 40t - 6 = 0$$
$$8t^2 - 20t - 3 = 0 \quad \text{Divide each side by 2.}$$
$$t = \frac{20 \pm \sqrt{(-20)^2 - 4(8)(-3)}}{2(8)}$$
$$= \frac{20 \pm \sqrt{496}}{16} = \frac{20 \pm 4\sqrt{31}}{16}$$
$$= \frac{5 \pm \sqrt{31}}{4}$$

Because the time must be positive, we have

$$t = \frac{5 + \sqrt{31}}{4} \approx 2.64 \text{ seconds.}$$

It takes the ball $\frac{5 + \sqrt{31}}{4}$ or 2.64 seconds to reach the earth.

Now do Exercises 13–16

Warm-Ups ▼

True or false? Explain your answer.

1. Two numbers that have a sum of 10 are represented by x and $x + 10$.
2. The area of a right triangle is one-half the product of the lengths of the legs.
3. If the speed of a boat in still water is x mph and the current is 5 mph, then the speed of the boat with the current is $5x$ mph.
4. If Boudreaux eats a 50-pound bag of crawfish in x hours, then his eating rate is $\frac{50}{x}$ bag/hr.
5. If the Concorde flew 1800 miles in $x + 2$ hours, then its average speed was $\frac{1800}{x+2}$ mph.
6. The quantity $\frac{7 - \sqrt{50}}{2}$ is negative.
7. The quantity $(-5 + \sqrt{27})$ is positive.
8. If the length of one side of a square is $x + 9$ meters, then the area of the square is $x^2 + 81$ square meters.
9. If Julia mows an entire lawn in x hours, then her mowing rate is $\frac{1}{x}$ lawn/hr.
10. If John's boat goes 20 miles per hour in still water, then against a 5-mph current it will go 15 miles per hour.

9.4 Exercises

Boost your GRADE at mathzone.com!

MathZone
- Practice Problems
- Net Tutor
- Self-Tests
- e-Professors
- Videos

Find the exact solution to each problem. See Example 1.

1. ***Length and width.*** The length of a rectangle is 2 meters longer than the width. If the area is 10 square meters, then what are the length and width?

2. ***Unequal legs.*** One leg of a right triangle is 4 centimeters longer than the other leg. If the area of the triangle is 8 square centimeters, then what are the lengths of the legs?

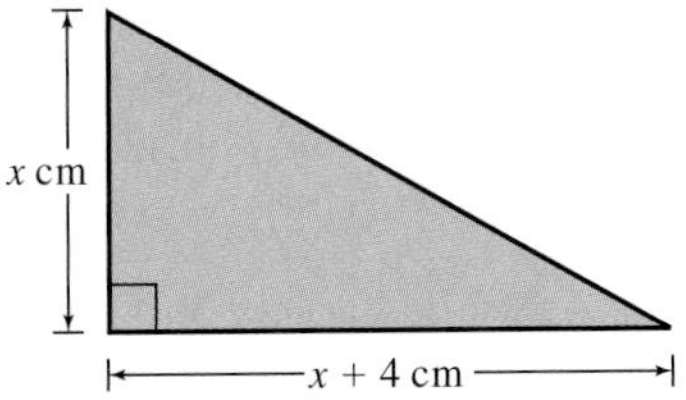

Figure for Exercise 2

3. ***Bracing a gate.*** If the diagonal brace of the square gate shown in the figure is 8 feet long, then what is the length of a side of the square gate?

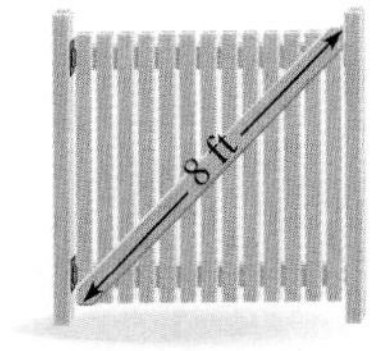

Figure for Exercise 3

4. ***Dimensions of a rectangle.*** If one side of a rectangle is 2 meters shorter than the other side and the diagonal is 10 meters long, then what are the dimensions of the rectangle?

5. ***Area of a parallelogram.*** The base of a parallelogram is 6 inches longer than its height. If the area of the parallelogram is 10 square inches, then what are the base and height?

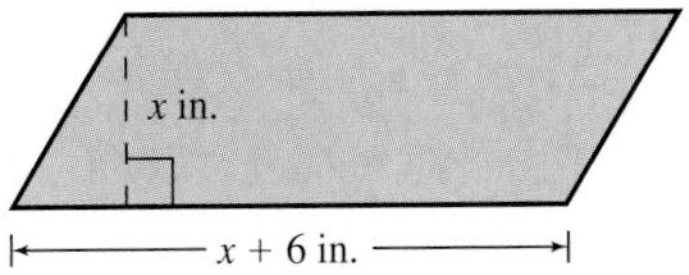

Figure for Exercise 5

6. ***Positive numbers.*** Find two positive real numbers that have a sum of 8 and a product of 4.

7. ***Rectangular frame.*** The area of a rectangular painting is 76 square inches and its perimeter is 36 inches. What are the length and width?

8. ***Rectangular cardboard.*** The area of a rectangular piece of cardboard is 47 square inches and its perimeter is 28 inches. What are the length and width?

Solve each problem. Give the exact answer and an approximate answer rounded to two decimal places. See Examples 2 and 3.

9. ***In the berries.*** On Monday, Alberta picked the strawberry patch, and Ernie sold the berries. On Tuesday, Ernie picked and Alberta sold, but it took him 2 hours longer to get the berries picked than it took Alberta. On Wednesday they worked together and got all of the berries picked in 2 hours. How long did it take Ernie to pick the berries by himself?

Photo for Exercise 9

10. ***Meter readers.*** Claude and Melvin read the water meters for the city of Ponchatoula. When Claude reads all of the meters by himself, it takes him a full day longer than it takes Melvin to read all of the meters by himself. If they can get the job done working together in 2 days, then how long does it take Claude by himself?

11. ***Hanging wallpaper.*** Working alone, Tasha can hang all of the paper in the McLendons' new house in 8 hours less time than it takes Tena working alone. Working together, they completed the job in 20 hours. How long would it take Tasha working alone?

12. ***Laying bricks.*** Chau's team of bricklayers can lay all of the bricks in the McLendons' new house in 3 working days less than Hong's team. To speed things up, the McLendons hire both teams and get the job done in 10 working days. How many working days do the McLendons save by using both teams rather than just the faster team?

13. ***Hang time.*** A punter kicks a football straight up from a height of 4 feet with an initial velocity of 60 feet per second. How long will it take the ball to reach the earth?

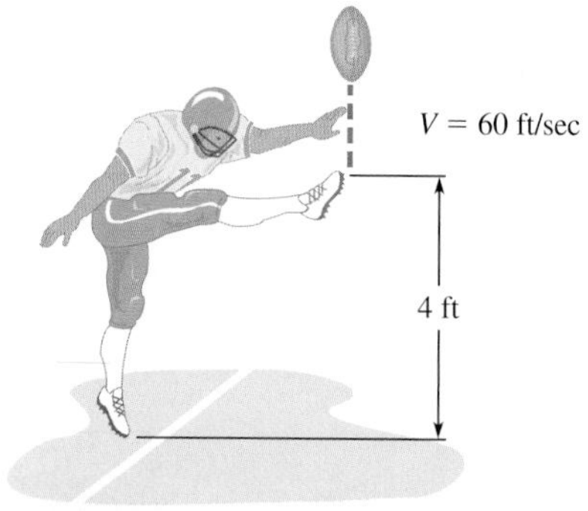

Figure for Exercise 13

14. ***Hunting accident.*** Dwight accidentally fired his rifle straight into the air while sitting in his deer stand 30 feet off the ground. If the bullet left the barrel with a velocity of 200 feet per second, then how long did it take the bullet to fall to the earth?

15. ***Going up.*** A ball is tossed into the air at 20 feet per second from a height of 5 feet. How long (to the nearest tenth of a second) will it take the ball to reach the ground?

16. ***Going down.*** A comedian throws a watermelon downward at 30 feet per second from a height of 200 feet. How long (to the nearest tenth of a second) will it take the watermelon to reach the ground? (The initial velocity of the watermelon is negative.)

17. ***Gone fishing.*** Nancy traveled 6 miles upstream to do some fly fishing. It took her 20 minutes longer to get there than to return. If the current in the river is 2 miles per hour, then how fast will her boat go in still water?

18. ***Commuting to work.*** Gladys and Bonita commute to work daily. Bonita drives 40 miles and averages 9 miles per hour more than Gladys. Gladys drives 50 miles, and she is on the road one-half hour longer than Bonita. How fast does each of them drive?

19. ***Expanding garden.*** Olin's garden is 5 feet wide and 8 feet long. He bought enough okra seed to plant 100 square feet in okra. If he wants to increase the width and the length by the same amount to plant all of his okra, then what should the increase be?

20. ***Spring flowers.*** Lillian has a 5-foot-square bed of tulips. She plans to surround this bed with a crocus bed of uniform width. If she has enough crocus bulbs to plant 100 square feet of crocus, then how wide should the crocus bed be?

Photo for Exercise 20

9.5 Complex Numbers

In this Section

- **Definition**
- **Operations with Complex Numbers**
- **Square Roots of Negative Numbers**
- **Complex Solutions to Quadratic Equations**

In this chapter we have seen quadratic equations that have no solution in the set of real numbers. In this section you will learn that the set of real numbers is contained in the set of complex numbers. *Quadratic equations that have no real number solutions have solutions that are complex numbers.*

Definition

If we try to solve the equation $x^2 + 1 = 0$ by the square root property, we get

$$x^2 = -1 \qquad \text{or} \qquad x = \pm\sqrt{-1}.$$

Because $\sqrt{-1}$ has no meaning in the real number system, the equation has no real solution. However, there is an extension of the real number system, called the *complex numbers,* in which $x^2 = -1$ has two solutions.

The complex numbers are formed by adding the **imaginary unit** i to the real number system. We make the definitions that

$$i = \sqrt{-1} \qquad \text{and} \qquad i^2 = -1.$$

In the complex number system, $x^2 = -1$ has two solutions: i and $-i$.

The set of complex numbers is defined as follows.

Complex Numbers

The set of complex numbers is the set of all numbers of the form

$$a + bi,$$

where a and b are real numbers, $i = \sqrt{-1}$, and $i^2 = -1$.

In the complex number $a + bi$, a is called the **real part** and b is called the **imaginary part.** If $b \neq 0$, the number $a + bi$ is called an **imaginary number.**

In dealing with complex numbers, we treat $a + bi$ as if it were a binomial with variable i. Thus we would write $2 + (-3)i$ as $2 - 3i$. We agree that $2 + i3$, $3i + 2$, and $i3 + 2$ are just different ways of writing $2 + 3i$. Some examples of complex numbers are

$$2 + 3i, \quad -2 - 5i, \quad 0 + 4i, \quad 9 + 0i, \quad \text{and} \quad 0 + 0i.$$

For simplicity we will write only $4i$ for $0 + 4i$. The complex number $9 + 0i$ is the real number 9, and $0 + 0i$ is the real number 0. Any complex number with $b = 0$, is a real number. The diagram in Fig. 9.3 shows the relationships between the complex numbers, the real numbers, and the imaginary numbers.

Complex numbers

Real numbers	Imaginary numbers
$-3, \pi, \frac{5}{2}, 0, -9, \sqrt{2}$	$i, 2 + 3i, \sqrt{-5}, -3 - 8i$

Figure 9.3

Helpful Hint

All numbers are ideas. They exist only in our minds. So in a sense we "imagine" all numbers. However, the imaginary numbers are a bit harder to imagine than the real numbers and so only they are called imaginary.

Operations with Complex Numbers

We define addition and subtraction of complex numbers as follows.

Addition and Subtraction of Complex Numbers

$$(a + bi) + (c + di) = (a + c) + (b + d)i$$
$$(a + bi) - (c + di) = (a - c) + (b - d)i$$

According to the definition, we perform these operations as if the complex numbers were binomials with i a variable.

EXAMPLE 1

Adding and subtracting complex numbers

Perform the indicated operations.

a) $(2 + 3i) + (4 + 5i)$ **b)** $(2 - 3i) + (-1 - i)$

c) $(3 + 4i) - (1 + 7i)$ **d)** $(2 - 3i) - (-2 - 5i)$

Study Tip

Stay calm and confident. Take breaks when you study. Get six to eight hours of sleep every night and keep reminding yourself that working hard all of the semester will really pay off.

Solution

a) $(2 + 3i) + (4 + 5i) = 6 + 8i$

b) $(2 - 3i) + (-1 - i) = 1 - 4i$

c) $(3 + 4i) - (1 + 7i) = 3 + 4i - 1 - 7i$
$= 2 - 3i$

d) $(2 - 3i) - (-2 - 5i) = 2 - 3i + 2 + 5i$
$= 4 + 2i$

Now do Exercises 7–18

We define multiplication of complex numbers as follows.

Multiplication of Complex Numbers

$$(a + bi)(c + di) = (ac - bd) + (ad + bc)i$$

There is no need to memorize the definition of multiplication of complex numbers. We multiply complex numbers in the same way we multiply polynomials: We use the distributive property. We can also use the FOIL method and the fact that $i^2 = -1$.

EXAMPLE 2

Multiplying complex numbers

Perform the indicated operations.

a) $2i(1 - 3i)$ **b)** $(-2 - 5i)(6 - 7i)$

c) $(5i)^2$ **d)** $(-5i)^2$

e) $(3 - 2i)(3 + 2i)$

Calculator Close-Up

Many calculators can perform operations with complex numbers.

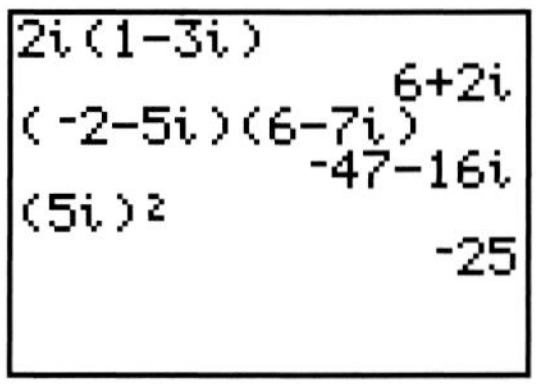

Solution

a) $2i(1 - 3i) = 2i - 6i^2$ Distributive property
$= 2i - 6(-1)$ $i^2 = -1$
$= 6 + 2i$

b) Use FOIL to multiply these complex numbers:

$(-2 - 5i)(6 - 7i) = -12 + 14i - 30i + 35i^2$
$= -12 - 16i + 35(-1)$ $i^2 = -1$
$= -12 - 16i - 35$
$= -47 - 16i$

c) $(5i)^2 = 25i^2 = 25(-1) = -25$

d) $(-5i)^2 = (-5)^2 i^2 = 25(-1) = -25$

e) $(3 - 2i)(3 + 2i) = 9 - 4i^2$
$= 9 - 4(-1)$
$= 9 + 4$
$= 13$

Now do Exercises 19–32

Notice that the product of the imaginary numbers $3 - 2i$ and $3 + 2i$ in Example 2(e) is a real number. We call $3 - 2i$ and $3 + 2i$ *complex conjugates* of each other.

Complex Conjugates

The complex numbers $a + bi$ and $a - bi$ are called **complex conjugates** of each other. Their product is the real number $a^2 + b^2$.

EXAMPLE 3

Complex conjugates

Find the product of the given complex number and its conjugate.

a) $4 - 3i$ **b)** $-2 + 5i$ **c)** $-i$

Helpful Hint

The last time we used "conjugate" was to refer to expressions such as $1 + \sqrt{3}$ and $1 - \sqrt{3}$ as conjugates. Note that we use the rule for the product of a sum and a difference to multiply the radical conjugates or the complex conjugates.

Solution

a) The complex conjugate of $4 - 3i$ is $4 + 3i$:

$$(4 - 3i)(4 + 3i) = 16 - 9i^2 = 16 + 9 = 25$$

b) The conjugate of $-2 + 5i$ is $-2 - 5i$:

$$(-2 + 5i)(-2 - 5i) = (-2)^2 + 5^2 = 4 + 25 = 29$$

c) The conjugate of $-i$ is i:

$$(-i)(i) = -i^2 = -(-1) = 1$$

Now do Exercises 33–40

To divide a complex number by a real number, we divide each part by the real number. For example,

$$\frac{4-6i}{2}=2-3i.$$

We use the idea of complex conjugates to divide by a complex number. The process is similar to rationalizing the denominator.

Dividing by a Complex Number

To divide by a complex number, multiply the numerator and denominator of the quotient by the complex conjugate of the denominator.

EXAMPLE 4

Dividing complex numbers

Perform the indicated operations.

a) $\frac{2}{3-4i}$ **b)** $\frac{6}{2+i}$ **c)** $\frac{3-2i}{i}$

Solution

a) Multiply the numerator and denominator by $3+4i$, the conjugate of $3-4i$:

$$\begin{aligned}\frac{2}{3-4i}&=\frac{2(3+4i)}{(3-4i)(3+4i)}\\&=\frac{6+8i}{9-16i^2}\\&=\frac{6+8i}{25}\qquad 9-16i^2=9-16(-1)=25\\&=\frac{6}{25}+\frac{8}{25}i\end{aligned}$$

b) Multiply the numerator and denominator by $2-i$, the conjugate of $2+i$:

$$\begin{aligned}\frac{6}{2+i}&=\frac{6(2-i)}{(2+i)(2-i)}\\&=\frac{12-6i}{4-i^2}\\&=\frac{12-6i}{5}\qquad 4-i^2=4-(-1)=5\\&=\frac{12}{5}-\frac{6}{5}i\end{aligned}$$

c) Multiply the numerator and denominator by $-i$, the conjugate of i:

$$\frac{3-2i}{i}=\frac{(3-2i)(-i)}{i(-i)}=\frac{-3i+2i^2}{-i^2}=\frac{-3i-2}{1}=-2-3i$$

Now do Exercises 41–52

Square Roots of Negative Numbers

In Example 2 we saw that both

$$(5i)^2 = -25 \quad \text{and} \quad (-5i)^2 = -25.$$

Because the square of each of these complex numbers is -25, both $5i$ and $-5i$ are square roots of -25. When we use the radical notation, we write

$$\sqrt{-25} = 5i.$$

The square root of a negative number is not a real number, it is a complex number.

Square Root of a Negative Number

For any positive number b,

$$\sqrt{-b} = i\sqrt{b}.$$

For example, $\sqrt{-9} = i\sqrt{9} = 3i$ and $\sqrt{-7} = i\sqrt{7}$. Note that the expression $\sqrt{7}i$ could easily be mistaken for the expression $\sqrt{7i}$, where i is under the radical. For this reason, when the coefficient of i contains a radical, we write i preceding the radical.

EXAMPLE 5

Square roots of negative numbers

Write each expression in the form $a + bi$, where a and b are real numbers.

a) $2 + \sqrt{-4}$ **b)** $\dfrac{2 + \sqrt{-12}}{2}$ **c)** $\dfrac{-2 - \sqrt{-18}}{3}$

Solution

a) $2 + \sqrt{-4} = 2 + i\sqrt{4}$

$= 2 + 2i$

b) $\dfrac{2 + \sqrt{-12}}{2} = \dfrac{2 + i\sqrt{12}}{2}$

$= \dfrac{2 + 2i\sqrt{3}}{2}$ $\quad \sqrt{12} = \sqrt{4} \cdot \sqrt{3} = 2\sqrt{3}$

$= 1 + i\sqrt{3}$

c) $\dfrac{-2 - \sqrt{-18}}{3} = \dfrac{-2 - i\sqrt{18}}{3}$

$= \dfrac{-2 - 3i\sqrt{2}}{3}$ $\quad \sqrt{18} = \sqrt{9}\sqrt{2} = 3\sqrt{2}$

$= -\dfrac{2}{3} - i\sqrt{2}$

Now do Exercises 53–64

Helpful Hint

The fundamental theorem of algebra says that an nth degree polynomial equation has exactly n solutions in the system of complex numbers. They don't all have to be different. For example, a quadratic equation such as $x^2 + 6x + 9 = 0$ has two solutions, both of which are -3.

Complex Solutions to Quadratic Equations

The equation $x^2 = -4$ has no real number solutions, but it has two complex solutions, which can be found as follows:

$$x^2 = -4$$

$$x = \pm\sqrt{-4} = \pm i\sqrt{4} = \pm 2i$$

Check:

$$(2i)^2 = 4i^2 = 4(-1) = -4$$
$$(-2i)^2 = 4i^2 = -4$$

Both $2i$ and $-2i$ are solutions to the equation.

Consider the general quadratic equation

$$ax^2 + bx + c = 0,$$

where a, b, and c are real numbers. If the discriminant $b^2 - 4ac$ is positive, then the quadratic equation has two real solutions. If the discriminant is 0, then the equation has one real solution. If the discriminant is negative, then the equation has two complex solutions. *In the complex number system, all quadratic equations have solutions.* We can use the quadratic formula to find them.

EXAMPLE 6

Quadratics with imaginary solutions

Find the complex solutions to the quadratic equations.

a) $x^2 - 2x + 5 = 0$

b) $2x^2 + 3x + 5 = 0$

Calculator Close-Up

The answers in Example 6(b) can be checked with a calculator as shown here.

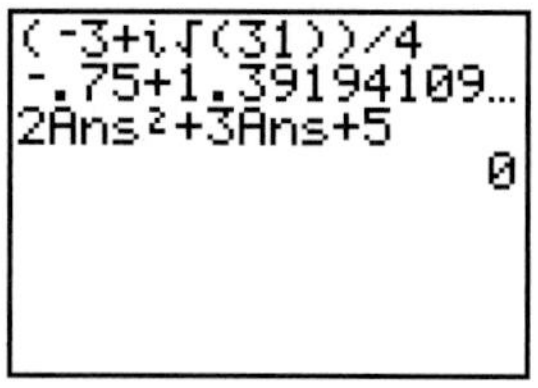

Solution

a) To solve $x^2 - 2x + 5 = 0$, use $a = 1$, $b = -2$, and $c = 5$ in the quadratic formula:

$$x = \frac{2 \pm \sqrt{(-2)^2 - 4(1)(5)}}{2(1)}$$
$$= \frac{2 \pm \sqrt{-16}}{2} = \frac{2 \pm 4i}{2} = 1 \pm 2i$$

We can use the operations with complex numbers to check these solutions:

$$(1 + 2i)^2 - 2(1 + 2i) + 5 = 1 + 4i + 4i^2 - 2 - 4i + 5$$
$$= 1 + 4i - 4 - 2 - 4i + 5 = 0$$

You should verify that $1 - 2i$ also satisfies the equation. The solutions are $1 - 2i$ and $1 + 2i$.

b) To solve $2x^2 + 3x + 5 = 0$, use $a = 2$, $b = 3$, and $c = 5$ in the quadratic formula:

$$x = \frac{-3 \pm \sqrt{3^2 - 4(2)(5)}}{2(2)}$$
$$= \frac{-3 \pm \sqrt{-31}}{4} = \frac{-3 \pm i\sqrt{31}}{4}$$

Check these answers. The solutions are $\frac{-3 + i\sqrt{31}}{4}$ and $\frac{-3 - i\sqrt{31}}{4}$.

Now do Exercises 65–84

The following box summarizes the basic facts about complex numbers.

Complex Numbers

1. $i = \sqrt{-1}$ and $i^2 = -1$.
2. A complex number has the form $a + bi$, where a and b are real numbers.
3. The complex number $a + 0i$ is the real number a.
4. If b is a positive real number, then $\sqrt{-b} = i\sqrt{b}$.
5. The complex conjugate of $a + bi$ is $a - bi$.
6. Add, subtract, and multiply complex numbers as if they were binomials with variable i.
7. Divide complex numbers by multiplying the numerator and denominator by the conjugate of the denominator.
8. In the complex number system, all quadratic equations have solutions.

Warm-Ups

True or false? Explain your answer.

1. $(3 + i) + (2 - 4i) = 5 - 3i$
2. $(4 - 2i)(3 - 5i) = 2 - 26i$
3. $(4 - i)(4 + i) = 17$
4. $i^4 = 1$
5. $\sqrt{-5} = 5i$
6. $\sqrt{-36} = \pm 6i$
7. The complex conjugate of $-2 + 3i$ is $2 - 3i$.
8. Zero is the only real number that is also a complex number.
9. Both $2i$ and $-2i$ are solutions to the equation $x^2 = 4$.
10. Every quadratic equation has at least one complex solution.

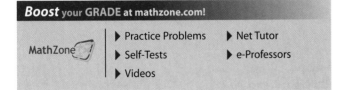

9.5 Exercises

Reading and Writing *After reading this section, write out the answers to these questions. Use complete sentences.*

1. What is a complex number?
2. What is the relationship between the complex numbers, the real numbers, and the imaginary numbers.
3. How do you add or subtract complex numbers?
4. What is a complex conjugate?
5. What is the square root of a negative number in the complex number system?

6. How many solutions do quadratic equations have in the complex number system?

Perform the indicated operations. See Example 1.

7. $(3 + 5i) + (2 + 4i)$

8. $(8 + 3i) + (1 + 2i)$

9. $(-1 + i) + (2 - i)$

10. $(-2 - i) + (-3 + 5i)$

11. $(4 - 5i) - (2 + 3i)$

12. $(3 - 2i) - (7 + 6i)$

13. $(-3 - 5i) - (-2 - i)$

14. $(-4 - 8i) - (-2 - 3i)$

15. $(8 - 3i) - (9 - 3i)$

16. $(5 + 6i) - (-3 + 6i)$

17. $\left(\frac{1}{2} + i\right) + \left(\frac{1}{4} - \frac{1}{2}i\right)$

18. $\left(\frac{2}{3} - i\right) - \left(\frac{1}{4} - \frac{1}{2}i\right)$

Perform the indicated operations. See Example 2.

19. $3(2 - 3i)$

20. $-4(3 - 2i)$

21. $(6i)^2$

22. $(3i)^2$

23. $(-6i)^2$

24. $(-3i)^2$

25. $(2 + 3i)(3 - 5i)$

26. $(4 - i)(3 - 6i)$

27. $(5 - 2i)^2$

28. $(3 + 4i)^2$

29. $(4 - 3i)(4 + 3i)$

30. $(-3 + 5i)(-3 - 5i)$

31. $(1 - i)(1 + i)$

32. $(3 - i)(3 + i)$

Find the product of the given complex number and its conjugate. See Example 3.

33. $2 + 5i$

34. $3 + 4i$

35. $4 - 6i$

36. $2 - 7i$

37. $-3 + 2i$

38. $-4 - i$

39. i

40. $-2i$

Perform the indicated operations. See Example 4.

41. $(2 - 6i) \div 2$

42. $(-3 + 6i) \div (-3)$

43. $\frac{-2 + 8i}{2}$

44. $\frac{6 - 9i}{-3}$

45. $\frac{4 + 6i}{-2i}$

46. $\frac{3 - 8i}{i}$

47. $\frac{4i}{3 + 2i}$

48. $\frac{5}{4 - 5i}$

49. $\frac{2 + i}{2 - i}$

50. $\frac{i - 5}{5 - i}$

51. $\frac{4 - 12i}{3 + i}$

52. $\frac{-4 + 10i}{5 - i}$

Write each expression in the form $a + bi$, where a and b are real numbers. See Example 5.

53. $5 + \sqrt{-9}$

54. $6 + \sqrt{-16}$

55. $-3 - \sqrt{-7}$

56. $2 - \sqrt{-3}$

57. $\frac{-2 + \sqrt{-12}}{2}$

58. $\frac{-6 + \sqrt{-18}}{3}$

59. $\frac{-8 - \sqrt{-20}}{-4}$

60. $\frac{6 + \sqrt{-24}}{-2}$

61. $\frac{-4 + \sqrt{-28}}{6}$

62. $\frac{6 - \sqrt{-45}}{6}$

63. $\frac{-2 + \sqrt{-100}}{-10}$

64. $\frac{-3 + \sqrt{-81}}{-9}$

Find the complex solutions to each quadratic equation. See Example 6.

65. $x^2 = -36$

66. $x^2 = -100$

67. $x^2 + 81 = 0$

68. $x^2 + 100 = 0$

69. $x^2 + 5 = 0$

70. $x^2 + 6 = 0$

71. $3y^2 + 2 = 0$

72. $5y^2 + 3 = 0$

73. $x^2 - 4x + 5 = 0$

74. $x^2 - 6x + 10 = 0$

75. $y^2 + 13 = 6y$

76. $y^2 + 29 = 4y$

77. $x^2 - 4x + 7 = 0$

78. $x^2 - 10x + 27 = 0$

79. $9y^2 - 12y + 5 = 0$

80. $2y^2 - 2y + 1 = 0$

81. $x^2 - x + 1 = 0$

82. $4x^2 - 20x + 27 = 0$

83. $-4x^2 + 8x - 9 = 0$

84. $-9x^2 + 12x - 10 = 0$

Solve each problem.

85. Evaluate $(2 - 3i)^2 + 4(2 - 3i) - 9$.

86. Evaluate $(3 + 5i)^2 - 2(3 + 5i) + 5$.

87. What is the value of $x^2 - 8x + 17$ if $x = 4 - i$?

88. What is the value of $x^2 - 6x + 34$ if $x = 3 + 5i$?

89. Find the product $[x - (6 - i)][x - (6 + i)]$.

90. Find the product $[x - (3 + 7i)][x - (3 - 7i)]$.

91. Write a quadratic equation that has $3i$ and $-3i$ as its solutions.

92. Write a quadratic equation that has $2 + 5i$ and $2 - 5i$ as its solutions.

Getting More Involved

93. *Discussion*

Determine whether each given number is in each of the following sets: the natural numbers, the integers, the rational numbers, the irrational numbers, the real numbers, the imaginary numbers, and the complex numbers.

a) 54 **b)** $-\frac{3}{8}$ **c)** $3\sqrt{5}$ **d)** $6i$ **e)** $\pi + i\sqrt{5}$

94. *Discussion*

Which of the following equations have real solutions? Imaginary solutions? Complex solutions?

a) $3x^2 - 2x + 9 = 0$
b) $5x^2 - 2x - 10 = 0$
c) $\frac{1}{2}x^2 - x + 3 = 0$
d) $7w^2 + 12 = 0$

9.6 Graphing Parabolas

In this Section

- **Finding Ordered Pairs**
- **Graphing Parabolas**
- **The Vertex and Intercepts**
- **Applications**

The graph of any equation of the form $y = mx + b$ is a straight line. In this section we will see that all equations of the form $y = ax^2 + bx + c$ have graphs that are in the shape of a *parabola*.

Finding Ordered Pairs

It is straightforward to calculate y when given x for an equation of the form $y = ax^2 + bx + c$. However, if we are given y and want to find x, then we must use methods for solving quadratic equations.

EXAMPLE 1

Finding ordered pairs

Complete each ordered pair so that it satisfies the given equation. For part (a) the pairs are of the form (x, y) and for part (b) they are of the form (t, s).

a) $y = x^2 - x - 6$; (2,), (, 0)
b) $s = -16t^2 + 48t + 84$; (0,), (, 20)

Study Tip

Although you should avoid cramming, there are times when you have no other choice. In this case concentrate on what is in your class notes and the homework assignments. Try to work one or two problems of each type. Instructors often ask some relatively easy questions on a test to see if you have understood the major ideas.

Solution

a) If $x = 2$, then $y = 2^2 - 2 - 6 = -4$. So the ordered pair is $(2, -4)$. To find x when $y = 0$, replace y by 0 and solve the resulting quadratic equation:

$$x^2 - x - 6 = 0$$
$$(x - 3)(x + 2) = 0$$
$$x - 3 = 0 \quad \text{or} \quad x + 2 = 0$$
$$x = 3 \quad \text{or} \quad x = -2$$

The ordered pairs are $(-2, 0)$ and $(3, 0)$.

b) If $t = 0$, then $s = -16 \cdot 0^2 + 48 \cdot 0 + 84 = 84$. The ordered pair is (0, 84). To find t when $s = 20$, replace s by 20 and solve the equation for t:

$$-16t^2 + 48t + 84 = 20$$

$$-16t^2 + 48t + 64 = 0 \quad \text{Subtract 20 from each side.}$$

$$t^2 - 3t - 4 = 0 \quad \text{Divide each side by } -16.$$

$$(t - 4)(t + 1) = 0 \quad \text{Factor.}$$

$$t - 4 = 0 \quad \text{or} \quad t + 1 = 0 \quad \text{Zero factor property}$$

$$t = 4 \quad \text{or} \quad t = -1$$

The ordered pairs are $(-1, 20)$ and $(4, 20)$.

Now do Exercises 7–10

CAUTION When variables other than x and y are used, the independent variable is the first coordinate of an ordered pair, and the dependent variable is the second coordinate. In Example 1(b), t is the independent variable and first coordinate because s depends on t by the formula $s = -16t^2 + 48t + 84$.

Graphing Parabolas

The graph of any equation of the form $y = ax^2 + bx + c$ (with $a \neq 0$) is called a **parabola.** To graph a parabola we simply plot points as we did when we first graphed lines in Section 3.1. Note that any real number may be used in place of x in the equation $y = ax^2 + bx + c$.

EXAMPLE 2

Calculator Close-Up

This close-up view of $y = x^2$ shows how rounded the curve is at the bottom. When drawing a parabola by hand, be sure to draw it smoothly.

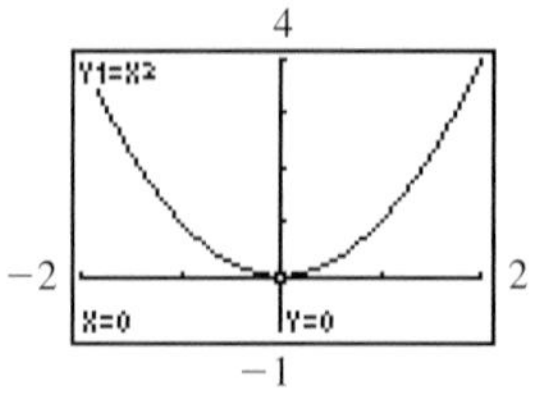

The simplest parabola

Graph $y = x^2$.

Solution

Arbitrarily select some values for the independent variable x and calculate the corresponding values for the dependent variable y:

If $x = -2$, then $y = (-2)^2 = 4$.
If $x = -1$, then $y = (-1)^2 = 1$.
If $x = 0$, then $y = 0^2 = 0$.
If $x = 1$, then $y = 1^2 = 1$.
If $x = 2$, then $y = 2^2 = 4$.

These values are displayed in the following table:

x	-2	-1	0	1	2
$y = x^2$	4	1	0	1	4

Plot $(-2, 4)$, $(-1, 1)$, $(0, 0)$, $(1, 1)$, and $(2, 4)$, as shown in Fig. 9.4, and draw a parabola through the points. Note that the x-coordinates can be any real numbers, but there are no negative y-coordinates. So $y \geq 0$ in every ordered pair on the graph.

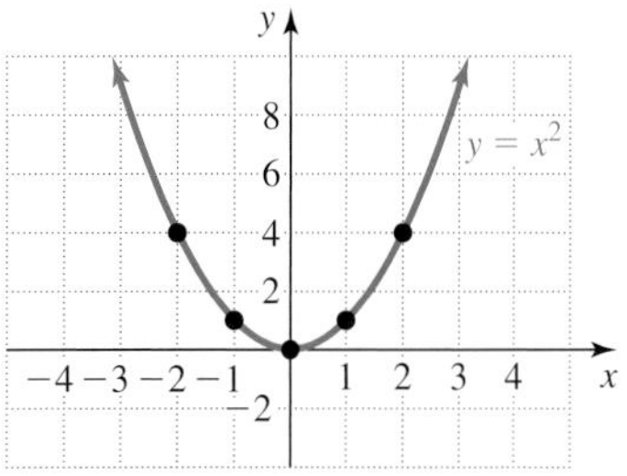

Figure 9.4

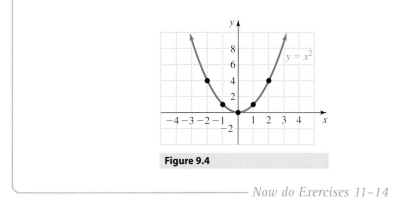

Now do Exercises 11–14

The parabola in Fig. 9.4 is said to **open upward.** In Example 3 we see a parabola that **opens downward.** If $a > 0$ in the equation $y = ax^2 + bx + c$, then the parabola opens upward. If $a < 0$, then the parabola opens downward.

EXAMPLE 3

A parabola that opens downward

Graph $y = 4 - x^2$.

Solution

Find some ordered pairs as follows:

x	-2	-1	0	1	2
$y = 4 - x^2$	0	3	4	3	0

Plot $(-2, 0)$, $(-1, 3)$, $(0, 4)$, $(1, 3)$, and $(2, 0)$, as shown in Fig. 9.5, and sketch a parabola through the points. Note that the largest y-coordinate on this graph is 4. So $y \leq 4$ in every ordered pair on the graph.

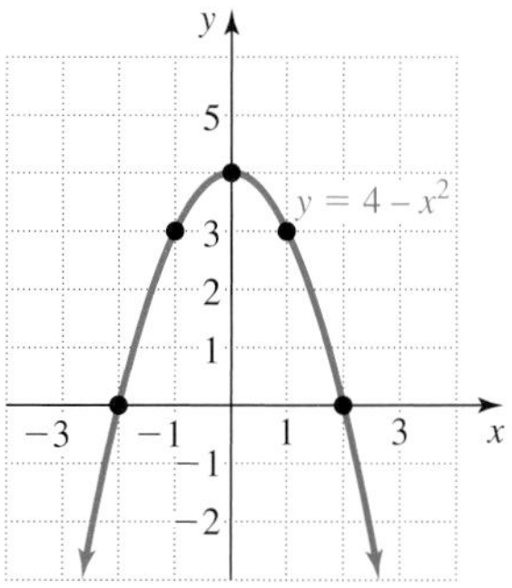

Figure 9.5

Now do Exercises 15–20

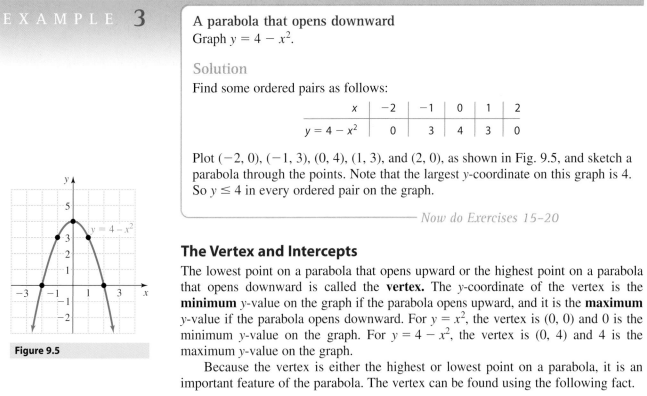

The Vertex and Intercepts

The lowest point on a parabola that opens upward or the highest point on a parabola that opens downward is called the **vertex.** The y-coordinate of the vertex is the **minimum** y-value on the graph if the parabola opens upward, and it is the **maximum** y-value if the parabola opens downward. For $y = x^2$, the vertex is $(0, 0)$ and 0 is the minimum y-value on the graph. For $y = 4 - x^2$, the vertex is $(0, 4)$ and 4 is the maximum y-value on the graph.

Because the vertex is either the highest or lowest point on a parabola, it is an important feature of the parabola. The vertex can be found using the following fact.

Vertex of a Parabola

The x-coordinate of the vertex of $y = ax^2 + bx + c$ is $\frac{-b}{2a}$, provided that $a \neq 0$.

Helpful Hint

To draw a parabola or any curve by hand, use your hand like a compass. The two halves of a parabola should be drawn in two steps. Position your paper so that your hand is approximately at the "center" of the arc you are trying to draw.

You can remember $\frac{-b}{2a}$ by observing that it is part of the quadratic formula:

$$x = \frac{-b \pm \sqrt{b^2 - 4ac}}{2a}$$

In Examples 2 and 3 we drew the graph by selecting five x-values and calculating y. But how did we know what to select for x? The best way to graph a parabola is to find the vertex first, then select two x-coordinates to the left of the vertex and two to the right of the vertex. This way you will always get a graph that shows the typical parabolic shape.

EXAMPLE 4

Using the vertex in graphing a parabola

Graph $y = -x^2 - x + 2$.

Solution

First find the x-coordinate of the vertex:

$$x = \frac{-b}{2a} = \frac{-(-1)}{2(-1)} = \frac{1}{-2} = -\frac{1}{2}$$

Now find y for $x = -\frac{1}{2}$:

$$y = -\left(-\frac{1}{2}\right)^2 - \left(-\frac{1}{2}\right) + 2 = -\frac{1}{4} + \frac{1}{2} + 2 = \frac{9}{4}$$

The vertex is $\left(-\frac{1}{2}, \frac{9}{4}\right)$. Now find a few points on either side of the vertex:

x	-2	-1	$-\frac{1}{2}$	0	1
$y = -x^2 - x + 2$	0	2	$\frac{9}{4}$	2	0

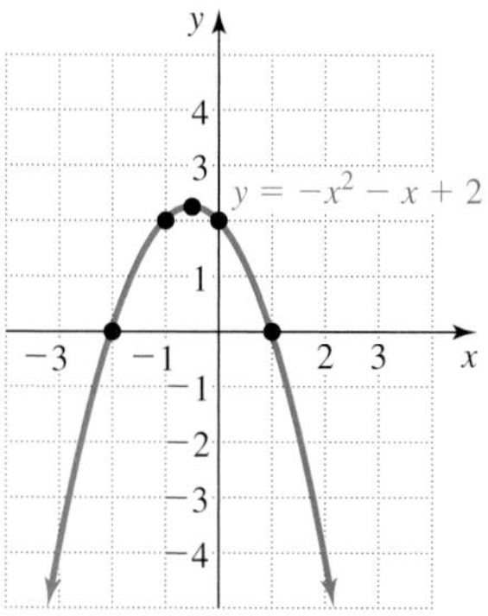

Figure 9.6

Sketch a parabola through these points as in Fig. 9.6.

Now do Exercises 21–26

The y-intercept of a parabola is the point that has 0 as the first coordinate. The x-intercepts are the points that have 0 as their second coordinates.

EXAMPLE 5

Using the intercepts in graphing a parabola

Find the vertex and intercepts, and sketch the graph of each equation.

a) $y = x^2 - 2x - 8$

b) $s = -16t^2 + 64t$

Solution

a) Use $x = \frac{-b}{2a}$ to get $x = 1$ as the x-coordinate of the vertex. If $x = 1$, then

$$\begin{aligned} y &= 1^2 - 2 \cdot 1 - 8 \\ &= -9. \end{aligned}$$

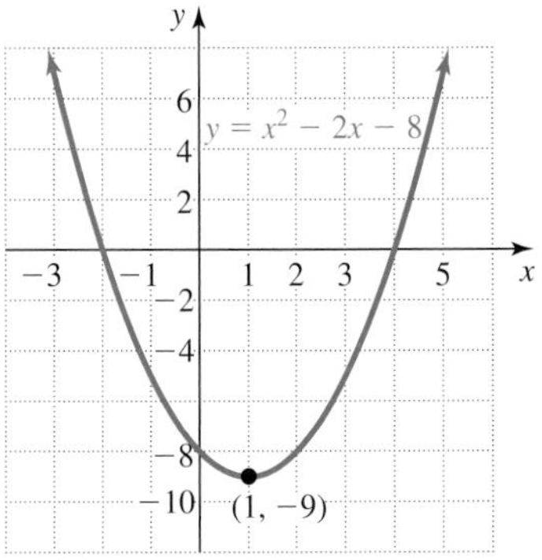

Figure 9.7

So the vertex is $(1, -9)$. If $x = 0$, then

$$y = 0^2 - 2 \cdot 0 - 8$$
$$= -8.$$

The y-intercept is $(0, -8)$. To find the x-intercepts, replace y by 0:

$$x^2 - 2x - 8 = 0$$
$$(x - 4)(x + 2) = 0$$
$$x - 4 = 0 \quad \text{or} \quad x + 2 = 0$$
$$x = 4 \quad \text{or} \quad x = -2$$

The x-intercepts are $(-2, 0)$ and $(4, 0)$. The graph is shown in Fig. 9.7.

b) Because s is expressed in terms of t, the first coordinate is t. Use $t = \frac{-b}{2a}$ to get

$$t = \frac{-64}{2(-16)} = 2.$$

If $t = 2$, then

$$s = -16 \cdot 2^2 + 64 \cdot 2$$
$$= 64.$$

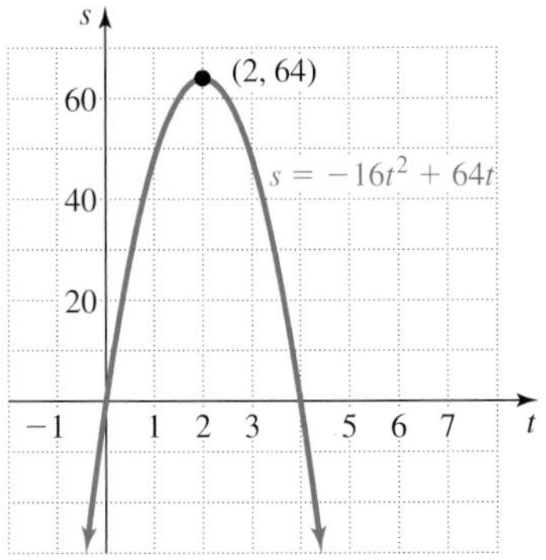

Figure 9.8

So the vertex is $(2, 64)$. If $t = 0$, then

$$s = -16 \cdot 0^2 + 64 \cdot 0$$
$$= 0.$$

So the s-intercept is $(0, 0)$. To find the t-intercepts, replace s by 0:

$$-16t^2 + 64t = 0$$
$$-16t(t - 4) = 0$$
$$-16t = 0 \quad \text{or} \quad t - 4 = 0$$
$$t = 0 \quad \text{or} \quad t = 4$$

The t-intercepts are $(0, 0)$ and $(4, 0)$. The graph is shown in Fig. 9.8.

Now do Exercises 27–30

Calculator Close-Up

You can find the vertex of a parabola with a calculator by using either the maximum or minimum feature. First graph the parabola as shown.

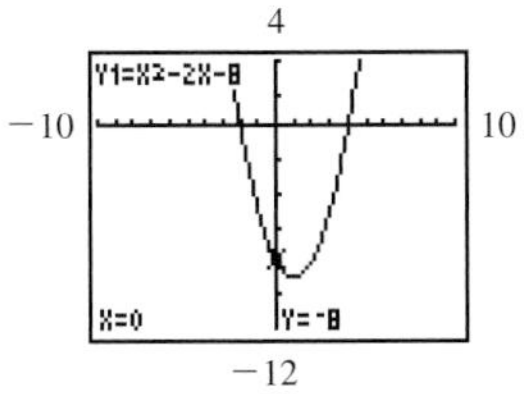

Because this parabola opens upward, the y-coordinate of the vertex is the minimum y-coordinate on the graph. Press CALC and choose minimum.

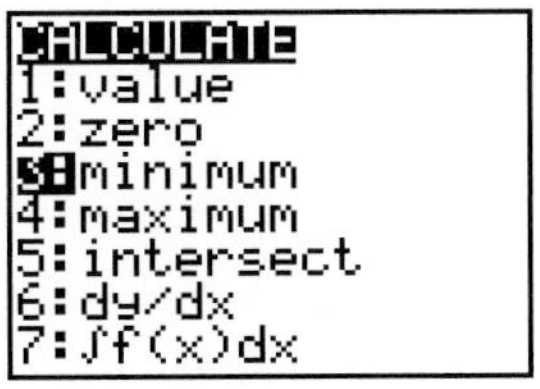

The calculator will ask for a left bound, a right bound, and a guess. For the left bound choose a point to the left of the vertex by moving the cursor to the point and pressing ENTER. For the right bound choose a point to the right of the vertex. For the guess choose a point close to the vertex.

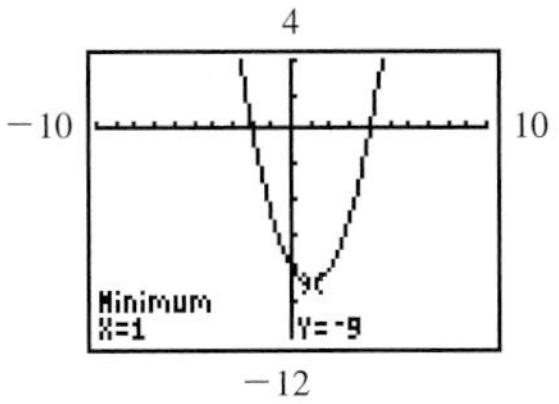

Applications

In applications we are often interested in finding the maximum or minimum value of a variable. If a parabola opens downward, then the maximum value of the dependent variable is the second coordinate of the vertex. If a parabola opens upward, then the minimum value of the dependent variable is the second coordinate of the vertex.

EXAMPLE 6

Finding the maximum height

A ball is tossed upward with a velocity of 64 feet per second from a height of 5 feet. What is the maximum height reached by the ball?

Solution

The height s of the ball for any time t is given by $s = -16t^2 + 64t + 5$. Because the maximum height occurs at the vertex of the parabola, we use $t = \frac{-b}{2a}$ to find the vertex:

$$t = \frac{-64}{2(-16)} = 2$$

Now use $t = 2$ to find the second coordinate of the vertex:

$$s = -16(2)^2 + 64(2) + 5 = 69$$

The maximum height reached by the ball is 69 feet. See Fig. 9.9.

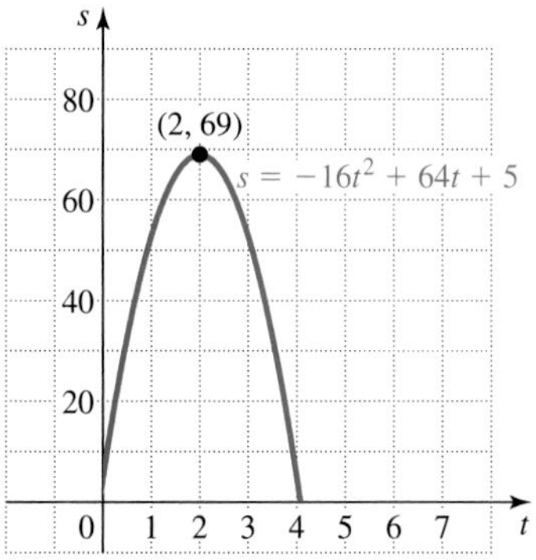

Figure 9.9

Now do Exercises 39–45

Warm-Ups ▼

True or false? Explain your answer.

1. The ordered pair $(-2, -1)$ satisfies $y = x^2 - 5$.
2. The y-intercept for $y = x^2 - 3x + 9$ is $(9, 0)$.
3. The x-intercepts for $y = x^2 - 5$ are $(\sqrt{5}, 0)$ and $(-\sqrt{5}, 0)$.
4. The graph of $y = x^2 - 12$ opens upward.
5. The graph of $y = 4 + x^2$ opens downward.
6. The vertex of $y = x^2 + 2x$ is $(-1, -1)$.
7. The parabola $y = x^2 + 1$ has no x-intercepts.
8. The y-intercept for $y = ax^2 + bx + c$ is $(0, c)$.
9. If $w = -2v^2 + 9$, then the maximum value of w is 9.
10. If $y = 3x^2 - 7x + 9$, then the maximum value of y occurs when $x = \frac{7}{6}$.

9.6 Exercises

Boost your GRADE at mathzone.com!

MathZone

- Practice Problems
- Self-Tests
- Videos
- Net Tutor
- e-Professors

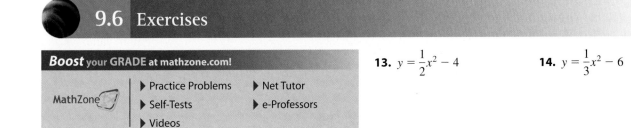

Reading and Writing *After reading this section, write out the answers to these questions. Use complete sentences.*

1. What type of equation has a straight line as its graph?

2. What type of equation has a parabola as its graph?

3. When does a parabola open upward?

4. What is the vertex of a parabola?

5. What is the first coordinate of the vertex for the parabola $y = ax^2 + bx + c$?

6. How do you find the second coordinate of the vertex?

Complete each ordered pair so that it satisfies the given equation. See Example 1.

7. $y = x^2 - x - 12$ $(3, \quad), (\quad, 0)$

8. $y = -\frac{1}{2}x^2 - x + 1$ $(0, \quad), (\quad, -3)$

9. $s = -16t^2 + 32t$ $(4, \quad), (\quad, 0)$

10. $a = b^2 + 4b + 5$ $(-2, \quad), (\quad, 2)$

Graph each equation. See Examples 2 and 3.

11. $y = x^2 + 2$

12. $y = x^2 - 4$

13. $y = \frac{1}{2}x^2 - 4$

14. $y = \frac{1}{3}x^2 - 6$

15. $y = -2x^2 + 5$

16. $y = -x^2 - 1$

17. $y = -\frac{1}{3}x^2 + 5$

18. $y = -\frac{1}{2}x^2 + 3$

19. $y = (x - 2)^2$

20. $y = (x + 3)^2$

Find the vertex and intercepts for each parabola. Then sketch the graph. See Examples 4 and 5.

21. $y = x^2 - x - 2$

22. $y = x^2 + 2x - 3$

23. $y = x^2 + 2x - 8$

24. $y = x^2 + x - 6$

25. $y = -x^2 - 4x - 3$

26. $y = -x^2 - 5x - 4$

27. $y = -x^2 + 3x + 4$

28. $y = -x^2 - 2x + 8$

29. $a = b^2 - 6b - 16$

30. $v = -u^2 - 8u + 9$

Find the maximum or minimum value for y for each equation.

31. $y = x^2 - 8$

32. $y = 33 - x^2$

33. $y = -3x^2 + 14$

34. $y = 6 + 5x^2$

35. $y = x^2 + 2x + 3$

36. $y = x^2 - 2x + 5$

37. $y = -2x^2 - 4x$

38. $y = -3x^2 + 24x$

Solve each problem. See Example 6.

39. ***Maximum height.*** If a baseball is projected upward from ground level with an initial velocity of 64 feet per second, then $s = -16t^2 + 64t$ gives its height in feet s in terms of time in seconds t. Graph this equation for $0 \le t \le 4$. What is the maximum height reached by the ball?

40. ***Maximum height.*** If a soccer ball is kicked straight up with an initial velocity of 32 feet per second, then $s = -16t^2 + 32t$ gives its height in feet s in terms of time in seconds t. Graph this equation for $0 \le t \le 2$. What is the maximum height reached by this ball?

41. ***Maximum area.*** Jason plans to fence a rectangular area with 100 meters of fencing. He has written the formula $A = w(50 - w)$ to express the area in square meters A in terms of the width in meters w. What is the maximum possible area that he can enclose with his fencing?

42. ***Minimizing cost.*** A company uses the formula $C = 0.02x^2 - 3.4x + 150$ to model the unit cost in dollars C for producing x stabilizer bars. For what number of bars is the unit cost at its minimum? What is the unit cost at that level of production?

43. ***Air pollution.*** The formula $A = -2t^2 + 32t + 12$ gives the amount of nitrogen dioxide A in parts per million (ppm) that was present in the air in the city of Homer on June 14, where t is the number of hours after 6:00 A.M. Use this equation to find the time at which the nitrogen dioxide level was at its maximum.

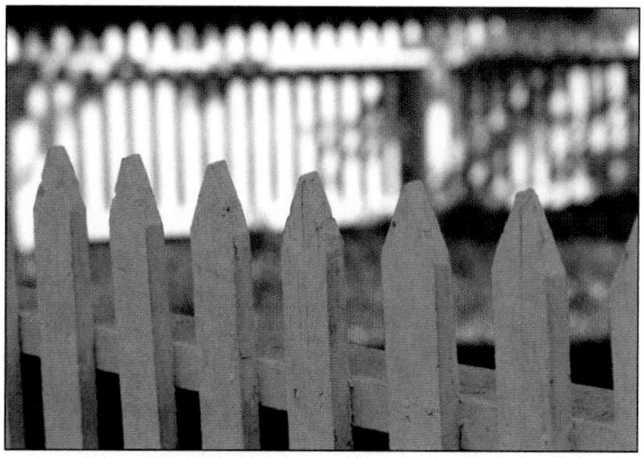

Photo for Exercise 41

44. ***Stabilization ratio.*** The stabilization ratio (births/deaths) for South and Central America can be modeled by the equation

$$y = -0.0012x^2 + 0.074x + 2.69,$$

where y is the number of births divided by the number of deaths in the year $1950 + x$ (World Resources Institute, www.wri.org).

a) Use the accompanying graph to estimate the year in which the stabilization ratio was at its maximum.

b) Use the equation to find the year in which the stabilization ratio was at its maximum.

c) What is the maximum stabilization ratio from part (b)?

d) What is the significance of a stabilization ratio of 1?

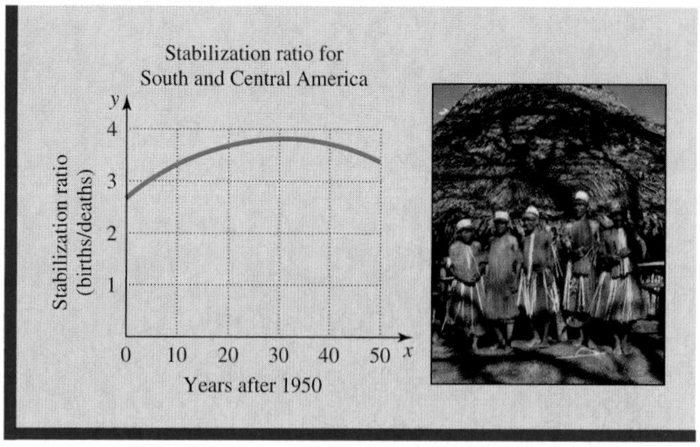

Figure for Exercise 44

45. ***Suspension bridge.*** The cable of the suspension bridge shown in the figure on the next page hangs in the shape of a parabola with equation $y = 0.0375x^2$, where x and y are in meters. What is the height of each tower above the roadway? What is the length z for the cable bracing the tower?

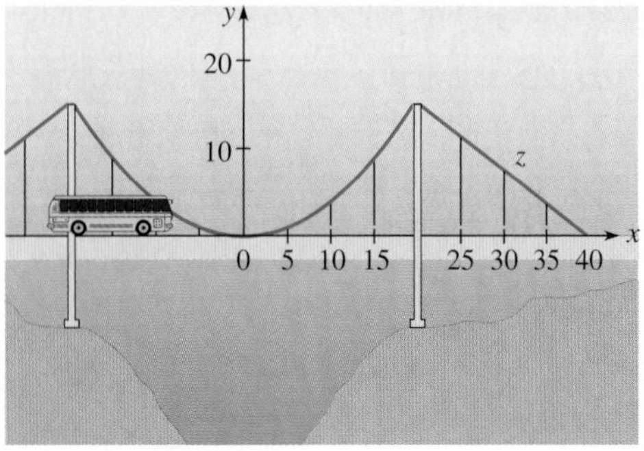

Figure for Exercise 45

Getting More Involved

46. *Exploration*

a) Write $y = 3(x - 2)^2 + 6$ in the form $y = ax^2 + bx + c$, and find the vertex of the parabola using the formula $x = \frac{-b}{2a}$.

b) Repeat part (a) with $y = -4(x - 5)^2 - 9$ and $y = 3(x + 2)^2 - 6$.

c) What is the vertex for a parabola that is written in the form $y = a(x - h)^2 + k$? Explain your answer.

Graphing Calculator Exercises

47. Graph $y = x^2$, $y = \frac{1}{2}x^2$, and $y = 2x^2$ on the same coordinate system. What can you say about the graph of $y = ax^2$?

48. Graph $y = x^2$, $y = (x - 3)^2$, and $y = (x + 3)^2$ on the same coordinate system. How does the graph of $y = (x - h)^2$ compare to the graph of $y = x^2$?

49. The equation $x = y^2$ is equivalent to $y = \pm\sqrt{x}$. Graph both $y = \sqrt{x}$ and $y = -\sqrt{x}$ on a graphing calculator. How does the graph of $x = y^2$ compare to the graph of $y = x^2$?

50. Graph each of the following equations by solving for y.

a) $x = y^2 - 1$

b) $x = -y^2$

c) $x^2 + y^2 = 4$

51. Determine the approximate vertex and x-intercepts for each parabola.

a) $y = 3.2x^2 - 5.4x + 1.6$

b) $y = -1.09x^2 + 13x + 7.5$

9.7 Introduction to Functions

In this Section

- Functions Expressed by Formulas
- Functions Expressed by Tables
- Functions Expressed by Ordered Pairs
- Graphs of Functions
- Domain and Range
- Function Notation

Even though we have not yet used the term *function,* we have used functions throughout this text. In this section we will study the concept of functions and see that they have been with us from the beginning.

Functions Expressed by Formulas

If you get a speeding ticket, then your speed determines the cost of the ticket. You may not know exactly how the judge determines the cost, but the judge is using some rule to determine a cost from knowing your speed. The cost of the ticket is a function of your speed.

Function (as a Rule)

A function is a rule by which any allowable value of one variable (the **independent variable**) determines a *unique* value of a second variable (the **dependent variable**).

Helpful Hint

According to the dictionary, "determine" means to settle conclusively. If the value of the dependent variable is inconclusive or there is more than one, then the rule is not a function.

One way to express a function is to use a formula. For example, the formula

$$A = \pi r^2$$

gives the area of a circle as a function of its radius. The formula gives us a rule for finding a *unique* area for any given radius. A is the dependent variable, and r is the independent variable. The formula

$$S = -16t^2 + v_0 t + s_0$$

expresses altitude S of a projectile as a function of time t, where v_0 is the initial velocity and s_0 is the initial altitude. S is the dependent variable, and t is the independent variable.

Since a formula can be used as a rule for obtaining the value of the dependent variable from the value of the independent variable, we say that the formula is a function. Formulas describe or **model** relationships between variables. In Example 1, we find a function in a real situation.

EXAMPLE 1

Writing a formula for a function

A carpet layer charges \$25 plus \$4 per square yard for installing carpet. Write the total charge C as a function of the number n of square yards of carpet installed.

Solution

At \$4 per square yard, n square yards installed cost $4n$ dollars. If we include the \$25 charge, then the total cost is $4n + 25$ dollars. Thus the equation

$$C = 4n + 25$$

expresses C as a function of n.

Now do Exercises 7–10

Any formula that has the form $y = mx + b$ is a **linear function.** So in Example 1, the charge is a linear function of the number of square yards installed and we say that $C = 4n + 25$ is a linear function.

EXAMPLE 2

A function in geometry

Express the area of a circle as a function of its diameter.

Solution

The area of a circle is given by $A = \pi r^2$. Because the radius of a circle is one-half of the diameter, we have $r = \frac{d}{2}$. Now replace r by $\frac{d}{2}$ in the

Math *at Work* Body Surface Area

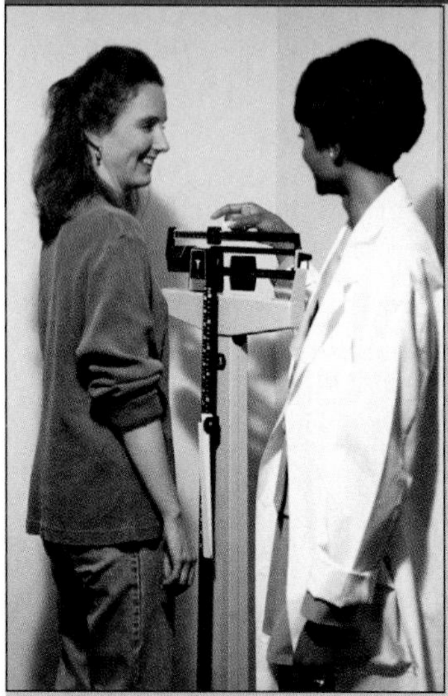

Body surface area (BSA) is difficult to determine but is essential for cancer chemotherapy dosage calculations. A recent study of 2819 chemotherapy orders found that 93 of them contained at least one error in the dose. Three of the errors were classified as potentially lethal. In chemotherapy the difference between an underdose and an overdose is small and the consequences of error can be fatal.

In 1916 Du Bois and Du Bois examined nine individuals of varying age, shape, and size. They measured their BSA directly using molds. From these measurements they derived the formula BSA $= 0.20247\,(h/100)^{0.725}\,w^{0.425}$ using height h (cm) and weight w (kg). The Du Bois formula was challenged in the 1970s by Gehan and George, who directly measured the skin-surface area of 401 individuals. They found that the Du Bois formula significantly overestimated BSA in about 15% of their cases. However, the Du Bois formula prevailed and is still a widely used method for finding BSA. More recently, Mosteller produced the formula BSA $= \sqrt{\frac{hw}{3600}}$, which is easier to remember and use. The Mosteller formula is now being promoted as the new standard for determining BSA.

To maintain consistency in chemotherapy a systematic approach to dosage calculations is needed. All BSA calculations should be based on the same formula. Current heights and weights should be used in BSA calculations and all dosage calculations should be checked. Verifying BSA and dose are needed to ensure optimal treatment and to prevent chemotherapy underdosing or overdosing.

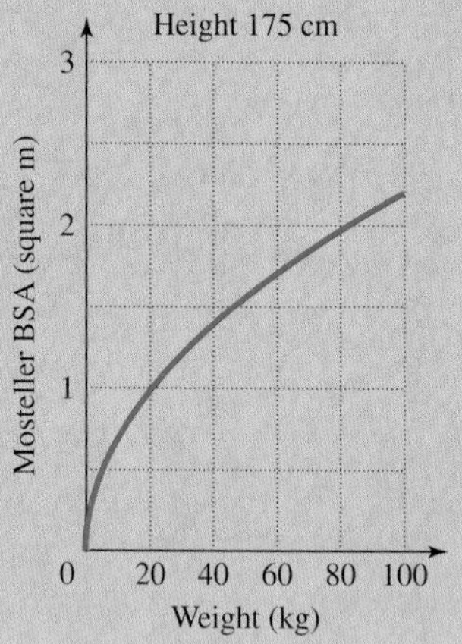

formula $A = \pi r^2$:

$$A = \pi\left(\frac{d}{2}\right)^2$$
$$= \frac{\pi}{4}d^2$$

So $A = \frac{\pi}{4}d^2$ expresses the area of a circle as a function of its diameter.

Now do Exercises 11–16

Any formula that has the form $y = ax^2 + bx + c$ (with $a \neq 0$) is a **quadratic function.** So in Example 2, the area is a quadratic function of the diameter and we say that $A = \frac{\pi}{4}d^2$ is a quadratic function.

Study Tip

When one student asks another for help, it is often the one who does the explaining that learns the most. When you work together, don't let one person do all of the talking. If you must work alone, then try explaining things to yourself. You might even try talking to yourself.

Functions Expressed by Tables

Another way to express a function is with a table. For example, Table 9.2 can be used to determine the cost at United Freight Service for shipping a package that weighs under 100 pounds. For any *allowable* weight, the table gives us a rule for finding the unique shipping cost. The weight is the independent variable, and the cost is the dependent variable.

Weight in Pounds	Cost
0 to 10	\$4.60
11 to 30	\$12.75
31 to 79	\$32.90
80 to 99	\$55.82

Table 9.2

Weight in Pounds	Cost
0 to 15	\$4.60
10 to 30	\$12.75
31 to 79	\$32.90
80 to 99	\$55.82

Table 9.3

Now consider Table 9.3. It does not look much different from Table 9.2, but there is an important difference. The cost for shipping a 12-pound package according to Table 9.3 is either \$4.60 or \$12.75. Either the table has an error or perhaps \$4.60 and \$12.75 are costs for shipping to different destinations. In any case the weight does not determine a unique cost. So Table 9.3 does not express the cost as a function of the weight.

EXAMPLE 3

Functions defined by tables

Which of the following tables expresses y as a function of x?

a)

x	y
1	3
2	6
3	9
4	12
5	15

b)

x	y
1	1
−1	1
2	2
−2	2
3	3
−3	3

c)

x	y
1988	27,000
1989	27,000
1990	28,500
1991	29,000
1992	30,000
1993	30,750

d)

x	y
23	48
35	27
19	28
23	37
41	56
22	34

Solution

In Tables a), b), and c), every value of x corresponds to only one value of y. Tables a), b), and c) each express y as a function of x. Notice that different values of x may correspond to the same value of y. In Table d), we have the value of 23 for x corresponding to two different values of y, 48 and 37. So Table d) does not express y as a function of x.

Now do Exercises 17–24

Helpful Hint

In a function, every value for the independent variable determines conclusively a corresponding value for the dependent variable. If there is more than one possible value for the dependent variable, then the set of ordered pairs is not a function.

Functions Expressed by Ordered Pairs

A computer at your grocery store determines the price of each item by searching a long list of ordered pairs in which the first coordinate is the universal product code and the second coordinate is the price of the item with that code. For each product code there is a unique price. This process certainly satisfies the rule definition of a function. Since the set of ordered pairs is the essential part of this rule we say that the set of ordered pairs is a function.

Function (as a Set of Ordered Pairs)

A function is a set of ordered pairs of real numbers such that no two ordered pairs have the same first coordinates and different second coordinates.

Note the importance of the phrase "no two ordered pairs have the same first coordinates and different second coordinates." Imagine the problems at the grocery store if the computer gave two different prices for the same universal product code. Note also that the product code is an identification number and it cannot be used in calculations. So the computer can use a function defined by a formula to determine the amount of tax, but it cannot use a formula to determine the price from the product code.

EXAMPLE 4

Functions expressed by a set of ordered pairs

Determine whether each set of ordered pairs is a function.

a) $\{(1, 2), (1, 5), (-4, 6)\}$ **b)** $\{(-1, 3), (0, 3), (6, 3), (-3, 2)\}$

Solution

a) This set of ordered pairs is not a function because $(1, 2)$ and $(1, 5)$ have the same first coordinates but different second coordinates.

b) This set of ordered pairs is a function. Note that the same second coordinate with different first coordinates is permitted in a function.

Now do Exercises 25–32

If there are infinitely many ordered pairs in a function, then we can use set-builder notation from Chapter 1 along with an equation to express the function. For example,

$$\{(x, y) \mid y = x^2\}$$

is the set of ordered pairs in which the y-coordinate is the square of the x-coordinate. Ordered pairs such as $(0, 0)$, $(2, 4)$, and $(-2, 4)$ belong to this set. This set is a function because every value of x determines only one value of y.

EXAMPLE 5

Functions expressed by set-builder notation

Determine whether each set of ordered pairs is a function.

a) $\{(x, y) \mid y = 3x^2 - 2x + 1\}$

b) $\{(x, y) \mid y^2 = x\}$

c) $\{(x, y) \mid x + y = 6\}$

Helpful Hint

Real-life variables are generally not as simple as the ones we consider. A student's college GPA is not a function of age, because many students with the same age have different GPAs. However, GPA is probably a function of a large number of variables: age, IQ, high school GPA, number of working hours, mother's IQ, and so on.

Solution

a) This set is a function because each value we select for x determines only one value for y.

b) If $x = 9$, then we have $y^2 = 9$. Because both 3 and -3 satisfy $y^2 = 9$, both $(9, 3)$ and $(9, -3)$ belong to this set. So the set is not a function.

c) If we solve $x + y = 6$ for y, we get $y = -x + 6$. Because each value of x determines only one value for y, this set is a function. In fact, this set is a linear function.

Now do Exercises 33–40

We often omit the set notation when discussing functions. For example, the equation

$$y = 3x^2 - 2x + 1$$

expresses y as a function of x because the set of ordered pairs determined by the equation is a function. However, the equation

$$y^2 = x$$

does not express y as a function of x because ordered pairs such as $(9, 3)$ and $(9, -3)$ satisfy the equation.

EXAMPLE 6

Functions expressed by equations

Determine whether each equation expresses y as a function of x.

a) $y = |x|$ **b)** $y = x^3$ **c)** $x = |y|$

Helpful Hint

To determine whether an equation expresses y as a function of x, always select a number for x (the independent variable) and then see if there is more than one corresponding value for y (the dependent variable). If there is more than one corresponding y-value, then y is not a function of x.

Solution

a) Because every number has a unique absolute value, $y = |x|$ is a function.

b) Because every number has a unique cube, $y = x^3$ is a function.

c) The equation $x = |y|$ does not express y as a function of x because both $(4, -4)$ and $(4, 4)$ satisfy this equation. These ordered pairs have the same first coordinate but different second coordinates.

Now do Exercises 41–48

Calculator Close-Up

Most calculators graph only functions. If you enter an equation using the Y= key and the calculator accepts it and draws a graph, the equation defines y as a function of x.

Graphs of Functions

Every function determines a set of ordered pairs, and any set of ordered pairs has a graph in the rectangular coordinate system. For example, the set of ordered pairs determined by the linear function $y = 2x - 1$ is shown in Fig. 9.10 on the next page.

Every graph illustrates a set of ordered pairs, but not every graph is a graph of a function. For example, the circle in Fig. 9.11 on the next page is not a graph of a function because the ordered pairs $(0, 4)$ and $(0, -4)$ are both on the graph, and these two ordered pairs have the same first coordinate and different second coordinates. Whether a graph has such ordered pairs can be determined by a simple visual test called the **vertical-line test.**

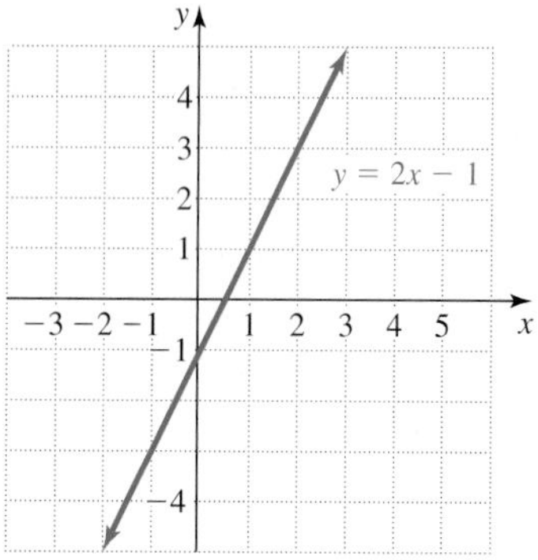

Figure 9.10

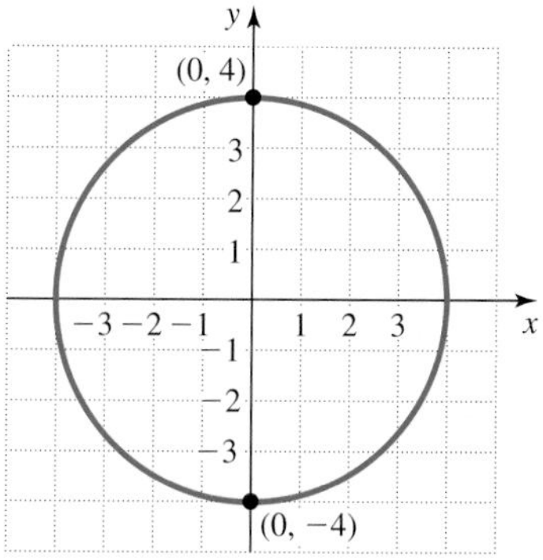

Figure 9.11

Vertical-Line Test

If it is possible to draw a vertical line that crosses a graph two or more times, then the graph is not the graph of a function.

If there is a vertical line that crosses a graph twice (or more), then we have two points (or more) with the same x-coordinate and different y-coordinates, and so the graph is not the graph of a function. If you mentally consider every possible vertical line and none of them cross the graph more than once, then you can conclude that the graph is the graph of a function.

EXAMPLE 7

Helpful Hint

Note that the vertical-line test works in theory, but it is certainly limited by the accuracy of the graph. Pictures are often deceptive. However, in spite of its limitations, the vertical-line test gives us a visual idea of what the graph of a function looks like.

Using the vertical-line test

Which of the following graphs are graphs of functions?

a)

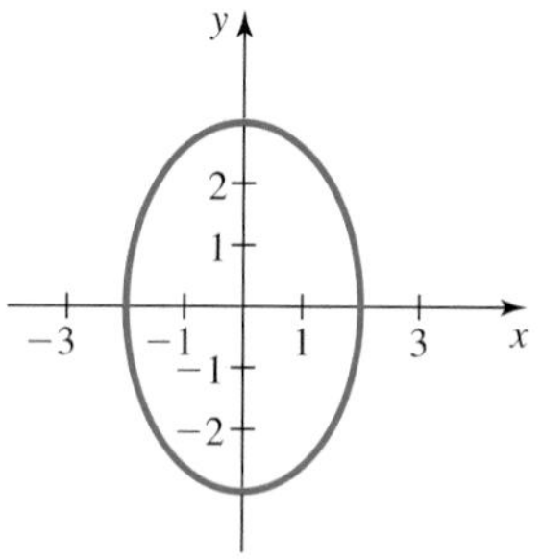

b)

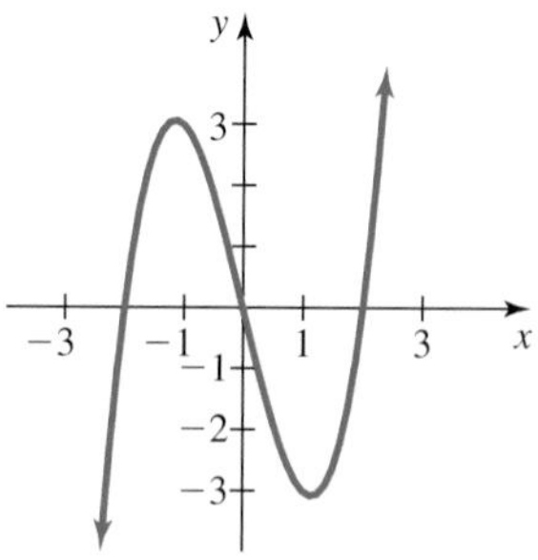

c)

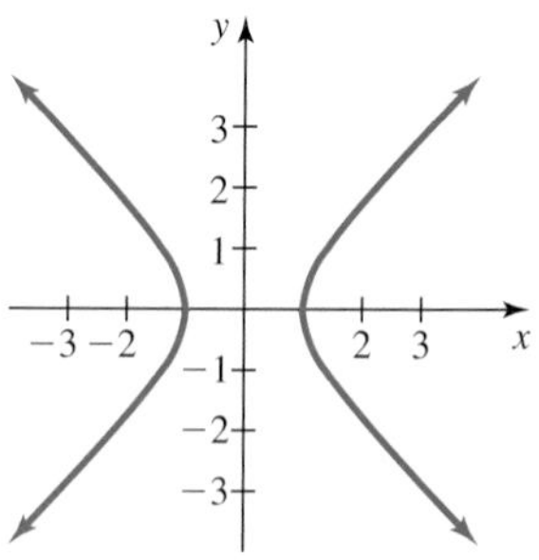

Solution

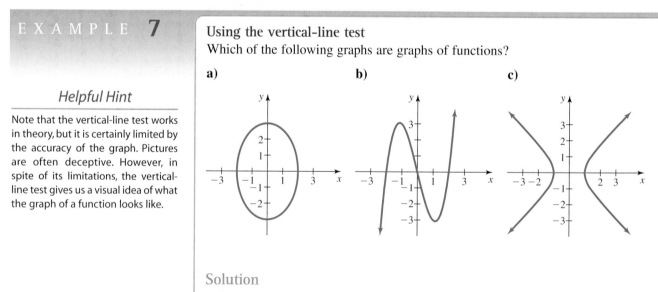

Neither a) nor c) is the graph of a function, since we can draw vertical lines that cross these graphs twice. Graph b) is the graph of a function, since no vertical line crosses it twice.

Now do Exercises 49–54

Domain and Range

A function is a set of ordered pairs. The set of all first coordinates of the ordered pairs is the **domain** of the function and the set of all second coordinates of the ordered pairs is the **range** of the function. In Example 8 we identify the domain and range for functions that are given as sets of ordered pairs.

EXAMPLE 8

Domain and range

Determine the domain and range of each function.

a) $\{(3, -1), (2, 5), (1, -4)\}$ **b)** $\{(2, 3), (4, 3), (6, 3), (8, 3)\}$

Solution

a) The domain is the set of numbers that occur as first coordinates, $\{1, 2, 3\}$. The range is the set of second coordinates, $\{-4, -1, 5\}$.

b) The domain is $\{2, 4, 6, 8\}$. Since 3 is the only number used as the second coordinate, the range is $\{3\}$.

Now do Exercises 55–58

If a function is defined by an equation, then the domain consists of those real numbers that can be used for the independent variable in the equation. For example, in $y = 3x$ we can use any real number for x. In $y = \frac{1}{x}$ we can use any nonzero real number for x. The domain is often the set of all real numbers, which is abbreviated as R.

The graph of a function is a picture of all ordered pairs of the function. So if a function is defined by an equation, it is usually helpful to graph the function and then use the graph as an aid in determining the domain and range.

EXAMPLE 9

Finding domain and range from a graph

Graph each function and then determine its domain and range.

a) $y = 2x - 1$ **b)** $y = 2x^2 - 8x + 5$ **c)** $y = -x^2 - 6x$

Solution

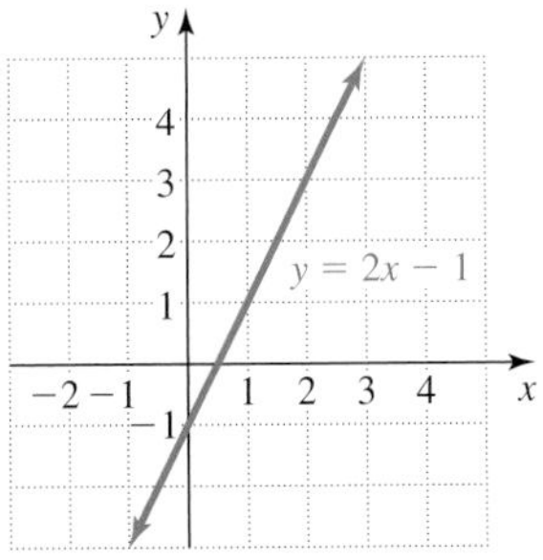

Figure 9.12

a) The graph of $y = 2x - 1$ is a line with y-intercept $(0, -1)$ and slope 2 as shown in Fig. 9.12. Any real number can be used for x in $y = 2x - 1$. So the domain is the set of real numbers, R. From the graph we can see that the line extends infinitely upward and downward. So the range is also R.

b) To graph the parabola $y = 2x^2 - 8x + 5$, first find the vertex:

$$x = \frac{-b}{2a} = \frac{8}{4} = 2 \qquad y = 2(2)^2 - 8(2) + 5 = -3$$

The vertex is $(2, -3)$. The parabola also goes through $(1, -1)$ and $(3, -1)$ as shown in Fig. 9.13 on the next page. Since any number can be used for x in $y = 2x^2 - 8x + 5$, the domain is R. Since the parabola opens upward, the smallest y-coordinate on the graph is -3. So the range is the set of real numbers that are greater than or equal to -3, which is written in set-builder notation as $\{y \mid y \geq -3\}$ or in interval notation as $[-3, \infty)$.

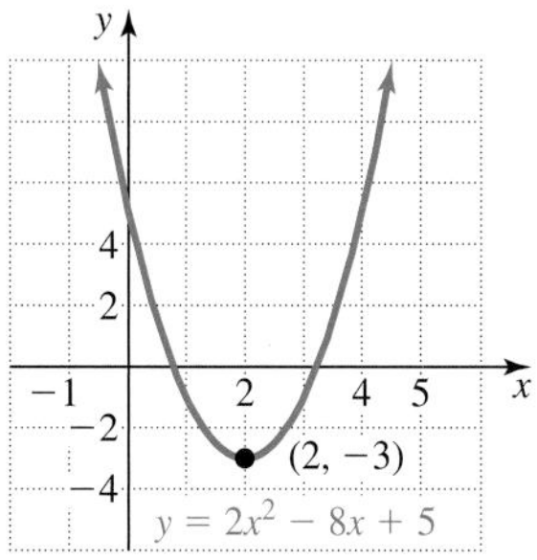

Figure 9.13

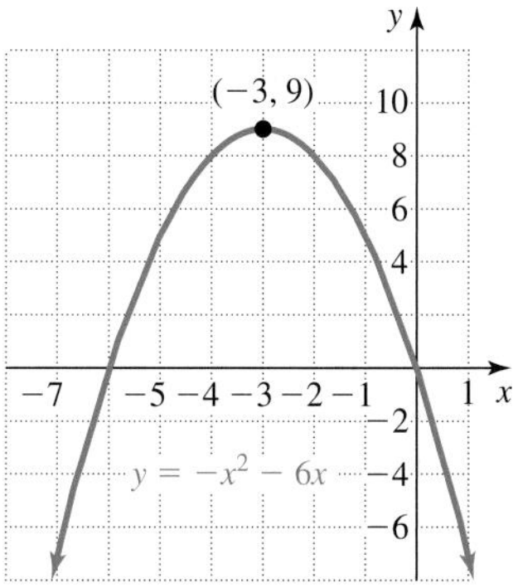

Figure 9.14

c) To graph the parabola $y = -x^2 - 6x$, first find the vertex:

$$x = \frac{-b}{2a} = \frac{6}{-2} = -3 \qquad y = -(-3)^2 - 6(-3) = 9$$

The vertex is $(-3, 9)$. The parabola also goes through $(-2, 8)$ and $(-4, 8)$ as shown in Fig. 9.14. Since any real number can be used in place of x in $y = -x^2 - 6x$, the domain is R. Since the parabola opens downward, the largest y-coordinate on the graph is 9. So the range is the set of real numbers that are less than or equal to 9, which is written in set notation as $\{y \mid y \leq 9\}$ or in interval notation as $(-\infty, 9]$.

Now do Exercises 59–66

Function Notation

When the variable y is a function of x, we may use the notation $f(x)$ to represent y. This **function notation** was first used in Section 5.1 with polynomial functions.

The symbol $f(x)$ is read as "f of x." So if x is the independent variable, we may use y or $f(x)$ to represent the dependent variable. For example, the function

$$y = 2x + 3$$

can also be written as

$$f(x) = 2x + 3.$$

We use y and $f(x)$ interchangeably. We think of f as the name of the function. We may use letters other than f. For example, the function $g(x) = 2x + 3$ is the same function as $f(x) = 2x + 3$.

The expression $f(x)$ represents the second coordinate when the first coordinate is x; it does not mean f times x. For example, if we replace x by 4 in $f(x) = 2x + 3$, we get

$$f(4) = 2 \cdot 4 + 3 = 11.$$

So if the first coordinate is 4, then the second coordinate is $f(4)$, or 11. The ordered pair (4, 11) belongs to the function f. This statement means that the function f pairs 4 with 11. We can use the diagram in Fig. 9.15 to picture this situation.

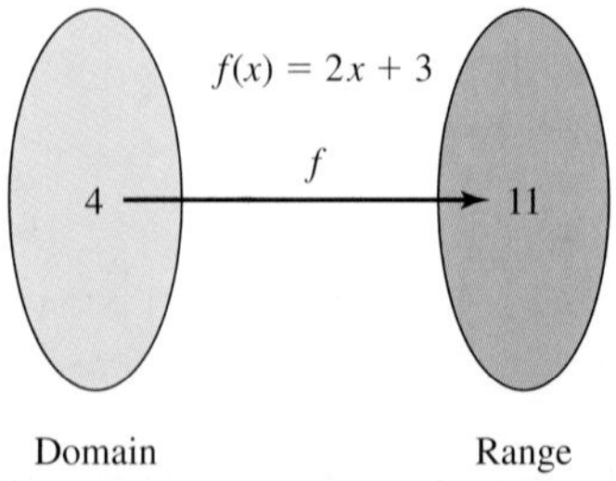

Figure 9.15

EXAMPLE 10

Using function notation

Suppose $f(x) = x^2 - 1$ and $g(x) = -3x + 2$. Find the following:

a) $f(-2)$ **b)** $f(-1)$ **c)** $g(0)$ **d)** $g(6)$

Study Tip

Everyone knows that you must practice to be successful with musical instruments, foreign languages, and sports. Success in algebra also requires regular practice. So budget your time so that you have a regular practice period for algebra.

Solution

a) Replace x by -2 in the formula $f(x) = x^2 - 1$:

$$\begin{aligned} f(-2) &= (-2)^2 - 1 \\ &= 4 - 1 \\ &= 3 \end{aligned}$$

So $f(-2) = 3$.

b) Replace x by -1 in the formula $f(x) = x^2 - 1$:

$$\begin{aligned} f(-1) &= (-1)^2 - 1 \\ &= 1 - 1 \\ &= 0 \end{aligned}$$

So $f(-1) = 0$.

c) Replace x by 0 in the formula $g(x) = -3x + 2$:

$$g(0) = -3 \cdot 0 + 2 = 2$$

So $g(0) = 2$.

d) Replace x by 6 in $g(x) = -3x + 2$ to get $g(6) = -16$.

Now do Exercises 67–78

EXAMPLE 11

Using function notation in an application

The formula $C(n) = 0.10n + 4.95$ gives the monthly cost in dollars for n minutes of long-distance calls. Find $C(40)$ and $C(100)$.

Solution

Replace n with 40 in the formula:

$$\begin{aligned} C(n) &= 0.10n + 4.95 \\ C(40) &= 0.10(40) + 4.95 \\ &= 8.95 \end{aligned}$$

So $C(40) = 8.95$. The cost for 40 minutes of calls is \$8.95. Now

$$C(100) = 0.10(100) + 4.95 = 14.95.$$

So $C(100) = 14.95$. The cost of 100 minutes of calls is \$14.95.

Now do Exercises 93–98

CAUTION $C(n)$ is not C times n. In the context of functions, $C(n)$ represents the value of C corresponding to a value of n.

Calculator Close-Up

A graphing calculator can be used to evaluate a formula in the same manner as in Example 11. To evaluate

$$C = 0.10n + 4.95$$

enter the formula into your calculator as $y_1 = 0.10x + 4.95$ using the Y = key:

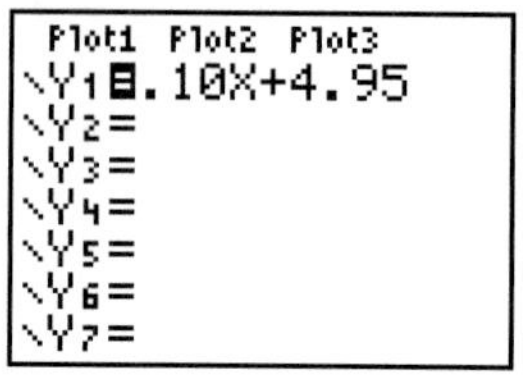

To find the cost of 40 minutes of calls, enter $y_1(40)$ on the home screen and press ENTER:

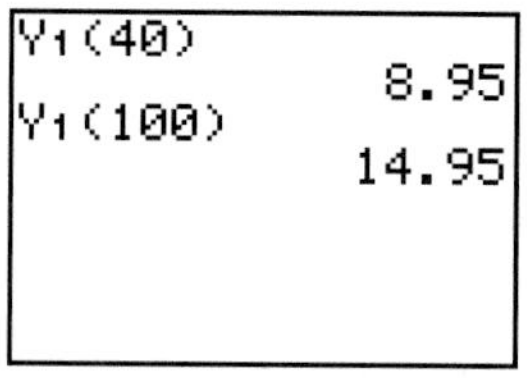

Warm-Ups ▼

True or false? Explain your answer.

1. Any set of ordered pairs is a function.
2. The area of a square is a function of the length of a side.
3. The set $\{(-1, 3), (-3, 1), (-1, -3)\}$ is a function.
4. The set $\{(1, 5), (3, 5), (7, 5)\}$ is a function.
5. The domain of $\{(1, 2), (3, 4)\}$, is $\{1, 3\}$.
6. The domain of $y = x^2$ is R.
7. The range of $y = x^2$ is $\{y \mid y \geq 0\}$.
8. The set $\{(x, y) \mid x = 2y\}$ is a function.
9. The set $\{(x, y) \mid x = y^2\}$ is a function.
10. If $f(x) = x^2 - 5$, then $f(-2) = -1$.

9.7 Exercises

Reading and Writing *After reading this section, write out the answers to these questions. Use complete sentences.*

1. What is a function?

2. What are the different ways to express functions?

3. What do all descriptions of functions have in common?

4. How can you tell at a glance if a graph is a graph of a function?

5. What is the domain of a function?

6. What is function notation?

Write a formula that describes the function for each of the following. See Examples 1 and 2.

7. A small pizza costs \$5.00 plus 50 cents for each topping. Express the total cost C as a function of the number of toppings t.

8. A developer prices condominiums in Florida at \$20,000 plus \$40 per square foot of living area. Express the cost C as a function of the number of square feet of living area s.

9. The sales tax rate on groceries in Mayberry is 9%. Express the total cost T (including tax) as a function of the total price of the groceries S.

10. With a GM MasterCard, 5% of the amount charged is credited toward a rebate on the purchase of a new car. Express the rebate R as a function of the amount charged A.

11. Express the circumference of a circle as a function of its radius.

12. Express the circumference of a circle as a function of its diameter.

13. Express the perimeter P of a square as a function of the length s of a side.

14. Express the perimeter P of a rectangle with width 10 ft as a function of its length L.

15. Express the area A of a triangle with a base of 10 m as a function of its height h.

16. Express the area A of a trapezoid with bases 12 cm and 10 cm as a function of its height h.

Determine whether each table expresses the second variable as a function of the first variable. See Example 3.

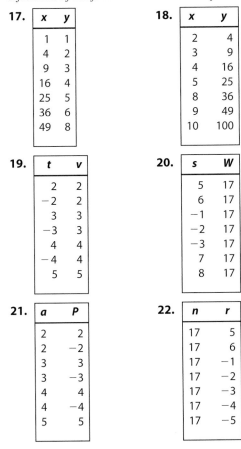

17.

x	y
1	1
4	2
9	3
16	4
25	5
36	6
49	8

18.

x	y
2	4
3	9
4	16
5	25
8	36
9	49
10	100

19.

t	v
2	2
−2	2
3	3
−3	3
4	4
−4	4
5	5

20.

s	W
5	17
6	17
−1	17
−2	17
−3	17
7	17
8	17

21.

a	P
2	2
2	−2
3	3
3	−3
4	4
4	−4
5	5

22.

n	r
17	5
17	6
17	−1
17	−2
17	−3
17	−4
17	−5

23.

b	q
1970	0.14
1972	0.18
1974	0.18
1976	0.22
1978	0.25
1980	0.28

24.

c	h
345	0.3
350	0.4
355	0.5
360	0.6
365	0.7
370	0.8
380	0.9

Determine whether each set of ordered pairs is a function. See Example 4.

25. $\{(1, 2), (2, 3), (3, 4)\}$

26. $\{(1, -3), (1, 3), (2, 12)\}$

27. $\{(-1, 4), (2, 4), (3, 4)\}$

28. $\{(1, 7), (7, 1)\}$

29. $\{(0, -1), (0, 1)\}$

30. $\{(1, 7), (-2, 7), (3, 7), (4, 7)\}$

31. $\{(50, 50)\}$

32. $\{(0, 0)\}$

Determine whether each set is a function. See Example 5.

33. $\{(x, y) \mid y = x - 3\}$

34. $\{(x, y) \mid y = x^2 - 2x - 1\}$

35. $\{(x, y) \mid x = |y|\}$

36. $\{(x, y) \mid x = y^2 + 1\}$

37. $\{(x, y) \mid x = y + 1\}$

38. $\left\{(x, y) \middle| y = \frac{1}{x}\right\}$

39. $\{(x, y) \mid x = y^2 - 1\}$

40. $\{(x, y) \mid x = 3y\}$

Determine whether each equation expresses y as a function of x. See Example 6.

41. $x = 4y$

42. $x = -3y$

43. $y = \frac{2}{x}$

44. $y = \frac{x}{2}$

45. $y = x^3 - 1$

46. $y = |x - 1|$

47. $x^2 + y^2 = 25$

48. $x^2 - y^2 = 9$

Which of the following graphs are graphs of functions? See Example 7.

49.

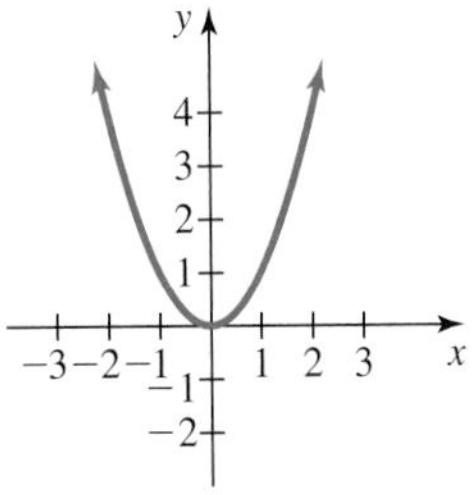

50.

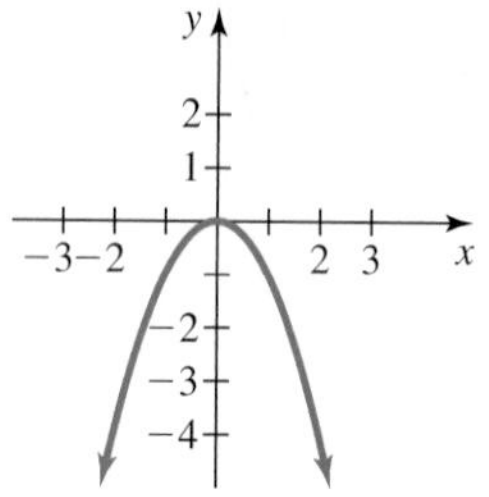

51.

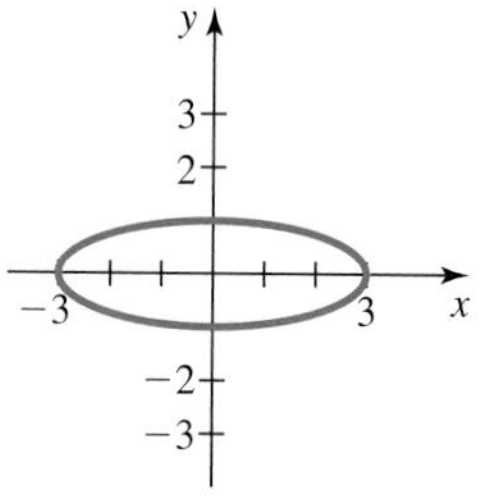

52.

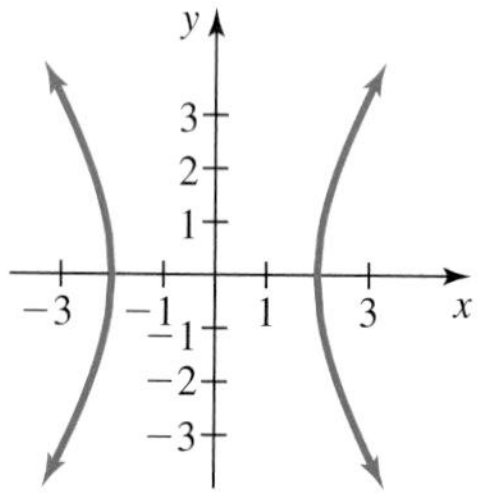

53.

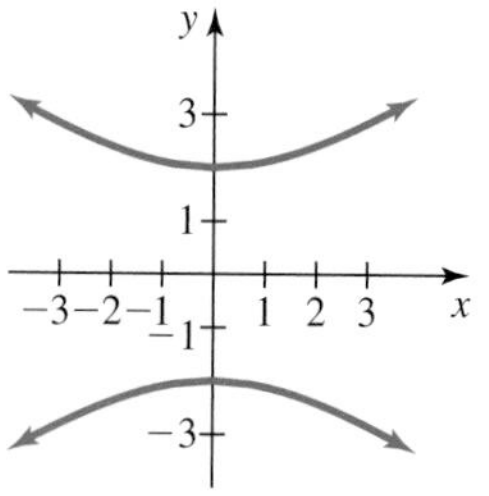

54.

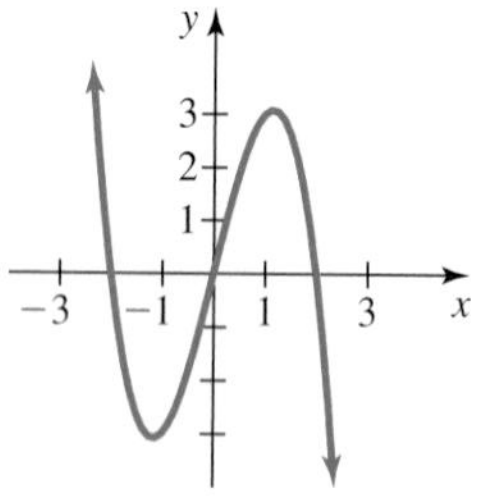

Determine the domain and range of each function. See Example 8.

55. $\{(3, 3), (2, 5), (1, 7)\}$

56. $\{(-1, 4), (3, 5)\}$

57. $\{(0, 1), (2, 1), (4, 1)\}$

58. $\{(4, -2), (5, -2), (7, -2), (8, -2)\}$

Graph each function and identify the domain and range. See Example 9.

59. $y = 2x - 6$

60. $y = 3x + 2$

61. $y = -x$

62. $y = 5 - x$

63. $y = x^2 - 4x + 2$

64. $y = x^2 + 6x + 5$

65. $y = -2x^2 - 4x + 5$

66. $y = -4x^2 + 8x$

Let $f(x) = 2x - 1$, $g(x) = x^2 - 3$, *and* $h(x) = |x - 1|$. *Find the following. See Example 10.*

67. $f(0)$ **68.** $f(-1)$

69. $f\left(\frac{1}{2}\right)$ **70.** $f\left(\frac{3}{4}\right)$

71. $g(4)$ **72.** $g(-4)$

73. $g(0.5)$ **74.** $g(-1.5)$

75. $h(3)$ **76.** $h(-1)$

77. $h(0)$ **78.** $h(1)$

Let $f(x) = \frac{2}{3}x - 4$, $g(x) = \sqrt{8 - x}$, *and* $h(x) = \sqrt[3]{x}$. *Find the following.*

79. $f(-3)$ **80.** $f(3)$

81. $f(6)$ **82.** $g(-1)$

83. $g(0)$ **84.** $g(-8)$

85. $g(7)$ **86.** $h(0)$

87. $h(-1)$ **88.** $h(8)$

Let $f(x) = x^3 - x^2$ *and* $g(x) = x^2 - 4.2x + 2.76$. *Find the following. Round each answer to three decimal places.*

89. $f(5.68)$ **90.** $g(-2.7)$

91. $g(3.5)$ **92.** $f(67.2)$

Solve each problem. See Example 11.

93. ***Velocity and time.*** If a ball is thrown straight upward into the air with a velocity of 100 ft/sec, then its velocity t seconds later is given by

$$v(t) = -32t + 100.$$

a) Find $v(0)$, $v(1)$, and $v(2)$.

b) Is the velocity increasing or decreasing as the time increases?

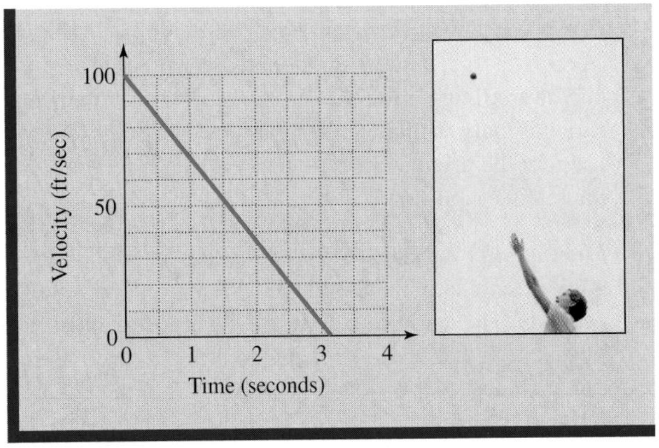

Figure for Exercise 93

94. ***Cost and toppings.*** The cost c in dollars for a pizza with n toppings is given by

$$c(n) = 0.75n + 6.99.$$

a) Find $c(2)$, $c(4)$, and $c(5)$.

b) Is the cost increasing or decreasing as the number of toppings increases?

95. ***Threshold weight.*** The threshold weight for an individual is the weight beyond which the risk of death increases significantly. For middle-aged males the function $W(h) = 0.000534h^3$ expresses the threshold weight in pounds as a function of the height h in inches. Find $W(70)$. Find the threshold weight for a 6′2″ middle-aged male.

96. ***Pole vaulting.*** The height a pole vaulter attains is a function of the vaulter's velocity on the runway. The function

$$h(v) = \frac{1}{64}v^2$$

gives the height in feet as a function of the velocity v in feet per second.

a) Find $h(35)$ to the nearest tenth of an inch.

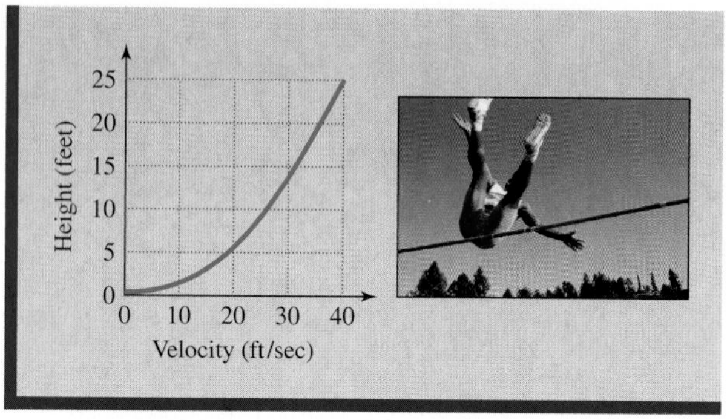

Figure for Exercise 96

b) Who gains more height from an increase of 1 ft/sec in velocity: a fast runner or a slow runner?

97. ***Credit card fees.*** A certain credit card company gets 4% of each charge, and the retailer receives the rest. At the end of a billing period the retailer receives a statement showing only the retailer's portion of each transaction. Express the original amount charged C as a function of the retailer's portion r.

98. ***More credit card fees.*** Suppose that the amount charged on the credit card in the previous exercise includes 8% sales tax. The credit card company does not get any of the sales tax. In this case the retailer's portion of each transaction includes sales tax on the original cost of the goods. Express the original amount charged C as a function of the retailer's portion r.

Getting More Involved

Discussion

In each situation determine whether a is a function of b, b is a function of a, or neither. Answers may vary depending on interpretations.

99. a = the price per gallon of regular unleaded.
b = the number of gallons that you get for $10.

100. a = the universal product code of an item at Sears.
b = the price of that item.

101. a = a student's score on the last test in this class.
b = the number of hours he/she spent studying.

102. a = a student's score on the last test in this class.
b = the IQ of the student's mother.

103. a = the weight of a package shipped by UPS.
b = the cost of shipping that package.

104. a = the Celsius temperature at any time.
b = the Fahrenheit temperature at the same time.

105. a = the weight of a letter.
b = the cost of mailing the letter.

106. a = the cost of a gallon of milk.
b = the amount of sales tax on that gallon.

9.8 Combining Functions

In this Section

- **Basic Operations with Functions**
- **Composition**

In this section you will learn how to combine functions to obtain new functions.

Basic Operations with Functions

An entrepreneur plans to rent a stand at a farmers market for $25 per day to sell strawberries. If she buys x flats of berries for $5 per flat and sells them for $9 per flat, then her daily cost in dollars can be written as a function of x:

$$C(x) = 5x + 25$$

Assuming she sells as many flats as she buys, her revenue in dollars is also a function of x:

$$R(x) = 9x$$

Because profit is revenue minus cost, we can find a function for the profit by subtracting the functions for cost and revenue:

$$\begin{aligned} P(x) &= R(x) - C(x) \\ &= 9x - (5x + 25) \\ &= 4x - 25 \end{aligned}$$

The function $P(x) = 4x - 25$ expresses the daily profit as a function of x. Since $P(6) = -1$ and $P(7) = 3$, the profit is negative if 6 or fewer flats are sold and positive if 7 or more flats are sold.

In the example of the entrepreneur we subtracted two functions to find a new function. In other cases we may use addition, multiplication, or division to combine two functions. For any two given functions we can define the sum, difference, product, and quotient functions as follows.

Sum, Difference, Product, and Quotient Functions

Given two functions f and g, the functions $f + g$, $f - g$, $f \cdot g$, and $\frac{f}{g}$ are defined as follows:

Sum function:	$(f + g)(x) = f(x) + g(x)$	
Difference function:	$(f - g)(x) = f(x) - g(x)$	
Product function:	$(f \cdot g)(x) = f(x) \cdot g(x)$	
Quotient function:	$\left(\frac{f}{g}\right)(x) = \frac{f(x)}{g(x)}$	provided that $g(x) \neq 0$

The domain of the function $f + g$, $f - g$, $f \cdot g$, or $\frac{f}{g}$ is the intersection of the domain of f and the domain of g. For the function $\frac{f}{g}$ we also rule out any values of x for which $g(x) = 0$.

EXAMPLE 1

Operations with functions

Let $f(x) = 4x - 12$ and $g(x) = x - 3$. Find the following.

a) $(f + g)(x)$

b) $(f - g)(x)$

c) $(f \cdot g)(x)$

d) $\left(\frac{f}{g}\right)(x)$

Helpful Hint

Note that we use $f + g$, $f - g$, $f \cdot g$, and f/g to name these functions only because there is no application in mind here. We generally use a single letter to name functions after they are combined as we did when using P for the profit function rather than $R - C$.

Solution

a) $(f + g)(x) = f(x) + g(x)$

$= 4x - 12 + x - 3$

$= 5x - 15$

b) $(f - g)(x) = f(x) - g(x)$

$= 4x - 12 - (x - 3)$

$= 3x - 9$

c) $(f \cdot g)(x) = f(x) \cdot g(x)$

$$= (4x - 12)(x - 3)$$
$$= 4x^2 - 24x + 36$$

d) $\left(\frac{f}{g}\right)(x) = \frac{f(x)}{g(x)} = \frac{4x - 12}{x - 3} = \frac{4(x - 3)}{x - 3} = 4 \quad \text{for } x \neq 3.$

Now do Exercises 5–8

EXAMPLE 2

Evaluating a sum function

Let $f(x) = 4x - 12$ and $g(x) = x - 3$. Find $(f + g)(2)$.

Solution

In Example 1(a) we found a general formula for the function $f + g$, namely, $(f + g)(x) = 5x - 15$. If we replace x by 2, we get

$$(f + g)(2) = 5(2) - 15$$
$$= -5.$$

We can also find $(f + g)(2)$ by evaluating each function separately and then adding the results. Because $f(2) = -4$ and $g(2) = -1$, we get

$$(f + g)(2) = f(2) + g(2)$$
$$= -4 + (-1)$$
$$= -5.$$

Now do Exercises 9–16

Helpful Hint

The difference between the first four operations with functions and composition is like the difference between parallel and series in electrical connections. Components connected in parallel operate simultaneously and separately. If components are connected in series, then electricity must pass through the first component to get to the second component.

Composition

A salesperson's monthly salary is a function of the number of cars he sells: \$1000 plus \$50 for each car sold. If we let S be his salary and n be the number of cars sold, then S in dollars is a function of n:

$$S = 1000 + 50n$$

Each month the dealer contributes \$100 plus 5% of his salary to a profit-sharing plan. If P represents the amount put into profit sharing, then P (in dollars) is a function of S:

$$P = 100 + 0.05S$$

Now P is a function of S, and S is a function of n. Is P a function of n? The value of n certainly determines the value of P. In fact, we can write a formula for P in terms of n by substituting one formula into the other:

$$P = 100 + 0.05S$$
$$= 100 + 0.05(1000 + 50n) \quad \text{Substitute } S = 1000 + 50n.$$
$$= 100 + 50 + 2.5n \quad \text{Distributive property}$$
$$= 150 + 2.5n$$

Now P is written as a function of n, bypassing S. We call this idea **composition of functions.**

EXAMPLE 3

The composition of two functions

Given that $y = x^2 - 2x + 3$ and $z = 2y - 5$, write z as a function of x.

Solution

Replace y in $z = 2y - 5$ by $x^2 - 2x + 3$:

$$\begin{aligned} z &= 2y - 5 \\ &= 2(x^2 - 2x + 3) - 5 \quad \text{Replace } y \text{ by } x^2 - 2x + 3. \\ &= 2x^2 - 4x + 1 \end{aligned}$$

The equation $z = 2x^2 - 4x + 1$ expresses z as a function of x.

Now do Exercises 17–24

The composition of two functions using function notation is defined as follows.

Composition of Functions

The **composition** of f and g is denoted $f \circ g$ and is defined by the equation

$$(f \circ g)(x) = f(g(x)),$$

provided that $g(x)$ is in the domain of f.

The notation $f \circ g$ is read as "the composition of f and g" or "f compose g." The diagram in Fig. 9.16 shows a function g pairing numbers in its domain with numbers in its range. If the range of g is contained in or equal to the domain of f, then f pairs the second coordinates of g with numbers in the range of f. The composition function $f \circ g$ is a rule for pairing numbers in the domain of g directly with numbers in the range of f, bypassing the middle set. The domain of the function $f \circ g$ is the domain of g (or a subset of it) and the range of $f \circ g$ is the range of f (or a subset of it).

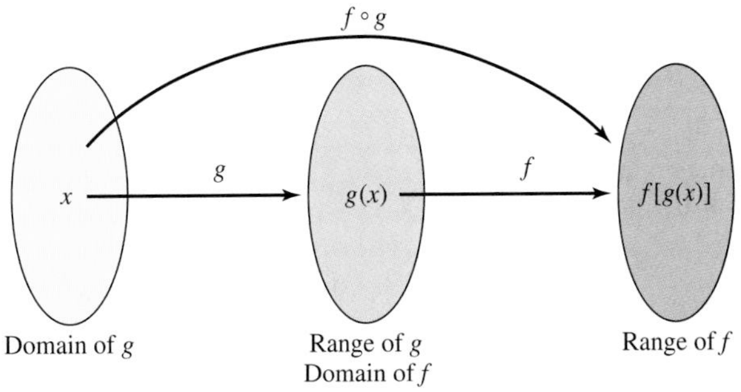

Figure 9.16

CAUTION The order in which functions are written is important in composition. For the function $f \circ g$ the function f is applied to $g(x)$. For the function $g \circ f$, the function g is applied to $f(x)$. The function closest to the variable x is applied first.

EXAMPLE 4

Composition of functions

Let $f(x) = 3x - 2$ and $g(x) = x^2 + 2x$. Find the following.

a) $(g \circ f)(2)$ **b)** $(f \circ g)(2)$ **c)** $(g \circ f)(x)$ **d)** $(f \circ g)(x)$

Calculator Close-Up

Set $y_1 = 3x - 2$ and $y_2 = x^2 + 2x$. You can find the composition for Examples 4(a) and 4(b) by evaluating $y_2(y_1(2))$ and $y_1(y_2(2))$. Note that the order in which you evaluate the functions is critical.

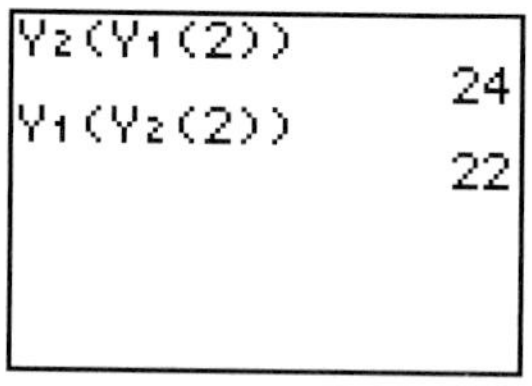

Solution

a) Because $(g \circ f)(2) = g(f(2))$, we first find $f(2)$:

$$f(2) = 3 \cdot 2 - 2 = 4$$

Because $f(2) = 4$, we have

$$(g \circ f)(2) = g(f(2)) = g(4) = 4^2 + 2 \cdot 4 = 24.$$

So $(g \circ f)(2) = 24$.

b) Because $(f \circ g)(2) = f(g(2))$, we first find $g(2)$:

$$g(2) = 2^2 + 2 \cdot 2 = 8$$

Because $g(2) = 8$, we have

$$(f \circ g)(2) = f(g(2)) = f(8) = 3 \cdot 8 - 2 = 22.$$

Thus $(f \circ g)(2) = 22$.

c)
$$\begin{aligned}(g \circ f)(x) &= g(f(x)) \\ &= g(3x - 2) \\ &= (3x - 2)^2 + 2(3x - 2) \\ &= 9x^2 - 12x + 4 + 6x - 4 = 9x^2 - 6x\end{aligned}$$

So $(g \circ f)(x) = 9x^2 - 6x$.

d)
$$\begin{aligned}(f \circ g)(x) &= f(g(x)) \\ &= f(x^2 + 2x) \\ &= 3(x^2 + 2x) - 2 = 3x^2 + 6x - 2\end{aligned}$$

So $(f \circ g)(x) = 3x^2 + 6x - 2$.

Now do Exercises 25–48

Helpful Hint

A composition of functions can be viewed as two function machines where the output of the first is the input of the second.

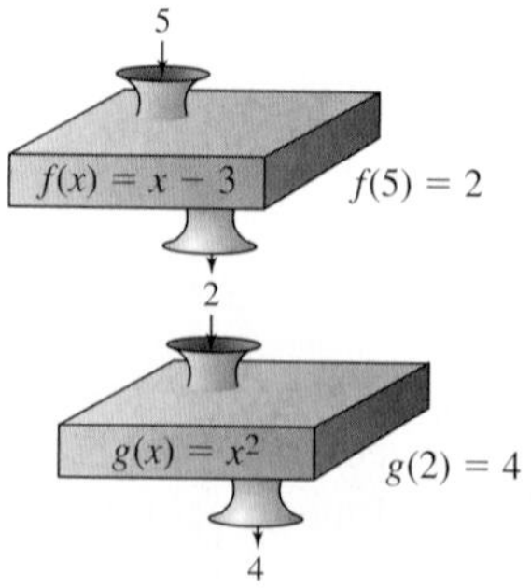

Notice that in Example 4(a) and (b), $(g \circ f)(2) \neq (f \circ g)(2)$. In Example 4(c) and (d) we see that $(g \circ f)(x)$ and $(f \circ g)(x)$ have different formulas defining them. In general, $f \circ g \neq g \circ f$.

It is often useful to view a complicated function as a composition of simpler functions. For example, the function $Q(x) = (x - 3)^2$ consists of two operations, subtracting 3 and squaring. So Q can be described as a composition of the functions $f(x) = x - 3$ and $g(x) = x^2$. To check this, we find $(g \circ f)(x)$:

$$\begin{aligned}(g \circ f)(x) &= g(f(x)) \\ &= g(x - 3) \\ &= (x - 3)^2\end{aligned}$$

We can express the fact that Q is the same as the composition function $g \circ f$ by writing $Q = g \circ f$ or $Q(x) = (g \circ f)(x)$.

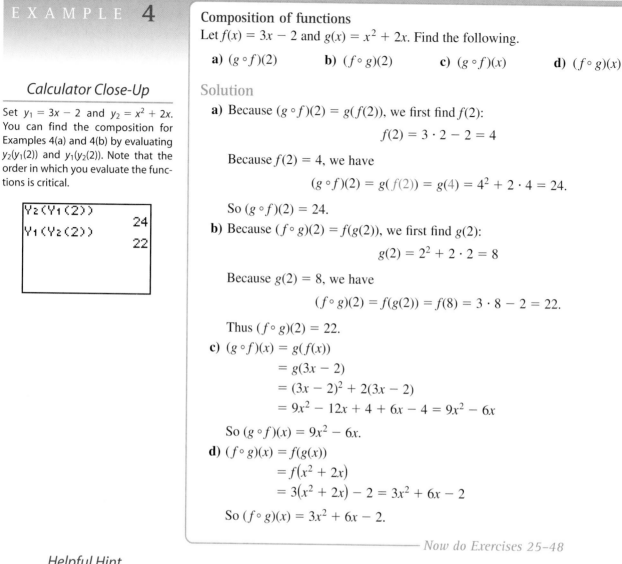

EXAMPLE 5

Expressing a function as a composition of simpler functions

Let $f(x) = x - 2$, $g(x) = 3x$, and $h(x) = \sqrt{x}$. Write each of the following functions as a composition, using f, g, and h.

a) $F(x) = \sqrt{x - 2}$

b) $H(x) = x - 4$

c) $K(x) = 3x - 6$

Solution

a) The function F consists of first subtracting 2 from x and then taking the square root of that result. So $F = h \circ f$. Check this result by finding $(h \circ f)(x)$:

$$(h \circ f)(x) = h(f(x)) = h(x - 2) = \sqrt{x - 2}$$

b) Subtracting 4 from x can be accomplished by subtracting 2 from x and then subtracting 2 from that result. So $H = f \circ f$. Check by finding $(f \circ f)(x)$:

$$(f \circ f)(x) = f(f(x)) = f(x - 2) = x - 2 - 2 = x - 4$$

c) Notice that $K(x) = 3(x - 2)$. The function K consists of subtracting 2 from x and then multiplying the result by 3. So $K = g \circ f$. Check by finding $(g \circ f)(x)$:

$$(g \circ f)(x) = g(f(x)) = g(x - 2) = 3(x - 2) = 3x - 6$$

Now do Exercises 49–58

CAUTION In Example 5(a) we have $F = h \circ f$ because in F we subtract 2 before taking the square root. If we had the function $G(x) = \sqrt{x} - 2$, we would take the square root before subtracting 2. So $G = f \circ h$. Notice how important the order of operations is here.

Warm-Ups ▼

True or false? Explain your answer.

1. If $f(x) = x - 2$ and $g(x) = x + 3$, then $(f - g)(x) = -5$.
2. If $f(x) = x + 4$ and $g(x) = 3x$, then $\left(\frac{f}{g}\right)(2) = 1$.
3. The functions $f \circ g$ and $g \circ f$ are always the same.
4. If $f(x) = x^2$ and $g(x) = x + 2$, then $(f \circ g)(x) = x^2 + 2$.
5. The functions $f \circ g$ and $f \cdot g$ are always the same.
6. If $f(x) = \sqrt{x}$ and $g(x) = x - 9$, then $g(f(x)) = f(g(x))$ for every x.
7. If $f(x) = 3x$ and $g(x) = \frac{x}{3}$, then $(f \circ g)(x) = x$.
8. If $a = 3b^2 - 7b$, and $c = a^2 + 3a$, then c is a function of b.
9. The function $F(x) = \sqrt{x - 5}$ is a composition of two functions.
10. If $F(x) = (x - 1)^2$, $h(x) = x - 1$, and $g(x) = x^2$, then $F = g \circ h$.

9.8 Exercises

Boost your GRADE at mathzone.com!

MathZone

- Practice Problems
- Self-Tests
- Videos
- Net Tutor
- e-Professors

Reading and Writing *After reading this section, write out the answers to these questions. Use complete sentences.*

1. What are the basic operations with functions?

2. How do we perform the basic operations with functions?

3. What is the composition of two functions?

4. How is the order of operations related to composition of functions?

Let $f(x) = 4x - 3$, *and* $g(x) = x^2 - 2x$. *Find the following. See Examples 1 and 2.*

5. $(f + g)(x)$

6. $(f - g)(x)$

7. $(f \cdot g)(x)$

8. $\left(\frac{f}{g}\right)(x)$

9. $(f + g)(3)$

10. $(f + g)(2)$

11. $(f - g)(-3)$

12. $(f - g)(-2)$

13. $(f \cdot g)(-1)$

14. $(f \cdot g)(-2)$

15. $\left(\frac{f}{g}\right)(4)$

16. $\left(\frac{f}{g}\right)(-2)$

For Exercises 17–24, use the two functions to write y as a function of x. See Example 3.

17. $y = 3a - 2, a = 2x - 6$

18. $y = 2c + 3, c = -3x + 4$

19. $y = 2d + 1, d = \dfrac{x + 1}{2}$

20. $y = -3d + 2, d = \dfrac{2 - x}{3}$

21. $y = m^2 - 1, m = x + 1$

22. $y = n^2 - 3n + 1, n = x + 2$

23. $y = \dfrac{a - 3}{a + 2}, a = \dfrac{2x + 3}{1 - x}$

24. $y = \dfrac{w + 2}{w - 5}, w = \dfrac{5x + 2}{x - 1}$

Let $f(x) = 2x - 3$, $g(x) = x^2 + 3x$, *and* $h(x) = \frac{x + 3}{2}$. *Find the following. See Example 4.*

25. $(g \circ f)(1)$

26. $(f \circ g)(-2)$

27. $(f \circ g)(1)$

28. $(g \circ f)(-2)$

29. $(f \circ f)(4)$

30. $(h \circ h)(3)$

31. $(h \circ f)(5)$

32. $(f \circ h)(0)$

33. $(f \circ h)(5)$

34. $(h \circ f)(0)$

35. $(g \circ h)(-1)$

36. $(h \circ g)(-1)$

37. $(f \circ g)(2.36)$

38. $(h \circ f)(23.761)$

39. $(g \circ f)(x)$

40. $(g \circ h)(x)$

41. $(f \circ g)(x)$

42. $(h \circ g)(x)$

43. $(h \circ f)(x)$

44. $(f \circ h)(x)$

45. $(f \circ f)(x)$

46. $(g \circ g)(x)$

47. $(h \circ h)(x)$

48. $(f \circ f \circ f)(x)$

Let $f(x) = \sqrt{x}$, $g(x) = x^2$, and $h(x) = x - 3$. Write each of the following functions as a composition using f, g, or h. See Example 5.

49. $F(x) = \sqrt{x - 3}$

50. $N(x) = \sqrt{x} - 3$

51. $G(x) = x^2 - 6x + 9$

52. $P(x) = x$ for $x \geq 0$

53. $H(x) = x^2 - 3$

54. $M(x) = x^{1/4}$

55. $J(x) = x - 6$

56. $R(x) = \sqrt{x^2 - 3}$

57. $K(x) = x^4$

58. $Q(x) = \sqrt{x^2 - 6x + 9}$

Solve each problem.

59. ***Color monitor.*** A color monitor has a square viewing area that has a diagonal measure of 15 inches. Find the area of the viewing area in square inches (in.^{2}). Write a formula for the area of a square as a function of the length of its diagonal.

60. ***Perimeter.*** Write a formula for the perimeter of a square as a function of its area.

61. ***Profit function.*** A plastic bag manufacturer has determined that the company can sell as many bags as it can produce each month. If it produces x thousand bags in a month, the revenue is $R(x) = x^2 - 10x + 30$ dollars, and the cost is $C(x) = 2x^2 - 30x + 200$ dollars. Use the fact that profit is revenue minus cost to write the profit as a function of x.

62. ***Area of a sign.*** A sign is in the shape of a square with a semicircle of radius x adjoining one side and a semicircle of diameter x removed from the opposite side. If the sides of the square are length $2x$, then write the area of the sign as a function of x.

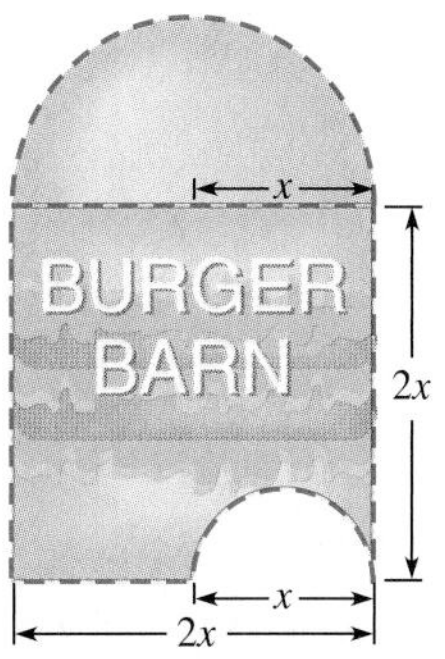

Figure for Exercise 62

63. ***Junk food expenditures.*** Suppose the average family spends 25% of its income on food, $F = 0.25I$, and 10% of each food dollar on junk food, $J = 0.10F$. Write J as a function of I.

64. ***Area of an inscribed circle.*** A pipe of radius r must pass through a square hole of area M as shown in the figure. Write the cross-sectional area of the pipe A as a function of M.

Figure for Exercise 64

65. ***Displacement-length ratio.*** To find the displacement-length ratio D for a sailboat, first find x, where $x = (L/100)^3$ and L is the length at the water line in feet. Next find D, where $D = (d/2240)/x$ and d is the displacement in pounds.

a) For the Pacific Seacraft 40, $L = 30$ ft 3 in. and $d = 24{,}665$ pounds. Find D.

b) For a boat with a displacement of 25,000 pounds, write D as a function of L.

c) The graph for the function in part (b) is shown in the figure on the next page. For a fixed displacement, does the displacement-length ratio increase or decrease as the length increases?

66. ***Sail area-displacement ratio.*** To find the sail area-displacement ratio S, first find y, where $y = (d/64)^{2/3}$ and d

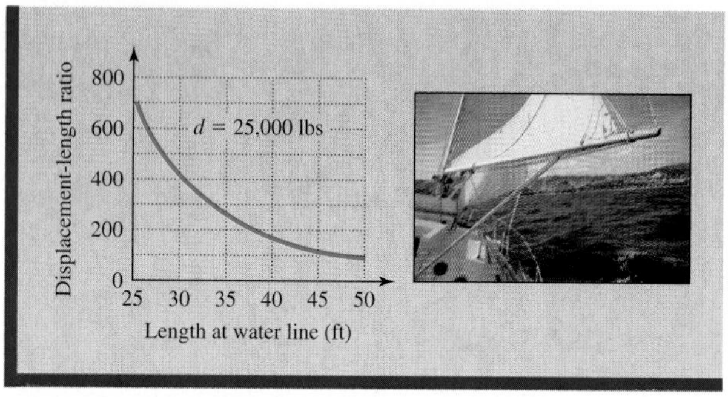

Figure for Exercise 65

is the displacement in pounds. Next find S, where $S = A/y$ and A is the sail area in square feet.

a) For the Pacific Seacraft 40, $A = 846$ square feet (ft^2) and $d = 24{,}665$ pounds. Find S.

b) For a boat with a sail area of 900 ft^2, write S as a function of d.

c) For a fixed sail area, does S increase or decrease as the displacement increases?

Collaborative Activities

Grouping: Four students

Topic: Solving quadratic equations

Designing a River Park

At the last election, voters approved the creation of a park along the river on the edge of the city. You are student representatives on the citizen's committee working with the park designer. The city is allocating 300 acres for the park. The park designer says that the committee needs to decide the dimensions of the park. In terms of the land available, the park dimensions can vary between a length that is twice the width to a length that is 10 times the width. The committee decides to consider the following three cases:

A. The length is twice the width.

B. The length is 10 times the width.

C. The length is 5 times the width.

Because you have all studied algebra, you offer to find the dimensions and draw sketches of the three cases.

The following suggestions will help you in solving this problem.

1. Determine what units of measurement you wish to use (feet, yards, miles, meters or kilometers).
2. A calculator with a conversion feature will make your work easier.
3. Assign roles. You might use the roles Moderator, Recorder, Calculator user, and Sketcher.

In your report to the committee, do the following:

1. For each case, include the equations used, the width and length you found (rounded to the nearest whole number), and your work showing how you found the length and width.
2. Include a small sketch of each case showing the given proportions.
3. Choose one of the three cases that your group feels gives the best dimensions for the park. Considering what features (flowerbeds, playground, soccer field, walking or bike paths, etc.) you would like to have in the park will help you decide which dimensions you would choose. Make a larger-scale drawing of this case and include the features you would want in a park.
4. Write a short paragraph explaining why you chose the case you did.

Chapter 9 Wrap-Up

Summary

Quadratic Equations		Examples
Quadratic equation	An equation of the form $ax^2 + bx + c = 0,$ where a, b, and c are real numbers with $a \neq 0$	$x^2 = 10$ $(x + 3)^2 = 8$ $x^2 + 5x - 7 = 0$
Methods for solving quadratic equations	Factoring	$x^2 + 5x + 6 = 0$ $(x + 3)(x + 2) = 0$
	Square root property	$(x - 3)^2 = 6$ $x - 3 = \pm\sqrt{6}$
	Completing the square (works on any quadratic): Take one-half of middle term, square it, then add it to each side.	$x^2 + 6x = 7$ $x^2 + 6x + 9 = 7 + 9$ $(x + 3)^2 = 16$
	Quadratic formula (works on any quadratic): $x = \frac{-b \pm \sqrt{b^2 - 4ac}}{2a}$	$2x^2 - 3x - 6 = 0$ $x = \frac{3 \pm \sqrt{9 - 4(2)(-6)}}{2(2)}$
Number and types of solutions	Determined by the discriminant $b^2 - 4ac$ $b^2 - 4ac > 0$: two real solutions	$x^2 + 5x - 9 = 0$ has two real solutions because $5^2 - 4(1)(-9) > 0.$
	$b^2 - 4ac = 0$: one real solution	$x^2 + 6x + 9 = 0$ has one real solution because $6^2 - 4(1)(9) = 0.$
	$b^2 - 4ac < 0$: no real solutions (two complex solutions)	$x^2 + 3x + 10 = 0$ has no real solutions because $3^2 - 4(1)(10) < 0.$

Complex Numbers		Examples
Complex numbers	Numbers of the form $a + bi$, where a and b are real numbers, $i = \sqrt{-1}$, and $i^2 = -1$.	$12, -3i, 5 + 4i, \sqrt{2} - i\sqrt{3}$
Imaginary numbers	Numbers of the form $a + bi$, where $b \neq 0$	$5i, 13 + i\sqrt{6}$
Square root of a negative number	If b is a positive real number, then $\sqrt{-b} = i\sqrt{b}$.	$\sqrt{-3} = i\sqrt{3}$ $\sqrt{-4} = i\sqrt{4} = 2i$

Complex conjugates	The complex numbers $a + bi$ and $a - bi$ are called complex conjugates of each other. Their product is a real number.	$(1 + 2i)(1 - 2i) = 1 + 4 = 5$
Complex number operations	Add, subtract, and multiply as if the complex numbers were binomials with variable i. Use the distributive property for multiplication. Remember that $i^2 = -1$. Divide complex numbers by multiplying the numerator and denominator by the conjugate of the denominator, then simplify.	$(2 + 3i) + (3 - 5i) = 5 - 2i$ $(2 - 5i) - (4 - 2i) = -2 - 3i$ $(3 - 4i)(2 + 5i) = 26 + 7i$ $\frac{4 - 6i}{5 + 2i} = \frac{(4 - 6i)(5 - 2i)}{(5 + 2i)(5 - 2i)}$
Parabolas		**Examples**
Opening	The parabola $y = ax^2 + bx + c$ opens upward if $a > 0$ or downward if $a < 0$.	$y = x^2 - 2x$ opens upward. $y = -x^2$ opens downward.
Vertex	The first coordinate of the vertex is $x = \frac{-b}{2a}$. The second coordinate of the vertex is the minimum value of y if $a > 0$ or the maximum value of y if $a < 0$.	$y = x^2 - 2x$ opens upward with vertex $(1, -1)$. Minimum y-value is -1.
Intercepts	The x-intercepts are found by solving $ax^2 + bx + c = 0$. The y-intercept is found by replacing x by 0 in $y = ax^2 + bx + c$.	$y = x^2 - 2x - 8$ has x-intercepts $(-2, 0)$ and $(4, 0)$ and y-intercept $(0, -8)$.
Functions		**Examples**
Definition of a function	A function is a rule by which any allowable value of one variable (the independent variable) determines a unique value of a second variable (the dependent variable).	$A = \pi r^2$
Equivalent definition of a function	A function is a set of ordered pairs such that no two ordered pairs have the same first coordinates and different second coordinates. To say that y is a function of x means that y is determined uniquely by x.	$\{(1, 0), (3, 8)\}$ $\{(x, y) \mid y = x^2\}$
Domain	The set of values of the independent variable, x	$y = x^2$ Domain: all real numbers, R
Range	The set of values of the dependent variable, y	$y = x^2$ Range: nonnegative real numbers, $\{y \mid y \geq 0\}$ or $[0, \infty)$
Linear function	If $y = mx + b$, we say that y is a linear function of x.	$F = \frac{9}{5}C + 32$

Quadratic function	A function of the form $y = ax^2 + bx + c$, where a, b, and c are real numbers and $a \neq 0$	$y = 3x^2 - 8x + 9$ $p = -3q^2 - 8q + 1$
Function notation	If x is the independent variable, then we use the notation $f(x)$ to represent the dependent variable.	$y = 2x + 3$ $f(x) = 2x + 3$
Combining Functions		**Examples**
Sum	$(f + g)(x) = f(x) + g(x)$	For $f(x) = x^2$ and $g(x) = x + 1$ $(f + g)(x) = x^2 + x + 1$
Difference	$(f - g)(x) = f(x) - g(x)$	$(f - g)(x) = x^2 - x - 1$
Product	$(f \cdot g)(x) = f(x) \cdot g(x)$	$(f \cdot g)(x) = x^3 + x^2$
Quotient	$\left(\frac{f}{g}\right)(x) = \frac{f(x)}{g(x)}$	$\left(\frac{f}{g}\right)(x) = \frac{x^2}{x + 1}$
Composition	$(g \circ f)(x) = g(f(x))$ $(f \circ g)(x) = f(g(x))$	$(g \circ f)(x) = g(x^2) = x^2 + 1$ $(f \circ g)(x) = f(x + 1)$ $= x^2 + 2x + 1$

Enriching Your Mathematical Word Power

For each mathematical term, choose the correct meaning.

1. quadratic equation
   a. $ax + b = c$ with $a \neq 0$
   b. $ax^2 + bx + c = 0$ with $a \neq 0$
   c. $ax + b = 0$ with $a \neq 0$
   d. $a/x^2 + b/x = c$ with $x \neq 0$

2. perfect square trinomial
   a. a trinomial of the form $a^2 + 2ab + b^2$
   b. a trinomial of the form $a^2 + b^2$
   c. a trinomial of the form $a^2 + ab + b^2$
   d. a trinomial of the form $a^2 - 2ab - b^2$

3. completing the square
   a. drawing a perfect square
   b. evaluating $(a + b)^2$
   c. drawing the fourth side when given three sides of a square
   d. finding the third term of a perfect square trinomial

4. quadratic formula
   a. $x = \dfrac{-b \pm \sqrt{b^2 - 4ac}}{2}$
   b. $x = -b \pm \dfrac{\sqrt{b^2 - 4ac}}{2a}$
   c. $x = \dfrac{-b \pm \sqrt{b^2 - 4ac}}{2a}$
   d. $x = \dfrac{b \pm \sqrt{b^2 - 4ac}}{2a}$

5. discriminant
   a. the vertex of a parabola
   b. the radicand in the quadratic formula
   c. the leading coefficient in $ax^2 + bx + c$
   d. to treat unfairly

6. complex numbers
   a. $a + bi$, where a and b are real
   b. irrational numbers
   c. imaginary numbers
   d. $\sqrt{-1}$

7. imaginary unit
   a. 1
   b. -1
   c. i
   d. $\sqrt{1}$

8. imaginary numbers
a. $a + bi$, where a and b are real and $b \neq 0$
b. i
c. a complex number
d. a complex number in which the real part is 0

9. complex conjugates
a. i and $\sqrt{-1}$
b. $a + bi$ and $a - bi$
c. $(a + b)(a - b)$
d. i and -1

10. function
a. domain and range
b. a set of ordered pairs
c. a rule by which any allowable value of one variable determines a unique value of a second variable
d. a graph

11. domain
a. the set of first coordinates of a function
b. the set of second coordinates of a function
c. the set of real numbers
d. the integers

12. range
a. all of the possibilities
b. the coordinates of a function
c. the entire set of numbers
d. the set of second coordinates of a function

13. quadratic function
a. $y = ax + b$ with $a \neq 0$
b. a parabola
c. $y = ax^2 + bx + c$ with $a \neq 0$
d. the quadratic formula

14. composition of *f* and *g*
a. the function $f \circ g$ where $(f \circ g)(x) = f(g(x))$
b. the function $f \circ g$ where $(f \circ g)(x) = g(f(x))$
c. the function $f \cdot g$ where $(f \cdot g)(x) = f(x) \cdot g(x)$
d. a diagram showing f and g

15. sum of *f* and *g*
a. the function $f \cdot g$ where $(f \cdot g)(x) = f(x) \cdot g(x)$
b. the function $f + g$ where $(f + g)(x) = f(x) + g(x)$
c. the function $f \circ g$ where $(f \circ g)(x) = g(f(x))$
d. the function obtained by adding the domains of f and g

Review Exercises

9.1 *Solve each equation.*

1. $x^2 - 9 = 0$

2. $x^2 - 1 = 0$

3. $x^2 - 9x = 0$

4. $x^2 - x = 0$

5. $x^2 - x = 2$

6. $x^2 - 9x = 10$

7. $(x - 9)^2 = 10$

8. $(x + 5)^2 = 14$

9. $4x^2 - 12x + 9 = 0$

10. $9x^2 + 6x + 1 = 0$

11. $t^2 - 9t + 20 = 0$

12. $s^2 - 4s + 3 = 0$

13. $\frac{x}{2} = \frac{7}{x + 5}$

14. $\sqrt{x + 4} = \frac{2x - 1}{3}$

15. $\frac{1}{2}x^2 + \frac{7}{4}x = 1$

16. $\frac{2}{3}x^2 - 1 = -\frac{1}{3}x$

9.2 *Solve each equation by completing the square.*

17. $x^2 + 4x - 7 = 0$

18. $x^2 + 6x - 3 = 0$

19. $x^2 + 3x - 28 = 0$

20. $x^2 - x - 6 = 0$

21. $x^2 + 3x - 5 = 0$

22. $x^2 + \frac{4}{3}x - \frac{1}{3} = 0$

23. $2x^2 + 9x - 5 = 0$

24. $2x^2 + 6x - 5 = 0$

9.3 *Find the value of the discriminant, and tell how many real solutions each equation has.*

25. $25t^2 - 10t + 1 = 0$

26. $3x^2 + 2 = 0$

27. $-3w^2 + 4w - 5 = 0$

28. $5x^2 - 7x = 0$

29. $-3v^2 + 4v = -5$

30. $49u^2 + 42u + 9 = 0$

Use the quadratic formula to solve each equation.

31. $6x^2 + x - 2 = 0$

32. $-6x^2 + 11x + 10 = 0$

33. $x^2 - x = 4$

34. $y^2 - 2y = 4$

35. $5x^2 - 6x - 1 = 0$

36. $t^2 - 6t + 4 = 0$

37. $3x^2 - 5x = 0$

38. $2w^2 - w = 15$

9.4 *For each problem, find the exact and approximate answers. Round the decimal answers to three decimal places.*

39. ***Bird watching.*** Chuck is standing 12 meters from a tree, watching a bird's nest that is 5 meters above eye level. Find the distance from Chuck's eyes to the nest.

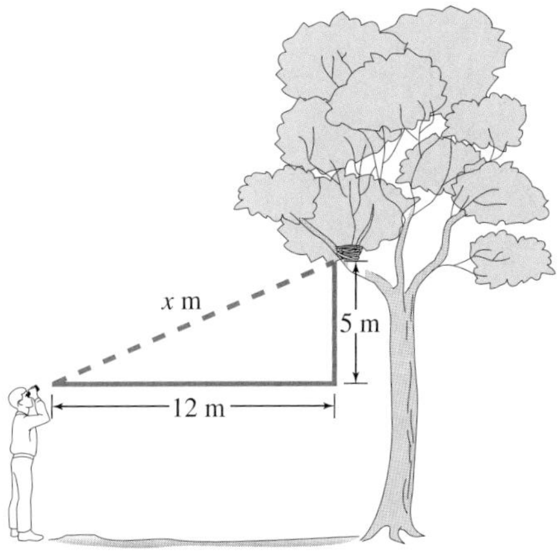

Figure for Exercise 39

40. ***Diagonal of a square.*** Find the diagonal of a square if the length of each side is 20 yards.

41. ***Lengthy legs.*** The hypotenuse of a right triangle measures 5 meters, and one leg is 2 meters longer than the other. Find the lengths of the legs.

42. ***Width and height.*** The width of a rectangular bookcase is 3 feet shorter than the height. If the diagonal is 7 feet, then what are the dimensions of the bookcase?

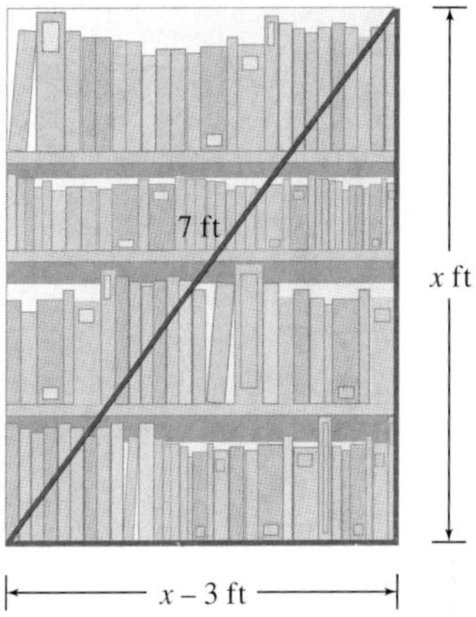

Figure for Exercise 42

43. ***Base and height.*** The base of a triangle is 4 inches longer than the height. If the area of the triangle is 20 square inches, then what are the lengths of the base and height?

44. ***Dimensions of a parallelogram.*** The base of a parallelogram is 1 meter longer than the height. If the area of the parallelogram is 8 square meters, then what are the lengths of the base and height?

45. ***Unknown numbers.*** Find two positive real numbers whose sum is 6 and whose product is 7.

46. ***Dimensions of a rectangle.*** The perimeter of a rectangle is 16 feet, and its area is 13 square feet. What are the dimensions of the rectangle?

47. ***Printing time.*** The old printer took 2 hours longer than the new printer to print 100,000 mailing labels. With both printers working on the job, the 100,000 labels can be printed in 8 hours. How long would it take each printer working alone to do the job?

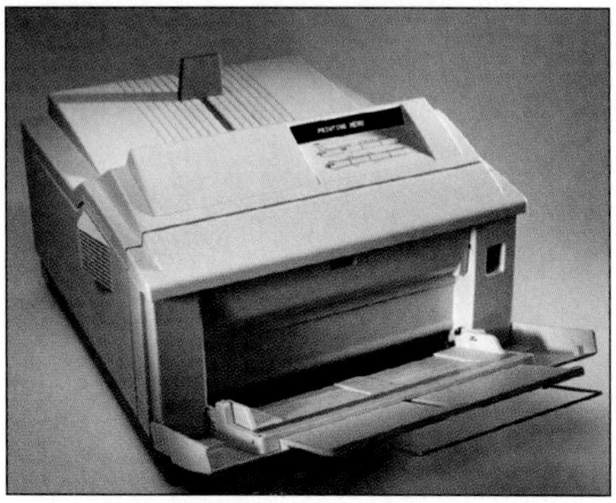

Photo for Exercise 47

48. ***Tilling the garden.*** When Blake uses his old tiller, it takes him 3 hours longer to till the garden than it takes Cassie using her new tiller. If Cassie will not let Blake use her new tiller and they can till the garden together in 6 hours then how long would it take each one working alone?

Photo for Exercise 48

9.5 *Perform the indicated operations. Write answers in the form $a + bi$.*

49. $(2 + 3i) + (5 - 6i)$

50. $(2 - 5i) + (-9 - 4i)$

51. $(-5 + 4i) - (-2 - 3i)$

52. $(1 - i) - (1 + i)$

53. $(2 - 9i)(3 + i)$

54. $2i - 3(6 - 2i)$

55. $(3 + 8i)^2$

56. $(-5 - 2i)(-5 + 2i)$

57. $\dfrac{-2 - \sqrt{-8}}{2}$

58. $\dfrac{-6 + \sqrt{-54}}{-3}$

59. $\dfrac{1 + 3i}{6 - i}$

60. $\dfrac{3i}{8 + 3i}$

61. $\dfrac{5 + i}{4 - i}$

62. $\dfrac{3 + 2i}{i}$

Find the complex solutions to the quadratic equations.

63. $x^2 + 121 = 0$

64. $x^2 + 120 = 0$

65. $x^2 - 16x + 65 = 0$

66. $x^2 - 10x + 28 = 0$

67. $2x^2 - 3x + 9 = 0$

68. $3x^2 - 6x + 4 = 0$

9.6 *Find the vertex and intercepts for each parabola and sketch its graph.*

69. $y = x^2 - 6x$

70. $y = x^2 + 4x$

71. $y = x^2 - 4x - 12$

72. $y = x^2 + 2x - 24$

73. $y = -2x^2 + 8x$

74. $y = -3x^2 + 6x$

75. $y = -x^2 + 2x + 3$

76. $y = -x^2 - 3x - 2$

Solve each problem.

77. ***Minimizing cost.*** The unit cost in dollars for manufacturing n starters is given by $C = 0.004n^2 - 3.2n + 660$. What is the unit cost when 390 starters are manufactured? For what number of starters is the unit cost at a minimum?

78. ***Maximizing profit.*** The total profit (in dollars) for sales of x rowing machines is given by $P = -0.2x^2 + 300x - 200$. What is the profit if 500 are sold? For what value of x will the profit be at a maximum?

9.7 *Determine whether each set of ordered pairs is a function.*

79. $\{(4, 3), (5, 3)\}$

80. $\{(0, 0), (0, 1), (0, 2)\}$

81. $\{(3, 4), (3, 5)\}$

82. $\{(1, 2), (2, 3), (3, 4)\}$

Determine whether each equation expresses y as a function of x.

83. $y = x^2 + 10$

84. $y = 2x - 7$

85. $x^2 + y^2 = 1$

86. $x^2 = y^2$

Determine the domain and range of each function.

87. $\{(1, 2), (2, 0), (3, 0)\}$

88. $\{(2, 3), (4, 3), (6, 3)\}$

Find the domain and range of each quadratic function.

89. $y = x^2 + 4x + 1$

90. $y = x^2 - 6x + 2$

91. $y = -2x^2 - x + 4$

92. $y = -3x^2 + 2x + 7$

9.8 *Let* $f(x) = \frac{1}{2}x + 1$, $g(x) = |5 - 2x|$, *and* $h(x) = \frac{1}{x - 1}$. *Find the following.*

93. $f(1/2)$

94. $f(-4/5)$

95. $f(-8)$

96. $g(5/2)$

97. $g(-1/2)$

98. $g(3)$

99. $g(0)$

100. $h(1/2)$

101. $h(-1)$

102. $h(5/4)$

Let $f(x) = 3x + 5$, $g(x) = x^2 - 2x$, *and* $h(x) = \frac{x - 5}{3}$. *Find the following.*

103. $f(-3)$

104. $h(-4)$

105. $(h \circ f)(\sqrt{2})$

106. $(f \circ h)(319)$

107. $(g \circ f)(2)$

108. $(g \circ f)(x)$

109. $(f + g)(3)$

110. $(f - g)(x)$

111. $(f \cdot g)(x)$

112. $\left(\frac{f}{g}\right)(1)$

113. $(f \circ f)(0)$

114. $(f \circ f)(x)$

Let $f(x) = |x|$, $g(x) = x + 2$, *and* $h(x) = x^2$. *Write each of the following functions as a composition of functions, using f, g, or h.*

115. $F(x) = |x + 2|$

116. $G(x) = |x| + 2$

117. $H(x) = x^2 + 2$

118. $K(x) = x^2 + 4x + 4$

119. $I(x) = x + 4$

120. $J(x) = x^4 + 2$

Chapter 9 Test

Calculate the value of $b^2 - 4ac$ and state how many real solutions each equation has.

1. $9x^2 - 12x + 4 = 0$

2. $-2x^2 + 3x - 5 = 0$

3. $-2x^2 + 5x - 1 = 0$

Solve by using the quadratic formula.

4. $5x^2 + 2x - 3 = 0$

5. $2x^2 - 4x - 3 = 0$

Solve by completing the square.

6. $x^2 + 4x - 21 = 0$

7. $x^2 + 3x - 5 = 0$

Solve by any method.

8. $x(x + 1) = 20$

9. $x^2 - 28x + 75 = 0$

10. $\dfrac{x - 1}{3} = \dfrac{x + 1}{2x}$

Perform the indicated operations. Write answers in the form $a + bi$.

11. $(2 - 3i) + (8 + 6i)$

12. $(-2 - 5i) - (4 - 12i)$

13. $(-6i)^2$

14. $(3 - 5i)(4 + 6i)$

15. $(8 - 2i)(8 + 2i)$

16. $(4 - 6i) \div 2$

17. $\dfrac{-2 + \sqrt{-12}}{2}$

18. $\dfrac{6 - \sqrt{-18}}{-3}$

19. $\dfrac{5i}{4 + 3i}$

Find the complex solutions to the quadratic equations.

20. $x^2 + 6x + 12 = 0$

21. $-5x^2 + 6x - 5 = 0$

Graph each quadratic function. State the domain and range.

22. $y = 16 - x^2$

23. $y = x^2 - 3x$

Let $f(x) = -2x + 5$ and $g(x) = x^2 + 4$. Find the following.

24. $f(-3)$

25. $(g \circ f)(-3)$

26. $(g + f)(x)$

27. $(f \cdot g)(1)$

28. $(f/g)(2)$

29. $(f \circ g)(x)$

30. $(g \circ f)(x)$

Let $f(x) = x - 7$ and $g(x) = x^2$. Write each of the following functions as a composition of functions using f and g.

31. $H(x) = x^2 - 7$

32. $W(x) = x^2 - 14x + 49$

Solve each problem.

33. Find the x-intercepts for the parabola $y = x^2 - 6x + 5$.

34. The height in feet for a ball thrown upward at 48 feet per second is given by $s = -16t^2 + 48t$, where t is the time in seconds after the ball is tossed. What is the maximum height that the ball will reach?

35. Find two positive numbers that have a sum of 10 and a product of 23. Give exact answers.

*Making*Connections | A Review of Chapters 1–9

Solve each equation.

1. $2x - 1 = 0$

2. $2(x - 1) = 0$

3. $2x^2 - 1 = 0$

4. $(2x - 1)^2 = 8$

5. $2x^2 - 4x - 1 = 0$

6. $2x^2 - 4x = 0$

7. $2x^2 + x = 1$

8. $x - 2 = \sqrt{2x - 1}$

9. $\frac{1}{x} = \frac{x}{2x - 15}$

10. $\frac{1}{x} - \frac{1}{x - 1} = -\frac{1}{2}$

Solve each equation for y.

11. $5x - 4y = 8$

12. $3x - y = 9$

13. $\frac{y - 4}{x + 2} = \frac{2}{3}$

14. $ay + b = 0$

15. $ay^2 + by + c = 0$

16. $y - 1 = -\frac{2}{3}(x - 9)$

17. $\frac{2}{3}x + \frac{1}{2}y = \frac{1}{9}$

18. $x^2 + y^2 = a^2$

Suppose that each side of a square has length s, the diagonal has length d, the area of the square is A, and its perimeter is P.

19. Write P in terms of s.

20. Write A in terms of s.

21. Write P in terms of d.

22. Write d in terms of A.

Solve each system of equations.

23. $3x - 2y = 12$
$2x + 5y = -11$

24. $y = 3x + 1$
$3x - 0.6y = 3$

Graph each function.

25. $y = x - 3$

26. $y = 2 - x$

27. $y = x^2 - 3$

28. $y = 2 - x^2$

29. $y = \frac{2}{3}x - 4$

30. $y = -\frac{4}{3}x + 5$

Solve the problem.

31. ***Maximizing revenue.*** For the last three years the Lakeland Air Show has raised the price of its tickets and has sold fewer and fewer tickets, as shown in the table.

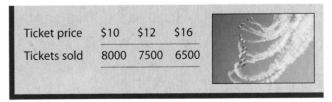

Ticket price	\$10	\$12	\$16
Tickets sold	8000	7500	6500

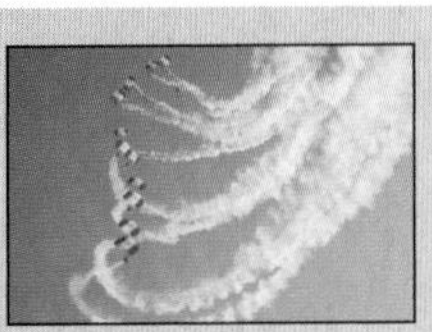

a) Use this information to write the number of tickets sold s as a linear function of the ticket price p.
b) Has the revenue from ticket sales increased or decreased as the ticket price was raised?
c) Write the revenue R as a function of the ticket price p.
d) What ticket price would produce the maximum revenue?

Critical Thinking | For Individual or Group Work | Chapter 9

These exercises can be solved by a variety of techniques, which may or may not require algebra. So be creative and think critically. Explain all answers. Answers are in the Instructor's Edition of this text.

1. ***Adjacent squares.*** A small square of area a is adjacent to a larger square of area b as shown in the accompanying figure. Find the distance between the centers of the two squares.

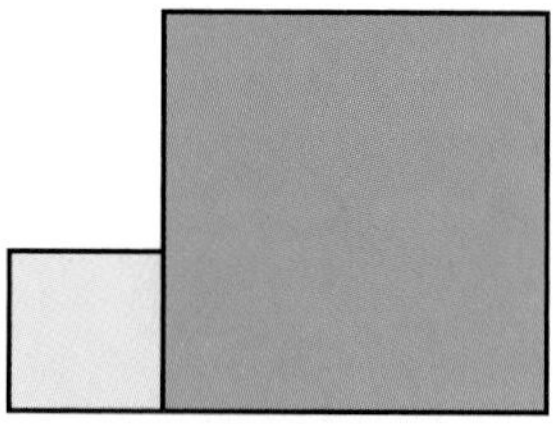

Figure for Exercise 1

2. ***Buried treasure.*** A pirate landed his boat on the northern most point of a circular island that had a diameter of 5 km. He walked 1 km due south and then walked due west until he reached the ocean. From there he walked 1 km due south and then walked due east until he reached the ocean. From there he walked 0.5 km due south and buried his treasure. As the crow flies, how far was it back to his boat?

3. ***Treadmill time.*** Angela starts running on a treadmill at 4 P.M. exactly. She quits running when as soon as the minute hand and hour hand of the clock coincide. Find the amount of time she spent on the treadmill to the nearest tenth of a second.

4. ***Perfect square.*** Find the smallest positive integer n such that $882n$ is a perfect square.

5. ***Pipeline problem.*** One-sixth of a gas pipeline is under water, two-fifths of the pipeline is in wetlands, and 78 kilometers is on dry land. How long is the pipeline?

6. ***Joni's rope.*** Joni positions her rope on the ground in the shape of an equilateral triangle. She then positions her rope on the ground in the shape of a regular hexagon. For each figure she uses the entire length of her rope.

a) Which figure has the greater area?

b) What is the ratio of the area of the hexagon to the area of the triangle?

7. ***Intersecting circles.*** What is the maximum number of times that five circles with the same radius but different centers could intersect? Six circles? What is an expression that gives the maximum number of intersections with n circles? How could you actually position 100 circles to intersect the maximum number of times?

8. ***Replacing football.*** A college has a football field that is surrounded by a 400-meter track. The track consists of two 100-meter straight sections and two 100-meter semicircles (measured on the inside edge) as shown in the accompanying figure. There is talk of replacing the football field with a soccer field. Will a soccer field fit inside the track? A college soccer field must be between 115 and 120 yards long and between 70 and 80 yards wide.

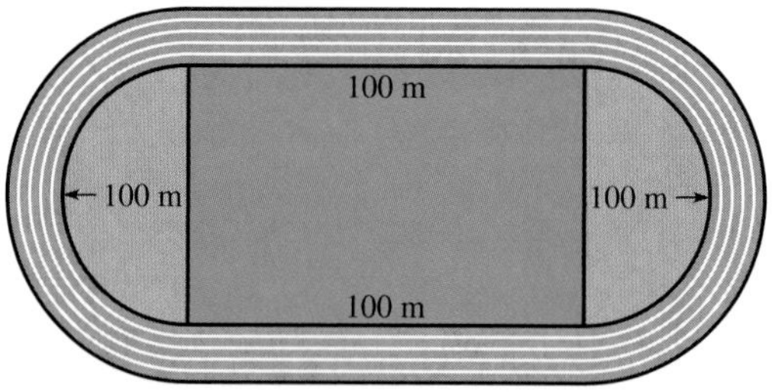

Figure for Exercise 8

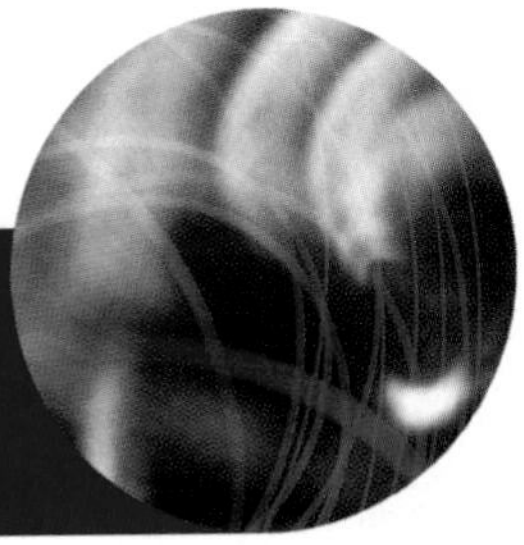

Appendix A

Geometry Review Exercises

(Answers are at the end of the answer section in this text.)

1. Find the perimeter of a triangle whose sides are 3 in., 4 in., and 5 in.
2. Find the area of a triangle whose base is 4 ft and height is 12 ft.
3. If two angles of a triangle are 30° and 90°, then what is the third angle?
4. If the area of a triangle is 36 ft^2 and the base is 12 ft, then what is the height?
5. If the side opposite 30° in a 30-60-90 right triangle is 10 cm, then what is the length of the hypotenuse?
6. Find the area of a trapezoid whose height is 12 cm and whose parallel sides are 4 cm and 20 cm.
7. Find the area of the right triangle that has sides of 6 ft, 8 ft, and 10 ft.
8. If a right triangle has sides of 5 ft, 12 ft, and 13 ft, then what is the length of the hypotenuse?
9. If the hypotenuse of a right triangle is 50 cm and the length of one leg is 40 cm, then what is the length of the other leg?
10. Is a triangle with sides of 5 ft, 10 ft, and 11 ft a right triangle?
11. What is the area of a triangle with sides of 7 yd, 24 yd, and 25 yd?
12. Find the perimeter of a parallelogram in which one side is 9 in. and another side is 6 in.
13. Find the area of a parallelogram which has a base of 8 ft and a height of 4 ft.
14. If one side of a rhombus is 5 km, then what is its perimeter.
15. Find the perimeter and area of a rectangle whose width is 18 in. and length is 2 ft.
16. If the width of a rectangle is 8 yd and its perimeter is 60 yd, then what is its length?
17. The radius of a circle is 4 ft. Find its area to the nearest tenth of a square foot.
18. The diameter of a circle is 12 ft. Find its circumference to the nearest tenth of a foot.
19. A right circular cone has radius 4 cm and height 9 cm. Find its volume to the nearest hundredth of a cubic centimeter.
20. A right circular cone has a radius 12 ft and a height of 20 ft. Find its lateral surface area to the nearest hundredth of a square foot.
21. A shoe box has a length of 12 in., a width of 6 in., and a height of 4 in. Find its volume and surface area.
22. The volume of a rectangular solid is 120 cm^3. If the area of its bottom is 30 cm^2, then what is its height?
23. What is the area and perimeter of a square in which one of the sides is 10 mi long?
24. Find the perimeter of a square whose area is 25 km^2.
25. Find the area of a square whose perimeter is 26 cm.
26. A sphere has a radius of 2 ft. Find its volume to the nearest thousandth of a cubic foot and its surface area to the nearest thousandth of a square foot.
27. A can of soup (right circular cylinder) has a radius of 2 in. and a height of 6 in. Find its volume to the nearest tenth of a cubic inch and total surface area to the nearest tenth of a square inch.
28. If one of two complementary angles is 34°, then what is the other angle?
29. If the perimeter of an isosceles triangle is 29 cm and one of the equal sides is 12 cm, then what is the length of the shortest side of the triangle?
30. A right triangle with sides of 6 in., 8 in., and 10 in., is similar to another right triangle that has a hypotenuse of 25 in. What are the lengths of the other two sides in the second triangle?
31. If one of two supplementary angles is 31°, then what is the other angle?
32. Find the perimeter of an equilateral triangle in which one of the sides is 4 km.
33. Find the length of a side of an equilateral triangle that has a perimeter of 30 yd.

Appendix B

Sets

Every subject has its own terminology, and **algebra** is no different. In this section we will learn the basic terms and facts about sets.

Set Notation

A **set** is a collection of objects. At home you may have a set of dishes and a set of steak knives. In algebra we generally discuss sets of numbers. For example, we refer to the numbers 1, 2, 3, 4, 5, and so on as the set of **counting numbers** or **natural numbers.** Of course, these are the numbers that we use for counting.

The objects or numbers in a set are called the **elements** or **members** of the set. To describe sets with a convenient notation, we use braces, { }, and name the sets with capital letters. For example,

$$A = \{1, 2, 3\}$$

means that set A is the set whose members are the natural numbers 1, 2, and 3. The letter N is used to represent the entire set of natural numbers.

A set that has a fixed number of elements such as {1, 2, 3} is a **finite** set, whereas a set without a fixed number of elements such as the natural numbers is an **infinite** set. When listing the elements of a set, we use a series of three dots to indicate a continuing pattern. For example, the set of natural numbers is written as

$$N = \{1, 2, 3, \ldots\}.$$

The set of natural numbers *between* 4 and 40 can be written

$$\{5, 6, 7, 8, \ldots, 39\}.$$

Note that since the members of this set are *between* 4 and 40, it does not include 4 or 40.

Set-builder notation is another method of describing sets. In this notation we use a variable to represent the numbers in the set. A **variable** is a letter that is used to stand for some numbers. The set is then built from the variable and a description of the numbers that the variable represents. For example, the set

$$B = \{1, 2, 3, \ldots, 49\}$$

is written in set-builder notation as

$$B = \{x \mid x \text{ is a natural number less than } 50\}.$$

↑ ↑ The set of numbers such that ↑ condition for membership

This notation is read as "B is the set of numbers x such that x is a natural number less than 50." Notice that the number 50 is not a member of set B.

The symbol $\in$ is used to indicate that a specific number is a member of a set, and $\notin$ indicates that a specific number is not a member of a set. For example, the statement $1 \in B$ is read as "1 is a member of B," "1 belongs to B," "1 is in B," or "1 is an element of B." The statement $0 \notin B$ is read as "0 is not a member of B," "0 does not belong to B," "0 is not in B," or "0 is not an element of B."

Two sets are **equal** if they contain exactly the same members. Otherwise, they are said to be not equal. To indicate equal sets, we use the symbol $=$. For sets that are not equal we use the symbol $\neq$. The elements in two equal sets do not need to be written in the same order. For example, $\{3, 4, 7\} = \{3, 4, 7\}$ and $\{2, 4, 1\} = \{1, 2, 4\}$, but $\{3, 5, 6\} \neq \{3, 5, 7\}$.

EXAMPLE 1

Set notation

Let $A = \{1, 2, 3, 5\}$ and $B = \{x \mid x \text{ is an even natural number less than } 10\}$. Determine whether each statement is true or false.

a) $3 \in A$ **b)** $5 \in B$ **c)** $4 \notin A$ **d)** $A = N$

e) $A = \{x \mid x \text{ is a natural number less than } 6\}$ **f)** $B = \{2, 4, 6, 8\}$

Solution

a) True, because 3 is a member of set A.

b) False, because 5 is not an even natural number.

c) True, because 4 is not a member of set A.

d) False, because A does not contain all of the natural numbers.

e) False, because 4 is a natural number less than 6, and $4 \notin A$.

f) True, because the even counting numbers less than 10 are 2, 4, 6, and 8.

Union of Sets

Any two sets A and B can be combined to form a new set called their union that consists of all elements of A together with all elements of B.

Union of Sets

If A and B are sets, the **union** of A and B, denoted $A \cup B$, is the set of all elements that are either in A, in B, or in both. In symbols,

$$A \cup B = \{x \mid x \in A \text{ or } x \in B\}.$$

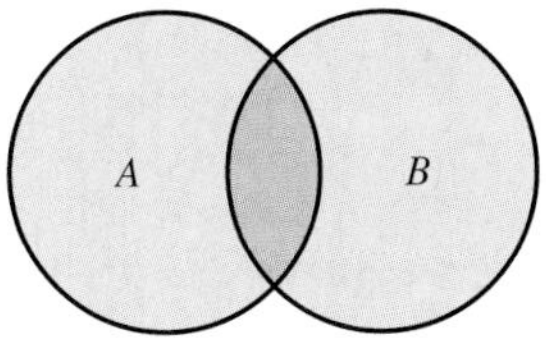

$A \cup B$

Figure B.1

In mathematics the word "or" is always used in an inclusive manner (allowing the possibility of both alternatives). The diagram in Fig. B.1 can be used to illustrate $A \cup B$. Any point that lies within circle A, circle B, or both is in $A \cup B$. Diagrams (like Fig. B.1) that are used to illustrate sets are called **Venn diagrams.**

EXAMPLE 2

Union of sets

Let $A = \{0, 2, 3\}$, $B = \{2, 3, 7\}$, and $C = \{7, 8\}$. List the elements in each of these sets.

a) $A \cup B$ **b)** $A \cup C$

Helpful Hint

To remember what "union" means think of a labor union, which is a group formed by joining together many individuals.

Solution

a) $A \cup B$ is the set of numbers that are in A, in B, or in both A and B.

$$A \cup B = \{0, 2, 3, 7\}$$

b) $A \cup C = \{0, 2, 3, 7, 8\}$

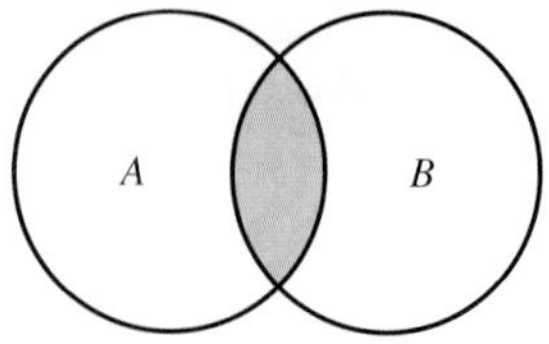

Figure B.2

Intersection of Sets

Another way to form a new set from two known sets is by considering only those elements that the two sets have in common. The diagram shown in Fig. B.2 illustrates the intersection of two sets A and B.

Intersection of Sets

If A and B are sets, the **intersection** of A and B, denoted $A \cap B$, is the set of all elements that are in both A and B. In symbols,

$$A \cap B = \{x \mid x \in A \text{ and } x \in B\}.$$

Helpful Hint

To remember the meaning of "intersection," think of the intersection of two roads. At the intersection you are on both roads.

It is possible for two sets to have no elements in common. A set with no members is called the **empty set** and is denoted by the symbol $\varnothing$. Note that $A \cup \varnothing = A$ and $A \cap \varnothing = \varnothing$ for any set A.

CAUTION The set $\{0\}$ is not the empty set. The set $\{0\}$ has one member, the number 0. Do not use the number 0 to represent the empty set.

EXAMPLE 3

Intersection of sets

Let $A = \{0, 2, 3\}$, $B = \{2, 3, 7\}$, and $C = \{7, 8\}$. List the elements in each of these sets.

a) $A \cap B$ **b)** $B \cap C$ **c)** $A \cap C$

Solution

a) $A \cap B$ is the set of all numbers that are in both A and B. So $A \cap B = \{2, 3\}$.

b) $B \cap C = \{7\}$ **c)** $A \cap C = \varnothing$

EXAMPLE 4

Membership and equality

Let $A = \{1, 2, 3, 5\}$, $B = \{2, 3, 7, 8\}$, and $C = \{6, 7, 8, 9\}$. Place one of the symbols $=$, $\neq$, $\in$, or $\notin$ in the blank to make each statement correct.

a) 5 _____ $A \cup B$ **b)** 5 _____ $A \cap B$

c) $A \cup B$ _____ $\{1, 2, 3, 5, 7, 8\}$ **d)** $A \cap B$ _____ $\{2\}$

Solution

a) $5 \in A \cup B$ because 5 is a member of A.

b) $5 \notin A \cap B$ because 5 must belong to *both* A and B to be a member of $A \cap B$.

c) $A \cup B = \{1, 2, 3, 5, 7, 8\}$ because the elements of A together with those of B are listed. Note that 2 and 3 are members of both sets but are listed only once.

d) $A \cap B \neq \{2\}$ because $A \cap B = \{2, 3\}$.

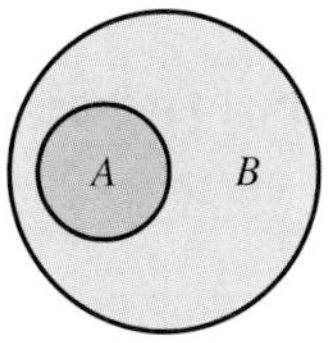

$A \subseteq B$

Figure B.3

Subsets

If every member of set A is also a member of set B, then we write $A \subseteq B$ and say that A is a **subset** of B. See Fig. B.3. For example,

$$\{2, 3\} \subseteq \{2, 3, 4\}$$

because $2 \in \{2, 3, 4\}$ and $3 \in \{2, 3, 4\}$. Note that the symbol for membership ($\in$) is used between a single element and a set, whereas the symbol for subset ($\subseteq$) is used between two sets. If A is not a subset of B, we write $A \not\subseteq B$.

CAUTION To claim that $A \not\subseteq B$, there *must* be an element of A that does *not* belong to B. For example,

$$\{1, 2\} \not\subseteq \{2, 3, 4\}$$

because 1 is a member of the first set but not of the second.

Is the empty set $\varnothing$ a subset of $\{2, 3, 4\}$? If we say that $\varnothing$ is *not* a subset of $\{2, 3, 4\}$, then there must be an element of $\varnothing$ that does not belong to $\{2, 3, 4\}$. But that cannot happen because $\varnothing$ is empty. So $\varnothing$ is a subset of $\{2, 3, 4\}$. In fact, by the same reasoning, *the empty set is a subset of every set.*

EXAMPLE 5

Subsets

Determine whether each statement is true or false.

a) $\{1, 2, 3\}$ is a subset of the set of natural numbers.

b) The set of natural numbers is not a subset of $\{1, 2, 3\}$.

c) $\{1, 2, 3\} \not\subseteq \{2, 4, 6, 8\}$

d) $\{2, 6\} \subseteq \{1, 2, 3, 4, 5\}$

e) $\varnothing \subseteq \{2, 4, 6\}$

Helpful Hint

The symbols $\subseteq$ and $\subset$ are often used interchangeably. The symbol $\subseteq$ combines the subset symbol $\subset$ and the equal symbol $=$. We use it when sets are equal, $\{1, 2\} \subseteq \{1, 2\}$, and when they are not, $\{1\} \subseteq \{1, 2\}$. When sets are not equal, we could simply use $\subset$, as in $\{1\} \subset \{1, 2\}$.

Solution

a) True, because 1, 2, and 3 are natural numbers.

b) True, because 5, for example, is a natural number and $5 \notin \{1, 2, 3\}$.

c) True, because 1 is in the first set but not in the second.

d) False, because 6 is in the first set but not in the second.

e) True, because we cannot find anything in $\varnothing$ that fails to be in $\{2, 4, 6\}$.

Combining Three or More Sets

We know how to find the union and intersection of two sets. For three or more sets we use parentheses to indicate which pair of sets to combine first. In Example 6, notice that different results are obtained from different placements of the parentheses.

EXAMPLE 6

Operations with three sets

Let $A = \{1, 2, 3, 4\}$, $B = \{2, 5, 6, 8\}$, and $C = \{4, 5, 7\}$. List the elements of each of these sets.

a) $(A \cup B) \cap C$ **b)** $A \cup (B \cap C)$

Solution

a) The parentheses indicate that the union of A and B is to be found first and then the result, $A \cup B$, is to be intersected with C.

$$A \cup B = \{1, 2, 3, 4, 5, 6, 8\}$$

Now examine $A \cup B$ and C to find the elements that belong to both sets:

$$A \cup B = \{1, 2, 3, 4, 5, 6, 8\}$$
$$C = \{4, 5, 7\}$$

The only numbers that are members of $A \cup B$ and C are 4 and 5. Thus

$$(A \cup B) \cap C = \{4, 5\}.$$

b) In $A \cup (B \cap C)$, first find $B \cap C$:

$$B \cap C = \{5\}$$

Now $A \cup (B \cap C)$ consist of all members of A together with 5 from $B \cap C$:

$$A \cup (B \cap C) = \{1, 2, 3, 4, 5\}$$

Exercises

Reading and Writing *After reading this section, write out the answers to these questions. Use complete sentences.*

1. What is a set?

2. What is the difference between a finite set and an infinite set?

3. What is a Venn diagram used for?

4. What is the difference between the intersection and the union of two sets?

5. What does it mean to say that set A is a subset of set B?

6. Which set is a subset of every set?

Using the sets A, B, C, and N, determine whether each statement is true or false. Explain. See Example 1.

$A = \{1, 3, 5, 7, 9\}$ $B = \{2, 4, 6, 8\}$
$C = \{1, 2, 3, 4, 5\}$ $N = \{1, 2, 3, \ldots\}$

7. $6 \in A$

8. $8 \in A$

9. $A \neq B$

10. $A = \{1, 3, 5, 7, \ldots\}$

11. $3 \in C$

12. $4 \notin B$

13. $A = \{1, 3, 7, 9\}$

14. $B \neq C$

15. $0 \in N$

16. $2.5 \in N$

17. $C = N$

18. $N = A$

Using the sets A, B, C, and N, list the elements in each set. If the set is empty write ∅. See Examples 2 and 3.

$A = \{1, 3, 5, 7, 9\}$ $B = \{2, 4, 6, 8\}$
$C = \{1, 2, 3, 4, 5\}$ $N = \{1, 2, 3, \ldots\}$

19. $A \cap B$

20. $A \cup B$

21. $A \cap C$

22. $A \cup C$

23. $B \cup C$

24. $B \cap C$

25. $A \cup \varnothing$

26. $B \cup \varnothing$

27. $A \cap \varnothing$

28. $B \cap \varnothing$

29. $A \cap N$

30. $A \cup N$

Use one of the symbols $\in$, $\notin$, $=$, $\neq$, $\cup$, *or* $\cap$ *in each blank to make a true statement. See Example 4.*

$A = \{1, 3, 5, 7, 9\}$ $B = \{2, 4, 6, 8\}$
$C = \{1, 2, 3, 4, 5\}$ $N = \{1, 2, 3, \ldots\}$

31. $A \cap B$ ____ $\varnothing$

32. $A \cap C$ ____ $\varnothing$

33. A ____ $B = \{1, 2, 3, 4, 5, 6, 7, 8, 9\}$

34. A ____ $B = \varnothing$

35. B ____ $C = \{2, 4\}$

36. B ____ $C = \{1, 2, 3, 4, 5, 6, 8\}$

37. 3 ____ $A \cap B$ **38.** 3 ____ $A \cap C$

39. 4 ____ $B \cap C$ **40.** 8 ____ $B \cup C$

Determine whether each statement is true or false. Explain your answer. See Example 5.

$A = \{1, 3, 5, 7, 9\}$ $B = \{2, 4, 6, 8\}$
$C = \{1, 2, 3, 4, 5\}$ $N = \{1, 2, 3, \ldots\}$

41. $A \subseteq N$ **42.** $B \subseteq N$

43. $\{2, 3\} \subseteq C$ **44.** $C \subseteq A$

45. $B \not\subseteq C$ **46.** $C \not\subseteq A$

47. $\varnothing \subseteq B$ **48.** $\varnothing \subseteq C$

49. $A \subseteq \varnothing$ **50.** $B \subseteq \varnothing$

51. $A \cap B \subseteq C$ **52.** $B \cap C \subseteq \{2, 4, 6, 8\}$

Using the sets D, E, and F, list the elements in each set. If the set is empty write $\varnothing$. *See Example 6.*

$D = \{3, 5, 7\}$ $E = \{2, 4, 6, 8\}$ $F = \{1, 2, 3, 4, 5\}$

53. $D \cup E$ **54.** $D \cap E$

55. $D \cap F$ **56.** $D \cup F$

57. $E \cup F$ **58.** $E \cap F$

59. $(D \cup E) \cap F$ **60.** $(D \cup F) \cap E$

61. $D \cup (E \cap F)$ **62.** $D \cup (F \cap E)$

63. $(D \cap F) \cup (E \cap F)$ **64.** $(D \cap E) \cup (F \cap E)$

65. $(D \cup E) \cap (D \cup F)$ **66.** $(D \cup F) \cap (D \cup E)$

Use one of the symbols $\in$, $\subseteq$, $=$, $\cup$, *or* $\cap$ *in each blank to make a true statement.*

$D = \{3, 5, 7\}$ $E = \{2, 4, 6, 8\}$ $F = \{1, 2, 3, 4, 5\}$

67. D ____ $\{x \mid x \text{ is an odd natural number}\}$

68. E ____ $\{x \mid x \text{ is an even natural number smaller than } 9\}$

69. 3 ____ D **70.** $\{3\}$ ____ D

71. D ____ $E = \varnothing$ **72.** $D \cap E$ ____ D

73. $D \cap F$ ____ F **74.** $3 \notin E$ ____ F

75. $E \not\subseteq E$ ____ F **76.** $E \subseteq E$ ____ F

77. D ____ $F = F \cup D$ **78.** E ____ $F = F \cap E$

List the elements in each set.

79. $\{x \mid x \text{ is an even natural number less than } 20\}$

80. $\{x \mid x \text{ is a natural number greater than } 6\}$

81. $\{x \mid x \text{ is an odd natural number greater than } 11\}$

82. $\{x \mid x \text{ is an odd natural number less than } 14\}$

83. $\{x \mid x \text{ is an even natural number between 4 and } 79\}$

84. $\{x \mid x \text{ is an odd natural number between 12 and } 57\}$

Write each set using set-builder notation. Answers may vary.

85. $\{3, 4, 5, 6\}$

86. $\{1, 3, 5, 7\}$

87. $\{5, 7, 9, 11, \ldots\}$

88. $\{4, 5, 6, 7, \ldots\}$

89. $\{6, 8, 10, 12, \ldots, 82\}$

90. $\{9, 11, 13, 15, \ldots, 51\}$

Determine whether each statement is true or false.

$A = \{1, 2, 3, 4\}$ $B = \{3, 4, 5\}$ $C = \{3, 4\}$

91. $A = \{x \mid x \text{ is a counting number}\}$

92. The set B has an infinite number of elements.

93. The set of counting numbers less than 50 million is an infinite set.

94. $1 \in A \cap B$ **95.** $3 \in A \cup B$

96. $A \cap B = C$ **97.** $C \subseteq B$

98. $A \subseteq B$ **99.** $\varnothing \subseteq C$

100. $A \not\subseteq C$

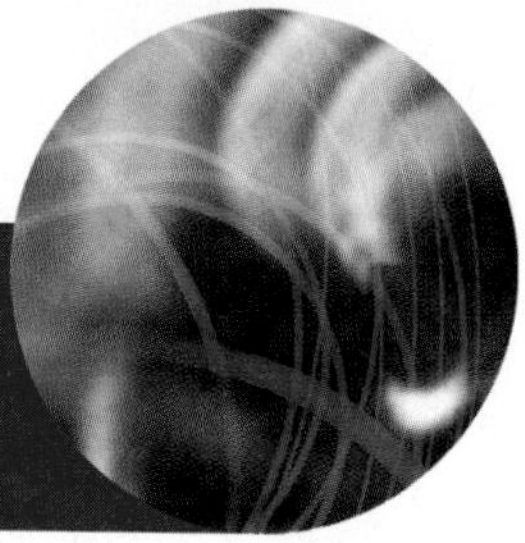

Appendix C

Final Exam Review

Note that this review does not cover every topic in this text. Use this review as a starting point for studying for your final exam. Check your tests, quizzes, and homework assignments to make sure that you have reviewed every topic covered in your course. The answers for all of these exercises can be found at the end of the answer section in this text.

Chapter 1

Evaluate each expression.

1. $\frac{3}{4} \cdot \frac{7}{9}$

2. $\frac{1}{4} + \frac{5}{6}$

3. $\frac{8}{9} \div 4$

4. $-4^2 - 3^3$

5. $|3 - 2^2| - |7 - 19|$

6. $\frac{-3 - 5}{-2 - (-1)}$

Name the property that justifies each equation.

7. $3(x + 4) = 3x + 12$

8. $x \cdot 7 = 7x$

9. $4 + (9 + y) = (4 + 9) + y$

10. $0 + 3 = 3$

Simplify each expression.

11. $5x - (3 - 8x)$

12. $x + 3 - 0.2(5x - 30)$

13. $(-3x)(-5x)$

14. $\frac{3x + 12}{-3}$

Chapter 2

Solve each equation and check your answer.

15. $11x - 2 = 3$

16. $4x - 5 = 12x + 11$

17. $3(x - 6) = 3x - 6$

18. $x - 0.1x = 0.9x$

Solve each equation for y.

19. $5x - 3y = 9$

20. $ay + b = 0$

21. $a = t - by$

22. $\frac{a}{2} + \frac{y}{3} = \frac{3a}{4}$

Solve each problem. Show all details.

23. The sum of three consecutive integers is 102. What are the integers?

24. The perimeter of a rectangular painting is 100 inches. If the width is 4 inches less than the length, then what is the width?

25. The area of a triangular piece of property is 44,000 square feet. If the base of the triangle is 400 feet, then what is the height?

26. Ivan has 400 pounds of mixed nuts that contain no peanuts. How many pounds of peanuts should he put into the mixed nuts so that 20% of the mixture is peanuts?

Solve each inequality. Express the solution set in interval notation.

27. $3x - 4 \le 11$

28. $5 - 7w > 26$

29. $-1 < 2a - 9 \le 7$

30. $5 < 6 - x < 6$

Chapter 3

Graph each equation in the coordinate plane and identify all intercepts.

31. $y = \frac{2}{3}x - 2$

32. $3x - 5y = 150$

33. $y = 2$

34. $x = 2$

Find the slope of each line.

35. The line passing through the points (1, 2) and (3, 6)

36. The line $y = \frac{1}{2}x - 4$

37. The line parallel to $2x + 3y = 9$

38. The line perpendicular to $y = -3x + 5$

Find the equation of each line in slope-intercept form when possible.

39. The line passing through the points (0, 3) and (2, 11)

40. The line passing through the points (−2, 4) and (1, −2)

41. The line through (3, 5) that is parallel to $x = 4$

42. The line through (0, 8) that is perpendicular to $y = \frac{1}{2}x$

Chapter 4

Solve the system by graphing.

43. $x + y = 4$
$y = 2x + 1$

44. $3x - 2y = 1$
$4x + y = 5$

45. $x - y = 3$
$y = -2$

Solve each system by substitution.

46. $y = 3x - 1$
$6x + 5y = 37$

47. $x + y = 2$
$-5x - y = 14$

48. $3x - 11y = 27$
$6x + y = -15$

Solve each system by the addition method.

49. $3x + 5y = 33$
$6x - 5y = -24$

50. $x - 3y = -1$
$5x + 6y = -12$

51. $4x - 3y = 14$
$6x - 2y = 11$

Determine whether each system is independent, inconsistent, or dependent.

52. $y = 2x + 2$
$y = 2x - 1$

53. $2x - 3y = 1$
$8x - 12y = 4$

54. $y = 6x$
$y = -4x$

For each problem, write a system of equations in two variables. Use the method of your choice to solve each system.

55. Bob and Carl studied a total of 72 hours for the final. If Bob studied only one-third as many hours as Carl, then how many hours did each of them study?

56. Hamburger Haven sold 15 singles and 22 doubles to one customer for a total of $51.75. The next customer bought 17 singles and 11 doubles for a total of $37.75. The third customer purchased only 1 single and 2 doubles. How much did the third customer spend?

Graph the solution set to each inequality in the coordinate plane.

57. $3x - 4y > 12$

58. $y \leq 3x + 2$

59. $x > -2$

60. $y \leq 4$

Chapter 5

Perform the indicated operations.

61. $(x^2 - 3x + 2) - (3x^2 + 9x - 4)$

62. $-3x^2(-2x^2 - 3)$

63. $(x + 7)(x - 9)$

64. $(x + 2)(x^2 - 2x + 4)$

65. $(4w^2 - 3)^2$

66. $(-8m^7) \div (2m^2)$

67. $(-9y^3 - 6y^2 + 3y) \div (3y)$

68. $(x^3 - 2x^2 - x - 6) \div (x - 3)$

Use the rules of exponents to simplify each expression. Write the answers without negative exponents.

69. $-8x^4 \cdot 4x^3$

70. $3x(5x^2)^3$

71. $\dfrac{-6x^2y^3}{-2x^{-3}y^4}$

72. $\left(\dfrac{2a^2}{a^{-3}}\right)^3$

Perform each operation without a calculator. Write the answer in scientific notation.

73. $400{,}000 \cdot 600$

74. $(9 \times 10^3)(2 \times 10^6)$

75. $(2 \times 10^{-3})^4$

76. $\dfrac{2 \times 10^{-9}}{2000}$

Chapter 6

Factor each polynomial completely.

77. $24x^2y^3 + 18xy^5$

78. $x^2 + 2x + ax + 2a$

79. $4m^2 - 49$

80. $x^2 - 3x - 54$

81. $6t^2 - 11t - 10$

82. $4w^2 - 36w + 81$

83. $2a^3 - 6a^2 - 108a$

84. $w^4 - 16$

Solve each equation.

85. $x^2 = x$

86. $2x^3 - 8x = 0$

87. $a^2 + a - 6 = 0$

88. $(b - 2)(b + 3) = 24$

Write a complete solution to each problem.

89. The sum of two numbers is 10 and their product is 21. Find the numbers.

90. The length of a new television screen is 14 inches larger than the width and the diagonal is 26 inches. What are the length and width?

Chapter 7

Solve each formula for y.

91. $\frac{3}{y} = \frac{5}{x}$

92. $a = \frac{1}{2}y(w - c)$

93. $\frac{y - 3}{x + 5} = -3$

94. $\frac{3}{y} + \frac{1}{2} = \frac{1}{t}$

Perform the indicated operations. Write answers in lowest terms.

95. $\frac{5}{2} - \frac{3}{5}$

96. $\frac{1}{x} + \frac{1}{2x}$

97. $\frac{5}{a - 7} - \frac{2}{7 - a}$

98. $\frac{5}{x^2 - 1} - \frac{3}{x^2 + 2x + 1}$

99. $\frac{12}{x^2 - 9} + \frac{2}{x + 3}$

100. $\frac{a^2 - 25}{(a - 5)^2} \cdot \frac{2a - 10}{4a + 20}$

101. $\frac{a - 9}{4} \div \frac{81 - a^2}{18}$

102. $\frac{3x^4y^3}{8x^3} \cdot \frac{10x^2y^8}{24xy^2}$

Simplify each complex fraction.

103. $\dfrac{\frac{1}{2} + \frac{2}{3}}{\frac{5}{6} - \frac{4}{9}}$

104. $\dfrac{\frac{2}{x} - \frac{1}{x - 3}}{\frac{3}{x - 3} - \frac{7}{x}}$

Solve each equation.

105. $\frac{5}{x} = \frac{2}{3}$

106. $\frac{1}{w - 3} = \frac{2}{w + 5}$

107. $\frac{x}{x - 1} - \frac{4}{x} = \frac{1}{6}$

108. $\frac{x}{3} - \frac{2}{3x} = \frac{1}{3}$

Chapter 8

Simplify each expression.

109. $\sqrt{8}$

110. $\sqrt{225}$

111. $\sqrt[3]{8}$

112. $\sqrt{\frac{3}{8}}$

113. $(\sqrt{3})^2$

114. $\sqrt{18} + \sqrt{2}$

115. $(1 + \sqrt{3})^2$

116. $(\sqrt{6} - \sqrt{5})(\sqrt{6} + \sqrt{5})$

117. $\sqrt{20} \div \sqrt{2}$

118. $\sqrt{36} \div \sqrt{3}$

119. $\frac{2 + \sqrt{8}}{2}$

120. $\sqrt{2}(\sqrt{8} - \sqrt{3})$

121. $(2\sqrt{5} - 3)(\sqrt{5} + 7)$

122. $\sqrt{\dfrac{18}{5}}$

123. $\dfrac{\sqrt{5} + 6}{\sqrt{5} - 1}$

Write each radical expression in simplified form. Assume that all variables represent nonnegative real numbers.

124. $\sqrt{x^6}$

125. $\sqrt[3]{x^6}$

126. $\sqrt{x^3}$

127. $\sqrt{18y^{10}}$

128. $\sqrt[3]{125y^{18}}$

129. $\sqrt{40t^{13}}$

130. $\sqrt{\dfrac{3x}{2y}}$

Solve each equation.

131. $x = \sqrt{2x + 3}$

132. $(x - 4)^2 = 12$

133. $2\sqrt{x} = 1$

Chapter 9

Solve by using the quadratic formula.

134. $x^2 + 2x - 5 = 0$

135. $6x^2 - 5x + 1 = 0$

136. $9x^2 - 6x + 1 = 0$

Solve by completing the square.

137. $x^2 + 4x - 1 = 0$

138. $x^2 + 3x - 1 = 0$

139. $2x^2 - 4x - 3 = 0$

Graph each parabola.

140. $y = 4 - x^2$

141. $y = x^2 - 5x$

142. $y = -x^2 + 2x + 1$

Solve each problem.

143. Find the intercepts and vertex for the parabola $y = x^2 - 6x + 8$.

144. Find the maximum value for y on the graph of $y = -x^2 + 4x - 3$.

145. Find the minimum value for y on the graph of $y = 6x^2 + 2x$.

146. The height in feet for a ball thrown upward at 16 feet per second is given by $h = -16t^2 + 16t$, where t is the time in seconds after the ball is tossed. What is the maximum height reached by the ball?

Let $f(x) = -2x - 1$, $g(x) = |3x - 1|$, *and* $h(x) = \frac{5}{x}$. *Find the following.*

147. $f(-1/2)$

148. $f(-3)$

149. $f(0)$

150. $g(1/3)$

151. $g(-5)$

152. $g(0)$

153. $h(0)$

154. $h(1/2)$

Answers to Selected Exercises

Chapter 1

Section 1.1 Warm-Ups T T F F T F T T F F

1. The integers are the numbers in the set $\{\ldots, -3, -2, -1, 0, 1, 2, 3, \ldots\}$.
3. A rational number is a ratio of integers and an irrational number is not.
5. The number a is larger than b if a lies to the right of b on the number line.
7. 6 **9.** 0 **11.** -2 **13.** -12 **15.** -2.1

17. 1, 2, 3, 4, 5
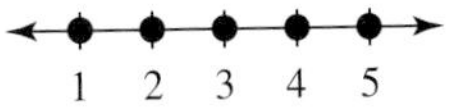

19. 0, 1, 2, 3, 4
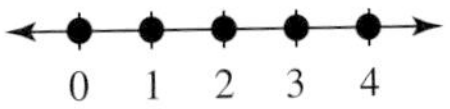

21. 0, 1, 2, 3, 4
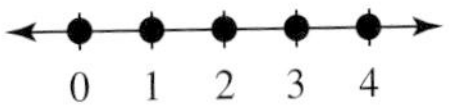

23. 1, 2, 3, 4, 5, . . .

25. 1, 2, 3, 4, 5, . . .
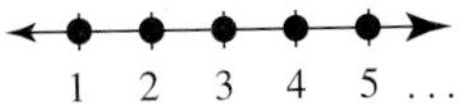

27. True **29.** False **31.** True **33.** True
35. True **37.** False

39. (0, 1)
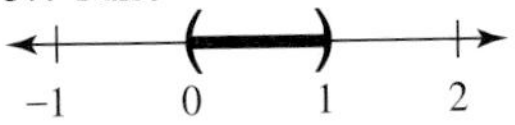

41. $[-2, 2]$

43. (0, 5]

45. $(4, \infty)$
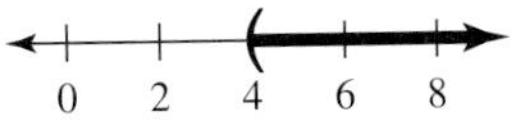

47. $(-\infty, -1]$

49. $[0, \infty)$
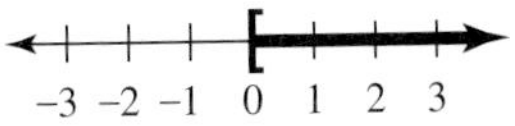

51. 6 **53.** 0 **55.** 7 **57.** 9 **59.** 45 **61.** $\frac{3}{4}$ **63.** 5.09
65. -16 **67.** $-\frac{5}{2}$ **69.** 2 **71.** 3 **73.** -9 **75.** 16
77. -4 **79.** -1.99 **81.** 74 **83.** 5.25 **85.** 40 **87.** $\frac{1}{2}$
89. -3 and 3 **91.** $-4, -3, 3, 4$ **93.** $-1, 0, 1$
95. [3, 8] **97.** $(-30, -20]$ **99.** $[30, \infty)$ **101.** True
103. True **105.** True
107. What is the probability that a tossed coin turns up heads?
109. If a is negative, then $-a$ and $|-a|$ are positive. The rest are negative.

Section 1.2 Warm-Ups T T F T T T T T F T

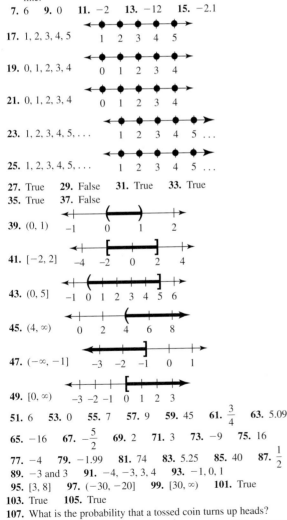

1. If two fractions are identical when reduced to lowest terms, then they are equivalent fractions.
3. To reduce a fraction to lowest terms means to find an equivalent fraction that has no factor common to the numerator and denominator.
5. Convert a fraction to a decimal by dividing the denominator into the numerator.
7. $\frac{6}{8}$ **9.** $\frac{32}{12}$ **11.** $\frac{10}{2}$ **13.** $\frac{75}{100}$ **15.** $\frac{30}{100}$ **17.** $\frac{70}{42}$ **19.** $\frac{1}{2}$
21. $\frac{2}{3}$ **23.** 3 **25.** $\frac{1}{2}$ **27.** 2 **29.** $\frac{3}{8}$ **31.** $\frac{13}{21}$ **33.** $\frac{12}{13}$
35. $\frac{10}{27}$ **37.** 5 **39.** $\frac{7}{10}$ **41.** $\frac{7}{13}$ **43.** $\frac{3}{5}$ **45.** $\frac{1}{6}$ **47.** 3
49. $\frac{1}{15}$ **51.** 4 **53.** $\frac{4}{5}$ **55.** $\frac{3}{40}$ **57.** $\frac{1}{2}$ **59.** $\frac{1}{3}$ **61.** $\frac{1}{4}$
63. $\frac{7}{12}$ **65.** $\frac{1}{12}$ **67.** $\frac{19}{24}$ **69.** $\frac{11}{72}$ **71.** $\frac{199}{48}$ **73.** 60%, 0.6
75. $\frac{9}{100}$, 0.09 **77.** 8%, $\frac{2}{25}$ **79.** 0.75, 75% **81.** $\frac{1}{50}$, 0.02
83. $\frac{1}{100}$, 1% **85.** 3 **87.** 1 **89.** $\frac{71}{96}$ **91.** $\frac{17}{120}$
93. $\frac{65}{16}$ **95.** $\frac{69}{4}$ **97.** $\frac{13}{12}$ **99.** $\frac{1}{8}$ **101.** $\frac{3}{8}$ **103.** $\frac{3}{16}$
105. $\frac{2}{3}$ **107.** $\frac{1}{2}$ **109.** $\frac{19}{96}$
111. a) 1.3 yd^3 b) $36\frac{11}{24}$ ft^3 or $1\frac{227}{648}$ yd^3
115. Each daughter gets 3 $km^2 \div 4$ or a $\frac{3}{4}$ km^2 piece of the farm. Divide the farm into 12 equal squares. Give each daughter an L-shaped piece consisting of 3 of those 12 squares.

Section 1.3 Warm-Ups T T T F F F T F T F

1. We studied addition and subtraction of signed numbers.
3. Two numbers are additive inverses of each other if their sum is zero.
5. To find the sum of two numbers with unlike signs, subtract their absolute values. The answer is given the sign of the number with the larger absolute value.
7. 13 **9.** -13 **11.** -1.15 **13.** $-\frac{1}{2}$ **15.** 0 **17.** 0
19. 2 **21.** -6 **23.** 5.6 **25.** -2.9 **27.** $-\frac{1}{4}$ **29.** $8 + (-2)$
31. $4 + (-12)$ **33.** $-3 + 8$ **35.** $8.3 + (1.5)$ **37.** -4
39. -10 **41.** 11 **43.** -11 **45.** $-\frac{1}{4}$ **47.** $\frac{3}{4}$ **49.** 7
51. 0.93 **53.** 9.3 **55.** -5.03 **57.** 3 **59.** -9 **61.** -120

63. 78 **65.** −27 **67.** −7 **69.** −201 **71.** −322

73. −15.97 **75.** −2.92 **77.** −3.73 **79.** 3.7 **81.** $\frac{3}{20}$ **83.** $\frac{7}{24}$

85. 13 **87.** −10 **89.** 14 **91.** −4 **93.** −3 **95.** −3.49

97. −0.3422 **99.** −48.84 **101.** −8.85 **103.** −\$8.85

105. −7°C

107. When adding signed numbers, we add or subtract only positive numbers which are the absolute values of the original numbers. We then determine the appropriate sign for the answer.

109. The distance between x and y is given by either $|x - y|$ or $|y - x|$.

Section 1.4 Warm-Ups T F T F T T T F T F

1. We learned to multiply and divide signed numbers.

3. To find the product of signed numbers, multiply their absolute values and then affix a negative sign if the two original numbers have opposite signs.

5. To find the quotient of nonzero numbers divide their absolute values and then affix a negative sign if the two original numbers have opposite signs.

7. −27 **9.** 132 **11.** $-\frac{1}{3}$ **13.** −0.3 **15.** 144 **17.** 0

19. −1 **21.** 3 **23.** $-\frac{2}{3}$ **25.** $\frac{5}{6}$ **27.** Undefined **29.** 0

31. −80 **33.** 0.25 **35.** −100 **37.** 27 **39.** −3 **41.** −4

43. −30 **45.** 19 **47.** −0.18 **49.** 0.3 **51.** −6

53. 1.5 **55.** 22 **57.** $-\frac{1}{3}$ **59.** −164.25 **61.** 1529.41

63. −12 **65.** −8 **67.** −6 **69.** −1 **71.** 5 **73.** 16

75. −8 **77.** 0 **79.** 0 **81.** −3.9 **83.** −40 **85.** 0.4

87. 0.4 **89.** −0.2 **91.** −7.5 **93.** $-\frac{1}{30}$ **95.** $-\frac{1}{10}$

97. 7.562 **99.** 19.35 **101.** 0 **103.** Undefined

Section 1.5 Warm-Ups F F T F F F F T F T

1. An arithmetic expression is the result of writing numbers in a meaningful combination with the ordinary operations of arithmetic.

3. An exponential expression is an expression of the form a^n.

5. The order of operations tells us the order in which to perform operations when grouping symbols are omitted.

7. −4 **9.** 1 **11.** −8 **13.** −7 **15.** −16 **17.** −4

19. 4^4 **21.** $(-5)^4$ **23.** $(-y)^3$ **25.** $\left(\frac{3}{7}\right)^5$ **27.** $5 \cdot 5 \cdot 5$

29. $b \cdot b$ **31.** $\left(-\frac{1}{2}\right)\left(-\frac{1}{2}\right)\left(-\frac{1}{2}\right)\left(-\frac{1}{2}\right)\left(-\frac{1}{2}\right)$

33. (0.22)(0.22)(0.22)(0.22) **35.** 81 **37.** 0 **39.** 625

41. −216 **43.** 100,000 **45.** −0.001 **47.** $\frac{1}{8}$ **49.** $\frac{1}{4}$

51. −64 **53.** −4096 **55.** 27 **57.** −13 **59.** 36 **61.** 18

63. −19 **65.** −17 **67.** −44 **69.** 18 **71.** −78 **73.** 0

75. 27 **77.** 1 **79.** 8 **81.** 7 **83.** 11 **85.** 111 **87.** 21

89. −1 **91.** −11 **93.** 9 **95.** 16 **97.** 28 **99.** 121

101. −73 **103.** 25 **105.** 0 **107.** −2 **109.** 12 **111.** 82

113. −54 **115.** −79 **117.** −24 **119.** 41.92

121. 184.643547 **123.** 8.0548

125. a) 330.2 million **b)** 2022

127. $(-5)^3 = -(5^3) = -5^3 = -1 \cdot 5^3$ and $-(-5)^3 = 5^3$

Section 1.6 Warm-Ups T F T F T F F F T F

1. An algebraic expression is the result of combining numbers and variables with the operations of arithmetic in some meaningful way.

3. An algebraic expression is named according to the last operation to be performed.

5. An equation is a sentence that expresses equality between two algebraic expressions.

7. Difference **9.** Cube **11.** Sum **13.** Difference

15. Product **17.** Square **19.** The difference of x^2 and a^2

21. The square of $x - a$ **23.** The quotient of $x - 4$ and 2

25. The difference of $\frac{x}{2}$ and 4 **27.** The cube of ab

29. $8 + y$ **31.** $5xz$ **33.** $8 - 7x$ **35.** $\frac{6}{x + 4}$

37. $(a + b)^2$ **39.** $x^3 + y^2$ **41.** $5m^2$ **43.** $(s + t)^2$

45. 3 **47.** 3 **49.** 16 **51.** −9 **53.** −3 **55.** −8

57. $-\frac{2}{3}$ **59.** 4 **61.** −1 **63.** 1 **65.** −4 **67.** 0

69. Yes **71.** No **73.** Yes **75.** Yes **77.** Yes **79.** Yes

81. No **83.** No **85.** $5x + 3x = 8x$ **87.** $3(x + 2) = 12$

89. $\frac{x}{3} = 5x$ **91.** $(a + b)^2 = 9$

93. −7, −5, −3, −1, 1

95. 4, 8, 16; $\frac{1}{4}, \frac{1}{8}, \frac{1}{16}$; 100, 1000, 10,000; 0.01, 0.001, 0.0001

97. 14.65 **99.** 37.12 **101.** 169.3 cm, 41 cm

103. 6, 15, 30, 38 **105.** 920 feet

107. For the square of the sum consider $(2 + 3)^2 = 5^2 = 25$. For the sum of the squares consider $2^2 + 3^2 = 4 + 9 = 13$. So $(2 + 3)^2 \neq 2^2 + 3^2$.

Section 1.7 Warm-Ups F F T F T T T T T T

1. The commutative property says that $a + b = b + a$ and the associative property says that $(a + b) + c = a + (b + c)$.

3. Factoring is the process of writing an expression or number as a product.

5. The properties help us to understand the operations and how they are related to each other.

7. $r + 9$ **9.** $3(x + 2)$ **11.** $-5x + 4$ **13.** $6x$

15. $-2(x - 4)$ **17.** $4 - 8y$ **19.** $4w^2$ **21.** $3a^2b$ **23.** $9x^3z$

25. −3 **27.** −10 **29.** −21 **31.** 0.6 **33.** −22.4

35. $3x - 15$ **37.** $2a + at$ **39.** $-3w + 18$ **41.** $-20 + 4y$

43. $-a + 7$ **45.** $-t - 4$ **47.** $2(m + 6)$ **49.** $4(x - 1)$

51. $4(y - 4)$ **53.** $4(a + 2)$ **55.** $x(1 + y)$ **57.** $2(3a - b)$

59. 2 **61.** $-\frac{1}{5}$ **63.** $\frac{1}{7}$ **65.** 1 **67.** −4 **69.** $\frac{2}{5}$

71. Commutative property of multiplication

73. Distributive property

75. Associative property of multiplication

77. Additive inverse property

79. Commutative property of multiplication

81. Multiplicative identity property

83. Distributive property

85. Additive inverse property

87. Multiplication property of 0

89. Distributive property

91. $y + a$ **93.** $(5a)w$ **95.** $\frac{1}{2}(x + 1)$ **97.** $3(2x + 5)$

99. 1 **101.** 0 **103.** $\frac{100}{33}$ **105. a)** 45 bricks/hour **b)** Bricklayer

107. a) 2.3213 people/second **b)** 1,403,900 people/week

109. The perimeter is twice the sum of the length and width.
111. **a)** Commutative **b)** Not commutative

Section 1.8 Warm-Ups T F T T F F F F F T

1. Like terms are terms with the same variables and exponents.
3. We can add or subtract like terms.
5. If a negative sign precedes a set of parentheses, then signs for all terms in the parentheses are changed when the parentheses are removed.
7. 7000 **9.** 1 **11.** 356 **13.** 350 **15.** 36 **17.** 36,000
19. 0 **21.** 98 **23.** $11w$ **25.** $3x$ **27.** $5x$ **29.** $-a$
31. $-2a$ **33.** $10 - 6t$ **35.** $8x^2$ **37.** $-4x + 2x^2$
39. $-7mw^2$ **41.** $\frac{5}{6}a$ **43.** $12h$ **45.** $-18b$ **47.** $-9m^2$
49. $12d^2$ **51.** y^2 **53.** $-15ab$ **55.** $-6a - 3ab$
57. $-k + k^2$ **59.** y **61.** $-3y$ **63.** y **65.** $2y^2$ **67.** $2a - 1$
69. $3x - 2$ **71.** $-2x + 1$ **73.** $8 - y$ **75.** $m - 6$
77. $w - 5$ **79.** $8x + 15$ **81.** $5x - 1$ **83.** $-2a - 1$
85. $5a - 2$ **87.** $6x^2 + x - 15$ **89.** $-2b^2 - 7b + 4$
91. $3m - 18$ **93.** $-3x - 7$ **95.** $0.95x - 0.5$
97. $4x - 4$ **99.** $2y + 4$ **101.** $2y + m - 1$ **103.** 3
105. $\frac{7}{6}a + \frac{13}{6}$ **107.** $0.15x - 0.4$ **109.** $-14k + 23$
111. 45 **113.** **a)** $0.25x - 6380$ **b)** \$13,620 **c)** \$48,000 **d)** \$300,000 **115.** $4x + 80$, 200 feet
117. If $x = 5$, then $1/2 \cdot 5 = \frac{1}{2} \cdot 5 = 2.5$ because we do division and multiplication from left to right.

Enriching Your Mathematical Word Power

1. c **2.** b **3.** a **4.** d **5.** b **6.** d **7.** a **8.** d
9. c **10.** a

Review Exercises

1. 0, 1, 2, 10 **3.** $-2, 0, 1, 2, 10$ **5.** $-\sqrt{5}, \pi$
7. True **9.** False **11.** False **13.** True
15. 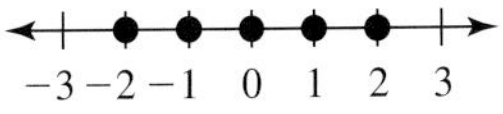

17. 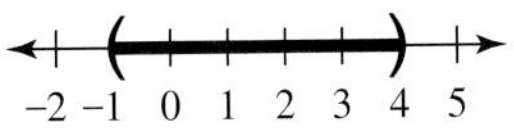

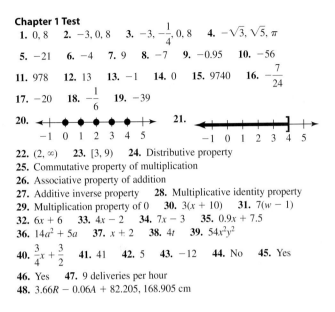

19. [4, 6] **21.** $[-30, \infty)$ **23.** $\frac{17}{24}$ **25.** 6 **27.** $\frac{3}{7}$ **29.** $\frac{14}{3}$
31. $\frac{13}{12}$ **33.** 2 **35.** -13 **37.** -7 **39.** -7 **41.** 11.95
43. -0.05 **45.** $-\frac{1}{6}$ **47.** $-\frac{11}{15}$ **49.** -15 **51.** 4 **53.** 5
55. $\frac{1}{6}$ **57.** -0.3 **59.** -0.24 **61.** 1 **63.** 66 **65.** 49
67. 41 **69.** 1 **71.** 50 **73.** -135 **75.** -2 **77.** -16
79. 16 **81.** 5 **83.** 9 **85.** 7 **87.** $-\frac{1}{3}$ **89.** 1 **91.** -9
93. Yes **95.** No **97.** Yes **99.** No
101. Distributive property
103. Multiplicative inverse property
105. Additive identity property
107. Associative property of addition
109. Commutative property of multiplication
111. Additive inverse property
113. Multiplicative identity property
115. $-a + 12$ **117.** $6a^2 - 6a$ **119.** $-12t + 39$
121. $-0.9a - 0.57$ **123.** $-0.05x - 4$ **125.** $27x^2 + 6x + 5$
127. $-2a$ **129.** $x^2 + 4x - 3$ **131.** 0 **133.** 8 **135.** -21
137. $\frac{1}{2}$ **139.** -0.5 **141.** -1 **143.** $x + 2$ **145.** $4 + 2x$
147. $2x$ **149.** $-4x + 8$ **151.** $6x$ **153.** x **155.** $8x$
157. $-x^2 + 6x - 8$ **159.** $\frac{1}{4}x - \frac{3}{2}$ **161.** 3, 2, 1, 0, −1
163. 25, 125, 625; 16, −64, 256 **165.** 18 memberships per hour

Chapter 1 Test

1. 0, 8 **2.** $-3, 0, 8$ **3.** $-3, -\frac{1}{4}, 0, 8$ **4.** $-\sqrt{3}, \sqrt{5}, \pi$
5. -21 **6.** -4 **7.** 9 **8.** -7 **9.** -0.95 **10.** -56
11. 978 **12.** 13 **13.** -1 **14.** 0 **15.** 9740 **16.** $-\frac{7}{24}$
17. -20 **18.** $-\frac{1}{6}$ **19.** -39
20. 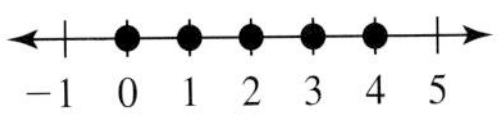

21. 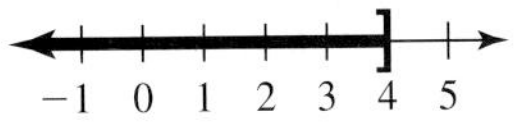

22. $(2, \infty)$ **23.** [3, 9) **24.** Distributive property
25. Commutative property of multiplication
26. Associative property of addition
27. Additive inverse property **28.** Multiplicative identity property
29. Multiplication property of 0 **30.** $3(x + 10)$ **31.** $7(w - 1)$
32. $6x + 6$ **33.** $4x - 2$ **34.** $7x - 3$ **35.** $0.9x + 7.5$
36. $14a^2 + 5a$ **37.** $x + 2$ **38.** $4t$ **39.** $54x^2y^2$
40. $\frac{3}{4}x + \frac{3}{2}$ **41.** 41 **42.** 5 **43.** -12 **44.** No **45.** Yes
46. Yes **47.** 9 deliveries per hour
48. $3.66R - 0.06A + 82.205$, 168.905 cm

Chapter 2

Section 2.1 Warm-Ups T T F T F T T T T T

1. The addition property of equality says that adding the same number to each side of an equation does not change the solution to the equation.
3. The multiplication property of equality says that multiplying both sides of an equation by the same nonzero number does not change the solution to the equation.
5. Replace the variable in the equation with your solution. If the resulting statement is correct, then the solution is correct.
7. $\{1\}$ **9.** $\{9\}$ **11.** $\{1\}$ **13.** $\left\{\frac{2}{3}\right\}$ **15.** $\{0.12\}$ **17.** $\{-9\}$
19. $\{-19\}$ **21.** $\left\{\frac{1}{4}\right\}$ **23.** $\{0\}$ **25.** $\{5.95\}$ **27.** $\{-5\}$ **29.** $\{-4\}$
31. $\{3\}$ **33.** $\left\{\frac{1}{4}\right\}$ **35.** $\{-8\}$ **37.** $\{1.8\}$ **39.** $\left\{\frac{2}{3}\right\}$ **41.** $\left\{\frac{1}{2}\right\}$
43. $\{-5\}$ **45.** $\{5\}$ **47.** $\{1.25\}$ **49.** $\left\{\frac{1}{4}\right\}$ **51.** $\left\{\frac{3}{20}\right\}$ **53.** $\{-2\}$
55. $\{120\}$ **57.** $\left\{\frac{5}{9}\right\}$ **59.** $\left\{-\frac{1}{2}\right\}$ **61.** $\{-8\}$ **63.** $\left\{\frac{1}{3}\right\}$ **65.** $\{-3.4\}$
67. $\{99\}$ **69.** $\{-7\}$ **71.** $\{9\}$ **73.** $\{8\}$ **75.** $\{5\}$ **77.** $\{-5\}$
79. $\{-8\}$ **81.** $\{2\}$ **83.** $\left\{\frac{1}{6}\right\}$ **85.** $\left\{-\frac{1}{3}\right\}$ **87.** $\{44\}$ **89.** $\left\{\frac{3}{4}\right\}$
91. $\{7\}$ **93.** $\{-14\}$ **95.** $\left\{\frac{3}{8}\right\}$
97. **a)** $\frac{4}{5}x = 48.5$, 60.6 births per 1000 females **b)** 54 births per 1000 females
99. 2877 stocks

Section 2.2 Warm-Ups T T T F T F T T T T

1. We can solve $ax + b = 0$ with the addition property and the multiplication property of equality.
3. Use the multiplication property of equality to solve $-x = 8$.
5. $\{2\}$ **7.** $\{-2\}$ **9.** $\left\{\frac{2}{3}\right\}$ **11.** $\left\{-\frac{5}{2}\right\}$ **13.** $\{6\}$ **15.** $\{12\}$
17. $\left\{\frac{1}{2}\right\}$ **19.** $\left\{-\frac{1}{6}\right\}$ **21.** $\{4\}$ **23.** $\left\{\frac{5}{6}\right\}$ **25.** $\{4\}$ **27.** $\{-5\}$
29. $\{34\}$ **31.** $\{9\}$ **33.** $\{1.2\}$ **35.** $\{3\}$ **37.** $\{4\}$ **39.** $\{-3\}$
41. $\left\{\frac{1}{2}\right\}$ **43.** $\{30\}$ **45.** $\{6\}$ **47.** $\{-2\}$ **49.** $\{18\}$ **51.** $\{0\}$
53. $\left\{\frac{1}{6}\right\}$ **55.** $\{-2\}$ **57.** $\left\{\frac{7}{3}\right\}$ **59.** $\{1\}$ **61.** $\{-6\}$ **63.** $\{-12\}$
65. $\{-4\}$ **67.** $\{-13\}$ **69.** $\{1.7\}$ **71.** $\{2\}$ **73.** $\{4.6\}$
75. $\{8\}$ **77.** $\{34\}$ **79.** $\{6\}$ **81.** $\{0\}$ **83.** $\{-10\}$ **85.** $\{18\}$
87. $\{-20\}$ **89.** $\{-3\}$ **91.** $\{-4.3\}$ **93.** 17 hr **95.** 20°C
97. 9 ft **99.** \$14,550

Section 2.3 Warm-Ups T T F F F T T F T T

1. If an equation involves fractions we usually multiply each side by the LCD of all of the fractions.
3. An identity is an equation that is satisfied by all numbers for which both sides are defined.
5. An inconsistent equation has no solutions.
7. $\left\{\frac{6}{5}\right\}$ **9.** $\left\{\frac{2}{9}\right\}$ **11.** $\{7\}$ **13.** $\{24\}$ **15.** $\{16\}$ **17.** $\{-12\}$
19. $\{60\}$ **21.** $\{24\}$ **23.** $\{90\}$ **25.** $\{6\}$ **27.** $\{-2\}$
29. $\{80\}$ **31.** $\{60\}$ **33.** $\{200\}$ **35.** $\{800\}$ **37.** $\left\{\frac{9}{2}\right\}$ **39.** $\{3\}$
41. $\{25\}$ **43.** $\{-2\}$ **45.** $\{-3\}$ **47.** $\{5\}$ **49.** $\{-10\}$ **51.** $\{2\}$
53. All real numbers, identity **55.** $\emptyset$, inconsistent
57. $\{0\}$, conditional **59.** $\emptyset$, inconsistent **61.** $\emptyset$, inconsistent
63. $\{1\}$, conditional **65.** $\emptyset$, inconsistent
67. All real numbers, identity **69.** All nonzero real numbers, identity
71. All real numbers, identity **73.** $\{-4\}$ **75.** R **77.** R **79.** $\{100\}$
81. $\left\{-\frac{3}{2}\right\}$ **83.** $\{30\}$ **85.** $\{6\}$ **87.** $\{0.5\}$ **89.** $\{19{,}608\}$
91. \$128,000 **93. a)** \$240,000 **b)** \$239,653

Section 2.4 Warm-Ups F F F F F T T T F T

1. A formula is an equation with two or more variables.
3. To solve for a variable means to find an equivalent equation in which the variable is isolated.
5. To find the value of a variable in a formula, we can solve for the variable and then insert values for the other variables, or insert values for the other variables and then solve for the variable.
7. $R = \frac{D}{T}$ **9.** $D = \frac{C}{\pi}$ **11.** $P = \frac{I}{rt}$ **13.** $C = \frac{5}{9}(F - 32)$
15. $h = \frac{2A}{b}$ **17.** $L = \frac{P - 2W}{2}$ **19.** $a = 2A - b$
21. $r = \frac{S - P}{Pt}$ **23.** $a = \frac{2A - bh}{h}$ **25.** $x = \frac{b - a}{2}$ **27.** $x = -7a$
29. $x = 12 - a$ **31.** $x = 7ab$ **33.** $y = -x - 9$ **35.** $y = -x + 6$
37. $y = 2x - 2$ **39.** $y = 3x + 4$ **41.** $y = -\frac{1}{2}x + 2$ **43.** $y = x - \frac{1}{2}$
45. $y = 3x - 14$ **47.** $y = \frac{1}{2}x$ **49.** $y = \frac{3}{2}x + 6$ **51.** $y = \frac{3}{2}x + \frac{13}{2}$
53. $y = -\frac{1}{4}x + \frac{5}{8}$ **55.** 60, 30, 0, −30, −60 **57.** 14, 23, 32, 104, 212
59. 40, 20, 10, 5, 4 **61.** 1, 3, 6, 10, 15 **63.** 2 **65.** 7 **67.** $-\frac{9}{5}$
69. 1 **71.** 1.33 **73.** 4% **75.** 4 years **77.** 7 yards **79.** 225 feet
81. \$60,500 **83.** \$300 **85.** 20% **87.** 160 feet **89.** 24 cubic feet
91. 4 inches **93.** 8 feet **95.** 12 inches **97.** 640 milligrams, age 13
99. 3.75 milliliters **101.** $L = F\sqrt{S} - 2D + 5.688$

Section 2.5 Warm-Ups T T T F T F F F T F

1. To express addition we use words such as plus, sum, increased by, and more than.
3. Complementary angles have degree measures with a sum of 90°.
5. Distance is the product of rate and time.
7. $x + 3$ **9.** $x - 3$ **11.** $5x$ **13.** $0.1x$ **15.** $\frac{x}{3}$ **17.** $\frac{1}{3}x$
19. x and $x + 15$ **21.** x and $6 - x$ **23.** x and $-4 - x$
25. x and $x + 3$ **27.** x and $0.05x$ **29.** x and $1.30x$
31. x and $90 - x$ **33.** x and $120 - x$
35. n and $n + 2$, where n is an even integer
37. x and $x + 1$, where x is an integer
39. x, $x + 2$, and $x + 4$, where x is an odd integer
41. x, $x + 2$, $x + 4$, and $x + 6$, where x is an even integer
43. $3x$ miles **45.** $0.25q$ dollars **47.** $\frac{x}{20}$ hour
49. $\frac{x - 100}{12}$ meters per second **51.** $5x$ square meters
53. $2w + 2(w + 3)$ inches **55.** $150 - x$ feet **57.** $2x + 1$ feet
59. $x(x + 5)$ square meters **61.** $0.18(x + 1000)$
63. $\frac{16.50}{x}$ dollars per pound **65.** $90 - x$ degrees
67. x is the smaller number, $x(x + 5) = 8$
69. x is the selling price, $x - 0.07x = 84{,}532$
71. x is the percent, $500x = 100$
73. x is the number of nickels, $0.05x + 0.10(x + 2) = 3.80$
75. x is the number, $x + 5 = 13$
77. x is the smallest integer, $x + (x + 1) + (x + 2) = 42$
79. x is the smaller integer, $x(x + 1) = 182$
81. x is Harriet's income, $0.12x = 3000$
83. x is the number, $0.05x = 13$
85. x is the width, $x(x + 5) = 126$
87. n is the number of nickels, $5n + 10(n - 1) = 95$
89. x is the measure of the larger angle, $x + x - 38 = 180$
91. a) $r + 0.6(220 - (30 + r)) = 144$, where r is the resting heart rate
b) Target heart rate increases as resting heart rate increases.
93. $6 + x$ **95.** $m + 9$ **97.** $11t$ **99.** $5(x - 2)$ **101.** $m - 3m$
103. $\frac{h + 8}{h}$ **105.** $\frac{5}{y - 9}$ **107.** $\frac{w - 8}{2w}$ **109.** $-3y - 9$
111. $x - \frac{x}{7}$ **113.** $m^2 - (m + 7)$ **115.** $x + (9x - 8)$ **117.** $13n - 9$
119. $6 + \frac{1}{3}(x + 2)$ **121.** $\frac{x}{2} + x$ **123.** $x(x + 3) = 24$
125. $w(w - 4) = 24$

Section 2.6 Warm-Ups F T T F F T T T F T

1. In this section we studied number, geometric, and uniform motion problems.
3. Uniform motion is motion at a constant rate of speed.
5. Complementary angles are angles whose degree measures have a sum of 90°.
7. 46, 47, 48 **9.** 75, 77 **11.** 47, 48, 49, 50
13. Length 50 meters, width 25 meters
15. Width 42 inches, length 46 inches

17. 13 inches **19.** 35° **21.** 1152 in. **23.** 22.88 km **25.** 5.31 in.
27. 402.57 g **29.** 58.67 ft/sec **31.** 548.53 km/hr
33. 65 miles per hour **35.** 55 miles per hour **37.** 4 hours, 2048 miles
39. Length 20 inches, width 12 inches **41.** 5 ft, 5 ft, 3 ft
43. 20°, 40°, 120° **45.** 20°, 80°, 80° **47.** Raiders 32, Vikings 14
49. 3 hours, 106 miles **51.** Crawford 1906, Wayne 1907, Stewart 1908
53. 7 ft, 7 ft, 16 ft

Section 2.7 Warm-Ups T F T F T T F T F T

1. We studied discount, investment, and mixture problems in this section.
3. The product of the rate and the original price gives the amount of discount. The original price minus the discount is the sale price.
5. A table helps us to organize the information given in a problem.
7. \$320 **9.** \$400 **11.** \$125,000 **13.** \$30.24
15. 100 Fund \$10,000, 101 Fund \$13,000
17. Fidelity \$14,000, Price \$11,000
19. 30 gallons **21.** 20 liters of 5% alcohol, 10 liters of 20% alcohol
23. 55,700 **25.** \$15,000 **27.** 75% **29.** 600
31. 42 private rooms, 30 semiprivate rooms **33.** 12 pounds
35. 4 nickels, 6 dimes **37.** 800 gallons **39.** $\frac{2}{3}$ gal
41. Shorts \$12, tops \$6

Section 2.8 Warm-Ups T T F T F T F F T F

1. The inequality symbols are $<$, $\le$, $>$, and $\ge$.
3. For $\le$ and $\ge$ use a bracket and for $<$ and $>$ use a parenthesis.
5. The compound inequality $a < b < c$ means $b > a$ and $b < c$, or b is between a and c.
7. True **9.** True **11.** False **13.** True **15.** True **17.** True
19. True

21. $(-\infty, 3]$
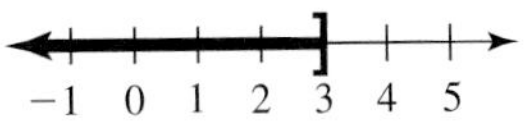

23. $(-2, \infty)$
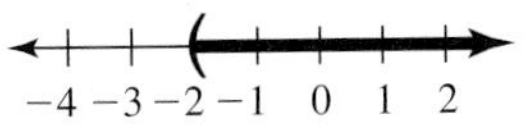

25. $(-\infty, -1)$
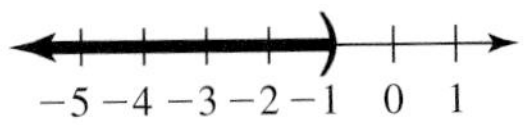

27. $[-2, \infty)$
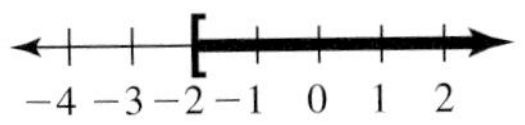

29. $\left[\frac{1}{2}, \infty\right)$
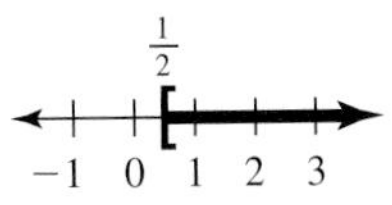

31. $(-\infty, 5.3]$
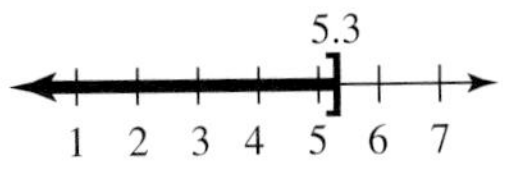

33. $(-3, 1)$
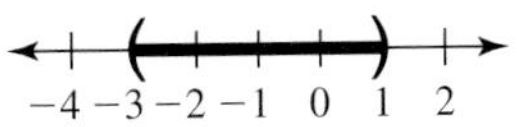

35. $[3, 7]$
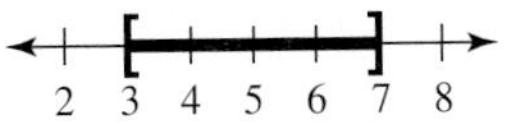

37. $[-5, 0)$
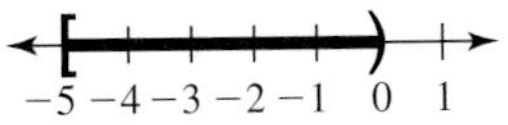

39. $(40, 100]$
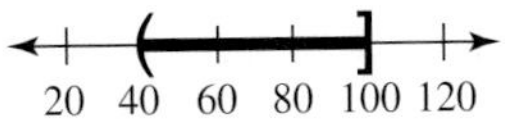

41. $x > 3$, $(3, \infty)$ **43.** $x \le 2$, $(-\infty, 2]$ **45.** $0 < x < 2$, $(0, 2)$
47. $-5 < x \le 7$, $(-5, 7]$ **49.** $x > -4$, $(-4, \infty)$
51. Yes **53.** No **55.** No **57.** Yes **59.** Yes **61.** Yes
63. No **65.** Yes **67.** No **69.** 0, 5.1 **71.** 5.1 **73.** 5.1
75. −5.1, 0, 5.1 **77.** $0.08p > 1500$ **79.** $p + 2p + p + 0.25 < 2.00$
81. $\frac{44 + 72 + s}{3} \ge 60$ **83.** $396 < 8R < 453$ **85.** $60 < 90 - x < 70$
87. **a)** $45 + 2(30) + 2h \le 130$ **b)** Approximately 12 in.
89. 79, moderate effort on level ground

Section 2.9 Warm-Ups T F F T F T F T T F

1. Equivalent inequalities are inequalities that have the same solutions.
3. According to the multiplication property of inequality, the inequality symbol is reversed when multiplying (or dividing) by a negative number and not reversed when multiplying (or dividing) by a positive number.
5. We solve compound inequalities using the properties of inequality as we do for simple inequalities.
7. $>$ **9.** $\ge$ **11.** $>$ **13.** $\le$

15. $(-3, \infty)$
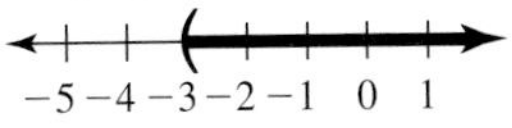

17. $(-2, \infty)$
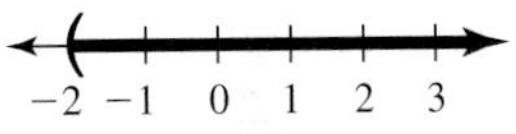

19. $(-\infty, 4)$
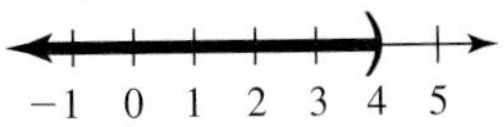

21. $\left[-\frac{1}{2}, \infty\right)$
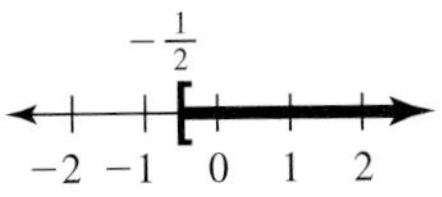

23. $(-\infty, 3)$
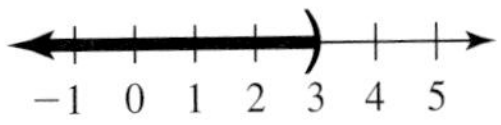

25. $\left[-\frac{1}{3}, \infty\right)$
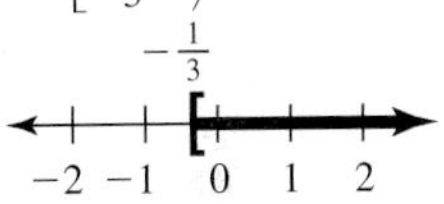

27. $(-3, \infty)$
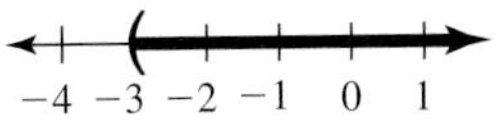

29. $(-\infty, 13)$
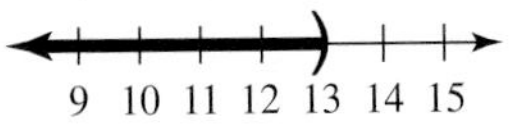

31. $(-\infty, 24]$
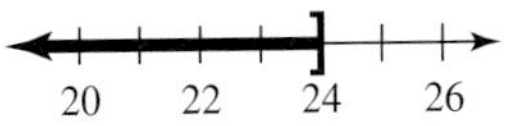

33. $\left(-\infty, \frac{7}{2}\right]$
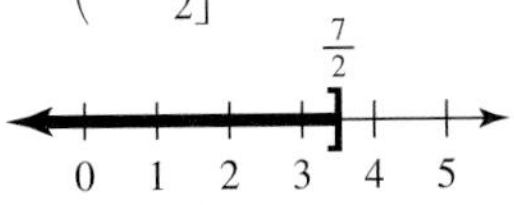

35. $(-1.5, \infty)$

37. $(-\infty, -11)$
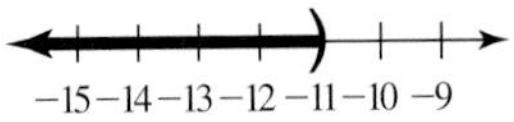

39. $(-10, \infty)$
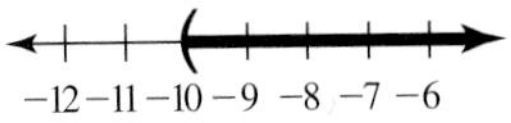

41. $(-\infty, 614.3)$
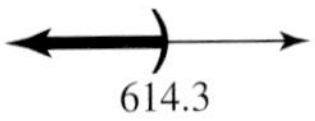

43. $(8, 10)$
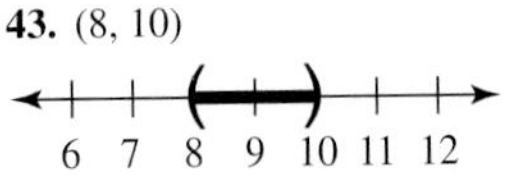

45. $\left(1, \frac{9}{2}\right)$
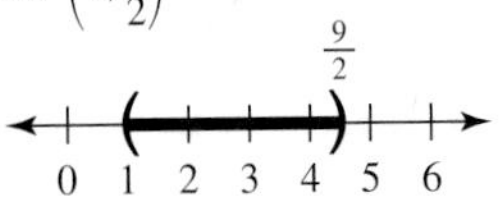

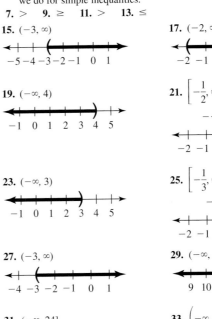

47. $[-2, 9]$

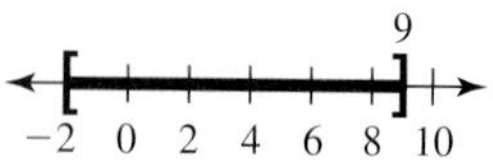

49. $[-5, 3)$

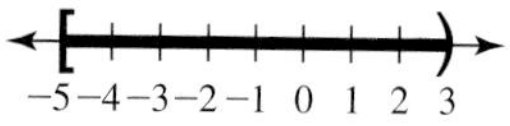

51. $(12, 30)$

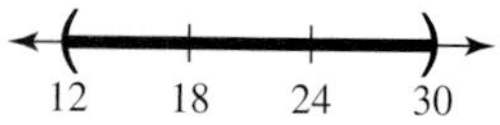

53. $\left(-\frac{1}{2}, \frac{3}{2}\right]$

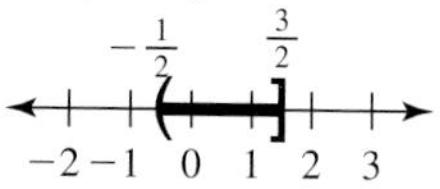

55. $(102.1, 108.3)$

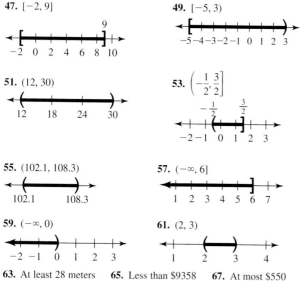

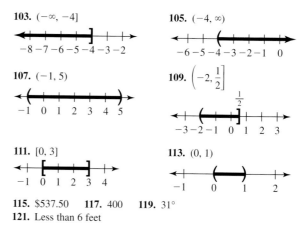

57. $(-\infty, 6]$

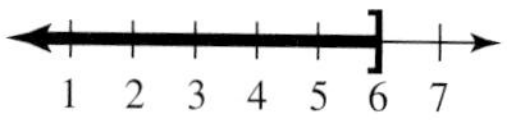

59. $(-\infty, 0)$

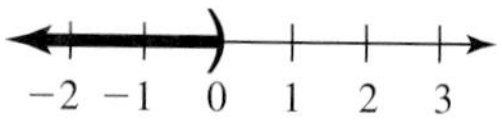

61. $(2, 3)$

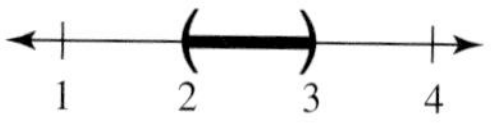

63. At least 28 meters **65.** Less than \$9358 **67.** At most \$550
69. At least 64 **71.** Between 81 and 94.5 inclusive
73. Between 49.5 and 56.625 miles per hour
75. Between 55° and 85°
77. a) Between 27 and 35 teeth inclusive **b)** Between 23.02 in. and 24.79 in. **c)** At least 14 teeth

Enriching Your Mathematical Word Power

1. b **2.** d **3.** c **4.** c **5.** d **6.** d **7.** a **8.** b **9.** c
10. d

Review Exercises

1. {35} **3.** {−6} **5.** {−7} **7.** {13} **9.** {7} **11.** {2} **13.** {7}
15. {0} **17.** {−8} **19.** ∅, inconsistent
21. All real numbers, identity **23.** All nonzero real numbers, identity
25. {24}, conditional **27.** {80}, conditional **29.** {1000}, conditional
31. $\left\{\frac{1}{4}\right\}$ **33.** $\left\{\frac{21}{8}\right\}$ **35.** $\left\{-\frac{4}{5}\right\}$ **37.** {4} **39.** {24} **41.** {−100}
43. $x = -\frac{b}{a}$ **45.** $x = \frac{b+2}{a}$ **47.** $x = \frac{V}{LW}$ **49.** $x = -\frac{b}{3}$
51. $y = -\frac{5}{2}x + 3$ **53.** $y = -\frac{1}{2}x + 4$ **55.** $y = -2x + 16$
57. −13 **59.** $-\frac{2}{5}$ **61.** 17 **63.** 15, 10, 5, 0, −5
65. −3, −1, 1, 3 **67.** $x + 9$, where x is the number
69. x and $x + 8$, where x is the smaller number
71. $0.65x$, where x is the number
73. $x(x + 5) = 98$, where x is the width
75. $2x = 3(x - 10)$, where x is Jim's rate
77. $x + x + 2 + x + 4 = 90$, where x is the smallest of the three even integers
79. $t + 2t + t - 10 = 180$, where t is the degree measure of an angle
81. 77, 79, 81 **83.** Betty 45 mph, Lawanda 60 mph
85. Wanda \$36,000, husband \$30,000 **87.** No **89.** No
91. $x > 1$, $(1, \infty)$ **93.** $x \ge 2$, $[2, \infty)$ **95.** $-3 \le x < 3$, $[-3, 3)$
97. $x < -1$, $(-\infty, -1)$

99. $(-1, \infty)$

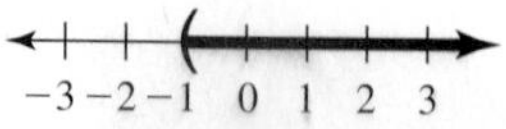

101. $(-\infty, 3)$

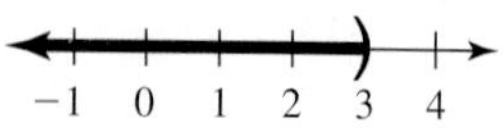

103. $(-\infty, -4]$

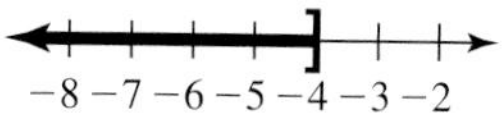

105. $(-4, \infty)$

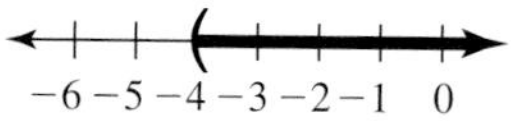

107. $(-1, 5)$

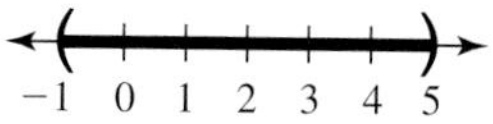

109. $\left(-2, \frac{1}{2}\right]$

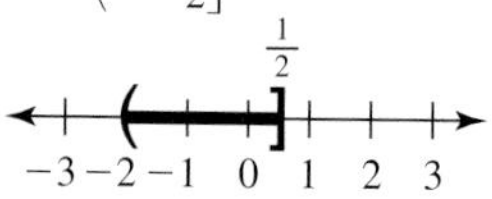

111. $[0, 3]$

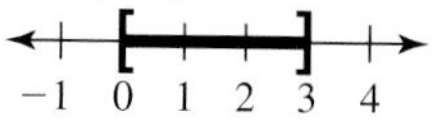

113. $(0, 1)$

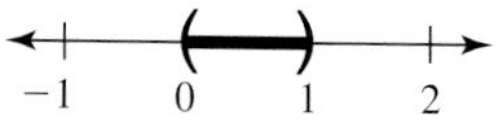

115. \$537.50 **117.** 400 **119.** 31°
121. Less than 6 feet

Chapter 2 Test

1. {−7} **2.** {2} **3.** {−9} **4.** {700} **5.** {1} **6.** $\left\{\frac{7}{6}\right\}$ **7.** {2}
8. ∅ **9.** All real numbers **10.** $y = \frac{2}{3}x - 3$ **11.** $a = \frac{m+w}{P}$
12. $-3 < x \le 2$, $(-3, 2]$ **13.** $x > 1$, $(1, \infty)$

14. $(19, \infty)$

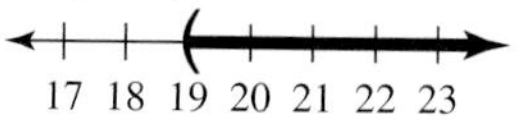

15. $(-7, -1)$

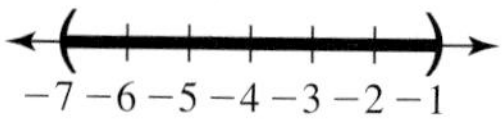

16. $(1, 3)$

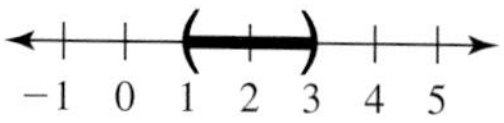

17. $(-6, \infty)$

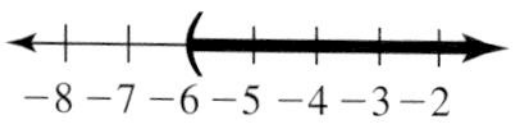

18. 14 meters **19.** 9 in. **20.** 150 liters **21.** At most \$2000
22. 30°, 60°, 90°

Making Connections Chapters 1–2

1. $8x$ **2.** $15x^2$ **3.** $2x + 1$ **4.** $4x - 7$ **5.** $-2x + 13$
6. 60 **7.** 72 **8.** −10 **9.** $-2x^3$ **10.** −1 **11.** $\frac{2}{3}$
12. $\frac{1}{6}$ **13.** $\frac{1}{9}$ **14.** $\frac{5}{9}$ **15.** 13 **16.** 8 **17.** $2x + 1$
18. $10x - 9$ **19.** $\left\{\frac{2}{3}\right\}$ **20.** $\left\{\frac{1}{6}\right\}$ **21.** $\left(\frac{2}{3}, \infty\right)$ **22.** $\left(-\infty, \frac{1}{6}\right]$
23. $\left\{\frac{1}{9}\right\}$ **24.** $\left\{\frac{5}{9}\right\}$ **25.** $\left[-\frac{1}{9}, \infty\right)$ **26.** $\left(-\infty, -\frac{5}{9}\right)$ **27.** $\left\{\frac{3}{10}\right\}$
28. $\left\{\frac{16}{5}\right\}$ **29.** $\left\{\frac{1}{2}\right\}$ **30.** $\left\{\frac{7}{5}\right\}$ **31.** {1} **32.** All real numbers
33. {0} **34.** {1} **35.** $(0, \infty)$ **36.** ∅ **37.** {2} **38.** {2}
39. $\left\{\frac{13}{2}\right\}$ **40.** {200} **41. a)** \$13,600 **b)** \$10,000 **c)** \$12,000

Chapter 3

Section 3.1 Warm-Ups F F F F T T T F F T

1. An ordered pair is a pair of numbers in which there is a first number and a second number, usually written as (a, b).
3. The origin is the point of intersection of the x-axis and y-axis.
5. A linear equation in two variables is an equation of the form $Ax + By = C$, where A and B are not both zero.
7. (0, 9), (5, 24), (2, 15) **9.** $(0, -7)$, $\left(\frac{1}{3}, -8\right)$, $\left(-\frac{2}{3}, -5\right)$
11. (0, 54.3), (10, 66.3), (0.5, 54.9) **13.** (3, 0), (0, −2), (12, 6)
15. (5, −3), (5, 5), (5, 0)

17–31 odd

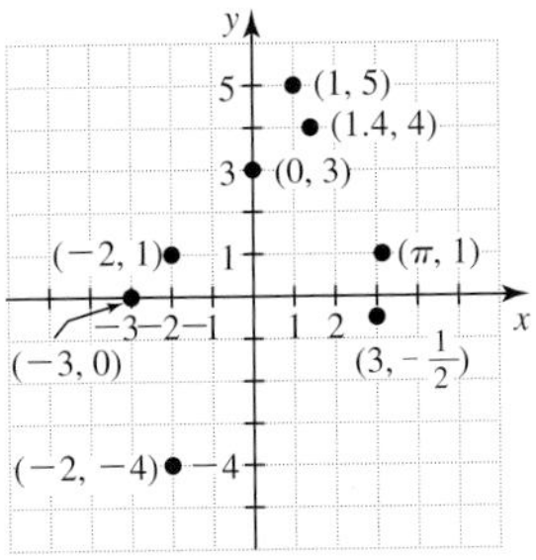

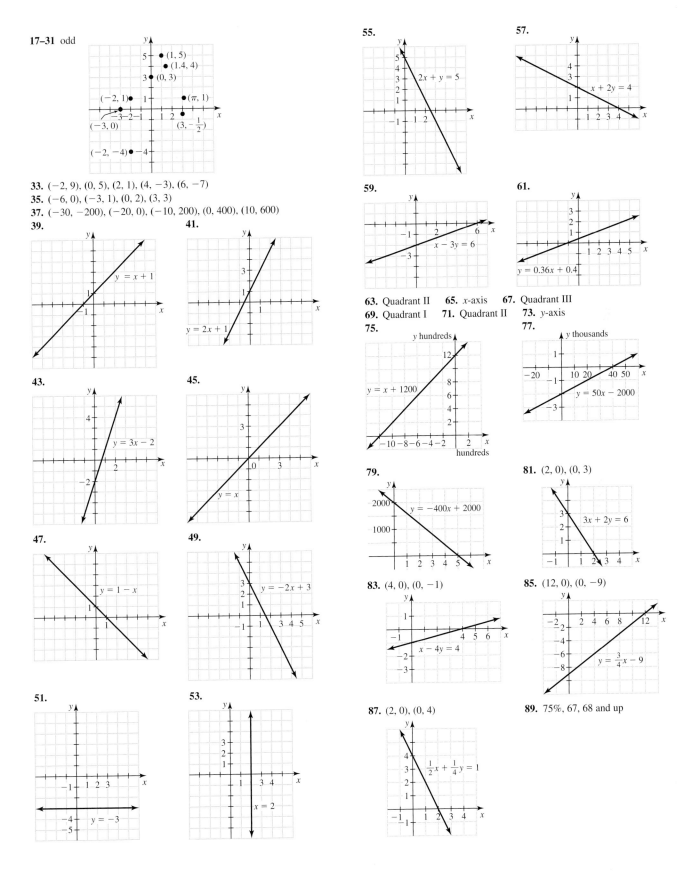

33. (−2, 9), (0, 5), (2, 1), (4, −3), (6, −7)
35. (−6, 0), (−3, 1), (0, 2), (3, 3)
37. (−30, −200), (−20, 0), (−10, 200), (0, 400), (10, 600)

39.

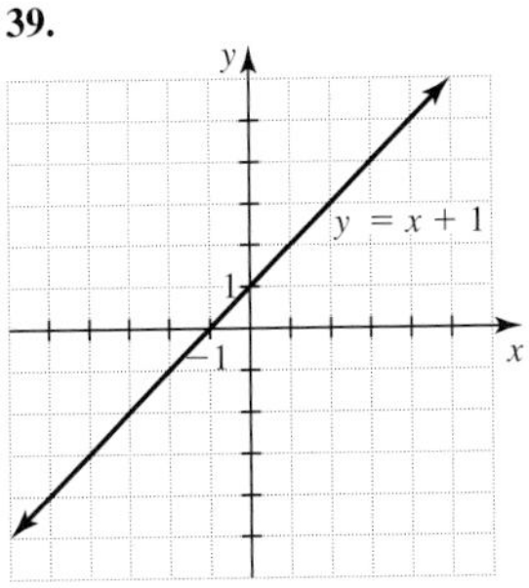

41.

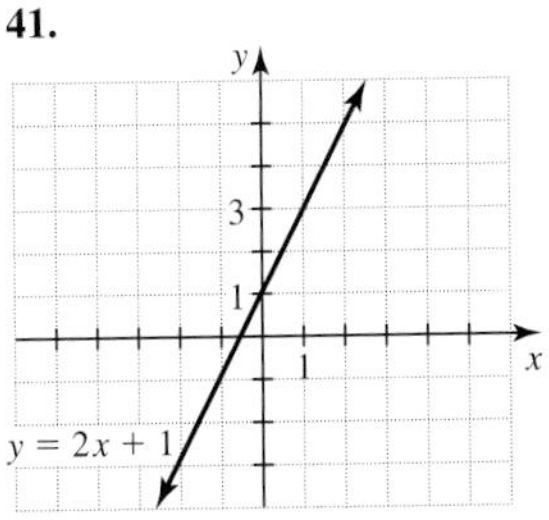

43.

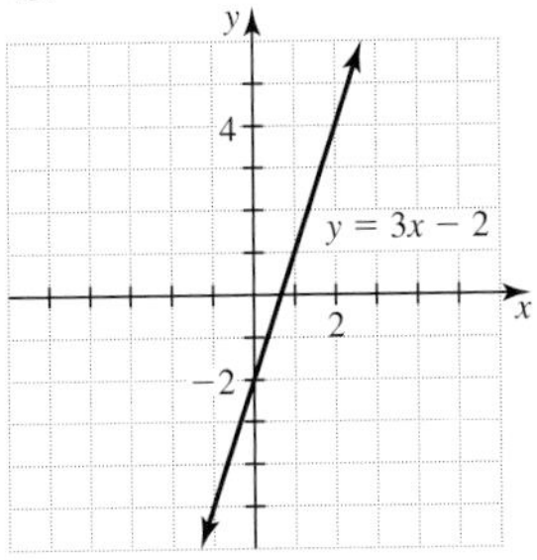

45.

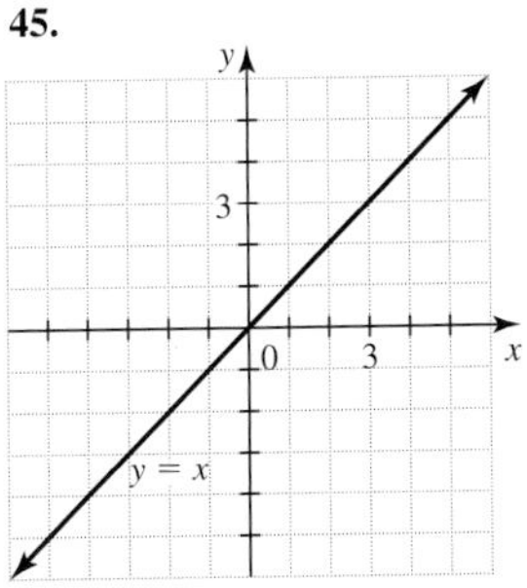

47.

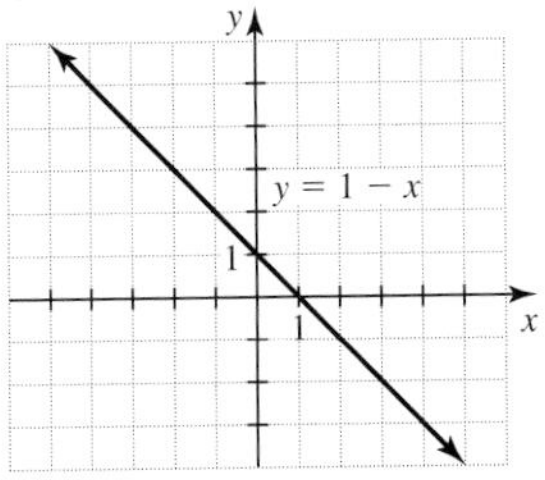

49.

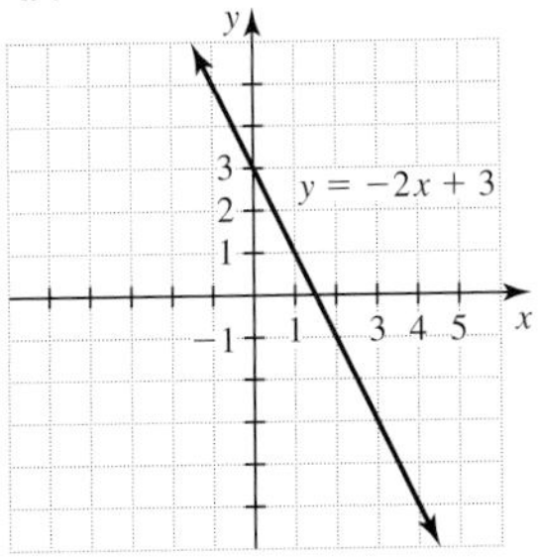

51.

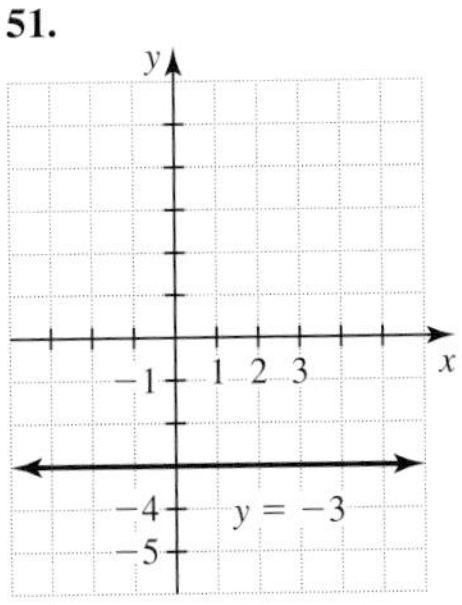

53.

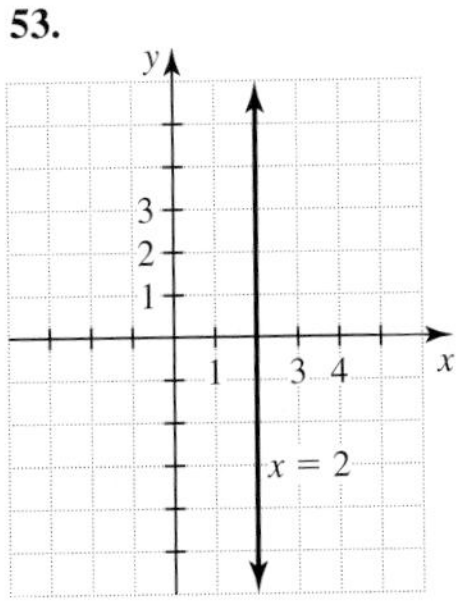

55.

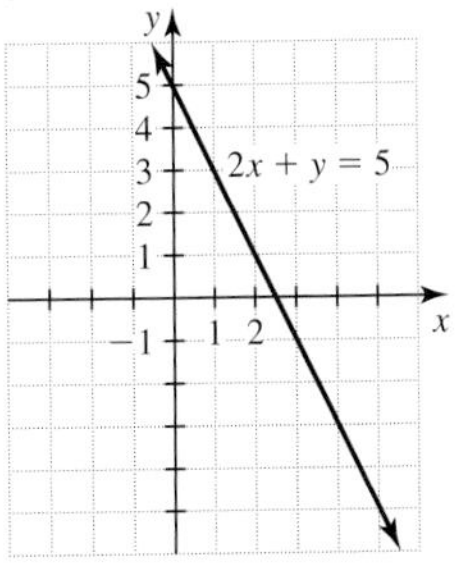

57.

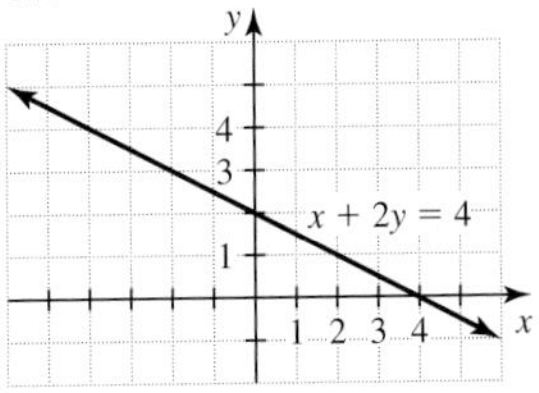

59.

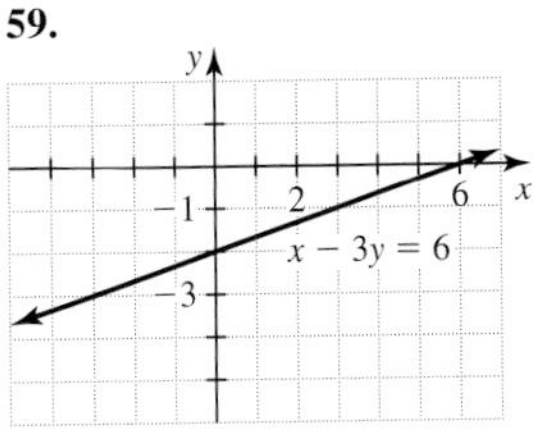

61.

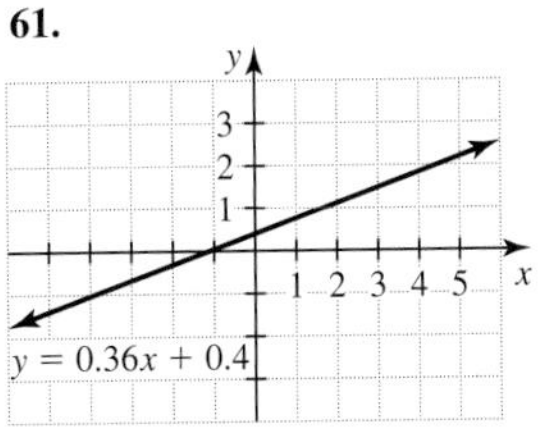

63. Quadrant II **65.** *x*-axis **67.** Quadrant III
69. Quadrant I **71.** Quadrant II **73.** *y*-axis

75.

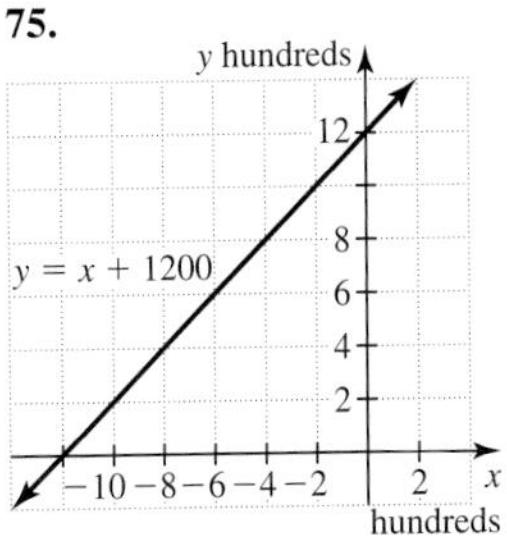

77.

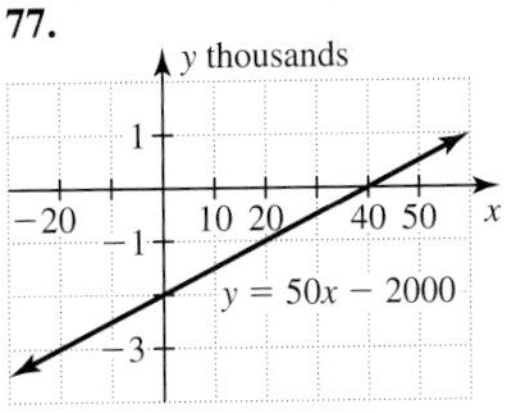

79.

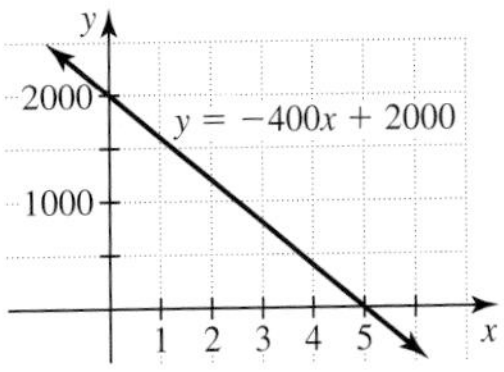

81. (2, 0), (0, 3)

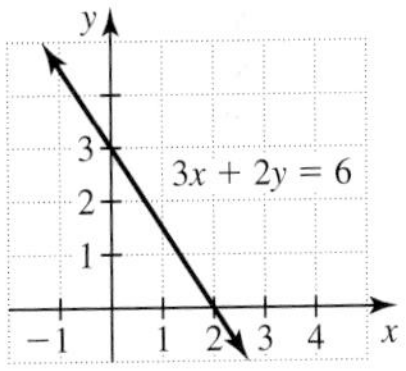

83. (4, 0), (0, −1)

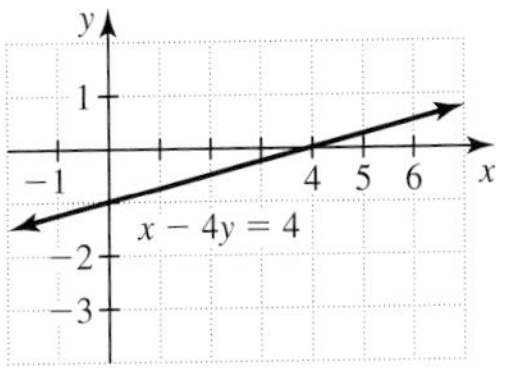

85. (12, 0), (0, −9)

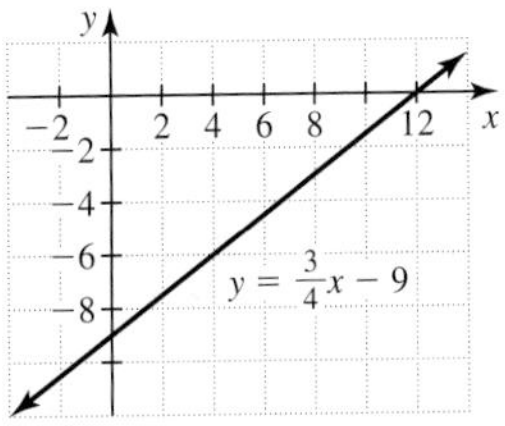

87. (2, 0), (0, 4)

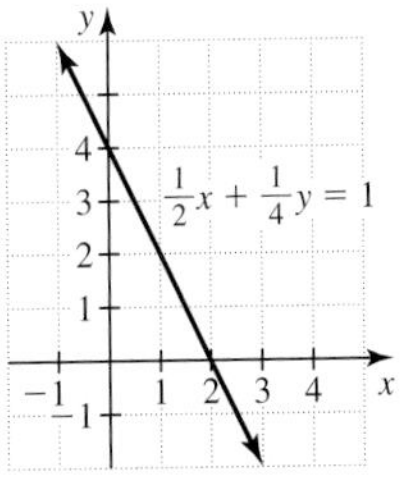

89. 75%, 67, 68 and up

91. **a)** \$97.3 billion **b)** 2016
c)

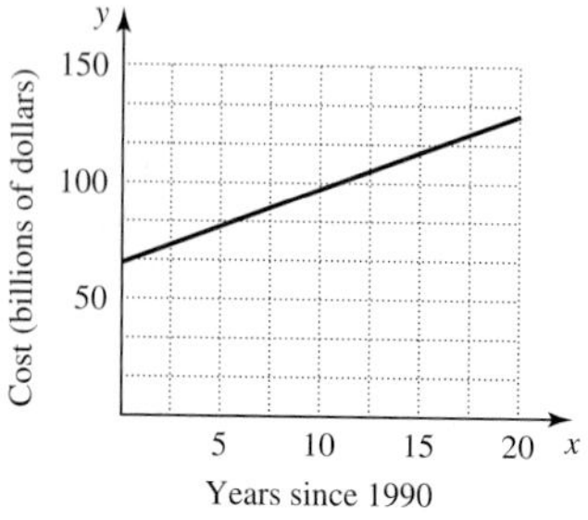

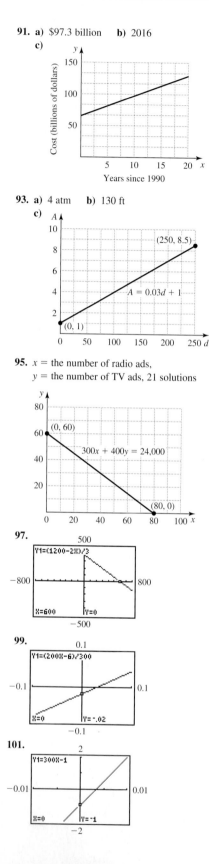

93. **a)** 4 atm **b)** 130 ft
c)

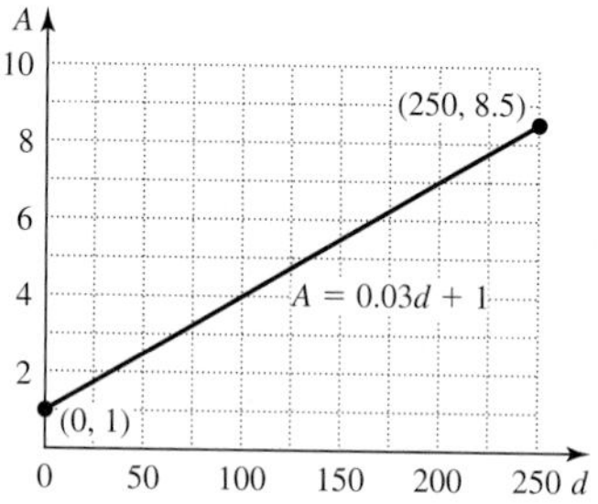

95. x = the number of radio ads,
y = the number of TV ads, 21 solutions

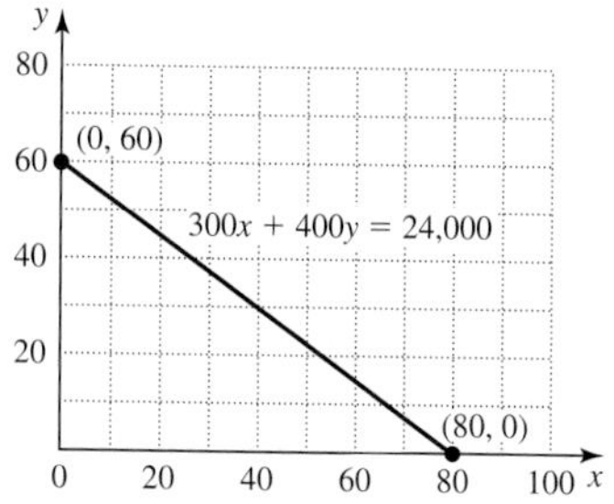

97.

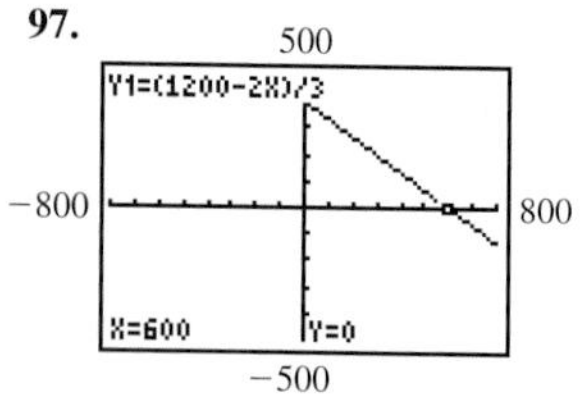

99.

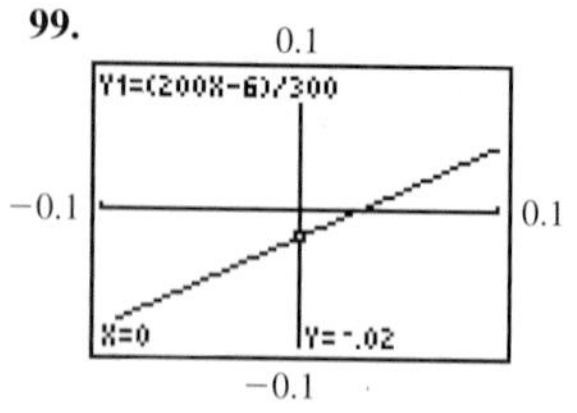

101.

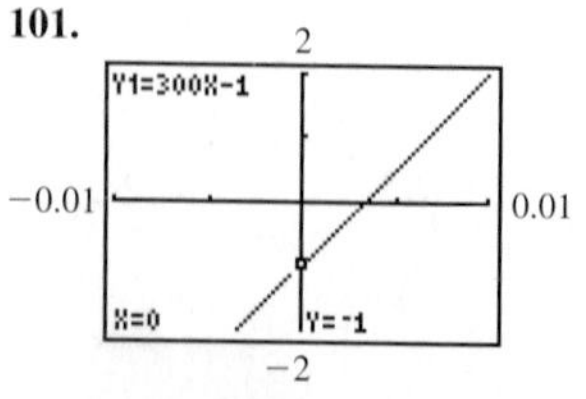

Section 3.2 Warm-Ups T T F T F F F F T T

1. The slope of a line is the ratio of its rise and run.
3. Slope is undefined for vertical lines.
5. Lines with positive slope are rising as you go from left to right, while lines with negative slope are falling as you go from left to right.
7. $-\frac{2}{3}$ **9.** $\frac{2}{3}$ **11.** $\frac{3}{2}$ **13.** 0 **15.** $\frac{2}{5}$ **17.** Undefined **19.** 2
21. $-\frac{5}{3}$ **23.** $\frac{5}{7}$ **25.** $-\frac{4}{3}$ **27.** -1 **29.** 1 **31.** Undefined
33. 0 **35.** 3
37.

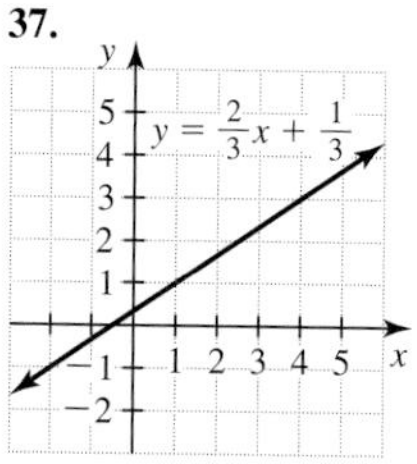

39.

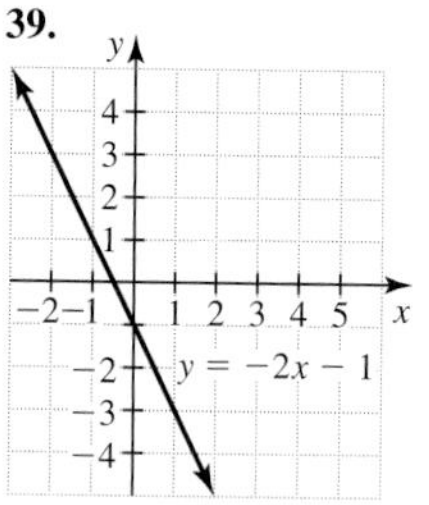

41.

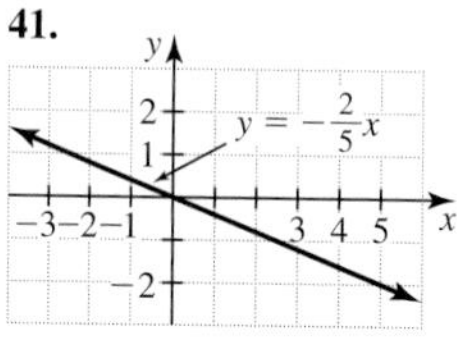

43.

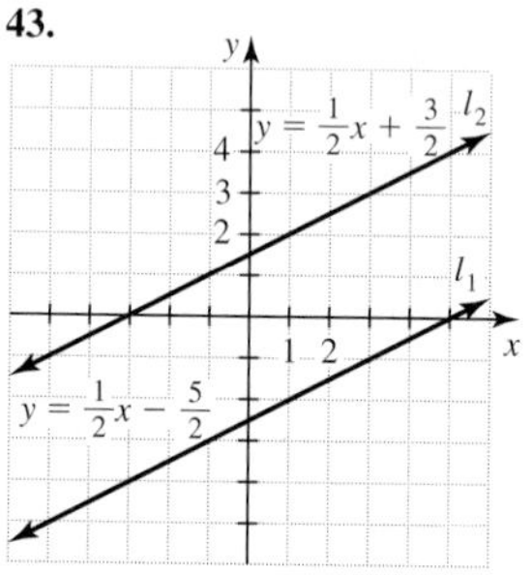

45.

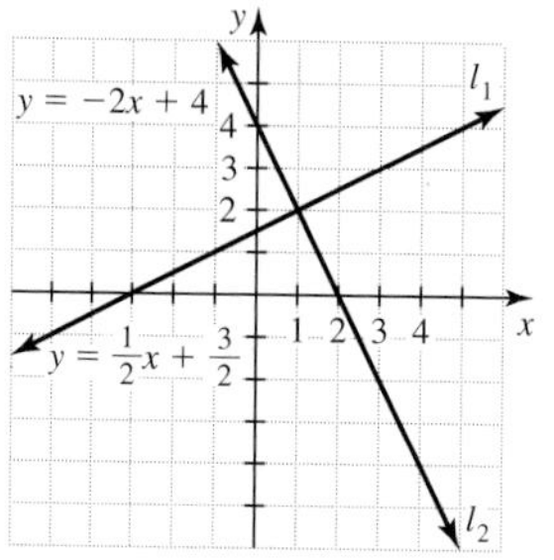

47. $-\frac{4}{3}$

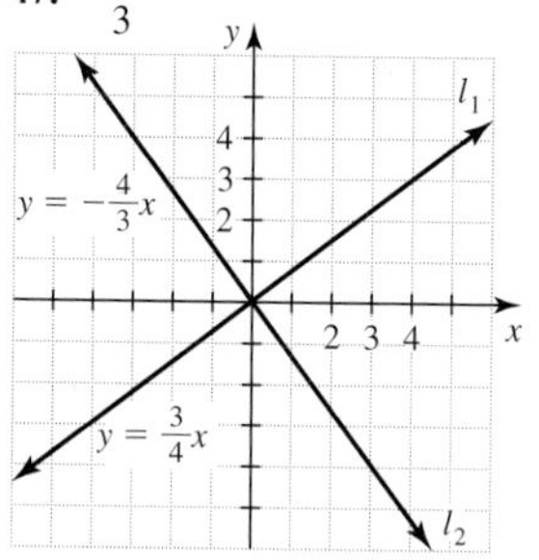

49. $\frac{1}{2}$

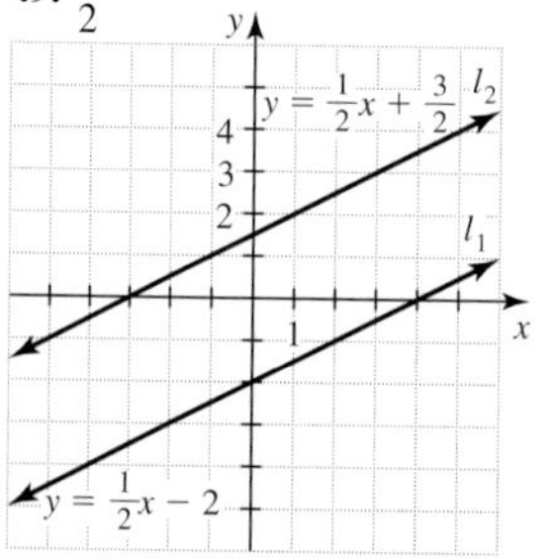

51. 1

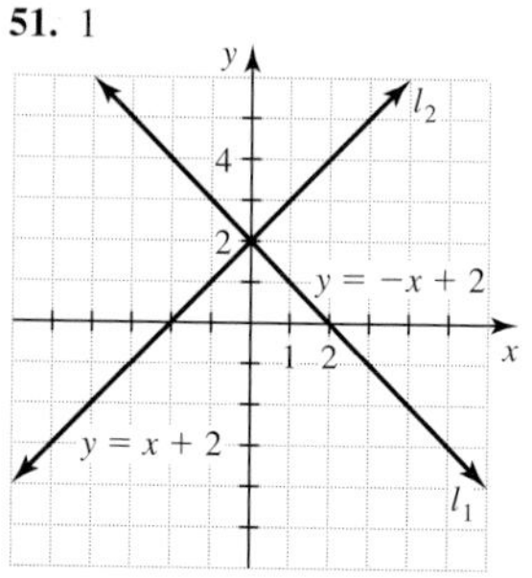

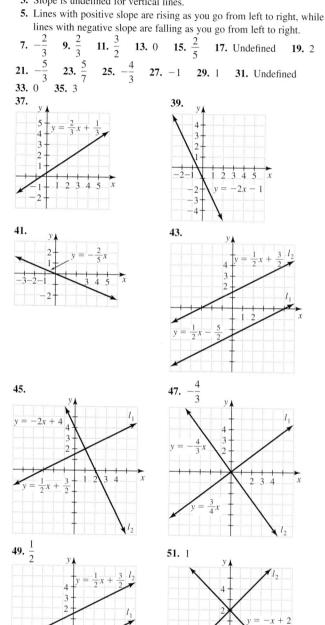

53. Parallel **55.** Neither **57.** Parallel **59.** Perpendicular
61. **a)** Approximately 0.183 slope; Cost is increasing about \$183,000 per year. **b)** \$2.03 million; yes **c)** \$3.13 million
63. 1 slope; The percentage increases 1% per year.
65. (2000, 28,100), (2003, 29,300), (2012, 32,900), (2015, 34,100)
67. Yes **69.** No

Section 3.3 Warm-Ups T F T T T F F T T F

1. Slope-intercept form is $y = mx + b$.
3. The standard form is $Ax + By = C$.
5. The slope-intercept form allows us to write the equation from the y-intercept and the slope.
7. $y = \frac{3}{2}x + 1$ **9.** $y = -2x + 2$ **11.** $y = x - 2$ **13.** $y = -x$
15. $y = -1$ **17.** $x = -2$ **19.** 3, (0, −9) **21.** $-\frac{1}{2}$, (0, 3)
23. 0, (0, 4) **25.** −3, (0, 0) **27.** −1, (0, 5) **29.** $\frac{1}{2}$, (0, −2)
31. $\frac{2}{5}$, (0, −2) **33.** 2, (0, 3) **35.** Undefined slope, no y-intercept
37. $x + y = 2$ **39.** $x - 2y = -6$ **41.** $9x - 6y = 2$
43. $6x + 10y = 7$ **45.** $x = -10$ **47.** $3y = 10$ **49.** $5x - 6y = 0$
51. $x - 50y = -25$

53.

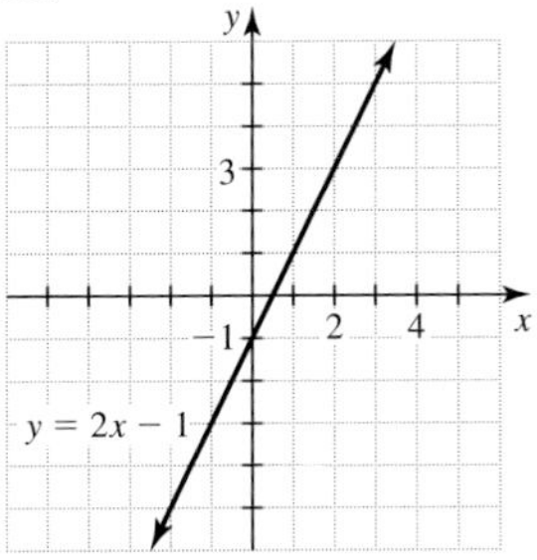

55.

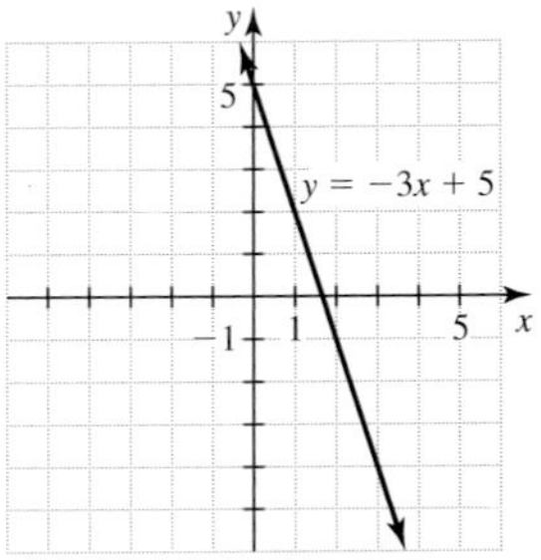

57.

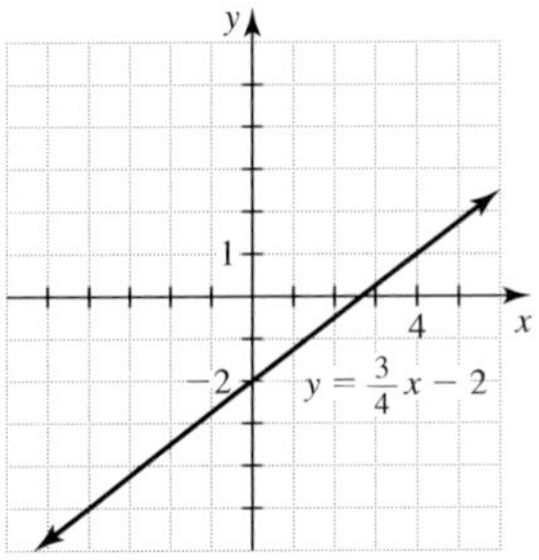

59.

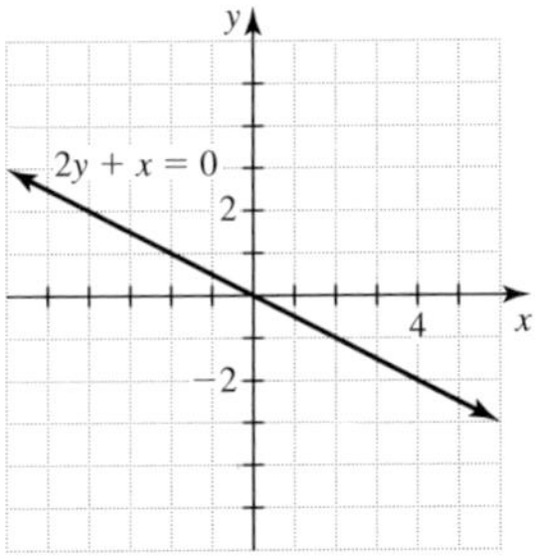

61.

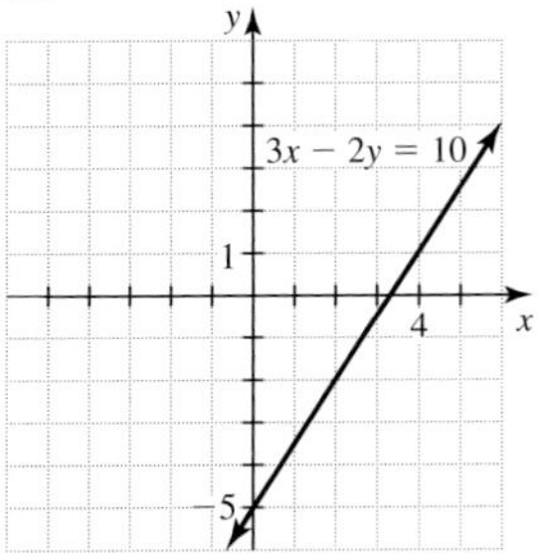

63.

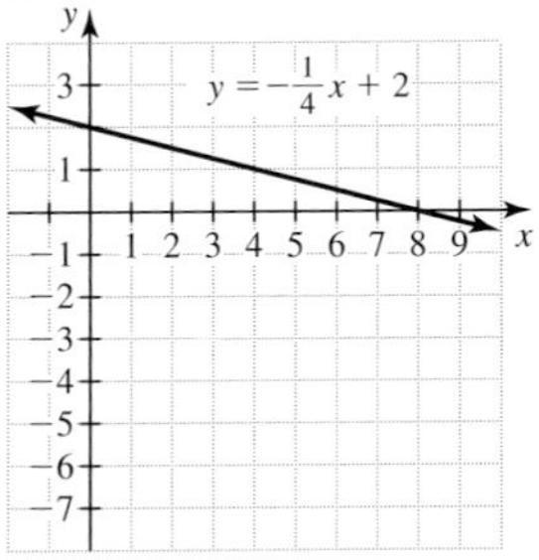

65.

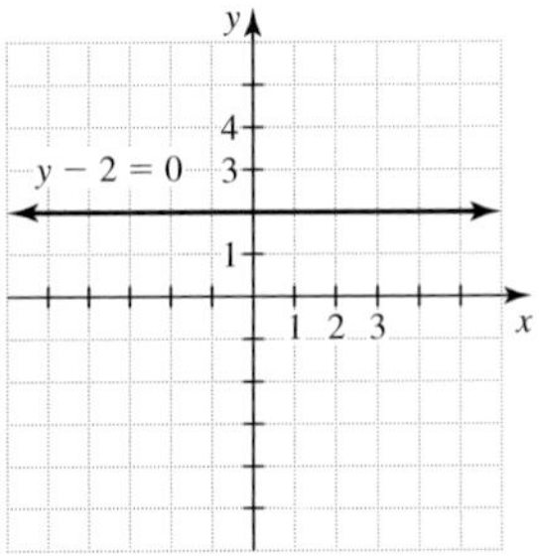

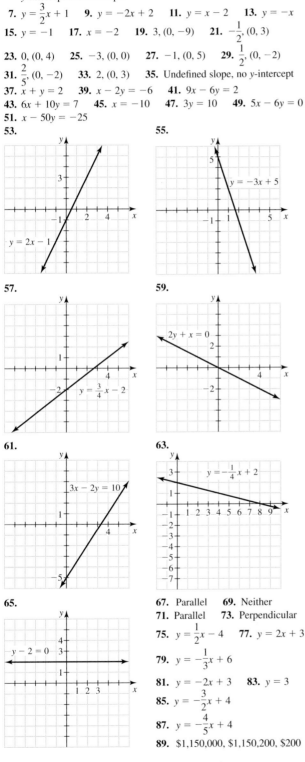

67. Parallel **69.** Neither
71. Parallel **73.** Perpendicular
75. $y = \frac{1}{2}x - 4$ **77.** $y = 2x + 3$
79. $y = -\frac{1}{3}x + 6$
81. $y = -2x + 3$ **83.** $y = 3$
85. $y = -\frac{3}{2}x + 4$
87. $y = -\frac{4}{5}x + 4$
89. \$1,150,000, \$1,150,200, \$200

91. a) A slope of 1 means that the percentage of workers receiving training is going up 1% per year. **b)** $y = x + 5$ where x is the number of years since 1982 **c)** The y-intercept (0, 5) means that 5% of the workers received training in 1982. **d)** 33%

93. a) x = the number of packs of pansies, y = the number of packs of snapdragons
b)

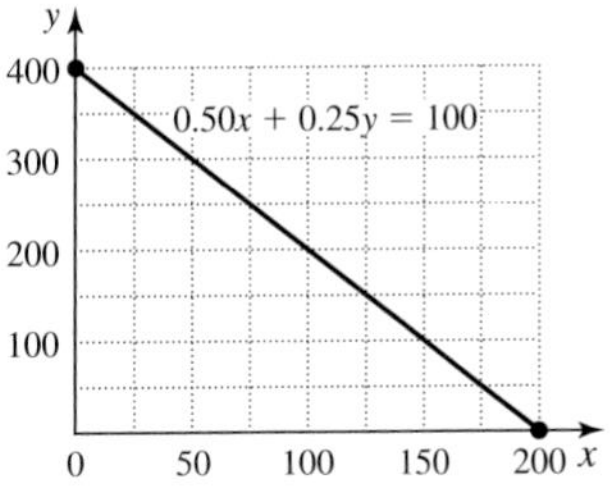

c) $y = -2x + 400$
d) −2
e) If the number of packs of pansies goes up by 1, then the number of packs of snapdragons goes down by 2.

95.

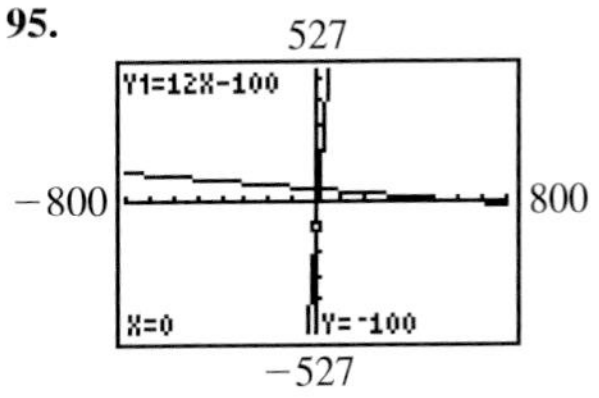

Section 3.4 Warm-Ups F F T T F T T T T T

1. Point-slope form is $y - y_1 = m(x - x_1)$.
3. If you know two points on a line, find the slope. Then use it along with a point in point-slope form to write the equation of the line.
5. Nonvertical parallel lines have equal slopes.
7. $y = 5x + 11$ **9.** $y = \frac{3}{4}x - 20$ **11.** $y = \frac{2}{3}x + \frac{1}{3}$ **13.** $y = 3x - 1$
15. $y = \frac{1}{2}x + 3$ **17.** $y = \frac{1}{3}x + \frac{7}{3}$ **19.** $y = -\frac{1}{2}x + 4$
21. $y = -6x - 13$ **23.** $2x - y = 7$ **25.** $x - 2y = 6$
27. $2x - 3y = 2$ **29.** $2x - y = -1$ **31.** $x - y = 0$
33. $3x - 2y = -1$ **35.** $3x + 5y = -11$ **37.** $x - y = -2$
39. $x = 2$ **41.** $y = 9$ **43.** $y = -x + 4$ **45.** $y = \frac{5}{3}x - 1$
47. $y = -\frac{1}{3}x + 5$ **49.** $y = x + 3$ **51.** $y = -\frac{2}{3}x + \frac{5}{3}$
53. $y = -2x - 5$ **55.** $y = \frac{1}{3}x + \frac{7}{3}$ **57.** $y = 2x - 1$ **59.** $y = 2$
61. $y = \frac{2}{3}x$ **63.** $y = -x$ **65.** $y = 50$ **67.** $y = -\frac{3}{5}x - 4$
69. e **71.** f **73.** h **75.** g
77. a) Slope 0.9 means that the number of ATM transactions is increasing by 0.9 billion per year. **b)** $y = 0.9x + 10.6$ **c)** 23.2 billion
79. a) $y = 1.5x + 53.8$ **b)** x = years since 1990, y = GDP in thousands of dollars **c)** \$83,800
d)

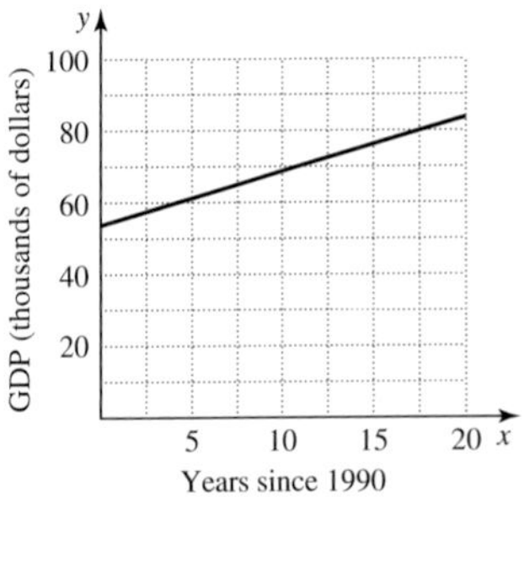

81. $C = 20n + 30$, \$170 **83.** $S = 3L - \frac{41}{4}$, 8.5

85. $v = 32t + 10$, 122 ft/sec

87. a) $w = -\frac{1}{120}t + \frac{3}{2}$ **b)** $\frac{5}{6}$ inch **c)** 60°F

89. $A = 0.6w$, 3.6 in. **91. a)** $a = 0.08c$ **b)** 0.24 **c)** 6.25 mg/ml

93. 2, 3, $-\frac{2}{3}$; 4, -5, $\frac{4}{5}$; $\frac{1}{2}$, 3, $-\frac{1}{6}$; 2, $-\frac{1}{3}$, 6

95. a)

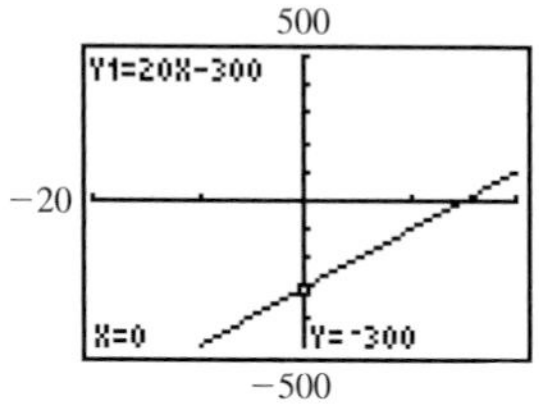

b)

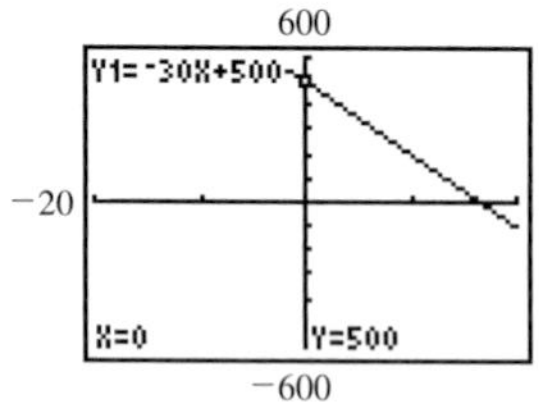

c)

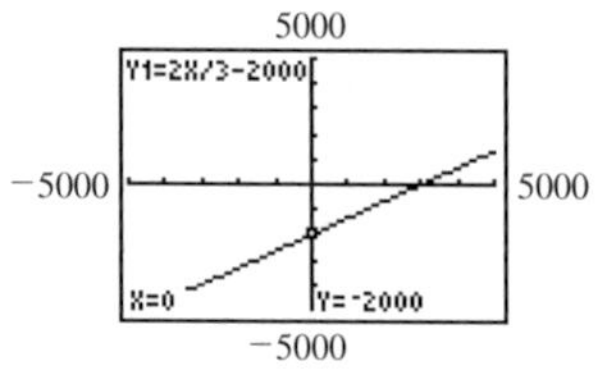

97. $-1 \le x \le 1$, $-1 \le y \le 1$

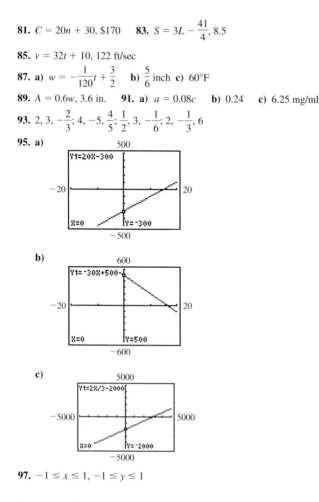

Section 3.5 Warm-Ups T F T F F T T T F F

1. If y varies directly as x, then there is a constant k such that $y = kx$.

3. If y is inversely proportional to x, then there is a constant k such that $y = \frac{k}{x}$.

5. $T = kh$ **7.** $y = \frac{k}{r}$ **9.** $R = kts$ **11.** $i = kb$ **13.** $A = kym$

15. $y = \frac{5}{3}x$ **17.** $A = \frac{6}{B}$ **19.** $m = \frac{198}{p}$ **21.** $A = 2tu$ **23.** $T = \frac{9}{2}u$

25. 25 **27.** 1 **29.** 105

31. $\left(\frac{1}{2}, 600\right)$, (1, 300), (30, 10), $\left(900, \frac{1}{3}\right)$, Inversely

33. $\left(\frac{1}{3}, \frac{1}{4}\right)$, (8, 6), (12, 9), (20, 15), Directly **35.** Directly, $y = 3.5x$

37. Inversely, $y = \frac{20}{x}$ **39.** (1, 65), (2, 130), (3, 195), (4, 260)

41. (20, 20), (40, 10), (50, 8), (200, 2)

43. 1600, 12, 12 **45.** 100.3 pounds

47. 50 minutes **49.** \$17.40 **51.** 80 mph **53.** 3 days

55. k, (0, 0), no, $y = kx$

Enriching Your Mathematical Word Power

1. d **2.** a **3.** b **4.** c **5.** b **6.** a **7.** c **8.** c **9.** d **10.** b **11.** c **12.** d

Review Exercises

1. Quadrant II **3.** x-axis **5.** y-axis **7.** Quadrant IV

9. $(0, -5)$, $(-3, -14)$, (4, 7)

11. $\left(0, -\frac{8}{3}\right)$, $\left(3, -\frac{2}{3}\right)$, $\left(-6, -\frac{20}{3}\right)$

13.

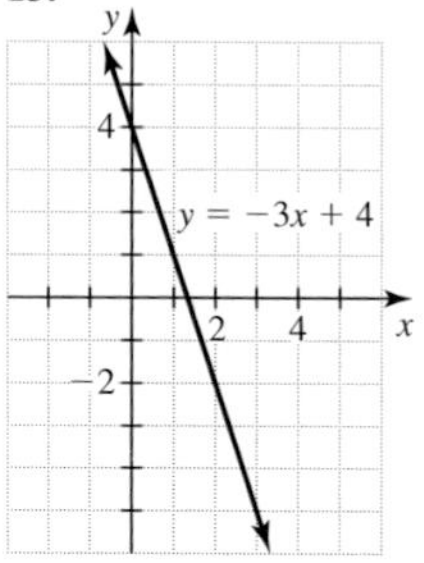

15.

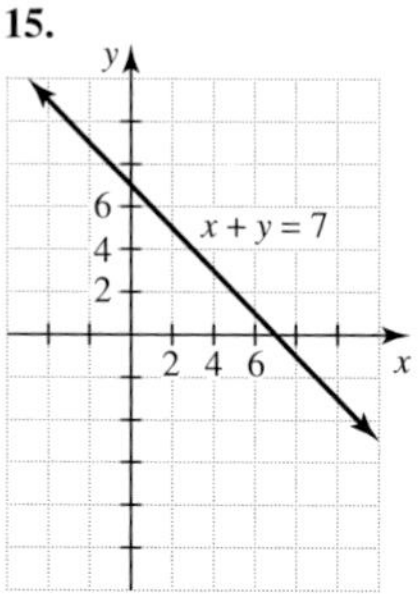

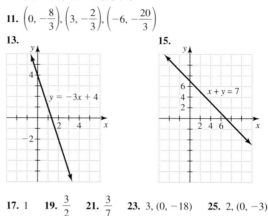

17. 1 **19.** $\frac{3}{2}$ **21.** $\frac{3}{7}$ **23.** 3, $(0, -18)$ **25.** 2, $(0, -3)$

27. 2, $(0, -4)$

29.

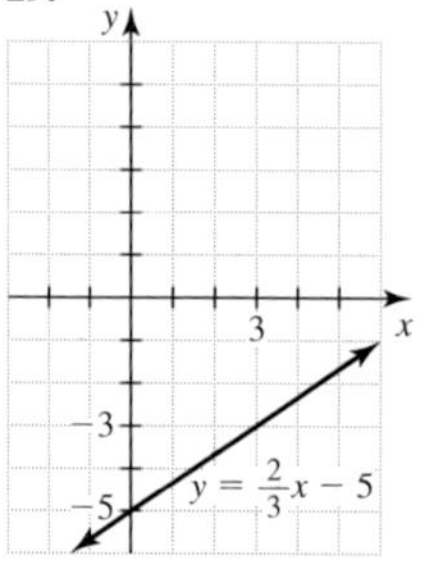

31.

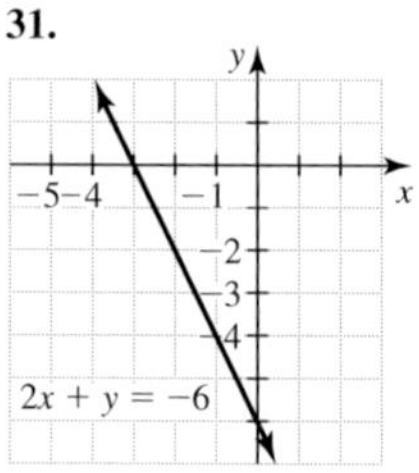

33.

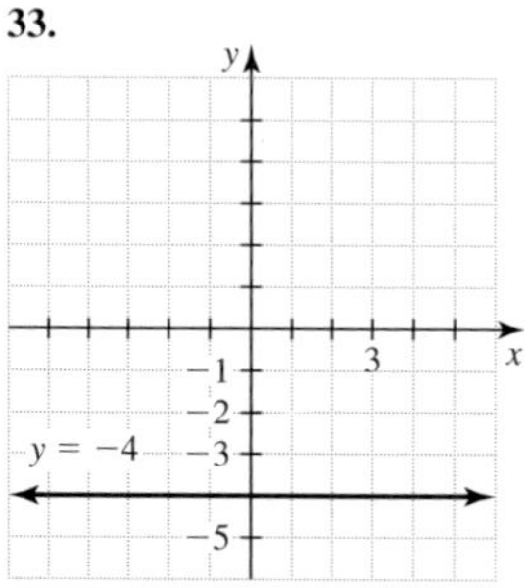

35. $x - 3y = -12$ **37.** $x + 2y = 0$ **39.** $y = 5$ **41.** $y = \frac{2}{3}x + 7$

43. $y = \frac{3}{7}x - 2$ **45.** $y = -\frac{3}{4}x + \frac{17}{4}$ **47.** $y = -2x - 1$

49. $y = \frac{6}{5}x + \frac{17}{5}$ **51.** $y = 3x - 14$ **53.** $C = 32n + 49$, \$177

55. a) $q = 1 - p$ **b)** 1 **57.** $y = 0.1x + 0.6$ **59.** 132 **61.** 2

63. 60 **65. a)** $C = 0.75T$ **b)** \$15 **c)** Increasing

Chapter 3 Test

1. Quadrant II **2.** x-axis **3.** Quadrant IV **4.** y-axis

5. 1 **6.** $-\frac{5}{6}$ **7.** 3 **8.** 0 **9.** Undefined **10.** $\frac{2}{3}$

11. $y = -\frac{1}{2}x + 3$ **12.** $y = \frac{3}{7}x - \frac{11}{7}$ **13.** $x - 3y = 11$

14. $5x + 3y = 27$

15.

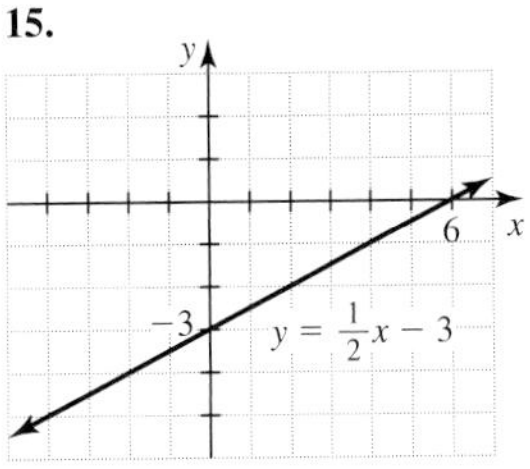

16.

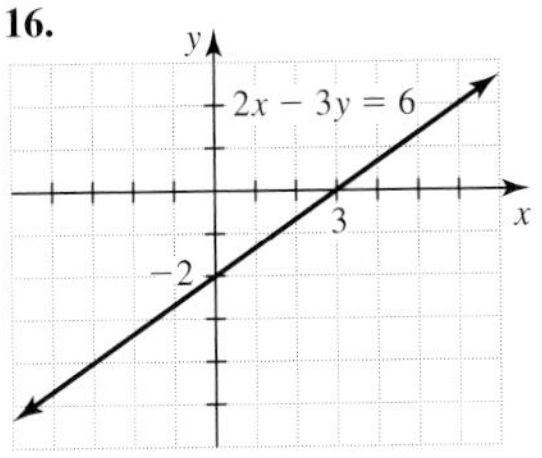

17.

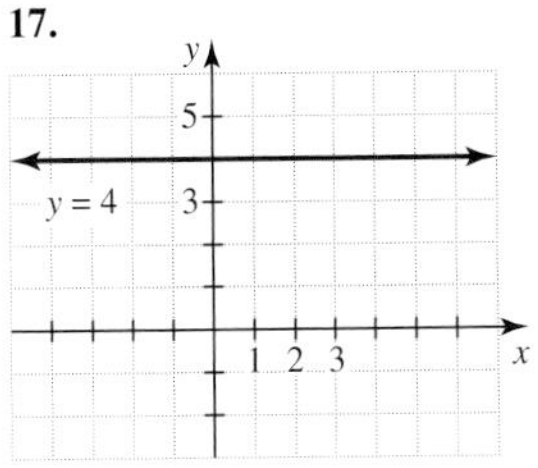

18.

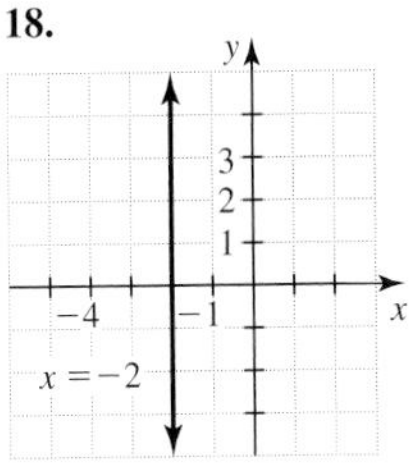

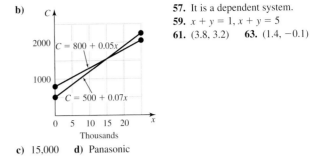

19. $S = 0.75n + 2.50$ **20.** $P = 3v + 20$, 80 cents
21. \$2.80 **22.** 18.75 days, decreases **23.** \$770

Making Connections Chapters 1–3

1. −1 **2.** −34 **3.** 1 **4.** 72 **5.** −4
6. −28 **7.** $-\frac{7}{2}$ **8.** 0.4 **9.** $\frac{1}{10}$ **10.** 15 **11.** $13x$
12. $3x - 36$ **13.** $\left\{\frac{5}{2}\right\}$ **14.** $\left\{\frac{7}{3}\right\}$ **15.** $\frac{1}{6}$ **16.** $\frac{5}{12}$
17. {2} **18.** {−4} **19.** $2x - 4$ **20.** $x + 2$
21. {5} **22.** {3} **23.** ∅ **24.** All real numbers
25. $(4.5, \infty)$ **26.** $\left(-\frac{2}{3}, \infty\right)$ **27.** $[10, \infty)$ **28.** $[20, \infty)$
29. $\left[-\frac{1}{2}, \frac{5}{2}\right)$ **30.** $\left[0, \frac{2}{3}\right)$ **31.** $y = \frac{t-2}{3\pi}$
32. $y = mx + b$ **33.** $y = x - 4$ **34.** $y = 6$
35. $y = \frac{4}{5}$ **36.** $y = 200$
37. a) $\frac{2}{15}$ **b)** $\frac{1}{5}$ **c)** About 13% per year
d) \$276,000 saved, \$12,000 per year

Chapter 4

Section 4.1 Warm-Ups T F F F T T T T T F
1. A system of equations is a pair of equations.
3. In this section systems of equations were solved by graphing.
5. A dependent system is one for which the two equations are equivalent.
7. (3, −2) **9.** All three **11.** None **13.** (−2, 3) **15.** (2, 4)
17. (1, 2) **19.** (0, −5) **21.** (−2, 3) **23.** (0, 0) **25.** (2, 3)
27. No solution, inconsistent **29.** (1, −2), independent
31. $\{(x, y) \mid 2x + y = 3\}$, dependent
33. $\{(x, y) \mid y = x\}$, dependent **35.** (−4, −3), independent
37. (−1, 0), independent **39.** Inconsistent **41.** Independent
43. Inconsistent **45.** Inconsistent **47.** Dependent
49. Independent **51.** Inconsistent
53. a) (5, 20)
b) For 5 toppings the cost is \$20 at both restaurants.
55. a) 800, 500; 1050, 850; 1300, 1200; 1800, 1900
b)

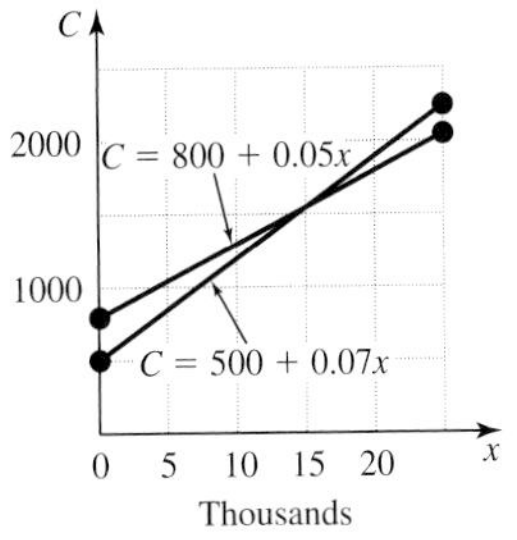

c) 15,000 **d)** Panasonic

57. It is a dependent system.
59. $x + y = 1, x + y = 5$
61. (3.8, 3.2) **63.** (1.4, −0.1)

Section 4.2 Warm-Ups T T T T T T F T T T
1. In this section we used the substitution method.
3. A dependent system is one in which the equations are equivalent.
5. Using substitution on a dependent system results in an equation that is always true.
7. (2, 5) **9.** (2, 3) **11.** (−2, 9) **13.** (−5, 5) **15.** $\left(\frac{1}{3}, \frac{2}{3}\right)$
17. $\left(\frac{1}{2}, \frac{1}{3}\right)$ **19.** $\left(3, \frac{5}{2}\right)$, independent **21.** No solution, inconsistent
23. $\{(x, y) \mid 3x - y = 5\}$, dependent **25.** No solution, inconsistent
27. $\left(\frac{11}{5}, \frac{3}{25}\right)$, independent **29.** (3, 2) **31.** (3, 1) **33.** No solution
35. Inconsistent **37.** Dependent **39.** Independent **41.** Inconsistent
43. Length 28 ft, width 14 ft **45.** \$12,000 at 10%, \$8000 at 5%
47. *Titanic* \$601 million, *Star Wars* \$461 million
49. Lawn \$12, sidewalk \$7
51. Left rear 288 pounds, left front 287 pounds, no
53. \$2.40 per pound
55. a) 69.2 years, 76.8 years
b)

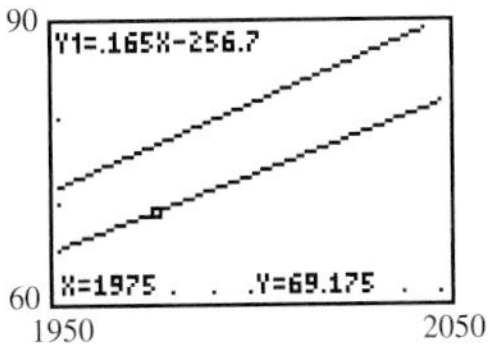

c) No **d)** 1614

Section 4.3 Warm-Ups F T T T T F F F F T
1. In this section we learned to solve systems by the addition method.
3. In addition and substitution we eliminate a variable and solve for the remaining variable.
5. Eliminate the variable that is easiest to eliminate.
7. (3, −1) **9.** (−3, 5) **11.** (−4, 3) **13.** $\left(\frac{1}{7}, \frac{9}{7}\right)$ **15.** (8, 31)
17. (−2, −3) **19.** (−1, 4) **21.** (−1, 2) **23.** (3, −1) **25.** (1, 2)
27. (12, 6) **29.** (24, 16) **31.** No solution, inconsistent
33. $\{(x, y) \mid x + y = 5\}$, dependent **35.** $\{(x, y) \mid 2x = y + 3\}$, dependent
37. $\{(x, y) \mid x + 3y = 3\}$, dependent **39.** No solution, inconsistent
41. (12, 18), independent **43.** (40, 60), independent
45. (0.1, 0.1), independent **47.** (1.5, −2.8) **49.** (4, 3)
51. (4, 1) **53.** No solution **55.** 150 cars, 100 trucks
57. 24 dimes, 16 nickels **59.** 24 dimes, 7 quarters
61. 6 adults, 24 children **63.** Coffee \$0.45, doughnut \$0.35
65. 300 men, 360 women **67.** 180 hours regular time, 30 hours overtime
69. \$3000 at 7%, \$5000 at 10% **71.** \$12,000 in stocks, \$9000 in bonds
73. \$1800 at 15%, \$3000 at 9%
75. 200 pounds of 20% synthetic, 600 pounds of 40% synthetic
77. 40 ounces of 10%, 80 ounces of 22%
79. 64 ounces of \$4.40 metal, 36 ounces of \$2.40 metal
81. 60 pounds of \$4.20 hamburger, 40 pounds of \$3.10 hamburger

21.

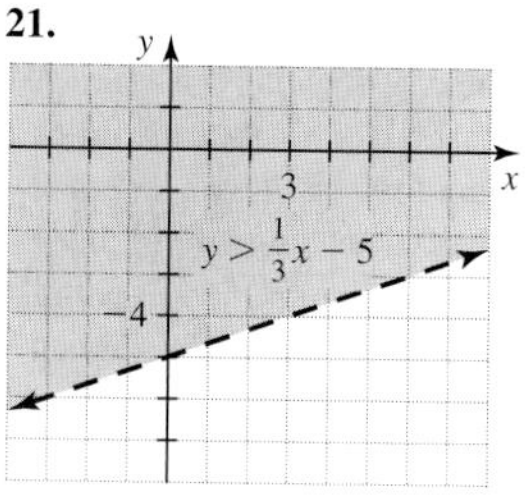

23.

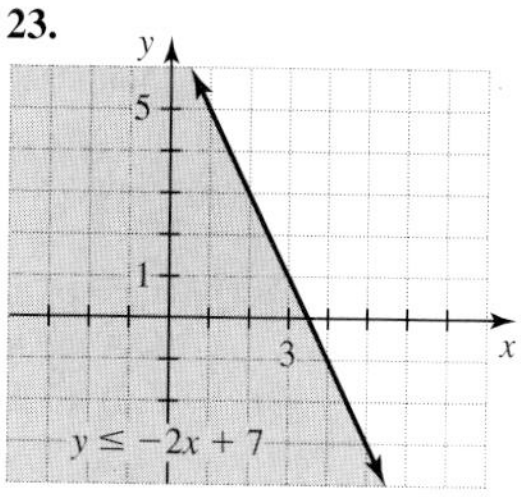

25.

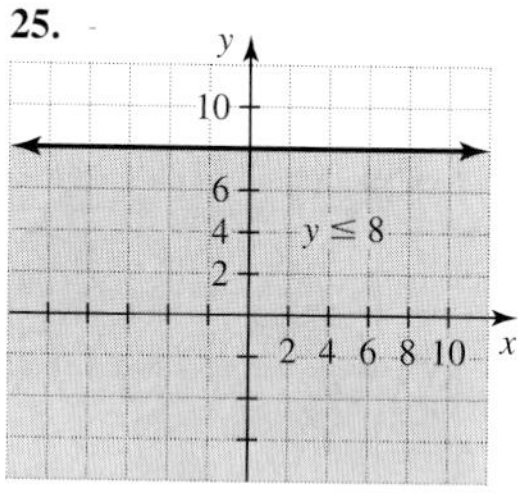

27.

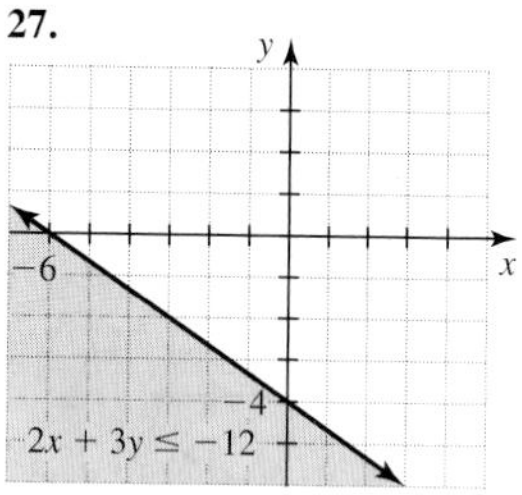

29.

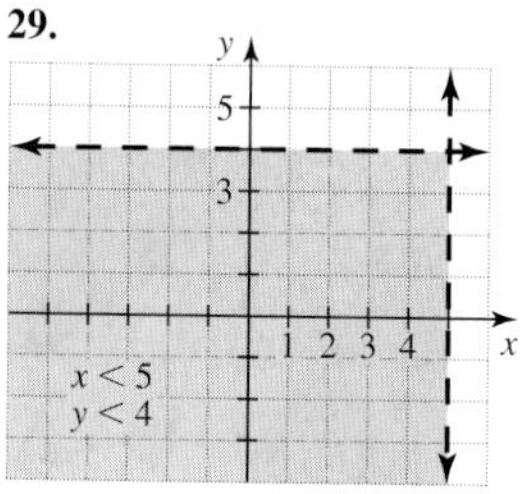

31.

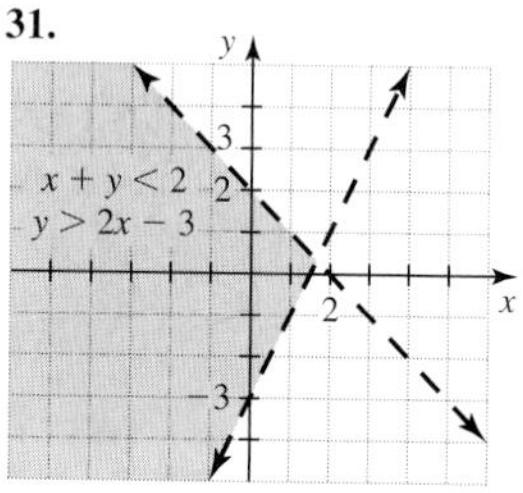

33.

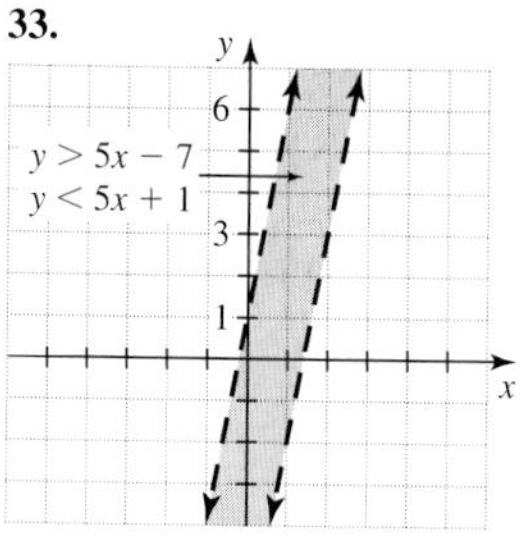

35.

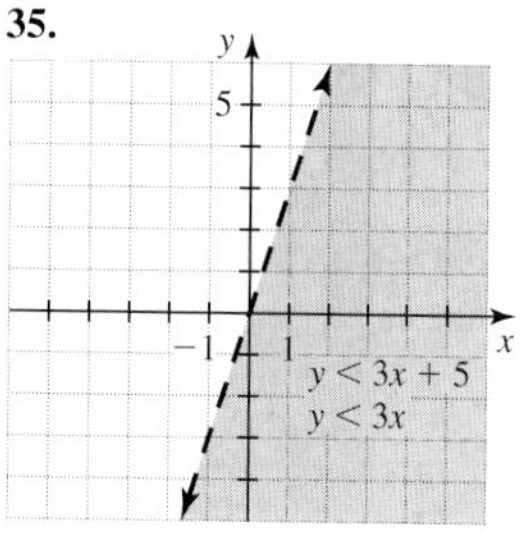

37. Apple \$0.45, orange \$0.35 **39.** 32 fives, 22 tens
41. 4 servings green beans, 3 servings chicken soup

Chapter 4 Test

1. $(-1, 3)$ **2.** $(2, 1)$ **3.** $(3, -1)$ **4.** $(2, 3)$ **5.** $(4, 1)$
6. Inconsistent **7.** Dependent
8. Independent

9.

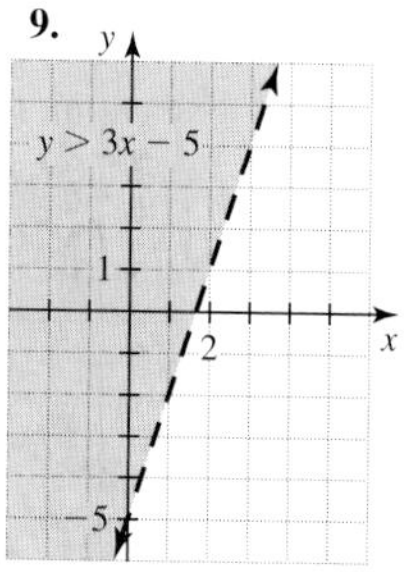

10.

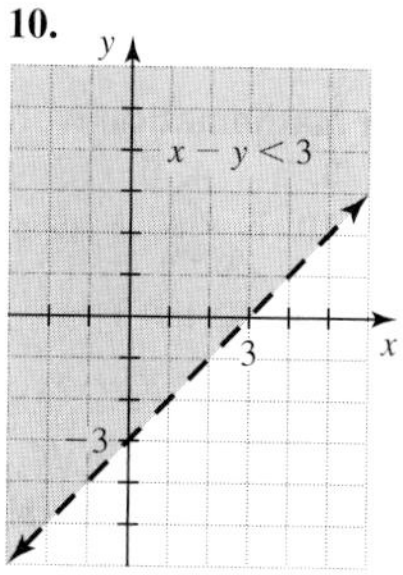

11.

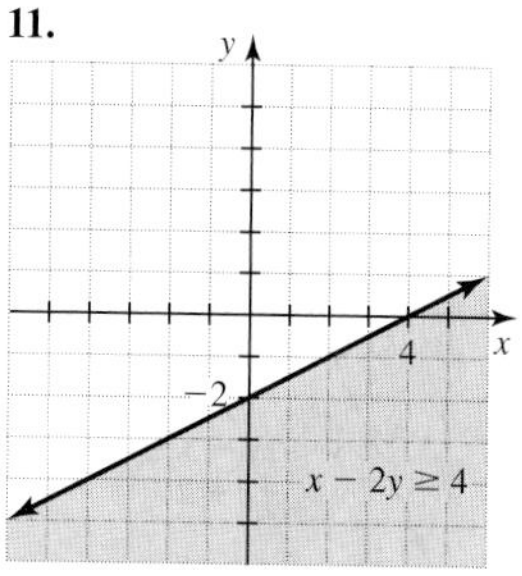

12.

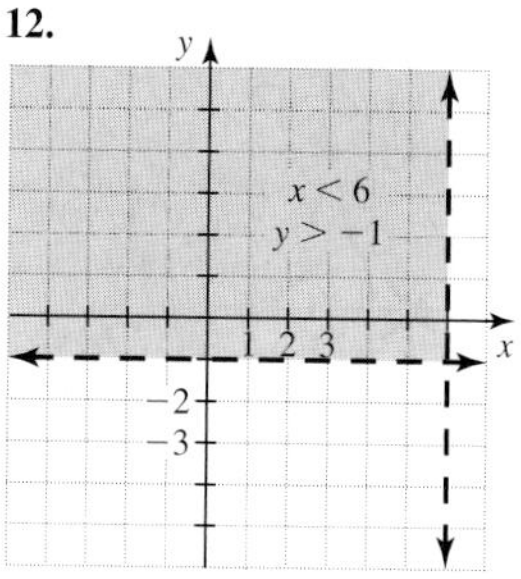

13.

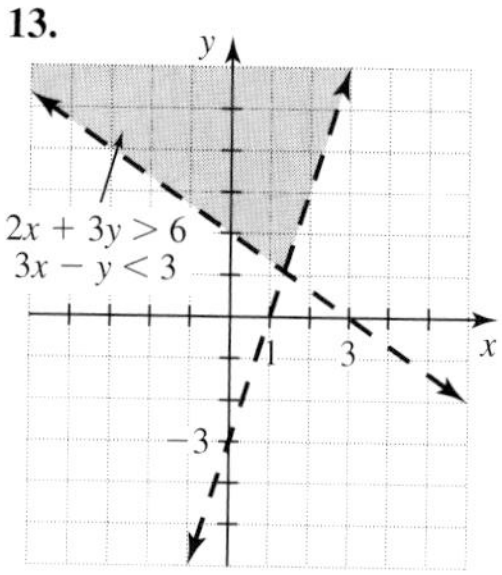

14.

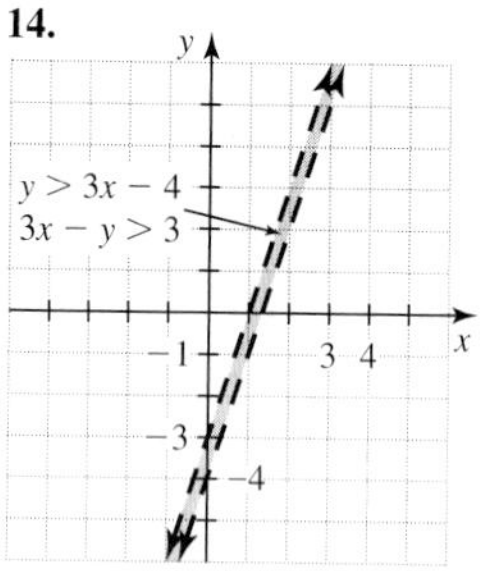

15. Kathy 36 hours, Chris 18 hours **16.** \$48

Making Connections A Review of Chapters 1–4

1. {7} **2.** $\left\{\frac{5}{3}\right\}$ **3.** {12} **4.** {1000} **5.** ∅ **6.** All real numbers

7. $(4, \infty)$

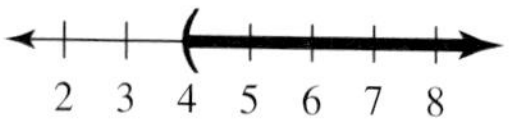

8. $\left[\frac{1}{2}, 5\right]$

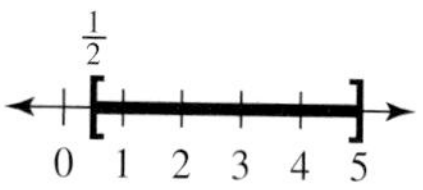

9. $[1, \infty)$

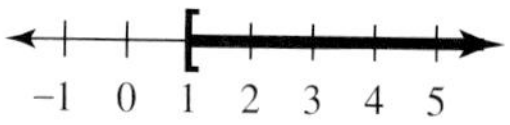

10.

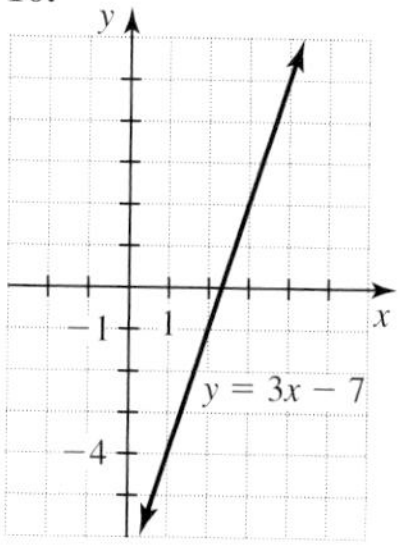

11.

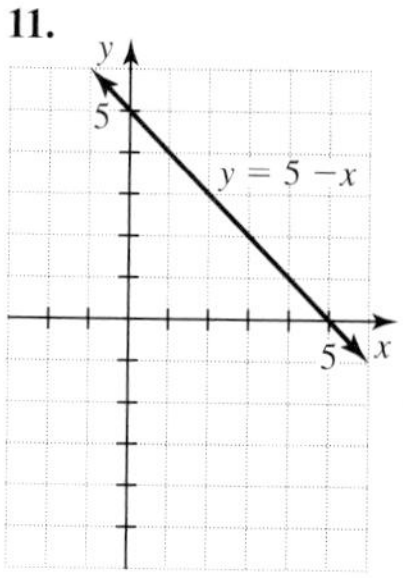

12.

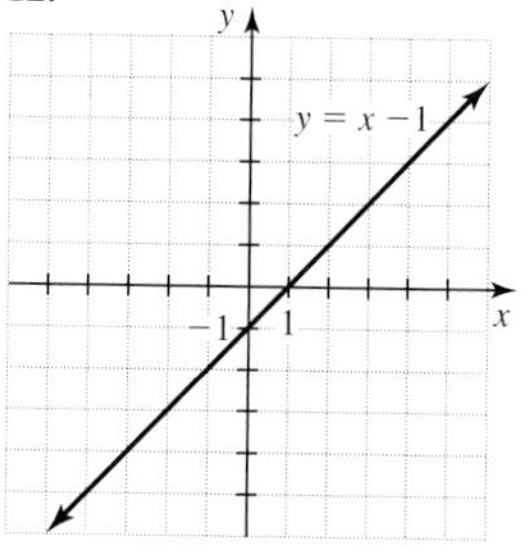

13.

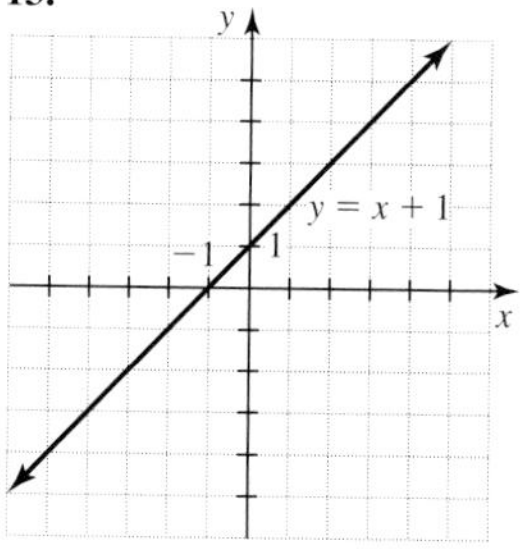

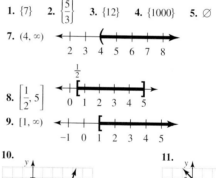

14.
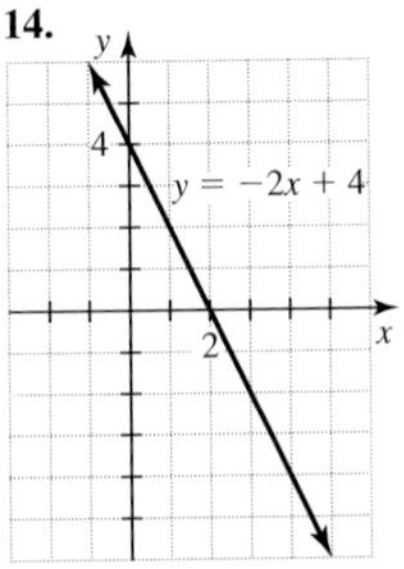

15.
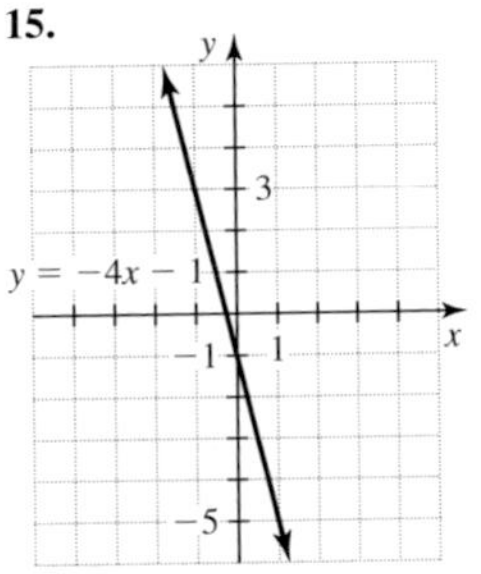

16.
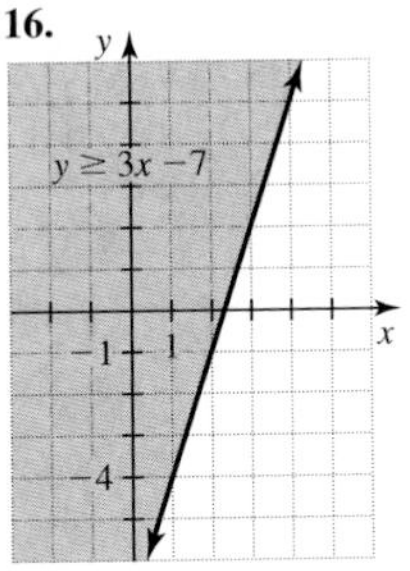

17.
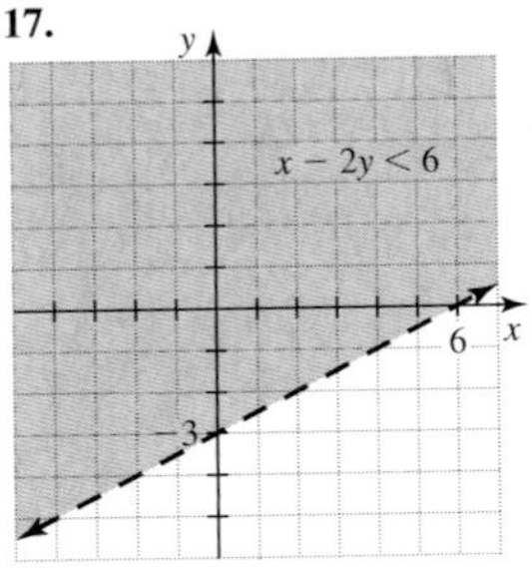

18.
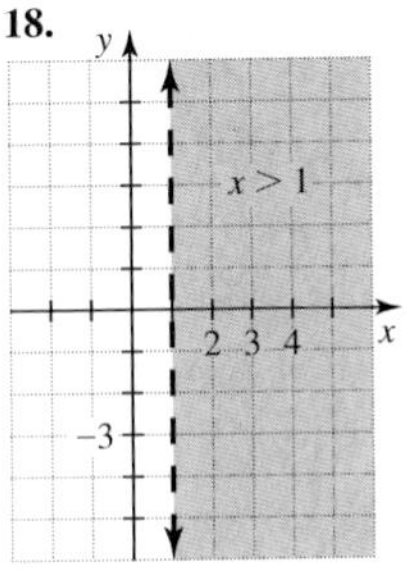

19. $y = 6x + 36$

20. $y = -7x + 95$

21. a) $p = -\frac{9}{10}x + 25$

b) $p = \frac{9}{10}x + 10$

c) $8\frac{1}{3}$ years after 1990 or 1998

Chapter 5

Section 5.1 Warm-Ups F F T T F T F T T F

1. A term is a single number or the product of a number and one or more variables raised to powers.
3. The degree of a polynomial in one variable is the highest power of the variable in the polynomial.
5. Polynomials are added by adding the like terms.

7. $-3, 7$ **9.** $0, 6$ **11.** $\frac{1}{3}, \frac{7}{2}$ **13.** Monomial, 0
15. Monomial, 3 **17.** Binomial, 1 **19.** Trinomial, 10
21. Binomial, 6 **23.** Trinomial, 3 **25.** 6 **27.** $\frac{5}{8}$ **29.** -85
31. 5 **33.** 71 **35.** -4.97665 **37.** $4x - 8$ **39.** $2q$
41. $x^2 + 3x - 2$ **43.** $x^3 + 9x - 7$ **45.** $3a^2 - 7a - 4$
47. $-3w^2 - 8w + 5$ **49.** $9.66x^2 - 1.93x - 1.49$
51. $-4x + 6$ **53.** -5 **55.** $-z^2 + 2z$ **57.** $w^5 + w^4 - w^3 - w^2$
59. $2t + 13$ **61.** $-8y + 7$ **63.** $-22.85x - 423.2$
65. $4a + 2$ **67.** $-2x + 4$ **69.** $2a$ **71.** $-5m + 7$
73. $4x^2 + 1$ **75.** $a^3 - 9a^2 + 2a + 7$ **77.** $-3x + 9$
79. $2y^3 + 7y^2 - 4y - 14$ **81.** $-3m + 3$ **83.** $-11y - 3$
85. $2x^2 - 6x + 12$ **87.** $-5z^4 - 8z^3 + 3z^2 + 7$
89. $P(x) = 100x + 500$ dollars, \$5500
91. $P(x) = 6x + 3$, $P(4) = 27$ meters
93. $5x + 40$ miles, 140 miles
95. 800 feet, 800 feet
97. $0.17x + 74.47$ dollars, \$244.47
99. 1321.39 calories
101. Yes, yes, yes
103. The highest power of x is 3.

Section 5.2 Warm-Ups F F T F T T T T T F

1. The product rule for exponents says that $a^m \cdot a^n = a^{m+n}$.
3. To multiply a monomial and a polynomial we use the distributive property.
5. To multiply any two polynomials we multiply each term of the first polynomial by every term of the second polynomial.

7. $27x^5$ **9.** $14a^{11}$ **11.** $-30x^4$ **13.** $27x^{17}$ **15.** $-54s^2t^2$
17. $24t^7w^8$ **19.** $25y^2$ **21.** $4x^6$ **23.** $x^2 + xy^2$ **25.** $4y^7 - 8y^3$
27. $-18y^2 + 12y$ **29.** $-3y^3 + 15y^2 - 18y$ **31.** $-xy^2 + x^3$
33. $15a^4b^3 - 5a^5b^2 - 10a^6b$ **35.** $-2t^5v^3 + 3t^3v^2 + 2t^2v^2$
37. $x^2 + 3x + 2$ **39.** $x^2 + 2x - 15$ **41.** $t^2 - 13t + 36$
43. $x^3 + 3x^2 + 4x + 2$ **45.** $6y^3 + y^2 + 7y + 6$
47. $2y^8 - 3y^6z - 5y^4z^2 + 3y^2z^3$ **49.** $2a^2 + 7a - 15$
51. $14x^2 + 95x + 150$ **53.** $20x^2 - 7x - 6$
55. $2am - 6an + bm - 3bn$ **57.** $x^3 + 9x^2 + 16x - 12$
59. $-4a^4 + 9a^2 - 8a - 12$ **61.** $x^2 - y^2$ **63.** $x^3 + y^3$
65. $u - 3t$ **67.** $-3x - y$ **69.** $3a^2 + a - 6$ **71.** $-3v^2 - v + 6$
73. $-6x^2 + 27x$ **75.** $-6x^2 + 27x + 2$ **77.** $-x - 7$ **79.** $36x^{12}$
81. $-6a^3b^{10}$ **83.** $25x^2 + 60x + 36$ **85.** $25x^2 - 36$
87. $6x^7 - 8x^4$ **89.** $m^3 - 1$ **91.** $3x^3 - 5x^2 - 25x + 18$
93. $x^2 + 4x$ square feet, 140 square feet
95. $A(x) = x^2 + \frac{1}{2}x$, $A(5) = 27.5$ square feet
97. $x^2 + 5x$ **99.** $8.05x^2 + 15.93x + 6.12$ square meters
101. 30,000, \$300,000, $40{,}000p - 1000p^2$
103. $10x^5 + 10x^4 + 10x^3 + 10x^2 + 10x$, \$67.16

Section 5.3 Warm-Ups F T T T T F T F F F

1. We use the distributive property to find the product of two binomials.
3. The purpose of FOIL is to provide a faster method for finding the product of two binomials.

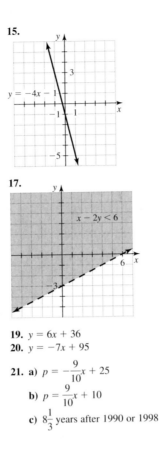

5. $x^2 + 6x + 8$ **7.** $a^2 + 5a + 4$ **9.** $x^2 + 19x + 90$
11. $2x^2 + 7x + 3$ **13.** $a^2 - a - 6$ **15.** $2x^2 - 5x + 2$
17. $2a^2 - a - 3$ **19.** $w^2 - 60w + 500$ **21.** $y^2 + 5y - ay - 5a$
23. $5w + 5m - w^2 - mw$ **25.** $10m^2 - 9mt - 9t^2$
27. $45a^2 + 53ab + 14b^2$ **29.** $x^4 - 3x^2 - 10$ **31.** $h^6 + 10h^3 + 25$
33. $3b^6 + 14b^3 + 8$ **35.** $y^3 - 2y^2 - 3y + 6$ **37.** $6m^6 + 7m^3n^2 - 3n^4$
39. $12u^4v^2 + 10u^2v - 12$ **41.** $b^2 + 9b + 20$ **43.** $x^2 + 6x - 27$
45. $a^2 + 10a + 25$ **47.** $4x^2 - 4x + 1$ **49.** $z^2 - 100$
51. $a^2 + 2ab + b^2$ **53.** $a^2 - 3a + 2$ **55.** $2x^2 + 5x - 3$
57. $5t^2 - 7t + 2$ **59.** $h^2 - 16h + 63$ **61.** $h^2 + 14hw + 49w^2$
63. $4h^4 - 4h^2 + 1$ **65.** $a^3 + 4a^2 - 7a - 10$
67. $h^3 + 9h^2 + 26h + 24$ **69.** $x^3 - 2x^2 - 64x + 128$
71. $x^3 + 8x^2 - \frac{1}{4}x - 2$ **73.** $x^2 + 15x + 50$ **75.** $x^2 + x + \frac{1}{4}$
77. $8x^2 + 2x + \frac{1}{8}$ **79.** $8a^2 + a - \frac{1}{4}$ **81.** $\frac{1}{8}x^2 + \frac{1}{6}x - \frac{1}{6}$
83. $a^3 + 7a^2 + 12a$ **85.** $x^5 + 13x^4 + 42x^3$
87. $-12x^6 - 26x^5 + 10x^4$ **89.** $x^3 + 3x^2 - x - 3$
91. $9x^3 + 45x^2 - 4x - 20$ **93.** $2x + 10$
95. $2x^2 + 5x - 3$ square feet
97. $5.2555x^2 + 0.41095x - 1.995$ square meters
99. 12 ft², $3h$ ft², $4h$ ft², h^2 ft², $h^2 + 7h + 12$ ft², $(h + 3)(h + 4) = h^2 + 7h + 12$

Section 5.4 Warm-Ups F T T T F T T T F F

1. The special products are $(a + b)^2$, $(a - b)^2$, and $(a + b)(a - b)$.
3. It is faster to do by the new rule than with FOIL.
5. $(a + b)(a - b) = a^2 - b^2$ **7.** $x^2 + 2x + 1$ **9.** $y^2 + 8y + 16$
11. $m^2 + 12m + 36$ **13.** $9x^2 + 48x + 64$ **15.** $s^2 + 2st + t^2$
17. $4x^2 + 4xy + y^2$ **19.** $4t^2 + 12ht + 9h^2$ **21.** $p^2 - 4p + 4$
23. $a^2 - 6a + 9$ **25.** $t^2 - 2t + 1$ **27.** $9t^2 - 12t + 4$
29. $s^2 - 2st + t^2$ **31.** $9a^2 - 6ab + b^2$ **33.** $9z^2 - 30yz + 25y^2$
35. $a^2 - 25$ **37.** $y^2 - 1$ **39.** $9x^2 - 64$ **41.** $r^2 - s^2$
43. $64y^2 - 9a^2$ **45.** $25x^4 - 4$
47. $x^3 + 3x^2 + 3x + 1$ **49.** $8a^3 - 36a^2 + 54a - 27$
51. $a^4 - 12a^3 + 54a^2 - 108a + 81$ **53.** $a^4 + 4a^3b + 6a^2b^2 + 4ab^3 + b^4$
55. $a^2 - 400$ **57.** $x^2 + 15x + 56$ **59.** $16x^2 - 1$
61. $81y^2 - 18y + 1$ **63.** $6t^2 - 7t - 20$ **65.** $4t^2 - 20t + 25$
67. $4t^2 - 25$ **69.** $x^4 - 1$ **71.** $4y^6 - 36y^3 + 81$
73. $4x^6 + 12x^3y^2 + 9y^4$ **75.** $\frac{1}{4}x^2 + \frac{1}{3}x + \frac{1}{9}$
77. $0.04x^2 - 0.04x + 0.01$ **79.** $a^3 + 3a^2b + 3ab^2 + b^3$
81. $2.25x^2 + 11.4x + 14.44$ **83.** $12.25t^2 - 6.25$
85. $x^2 - 25$ square feet, 25 square feet smaller
87. $3.14b^2 + 6.28b + 3.14$ square meters
89. $v = k(R^2 - r^2)$ **91.** $P + 2Pr + Pr^2$, \$242 **93.** \$20,230.06
95. The first is an identity and the second is a conditional equation.

Section 5.5 Warm-Ups F F T F T F T T T T

1. The quotient rule is used for dividing monomials.
3. When dividing a polynomial by a monomial the quotient should have the same number of terms as the polynomial.
5. The long division process stops when the degree of the remainder is less than the degree of the divisor.
7. 1 **9.** 1 **11.** 1 **13.** 1 **15.** x^6 **17.** $\frac{1}{a^9}$ **19.** $\frac{3}{a^5}$ **21.** a^6
23. $\frac{-4}{x^4}$ **25.** $-y$ **27.** $-3x$ **29.** $\frac{-3}{x^3}$ **31.** $x - 2$
33. $x^3 + 3x^2 - x$ **35.** $4xy - 2x + y$ **37.** $y^2 - 3xy$ **39.** 2, -1
41. $x + 5$, 16 **43.** $x + 2$, 7 **45.** 2, -10 **47.** $a^2 + 2a + 8$, 13
49. $x - 4$, 4 **51.** $h^2 + 3h + 9$, 0 **53.** $2x - 3$, 1 **55.** $x^2 + 1$, -1
57. $3 + \frac{15}{x - 5}$ **59.** $-1 + \frac{3}{x + 3}$ **61.** $1 - \frac{1}{x}$ **63.** $3 + \frac{1}{x}$
65. $x - 1 + \frac{1}{x + 1}$ **67.** $x - 2 + \frac{8}{x + 2}$ **69.** $x^2 + 2x + 4 + \frac{8}{x - 2}$
71. $x^2 + \frac{3}{x}$ **73.** $-3a$ **75.** $\frac{4t^4}{w^5}$ **77.** $-a + 4$ **79.** $x - 3$
81. $-6x^2 + 2x - 3$ **83.** $t + 4$ **85.** $2w + 1$ **87.** $4x^2 - 6x + 9$
89. $t^2 - t + 3$ **91.** $v^2 - 2v + 1$ **93.** $x - 5$ meters
95. $x^8 + x^7 + x^6 + x^5 + x^4 + x^3 + x^2 + x + 1$
97. $10x \div 5x$ is not equivalent to the other two.

Section 5.6 Warm-Ups F F F F F F T F F T

1. The product rule says that $a^m a^n = a^{m+n}$.
3. These rules do not make sense without identical bases.
5. The power of a product rule says that $(ab)^n = a^n b^n$.
7. 128 **9.** $6u^{10}$ **11.** a^4b^{10} **13.** $\frac{-1}{2a^4}$ **15.** $\frac{2a^6}{5b^4}$ **17.** 200
19. x^6 **21.** $2x^{12}$ **23.** $\frac{1}{t^2}$ **25.** $\frac{1}{2}$ **27.** x^3y^6 **29.** $-8t^{15}$
31. $-8x^6y^{15}$ **33.** $a^9b^2c^{14}$ **35.** $\frac{x^{12}}{64}$ **37.** $\frac{16a^8}{b^{12}}$ **39.** $-\frac{x^6}{8y^3}$
41. $\frac{4z^{12}}{x^8}$ **43.** 45 **45.** 81 **47.** -19 **49.** -1 **51.** $\frac{8}{125}$
53. 200 **55.** 128 **57.** $\frac{1}{16}$ **59.** x^7 **61.** a^{32} **63.** $a^{12}b^6$
65. $\frac{1}{x^3}$ **67.** a^4 **69.** $\frac{a^9}{b^{12}}$ **71.** $\frac{1}{x^5}$ **73.** $15x^{11}$ **75.** $-125x^{12}$
77. $-27y^6z^{19}$ **79.** $\frac{3v}{u}$ **81.** $-16x^9t^6$ **83.** $\frac{8}{x^6}$ **85.** $\frac{-32a^{15}b^{20}}{c^{25}}$
87. $\frac{y^5}{32x^5}$ **89.** Product rule, $P(1 + r)^{15}$

Section 5.7 Warm-Ups T F F T F T T T T F

1. A negative exponent means "reciprocal," as in $a^{-n} = \frac{1}{a^n}$.
3. The new quotient rule is $a^m/a^n = a^{m-n}$ for any integers m and n.
5. Convert from standard notation by counting the number of places the decimal must move so that there is one nonzero digit to the left of the decimal point.
7. $\frac{1}{3}$ **9.** $\frac{1}{16}$ **11.** $-\frac{1}{16}$ **13.** 4 **15.** $\frac{8}{125}$ **17.** $\frac{1}{3}$ **19.** 1250
21. 82 **23.** x **25.** $-\frac{16}{x^4}$ **27.** $\frac{6}{a^5}$ **29.** $\frac{1}{u^8}$ **31.** $-4t^2$
33. $2x^{11}$ **35.** $\frac{1}{x^{10}}$ **37.** a^9 **39.** $\frac{x^{12}}{16}$ **41.** $\frac{y^6}{16x^4}$ **43.** $\frac{x^2}{4y^6}$
45. $\frac{a^{16}}{16c^8}$ **47.** $\frac{6}{w^5}$ **49.** $\frac{1}{8h^3}$ **51.** $\frac{1}{x^{18}}$ **53.** b^{13} **55.** v^{23}
57. $\frac{1}{c^9}$ **59.** $\frac{1}{6}$ **61.** $\frac{3}{2}$ **63.** $-14x^6$ **65.** $\frac{2a^4}{b^2}$
67. 9,860,000,000 **69.** 0.00137 **71.** 0.000001 **73.** 600,000
75. 9×10^3 **77.** 7.8×10^{-4} **79.** 8.5×10^{-6} **81.** 5.25×10^{11}
83. 6×10^{-10} **85.** 2×10^{-38} **87.** 5×10^{27} **89.** 9×10^{24}
91. 1.25×10^{14} **93.** 2.5×10^{-33} **95.** 8.6×10^9
97. 2.1×10^2 **99.** 2.7×10^{-23} **101.** 3×10^{15}
103. 9.135×10^2 **105.** 5.715×10^{-4} **107.** 4.426×10^7
109. 1.577×10^{182} **111.** 4.910×10^{11} feet
113. 4.65×10^{-28} hours **115.** 9.040×10^8 feet **117.** \$10,727.41
119. a) $w < 0$ b) m is odd c) $w < 0$ and m odd

Enriching Your Mathematical Word Power

1. a **2.** d **3.** b **4.** c **5.** d **6.** b **7.** a **8.** b
9. c **10.** a **11.** a **12.** c

Review Exercises

1. $5w - 2$ **3.** $-6x + 4$ **5.** $2w^2 - 7w - 4$ **7.** $-2m^2 + 3m - 1$
9. $-50x^{11}$ **11.** $121a^{14}$ **13.** $-4x + 15$ **15.** $3x^2 - 10x + 12$
17. $15m^5 - 3m^3 + 6m^2$ **19.** $x^3 - 7x^2 + 20x - 50$
21. $3x^3 - 8x^2 + 16x - 8$ **23.** $q^2 + 2q - 48$ **25.** $2t^2 - 21t + 27$
27. $20y^2 - 7y - 6$ **29.** $6x^4 + 13x^2 + 5$ **31.** $z^2 - 49$
33. $y^2 + 14y + 49$ **35.** $w^2 - 6w + 9$ **37.** $x^4 - 9$
39. $9a^2 + 6a + 1$ **41.** $16 - 8y + y^2$ **43.** $-5x^2$ **45.** $\frac{-2a^2}{b^2}$
47. $-x + 3$ **49.** $-3x^2 + 2x - 1$ **51.** -1
53. $m^3 + 2m^2 + 4m + 8$ **55.** $m^2 - 3m + 6$, 0 **57.** $b - 5$, 15
59. $2x - 1$, -8 **61.** $x^2 + 2x - 9$, 1 **63.** $2 + \frac{6}{x - 3}$
65. $-2 + \frac{2}{1 - x}$ **67.** $x - 1 - \frac{2}{x + 1}$ **69.** $x - 1 + \frac{1}{x + 1}$
71. $6y^{30}$ **73.** $\frac{-5}{c^6}$ **75.** b^{30} **77.** $-8x^9y^6$ **79.** $\frac{8a^3}{b^3}$
81. $\frac{8x^6y^{15}}{z^{18}}$ **83.** $\frac{1}{32}$ **85.** $\frac{1}{1000}$ **87.** $\frac{1}{x^3}$ **89.** a^4 **91.** a^{10}
93. $\frac{1}{x^{12}}$ **95.** $\frac{x^9}{8}$ **97.** $\frac{9}{a^2b^6}$ **99.** 5×10^3 **101.** 340,000
103. 4.61×10^{-5} **105.** 0.00000569 **107.** 7×10^{-4}
109. 1.6×10^{-15} **111.** 8×10^1 **113.** 3.2×10^{-34}
115. $x^2 + 10x + 21$ **117.** $t^2 - 7ty + 12y^2$ **119.** 2

121. $-27h^3t^{18}$ **123.** $2w^2 - 9w - 18$ **125.** $9u^2 - 25v^2$
127. $9h^2 + 30h + 25$ **129.** $x^3 + 9x^2 + 27x + 27$ **131.** $14s^5t^6$
133. $\frac{k^8}{16}$ **135.** $x^2 - 9x - 5$ **137.** $5x^2 - x - 12$
139. $x^3 - x^2 - 19x + 4$ **141.** $x + 6$
143. $P(w) = 4w + 88, A(w) = w^2 + 44w, P(50) = 288$ ft, $A(50) = 4700$ ft^2
145. $R = -15p^2 + 600p$, \$5040, \$20

Chapter 5 Test

1. $7x^3 + 4x^2 + 2x - 11$ **2.** $-x^2 - 9x + 2$ **3.** $-2y^2 + 3y$
4. -1 **5.** $x^2 + x - 1$ **6.** $15x^5 - 21x^4 + 12x^3 - 3x^2$
7. $x^2 + 3x - 10$ **8.** $6a^2 + a - 35$ **9.** $a^2 - 14a + 49$
10. $16x^2 + 24xy + 9y^2$ **11.** $b^2 - 9$ **12.** $9t^4 - 49$
13. $4x^4 + 5x^2 - 6$ **14.** $x^3 - 3x^2 - 10x + 24$ **15.** $2 + \frac{6}{x - 3}$
16. $x - 5 + \frac{15}{x + 2}$ **17.** $-35x^8$ **18.** $12x^5y^9$ **19.** $-2ab^4$
20. $15x^5$ **21.** $\frac{-32a^5}{b^{10}}$ **22.** $\frac{3a^4}{b^2}$ **23.** $\frac{3}{t^{16}}$ **24.** $\frac{1}{w^2}$
25. $\frac{s^6}{9t^4}$ **26.** $\frac{-8y^3}{x^{18}}$ **27.** 5.433×10^6 **28.** 6.5×10^{-6}
29. 4.8×10^{-1} **30.** 8.1×10^{-27} **31.** $x - 2, 3$ **32.** $-2x^2 + x + 15$
33. $A(x) = x^2 + 4x, P(x) = 4x + 8, A(4) = 32$ ft^2, $P(4) = 24$ ft
34. $R = -150q^2 + 3000q$, \$14,400

Making Connections A Review of Chapters 1–5

1. 8 **2.** 32 **3.** 41 **4.** -2 **5.** 32 **6.** 32 **7.** -144
8. 144 **9.** $\frac{5}{8}$ **10.** $\frac{1}{9}$ **11.** 64 **12.** 34 **13.** $\frac{5}{6}$ **14.** $\frac{5}{36}$
15. 899 **16.** -1 **17.** $x^2 + 8x + 15$ **18.** $4x + 15$ **19.** $-15t^5v^7$
20. $5tv$ **21.** $x^2 + 9x + 20$ **22.** $x^2 + 7x + 10$ **23.** $x + 3$
24. $x^3 + 13x^2 + 55x + 75$ **25.** $3y - 4$ **26.** $6y^2 - 4y + 1$
27. $\left\{-\frac{1}{2}\right\}$ **28.** $\{7\}$ **29.** $\left\{\frac{14}{3}\right\}$ **30.** $\left\{\frac{7}{4}\right\}$ **31.** $\{0\}$
32. All real numbers **33.** $\left(-\frac{1}{2}, 0\right)$ **34.** $(0, -7)$ **35.** 2
36. $\frac{2}{3}$ **37.** $\frac{14}{3}$ **38.** $-\frac{1}{2}$
39. $\frac{2.25n + 100{,}000}{n}$, \$102.25, \$3.25, \$2.35, It averages out to 10 cents per disk.

Chapter 6

Section 6.1 Warm-Ups F F F T T T T F F T

1. To factor means to write as a product.
3. You can find the prime factorization by dividing by prime factors until the result is prime.
5. The GCF for two monomials consists of the GCF of their coefficients and every variable that they have in common raised to the lowest power that appears on the variable.
7. $2 \cdot 3^2$ **9.** $2^2 \cdot 13$ **11.** $2 \cdot 7^2$ **13.** $2^2 \cdot 5 \cdot 23$ **15.** $2^2 \cdot 3 \cdot 7 \cdot 11$
17. 4 **19.** 12 **21.** 8 **23.** 4 **25.** 1 **27.** $2x$ **29.** $2x$ **31.** xy
33. $12ab$ **35.** 1 **37.** $6ab$ **39.** $3x$ **41.** $3t$ **43.** $9y^3$ **45.** u^3v^2
47. $-7n^3$ **49.** $11xy^2z$ **51.** $2(w + 2t)$ **53.** $6(2x - 3y)$
55. $x(x^2 - 6)$ **57.** $5a(x + y)$ **59.** $h^3(h^2 + 1)$
61. $2k^3m^4(-k^4 + 2m^2)$ **63.** $2x(x^2 - 3x + 4)$ **65.** $6x^2t(2x^2 + 5x - 4t)$
67. $(x - 3)(a + b)$ **69.** $(x - 5)(x - 1)$ **71.** $(m + 1)(m + 9)$
73. $(a + b)(y + 1)^2$ **75.** $8(x - y), -8(-x + y)$
77. $4x(-1 + 2x), -4x(1 - 2x)$ **79.** $1(x - 5), -1(-x + 5)$
81. $1(4 - 7a), -1(-4 + 7a)$ **83.** $8a^2(-3a + 2), -8a^2(3a - 2)$
85. $6x(-2x - 3), -6x(2x + 3)$
87. $2x(-x^2 - 3x + 7), -2x(x^2 + 3x - 7)$
89. $2ab(2a^2 - 3ab - 2b^2), -2ab(-2a^2 + 3ab + 2b^2)$ **91.** $x + 2$ hours
93. a) $S = 2\pi r(r + h)$ **b)** $S = 2\pi r^2 + 10\pi r$ **c)** 3 in.
95. The GCF is an algebraic expression.

Section 6.2 Warm-Ups F T F F T T F F T T

1. A perfect square is a square of an integer or an algebraic expression.
3. A perfect square trinomial is of the form $a^2 + 2ab + b^2$ or $a^2 - 2ab + b^2$.
5. A polynomial is factored completely when it is a product of prime polynomials.
7. $(a - 2)(a + 2)$ **9.** $(x - 7)(x + 7)$ **11.** $(y + 3x)(y - 3x)$
13. $(5a + 7b)(5a - 7b)$ **15.** $(11m + 1)(11m - 1)$
17. $(3w - 5c)(3w + 5c)$ **19.** Perfect square trinomial
21. Neither **23.** Perfect square trinomial **25.** Neither
27. Difference of two squares **29.** Perfect square trinomial
31. $(x + 1)^2$ **33.** $(a + 3)^2$ **35.** $(x + 6)^2$ **37.** $(a - 2)^2$
39. $(2w + 1)^2$ **41.** $(4x - 1)^2$ **43.** $(2t + 5)^2$ **45.** $(3w + 7)^2$
47. $(n + t)^2$ **49.** $5(x - 5)(x + 5)$ **51.** $-2(x - 3)(x + 3)$
53. $a(a - b)(a + b)$ **55.** $3(x + 1)^2$ **57.** $-5(y - 5)^2$
59. $x(x - y)^2$ **61.** $-3(x - y)(x + y)$ **63.** $2a(x - 7)(x + 7)$
65. $3a(b - 3)^2$ **67.** $-4m(m - 3n)^2$ **69.** $(b + c)(x + y)$
71. $(x - 2)(x + 2)(x + 1)$ **73.** $(3 - x)(a - b)$ **75.** $(a^2 + 1)(a + 3)$
77. $(a + 3)(x + y)$ **79.** $(c - 3)(ab + 1)$ **81.** $(a + b)(x - 1)(x + 1)$
83. $(y + b)(y + 1)$ **85.** $6ay(a + 2y)^2$ **87.** $6ay(2a - y)(2a + y)$
89. $2a^2y(ay - 3)$ **91.** $(b - 4w)(a + 2w)$
93. $h = -16(t - 20)(t + 20)$, 6336 feet **95.** $y - 3$ inches

Section 6.3 Warm-Ups T T F F T F T F F F

1. We factored $ax^2 + bx + c$ with $a = 1$.
3. If there are no two integers that have a product of c and a sum of b, then $x^2 + bx + c$ is prime.
5. A polynomial is factored completely when all of the factors are prime polynomials.
7. $(x + 3)(x + 1)$ **9.** $(x + 3)(x + 6)$ **11.** $(a + 2)(a + 5)$
13. $(a - 3)(a - 4)$ **15.** $(b - 6)(b + 1)$ **17.** $(x - 2)(x + 5)$
19. $(y + 2)(y + 5)$ **21.** $(a - 2)(a - 4)$ **23.** $(m - 8)(m - 2)$
25. $(w + 10)(w - 1)$ **27.** $(w - 4)(w + 2)$ **29.** Prime
31. $(m + 16)(m - 1)$ **33.** Prime **35.** $(z - 5)(z + 5)$ **37.** Prime
39. $(m + 2)(m + 10)$ **41.** Prime **43.** $(m - 18)(m + 1)$ **45.** Prime
47. $(t + 8)(t - 3)$ **49.** $(t - 6)(t + 4)$ **51.** $(t - 20)(t + 10)$
53. $(x - 15)(x + 10)$ **55.** $(y + 3)(y + 10)$ **57.** $(x + 3a)(x + 2a)$
59. $(x - 6y)(x + 2y)$ **61.** $(x - 12y)(x - y)$ **63.** Prime
65. $5x(x^2 + 1)$ **67.** $w(w - 8)$ **69.** $2(w - 9)(w + 9)$
71. $-2(b^2 + 49)$ **73.** $(x + 3)(x - 3)(x - 2)$ **75.** Prime
77. $x^2(w^2 + 9)$ **79.** $(w - 9)^2$ **81.** $6(w - 3)(w + 1)$
83. $3(y^2 + 25)$ **85.** $(a + c)(x + y)$ **87.** $-2(x + 2)(x + 3)$
89. $2x^2(4 - x)(4 + x)$ **91.** $3(w + 3)(w + 6)$ **93.** $w(w^2 + 18w + 36)$
95. $(3y + 1)^2$ **97.** $8v(w + 2)^2$ **99.** $6xy(x + 3y)(x + 2y)$
101. $(3w + 5)(w + 1)$ **103.** $-3y(y - 1)^2$ **105.** $(a + 3)(a^2 + b)$
107. $x + 4$ feet **109.** 3 feet and 5 feet **111.** d

Section 6.4 Warm-Ups T F T F T F F F F T

1. We factored $ax^2 + bx + c$ with $a \neq 1$.
3. If there are no two integers whose product is ac and whose sum is b, then $ax^2 + bx + c$ is prime.
5. 2 and 10 **7.** -6 and 2 **9.** 3 and 4 **11.** -2 and -9
13. -3 and 4 **15.** $(2x + 1)(x + 1)$ **17.** $(2x + 1)(x + 4)$
19. $(3t + 1)(t + 2)$ **21.** $(2x - 1)(x + 3)$ **23.** $(3x - 1)(2x + 3)$
25. Prime **27.** $(2x - 3)(x - 2)$ **29.** $(5b - 3)(b - 2)$

31. $(4y + 1)(y - 3)$ **33.** Prime **35.** $(4x + 1)(2x - 1)$
37. $(3t - 1)(3t - 2)$ **39.** $(5x + 1)(3x + 2)$ **41.** $(2a + 3b)(2a + 5b)$
43. $(3m - 5n)(2m + n)$ **45.** $(x - y)(3x - 5y)$ **47.** $(5a + 1)(a + 1)$
49. $(2x + 1)(3x + 1)$ **51.** $(5a + 1)(a + 2)$ **53.** $(2w + 3)(2w + 1)$
55. $(5x - 2)(3x + 1)$ **57.** $(4x - 1)(2x - 1)$ **59.** $(15x - 1)(x - 2)$
61. Prime **63.** $2(x^2 + 9x - 45)$ **65.** $(3x - 5)(x + 2)$
67. $(5x + y)(2x - y)$ **69.** $(6a - b)(7a - b)$ **71.** $3x + 1$
73. $x + 2$ **75.** $2a - 5$ **77.** $w(9w - 1)(9w + 1)$
79. $2(2w - 5)(w + 3)$ **81.** $3(2x + 3)^2$ **83.** $(3w + 5)(2w - 7)$
85. $3z(x - 3)(x + 2)$ **87.** $3x(3x^2 - 7x + 6)$ **89.** $(a + 5b)(a - 3b)$
91. $y^2(2x^2 + x + 3)$ **93.** $-t(3t + 2)(2t - 1)$ **95.** $2t^2(3t - 2)(2t + 1)$
97. $y(2x - y)(2x - 3y)$ **99.** $-1(w - 1)(4w - 3)$
101. $-2a(2a - 3b)(3a - b)$ **103.** $h = -8(2t + 1)(t - 3)$, 0 feet
105. **a)** ± 4 **b)** $\pm 8, \pm 16$ **c)** $\pm 1, \pm 7, \pm 13, \pm 29$

Section 6.5 Warm-Ups F F T T F T F T T F

1. If there is no remainder, then the dividend factors as the divisor times the quotient.
3. If you divide $a^3 + b^3$ by $a + b$ there will be no remainder.
5. $a^3 + b^3 = (a + b)(a^2 - ab + b^2)$
7. $(x + 4)(x - 3)(x + 2)$ **9.** $(x - 1)(x + 3)(x + 2)$
11. $(x - 2)(x^2 + 2x + 4)$ **13.** $(x + 5)(x^2 - x + 2)$
15. $(x + 1)(x^2 + x + 1)$ **17.** $(m - 1)(m^2 + m + 1)$
19. $(x + 2)(x^2 - 2x + 4)$ **21.** $(a + 5)(a^2 - 5a + 25)$
23. $(c - 7)(c^2 + 7c + 49)$ **25.** $(2w + 1)(4w^2 - 2w + 1)$
27. $(2t - 3)(4t^2 + 6t + 9)$ **29.** $(x - y)(x^2 + xy + y^2)$
31. $(2t + y)(4t^2 - 2ty + y^2)$ **33.** $(x - y)(x + y)(x^2 + y^2)$
35. $(x - 1)(x + 1)(x^2 + 1)$ **37.** $(2b - 1)(2b + 1)(4b^2 + 1)$
39. $(a - 3b)(a + 3b)(a^2 + 9b^2)$ **41.** $2(x - 3)(x + 3)$
43. Prime **45.** $4(x + 5)(x - 3)$ **47.** $x(x + 2)^2$
49. $5am(x^2 + 4)$ **51.** Prime **53.** $(3x + 1)^2$ **55.** Prime
57. $(w - z)(w + z)(w^2 + z^2)$ **59.** $y(3x + 2)(2x - 1)$
61. Prime **63.** $3(4a - 1)^2$ **65.** $2(4m + 1)(2m - 1)$
67. $(s - 2t)(s + 2t)(s^2 + 4t^2)$ **69.** $(3a + 4)^2$ **71.** $2(3x - 1)(4x - 3)$
73. $3(m^2 + 9)$ **75.** $3a(a - 9)$ **77.** $2(2 - x)(2 + x)$ **79.** Prime
81. $x(6x^2 - 5x + 12)$ **83.** $ab(a - 2)(a + 2)$ **85.** $(x - 2)(x + 2)^2$
87. $-7mn(m^2 + 4n^2)$ **89.** $2(x + 2)(x^2 - 2x + 4)$
91. $2w(w - 2)(w^2 + 2w + 4)$ **93.** $3w(a - 3)^2$
95. $5(x - 10)(x + 10)$ **97.** $(2 - w)(m + n)$
99. $3x(x + 1)(x^2 - x + 1)$ **101.** $4(w^2 + w - 1)$
103. $a^2(a + 10)(a - 3)$ **105.** $aw(2w - 3)^2$ **107.** $(t + 3)^2$
109. Length $x + 5$ cm, width $x + 3$ cm
111. $(-1 + 1)^3 = (-1)^3 + 1^3$, $(1 + 2)^3 \neq 1^3 + 2^3$

Section 6.6 Warm-Ups F F T T T F T T T F

1. A quadratic equation has the form $ax^2 + bx + c = 0$ with $a \neq 0$.
3. The zero factor property says that if $ab = 0$ then $a = 0$ or $b = 0$.
5. Dividing each side by a variable is not usually done because the variable might have a value of zero.
7. $-4, -5$ **9.** $-\frac{5}{2}, \frac{4}{3}$ **11.** $-2, -1$ **13.** $2, 7$ **15.** $-4, 6$
17. $-1, \frac{1}{2}$ **19.** $0, 1$ **21.** $0, -7$ **23.** $-5, 4$ **25.** $\frac{1}{2}, -3$
27. $0, -8$ **29.** $-\frac{9}{2}, 2$ **31.** $\frac{2}{3}, -4$ **33.** 5 **35.** $\frac{3}{2}$ **37.** $0, -3, 3$
39. $-4, -2, 2$ **41.** $-1, 1, 3$ **43.** $0, 4, 5$ **45.** $-4, 4$ **47.** $-3, 3$
49. $0, -1, 1$ **51.** $-3, -2$ **53.** $-\frac{3}{2}, -4$ **55.** $-6, 4$ **57.** $-1, 3$
59. $-4, 2$ **61.** $-5, -3, 5$ **63.** Length 12 ft, width 5 ft
65. Width 5 ft, length 12 ft **67.** 2 and 3, or -3 and -2 **69.** 5 and 6
71. $-8, -6, -4$, or 4, 6, 8 **73.** -2 and -1, or 3 and 4
75. -7 and -2, or 2 and 7 **77.** Length 12 feet, width 6 feet
79. 9 meters and 12 meters
81. **a)** 25 sec **b)** last 5 sec **c)** increasing **83.** 6 sec
85. Base 6 in., height 13 in. **87.** 20 ft by 20 ft **89.** 80 ft
91. 3 yd by 3 yd, 6 yd by 6 yd **93.** 12 mi **95.** 25%

Enriching Your Mathematical Word Power

1. a **2.** d **3.** c **4.** a **5.** c **6.** b **7.** c **8.** a
9. d **10.** c

Review Exercises

1. $2^4 \cdot 3^2$ **3.** $2 \cdot 29$ **5.** $2 \cdot 3 \cdot 5^2$ **7.** 18 **9.** $4x$ **11.** $x + 2$
13. $-a + 10$ **15.** $a(2 - a)$ **17.** $3x^2y(2y - 3x^3)$
19. $3y(x^2 - 4x - 3y)$ **21.** $(y - 20)(y + 20)$ **23.** $(w - 4)^2$
25. $(2y + 5)^2$ **27.** $(r - 2)^2$ **29.** $2t(2t - 3)^2$ **31.** $(x + 6y)^2$
33. $(x - y)(x + 5)$ **35.** $(b + 8)(b - 3)$ **37.** $(r - 10)(r + 6)$
39. $(y - 11)(y + 5)$ **41.** $(u + 20)(u + 6)$ **43.** $3t^2(t + 4)$
45. $5w(w^2 + 5w + 5)$ **47.** $ab(2a + b)(a + b)$
49. $x(3x - y)(3x + y)$ **51.** $(7t - 3)(2t + 1)$ **53.** $(3x + 1)(2x - 7)$
55. $(3p + 4)(2p - 1)$ **57.** $-2p(5p + 2)(3p - 2)$
59. $(6x + y)(x - 5y)$ **61.** $2(4x + y)^2$ **63.** $5x(x^2 + 8)$
65. $(3x - 1)(3x + 2)$ **67.** Prime **69.** $(x + 2)(x - 1)(x + 1)$
71. $xy(x - 16y)$ **73.** Prime **75.** $(a + 1)^2$ **77.** $(x^2 + 1)(x - 1)$
79. $(a + 2)(a + b)$ **81.** $-2(x - 6)(x - 2)$
83. $(m - 10)(m^2 + 10m + 100)$ **85.** $(p - q)(p + q)(p^2 + q^2)$
87. $(x + 2)(x^2 - 2x + 5)$ **89.** $(x + 4)(x + 5)(x - 3)$
91. 0, 5 **93.** 0, 5 **95.** $-\frac{1}{2}, 5$ **97.** $-4, -3, 3$ **99.** $-2, -1$
101. $-\frac{1}{2}, \frac{1}{4}$ **103.** 5, 11 **105.** 6 in. by 8 in.
107. $v = k(R - r)(R + r)$ **109.** 6 ft

Chapter 6 Test

1. $2 \cdot 3 \cdot 11$ **2.** $2^4 \cdot 3 \cdot 7$ **3.** 16 **4.** 6 **5.** $3y^2$ **6.** $6ab$
7. $5x(x - 2)$ **8.** $6y^2(x^2 + 2x + 2)$ **9.** $3ab(a - b)(a + b)$
10. $(a + 6)(a - 4)$ **11.** $(2b - 7)^2$ **12.** $3m(m^2 + 9)$
13. $(a + b)(x - y)$ **14.** $(a - 5)(x - 2)$ **15.** $(3b - 5)(2b + 1)$
16. $(m + 2n)^2$ **17.** $(2a - 3)(a - 5)$ **18.** $z(z + 3)(z + 6)$
19. $(x + 5)(x^2 - 5x + 25)$ **20.** $a(a - b)(a^2 + ab + b^2)$
21. $(x - 1)(x - 2)(x - 3)$ **22.** -3 **23.** $\frac{3}{2}, -4$ **24.** $0, -2, 2$
25. $-2, \frac{5}{6}$ **26.** Length 12 ft, width 9 ft **27.** -4 and 8

Making Connections A Review of Chapters 1–6

1. -1 **2.** 2 **3.** -3 **4.** 57 **5.** 16 **6.** 7 **7.** $2x^2$ **8.** $3x$
9. $3 + x$ **10.** $6x$ **11.** $24yz$ **12.** $6y + 8z$ **13.** $4z - 1$ **14.** t^6
15. t^{10} **16.** $4t^6$
17. $(-\infty, -9)$

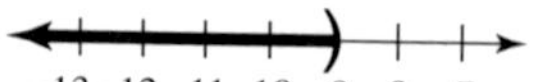

18. $[3, \infty)$

19. $(12, \infty)$

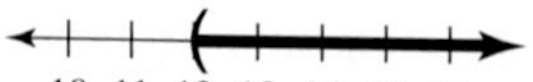

20. $(-\infty, 600)$

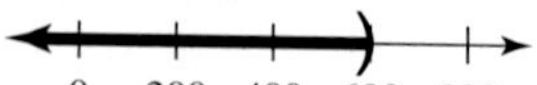

21. $\left\{\frac{3}{2}\right\}$ **22.** $\left\{-\frac{1}{2}\right\}$ **23.** $\{3, -5\}$ **24.** $\left\{\frac{3}{2}, -\frac{1}{2}\right\}$ **25.** $\{0, 3\}$
26. $\{0, 1\}$ **27.** R **28.** No solution or $\varnothing$ **29.** $\{10\}$
30. $\{40\}$ **31.** $\{-3, 3\}$ **32.** $\left\{-5, \frac{3}{2}\right\}$
33. Length 21 ft, width 13.5 ft

Chapter 7

Section 7.1 Warm-Ups F T T F F T T F F T

1. A rational number is a ratio of two integers with the denominator not 0.
3. A rational number is reduced to lowest terms by dividing the numerator and denominator by the GCF.
5. The quotient rule is used in reducing ratios of monomials.
7. -3 **9.** 5 **11.** $-0.6, 9, 401, -199$ **13.** -1
15. $\frac{5}{3}$ **17.** 4, -4 **19.** Any number can be used. **21.** $\frac{2}{9}$
23. $\frac{7}{15}$ **25.** $\frac{2a}{5}$ **27.** $\frac{13}{5w}$ **29.** $\frac{3x+1}{3}$ **31.** $\frac{2}{3}$ **33.** $w-7$
35. $\frac{a-1}{a+1}$ **37.** $\frac{x+1}{2(x-1)}$ **39.** $\frac{x+3}{7}$ **41.** x^3 **43.** $\frac{1}{z^5}$
45. $-2x^2$ **47.** $\frac{-3m^3n^2}{2}$ **49.** $\frac{-3}{4c^3}$ **51.** $\frac{5c}{3a^4b^{16}}$ **53.** $\frac{35}{44}$
55. $\frac{11}{8}$ **57.** $\frac{21}{10x^4}$ **59.** $\frac{33a^4}{16}$ **61.** -1 **63.** $-h-t$
65. $\frac{-2}{3h+g}$ **67.** $\frac{-x-2}{x+3}$ **69.** -1 **71.** $\frac{-2y}{3}$ **73.** $\frac{x+2}{2-x}$
75. $\frac{-6}{a+3}$ **77.** $\frac{x^4}{2}$ **79.** $\frac{x+2}{2x}$ **81.** -1 **83.** $\frac{-2}{c+2}$
85. $\frac{x+2}{x-2}$ **87.** $\frac{-2}{x+3}$ **89.** q^2 **91.** $\frac{u+2}{u-8}$
93. $\frac{a^2+2a+4}{2}$ **95.** $y+2$ **97.** $\frac{300}{x+10}$ hr
99. $\frac{4.50}{x+4}$ dollars/lb **101.** $\frac{1}{x}$ pool/hr
103. **a)** \$0.75 **b)** \$0.75, \$0.63, \$0.615 **c)** Approaches \$0.60

Section 7.2 Warm-Ups T T T F T F F T T T

1. Rational numbers are multiplied by multiplying their numerators and their denominators.
3. Reducing can be done before multiplying rational numbers or expressions.
5. $\frac{5}{9}$ **7.** $\frac{7}{9}$ **9.** $\frac{18}{5}$ **11.** $\frac{42}{5}$ **13.** $\frac{5}{6}$ **15.** $\frac{a}{44}$ **17.** $\frac{-x^5}{a^3}$
19. $\frac{18t^8y^7}{w^4}$ **21.** $\frac{5}{7}$ **23.** $\frac{2a}{a-b}$ **25.** $3x-9$ **27.** $\frac{8a+8}{5(a^2+1)}$
29. $\frac{1}{2}$ **31.** 30 **33.** $\frac{2}{3}$ **35.** $\frac{10}{9}$ **37.** $\frac{x}{2}$ **39.** $\frac{7x}{2}$ **41.** $\frac{2m^2}{3n^6}$
43. -3 **45.** $\frac{2}{x+2}$ **47.** $\frac{1}{4(t-5)}$ **49.** x^2-1 **51.** $2x-4y$
53. $\frac{x+2}{2}$ **55.** $\frac{x^2+9}{15}$ **57.** $9x+9y$ **59.** -3 **61.** $\frac{a+b}{a}$
63. $\frac{2b}{a}$ **65.** $\frac{y}{x}$ **67.** $\frac{-a^6b^8}{2}$ **69.** $\frac{1}{9m^3n}$ **71.** $\frac{x^2+5x}{3x-1}$
73. $\frac{a^3+8}{2(a-2)}$ **75.** 1 **77.** $\frac{m^2+6m+9}{(m-3)(m+k)}$ **79.** $\frac{13.1}{x}$ mi
81. 5 square meters **83.** **a)** $\frac{1}{8}$ **b)** $\frac{4}{3}$ **c)** $\frac{2x}{3}$ **d)** $\frac{3x}{4}$

Section 7.3 Warm-Ups F F T T F F F F T T

1. We can build up a denominator by multiplying the numerator and denominator of a fraction by the same nonzero number.
3. For fractions, the LCD is the smallest number that is a multiple of all of the denominators.
5. $\frac{9}{27}$ **7.** $\frac{12}{16}$ **9.** $\frac{12}{6}$ **11.** $\frac{5a}{ax}$ **13.** $\frac{14x}{2x}$ **15.** $\frac{15t}{3bt}$
17. $\frac{-36z^2}{8awz}$ **19.** $\frac{10a^2}{15a^3}$ **21.** $\frac{8xy^3}{10x^2y^5}$ **23.** $\frac{10}{2x+6}$ **25.** $\frac{-20}{-8x-8}$
27. $\frac{-32ab}{20b^2-20b^3}$ **29.** $\frac{3x-6}{x^2-4}$ **31.** $\frac{3x^2+3x}{x^2+2x+1}$ **33.** $\frac{y^2-y-30}{y^2+y-20}$
35. 48 **37.** 180 **39.** $30a^2$ **41.** $12a^4b^6$
43. $(x-4)(x+4)^2$ **45.** $x(x+2)(x-2)$ **47.** $2x(x-4)(x+4)$
49. $\frac{4}{24}, \frac{9}{24}$ **51.** $\frac{3}{6x}, \frac{5}{6x}$ **53.** $\frac{4b}{6ab}, \frac{3a}{6ab}$ **55.** $\frac{9b}{252ab}, \frac{20a}{252ab}$
57. $\frac{2x^3}{6x^5}, \frac{9}{6x^5}$ **59.** $\frac{4x^4}{36x^3y^5z}, \frac{3y^6z}{36x^3y^5z}, \frac{6xy^4z}{36x^3y^5z}$
61. $\frac{2x^2+4x}{(x-3)(x+2)}, \frac{5x^2-15x}{(x-3)(x+2)}$ **63.** $\frac{4}{a-6}, \frac{-5}{a-6}$
65. $\frac{x^2-3x}{(x-3)^2(x+3)}, \frac{5x^2+15x}{(x-3)^2(x+3)}$
67. $\frac{w^2+3w+2}{(w-5)(w+3)(w+1)}, \frac{-2w^2-6w}{(w-5)(w+3)(w+1)}$
69. $\frac{-5x-10}{6(x-2)(x+2)}, \frac{6x}{6(x-2)(x+2)}, \frac{9x-18}{6(x-2)(x+2)}$
71. $\frac{2q+8}{(2q+1)(q-3)(q+4)}, \frac{3q-9}{(2q+1)(q-3)(q+4)}, \frac{8q+4}{(2q+1)(q-3)(q+4)}$
73. Identical denominators are needed for addition and subtraction.

Section 7.4 Warm-Ups F T T T T F T F T F

1. We can add rational numbers with identical denominators as follows: $\frac{a}{c}+\frac{b}{c}=\frac{a+b}{c}$.
3. The LCD is the smallest number that is a multiple of all denominators.
5. $\frac{1}{5}$ **7.** $\frac{3}{4}$ **9.** $-\frac{2}{3}$ **11.** $-\frac{3}{4}$ **13.** $\frac{5}{9}$ **15.** $\frac{103}{144}$ **17.** $-\frac{31}{40}$
19. $\frac{5}{24}$ **21.** $\frac{1}{x}$ **23.** $\frac{5}{w}$ **25.** 3 **27.** -2 **29.** $\frac{3}{h}$
31. $\frac{x-4}{x+2}$ **33.** $\frac{3}{2a}$ **35.** $\frac{5x}{6}$ **37.** $\frac{6m}{5}$ **39.** $\frac{2x+y}{xy}$
41. $\frac{17}{10a}$ **43.** $\frac{w}{36}$ **45.** $\frac{b^2-4ac}{4a}$ **47.** $\frac{2w+3z}{w^2z^2}$
49. $\frac{2x+2}{x(x+2)}$ **51.** $\frac{-x-3}{x(x+1)}$ **53.** $\frac{3a+b}{(a-b)(a+b)}$
55. $\frac{15-4x}{5x(x+1)}$ **57.** $\frac{a^2+5a}{(a-3)(a+3)}$ **59.** 0 **61.** $\frac{7}{2(a-1)}$
63. $\frac{-2x+1}{(x-5)(x+2)(x-2)}$ **65.** $\frac{7x+17}{(x+2)(x-1)(x+3)}$
67. $\frac{bc+ac+ab}{abc}$ **69.** $\frac{2x^2-x-4}{x(x-1)(x+2)}$ **71.** $\frac{a+51}{6a(a-3)}$
73. **a)** F **b)** A **c)** E **d)** B **e)** D **f)** C **75.** $\frac{p+6}{p(p+4)}$
77. $\frac{6}{(a+1)(a+3)}$ **79.** $\frac{1}{(b+1)(b+2)}$ **81.** $\frac{-1}{2(t+2)}$
83. $\frac{11}{x}$ feet **85.** $\frac{315x+600}{x(x+5)}$ hours, 5 hours **87.** $\frac{4x+6}{x(x+3)}$ job, $\frac{5}{9}$ job

Section 7.5 Warm-Ups F T F F F F F T T T

1. A complex fraction is a fraction that has fractions in its numerator, denominator, or both.
3. $\frac{3}{5}$ **5.** $-\frac{10}{3}$ **7.** $\frac{22}{7}$ **9.** $\frac{2}{3}$ **11.** $\frac{14}{17}$ **13.** $\frac{45}{23}$ **15.** $\frac{1}{2}$
17. $\frac{3a+b}{a-3b}$ **19.** $\frac{5a-3}{3a+1}$ **21.** $\frac{x^2-4x}{2(3x^2-1)}$ **23.** $\frac{10b}{3b^2-4}$ **25.** $\frac{1}{3}$

27. $\frac{y-2}{3y+4}$ **29.** $\frac{x^2-2x+4}{x^2-3x-1}$ **31.** $\frac{5x-14}{2x-7}$ **33.** $\frac{a-6}{3a-1}$ **35.** $\frac{-3m+12}{4m-3}$ **37.** $\frac{-w+5}{9w+1}$ **39.** -1 **41.** $\frac{a+2}{a+4}$ **43.** $\frac{3}{2x-1}$ **45.** $\frac{x-2}{x+3}$ **47.** $\frac{6x-27}{2(2x-3)}$ **49.** $\frac{2x^2}{3y}$ **51.** $\frac{a^2+7a+6}{a+3}$ **53.** $1-x$ **55.** $\frac{32}{95}, \frac{11}{35}$ **57. a)** Neither **b)** $\frac{8}{13}, \frac{13}{21}$ **c)** Converging to 0.61803

Section 7.6 Warm-Ups F F F F F T T T T T

1. The first step is usually to multiply each side by the LCD.
3. An extraneous solution is a number that appears to be a solution when we solve an equation, but it does not check in the original equation.

5. -4 **7.** 12 **9.** 30 **11.** 5 **13.** 4 **15.** $\frac{2}{5}$ **17.** $\frac{3}{7}$ **19.** 4 **21.** 4 **23.** 3 **25.** 2 **27.** -5, 2 **29.** -3, 2 **31.** 2, 3 **33.** -3, 3 **35.** 2 **37.** No solution **39.** No solution **41.** 3 **43.** 10 **45.** 0 **47.** -5, 5 **49.** 3, 5 **51.** 1 **53.** 3 **55.** 0 **57.** 4 **59.** -20 **61.** 3 **63.** 3 **65.** $54\frac{6}{11}$ mm

Section 7.7 Warm-Ups T F F T T T T F F F T

1. A ratio is a comparison of two numbers.
3. Equivalent ratios are ratios that are equivalent as fractions.
5. In the proportion $\frac{a}{b} = \frac{c}{d}$ the means are b and c and the extremes are a and d.

7. $\frac{2}{3}$ **9.** $\frac{4}{3}$ **11.** $\frac{5}{7}$ **13.** $\frac{8}{15}$ **15.** $\frac{7}{2}$ **17.** $\frac{9}{14}$ **19.** $\frac{5}{2}$ **21.** $\frac{15}{1}$ **23.** 3 to 2 **25.** 9 to 16 **27.** 31 to 1 **29.** 2 to 3 **31.** 6 **33.** $-\frac{2}{5}$ **35.** $-\frac{27}{5}$ **37.** 5 **39.** $-\frac{3}{4}$ **41.** $\frac{5}{4}$ **43.** 108 **45.** 176,000 **47.** Lions 85, Tigers 51 **49.** 40 luxury cars, 60 sports cars **51.** 84 in. **53.** 15 min **55.** $\frac{1610}{3}$ or 536.7 mi **57.** 3920 lbs, 2000 lbs **59.** 6000 **61. a)** 3 to 17 **b)** $\frac{201}{14}$ or 14.4 lbs **63.** 4074

Section 7.8 Warm-Ups T T F T T F F T F T

1. $y = 2x - 5$ **3.** $y = -\frac{1}{2}x - 2$ **5.** $y = mx - mb - a$ **7.** $y = -\frac{1}{3}x - \frac{1}{3}$ **9.** $C = \frac{B}{A}$ **11.** $p = \frac{a}{1 + am}$ **13.** $m_1 = \frac{r^2F}{km_2}$ **15.** $a = \frac{bf}{b-f}$ **17.** $r = \frac{S-a}{S}$ **19.** $P_2 = \frac{P_1V_1T_2}{T_1V_2}$ **21.** $h = \frac{3V}{4\pi r^2}$ **23.** $\frac{5}{12}$ **25.** $-\frac{6}{23}$ **27.** $\frac{128}{3}$ **29.** -6 **31.** $\frac{6}{5}$ **33.** Marcie 4 mph, Frank 3 mph **35.** Bob 25 mph, Pat 20 mph **37.** 5 mph **39.** 6 hours **41.** 40 minutes **43.** 1 hour 36 minutes **45.** Bananas 8 pounds, apples 10 pounds **47.** 80 gallons **49.** 140 mph **51.** 10 mph **53.** Ben 15 mph, Jerry 7.5 mph **55.** 1800 miles **57.** 4 hours **59.** 1.2 hours or 1 hour 12 minutes **61.** 24 minutes

Enriching Your Mathematical Word Power

1. b **2.** a **3.** a **4.** d **5.** a **6.** b **7.** d **8.** a **9.** c **10.** d

Review Exercises

1. $\frac{6}{7}$ **3.** $\frac{c^2}{4a^2}$ **5.** $\frac{2w-3}{3w-4}$ **7.** $-\frac{x+1}{3}$ **9.** $\frac{1}{2}k$ **11.** $\frac{2x}{3y}$ **13.** $a^2 - a - 6$ **15.** $\frac{1}{2}$ **17.** 108 **19.** $24a^7b^3$ **21.** $12x(x-1)$ **23.** $(x+1)(x-2)(x+2)$ **25.** $\frac{15}{36}$ **27.** $\frac{10x}{15x^2y}$ **29.** $\frac{-10}{12-2y}$ **31.** $\frac{x^2+x}{x^2-1}$ **33.** $\frac{29}{63}$ **35.** $\frac{3x-4}{x}$ **37.** $\frac{2a-b}{a^2b^2}$ **39.** $\frac{27a^2-8a-15}{(2a-3)(3a-2)}$ **41.** $\frac{3}{a-8}$ **43.** $\frac{3x+8}{2(x+2)(x-2)}$ **45.** $-\frac{3}{14}$ **47.** $\frac{6b+4a}{3(a-6b)}$ **49.** $\frac{-2x+9}{3x-1}$ **51.** $\frac{x^2+x-2}{-4x+13}$ **53.** $-\frac{15}{2}$ **55.** 9 **57.** -3 **59.** $\frac{21}{2}$ **61.** 5 **63.** 8 **65.** 56 cups water, 28 cups rice **67.** $y = mx + b$ **69.** $m = \frac{1}{F-v}$ **71.** $y = 4x - 13$ **73.** 200 hours **75.** Bert 60 cars, Ernie 50 cars **77.** 27.83 million tons **79.** $\frac{10}{2x}$ **81.** $\frac{-2}{5-a}$ **83.** $\frac{3x}{x}$ **85.** $2m$ **87.** $\frac{1}{6}$ **89.** $\frac{1}{a+1}$ **91.** $\frac{5-a}{5a}$ **93.** $\frac{a-2}{2}$ **95.** $b-a$ **97.** $\frac{1}{10a}$ **99.** $\frac{3}{2x}$ **101.** $\frac{4+y}{6xy}$ **103.** $\frac{8}{a-5}$ **105.** -1, 2 **107.** $-\frac{5}{3}$ **109.** 6 **111.** $\frac{1}{2}$ **113.** $\frac{3x+7}{(x-5)(x+5)(x+1)}$ **115.** $\frac{-5a}{(a-3)(a+3)(a+2)}$ **117.** $\frac{2}{5}$

Chapter 7 Test

1. -1, 1 **2.** $\frac{2}{3}$ **3.** 0 **4.** $-\frac{14}{45}$ **5.** $\frac{1+3y}{y}$ **6.** $\frac{4}{a-2}$ **7.** $\frac{-x+4}{(x+2)(x-2)(x-1)}$ **8.** $\frac{2}{3}$ **9.** $\frac{-2}{a+b}$ **10.** $\frac{a^3}{18b^4}$ **11.** $-\frac{4}{3}$ **12.** $\frac{-3x+4}{2(x-3)}$ **13.** $\frac{15}{7}$ **14.** 2, 3 **15.** 12 **16.** $y = -\frac{1}{5}x + \frac{13}{5}$ **17.** $c = \frac{3M - bd}{b}$ **18.** 29 **19.** 7.2 minutes
20. Brenda 15 mph and Randy 20 mph, or Brenda 10 mph and Randy 15 mph
21. $72 billion

Making Connections A Review of Chapters 1–7

1. $\frac{7}{3}$ **2.** $-\frac{10}{3}$ **3.** -2 **4.** No solution **5.** 0 **6.** -4, -2 **7.** -1, 0, 1 **8.** $-\frac{15}{2}$ **9.** -6, 6 **10.** -2, 4 **11.** 5 **12.** 3 **13.** $y = \frac{c-2x}{3}$ **14.** $y = \frac{1}{2}x + \frac{1}{2}$ **15.** $y = \frac{c}{2-a}$ **16.** $y = \frac{AB}{C}$ **17.** $y = 3B - 3A$ **18.** $y = \frac{6A}{5}$ **19.** $y = \frac{8}{3-5a}$ **20.** $y = 0$ or $y = B$ **21.** $y = \frac{2A - hb}{h}$ **22.** $y = -\frac{b}{2}$ **23.** 64 **24.** 16 **25.** 49 **26.** 121 **27.** $-2x-2$ **28.** $2a^2 - 11a + 15$ **29.** x^4 **30.** $\frac{2x+1}{5}$ **31.** $\frac{1}{2x}$ **32.** $\frac{x+2}{2x}$ **33.** $\frac{x}{2}$ **34.** $\frac{x-2}{2x}$ **35.** $-\frac{7}{5}$ **36.** $\frac{3a}{4}$ **37.** $x^2 - 64$ **38.** $3x^3 - 21x$ **39.** $10a^{14}$ **40.** x^{10} **41.** $k^2 - 12k + 36$ **42.** $j^2 + 10j + 25$ **43.** -1 **44.** $3x^2 - 4x$ **45.** $P = \frac{r+2}{(1+r)^2}$, $1.81, $7.72

Chapter 8

Section 8.1 Warm-Ups T F T F T T T F F T T

1. If $b^n = a$, then b is an nth root of a. **3.** If $b^n = a$, then b is an even root provided n is even or an odd root provided n is odd. **5.** The product rule for radicals says that $\sqrt[n]{a} \cdot \sqrt[n]{b} = \sqrt[n]{ab}$ provided all of these roots are real. **7.** 6 **9.** 2 **11.** 10 **13.** Not a real number **15.** 0 **17.** -1 **19.** 1 **21.** Not a real number **23.** 2 **25.** -2 **27.** -10 **29.** Not a real number **31.** 5 **33.** 6 **35.** 3 **37.** 10 **39.** 3 **41.** m **43.** y^3 **45.** y^5 **47.** m **49.** 27 **51.** 32 **53.** 125 **55.** 10^{10} **57.** $3\sqrt{y}$ **59.** $2a$ **61.** x^2y **63.** $m^6\sqrt{5}$ **65.** $2\sqrt[3]{y}$ **67.** $-3w$ **69.** $2\sqrt[4]{s}$ **71.** $-5a^3y^2$ **73.** $\frac{\sqrt{t}}{2}$ **75.** $\frac{25}{4}$ **77.** $\frac{\sqrt[3]{t}}{2}$ **79.** $\frac{-2x^2}{y}$ **81.** $\frac{2a^3}{3}$ **83.** $\frac{\sqrt[4]{y}}{2}$ **85.** 3.968 **87.** -1.610 **89.** 6.001 **91.** 0.769 **93.** 46 **95. a)** 228 mi **b)** 7000 ft **97. a)** No **b)** Yes **c)** No **d)** Yes

Section 8.2 Warm-Ups T F T F T F T F F F

1. We use the product rule to factor out a perfect square from inside a square root. **3.** To rationalize a denominator means to rewrite the expression so that the denominator is a rational number. **5.** To simplify square roots containing variables use the same techniques as we use on square roots of numbers. **7.** $2\sqrt{2}$ **9.** $2\sqrt{6}$ **11.** $2\sqrt{7}$ **13.** $3\sqrt{10}$ **15.** $10\sqrt{5}$ **17.** $5\sqrt{6}$ **19.** $\frac{\sqrt{5}}{5}$ **21.** $\frac{3\sqrt{2}}{2}$ **23.** $\frac{\sqrt{6}}{2}$ **25.** $\frac{-3\sqrt{10}}{10}$ **27.** $\frac{-10\sqrt{17}}{17}$ **29.** $\frac{\sqrt{77}}{7}$ **31.** $3\sqrt{7}$ **33.** $\frac{\sqrt{6}}{2}$ **35.** $\frac{\sqrt{10}}{4}$ **37.** $\frac{\sqrt{15}}{5}$ **39.** 5 **41.** $\frac{\sqrt{6}}{2}$ **43.** a^4 **45.** $a^4\sqrt{a}$ **47.** $2a^3\sqrt{2}$ **49.** $2a^2b^4\sqrt{5b}$ **51.** $3xy\sqrt{3xy}$ **53.** $3ab^4c\sqrt{3a}$ **55.** $\frac{\sqrt{x}}{x}$ **57.** $\frac{\sqrt{6a}}{3a}$ **59.** $\frac{\sqrt{5y}}{5y}$ **61.** $\frac{\sqrt{6xy}}{2y}$ **63.** $\frac{\sqrt{6xy}}{3x}$ **65.** $\frac{2x\sqrt{2xy}}{y}$ **67.** $2\sqrt{2a}$ **69.** w^2 **71.** $z^2\sqrt{z}$ **73.** $\frac{\sqrt{3p}}{3}$ **75.** $\frac{\sqrt{ab}}{a}$ **77.** $b\sqrt{a}$ **79.** $4x\sqrt{5x}$ **81.** $3y^4x^7\sqrt{yx}$ **83.** $4x^3\sqrt{5x}$ **85.** $\frac{-11\sqrt{6pq}}{3q}$ **87.** $a^3b^8c\sqrt{ac}$ **89.** $\frac{\sqrt{6y}}{3x^9y^4}$ **91.** 0 **93.** 0 **95. a)** $E = \frac{\sqrt{2AIS}}{I}$ **b)** 51.2

Section 8.3 Warm-Ups F T F F T F F F T T

1. Like radicals have the same index and same radicand. **3.** Radicals can be added, subtracted, multiplied, and divided. **5.** Radical expressions such as $\sqrt{a} + \sqrt{b}$ and $\sqrt{a} - \sqrt{b}$ are conjugates. **7.** $7\sqrt{5}$ **9.** $2\sqrt{2}$ **11.** $8u\sqrt{11}$ **13.** $4\sqrt{3} - 4\sqrt{2}$ **15.** $-4\sqrt{x} - \sqrt{y}$ **17.** $5x\sqrt{y} + 2\sqrt{a}$ **19.** $5\sqrt{6}$ **21.** $-14\sqrt{3}$ **23.** $-\sqrt{3a}$ **25.** $3x\sqrt{x}$ **27.** $\frac{2\sqrt{3}}{3}$ **29.** $\frac{7\sqrt{3}}{3}$ **31.** $\sqrt{77}$ **33.** 36 **35.** $-12\sqrt{10}$ **37.** $2a^4\sqrt{3}$ **39.** 3 **41.** $-2m$ **43.** $2 + \sqrt{6}$ **45.** $12\sqrt{3} + 6\sqrt{5}$ **47.** $10 - 30\sqrt{2}$ **49.** $-7 - \sqrt{5}$ **51.** 2 **53.** 3 **55.** $28 - \sqrt{5}$ **57.** $-42 - \sqrt{15}$ **59.** $37 + 20\sqrt{3}$ **61.** $\sqrt{2}$ **63.** $\frac{\sqrt{15}}{3}$ **65.** $\frac{2\sqrt{30}}{9}$ **67.** $\frac{5\sqrt{7}}{3}$ **69.** $1 + \sqrt{2}$ **71.** $-2 + \sqrt{5}$ **73.** $\frac{2 - \sqrt{5}}{3}$ **75.** $\frac{2 + \sqrt{6}}{3}$ **77.** $\frac{3 + 2\sqrt{3}}{6}$ **79.** $5\sqrt{3} + 5\sqrt{2}$ **81.** $\frac{\sqrt{15} + 3}{2}$ **83.** $\frac{13 + 7\sqrt{3}}{22}$ **85.** $\frac{19 - 11\sqrt{7}}{27}$ **87.** $3\sqrt{5a}$ **89.** $\frac{5\sqrt{2}}{2}$ **91.** $70 + 30\sqrt{5}$ **93.** $\frac{5\sqrt{5}}{3}$ **95.** 1.866 **97.** -1.974 **99.** 12 ft^2, $10\sqrt{2}$ ft **101.** 6 m^3

Section 8.4 Warm-Ups F T F F T F F T T T

1. The square root property says that $x^2 = k$ for $k > 0$ is equivalent to $x = \pm\sqrt{k}$. **3.** The square root property and squaring each side are two new techniques used for solving equations. **5.** Squaring each side can produce extraneous roots. **7.** $-4, 4$ **9.** $-2\sqrt{10}, 2\sqrt{10}$ **11.** $-\frac{\sqrt{6}}{3}, \frac{\sqrt{6}}{3}$ **13.** No solution **15.** $-1, 3$ **17.** $5 - \sqrt{3}, 5 + \sqrt{3}$ **19.** -19 **21.** 90 **23.** No solution **25.** $-5, 5$ **27.** 3 **29.** 0, 1 **31.** 6 **33.** No solution **35.** $\frac{13}{18}$ **37.** 3, 5 **39.** 3 **41.** 6, 8 **43.** $r = \pm\sqrt{\frac{V}{\pi h}}$ **45.** $b = \pm\sqrt{c^2 - a^2}$ **47.** $b = \pm 2\sqrt{ac}$ **49.** $t = \frac{v^2}{2p}$ **51.** $-\frac{1}{2}, \frac{1}{2}$ **53.** No solution **55.** $\frac{1}{81}$ **57.** $-8, 8$ **59.** $0, \frac{1}{4}$ **61.** $-3, 1$ **63.** $-\sqrt{2}, \sqrt{2}$ **65.** No solution **67.** $\frac{1 + 2\sqrt{2}}{2}, \frac{1 - 2\sqrt{2}}{2}$ **69.** $\frac{9}{2}$ **71.** 3 **73.** $-4, 2$ **75.** $\frac{5}{2}$ **77.** $-1.803, 1.803$ **79.** 0.993 **81.** 0.362, 4.438 **83.** $3\sqrt{2}$ or 4.243 ft **85.** $3\sqrt{2}$ or 4.243 ft **87.** $\sqrt{2}$ or 1.414 ft **89.** 5 ft **91.** 2.5 sec **93.** $\frac{200\sqrt{13}}{3}$ or 240.370 ft

Section 8.5 Warm-Ups T F F T T T T F T T

1. The nth root of a is $a^{1/n}$. **3.** The expression $a^{-m/n}$ means $\frac{1}{a^{m/n}}$. **5.** The operations can be performed in any order, but the easiest is usually root, power, and then reciprocal. **7.** $7^{1/4}$ **9.** $\sqrt[5]{9}$ **11.** $(5x)^{1/2}$ **13.** $\sqrt{-a}$ **15.** $-x^{1/2}$ **17.** $-\sqrt[4]{6}$ **19.** 5 **21.** 5 **23.** -5 **25.** -2 **27.** Not a real number **29.** -2 **31.** -4 **33.** $w^{7/3}$ **35.** $2^{-10/3}$ **37.** $\frac{1}{\sqrt[4]{w^3}}$ **39.** $\sqrt{(ab)^3}$ **41.** $-t^{3/4}$ **43.** $-\frac{1}{\sqrt{a^3}}$ **45.** 25 **47.** 125 **49.** $\frac{1}{81}$ **51.** $\frac{1}{8}$ **53.** $-\frac{1}{3}$ **55.** Not a real number **57.** -8 **59.** $-\frac{1}{4}$ **61.** $x^{1/2}$ **63.** $n^{1/6}$ **65.** $x^{3/2}$ **67.** $2t^{1/4}$ **69.** x^2 **71.** $5^{1/8}$ **73.** xy^3 **75.** $\frac{x}{3y^4}$ **77.** $\frac{3}{4}$ **79.** $\frac{1}{9}$ **81.** 27 **83.** $\frac{1}{12}$ **85.** Yes, 9.2 m^2 **87.** 11.8% **89. a)** 15.9 **b)** Decreasing **c)** 28,000 pounds **91.** $a < 0$ and m and n are odd **93.** $3^{-8/9} < 3^{-2/3} < 3^{-1/2} < 3^0 < 3^{1/2} < 3^{2/3} < 3^{8/9}$, $7^{-9/10} < 7^{-3/4} < 7^{-1/3} < 7^0 < 7^{1/3} < 7^{3/4} < 7^{9/10}$, $m < n$

Enriching Your Mathematical Word Power

1. d **2.** b **3.** b **4.** b **5.** d **6.** b **7.** c **8.** d

Review Exercises

1. 2 **3.** 10 **5.** x^6 **7.** x^2 **9.** $2x$ **11.** $5x^2$ **13.** $\frac{2x^8}{y^7}$ **15.** $\frac{w}{4}$
17. $6\sqrt{2}$ **19.** $\frac{\sqrt{3}}{3}$ **21.** $\frac{\sqrt{15}}{5}$ **23.** $\sqrt{11}$ **25.** $\frac{\sqrt{6}}{4}$ **27.** y^3
29. $2t^4\sqrt{6t}$ **31.** $2m^2t\sqrt{3mt}$ **33.** $\frac{\sqrt{2x}}{x}$ **35.** $\frac{a^2\sqrt{6as}}{2s}$ **37.** $10\sqrt{7}$
39. $-\sqrt{3}$ **41.** 30 **43.** $-45\sqrt{2}$ **45.** $-15\sqrt{3} - 9$
47. $-3\sqrt{2} + 3\sqrt{5}$ **49.** -22 **51.** $26 - 4\sqrt{30}$ **53.** $38 - 19\sqrt{6}$
55. $\frac{\sqrt{10}}{4}$ **57.** $\frac{1 - \sqrt{5}}{5}$ **59.** $-\frac{3 + 3\sqrt{5}}{4}$ **61.** $-20, 20$
63. $-\frac{\sqrt{21}}{7}, \frac{\sqrt{21}}{7}$ **65.** $4 - 3\sqrt{2}, 4 + 3\sqrt{2}$ **67.** 81 **69.** 4
71. 6 **73.** $t = \pm 2\sqrt{2sw}$ **75.** $t = \frac{9a^2}{b}$ **77.** $\frac{1}{125}$ **79.** 5
81. $\frac{1}{8}$ **83.** $-\frac{1}{125}$ **85.** Not a real number **87.** $-\frac{1}{2}$ **89.** $\frac{1}{x}$
91. $-\frac{1}{2}x^2$ **93.** w^2 **95.** $\frac{t^3}{3s^2}$ **97.** $\frac{4}{x^8y^{20}}$
99. a) 17.2% **b)** First two years
101. 0.4 cm **103.** $4\pi\sqrt{10}$ or 40 in.^{2} **105.** 22 in.

Chapter 8 Test

1. 6 **2.** 12 **3.** -3 **4.** 2 **5.** 2 **6.** $2\sqrt{6}$ **7.** $\frac{\sqrt{6}}{4}$
8. Not a real number **9.** 81 **10.** $-\frac{1}{3}$ **11.** -16 **12.** $3\sqrt{2}$
13. $7 + 4\sqrt{3}$ **14.** 11 **15.** $\sqrt{7}$ **16.** $\frac{2\sqrt{15}}{3}$ **17.** $1 + \sqrt{2}$
18. $3\sqrt{2} - 3$ **19.** $y^{3/4}$ **20.** $3x^{2/3}$ **21.** xy^3 **22.** $\frac{1}{5u^4w}$
23. $\frac{\sqrt{3t}}{t}$ **24.** $2y^3$ **25.** $2y^4$ **26.** $3t^3\sqrt{2t}$ **27.** $-9, 3$ **28.** 18
29. $-\frac{\sqrt{10}}{5}, \frac{\sqrt{10}}{5}$ **30.** $\frac{4}{3}$ **31.** 11 **32.** $r = \pm\sqrt{\frac{S}{\pi h}}$
33. $b = \pm\sqrt{c^2 - a^2}$ **34.** $\frac{5\sqrt{2}}{2}$ meters **35.** 178.4 meters

Making Connections A Review of Chapters 1–8

1. $\left\{-\frac{3}{2}\right\}$ **2.** $\left\{\frac{3}{2}\right\}$

3. $\left(-\frac{3}{2}, \infty\right)$ 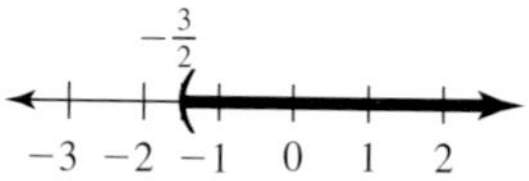

4. $\left(-\infty, \frac{3}{2}\right)$ 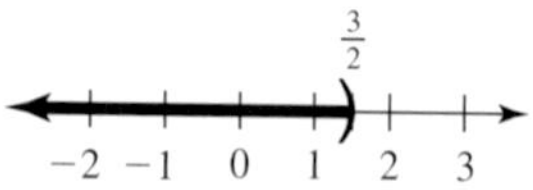

5. $\{-3\}$ **6.** $\left\{-\frac{\sqrt{6}}{2}, \frac{\sqrt{6}}{2}\right\}$ **7.** $\{-\sqrt{6}, \sqrt{6}\}$ **8.** $\left\{\frac{2}{3}\right\}$ **9.** $\left\{-\frac{3}{2}\right\}$
10. $\left\{-\frac{3}{2}, 3\right\}$ **11.** No solution, $\varnothing$ **12.** $\{-2, -1\}$
13. No solution, $\varnothing$ **14.** $\{3\}$ **15.** 9 **16.** 72 **17.** 81 **18.** 9
19. 12 **20.** -6 **21.** 3 **22.** $-\frac{3}{2}$ **23.** $(x - 3)^2$ **24.** $(x + 5)^2$
25. $(x + 6)^2$ **26.** $(x - 10)^2$ **27.** $2(x - 2)^2$ **28.** $3(x + 1)^2$
29. $7x - 3$ **30.** $-15t^2 - 13t + 20$ **31.** $-24j^2 + 14j + 24$
32. $6u + 6$ **33.** $v + 1$ **34.** $t^2 + 4t + 4$ **35.** $t^2 - 49$
36. $-4n^2 + 9$ **37.** $m^2 - 2m + 1$ **38.** $2t - 1$ **39.** $-4r^2 - r + 3$
40. $-18y^2 + 2$ **41.** $3j - 5$ **42.** $-6j + 2$ **43.** $2 - 3x$
44. $-1 - 3p$ **45.** $-2 + 3q$ **46.** $-4 + z$
47. a) 16.788 kilocalories per minute **b)** 275.24 m/min **c)** Increasing

Chapter 9

Section 9.1 Warm-Ups T F F T T F F T F F

1. A quadratic equation is an equation of the form $ax^2 + bx + c = 0$, where $a \neq 0$.
3. If $b = 0$, we can get the square root of a negative number and no real solution.
5. After applying the zero factor property we will have linear equations to solve.

7. $-8, 8$ **9.** $-\frac{3}{2}, \frac{3}{2}$ **11.** $-6, 6$ **13.** No real solution
15. $-\sqrt{10}, \sqrt{10}$ **17.** $\frac{-\sqrt{15}}{3}, \frac{\sqrt{15}}{3}$ **19.** $-\frac{2\sqrt{6}}{3}, \frac{2\sqrt{6}}{3}$ **21.** 1, 5
23. $2 - 3\sqrt{2}, 2 + 3\sqrt{2}$ **25.** $-\frac{3}{2}, -\frac{1}{2}$ **27.** $\frac{2 - \sqrt{2}}{2}, \frac{2 + \sqrt{2}}{2}$
29. $\frac{-1 - \sqrt{2}}{2}, \frac{-1 + \sqrt{2}}{2}$ **31.** 11 **33.** $-2, -1$ **35.** $-5, 6$
37. $-3, 5$ **39.** -3 **41.** $-1, 2$ **43.** 0, 2 **45.** $0, \frac{3}{2}$ **47.** $-7, \frac{3}{2}$
49. 5 **51.** $-\frac{5}{2}, 6$ **53.** 4, 6 **55.** $-\frac{1}{2}, \frac{1}{2}$ **57.** $-6, 4$ **59.** $-6, 4$
61. $-\frac{9}{2}$ **63.** $-\sqrt{6}, \sqrt{6}$ **65.** 3, 9 **67.** 3, 4 **69.** $\frac{5\sqrt{2}}{2}$ meters
71. $4\sqrt{5}$ blocks **73.** $2\sqrt{61}$ feet **75.** 6.3%
77. a) 2 sec and 3 sec **b)** 5 sec **79.** 9 **81.** c

Section 9.2 Warm-Ups F F F F T T T T T T

1. Not every quadratic can be solved by factoring or the square root property.
3. The last term is the square of one-half of the coefficient of the middle term.
5. $x^2 + 6x + 9 = (x + 3)^2$ **7.** $x^2 + 14x + 49 = (x + 7)^2$
9. $x^2 - 16x + 64 = (x - 8)^2$ **11.** $t^2 - 18t + 81 = (t - 9)^2$
13. $m^2 + 3m + \frac{9}{4} = \left(m + \frac{3}{2}\right)^2$ **15.** $z^2 + z + \frac{1}{4} = \left(z + \frac{1}{2}\right)^2$
17. $x^2 - \frac{1}{2}x + \frac{1}{16} = \left(x - \frac{1}{4}\right)^2$ **19.** $y^2 + \frac{1}{4}y + \frac{1}{64} = \left(y + \frac{1}{8}\right)^2$
21. $(x + 5)^2$ **23.** $(m - 1)^2$ **25.** $\left(x + \frac{1}{2}\right)^2$ **27.** $\left(t + \frac{1}{6}\right)^2$
29. $\left(x + \frac{1}{5}\right)^2$ **31.** $-5, 3$ **33.** $-3, 7$ **35.** -3 **37.** $\frac{1}{2}, 1$
39. $\frac{3}{2}, 2$ **41.** $-1, \frac{1}{3}$ **43.** $-1 - \sqrt{7}, -1 + \sqrt{7}$
45. $-3 - 2\sqrt{2}, -3 + 2\sqrt{2}$ **47.** $\frac{1 - \sqrt{13}}{2}, \frac{1 + \sqrt{13}}{2}$
49. $\frac{-3 - \sqrt{21}}{2}, \frac{-3 + \sqrt{21}}{2}$ **51.** $\frac{1 - \sqrt{33}}{4}, \frac{1 + \sqrt{33}}{4}$
53. $\frac{-3 - \sqrt{15}}{2}, \frac{-3 + \sqrt{15}}{2}$ **55.** $\frac{3 - \sqrt{5}}{4}, \frac{3 + \sqrt{5}}{4}$
57. $5 - \sqrt{7}, 5 + \sqrt{7}$ **59.** $-\frac{\sqrt{15}}{3}, \frac{\sqrt{15}}{3}$ **61.** $\frac{-2 - \sqrt{5}}{2}, \frac{-2 + \sqrt{5}}{2}$
63. No real solution **65.** $4 - 2\sqrt{2}, 4 + 2\sqrt{2}$ **67.** $\frac{7}{2}$
69. $-3 - 2\sqrt{5}, -3 + 2\sqrt{5}$ **71.** $2 - \sqrt{2}, 2 + \sqrt{2}$ **73.** $-0.6, 0.4$
75. $\frac{-1 - 2\sqrt{2}}{2}, \frac{-1 + 2\sqrt{2}}{2}$ **77.** $5 - \sqrt{3}$ in. and $5 + \sqrt{3}$ in.

79. $6 - \sqrt{2}$ and $6 + \sqrt{2}$ **81.** 12 years old

83. a) $x^2 - 10x + 22$ **b)** $5 \pm \sqrt{3}$ **c)** $x^2 - 3x + \frac{7}{4} = 0$

Section 9.3 Warm-Ups T F F F T F F T T T

1. The quadratic formula solves any quadratic equation.
3. The quadratic formula is used to solve $ax^2 + bx + c = 0$ where $a \neq 0$.
5. If $b^2 - 4ac < 0$, then there are no real solutions.
7. $-5, 3$ **9.** -5 **11.** $-2, \frac{3}{2}$ **13.** $-\frac{3}{2}, \frac{1}{2}$ **15.** $3 - \sqrt{5}, 3 + \sqrt{5}$
17. $\frac{3 - \sqrt{3}}{2}, \frac{3 + \sqrt{3}}{2}$ **19.** $\frac{-2 - \sqrt{2}}{2}, \frac{-2 + \sqrt{2}}{2}$ **21.** No real solution
23. $0, \frac{1}{2}$ **25.** $-\frac{\sqrt{15}}{5}, \frac{\sqrt{15}}{5}$ **27.** $-7 + 4\sqrt{3}, -7 - 4\sqrt{3}$
29. 0, one **31.** -47, none **33.** 181, two **35.** 0, one
37. -15, none **39.** -59, none **41.** $-2, \frac{1}{2}$ **43.** 0, 3 **45.** $-\frac{6}{5}, 4$
47. $-4, -2$ **49.** 0, 9 **51.** $-8, -2$ **53.** No real solution
55. $\frac{3 - \sqrt{33}}{2}, \frac{3 + \sqrt{33}}{2}$ **57.** $0, \frac{3}{2}$ **59.** $-0.79, 3.79$
61. $-1.36, 2.36$ **63.** 0.30
65. a) \$150,000 **b)** \$0 and \$40 **c)** \$17.50 or \$22.50 **d)** \$20

Section 9.4 Warm-Ups F T F F T T T F T T

1. Width $-1 + \sqrt{11}$ meters, length $1 + \sqrt{11}$ meters
3. $4\sqrt{2}$ feet **5.** Height $-3 + \sqrt{19}$ inches, base $3 + \sqrt{19}$ inches
7. Length $9 + \sqrt{5}$ in., width $9 - \sqrt{5}$ in. **9.** $3 + \sqrt{5}$ or 5.24 hours
11. $16 + 4\sqrt{26}$ or 36.40 hours **13.** $\frac{15 + \sqrt{241}}{8}$ or 3.82 seconds
15. 1.5 seconds **17.** $2\sqrt{19}$ or 8.72 mph
19. $\frac{-13 + \sqrt{409}}{2}$ or 3.61 feet

Section 9.5 Warm-Ups T T T T F F F F F T

1. A complex number is a number of the form $a + bi$, where a and b are real and $i = \sqrt{-1}$.
3. Complex numbers are added or subtracted like binomials in which i is the variable.
5. If $b > 0$, then $\sqrt{-b} = i\sqrt{b}$.
7. $5 + 9i$ **9.** 1 **11.** $2 - 8i$ **13.** $-1 - 4i$ **15.** -1 **17.** $\frac{3}{4} + \frac{1}{2}i$
19. $6 - 9i$ **21.** -36 **23.** -36 **25.** $21 - i$ **27.** $21 - 20i$
29. 25 **31.** 2 **33.** 29 **35.** 52 **37.** 13 **39.** 1 **41.** $1 - 3i$
43. $-1 + 4i$ **45.** $-3 + 2i$ **47.** $\frac{8}{13} + \frac{12}{13}i$ **49.** $\frac{3}{5} + \frac{4}{5}i$ **51.** $-4i$
53. $5 + 3i$ **55.** $-3 - i\sqrt{7}$ **57.** $-1 + i\sqrt{3}$ **59.** $2 + \frac{1}{2}i\sqrt{5}$
61. $-\frac{2}{3} + \frac{1}{3}i\sqrt{7}$ **63.** $\frac{1}{5} - i$ **65.** $-6i, 6i$ **67.** $-9i, 9i$
69. $-i\sqrt{5}, i\sqrt{5}$ **71.** $-i\frac{\sqrt{6}}{3}, i\frac{\sqrt{6}}{3}$ **73.** $2 - i, 2 + i$
75. $3 - 2i, 3 + 2i$ **77.** $2 - i\sqrt{3}, 2 + i\sqrt{3}$ **79.** $\frac{2 - i}{3}, \frac{2 + i}{3}$
81. $\frac{1 - i\sqrt{3}}{2}, \frac{1 + i\sqrt{3}}{2}$ **83.** $\frac{2 - i\sqrt{5}}{2}, \frac{2 + i\sqrt{5}}{2}$ **85.** $-6 - 24i$
87. 0 **89.** $x^2 - 12x + 37$ **91.** $x^2 + 9 = 0$
93. Natural: 54, integers: 54, rational: 54, $-3/8$, irrational: $3\sqrt{5}$, real: 54, $-3/8$, $3\sqrt{5}$, imaginary: $6i$, $\pi + i\sqrt{5}$, complex: all

Section 9.6 Warm-Ups T F T T F T T T T F

1. The graph of $y = mx + b$ is a straight line.
3. The graph of $y = ax^2 + bx + c$ opens upward if $a > 0$.
5. The x-coordinate of the vertex is $-b/(2a)$.
7. $(3, -6), (4, 0), (-3, 0)$ **9.** $(4, -128), (0, 0), (2, 0)$

11.
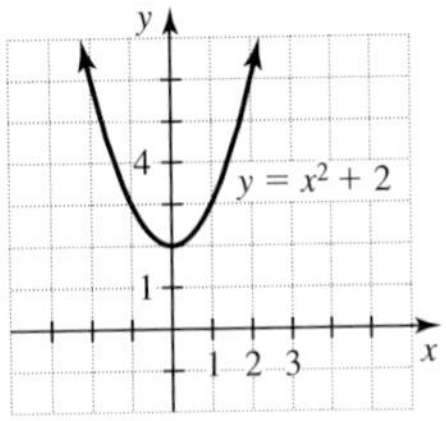

13.
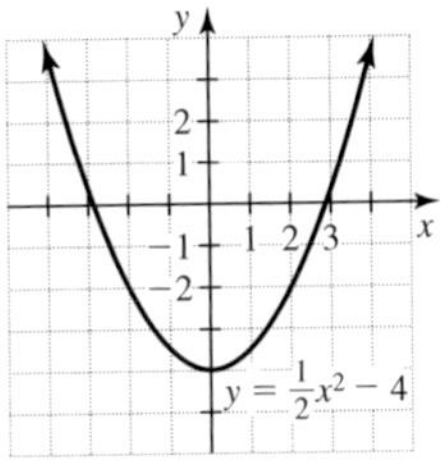

15.
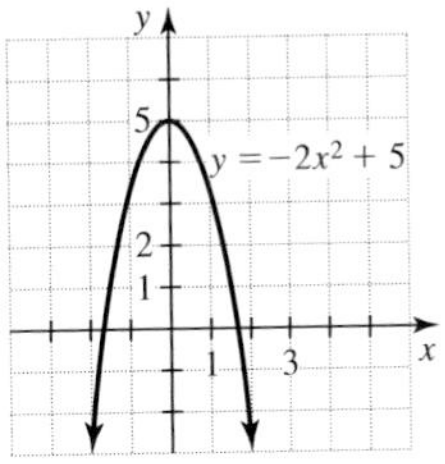

17.
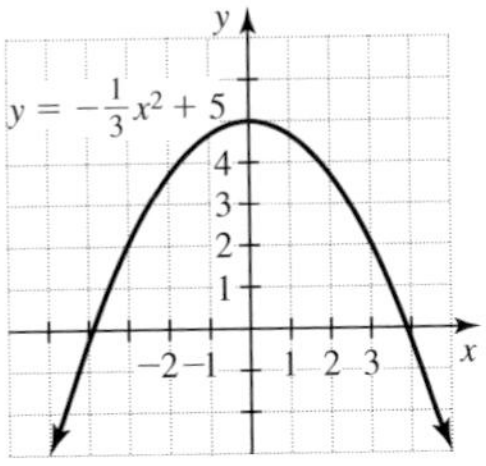

19.
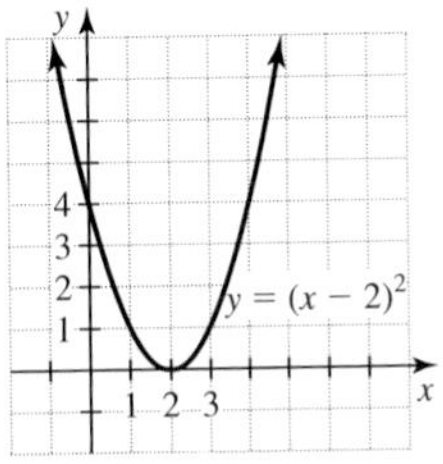

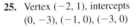

21. Vertex $\left(\frac{1}{2}, -\frac{9}{4}\right)$, intercepts $(0, -2), (-1, 0), (2, 0)$
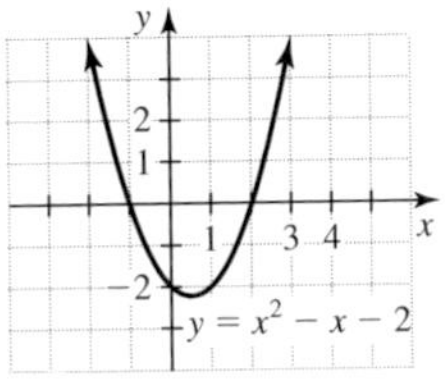

23. Vertex $(-1, -9)$, intercepts $(0, -8), (-4, 0), (2, 0)$
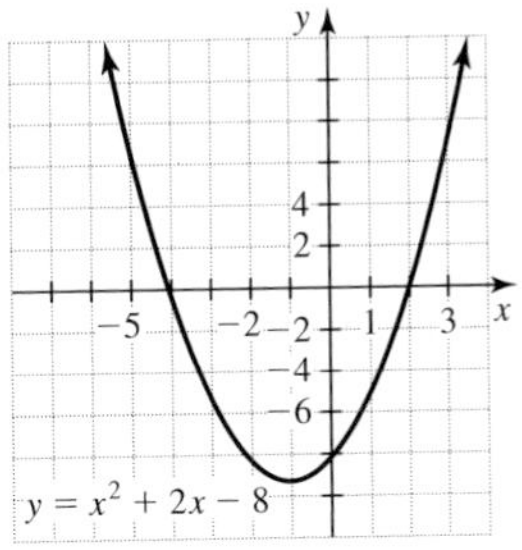

25. Vertex $(-2, 1)$, intercepts $(0, -3), (-1, 0), (-3, 0)$
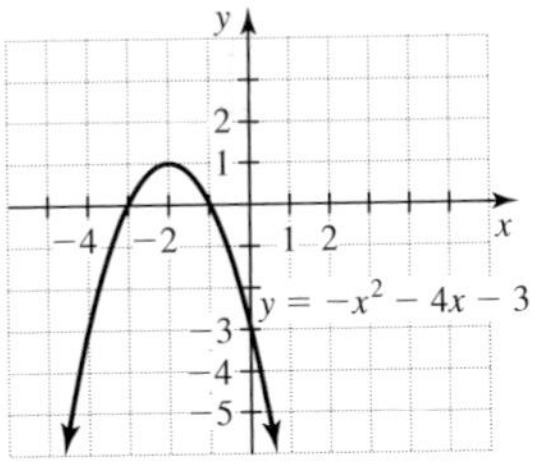

27. Vertex $\left(\frac{3}{2}, \frac{25}{4}\right)$, intercepts $(0, 4), (4, 0), (-1, 0)$
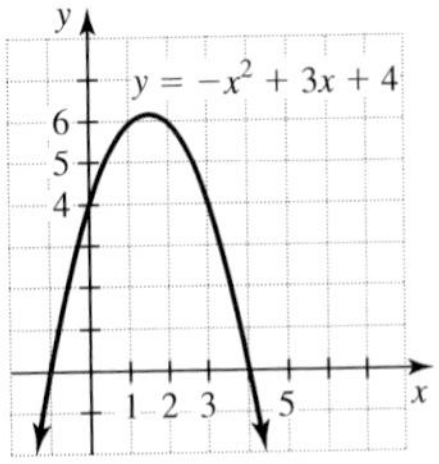

29. Vertex $(3, -25)$, intercepts $(0, -16), (8, 0), (-2, 0)$
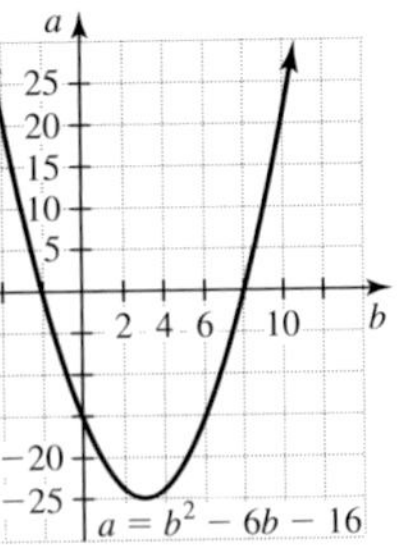

31. Minimum -8 **33.** Maximum 14 **35.** Minimum 2
37. Maximum 2

39. Maximum 64 feet

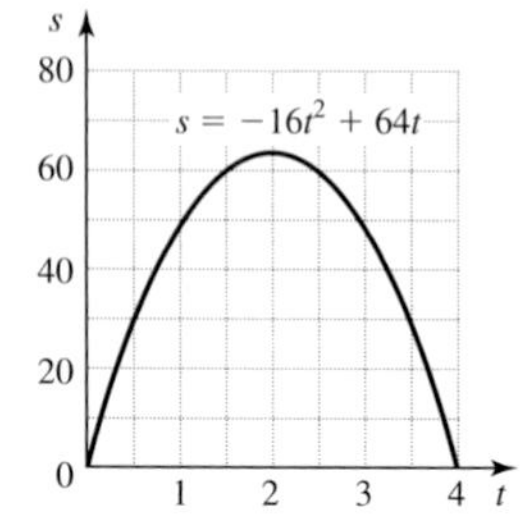

41. 625 square meters **43.** 2 P.M.
45. 15 meters, 25 meters
47. The graph of $y = kx^2$ gets narrower as k gets larger.
49. The graph of $y = x^2$ has the same shape as $x = y^2$.
51. a) Vertex $(0.84, -0.68)$, x-intercepts $(1.30, 0)$, $(0.38, 0)$
b) Vertex $(5.96, 46.26)$, x-intercepts $(12.48, 0)$, $(-0.55, 0)$

Section 9.7 Warm-Ups F T F T T T T T F T

1. A function is a set of ordered pairs in which no two have the same first coordinate and different second coordinates.
3. All descriptions of functions involve ordered pairs that satisfy the definition.
5. The domain is the set of all first coordinates of the ordered pairs.
7. $C = 0.50t + 5$ **9.** $T = 1.09S$ **11.** $C = 2\pi r$ **13.** $P = 4s$
15. $A = 5h$ **17.** Yes **19.** Yes **21.** No **23.** Yes **25.** Yes
27. Yes **29.** No **31.** Yes **33.** Yes **35.** No **37.** Yes
39. No **41.** Yes **43.** Yes **45.** Yes **47.** No **49.** Yes
51. No **53.** No **55.** $\{1, 2, 3\}, \{3, 5, 7\}$ **57.** $\{0, 2, 4\}, \{1\}$
59. R (real numbers), R **61.** R, R

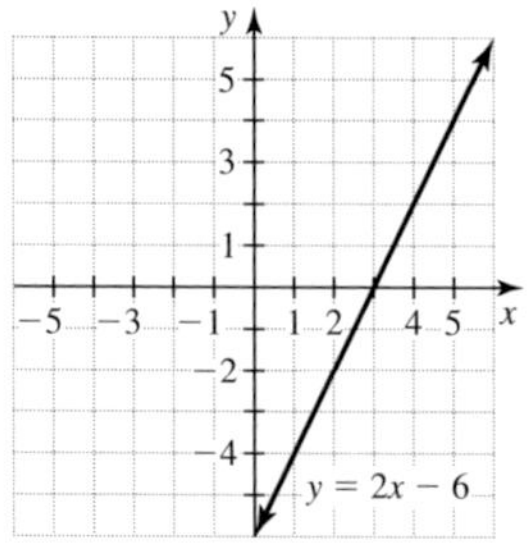

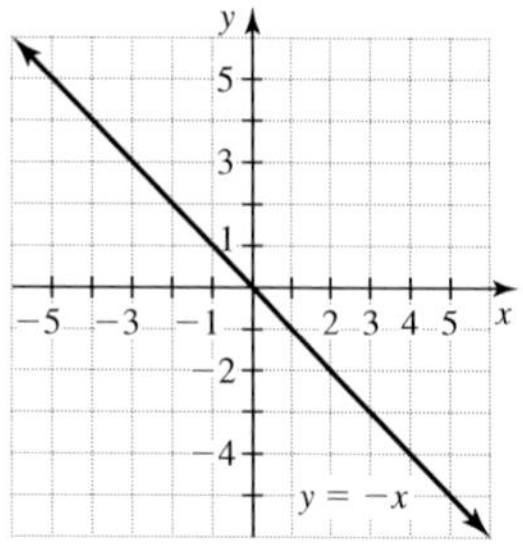

63. R, $\{y \mid y \geq -2\}$ or $[-2, \infty)$ **65.** R, $\{y \mid y \leq 7\}$ or $(-\infty, 7]$

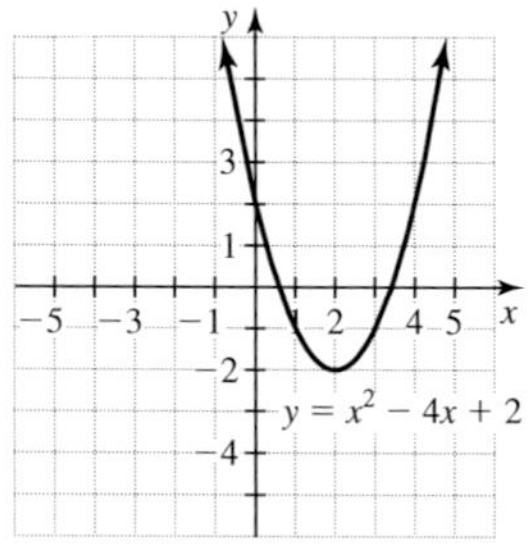

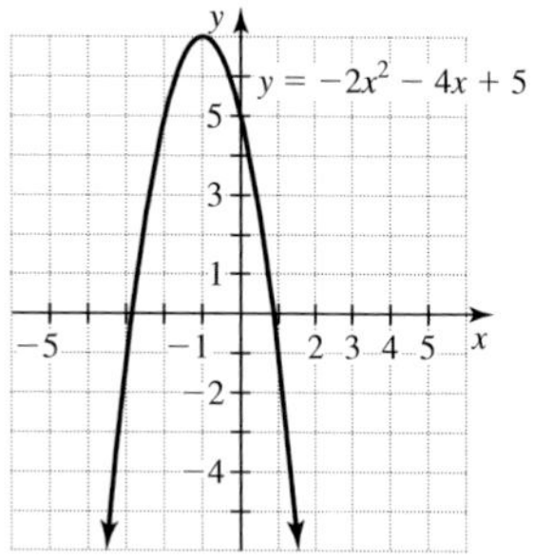

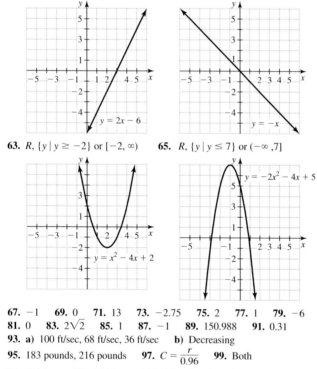

67. -1 **69.** 0 **71.** 13 **73.** -2.75 **75.** 2 **77.** 1 **79.** -6
81. 0 **83.** $2\sqrt{2}$ **85.** 1 **87.** -1 **89.** 150.988 **91.** 0.31
93. a) 100 ft/sec, 68 ft/sec, 36 ft/sec **b)** Decreasing
95. 183 pounds, 216 pounds **97.** $C = \dfrac{r}{0.96}$ **99.** Both
101. Neither **103.** Neither **105.** b is a function of a

Section 9.8 Warm-Ups T T F F F F T T T T

1. The basic operations of functions are addition, subtraction, multiplication, and division.
3. In the composition function the second function is evaluated on the result of the first function.
5. $x^2 + 2x - 3$ **7.** $4x^3 - 11x^2 + 6x$ **9.** 12 **11.** -30 **13.** -21
15. $\dfrac{13}{8}$ **17.** $y = 6x - 20$ **19.** $y = x + 2$ **21.** $y = x^2 + 2x$
23. $y = x$ **25.** -2 **27.** 5 **29.** 7 **31.** 5 **33.** 5 **35.** 4
37. 22.2992 **39.** $4x^2 - 6x$ **41.** $2x^2 + 6x - 3$ **43.** x **45.** $4x - 9$
47. $\dfrac{x + 9}{4}$ **49.** $F = f \circ h$ **51.** $G = g \circ h$ **53.** $H = h \circ g$
55. $J = h \circ h$ **57.** $K = g \circ g$ **59.** 112.5 in.^{2}, $A = \dfrac{d^2}{2}$
61. $P(x) = -x^2 + 20x - 170$ **63.** $J = 0.025I$
65. a) 397.8 **b)** $D = \dfrac{1.116 \times 10^7}{L^3}$ **c)** decreases

Enriching Your Mathematical Word Power

1. b **2.** a **3.** d **4.** c **5.** b **6.** a **7.** c **8.** a **9.** b
10. c **11.** a **12.** d **13.** c **14.** a **15.** b

Review Exercises

1. $-3, 3$ **3.** $0, 9$ **5.** $-1, 2$ **7.** $9 - \sqrt{10}, 9 + \sqrt{10}$ **9.** $\dfrac{3}{2}$
11. $4, 5$ **13.** $-7, 2$ **15.** $-4, \dfrac{1}{2}$ **17.** $-2 - \sqrt{11}, -2 + \sqrt{11}$
19. $-7, 4$ **21.** $\dfrac{-3 - \sqrt{29}}{2}, \dfrac{-3 + \sqrt{29}}{2}$ **23.** $-5, \dfrac{1}{2}$ **25.** 0, one
27. -44, none **29.** 76, two **31.** $-\dfrac{2}{3}, \dfrac{1}{2}$
33. $\dfrac{1 - \sqrt{17}}{2}, \dfrac{1 + \sqrt{17}}{2}$ **35.** $\dfrac{3 - \sqrt{14}}{5}, \dfrac{3 + \sqrt{14}}{5}$
37. $0, \dfrac{5}{3}$ **39.** 13 meters
41. $\dfrac{-2 + \sqrt{46}}{2}$ or 2.391 meters and $\dfrac{2 + \sqrt{46}}{2}$ or 4.391 meters
43. Height $-2 + 2\sqrt{11}$ or 4.633 inches, base $2 + 2\sqrt{11}$ or 8.633 inches
45. $3 + \sqrt{2}$ or 4.414 and $3 - \sqrt{2}$ or 1.586
47. New printer $7 + \sqrt{65}$ or 15.062 hours, old printer $9 + \sqrt{65}$ or 17.062 hours
49. $7 - 3i$ **51.** $-3 + 7i$ **53.** $15 - 25i$ **55.** $-55 + 48i$
57. $-1 - i\sqrt{2}$ **59.** $\dfrac{3}{37} + \dfrac{19}{37}i$ **61.** $\dfrac{19}{17} + \dfrac{9}{17}i$ **63.** $-11i, 11i$
65. $8 - i, 8 + i$ **67.** $\dfrac{3 - 3i\sqrt{7}}{4}, \dfrac{3 + 3i\sqrt{7}}{4}$

69. Vertex $(3, -9)$, intercepts $(0, 0)$, $(6, 0)$

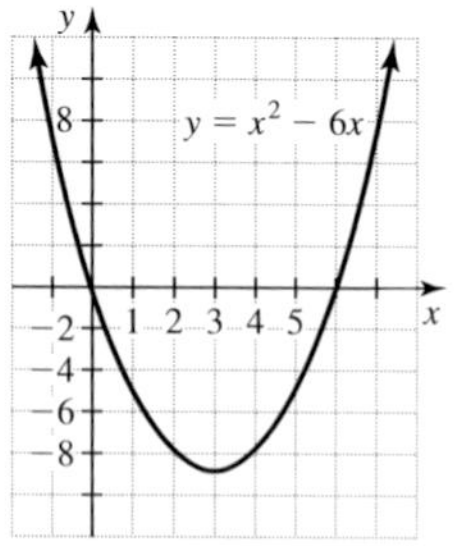

71. Vertex $(2, -16)$, intercepts $(0, -12)$, $(-2, 0)$, and $(6, 0)$

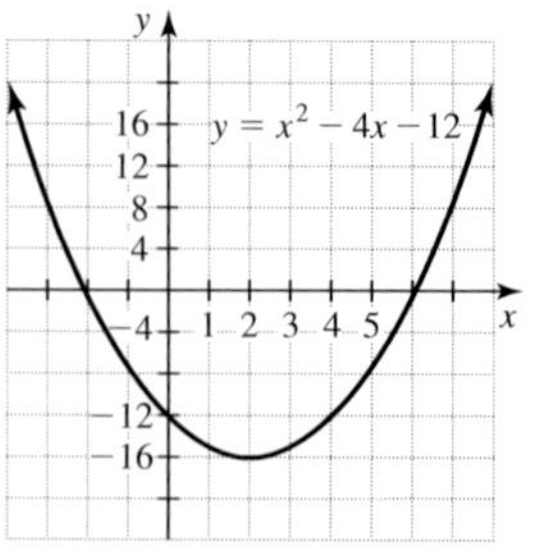

73. Vertex $(2, 8)$, intercepts $(0, 0)$, $(4, 0)$

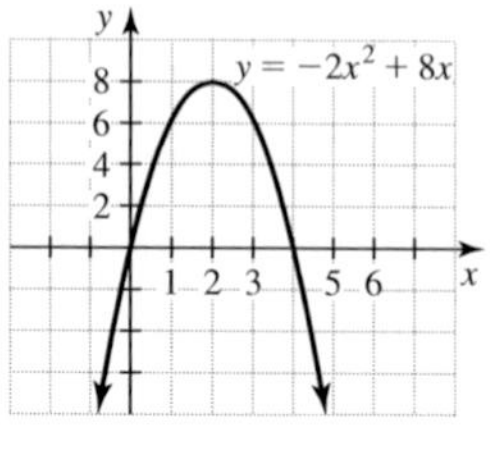

75. Vertex $(1, 4)$, intercepts $(0, 3)$, $(-1, 0)$, $(3, 0)$

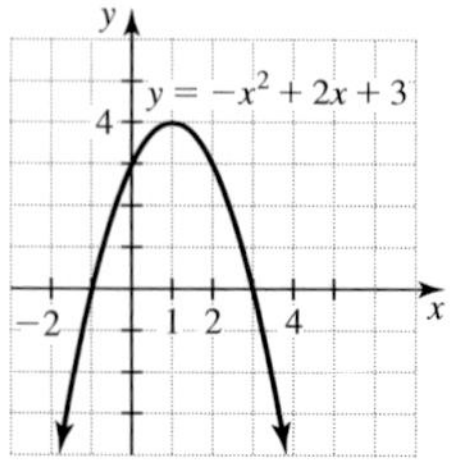

77. \$20.40, 400 **79.** Yes **81.** No **83.** Yes **85.** No
87. $\{1, 2, 3\}, \{0, 2\}$ **89.** Domain R, range $\{y \mid y \geq -3\}$ or $[-3, \infty)$
91. Domain R, range $\{y \mid y \leq 4.125\}$ or $(-\infty, 4.125]$ **93.** $5/4$
95. -3 **97.** 6 **99.** 5 **101.** $-1/2$ **103.** -4 **105.** $\sqrt{2}$
107. 99 **109.** 17 **111.** $3x^3 - x^2 - 10x$ **113.** 20 **115.** $F = f \circ g$
117. $H = g \circ h$ **119.** $I = g \circ g$

Chapter 9 Test

1. 0, one **2.** -31, none **3.** 17, two **4.** $-1, \frac{3}{5}$

5. $\frac{2-\sqrt{10}}{2}, \frac{2+\sqrt{10}}{2}$ **6.** $-7, 3$ **7.** $\frac{-3-\sqrt{29}}{2}, \frac{-3+\sqrt{29}}{2}$

8. $-5, 4$ **9.** 3, 25 **10.** $-\frac{1}{2}, 3$ **11.** $10 + 3i$ **12.** $-6 + 7i$

13. -36 **14.** $42 - 2i$ **15.** 68 **16.** $2 - 3i$ **17.** $-1 + i\sqrt{3}$

18. $-2 + i\sqrt{2}$ **19.** $\frac{3}{5} + \frac{4}{5}i$ **20.** $-3 - i\sqrt{3}, -3 + i\sqrt{3}$

21. $\frac{3}{5} - \frac{4}{5}i, \frac{3}{5} + \frac{4}{5}i$

22. Domain R, range $\{y \mid y \leq 16\}$ or $(-\infty, 16]$

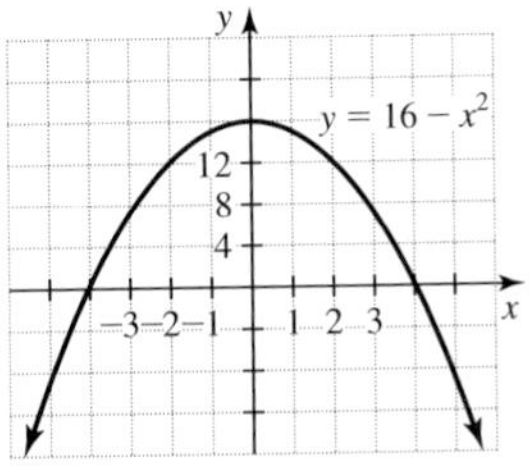

23. Domain R, range $\left\{y \mid y \geq -\frac{9}{4}\right\}$ or $\left[-\frac{9}{4}, \infty\right)$

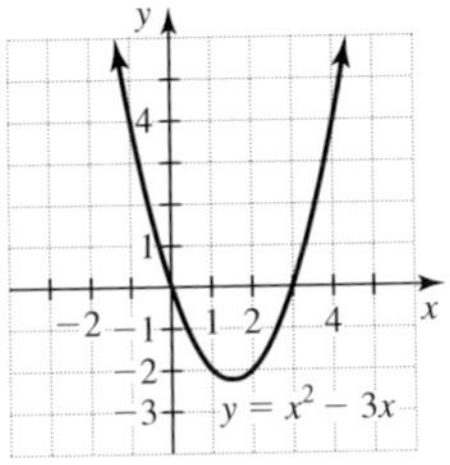

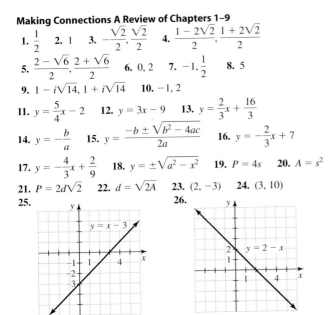

24. 11 **25.** 125 **26.** $x^2 - 2x + 9$ **27.** 15 **28.** $\frac{1}{8}$
29. $-2x^2 - 3$ **30.** $4x^2 - 20x + 29$ **31.** $H = f \circ g$ **32.** $W = g \circ f$
33. (1, 0), (5, 0) **34.** 36 feet **35.** $5 - \sqrt{2}$ and $5 + \sqrt{2}$

Making Connections A Review of Chapters 1–9

1. $\frac{1}{2}$ **2.** 1 **3.** $-\frac{\sqrt{2}}{2}, \frac{\sqrt{2}}{2}$ **4.** $\frac{1-2\sqrt{2}}{2}, \frac{1+2\sqrt{2}}{2}$

5. $\frac{2-\sqrt{6}}{2}, \frac{2+\sqrt{6}}{2}$ **6.** 0, 2 **7.** $-1, \frac{1}{2}$ **8.** 5

9. $1 - i\sqrt{14}, 1 + i\sqrt{14}$ **10.** $-1, 2$

11. $y = \frac{5}{4}x - 2$ **12.** $y = 3x - 9$ **13.** $y = \frac{2}{3}x + \frac{16}{3}$

14. $y = -\frac{b}{a}$ **15.** $y = \frac{-b \pm \sqrt{b^2 - 4ac}}{2a}$ **16.** $y = -\frac{2}{3}x + 7$

17. $y = -\frac{4}{3}x + \frac{2}{9}$ **18.** $y = \pm\sqrt{a^2 - x^2}$ **19.** $P = 4s$ **20.** $A = s^2$

21. $P = 2d\sqrt{2}$ **22.** $d = \sqrt{2A}$ **23.** $(2, -3)$ **24.** $(3, 10)$

25.

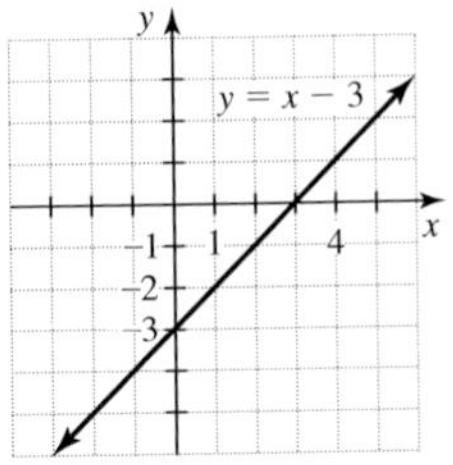

26.

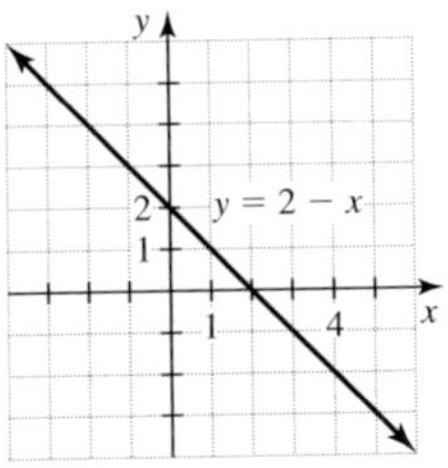

27.

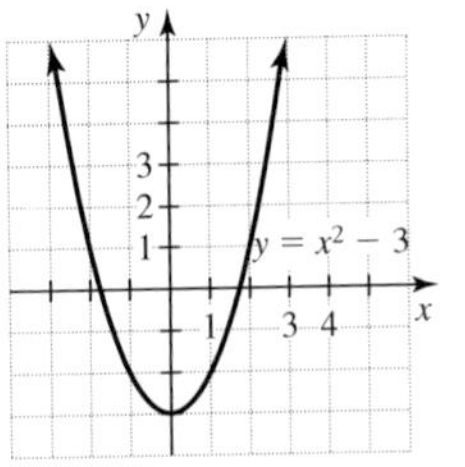

28.

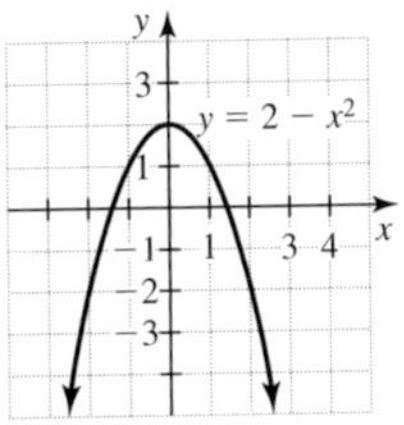

29.

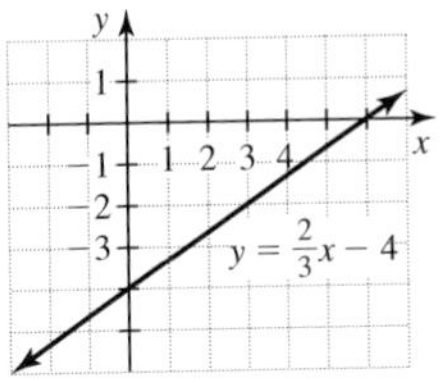

30.

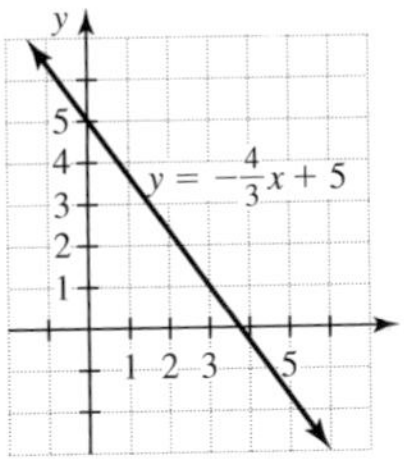

31. a) $s = -250p + 10{,}500$ **b)** Increased
c) $R = -250p^2 + 10{,}500p$ **d)** \$21

Appendix A

Geometry Review Exercises

1. 12 in. **2.** 24 ft^2 **3.** 60° **4.** 6 ft **5.** 20 cm **6.** 144 cm^2
7. 24 ft^2 **8.** 13 ft **9.** 30 cm **10.** No **11.** 84 yd^2 **12.** 30 in.
13. 32 ft^2 **14.** 20 km **15.** 7 ft, 3 ft^2 **16.** 22 yd **17.** 50.3 ft^2
18. 37.7 ft **19.** 150.80 cm^3 **20.** 879.29 ft^2 **21.** 288 $in.^3$, 288 $in.^2$
22. 4 cm **23.** 100 mi^2, 40 mi **24.** 20 km **25.** 42.25 cm^2
26. 33.510 ft^3, 50.265 ft^2 **27.** 75.4 $in.^3$, 100.5 $in.^2$ **28.** 56°
29. 5 cm **30.** 15 in. and 20 in. **31.** 149° **32.** 12 km **33.** 10 yd

Appendix B

Sets

1. A set is a collection of objects.
2. A finite set has a fixed number of elements and an infinite set does not.
3. A Venn diagram is used to illustrate relationships between sets.
4. The intersection of two sets consists of elements that are in both sets, whereas the union of two sets consists of elements that are in one, in the other, or in both sets.
5. Every member of set A is also a member of set B.
6. The empty set is a subset of every set.
7. False **8.** False **9.** True **10.** False **11.** True **12.** False
13. False **14.** True **15.** False **16.** False **17.** False
18. False **19.** $\varnothing$ **20.** $\{1, 2, 3, 4, 5, 6, 7, 8, 9\}$ **21.** $\{1, 3, 5\}$
22. $\{1, 2, 3, 4, 5, 7, 9\}$ **23.** $\{1, 2, 3, 4, 5, 6, 8\}$ **24.** $\{2, 4\}$ **25.** A
26. B **27.** $\varnothing$ **28.** $\varnothing$ **29.** A **30.** N **31.** $=$ **32.** $\neq$
33. $\cup$ **34.** $\cap$ **35.** $\cap$ **36.** $\cup$ **37.** $\notin$ **38.** $\in$ **39.** $\in$
40. $\in$ **41.** True **42.** True **43.** True **44.** False **45.** True
46. True **47.** True **48.** True **49.** False **50.** False
51. True **52.** True **53.** $\{2, 3, 4, 5, 6, 7, 8\}$ **54.** $\varnothing$ **55.** $\{3, 5\}$
56. $\{1, 2, 3, 4, 5, 7\}$ **57.** $\{1, 2, 3, 4, 5, 6, 8\}$ **58.** $\{2, 4\}$
59. $\{2, 3, 4, 5\}$ **60.** $\{2, 4\}$ **61.** $\{2, 3, 4, 5, 7\}$ **62.** $\{2, 3, 4, 5, 7\}$
63. $\{2, 3, 4, 5\}$ **64.** $\{2, 4\}$ **65.** $\{2, 3, 4, 5, 7\}$ **66.** $\{2, 3, 4, 5, 7\}$
67. $\subseteq$ **68.** $=$ **69.** $\in$ **70.** $\subseteq$ **71.** $\cap$ **72.** $\subseteq$ **73.** $\subseteq$
74. $\cap$ **75.** $\cap$ **76.** $\cup$ **77.** $\cup$ **78.** $\cap$ **79.** $\{2, 4, 6, \ldots, 18\}$
80. $\{7, 8, 9, \ldots\}$ **81.** $\{13, 15, 17, \ldots\}$ **82.** $\{1, 3, 5, \ldots, 13\}$
83. $\{6, 8, 10, \ldots, 78\}$ **84.** $\{13, 15, 17, \ldots, 55\}$
85. $\{x \mid x$ is a natural number between 2 and 7$\}$
86. $\{x \mid x$ is an odd natural number less than 8$\}$
87. $\{x \mid x$ is an odd natural number greater than 4$\}$

88. $\{x \mid x$ is a natural number greater than $3\}$
89. $\{x \mid x$ is an even natural number between 5 and $83\}$
90. $\{x \mid x$ is an odd natural number between 8 and $52\}$
91. False **92.** False **93.** False **94.** False **95.** True
96. True **97.** True **98.** False **99.** True **100.** True

Appendix C

Final Exam Review

1. $\frac{7}{12}$ **2.** $\frac{13}{12}$ **3.** $\frac{2}{9}$ **4.** -43 **5.** -11 **6.** 8
7. Distributive property **8.** Commutative property of multiplication
9. Associative property of addition **10.** Additive identity
11. $13x - 3$ **12.** 9 **13.** $15x^2$ **14.** $-x - 4$ **15.** $\left\{\frac{5}{11}\right\}$
16. $\{-2\}$ **17.** No solution, $\varnothing$ **18.** All real numbers
19. $y = \frac{5}{3}x - 3$ **20.** $y = -\frac{b}{a}$ **21.** $y = \frac{t - a}{b}$ **22.** $y = \frac{3}{4}a$
23. 33, 34, 35 **24.** 23 in. **25.** 220 ft **26.** 100 pounds
27. $(-\infty, 5]$ **28.** $(-\infty, -3)$ **29.** $(4, 8]$ **30.** $(0, 1)$
31. $(0, -2), (3, 0)$ **32.** $(0, -30), (50, 0)$

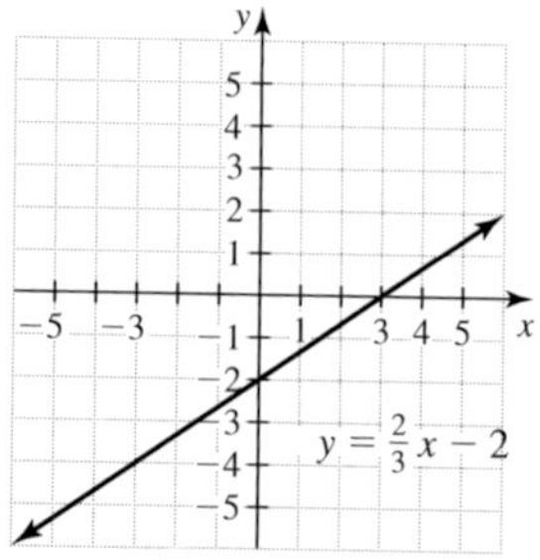

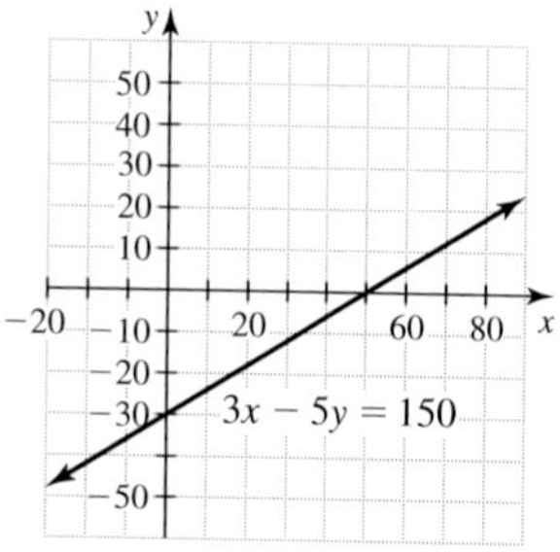

33. $(0, 2)$ **34.** $(2, 0)$

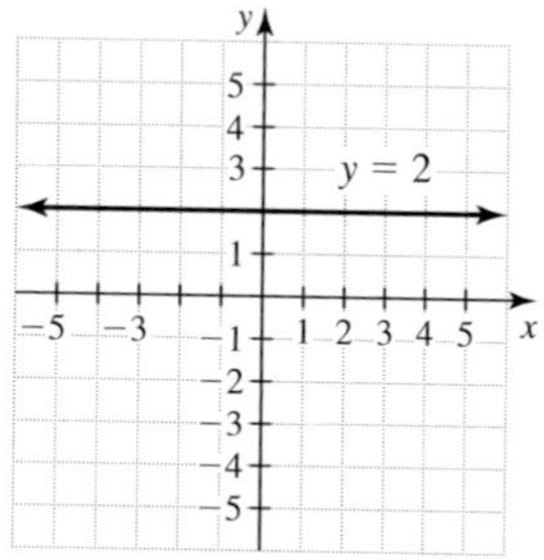

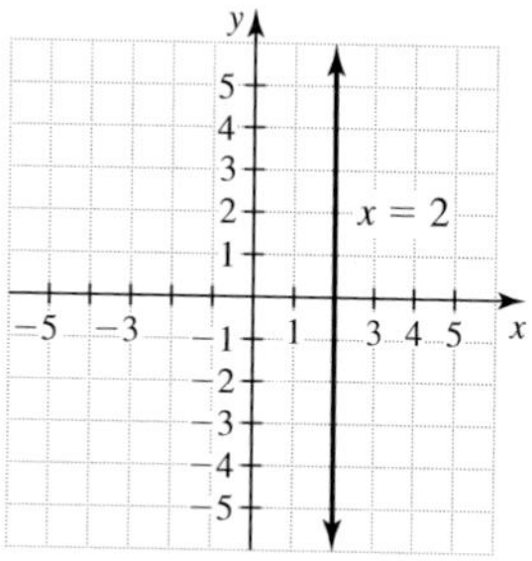

35. 2 **36.** $\frac{1}{2}$ **37.** $-\frac{2}{3}$ **38.** $\frac{1}{3}$ **39.** $y = 4x + 3$ **40.** $y = -2x$
41. $x = 3$ **42.** $y = -2x + 8$ **43.** $(1, 3)$ **44.** $(1, 1)$ **45.** $(1, -2)$
46. $(2, 5)$ **47.** $(-4, 6)$ **48.** $(-2, -3)$ **49.** $(1, 6)$
50. $(-2, -1/3)$ **51.** $(1/2, -4)$ **52.** Inconsistent **53.** Dependent
54. Independent **55.** Bob 18 hours and Carl 54 hours **56.** \$4.25

57.

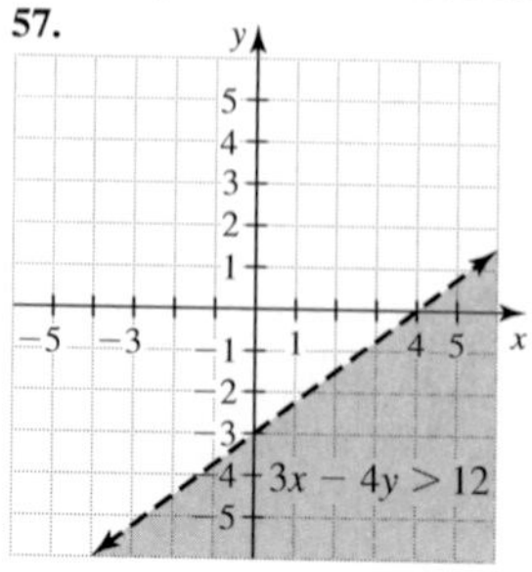

58.

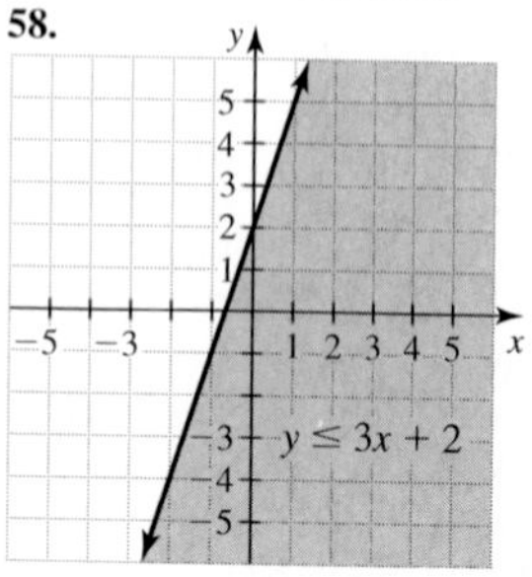

59.

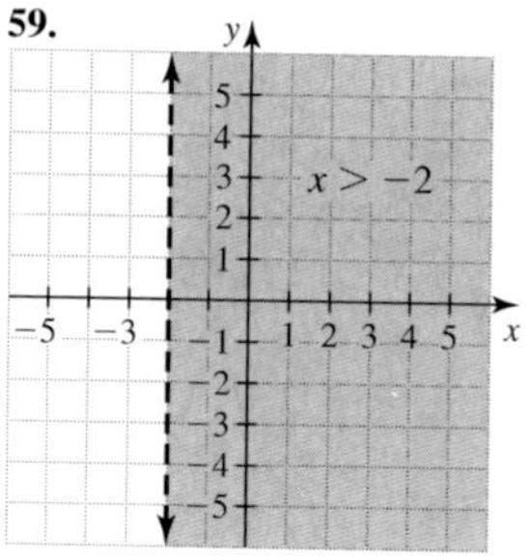

60.

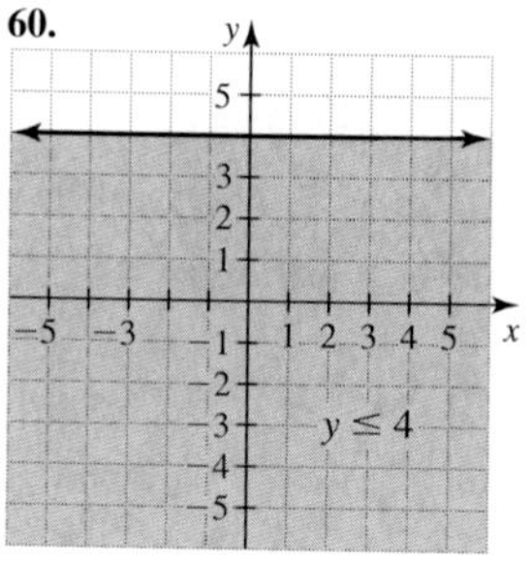

61. $-2x^2 - 12x + 6$ **62.** $6x^4 + 9x^2$ **63.** $x^2 - 2x - 63$
64. $x^3 + 8$ **65.** $16w^4 - 24w^2 + 9$ **66.** $-4m^5$ **67.** $-3y^2 - 2y + 1$
68. $x^2 + x + 2$ **69.** $-32x^7$ **70.** $375x^7$ **71.** $\frac{3x^5}{y}$ **72.** $8a^{15}$
73. 2.4×10^8 **74.** 1.8×10^{10} **75.** 1.6×10^{-11} **76.** 1×10^{-12}
77. $6xy^3(4x + 3y^2)$ **78.** $(x + a)(x + 2)$ **79.** $(2m - 7)(2m + 7)$
80. $(x - 9)(x + 6)$ **81.** $(2t - 5)(3t + 2)$ **82.** $(2w - 9)^2$
83. $2a(a - 9)(a + 6)$ **84.** $(w^2 + 4)(w + 2)(w - 2)$ **85.** $\{0, 1\}$
86. $\{-2, 0, 2\}$ **87.** $\{-3, 2\}$ **88.** $\{-6, 5\}$ **89.** 3 and 7
90. Length 24 in., width 10 in. **91.** $y = \frac{3}{5}x$ **92.** $y = \frac{2a}{w - c}$
93. $y = -3x - 12$ **94.** $y = \frac{6t}{2 - t}$ **95.** $\frac{19}{10}$ **96.** $\frac{3}{2x}$ **97.** $\frac{7}{a - 7}$
98. $\frac{2x + 8}{(x - 1)(x + 1)^2}$ **99.** $\frac{2}{x - 3}$ **100.** $\frac{1}{2}$ **101.** $\frac{-9}{2(9 + a)}$
102. $\frac{5x^2y^9}{32}$ **103.** 3 **104.** $\frac{x - 6}{-4x + 21}$ **105.** $\left\{\frac{15}{2}\right\}$ **106.** $\{11\}$
107. $\left\{\frac{8}{5}, 3\right\}$ **108.** $\{2, -1\}$ **109.** $2\sqrt{2}$ **110.** 15 **111.** 2
112. $\frac{\sqrt{6}}{4}$ **113.** 3 **114.** $4\sqrt{2}$ **115.** $4 + 2\sqrt{3}$ **116.** 1 **117.** $\sqrt{10}$
118. $2\sqrt{3}$ **119.** $1 + \sqrt{2}$ **120.** $4 - \sqrt{6}$ **121.** $-11 + 11\sqrt{5}$
122. $\frac{3\sqrt{10}}{5}$ **123.** $\frac{11 + 7\sqrt{5}}{4}$ **124.** x^3 **125.** x^2 **126.** $x\sqrt{x}$
127. $3y^5\sqrt{2}$ **128.** $5y^6$ **129.** $2t^6\sqrt{10t}$ **130.** $\frac{\sqrt{6xy}}{2y}$ **131.** $\{3\}$
132. $\{4 \pm 2\sqrt{3}\}$ **133.** $\left\{\frac{1}{4}\right\}$ **134.** $\{-1 \pm \sqrt{6}\}$ **135.** $\left\{\frac{1}{2}, \frac{1}{3}\right\}$
136. $\left\{\frac{1}{3}\right\}$ **137.** $\{-2 \pm \sqrt{5}\}$ **138.** $\left\{\frac{-3 \pm \sqrt{13}}{2}\right\}$ **139.** $\left\{\frac{2 \pm \sqrt{10}}{2}\right\}$

140.

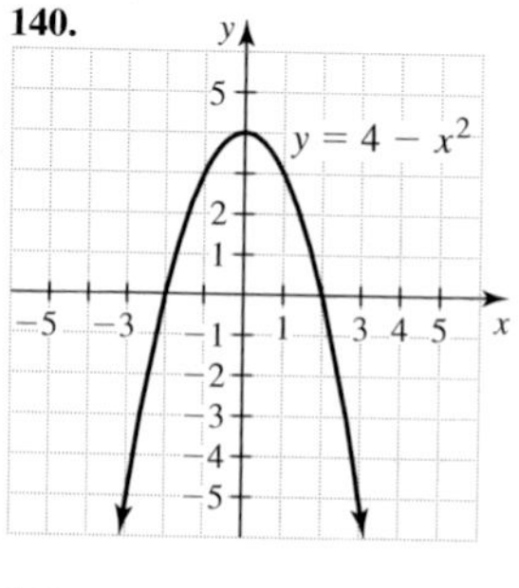

141.

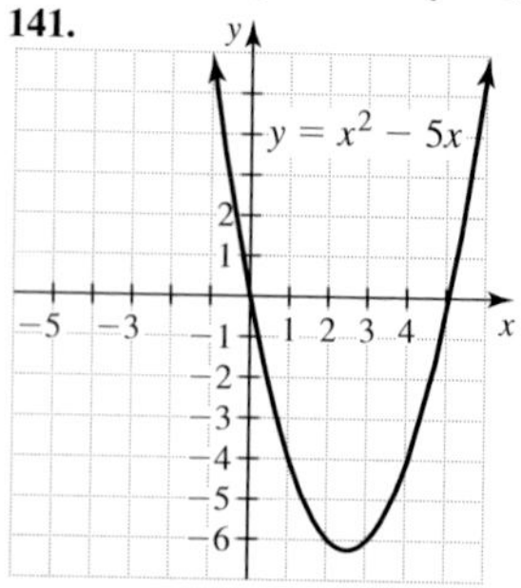

142.

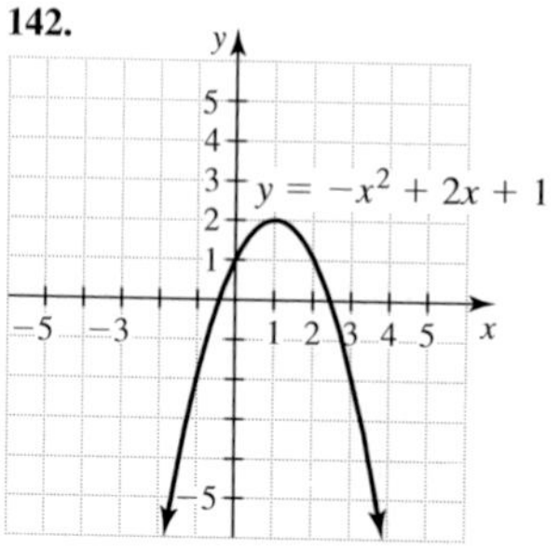

143. $(2, 0), (4, 0), (0, 8), (3, -1)$
144. 1 **145.** $-\frac{1}{6}$ **146.** 4 feet
147. 0 **148.** 5 **149.** -1
150. 0 **151.** 16 **152.** 1
153. Undefined **154.** 10

Index

F

G

Q

R

DEFINITIONS, RULES, AND FORMULAS

Subsets of the Real Numbers

Natural Numbers = $\{1, 2, 3, \ldots\}$

Whole Numbers = $\{0, 1, 2, 3, \ldots\}$

Integers = $\{\ldots -3, -2, -1, 0, 1, 2, 3, \ldots\}$

Rational = $\left\{\frac{a}{b} \,\middle|\, a \text{ and } b \text{ are integers with } b \neq 0\right\}$

Irrational = $\{x \mid x \text{ is not rational}\}$

Properties of the Real Numbers

For all real numbers a, b, and c

$a + b = b + a$; $a \cdot b = b \cdot a$ Commutative

$(a + b) + c = a + (b + c)$; $(ab)c = a(bc)$ Associative

$a(b + c) = ab + ac$; $a(b - c) = ab - ac$ Distributive

$a + 0 = a$; $1 \cdot a = a$ Identity

$a + (-a) = 0$; $a \cdot \frac{1}{a} = 1\ (a \neq 0)$ Inverse

$a \cdot 0 = 0$ Multiplication property of 0

Absolute Value

$$|a| = \begin{cases} a & \text{for } a \geq 0 \\ -a & \text{for } a < 0 \end{cases}$$

Order of Operations

No parentheses or absolute value present:

1. Exponential expressions
2. Multiplication and division
3. Addition and subtraction

With parentheses or absolute value:

First evaluate within each set of parentheses or absolute value, using the order of operations.

Exponents

$a^0 = 1$ $\qquad a^{-1} = \frac{1}{a}$

$a^{-r} = \frac{1}{a^r} = \left(\frac{1}{a}\right)^r$ $\qquad \frac{1}{a^{-r}} = a^r$

$a^r a^s = a^{r+s}$ $\qquad \frac{a^r}{a^s} = a^{r-s}$

$(a^r)^s = a^{rs}$ $\qquad (ab)^r = a^r b^r$

$\left(\frac{a}{b}\right)^r = \frac{a^r}{b^r}$ $\qquad \left(\frac{a}{b}\right)^{-r} = \left(\frac{b}{a}\right)^r$

Roots and Radicals

$a^{1/n} = \sqrt[n]{a}$ $\qquad a^{m/n} = \left(\sqrt[n]{a}\right)^m = \sqrt[n]{a^m}$

$\sqrt[n]{ab} = \sqrt[n]{a} \cdot \sqrt[n]{b}$ $\qquad \sqrt[n]{\frac{a}{b}} = \frac{\sqrt[n]{a}}{\sqrt[n]{b}}$

Factoring

$a^2 + 2ab + b^2 = (a + b)^2$

$a^2 - 2ab + b^2 = (a - b)^2$

$a^2 - b^2 = (a + b)(a - b)$

$a^3 - b^3 = (a - b)(a^2 + ab + b^2)$

$a^3 + b^3 = (a + b)(a^2 - ab + b^2)$

Rational Expressions

$\frac{a}{b} + \frac{c}{b} = \frac{a + c}{b}$ $\qquad \frac{a}{b} - \frac{c}{b} = \frac{a - c}{b}$

$\frac{ac}{bc} = \frac{a}{b}$ $\qquad \frac{a}{b} + \frac{c}{d} = \frac{ad + bc}{bd}$

$\frac{a}{b} \cdot \frac{c}{d} = \frac{ac}{bd}$ $\qquad \frac{a}{b} \div \frac{c}{d} = \frac{a}{b} \cdot \frac{d}{c}$

If $\frac{a}{b} = \frac{c}{d}$, then $ad = bc$.